THEORY OF ELASTICITY

By the Same Author :

1. Applied Stress Analysis
2. Computer Aided Design and Manufacturing
3. Engineering Mechanics
4. Experimental Stress Analysis
5. Machine Design
6. Mechanical Engineer's Handbook
7. Strength of Materials
8. Theory of Plasticity and Metal Forming Processes

Other Related Books :

1. Plate Analysis *by* Dr. N.K. Bairagi
2. Shell Analysis *by* Dr. N.K. Bairagi

THEORY OF ELASTICITY

Dr. Sadhu Singh

Ex-Director (Colleges)

Punjab Technical University, Jalandhar - 144 011

Formerly

Professor and Head, Mechanical Engineering Department

&

Dean, Faculty of Engineering and Technology

G.B. Pant University of Agriculture & Technology,

Pantnagar-263145

KHANNA PUBLISHERS®

Investing in Learning®

Operational Office :

4575/15, Onkar House, Opp. Happy School,

Ground Floor, Daryaganj, New Delhi 110 002

Phones : 011-45033819 • *Mob.* 09811541460

email : contactus@khannapublishers.in

Published by :
Romesh Chander Khanna & Vineet Khanna
for **KHANNA PUBLISHERS**
2-B, Nath Market, Nai Sarak
Delhi- 110 006 (India)

Website : www.khannapublishers.in

Hologram & Description

To all readers of our books, to prevent yourself from being defrauded by piraters, please make sure that there is an Hologram on the cover of our books with the below specifications. If you find any book without Hologram and Description, please mail us at **contactus@khannapublishers.in**

Thanking you

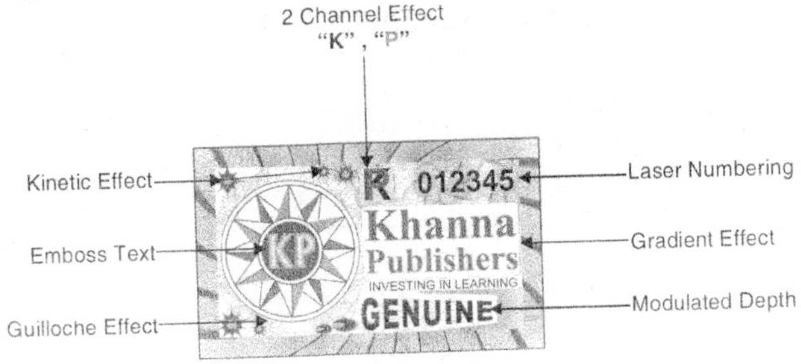

ISBN No. 978-81-7409-060-6

First Edition : 1978
Fourth Edition
Fourteenth Reprint : 2024

***Price :* ₹ 549.00**

Dedicated to the
Loving Memory
of my Parents

Preface to the Fourth Edition

The fourth edition of the book is being presented to the readers after a careful revision. The suggestions received from the readers have been incorporated. Many more numerical examples have been included to cope up with the needs of the readers. A large number of exercises have been added to present varied types of problems.

Further suggestions for the improvement of the text shall be welcome. I express my sincere gratitude to all those readers who sent their valuable suggestions. The continued support from my family is highly appreciated.

Pantnagar SADHU SINGH

Preface to the First Edition

Elasticity may be defined as the property of a deformable body to recover its original size and shape when the forces producing deformation are removed. The Theory of Elasticity, in its broad aspects, deals with the study of behaviour of such deformable bodies. It is further assumed that the body is homogeneous, perfectly elastic and isotropic.

When the elementary methods of strength of materials are inadequate to furnish satisfactory information regarding stress distribution in engineering structures, then a resort is made to the more powerful methods of the theory of elasticity. However, sometimes help is taken of the experimental and numerical methods also.

During the recent past the subject of theory of elasticity has developed tremendously. This subject is now being offered as a full course of study both at the undergraduate and postgraduate levels in almost all the universities and institutions. Most of the books available to the students and teachers do not fulfil their requirements. This book has been written with a view to present to the readers the requisite subject matter in a concise form. A number of solved and unsolved problems have been included to facilitate easy understanding of the subject.

This book in total consists of sixteen chapters. The first chapter deals with the basic equations of the theory of elasticity. Cartesian, cylindrical and spherical co-ordinates have been used to list these equations. Chapter 2 to 5 deal with the solution of two-dimensional problems in Cartesian, polar and curvilinear co-ordinates. Both the plane stress and plane strain problems have been dealt with. Airy's stress function method, complex potentials, Cauchy's integral method and Fourier transform method in general have been used to solve the two-dimensional problems. St. Venant's principle has also been illustrated by solved examples.

Chapters 5 and 6 deal with the torsion and bending of prismatic bars respectively. St. Venant's theory of torsion and complex torsion functions have been utilized to solve torsion problems. Harmonic functions have been utilized to solve bending problems. Analogous methods like the membrane analogy, hydrodynamic analogy, conducting sheet analogy and the soap film method have also been discussed. In chapter 7, solution to some simple three dimensional problems have been presented. Kelvin's, Betti's and integral transform methods have been illustrated by solved examples. In chapter 8, some axisymmetric problems have been solved by stress function and integral transforms methods.

Chapters 9 and 10 deals with the bending plates and thin shells respectively. Solution to plates of various profiles and end conditions have been presented in chapter 9. In chapter 10, the equilibrium equations for shells of general shape have been derived and then these equations have been reduced for shells of revolution, cylindrical and spherical shells. Membrane solution has also been presented for these shells. Donnell's equations for cylindrical shells have been derived and discussed.

Chapter 11 deals with the bending of beams resting on an elastic foundation. Both the foundations of the Winkler type and abstract type have been discussed. Some problems of cylindrical shells, curved beams, arches and plates resting on an elastic foundation have been attempted in this chapter. Chapter 12 deals with some of the beam column problems.

In chapter 13, the most commonly used numerical and energy methods have been discussed. The foundation theorems, like the Castigliano and Maxwell-Betti theorem and the principles of virtual work, potential energy and complementary energy have been discussed. The Dummy load method, Rayleigh's method and the Rayleigh-Ritz method have also been dealt with in this chapter. An introduction to the finite-difference and finite-element methods have been included in this chapter.

Chapter 14 and 15 deal with the solution to problems arising due to thermal effects and contact pressure respectively. Stresses due to thermal effects have been investigated by assuming one-dimensional variation of temperature only and the contact stresses have been determined by assuming prefect contact.

Chapters 16 deals with Stress Concentration Problems.

During the preparation of this book a number of books and journals were consulted. To that effect a brief bibliography is included at the end of the book. Some of the original articles published by various researchers are reproduced in this book, references to which have been cited as footnotes. Although utmost care has been taken to use the same notations throughout the book, deviations if any, if brought to the notice of the author, shall be highly acknowledged and incorporated in future.

I acknowledge with thanks the encouragement given to me by Principal, D.M. Sen and Prof. S.V. Kulkarni, whose association and guidance for the past many years have gone a long way in writing this book I also sincerely thank M/s Khanna Publishers for providing whole hearted co-operation and bringing out this book in a short period.

I am especially indebted to my wife Mrs. Manjit, for her patience and supporting nature without which this work would not have been possible.

Dr. SADHU SINGH

CONTENTS

NOTATIONS

a_{ij} = direction cosines

B_x, B_y, B_z = body force components in Cartesian coordinates

B_r, B_θ, B_z = body force components in cylindrical coordinates

B_r, B_θ, B_ϕ = body force components in spherical coordinates

$D = EI$ = flexural rigidity

$D = u_x + iu_y$, complex displacement

E = modulus of elasticity

F_x, F_y = resultant force components

G = modulus of rigidity

g = acceleration due to gravity, 9.81 m/s^2

h = shell or plate thickness

I = moment of inertia

I_0, I_1 = modified Bessel functions of first kind

$i = \sqrt{-1}$

J = polar moment of inertia

k = modulus of elastic foundation

K = stiffness matrix

l = length

M = bending moment

m, n = integers

P = single load

p = pressure

p_0 = contract pressure

Q = shear force

q = intensity of distributed load

R = radius

Re = real part of

r_x, r_y = radii of curvature of the middle surface of the plate in the xz and yz planes respectively

S = uniform tension

T = torque

T = temperature

t = thickness

U = strain energy

u = strain energy per unit volume

V = volume

V = body force potential

$w = \alpha + i\beta$

$z = x + iy$

$z = x - iy$

x, y, z = Cartesian co-ordinates

r, θ, z = cylindrical co-ordinates

r, θ, ϕ = spherical co-ordinates

X_n, Y_n, Z_n = boundary stresses

u_x, u_y, u_z = Cartesian components of displacement field

u_r, u_θ = components of displacement field in cylindrical co-ordinates

u_r, u_θ, u_ϕ = components of displacement field in spherical co-ordinates

M_x, M_y = bending moment per unit length of section of a plate perpendicular to x and y-axes respectively

M_{xy} = twisting moment per unit length of section of a plate perpendicular to x-axis

M_n, M_{nt} = bending and twisting moments per unit length of a section of a plate perpendicular to n-direction

Q_x, Q_y = shearing forces parallel to z-axis per unit length of section of a plate perpendicular to x and y axes respectively

N_x, N_y = normal forces per unit length of sections of a plate perpendicular to x and y-axes respectively.

N_{xy} = shearing force in direction of y-axis per unit length of section of a plate perpendicular to x-axis

$M_r, M_\theta, M_{r\theta}$ = radial, tangential and twisting moments in polar co-ordinates

Q_r, Q_θ = radial and tangential shearing force

N_r, N_θ = normal forces per unit length in radial and tangential directions

$N_\phi, N_\theta, N_{\phi\theta}$ = membrane forces per unit length of principal normal sections of a shell

M_θ, M_ϕ = bending moments in a shell per unit length of meridional section and a section perpendicular to meridian respectively

$N_\phi, N_x, N_{\phi x}$ = membrane force per unit length of axial section and a section perpendicular to the axis of a cylindrical shell respectively

M_ϕ, M_x = bending moment per unit length of an axial section and a section perpendicular to the axis of a cylindrical shell respectively

$M_{x\phi}$ = twisting moment per unit length of an axial section of a cylindrical shell

Q_ϕ, Q_x = Shearing forces parallel to z-axis per unit length of an axial section and a section perpendicular to the axis of a cylindrical shell respectively

σ = normal stress

τ = shear stress

ε = normal strain

γ = shear strain

τ_{ij} = stress tensor

ε_{ij} = strain tensor

δ_{ij} = Kronecker delta

$\sigma_x, \sigma_y, \sigma_z$ = normal stresses in Cartesian co-ordinates

$\tau_{xy}, \tau_{yz}, \tau_{zx}$ = shear stresses in Cartesian co-ordinates

$\sigma_r, \sigma_\theta, \sigma_z$ = normal stresses in cylindrical co-ordinates

$\tau_{\theta r}, \tau_{\theta z}, \tau_{rz}$ = shear stresses in cylindrical co-ordinates

$\sigma_r, \sigma_\theta, \sigma_\phi$ = normal stresses in spherical co-ordinates

$\tau_{r\theta}, \tau_{\phi\theta}, \tau_{r\phi}$ = shear stress in spherical co-ordinates

$\varepsilon_x, \varepsilon_y, \varepsilon_z$ = normal strains in Cartesian co-ordinates

$\gamma_{xy}, \gamma_{yz}, \gamma_{xz}$ = shear strains in Cartesian co-ordinates

$\varepsilon_r, \varepsilon_\theta, \varepsilon_z$ = normal strains in cylindrical co-ordinates

$\gamma_{\theta r}, \gamma_{\theta z}, \gamma_{rz}$ = shear strains in cylindrical co-ordinates

$\varepsilon_r, \varepsilon_\theta, \varepsilon_\phi$ = normal strains in cylindrical co-ordinates

$\gamma_{r\theta}, \gamma_{\theta\phi}, \gamma_{r\phi}$ = shear strains in spherical co-ordinates

$\sigma_m = \frac{1}{3}(\sigma_x + \sigma_y + \sigma_z), \frac{1}{3}(\sigma_r + \sigma_\theta + \sigma_z),$

$\frac{1}{3}(\sigma_r + \sigma_\theta + \sigma_\phi)$, mean stress

$\varepsilon_m = \frac{1}{3}(\varepsilon_x + \varepsilon_y + \varepsilon_z), \frac{1}{3}(\varepsilon_r + \varepsilon_\theta + \varepsilon_z),$

$\frac{1}{3}(\varepsilon_r + \varepsilon_\theta + \varepsilon_\phi)$, mean strain,

$\Theta = 3\,\sigma_m$, first invariant of stress,

$e = 3\,\varepsilon_m$, dilatation

ξ, η, ζ = curvilinear co-ordinates

$\sigma_\xi, \sigma_\eta, \tau_{\xi\eta}$ = stresses in curvilinear co-ordinates

γ = specific weight

θ = angle of twist

σ = density

ν = Poisson's ratio

α = coefficient of linear expansion

λ = Lame's constant

λ = characteristic for foundation

ω = angular speed

Φ = Airy's stress function

$\phi(z), \psi(z), \chi(x)$ = complex potentials

ϕ = stress function for torsion

φ = warping function

ψ = Prandtl's torsion function

Δ = area of triangle

π = strain energy

π^* = complementary strain energy

∇^2 = Laplacian operator

$\chi = 3 - 4\nu$ for plane strain and

$\frac{3-\nu}{1+\nu}$ for plane stress

$\kappa = \frac{1-2\nu}{1-\nu}$

1

Basic Equations of Elasticity

1.1 Introduction

Elasticity is the property of a deformable body due to which the body recovers its original shape upon the removal of the forces causing the deformation. Almost all engineering materials possess to a certain extent the property of elasticity. In the Theory of Elasticity, we assume the body to be perfectly elastic, homogeneous and isotropic. The first discussions of elastic phenomena occur in the writings of Hooke (1676) but the first real attempts to construct a theory of elasticity using the continuum approach, date from the first half of the eighteen century*. Since that time, the theory of elasticity has developed tremendously. In this chapter, we will study the basic equations of the Theory of Elasticity without giving their proof or deriving them†.

1.2 The State of Stress at a Point

There are two types of forces acting on a material body to produce stresses : Surface forces acting on the boundaries of the body, and body forces distributed throughout the volume of the body. These applied forces are resisted by a system of stresses throughout the body.

The state of stress at any given point of a continuous body is determined entirely by the components of stress in three mutually perpendicular planes which pass through the chosen point. The planes are usually taken perpendicular to the co-ordinate directions of some orthogonal co-ordinate system. The stress at a point on a plane section through the point is defined as the force per unit area transmitted through that section at the point. The stress components are a function of both the position of the point in the body

*For the early history of the theory of elasticity, the reader may refer to I. Todhunter and K. Pearson : A History of the Theory of Elasticity and of the Strength of Materials, 2 vols. Cambridge, 1893.

†For the derivation of these equations the reader may refer to : Singh, S., "Applied Stress Analysis", Khanna Publishers.

and the orientation of the plane passed through that point. Each of the above stress components, being a vector, can be resolved further into three components along three co-ordinate axes.

Let us refer the continuous body in a state of stress under the influence of some external forces to a Cartesian co-ordinate system x, y, z and at a given point let us consider three planes which are normal to the axes of the co-ordinates. The stress which acts on each area is resolved into three components along the axes. We denote the normal components (the normal stresses) by the symbol σ with an index which indicates the direction of the normal to the area. We denote the tangential components (the tangential stresses) by τ with two indices. The components of stress acting on an area normal to the x-axis are σ_x, τ_{xy}, τ_{xz} ; the components of stress acting on an area normal to the y-axis are τ_{yz}, σ_y, τ_{yz} ; the components of stress acting on an area normal to the z-axis are τ_{zx}, τ_{zy}, σ_z (Fig. 1.1). These nine components define the state of stress at a point given by the stress tensor :

$$[\tau_{ij}] = \begin{bmatrix} \sigma_x & \tau_{xy} & \tau_{xz} \\ \tau_{yx} & \sigma_y & \tau_{yz} \\ \tau_{zx} & \tau_{zy} & \sigma_z \end{bmatrix}$$

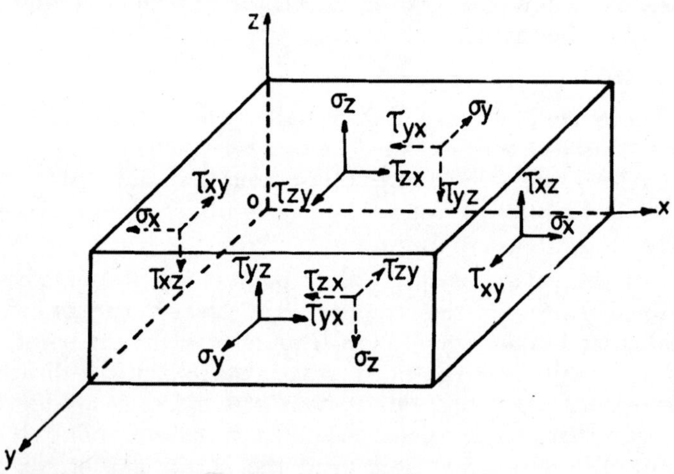

Fig. 1.1 Stress components on a rectangular parallelopiped.

The stress tensor is symmetric because $\tau_{yx} = \tau_{xy}$; $\tau_{zy} = \tau_{yz}$; $\tau_{zx} = \tau_{xz}$ (generally, $\tau_{ij} = \tau_{ji}$, $i \neq j$, where i and j denote mutually perpendicular directions).

Similarly the stress tensor in cylindrical co-ordinate system r, θ, z is given by :

$$[\tau_{ij}] = \begin{bmatrix} \sigma_r & \tau_{r\theta} & \tau_{rz} \\ \tau_{\theta r} & \sigma_\theta & \tau_{\theta z} \\ \tau_{zr} & \tau_{z\theta} & \sigma_z \end{bmatrix}$$

where $\tau_{\theta r} = \tau_{r\theta}$; $\tau_{zr} = \tau_{rz}$; $\tau_{z\theta} = \tau_{\theta z}$, and in spherical co-ordinates r, θ, ϕ by,

$$[\tau_{ij}] = \begin{bmatrix} \sigma_r & \tau_{r\theta} & \tau_{r\phi} \\ \tau_{\theta r} & \sigma_\theta & \tau_{\theta\phi} \\ \tau_{\phi r} & \tau_{\phi\theta} & \sigma_\phi \end{bmatrix}$$

where $\tau_{\theta r} = \tau_{r\theta}$; $\tau_{\phi r} = \tau_{r\phi}$; $\tau_{\phi\theta} = \tau_{\theta\phi}$.

If we know the stresses in three mutually perpendicular areas we can always determine the stress which acts on any area passing through the same point. We have the formulae :

$$\left.\begin{aligned} X_n &= \sigma_x \cos(n, x) + \tau_{xy} \cos(n, y) + \tau_{xz} \cos(n, z) \\ Y_n &= \tau_{xy} \cos(n, x) + \sigma_y \cos(n, y) + \tau_{yz} \cos(n, z) \\ Z_n &= \tau_{xz} \cos(n, x) + \tau_{yz} \cos(n, y) + \sigma_z \cos(n, z) \end{aligned}\right\} \quad ...(1.1)$$

where X_n, Y_n, Z_n are components of stress which act on an area with the arbitrary normal direction $\overline{n}$.

If it is desired to express the stress components referred to co-ordinate system $0, x', y', z'$ in terms of the components of system $0, x, y, z$, then the direction cosines are determined by :

	x	y	z
x'	$a_{x'x}$	$a_{x'y}$	$a_{x'z}$
y'	$a_{y'x}$	$a_{y'y}$	$a_{y'z}$
z'	$a_{z'x}$	$a_{z'y}$	$a_{z'z}$

where $a_{x'x} = \cos(x', x)$; $a_{y'z} = \cos(y', z)$ and so on.

The stress components in the new co-ordinate system may be obtained from :

$$\tau_{ij}' = a_{il}\, a_{jm}\, \tau_{lm} \quad ...(1.2)$$

Thereby, we get

$$\begin{aligned} \sigma_x' &= \sigma_x a^2_{x'x} + \sigma_y a^2_{x'y} + \sigma_z\, a^2_{x'z} \\ &\quad + 2(\tau_{xy}\, a_{x'x} \,.\, a_{x'y} + \tau_{yz}\, a_{x'y}\, a_{x'z} + \tau_{xz}\, a_{x'x}\, a_{x'z}) \quad ...(1.3) \\ \tau_{x'y'} &= \sigma_x\, a_{x'x} \,.\, a_{y'x} + \sigma_y\, a_{x'y}\, a_{y'y} + \sigma_z\, a_{x'z}\, a_{y'z} \\ &\quad + \tau_{xy}(a_{x'x}\, a_{y'y} + a_{x'y}\, a_{y'x}) \\ &\quad + \tau_{yz}(a_{x'y}\, a_{y'z} + a_{x'z}\, a_{y'y}) \\ &\quad + \tau_{xz}(a_{x'x}\, a_{y'z} + a_{x'z}\, a_{y'x}) \end{aligned}$$

The rest of the stress components can be found by cyclic permutation of symbols x, y and z. The formulas for the transformation from the system x', y', z' to x, y, z, can be written as :

$$\left.\begin{aligned} \sigma_x &= \sigma_x'\, a^2_{x'x} + \sigma_y'\, a^2_{y'x} + \sigma_z'\, a^2_{z'x} \\ &\quad + 2\,(\tau_{x'y'}\, a_{x'x}\, a_{y'x} + \tau_{y'z'}\, a_{y'x}\, a_{z'x} \\ &\quad + \tau_{x'z'}\, a_{x'x}\, a_{x'x}) \\ \tau_{xy} &= \sigma_x'\, a_{x'x}\, a_{x'y} + \sigma_{y'}\, a_{y'x}\, a_{y'y} \\ &\quad + \sigma_{z'}\, a_{z'x}\, a_{z'y} \\ &\quad + \tau_{x'y'}\,(a_{x'x}\, a_{y'y} + a_{x'y}\, a_{y'x}) \\ &\quad + \tau_{y'z'}\,(a_{y'x}\, a_{z'y} + a_{y'y}\, a_{z'x}) \\ &\quad + \tau_{x'z'}\,(a_{x'x}\, a_{z'y} + a_{x'y}\, a_{z'x}) \end{aligned}\right\} \quad \text{...(1.4)}$$

The expressions for the remaining stress components can be obtained by a cyclic permutation of the symbols x', y', z'.

Analogous formulas hold for other orthogonal co-ordinate systems. For example the relations between the stresses in the Cartesian and cylindrical co-ordinate systems with the same z-axis are :

		x	y	z
r	x'	$\cos\theta$	$\sin\theta$	0
θ	y'	$-\sin\theta$	$\cos\theta$	0
z	z'	0	0	1

$$\left.\begin{aligned} \sigma_r &= \sigma_x \cos^2\theta + \sigma_y \sin^2\theta + 2\tau_{xy} \sin\theta\cos\theta \\ \sigma_\theta &= \sigma_x \sin^2\theta + \sigma_y \cos^2\theta - 2\tau_{xy} \sin\theta\cos\theta \\ \sigma_z &= \sigma_z \\ \tau_{r\theta} &= -\sigma_x \cos\theta\sin\theta + \sigma_y \sin\theta\cos\theta \\ &\qquad + \tau_{xy}(\cos^2\theta - \sin^2\theta) \\ \tau_{\theta z} &= \tau_{yz}\,.\,\cos\theta - \tau_{xz}\sin\theta \\ \tau_{rz} &= \tau_{yz}\sin\theta + \tau_{xz}\cos\theta \end{aligned}\right\} \quad \text{...(1.5)}$$

and

$$\left.\begin{aligned} \sigma_x &= \sigma_r \cos^2\theta + \sigma_\theta \sin^2\theta - 2\tau_{r\theta} \sin\theta\cos\theta \\ \sigma_y &= \sigma_r \sin^2\theta + \sigma_\theta \cos^2\theta + 2\tau_{r\theta} \sin\theta\cos\theta \\ \sigma_z &= \sigma_z \\ \tau_{xy} &= (\sigma_r - \sigma_\theta)\sin\theta\cos\theta + \tau_{r\theta}(\cos^2\theta - \sin^2\theta) \\ \tau_{yz} &= \tau_{\theta z}\cos\theta + \tau_{rz}\sin\theta \\ \tau_{xz} &= -\tau_{\theta z}\sin\theta + \tau_{rz}\cos\theta \end{aligned}\right\} \quad \text{...(1.6)}$$

The importance of these stress-transformation equations is that they can be used to determine the planes on which the shear stress is zero. It can be shown that, for the most general state of stress, three such planes exist. These planes, referred to as principal planes, are mutually perpendicular. The stresses on these planes, σ_1, σ_2 and σ_3 are called principal stresses. The principal stresses are the roots of the cubic equation :

$$\sigma^3 - I_1\sigma^2 + I_2\sigma - I_3 = 0 \qquad \text{...(1.7)}$$

where

$$\left.\begin{aligned} I_1 &= \sigma_x + \sigma_y + \sigma_z \\ I_2 &= \sigma_x\sigma_y + \sigma_y\sigma_z + \sigma_x\sigma_z - \tau_{xy}^2 - \tau_{yz}^2 - \tau_{xz}^2 \\ I_3 &= \sigma_x\sigma_y\sigma_z - \sigma_x\tau_{yz}^2 - \sigma_y\tau_{xz}^2 - \sigma_z\tau_{xy}^2 + 2\tau_{xy}\tau_{yz}\tau_{xz} \end{aligned}\right\} \qquad \text{...(1.8)}$$

are three stress invariants.

Eq. (1.7) may be solved by hit and trial, Newton-Raphson method or any other numerical method and the principal stresses and principal directions may be determined.

1.3 The State of Strain at a Point

The displacements of the different points of a material body are resolved into two parts, rigid-body displacements and deformation. The first corresponds to a translation or rotation of the body as a whole, and the second corresponds to displacements of points relative to each other. It is only this later part that enters the definition of strain. In the linear theory or small deformation theory of elasticity, it is assumed that the deformations are infinitesimal and it applies to a small neighbourhood about the point in question.

The state of deformation in the neighbourhood of a given point of a continuous body is determined by six components of strain ; three components of normal strain, which we shall denote by ε, with one index, which indicates the primary direction of the segment ; three components of shear strain which we shall denote by γ with two indices. Physically, the normal strain is defined as the change in length of a small line segment divided by the length of the segment. The shearing strain is visualized as the change in angle between two initially perpendicular line segments. The components of strain determine the strain tensor,

$$[\varepsilon_{ij}] = \begin{bmatrix} \varepsilon_x & \dfrac{\gamma_{xy}}{2} & \dfrac{\gamma_{xz}}{2} \\ \dfrac{\gamma_{xy}}{2} & \varepsilon_y & \dfrac{\gamma_{yz}}{2} \\ \dfrac{\gamma_{xz}}{2} & \dfrac{\gamma_{yz}}{2} & \varepsilon_z \end{bmatrix}$$

which, as in the case of stress tensor, is symmetric, *i.e.* $\varepsilon_{ij} = \varepsilon_{ji} : i \neq j$, and $\gamma_{ij} = 2\,\varepsilon_{ij}\,;\, i \neq j$.

The components of rotation tensor are :

$$[\omega_{ij}] = \begin{bmatrix} 0 & \omega_{xy} & \omega_{xz} \\ -\omega_{xy} & 0 & \omega_{yz} \\ -\omega_{xz} & -\omega_{yz} & 0 \end{bmatrix}$$

which is skew-symmetric.

If u_x, u_y, u_z are the components of the displacement field u, then

$$\omega_{xy} = \frac{1}{2}\left(\frac{\partial u_x}{\partial y} - \frac{\partial u_y}{\partial x}\right)$$

$$\omega_{xz} = \frac{1}{2}\left(\frac{\partial u_x}{\partial z} - \frac{\partial u_z}{\partial x}\right)$$

$$\omega_{yz} = \frac{1}{2}\left(\frac{\partial u_y}{\partial z} - \frac{\partial u_z}{\partial y}\right)$$

so that the components of rigid body rotations are :

$$\left.\begin{aligned} \delta\phi_x &= \omega_{zy} = \frac{1}{2}\left(\frac{\partial u_z}{\partial y} - \frac{\partial u_y}{\partial z}\right) \\ \delta\phi_y &= \omega_{xz} = \frac{1}{2}\left(\frac{\partial u_x}{\partial z} - \frac{\partial u_z}{\partial x}\right) \\ \delta\phi_z &= \omega_{yx} = \frac{1}{2}\left(\frac{\partial u_y}{\partial x} - \frac{\partial u_x}{\partial y}\right) \end{aligned}\right\} \quad \text{...(1.9)}$$

and the components of strain are :

$$\left.\begin{aligned} \varepsilon_x &= \frac{\partial u_x}{\partial x} \\ \varepsilon_y &= \frac{\partial u_y}{\partial y} \\ \varepsilon_z &= \frac{\partial u_z}{\partial z} \\ \frac{\gamma_{xy}}{2} &= \varepsilon_{xy} = \frac{1}{2}\left(\frac{\partial u_x}{\partial y} + \frac{\partial u_y}{\partial x}\right) \\ \frac{\gamma_{yz}}{2} &= \varepsilon_{yz} = \frac{1}{2}\left(\frac{\partial u_y}{\partial z} + \frac{\partial u_z}{\partial y}\right) \\ \frac{\gamma_{xz}}{2} &= \varepsilon_{xz} = \frac{1}{2}\left(\frac{\partial u_x}{\partial z} + \frac{\partial u_z}{\partial x}\right) \end{aligned}\right\} \quad \text{...(1.10)}$$

For the transformation of strain components from one orthogonal co-ordinate system x, y, z to another orthogonal co-ordinate system x', y', z', the same laws of transformation as for the stress components hold good, *i.e.*,

$$\varepsilon'_{ij} = a_{il}\, a_{jm}\, \varepsilon_{lm} \qquad \text{...(1.11)}$$

Therefore, it is necessary to substitute ε in place of σ and $\frac{\gamma}{2}$ in place of τ in Eq. (1.3). Thus, for example

$$\left.\begin{aligned}
\varepsilon_x' &= \varepsilon_x\, a^2_{x'x} + \varepsilon_y\, a^2_{x'y} + \varepsilon_z\, a^2_{x'z} \\
&\quad + \gamma_{xy}\, a_{x'x}\, a_{x'y} + \gamma_{yz}\, a_{x'y}\, a_{x'z} \\
&\quad + \gamma_{xz}\, a_{x'x}\, a_{x'z} \\
\frac{\gamma_{x'y'}}{2} &= \varepsilon_x\, a_{x'x} \,.\, a_{y'x} + \varepsilon_y \,.\, a_{x'y}\, a_{y'y} + \varepsilon_z\, a_{x'z}\, a_{y'z} \\
&\quad + \frac{\gamma_{xy}}{2}\,(a_{x'x}\, a_{y'y} + a_{x'y}\, a_{y'x}) \\
&\quad + \frac{\gamma_{yz}}{2}\,(a_{x'y}\, a_{y'z} + a_{x'z}\, a_{y'y}) \\
&\quad + \frac{\gamma_{xz}}{2}\,(a_{x'x}\, a_{y'z} + a_{x'z}\, a_{y'x})
\end{aligned}\right\} \qquad \text{...(1.12)}$$

and conversely,

$$\left.\begin{aligned}
\varepsilon_x &= \varepsilon_x'\, a^2_{x'x} + \varepsilon_y'\, a^2_{y'x} + \varepsilon_z a^2_{z'x} \\
&\quad + \gamma_{x'y'} + a_{x'x}\, a_{y'x} + \gamma_{y'z'}\, a_{y'z'}\, a_{y'x}\, a_{z'y} \\
&\quad + \gamma_{x'z'}\, a_{x'x}\, a_{z'x} \\
\frac{\gamma_{xy}}{2} &= \varepsilon_x'\, a_{x'x}\, a_{x'y} + \varepsilon_y'\, a_{y'x}\, a_{y'y} \\
&\quad + \varepsilon_z'\, a_{z'x} + a_{z'y} \\
&\quad + \frac{\gamma_{x'y'}}{2}\,(a_{x'x}\, a_{y'y} + a_{x'y}\, a_{y'x}) \\
&\quad + \frac{\gamma_{y'z'}}{2}\,(a_{y'x}\, a_{z'y} + a_{y'y}\, a_{z'x}) \\
&\quad + \frac{\gamma_{x'z'}}{2}\,(a_{x'x}\, a_{z'y} + a_{x'y}\, a_{z'x})
\end{aligned}\right\} \qquad \text{...(1.13)}$$

other components may be written by cyclic permutation of the symbols x, y, z or x', y', z'.

As in the case of stresses, there exist three mutually perpendicular planes on which the shear strains are zero. The normal components in these directions ε_1, ε_2 and ε_3, which include the algebraically largest and smallest of all normal strains at the point, are called principal strains and the directions, principal strain directions. The principal strains are the roots of the cubic equation :

$$\varepsilon^3 - J_1\varepsilon^2 + J_2\varepsilon - J_3 = 0 \qquad ...(1.14)$$

where

$$\left.\begin{aligned} J_1 &= \varepsilon_x + \varepsilon_y + \varepsilon_z \\ J_2 &= \varepsilon_x\,\varepsilon_y + \varepsilon_y\,\varepsilon_z + \varepsilon_x\,\varepsilon_z \\ &\quad - \frac{\gamma_{xy}^2}{4} - \frac{\gamma_{yz}^2}{4} - \frac{\gamma_{xz}^2}{4} \\ J_3 &= \varepsilon_x\varepsilon_y\varepsilon_z - \varepsilon_x\frac{\gamma_{xy}^2}{4} - \varepsilon_y\frac{\gamma_{xz}^2}{4} \\ &\quad - \varepsilon_z\frac{\gamma_{xy}^2}{4} + \frac{1}{4}\gamma_{xy}\,\gamma_{yz}\,\gamma_{xz} \end{aligned}\right\} \qquad ...(1.15)$$

are the three strain invariants.

Eq. (1.14) may be solved by hit and trial, Newton-Raphson method or any other numerical method to determine the principal strains and principal directions. The first invariant of strain J_1 is called dilatation and has the physical significance of representing the change of a unit volume of the material.

1.4 Basic Equations of Elasticity

1.4.1 Cartesian Co-ordinates

(i) Equations of equilibrium (3)

Consideration of the variation of the state of stress from point to point leads to the equilibrium equations given by

$$\left.\begin{aligned} \frac{\partial\sigma_x}{\partial x} + \frac{\partial\tau_{xy}}{\partial y} + \frac{\partial\tau_{xz}}{\partial z} + B_x &= 0 \\ \frac{\partial\tau_{xy}}{\partial x} + \frac{\partial\sigma_y}{\partial y} + \frac{\partial\tau_{yz}}{\partial z} + B_y &= 0 \\ \frac{\partial\tau_{xz}}{\partial x} + \frac{\partial\tau_{yz}}{\partial y} + \frac{\partial\sigma_z}{\partial z} + B_z &= 0 \end{aligned}\right\} \qquad ...(1.16)$$

where B_x, B_y and B_z are the components of the body force (in N/m^3), such as gravitational, centrifugal, or other inertia forces.

(*ii*) *Strain-displacement relations (6)*

$$\left.\begin{aligned}
\varepsilon_x &= \frac{\partial u_x}{\partial x} \\
\varepsilon_y &= \frac{\partial u_y}{\partial y} \\
\varepsilon_z &= \frac{\partial u_z}{\partial z} \\
\gamma_{xy} &= \frac{\partial u_x}{\partial y} + \frac{\partial u_y}{\partial x} \\
\gamma_{yz} &= \frac{\partial u_y}{\partial z} + \frac{\partial u_z}{\partial y} \\
\gamma_{xz} &= \frac{\partial u_x}{\partial z} + \frac{\partial u_z}{\partial x}
\end{aligned}\right\} \quad \text{...(1.17)}$$

(*iii*) *Stress-strain relations (6)*

For an isotropic material, we have only two independent elastic constants λ and G, known as Lame's constants and the generalised Hooke's law gives the following stress-strain relations.

$$\left.\begin{aligned}
\varepsilon_x &= \frac{1}{E}[\sigma_x - \nu(\sigma_y + \sigma_z)] \\
\varepsilon_y &= \frac{1}{E}[\sigma_y - \nu(\sigma_x + \sigma_z)] \\
\varepsilon_z &= \frac{1}{E}[\sigma_z - \nu(\sigma_x + \sigma_y)] \\
\gamma_{xy} &= \frac{\tau_{xy}}{G} \\
\gamma_{yz} &= \frac{\tau_{yz}}{G} \\
\gamma_{xz} &= \frac{\tau_{xz}}{G}
\end{aligned}\right\} \quad \text{...(1.18)}$$

where ν is Poisson's ratio and E is the modulus of elasticity

$$G = \frac{E}{2(1 + \nu)} = \text{Modulus of rigidity}$$

(*iv*) *Strain compatibility equations (6)*

Given a displacement field, a unique strain field can be computed. However, given a strain field, the corresponding displacements are not necessarily unique and continuous, unless certain conditions are satisfied. These conditions, known as the compatibility equations, are expressed as,

$$\frac{\partial^2 \gamma_{xy}}{\partial x \partial y} = \frac{\partial^2 \varepsilon_x}{\partial y^2} + \frac{\partial^2 \varepsilon_y}{\partial x^2}$$

$$\frac{\partial^2 \gamma_{yz}}{\partial y \partial z} = \frac{\partial^2 \varepsilon_y}{\partial z^2} + \frac{\partial^2 \varepsilon_z}{\partial y^2}$$

$$\frac{\partial^2 \gamma_{zx}}{\partial z \partial x} = \frac{\partial^2 \varepsilon_z}{\partial x^2} + \frac{\partial^2 \varepsilon_x}{\partial z^2} \qquad \text{...(1.19)}$$

$$2\frac{\partial^2 \varepsilon_x}{\partial y \partial z} = \frac{\partial}{\partial x}\left(-\frac{\partial \gamma_{yz}}{\partial x} + \frac{\partial \gamma_{zx}}{\partial y} + \frac{\partial \gamma_{xy}}{\partial z}\right)$$

$$2\frac{\partial^2 \varepsilon_y}{\partial z \partial x} = \frac{\partial}{\partial y}\left(-\frac{\partial \gamma_{zx}}{\partial y} + \frac{\partial \gamma_{xy}}{\partial z} + \frac{\partial \gamma_{yz}}{\partial x}\right)$$

$$2\frac{\partial^2 \varepsilon_z}{\partial x \partial y} = \frac{\partial}{\partial z}\left(-\frac{\partial \gamma_{xy}}{\partial z} + \frac{\partial \gamma_{yz}}{\partial x} + \frac{\partial \gamma_{zx}}{\partial y}\right)$$

(*v*) *Stress-displacement relations (6)*

The stress-strain relations using Eqs. (1.17) may be written as,

$$\sigma_x = 2G\frac{\partial u_x}{\partial x} + \lambda\left(\frac{\partial u_x}{\partial x} + \frac{\partial u_y}{\partial y} + \frac{\partial u_z}{\partial z}\right)$$

$$\sigma_y = 2G\frac{\partial u_y}{\partial y} + \lambda\left(\frac{\partial u_x}{\partial x} + \frac{\partial u_y}{\partial y} + \frac{\partial u_z}{\partial z}\right)$$

$$\sigma_z = 2G\frac{\partial u_z}{\partial z} + \lambda\left(\frac{\partial u_x}{\partial x} + \frac{\partial u_y}{\partial y} + \frac{\partial u_z}{\partial z}\right) \qquad \text{...(1.20)}$$

$$\tau_{xy} = G\left(\frac{\partial u_x}{\partial y} + \frac{\partial u_y}{\partial x}\right)$$

$$\tau_{yz} = G\left(\frac{\partial u_y}{\partial z} + \frac{\partial u_z}{\partial y}\right)$$

$$\tau_{zx} = G\left(\frac{\partial u_z}{\partial x} + \frac{\partial u_x}{\partial z}\right)$$

where $\lambda = \dfrac{\nu E}{(1-\nu)(1-2\nu)}$ = Lame's constant

(*vi*) *Equilibrium equations in terms of displacements (3)*

Using the stress-displacement relations Equations (1.20), the equilibrium Equations (1.16) can be expressed in terms of displacements. These equations are known as Navier equations and are given by,

$$\nabla^2 u_x + \left(\frac{3}{1-2\nu}\right)\frac{\partial \varepsilon_m}{\partial x} + \frac{B_x}{G} = 0$$

$$\nabla^2 u_y + \left(\frac{3}{1-2\nu}\right)\frac{\partial \varepsilon_m}{\partial y} + \frac{B_y}{G} = 0 \qquad \text{...(1.21)}$$

$$\nabla^2 u_z + \left(\frac{3}{1-2\nu}\right)\frac{\partial \varepsilon_m}{\partial z} + \frac{B_z}{G} = 0$$

where $\nabla^2 = \frac{\partial^2}{\partial x^2} + \frac{\partial^2}{\partial y^2} + \frac{\partial^2}{\partial z^2}$

$$\varepsilon_m = \frac{1}{3}\left(\frac{\partial u_x}{\partial x} + \frac{\partial u_y}{\partial y} + \frac{\partial u_z}{\partial z}\right).$$

(vii) Stress-compatibility equations (6)

Using Eqs. (1.18) and (1.16), the compatibility equations (1.19) may be expressed in terms of stress components. These equations are known as Beltrami-Michell equations and are given by :

$$\nabla^2 \sigma_x + \left(\frac{3}{1+\nu}\right)\frac{\partial^2 \sigma_m}{\partial x^2} = -\left(\frac{\nu}{1-\nu}\right)\nabla.\overline{B} - 2\frac{\partial B_x}{\partial x}$$

$$\nabla^2 \sigma_y + \left(\frac{3}{1+\nu}\right)\frac{\partial^2 \sigma_m}{\partial y^2} = -\left(\frac{\nu}{1-\nu}\right)\nabla.\overline{B} - 2\frac{\partial B_y}{\partial y}$$

$$\nabla^2 \sigma_z + \left(\frac{3}{1+\nu}\right)\frac{\partial^2 \sigma_m}{\partial z^2} = -\left(\frac{\nu}{1-\nu}\right)\nabla.\overline{B} - 2\frac{\partial B_z}{\partial z} \qquad \text{...(1.22)}$$

$$\nabla^2 \tau_{xy} + \left(\frac{3}{1+\nu}\right)\frac{\partial^2 \sigma_m}{\partial x \partial y} = -\left(\frac{\partial B_x}{\partial y} + \frac{\partial B_y}{\partial x}\right)$$

$$\nabla^2 \tau_{yz} + \left(\frac{3}{1+\nu}\right)\frac{\partial^2 \sigma_m}{\partial y \partial z} = -\left(\frac{\partial B_y}{\partial z} + \frac{\partial B_z}{\partial y}\right)$$

$$\nabla^2 \tau_{zx} + \left(\frac{3}{1+\nu}\right)\frac{\partial^2 \sigma_m}{\partial z \partial x} = -\left(\frac{\partial B_z}{\partial x} + \frac{\partial B_x}{\partial z}\right)$$

where $\nabla = \frac{\partial}{\partial x} + \frac{\partial}{\partial y} + \frac{\partial}{\partial z}$

$$\sigma_m = \frac{1}{3}(\sigma_x + \sigma_y + \sigma_z)$$

$$\overline{B} = B_x\hat{i} + B_y\hat{j} + B_z\hat{k}.$$

(viii) Boundary conditions in terms of stresses (3)

When the stresses are prescribed over the entire boundary, then the boundary conditions can be expressed as :

$$\begin{aligned} X_n &= \sigma_x \cdot a_{nx} + \tau_{xy} \cdot a_{ny} + \tau_{xz}\, a_{nz} \\ Y_n &= \tau_{xy} \cdot a_{nx} + \sigma_y \cdot a_{ny} + \tau_{yz} \cdot a_{nz} \\ Z_n &= \tau_{zx} \cdot a_{nz} + \tau_{zy}\, a_{ny} + \sigma_z \cdot a_{nz} \end{aligned} \qquad \text{...(1.23)}$$

(ix) Boundary conditions in terms of displacements (3)

Using Eqs. (1.20), the Eqs. (1.23) can be expressed in terms of displacements as given below :

$$\begin{aligned} X_n &= 3\lambda\varepsilon_m \cdot a_{nx} + G\left[2\frac{\partial u_x}{\partial x} a_{nx} + \left(\frac{\partial u_x}{\partial y} + \frac{\partial u_y}{\partial x} \right) a_{ny} + \left(\frac{\partial u_x}{\partial z} + \frac{\partial u_z}{\partial x} \right) a_{nz} \right] \\ Y_n &= 3\lambda\varepsilon_m \cdot a_{ny} + G\left[2\frac{\partial u_y}{\partial y} \cdot a_{ny} + \left(\frac{\partial u_x}{\partial y} + \frac{\partial u_y}{\partial x} \right) a_{nx} + \left(\frac{\partial u_y}{\partial z} + \frac{\partial u_z}{\partial y} \right) a_{nz} \right] \\ Z_n &= 3\lambda\varepsilon_m \cdot a_{nz} + G\left[2\frac{\partial u_z}{\partial z} a_{nz} + \left(\frac{\partial u_z}{\partial x} + \frac{\partial u_x}{\partial z} \right) a_{nx} + \left(\frac{\partial u_y}{\partial z} + \frac{\partial u_z}{\partial y} \right) a_{ny} \right] \end{aligned} \qquad \text{...(1.24)}$$

where $\varepsilon_m = \frac{1}{3}\left(\frac{\partial u_x}{\partial x} + \frac{\partial u_y}{\partial y} + \frac{\partial u_z}{\partial z} \right)$.

1.4.2 Cylindrical Co-ordinates

(i) Equations of equilibrium (3) (Fig. 1.2)

$$\begin{aligned} &\frac{\partial \sigma_r}{\partial r} + \frac{1}{r}\frac{\partial \tau_{r\theta}}{\partial \theta} + \frac{\partial \tau_{rz}}{\partial z} + \frac{\sigma_r - \sigma_\theta}{r} + B_r = 0 \\ &\frac{\partial \tau_{r\theta}}{\partial r} + \frac{1}{r}\frac{\partial \sigma_\theta}{\partial \theta} + \frac{\partial \tau_{\theta z}}{\partial z} + \frac{2\tau_{r\theta}}{r} + B_\theta = 0 \\ &\frac{\partial \tau_{rz}}{\partial r} + \frac{1}{r}\frac{\partial \tau_{\theta z}}{\partial \theta} + \frac{\partial \sigma_z}{\partial z} + \frac{\tau_{rz}}{r} + B_z = 0 \end{aligned} \qquad \text{...(1.25)}$$

(ii) Strain-displacement relations (6)

$$\varepsilon_r = \frac{\partial u_r}{\partial r}$$

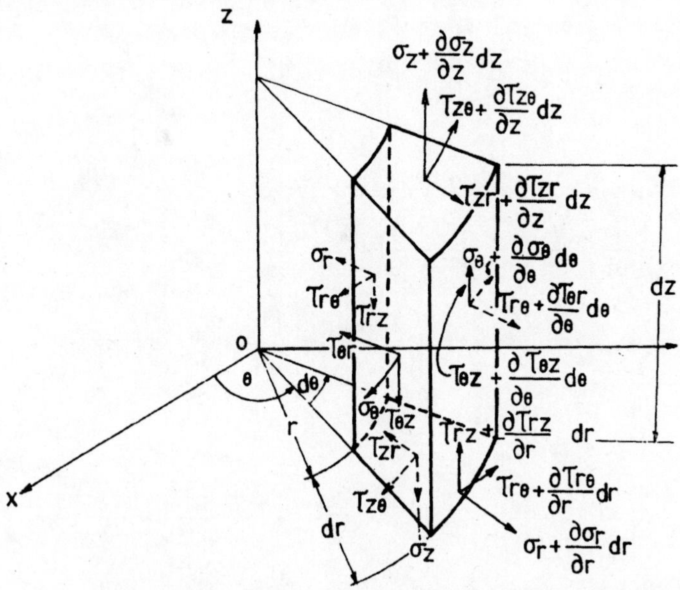

Fig. 1.2 Stresses on a cylindrical parallelopiped.

$$\varepsilon_\theta = \frac{1}{r}\frac{\partial u_\theta}{\partial \theta} + \frac{u_r}{r}$$

$$\varepsilon_z = \frac{\partial u_z}{\partial z}$$

$$\gamma_{r\theta} = \frac{1}{r}\frac{\partial u_r}{\partial \theta} + \frac{\partial u_\theta}{\partial r} - \frac{u_\theta}{r} \qquad \text{...(1.26)}$$

$$\gamma_{\theta z} = \frac{\partial u_\theta}{\partial z} + \frac{1}{r}\frac{\partial u_z}{\partial \theta}$$

$$\gamma_{zr} = \frac{\partial u_z}{\partial r} + \frac{\partial u_r}{\partial z}$$

(iii) Stress-strain relations (6)

$$\varepsilon_r = \frac{1}{E}[\sigma_r - \nu(\sigma_\theta + \sigma_z)]$$

$$\varepsilon_\theta = \frac{1}{E}[\sigma_\theta - \nu(\sigma_r + \sigma_z)]$$

$$\varepsilon_z = \frac{1}{E}[\sigma_z - \nu(\sigma_\theta + \sigma_r)]$$

$$\gamma_{r\theta} = \frac{\tau_{r\theta}}{G} \qquad \text{...(1.27)}$$

$$\gamma_{\theta z} = \frac{\tau_{\theta z}}{G}$$

$$\gamma_{zr} = \frac{\tau_{zr}}{G}$$

(iv) Strain-compatibility equations (6)

$$\frac{1}{r}\frac{\partial}{\partial z}\left(\frac{\partial \gamma_{\theta z}}{\partial \theta} + \gamma_{zr}\right) = \frac{1}{r}\frac{\partial \varepsilon_z}{\partial r} + \frac{1}{r^2}\frac{\partial^2 \varepsilon_z}{\partial \theta^2} + \frac{d^2 \varepsilon_\theta}{\partial z^2}$$

$$\frac{\partial^2 \gamma_{zr}}{2r\partial\theta\partial z} + \frac{r}{2}\frac{\partial}{\partial r}\left(\frac{1}{r}\frac{\partial \gamma_{z\theta}}{\partial z^2}\right) - \frac{\partial^2 \gamma_{r\theta}}{2\partial z^2} = \frac{\partial}{\partial r}\left(\frac{\partial \varepsilon_z}{r\partial\theta}\right)$$

$$\frac{\partial^2 \gamma_{zr}}{\partial r \partial z} = \frac{\partial^2 \varepsilon_z}{\partial r^2} + \frac{\partial^2 \varepsilon_r}{\partial z^2} \qquad ...(1.28)$$

$$\frac{1}{r^2}\frac{\partial}{\partial r}\left(r\frac{\partial \gamma_{r\theta}}{\partial \theta}\right) = \frac{1}{r^2}\frac{\partial}{\partial r}\left(r^2\frac{\partial \varepsilon_\theta}{\partial r}\right) + \left(\frac{1}{r^2}\frac{\partial^2}{\partial \theta^2} - \frac{1}{r^2}\frac{\partial}{\partial r}\right)\varepsilon_r$$

$$\frac{1}{2}\cdot\frac{1}{r^2}\frac{\partial}{\partial r}\left(r\frac{\partial \gamma_{z\theta}}{\partial \theta}\right) + \frac{1}{r}\frac{\partial^2 \gamma_{\theta z}}{\partial\theta\partial z} - \frac{\partial^2 \gamma_{zr}}{2r^2\partial\theta^2} = \frac{1}{r}\frac{\partial}{\partial r}\left(r\frac{\partial \varepsilon_\theta}{\partial z}\right) - \frac{1}{r}\frac{\partial \varepsilon_r}{\partial z}$$

$$-\frac{1}{2}\frac{\partial}{\partial r}\left(\frac{1}{r}\frac{\partial}{\partial r} r\gamma_{\theta z}\right) + \frac{1}{2r^2}\frac{\partial}{\partial r}\left(r^2\frac{\partial \gamma_{r\theta}}{\partial \theta}\right) + \frac{\partial}{\partial r}\left(\frac{1}{\partial r}\frac{\partial \gamma_{zr}}{\partial \theta}\right)$$

$$= \frac{\partial^2 \varepsilon_r}{r\partial\theta\partial z}$$

(v) Stress-displacement relations (6)

$$\sigma_r = 2G\frac{\partial u_r}{\partial r} + \lambda\left(\frac{\partial u_r}{\partial r} + \frac{1}{r}\frac{\partial u_\theta}{\partial \theta} + \frac{u_r}{r} + \frac{\partial u_z}{\partial z}\right)$$

$$\sigma_\theta = 2G\left(\frac{1}{r}\frac{\partial u_\theta}{\partial \theta} + \frac{u_r}{r}\right) + \lambda\left(\frac{\partial u_r}{\partial r} + \frac{1}{r}\frac{\partial u_\theta}{\partial \theta} + \frac{u_r}{r} + \frac{\partial u_z}{\partial_z}\right)$$

$$\sigma_z = 2G\frac{\partial u_z}{\partial z} + \lambda\left(\frac{\partial u_r}{\partial r} + \frac{1}{r}\frac{\partial u_\theta}{\partial \theta} + \frac{u_r}{r} + \frac{\partial u_z}{\partial z}\right)$$

$$\tau_{r\theta} = G\left(\frac{1}{r}\frac{\partial u_r}{\partial \theta} + \frac{\partial u_\theta}{\partial r} - \frac{u_\theta}{r}\right) \qquad ...(1.29)$$

$$\tau_{\theta z} = G\left(\frac{\partial u_\theta}{\partial z} + \frac{1}{r}\frac{\partial u_z}{\partial \theta}\right)$$

$$\tau_{zr} = G\left(\frac{\partial u_z}{\partial r} + \frac{\partial u_r}{\partial z}\right)$$

(vi) Equilibrium equations in terms of displacements (3)

$$\nabla^2 u_r + \left(\frac{3}{1-2\nu}\right)\frac{\partial \varepsilon_m}{\partial r} - \frac{u_r}{r^2} - \frac{2}{r^2}\frac{\partial u_\theta}{\partial \theta} + \frac{B_r}{G} = 0$$

$$\nabla^2 u_\theta \left(\frac{3}{1-2\nu}\right)\frac{\partial \varepsilon_m}{r\partial \theta} - \frac{u_\theta}{r^2} + \frac{2}{r^2}\frac{\partial u_r}{\partial \theta} + \frac{B_\theta}{G} = 0 \quad \text{...(1.30)}$$

$$\nabla^2 u_z + \left(\frac{3}{1-2\nu}\right)\frac{\partial \varepsilon_m}{\partial z} + \frac{B_z}{G} = 0$$

where $\varepsilon_m = \frac{1}{3}\left(\frac{\partial u_r}{\partial r} + \frac{1}{r}\frac{\partial u_\theta}{\partial \theta} + \frac{u_r}{r} + \frac{\partial u_z}{\partial z}\right)$

$$\nabla^2 = \frac{\partial^2}{\partial r^2} + \frac{1}{r}\frac{\partial}{\partial r} + \frac{1}{r^2}\frac{\partial^2}{\partial \theta^2} + \frac{\partial^2}{\partial z^2}.$$

(vii) Stress-compatibility equations (6)

In the absence of body forces, we have

$$\nabla^2 \sigma_r + \frac{2(\sigma_\theta - \sigma_r)}{r^2} - \frac{4}{r^2}\frac{\partial \tau_{r\theta}}{\partial \theta} + \left(\frac{3}{1+\nu}\right)\frac{\partial^2 \sigma_m}{\partial r^2} = 0$$

$$\nabla^2 \sigma_\theta - \frac{2(\sigma_\theta - \sigma_r)}{r^2} + \frac{4}{r^2}\frac{\partial \tau_{r\theta}}{\partial \theta} + \left(\frac{3}{1+\nu}\right)\left(\frac{1}{r}\frac{\partial \sigma_m}{\partial r} + \frac{\partial^2 \sigma_m}{r^2 \partial \theta^2}\right) = 0$$

$$\nabla^2 \sigma_z + \left(\frac{3}{1+\nu}\right)\frac{\partial^2 \sigma_m}{\partial z^2} = 0 \quad \text{...(1.31)}$$

$$\nabla^2 \tau_{r\theta} - \frac{2}{r^2}\frac{\partial}{\partial \theta}(\sigma_\theta - \sigma_r) - \frac{4}{r^2}\tau_{r\theta} + \left(\frac{3}{1+\nu}\right)\frac{\partial}{\partial r}\left(\frac{1}{r}\frac{\partial \sigma_m}{\partial \theta}\right) = 0$$

$$\nabla^2 \tau_{\theta z} - \frac{\tau_{\theta z}}{r^2} + \frac{2}{r^2}\frac{\partial \tau_{rz}}{\partial \theta} + \left(\frac{3}{1+\nu}\right)\frac{\partial^2 \sigma_m}{r\partial \theta \partial z} = 0$$

$$\nabla^2 \tau_{zr} - \frac{\tau_{zr}}{r^2} - \frac{2}{r^2}\frac{\partial \tau_{\theta z}}{\partial \theta} + \left(\frac{3}{1+\nu}\right)\frac{\partial^2 \sigma_m}{\partial r \partial z} = 0$$

where $\sigma_m = \frac{1}{3}(\sigma_r + \sigma_\theta + \sigma_z)$

(viii) Boundary conditions in terms of stresses (3)

$$r_n = \sigma_r \,.\, a_{nr} + \tau_{r\theta}\, a_{n\theta} + \tau_{rz}\, a_{nz}$$

$$\theta_n = \tau_{r\theta} \cdot a_{nr} + \sigma_\theta \cdot a_{n\theta} + \tau_{\theta z}\, a_{nz} \quad \text{...(1.32)}$$

$$z_n = \tau_{zr}\, a_{nr} + \tau_{\theta z}\, a_{n\theta} + \sigma_z\, a_{nz}$$

(ix) Boundary conditions in terms of displacements (3)

$$r_n = 3\lambda\varepsilon_m \,.\, a_{nr} + G\,[2\,\varepsilon_r\, a_{nr} + \gamma_{r\theta}\, a_{n\theta} + \gamma_{rz}\, a_{nz}]$$

$$\theta_n = 3\lambda\varepsilon_m \cdot a_{n\theta} + G\,[2\varepsilon_\theta \cdot a_{n\theta} + \gamma_{r\theta}\, a_{nr} + \gamma_{\theta z}\, a_{n\theta}] \quad \text{...(1.33)}$$

$$z_n = 3\lambda\varepsilon_m . a_{nz} + G\,[2\varepsilon_z . a_{nz} + \gamma_{zr}\,a_{nr} + \gamma_{\theta z}\,a_{ny}]$$

For axis symmetric bodies, if z is the axis of revolution, then all stresses and strains are independent of θ.

1.5 Methods of Solution of Elasticity Problems

Given the geometry of the body, the manner in which it is loaded, *i.e.* specified boundary conditions, the distribution of body forces, and the elastic constants, it is required to determine stresses, strains and displacements as a function of location. Depending on the form of the boundary conditions, problems, can be classified in three groups :

1. *First Boundary*-Value Problem : where normal and shear stresses are prescribed on the whole boundary of the body.

2. *Second Boundary*-Value Problem : where only displacements are prescribed on the boundary.

3. *Mixed Boundary*-Value Problem : where stresses are prescribed on some parts of the boundary and displacements on other parts.

In the most general case, it is required to solve 15 equations : 3 equilibrium equations (1.16) or (1.25), 6 strain-displacement relations Eqs. (1.17) or (1.26), 6 stress-strain relations Eqs. (1.18) or (1.27)] for 15 unknowns [6 stress components, 6 strains components and 3 displacement components]. With the requirement that compatibility equations (1.19) or (1.28) and boundary conditions (1.23) or (1.32) be satisfied. The boundary conditions serve to solve for the integration constants that arise in the solution of the system of differential equations. Such a formulation of elasticity problems is known as *stress formulation.* Another approach suitable for the second boundary value problem is to solve the three equations of equilibrium in terms of displacements. Eqs. (1.21) or (1.30) for the three displacement components subject to the displacement boundary conditions, Eqs. (1.24) or (1.33). This formulation of elasticity problems is called *displacement formulation.* However, these direct methods of solving elasticity problems are too complicated to yield practical results. In order to simplify the solution methods, inverse and semi-inverse methods can be used. In the inverse method, a judicious guess of the solution is made and is checked against the basic equations and boundary conditions. If all conditions are satisfied the solution is the correct one ; otherwise, corrections are made and a revised solution is checked in the same manner. In the semi-inverse method, some of the unknowns are guessed and a simplified group of equations is solved for the remaining unknowns.

These educated guesses are based on experience, intuition, solutions of similar problems, or experiments.

1.6 Plane Stress

If a thin plate is loaded by forces applied at the boundary, parallel to the plane of the plate and distributed uniformly over the thickness (Fig. 1.3), the stress components σ_z, τ_{xz}, τ_{yz} are zero on both faces of the plate, and if also assumed to be zero within the plate, then the state of stress is called plane stress.

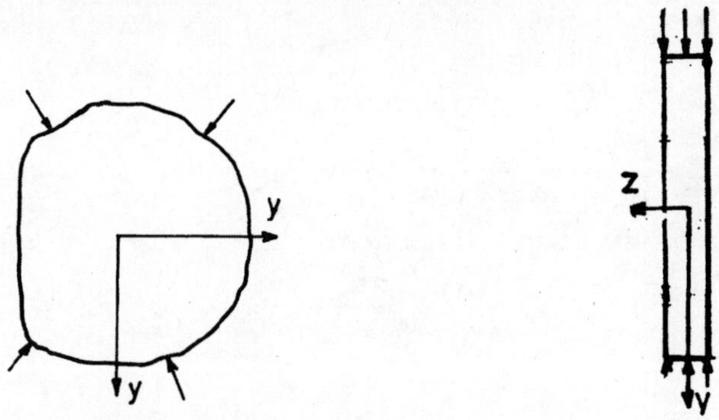

Fig. 1.3 State of plane stress.

For plane stress case, $\sigma_x, \sigma_y, \tau_{xy} = f(x, y)$ ...(1.34)

and governing equations become.

(i) Equilibrium equations

$$\left.\begin{aligned} \frac{\partial \sigma_x}{\partial x} + \frac{\partial \tau_{xy}}{\partial y} + B_x &= 0 \\ \frac{\partial \tau_{xy}}{\partial x} + \frac{\partial \sigma_y}{\partial y} + B_y &= 0 \\ B_z &= 0 \end{aligned}\right\} \quad \text{...(1.35)}$$

(ii) Strain-displacement relations

$$\left.\begin{aligned} \varepsilon_x &= \frac{\partial u_x}{\partial x} \\ \varepsilon_y &= \frac{\partial u_y}{\partial y} \\ \varepsilon_z &= \frac{\partial u_z}{\partial z} \\ \gamma_{xy} &= \frac{\partial u_x}{\partial y} + \frac{\partial u_y}{\partial x} \end{aligned}\right\} \quad \text{...(1.36)}$$

(iii) Stress-strain relations

$$\left.\begin{aligned}
\varepsilon_x &= \frac{1}{E}[\sigma_x - \nu\sigma_y] \\
\varepsilon_y &= \frac{1}{E}[\sigma_y - \nu\sigma_x] \\
\varepsilon_z &= -\frac{\nu}{E}[\sigma_x + \sigma_y \\
\gamma_{xy} &= \frac{\tau_{xy}}{G} \\
\gamma_{yz} &= \gamma_{zx} = 0
\end{aligned}\right\} \quad \ldots(1.37)$$

(iv) Strain-compatibility equations

$$\frac{\partial^2 \varepsilon_x}{\partial y^2} + \frac{\partial^2 \varepsilon_y}{\partial x^2} = \frac{\partial^2 \gamma_{xy}}{\partial x \partial y} \quad \ldots(1.38)$$

(v) Stress-displacement relations

$$\left.\begin{aligned}
\sigma_x &= 2G\frac{\partial u_x}{\partial x} + \lambda\left(\frac{\partial u_x}{\partial x} + \frac{\partial u_y}{\partial y}\right) \\
\sigma_y &= 2G\frac{\partial u_y}{\partial y} + \lambda\left(\frac{\partial u_y}{\partial y} + \frac{\partial u_x}{\partial x}\right) \\
\tau_{xy} &= G\left(\frac{\partial u_x}{\partial y} + \frac{\partial u_y}{\partial x}\right)
\end{aligned}\right\} \quad \ldots(1.39)$$

(vi) Navier equations

$$\left.\begin{aligned}
\left(\frac{\partial^2}{\partial x^2} + \frac{\partial^2}{\partial y^2}\right)u_x + \left(\frac{1}{1-2\nu}\right)\frac{\partial}{\partial x}(\varepsilon_x + \varepsilon_y) + \frac{B_x}{G} = 0 \\
\left(\frac{\partial^2}{\partial x^2} + \frac{\partial^2}{\partial y^2}\right)u_y + \left(\frac{1}{1-2\nu}\right)\frac{\partial}{\partial y}(\varepsilon_x + \varepsilon_y) + \frac{B_y}{G} = 0
\end{aligned}\right\} \quad \ldots(1.40)$$

(vii) Stress-compatibility Equations

$$\left(\frac{\partial^2}{\partial x^2} + \frac{\partial^2}{\partial y^2}\right)(\sigma_x + \sigma_y) = -(1+\nu)\left(\frac{\partial B_x}{\partial x} + \frac{\partial B_y}{\partial y}\right)$$

or

$$\nabla^2(\sigma_x + \sigma_y) = -(1+\nu)\left(\frac{\partial B_x}{\partial x} + \frac{\partial B_y}{\partial y}\right) \quad \ldots(1.41)$$

(viii) Stress-boundary conditions

$$\left.\begin{aligned}
X_n &= \sigma_x \,.\, a_{nx} + \tau_{xy}\, a_{ny} \\
Y_n &= \tau_{xy} \,.\, a_{nx} + \sigma_y \,.\, a_{ny}
\end{aligned}\right\} \quad \ldots(1.42)$$

(ix) Boundary conditions in terms of displacements

$$\left.\begin{aligned} X_n &= \lambda(\varepsilon_x + \varepsilon_y)\, a_{nx} \\ &\quad + G\left[2\frac{\partial u_x}{\partial x}\,.\,a_{nx} + \left(\frac{\partial u_y}{\partial x} + \frac{\partial u_x}{\partial y}\right)a_{ny}\right] \\ Y_n &= \lambda(\varepsilon_x + \varepsilon_y)\, a_{ny} \\ &\quad + G\left[2\frac{\partial u_y}{\partial y}\,.\,a_{ny} + \left(\frac{\partial u_y}{\partial x} + \frac{\partial u_x}{\partial y}\right)a_{nx}\right] \end{aligned}\right\} \quad ...(1.43)$$

(x) Biharmonic equation

Let there exist a body force potential V such that,

$$B_x = \frac{\partial V}{\partial x}$$

$$B_y = \frac{\partial V}{\partial y}$$

and define

$$\left.\begin{aligned} \sigma_x &= \frac{\partial^2 \phi}{\partial y^2} - V \\ \sigma_y &= \frac{\partial^2 \phi}{\partial x^2} - V \\ \tau_{xy} &= -\frac{\partial^2 \phi}{\partial x \partial y} \end{aligned}\right\} \quad ...(1.44)$$

where ϕ = Airy's stress function.

Substituting Eq. (1.44) in Eq. (1.35), we get

$$\frac{\partial}{\partial x}\left[\frac{\partial^2 \phi}{\partial y^2} - V\right] + \frac{\partial}{\partial y}\left[-\frac{\partial^2 \phi}{\partial x \partial y}\right] + B_x = 0$$

$$\frac{\partial^3 \phi}{\partial x \partial y^2} - \frac{\partial V}{\partial x} - \frac{\partial^3 \phi}{\partial x \partial y^2} + B_x = 0$$

and

$$\frac{\partial}{\partial x}\left[-\frac{\partial^2 \phi}{\partial x \partial y}\right] + \frac{\partial}{\partial y}\left[\frac{\partial^2 \phi}{\partial x^2} - V\right] + B_y = 0$$

$$-\frac{\partial^3 \phi}{\partial x^2 \partial y} + \frac{\partial^3 \phi}{\partial y \partial x^2} - \frac{\partial V}{\partial y} + B_y = 0.$$

Hence, we see that the stress components given by Eq. (1.44) satisfy the equilibrium equations.

Now substituting Eq. (1.44) in Eq. (1.41), we get

$$\nabla^2\left[\frac{\partial^2 \phi}{\partial x^2} + \frac{\partial^2 \phi}{\partial y^2} - 2V\right] = -(1+\nu)\left(\frac{\partial B_x}{\partial x} + \frac{\partial B_y}{\partial y}\right)$$

$$\nabla^2.\nabla^2\phi - 2\nabla^2 V = -(1+\nu)\left(\frac{\partial B_x}{\partial x} + \frac{\partial B_y}{\partial y}\right)$$

$$\nabla^4\phi = 2\,\nabla^2 V - (1+\nu)\left(\frac{\partial B_x}{\partial x} + \frac{\partial B_y}{\partial y}\right)$$

$$= 2\left(\frac{\partial^2 V}{\partial x^2} + \frac{\partial^2 V}{\partial y^2}\right) - (1+\nu)\left(\frac{\partial B_x}{\partial x} + \frac{\partial B_y}{\partial y}\right)$$

$$= 2\left(\frac{\partial B_x}{\partial x} + \frac{\partial B_y}{\partial y}\right) - (1+\nu)\left(\frac{\partial B_x}{\partial x} + \frac{\partial B_y}{\partial y}\right)$$

$$\nabla^4\phi = (1-\nu)\left(\frac{\partial B_x}{\partial x} + \frac{\partial B_y}{\partial y}\right) \qquad ...(1.45)$$

Eq. (1.45) is known as the *Biharmonic equation* for plane stress problems.

where $$\nabla^4 = \frac{\partial^4}{\partial x^4} + 2\frac{\partial^4}{\partial x^2 \partial y^2} + \frac{\partial^4}{\partial y^4}$$

If body forces are zero, then Eq. (1.45) becomes

$$\nabla^4\phi = 0 \qquad ...(1.46)$$

1.7 Plane Strain

When the length of the member in the z-direction is either very large so that no displacement is possible or the movements along the z-axis are otherwise prevented so that $\varepsilon_z = \gamma_{yz} = \gamma_{zx} = 0$, then the state of strain is said to be plane strain. A retaining wall with internal pressure, a culvert or tunnel and a cylindrical tube with internal pressure are some of the important problems of this kind.

The governing equations for plane strain case are :

(*i*) *Equilibrium equations*

$$\left.\begin{aligned} \frac{\partial \sigma_x}{\partial x} + \frac{\partial \tau_{xy}}{\partial y} + B_x &= 0 \\ \frac{\partial \tau_{xy}}{\partial x} + \frac{\partial \sigma_y}{\partial y} + B_y &= 0 \\ \frac{\partial \sigma_z}{\partial z} + B_z &= 0 \end{aligned}\right\} \qquad ...(1.47)$$

(*ii*) *Strain-displacement relations*

$$\left.\begin{aligned} \varepsilon_x &= \frac{\partial u_x}{\partial x} \\ \varepsilon_y &= \frac{\partial u_y}{\partial y} \\ \gamma_{xy} &= \frac{\partial u_x}{\partial y} + \frac{\partial u_y}{\partial x} \\ \varepsilon_z &= \gamma_{yz} = \gamma_{zx} = 0 \end{aligned}\right\} \quad \text{...(1.48)}$$

(*iii*) *Stress-strain relations*

$$\left.\begin{aligned} \varepsilon_z &= \frac{1}{E}\,[\sigma_z - \nu(\sigma_x + \sigma_y)] = 0 \\ \therefore \quad \sigma_z &= \nu(\sigma_x + \sigma_y) \\ \text{Hence} \quad \varepsilon_x &= \frac{1}{E}\,[\sigma_x - \nu(\sigma_y + \sigma_z)] \\ \varepsilon_x &= \left(\frac{1+\nu}{E}\right)[\sigma_x(1-\nu) - \upsilon\sigma_y] \\ \varepsilon_y &= \left(\frac{1+\nu}{E}\right)[\sigma_y(1-\nu) - \upsilon\sigma_x] \\ \gamma_{xy} &= \frac{\tau_{xy}}{G} \end{aligned}\right\} \quad \text{...(1.49)}$$

(*iv*) *Strain compatibility equation*

$$\frac{\partial^2 \varepsilon_x}{\partial y^2} + \frac{\partial^2 \varepsilon_y}{\partial x^2} = \frac{\partial^2 \gamma_{xy}}{\partial x\,\partial y} \quad \text{...(1.50)}$$

(*v*) *Stress displacement relations*

$$\left.\begin{aligned} \sigma_x &= 2G\frac{\partial u_x}{\partial x} + \lambda\left(\frac{\partial u_x}{\partial x} + \frac{\partial u_y}{\partial y}\right) \\ \sigma_y &= 2G\frac{\partial u_y}{\partial y} + \lambda\left(\frac{\partial u_y}{\partial y} + \frac{\partial u_x}{\partial x}\right) \\ \sigma_z &= 2\upsilon(\lambda + G)\left(\frac{\partial u_x}{\partial x} + \frac{\partial u_y}{\partial y}\right) \\ \tau_{xy} &= G\left(\frac{\partial u_x}{\partial y} + \frac{\partial u_y}{\partial x}\right) \end{aligned}\right\} \quad \text{...(1.51)}$$

(*vi*) *Navier equations*

$$\left(\frac{\partial^2}{\partial x^2} + \frac{\partial^2}{\partial y^2}\right)u_x + \left(\frac{1}{1-2\upsilon}\right)\frac{\partial}{\partial x}\left(\frac{\partial u_x}{\partial x} + \frac{\partial u_y}{\partial y}\right) + \frac{B_x}{G} = 0 \quad \text{...(1.52 a)}$$

$$\left(\frac{\partial^2}{\partial x^2}+\frac{\partial^2}{\partial y^2}\right)u_y+\left(\frac{1}{1-2\upsilon}\right)\frac{\partial}{\partial x}\left(\frac{\partial u_y}{\partial y}+\frac{\partial u_x}{\partial x}\right)+\frac{B_y}{G}=0 \quad \text{...(1.52 } b\text{)}$$

(*vii*) *Stress compatibility equation*

$$\nabla^2(\sigma_x+\sigma_y)=-\left(\frac{1}{1-\nu}\right)\left(\frac{\partial B_x}{\partial x}+\frac{\partial B_y}{\partial y}\right) \quad \text{...(1.53)}$$

(*viii*) *Biharmonic equation*

Let $\sigma_x=\dfrac{\partial^2\phi}{\partial y^2}-V,\quad \sigma_y=\dfrac{\partial^2\phi}{\partial x^2}-V$ and $\tau_{xy}=\dfrac{-\partial^2\phi}{\partial x\,\partial y}$.

Substituting in Eq. (1.53), we get

$$\nabla^2\left(\frac{\partial^2\phi}{\partial x^2}+\frac{\partial^2\phi}{\partial y^2}-2V\right)=-\left(\frac{1}{1-\nu}\right)\left(\frac{\partial B_z}{\partial x}+\frac{\partial B_y}{\partial y}\right)$$

$$\nabla^4\phi=2\nabla^2V-\left(\frac{1}{1-\nu}\right)\left(\frac{\partial B_x}{\partial x}+\frac{\partial B_y}{\partial y}\right)$$

$$=2\left(\frac{\partial B_x}{\partial x}+\frac{\partial B_y}{\partial y}\right)-\left(\frac{1}{1-\nu}\right)\left(\frac{\partial B_x}{\partial x}+\frac{\partial B_y}{\partial y}\right)$$

$$\nabla^4\phi=\left(\frac{1-2\nu}{1-\nu}\right)\left(\frac{\partial B_x}{\partial x}+\frac{\partial B_y}{\partial y}\right) \quad \text{...(1.54)}$$

Eq. (1.54) is the *Biharmonic equation* for plane strain case. When body forces are zero, then

$$\nabla^4\phi=0 \quad \text{...(1.55)}$$

1.8 Spherical Coordinates

(*i*) *Equilibrium equations*

With reference to Fig. 1.4, the equilibrium equations can be written as :

$$\frac{\partial\sigma_r}{\partial r}+\frac{1}{r}\frac{\partial\tau_{r\theta}}{\partial\theta}+\frac{1}{r\sin\theta}\frac{\partial\tau_{r\phi}}{\partial\phi}+\frac{1}{r}[2\sigma_r-\sigma_\theta-\sigma_\phi+\tau_{r\theta}\cot\theta]+B_r=0$$

$$\frac{\partial\tau_{r\theta}}{\partial r}+\frac{1}{r}\frac{\partial\sigma_\theta}{\partial\theta}+\frac{1}{r\sin\theta}\frac{\partial\tau_{\theta\phi}}{\partial\phi}+\frac{1}{r}[(\sigma_\theta-\sigma_\phi)\cot\theta+3\tau_{r\theta}]+B_\theta=0$$

$$\frac{\partial\tau_{r\phi}}{\partial r}+\frac{1}{r}\frac{\partial\tau_{\theta\phi}}{\partial\theta}+\frac{1}{r\sin\theta}\frac{\partial\sigma_\phi}{\partial_\phi}+\frac{1}{r}[3\tau_{r\phi}+2\tau_{\theta\phi}\cot\theta]+B_\phi=0 \quad \text{...(1.56)}$$

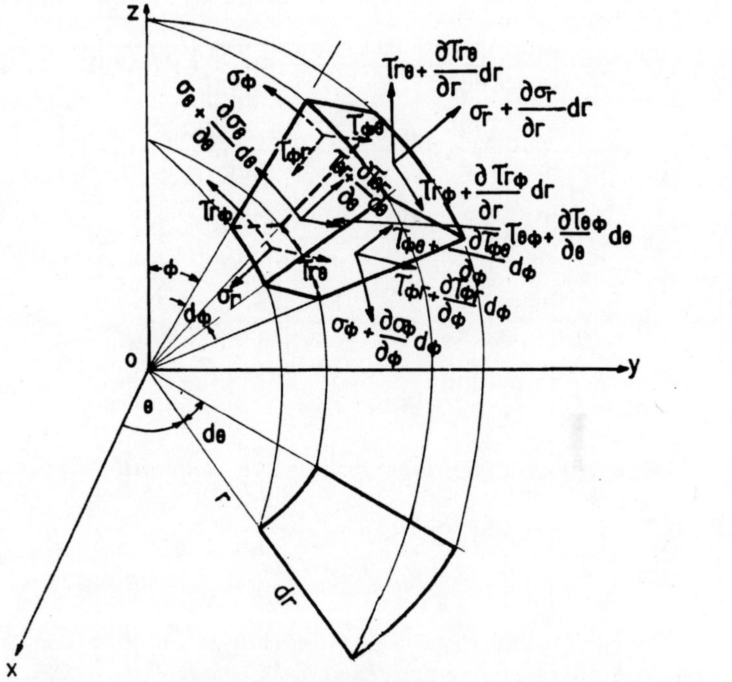

Fig. 1.4 Stress components on a spherical parallelopiped.

(ii) Strain-displacement relations

$$\left.\begin{aligned}
\varepsilon_r &= \frac{\partial u_r}{\partial} \quad \varepsilon_\theta = \frac{u_r}{r} + \frac{1}{r}\frac{\partial u_\theta}{\partial \theta} \\
\varepsilon_\phi &= \frac{u_r}{r} + \frac{\partial \theta}{r}\cot\theta + \frac{1}{r\sin\theta}\cdot\frac{\partial u_\phi}{\partial \phi} \\
\gamma_{r\theta} &= \frac{1}{r}\frac{\partial u_r}{\partial \theta} + \frac{\partial u_\theta}{\partial r} - \frac{u_\theta}{r} \\
\gamma_{r\phi} &= \frac{1}{r\sin\theta}\frac{\partial u_r}{\partial \phi} = \frac{\partial u_\phi}{\partial r} - \frac{u_\phi}{\partial r} \\
\gamma_{\theta\phi} &= \frac{1}{r}\frac{\partial u_\phi}{\partial \theta} - \frac{\cot\theta}{r}u_\phi + \frac{1}{r\sin\theta}\cdot\frac{\partial u_\theta}{\partial \phi}
\end{aligned}\right\} \quad \text{...(1.57)}$$

(iii) Equilibrium equations in term of displacement components

$$(\lambda + 2G)\frac{1}{r}\frac{\partial e}{\partial \theta} - \frac{2G}{r\sin\theta}\left[\frac{\partial \omega_r}{\partial \phi} - \sin\theta\,\frac{\partial (r\omega_\phi)}{\partial r}\right] + B_\theta = 0$$

$$(\lambda + 2G)\frac{1}{r\sin\theta}\frac{\partial e}{\partial \phi} - \frac{2G}{r}\left[\frac{\partial (r\omega_\theta)}{\partial r} - \frac{\partial \omega_r}{\partial \theta}\right] + B_\phi = 0$$

$$(\lambda + 2G)\frac{\partial e}{\partial r} - \frac{2G}{r \sin\theta}\left[\frac{\partial(\omega_\phi \sin\theta)}{\partial\theta} - \frac{\partial\omega_\theta}{\partial\phi}\right] + B_r = 0 \qquad ...(1.58)$$

where $e = \dfrac{1}{r^2 \sin\theta}\left[\sin\theta\,\dfrac{\partial(r^2 u_r)}{\partial r} + r\,\dfrac{\partial(u_\theta \sin\theta)}{\partial\theta} + r\,\dfrac{\partial u_\phi}{\partial\phi}\right]$

$$\omega_r = \frac{1}{2r\sin\theta}\left[\frac{\partial(u_\varphi \sin\theta)}{\partial\theta} - \frac{\partial u_\theta}{\partial\phi}\right]$$

$$\omega_\theta = \frac{1}{2r\sin\theta}\left[\frac{\partial u_r}{\partial\phi} - \sin\theta\,\frac{\partial(ru_\phi)}{\partial r}\right]$$

$$\omega_\phi = \frac{1}{2r}\left[\frac{\partial(ru_\theta)}{\partial r} - \frac{\partial u_r}{\partial\theta}\right].$$

Example 1.1 *The state of stress at a point with respect to the xyz system is*

$$\begin{bmatrix} 3 & 2 & -2 \\ 2 & 0 & -1 \\ -2 & -1 & 2 \end{bmatrix} kN/m^2$$

Determine the stress tensor relative to the x'y'z' coordinate system obtained by a rotation through 30° about the z-axis.

Solution. The direction cosines are,

$$a_{x'x} = \frac{\sqrt{3}}{2},\; a_{x'y} = \frac{1}{2},\; a_{x'z} = 0$$

$$a_{y'x} = -\frac{1}{2},\; a_{y'y} = \frac{\sqrt{3}}{2},\; a_{y'z} = 0$$

$$a_{z'x} = 0,\; a_{z'y} = 0,\; a_{z'z} = 1.$$

Using Eqs. (1.2), we get

$$\sigma_{x'} = \sigma_x a^2_{x'x} + \sigma_y a^2_{x'y} + \sigma_z a^2_{x'z}$$
$$+ 2(\tau_{xy} a_{x'x} a_{x'y} + \tau_{yz} a_{x'y} a_{x'z} + \tau_{xz} a_{x'x} a_{x'z})$$
$$= 1000\left[3 \times \frac{3}{4} + 0 + 0 + 2\left(2 \times \frac{\sqrt{3}}{2} \times \frac{1}{2} + 0 + 0\right)\right]$$
$$= 1000\left[\frac{9}{4} + \sqrt{3}\right] = 3982 \text{ N/m}^2$$

Similarly

$$\sigma_{y'} = -982 \text{ N/m}^2,\; \sigma_{z'} = 1000 \text{ N/m}^2$$

$$\tau_{x'y'} = \sigma_x \,.\, a_{x'x} a_{y'x} + \sigma_y \,.\, a_{x'y} \,.\, a_{y'y} + \sigma_z a_{x'z} a_{y'z}$$
$$+ \tau_{xy}(a_{x'x} a_{y'y} + a_{x'y} a_{y'x})$$

$$+ \tau_{yz} (a_{x'y} a_{y'z} + a_{x'z} a_{y'y})$$
$$+ \tau_{xz} (a_{x'x} a_{y'z} + a_{x'z} a_{y'x})$$

$$\tau_{x'y'} = 1000 \left[3 \times \frac{\sqrt{3}}{2}\left(-\frac{1}{2}\right) + 0 + 0 + 2 \left\{ \frac{\sqrt{3}}{2} \times \frac{\sqrt{3}}{2} + \frac{1}{2}\left(-\frac{1}{2}\right) + 0 + 0 \right]$$

$$= 1000 \left[-\frac{3\sqrt{3}}{4} + 1 \right] = -299 \text{ N/m}^2.$$

Similarly,

$$a_{y'z'} = 134 \text{ N/m}^2 \; \tau_{x'z'} = -2232 \text{ N/m}^2.$$

Example 1.2 *The state of stress at a particular point relative to the xyz coordinate system is given by the stress matrix*

$$\begin{bmatrix} 15 & 10 & -10 \\ 10 & 10 & 0 \\ -10 & 0 & 40 \end{bmatrix} MPa$$

Determine the normal stress and the magnitude and direction of the shear stress on a surface intersecting the point and parallel to the plane given by the equation

$$2x - y + 3z = 9$$

Solution. The direction cosines of the normal to this plane are

$$a_{nx} = \frac{2}{\sqrt{14}},\; a_{ny} = \frac{-1}{\sqrt{14}},\; a_{nz} = \frac{3}{\sqrt{14}}$$

The normal stress is,

$$\sigma_n = \sigma_x . a^2_{nx} + \sigma_y a^2_{ny} + \sigma_z . a^2_{nz} + 2(\tau_{xy}\, a_{nx}\, a_{ny} + \tau_{yz} . a_{ny}\, a_{nz} + \tau_{xz} . a_{nx} . a_{nz})$$

$$\sigma_n = 15 \times \frac{4}{14} + 10 \times \frac{1}{14} + 40 \times \frac{9}{14} + 2\left(10 \times \frac{-3}{14} + 0 - 10 \times \frac{6}{14}\right)$$

$$= \frac{1}{14} [90 + 10 + 360 - 60 - 60]$$

$$= \frac{310}{14} = 22.1428 \text{ MPa}$$

The resultant stress σ_R is

$$\sigma_R\ [(\sigma_x.a_{nx} + \tau_{xy}.a_{ny} + \tau_{xz}.a_{nz})^2 + (\tau_{xy}.a_{nx} + \sigma_y.a_{ny} + \tau_{yz}.a_{nz})^2 + (\tau_{xz}.a_{nx} + \tau_{yz}.a_{ny} + \sigma_z.a_{nz})^2]^{1/2}$$

$$= \left[\left(15 \times \frac{2}{\sqrt{14}} - \frac{10}{\sqrt{14}} - \frac{30}{\sqrt{14}}\right)^2 + \left(10 \times \frac{2}{\sqrt{14}} - 10 \times \frac{1}{\sqrt{14}} + 0\right)^2 + \left(-10 \times \frac{2}{\sqrt{14}} + 0 + 40 \times \frac{3}{\sqrt{14}}\right)^3\right]^{1/2}$$

$$= \left[\frac{100}{14} + \frac{100}{14} + \frac{10000}{14}\right)^{1/2} = \left(\frac{10200}{14}\right)^{1/2}$$

$$= (728.57)^{1/2} = 26.992 \text{ MPa}$$

Hence shear stress is

$$\tau = \sqrt{\sigma_R^2 - \sigma_n^2} = \sqrt{728.57 - 490.30}$$

$$= \sqrt{238.27} = 15.436 \text{ MPa} .$$

To determine the direction cosines for τ, let a_{sx}, a_{sy} and a_{sz} be the direction cosines of τ. For equilibrium of forces in the x-direction

$$\sigma_n.A.a_{nx} + \tau.A.a_{sx} = X_n.A$$

where A is the area over which these stresses act.

$$\therefore \quad a_{sz} = \frac{1}{\tau}[X_n - \sigma_n \,.\, a_{nx}]$$

$$= \frac{1}{\tau}[\sigma_x.a_{nx} + \tau_{xy}a_{ny} + \tau_{xz}.a_{nz} - \sigma_n.a_{nx}]$$

$$= \frac{1}{\tau}[(\sigma_x - \sigma_n)a_{nx} + \tau_{xy}a_{ny} + \tau_{xz}a_{nz}]$$

$$= \frac{1}{15.436}\left[(15 - 22.1428)\frac{2}{\sqrt{14}} - 10 \times \frac{1}{\sqrt{14}} - 10 \times \frac{3}{\sqrt{14}}\right]$$

$$= \frac{1}{15.436}\left[-\frac{54.2856}{\sqrt{14}}\right] = -0.9399$$

Similarly

$$a_{sy} = \frac{1}{\tau}[\tau_{xy}.a_{nx} + (\sigma_y - \sigma_n)a_{ny} + \tau_{yz}.a_{nz}]$$

$$= \frac{1}{15.436}\left[10 \times \frac{2}{\sqrt{14}} + (10 - 22.1428) \,.\left(-\frac{1}{\sqrt{14}}\right) + 0\right]$$

$$= \frac{1}{15.436}\left[\frac{20}{\sqrt{14}} + \frac{12.1428}{\sqrt{14}}\right] = \frac{32.1428}{15.436\sqrt{14}} = 0.5565$$

$$a_{sz} = \frac{1}{15.436}\left[\tau_{xz}.a_{nx} + \tau_{yz}a_{ny} + (\sigma_z - \sigma_n)a_{nz}\right]$$

$$= \frac{1}{15.436}\left[-10 \times \frac{2}{\sqrt{14}} + 0 + (40 - 22.1428)\,.\,\frac{3}{\sqrt{14}}\right]$$

$$= \frac{1}{15.436}\left[-\frac{20}{\sqrt{14}} + \frac{53.5716}{\sqrt{14}}\right]$$

$$= \frac{33.5716}{15.436\sqrt{14}} = 0.5812$$

Example 1.3 *For the stress tensor given below, determine the principal stresses and the direction cosines associated with the normals to the surface of each principal stress.*

$$\begin{bmatrix} 3000 & 1000 & 1000 \\ 1000 & 0 & 2000 \\ 1000 & 2000 & 0 \end{bmatrix} N/m^2.$$

Solution. The stress invariants are

$$I_1 = 3000$$

$$I_2 = -(1000)^2 - (2000)^2 - (1000)^2$$

$$= -[10^6 + 4 \times 10^6 + 10^6] = -6 \times 10^6$$

$$I_3 = 3000\,(-4 \times 10^6) - 1000\,(-2 \times 10^6) + 1000(2 \times 10^6)$$

$$= -12 \times 10^9 + 2 \times 10^9 + 2 \times 10^9 = -8 \times 10^9$$

The cubic equation is

$$\sigma^3 - 3000\,\sigma^2 - 6 \times 10^6\sigma + 8 \times 10^9 = 0$$

Its solution is

$$\sigma_1 = 4000 \text{ N/m}^2$$

$$\sigma_2 = 1000 \text{ N/m}^2$$

$$\sigma_3 = -2000 \text{ N/m}^2$$

To determine the direction cosines for σ_1, we have

$$[\tau_{ij} - \sigma_1 I] = \begin{bmatrix} -1000 & 1000 & 1000 \\ 1000 & -4000 & 2000 \\ 1000 & 2000 & -4000 \end{bmatrix}$$

where I is the unit matrix.

$$\therefore \qquad A_1 = (16 \times 10^6 - 4 \times 10^6) = 12 \times 10^6$$

$$B_1 = -(-4 \times 10^6 - 2 \times 10^6) = 6 \times 10^6$$

$$C_1 = 2 \times 10^6 + 4 \times 10^6 = 6 \times 10^6$$

$$\therefore \quad k_1 = \sqrt{A_1^2 + B_1^2 + C_1^2} = 10^6\sqrt{144 + 36 + 36} = 10^6\sqrt{216}$$

$$\therefore \quad a_{nx_1} = \frac{A_1}{k_1} = \frac{12 \times 10^6}{10^6 \times \sqrt{216}} = \frac{2}{\sqrt{6}}$$

$$a_{ny_1} = \frac{6 \times 10^6}{\sqrt{216} \times 10^6} = \frac{1}{\sqrt{6}}$$

$$a_{nz_1} = \frac{6 \times 10^6}{\sqrt{216} \times 10^6} = \frac{1}{\sqrt{6}}$$

Similarly,

$$a_{nx_2} = \frac{1}{\sqrt{3}}, \; a_{ny_2} = \frac{-1}{\sqrt{3}}, \; a_{nz_2} = \frac{-1}{\sqrt{3}}$$

$$a_{nx_3} = 0, \; a_{ny_3} = \frac{1}{\sqrt{2}}, \; a_{nz_3} = \frac{-1}{\sqrt{2}}$$

Example 1.4 *The stress components at a point are given by the following array :*

$$\begin{bmatrix} 10 & 5 & 6 \\ 5 & 8 & 10 \\ 6 & 10 & 6 \end{bmatrix} MPa$$

Calculate the principal stresses and principal planes.

Solution.

$$I_1 = \sigma_x + \sigma_y + \sigma_z$$
$$= 10 + 8 + 6 = 24$$
$$I_2 = \sigma_x\sigma_y + \sigma_y\sigma_z + \sigma_z\sigma_x - \tau_{xy}^2 - \tau_{yz}^2 - \tau_{xz}^2$$
$$= 10 \times 8 + 8 \times 6 + 6 \times 10 - 5^2 - 10^2 - 6^2$$
$$= 80 + 48 + 60 - 25 - 100 - 36 = 27$$
$$I_3 = \sigma_x\sigma_{yz} - \sigma_x\tau_{yz}^2 - \sigma_y\tau_{xz}^2 - \sigma_z\tau_{xy}^2 \; 2\tau_{xy}\tau_{yz}\tau_{xz}$$
$$= 10 \times 8 \times 6 - 10 \times 100 - 8 \times 36 - 6 \times 25$$
$$+ 2 \times 5 \times 10 \times 6$$
$$= 480 - 1000 - 288 - 150 + 600$$
$$= -358$$

$$\sigma^3 - 24\sigma^2 + 27\sigma + 358 = 0 \qquad ...(1)$$

Now $\cos 3\theta = 4\cos^3\theta - 4\cos\theta$

or $$\cos^3\theta - \frac{3}{4}\cos\theta - \frac{1}{4}\cos 3\theta = 0 \qquad ...(2)$$

Put $$\sigma = r\cos\theta + \frac{I_1}{3}$$
$$= r\cos\theta + 8$$

Eq. (1) becomes,

$$r^3 \cos^3 \theta + 512 + 24 r^2 \cos^2 \theta + 192 r \cos \theta - 24 r^2 \cos^2 \theta - 1536$$
$$- 384 r \cos \theta + 27 r \cos \theta + 216 + 358 = 0$$
$$r^3 \cos^3 \theta - 165 r \cos \theta - 450 = 0$$

or
$$\cos^3 \theta - \frac{165}{r^2} \cos \theta - \frac{450}{r^3} = 0 \quad \text{...(3)}$$

Comparing Eqs. (2) and (3), we have

$$\frac{165}{r^2} = \frac{3}{4}$$

or
$$r = 14.8324$$

and
$$\frac{450}{r^3} = \frac{\cos 3\theta}{4}$$

$$\cos 3\theta = \frac{450 \times 4}{(14.8324)^3} = 0.551618$$

$$3\theta = 56.5219$$
$$\theta_1 = 18.84°$$
$$\theta_2 = 138.84°, \quad \theta_3 = 258.84°$$
$$\sigma_1 = r \cos \theta_1 + 8$$
$$= 14.8324 \cos 18.84° + 8$$
$$= 22.04 \text{ MPa}$$
$$\sigma_2 = r \cos \theta_2 + 8$$
$$= 14.8324 \cos 138.84° + 8$$
$$= -3.17 \text{ MPa}$$
$$\sigma_3 = r \cos \theta_3 + 8$$
$$= 14.8324 \cos 258.84° + 8$$
$$= 5.13 \text{ MPa.}$$

Example 1.5 *Calculate the principal stresses for the stress components given below :*

$$\begin{bmatrix} 100 & 200 & 100 \\ 200 & -200 & -300 \\ 100 & -300 & 400 \end{bmatrix} MPa.$$

Solution. $I_1 = 100 - 200 + 400 = 300$

$$I_2 = -100 \times 200 - 200 \times 400 + 100 \times 400$$
$$- 200^2 - 300^2 - 100^2$$
$$= (-2 - 8 + 4 - 4 - 9 - 1) \times 10^4$$
$$= -20 \times 10^4$$

$$I_3 = [1 \times (-8-9) - 2 \times (8+3) + 1 \times (-6+2)] \times 10^4$$
$$= [-17 - 22 - 4] \times 10^4$$
$$= -43 \times 10^4$$

$\therefore \quad f(\sigma) = \sigma^3 - 300\,\sigma^2 - 20 \times 10^4\,\sigma + 43 \times 10^4$

Consider the cubic equation :

$$y^3 + py^2 + qy + r = 0$$

Substitute $y = x - \frac{p}{3}$

$$\left(x - \frac{p}{3}\right)^3 + p\left(x - \frac{p}{3}\right)^2 + q\left(x - \frac{p}{3}\right) r = 0$$

$$x^3 - x^2p + \frac{xp^2}{3} - \frac{p^3}{27} + px^2 + \frac{p^3}{9} - \frac{2}{3}xp^2 + qx - \frac{pq}{3} + r = 0$$

$$x^3 + x\left(\frac{p^2}{3} - \frac{2}{3}p^2 + q\right) - \frac{p^3}{27} + \frac{p^3}{9} - \frac{pq}{3} + r = 0$$

$$x^3 + \left(\frac{3q - p^2}{3}\right)x + \frac{2}{27}p^3 - \frac{pq}{3} + r = 0$$

$$x^3 + ax + b = 0$$

where $a = \frac{1}{3}(3q - p^2)$, $b = \frac{1}{27}(2p^3 - 9pq + 27r)$

Put $\cos\theta = \dfrac{-b}{2\left(-\dfrac{a^3}{27}\right)^{1/2}}$

$$g = 2\sqrt{-\frac{a}{3}}$$

then

$$y_1 = g\cos\frac{\theta}{3} - \frac{p}{3}$$

$$y_2 = g\cos\left(120° + \frac{\theta}{3}\right) - \frac{p}{3}$$

$$y_3 = g\cos\left(240° + \frac{\theta}{3}\right) - \frac{p}{3}$$

Here $p = -300$, $q = -20 \times 10^4$, $r = 43 \times 10^4$

$$a = \frac{1}{3}[3 \times (-20 \times 10^4) - (300)^2]$$

$$= \frac{1}{3}[-60 - 9] \times 10^4 = -23 \times 10^4$$

$$b = \frac{1}{27}[2 \times (-300)^3 - 9(-300)(-20 \times 10^4) + 27 \times 43 \times 10^4]$$

$$= \frac{1}{27}[-54 - 540 + 1161] \times 10^6$$

$$= 21 \times 10^6$$

$$\cos\theta = \frac{-21 \times 10^6}{2\left(\frac{23^3}{27}\right)^{1/2} \times 10^6} = -0.49462$$

$$\theta = 119.645°$$

$$g = 2\sqrt{\frac{23 \times 10^4}{3}} = 553.775$$

$$\sigma_1 = y_1 = 553.775 \cos\left(\frac{119.645}{3}\right)° + 100$$

$$= 524.95 \text{ MPa}$$

$$\sigma_2 = y_2 = 553.775 \cos(159.8817)° + 100$$

$$= -419.986 \text{ MPa}$$

$$\sigma_3 = y_3 = 553.775 \cos(279.8817)^0 + 100$$

$$= 195.035 \text{ MPa}$$

1.9 Computer Program for Principal Stresses and Principal Planes

Nomenclature

$SX = \sigma_x$, $SY = \sigma_y$, $SZ = \sigma_z$

$SXY = \tau_{xy}$, $SYZ = \tau_{yz}$, $SXZ = \tau_{xz}$

$a_1 = 1$, $a_2 = I_2$, $a_3 = I_3$

$S(1) = \sigma_1$, $S(2) = \sigma_2$, $S(3) = \sigma_3$

$l(j) = a_{nxj}$, $m(j) = a_{nyj}$, $n(j) = a_{nzj}$

```
C     Program for principal stresses and planes
C     developed by Dr. Sadhu singh
C     Mechanical Engineering Department
C     Pantnagar University
      Dimension a(3),b(3),c(3),s(3)
      Integer x,y
      Real l(3),m(3),n(3),k(3)
      Open(unit=7,file='ps.in',status='old)
      Open(unit=8,file='ps.out',status='new')
      Write(8,5)
5     Format(1x,'Input',11,'Given Stress Components are :', //)
```

```
      Read(7,*)SX,SY,SZ,SXY,SYZ,SXZ
      Write(8,10)SX,SY,SZ
10    Format(1x,'SX=',F7.3,5X,'SY=',f7.3,5X,'SZ=',f7.3,//)
      Write(8,15)SXY,SYZ,SXZ
15    Format (1x,'SXY=',f7.3,5x,'SYZ=',F7.3,5x,'SXZ=',f7.3,//)
      a1=SX+SY+SZ
      a2 = SX*SY+XY*SZ+SX*SZ-SXY**2-SYZ**2-SXZ**2
      a3 = SX*SY*SZ+2SZY*SYZ*SXZ-SX*SYZ**2
      1-SY*SXZ**2-SZ*SXY**2
20    Write(8,20)
      Format (1x,'Ouput',//,'Principal stresses are :')
      r = (1./3.)*a1**2-a2
      t=SQRT[(1./27.)*r**3]
      q = (1./3.)*a1*a2-a3-(2./27.)*a1**3
      st=SQRT(1./3.)*r)
      alpha=acos(-q/(2.*t))
      S(1)=2.*st*(cos(alpha/3.)+(1./3.)*a1
      S(2)=2.*st*cos((alpha/3.)+2.0944)+(1./3.)*a1
      S(3)=2.*St*cos((alpha/3.)+4.1888)+(1./3.)*a1
      Do50x=1,2
      Do40y=x,3
      If(S(X).lt.S(Y))Go To 30
      Go To 40
30    temp=S(X)
      S(X) = S(Y)
      S(Y) = temp
40    Continue
50    Continue
      Write(8,60) S(1),S(2),S(3)
60    FORMAT(/,1x,'S1=',f7.3,5x,'S2=',f7.3,5x,'S3=',f7.3,/)
      Write(8,65)
65    FORMAT (/, 'Direction cosines are :', //)
      Do 80j=1,3
      a(j)=(SY-S(j))*(SZ-S(j))-SYZ**2
      b(j)= -(SXY*(SZ-S(j))-SXZ*SYZ)
      C(j)=SXY*SYZ-SXZ*(SY-S(j))
      k(j)=1./(SQRT(a(j)**2+b(j)**2+c(j)**2)
      l(j)=a(j)*k(j)
      m(j)=b(j)*k(j)
      n(j)=c(j)*k(j)
      Write(8,70)j,l(j),m(j),n(j)
70    Format(1x,'l(',i1,')=',f7.4,10x,'m(',i1,')=',
      1f7.4,10x,'n(',i1,')=',f7.4)
80    Continue
      Close(7)
      Close(8)
      Stop
```

```
END
Input
Given stress components are :
SX=21.450       SY=5.740        SZ= -6.850
SXY= -5.720     SYZ=11.750      SXZ=5.500
Output
Principal stresses are :
S1=23.428       S2=12.391       S3= -15.479
Direction Cosines are :
l(1)=0.9621     m(1)= -0.2628   n(1)=0.0728
l(2)= -0.1810   m(2)= -0.8155   n(2)= -0.5497
l(3)=0.2038     m(3)=0.5157     n(3)= -0.8321
```

EXERCISES

1.1 **The state of stress at a point relative to an *xyz* coordinate system is**

$$\begin{bmatrix} 400 & 100 & -100 \\ 100 & 0 & 200 \\ -100 & 200 & 0 \end{bmatrix} \text{N/mm}^2$$

Determine the complete state of stress relative to an x' y' z' co-ordinate system if

	x	y	z
x'	30°	60°	90°
y'	120°	30°	90°
z'	90°	90°	90°

1.2 **At point *Q* in a body the state of stress relative to a *xyz* co-ordinate system is**

$$\begin{bmatrix} 500 & 200 & -200 \\ 200 & 0 & 400 \\ -200 & 400 & 300 \end{bmatrix} \text{MPa}$$

Using the cube shown in Fig. 1.5, determine the normal and shear stress at point *Q* for surface parallel to the following planes.

(*i*) *BCGF*, (*ii*) *ABEF*, and (*iii*) *BGE*.

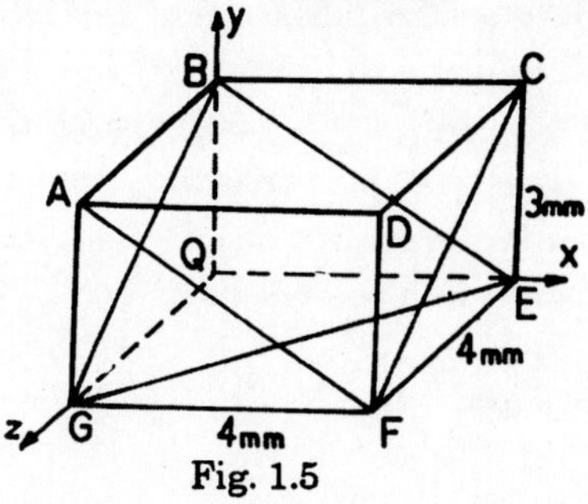

Fig. 1.5

1.3 If $\sigma_x = 4, \sigma_y = 2, \sigma_z = -2,$
$\tau_{xy} = 3, \tau_{yz} = 8, \tau_{xz} = -2$, MPa,
compute the stress vectors on planes with unit normals $\left(\frac{2}{3}, \frac{1}{3}, \frac{2}{3}\right)$ and $\left(\frac{1}{\sqrt{14}}, \frac{2}{\sqrt{14}}, \frac{3}{\sqrt{14}}\right)$. Compute the normal and shearing stresses on these planes.

1.4 Determine the magnitude and the direction of the principal stresses and the maximum shearing stress when,

$$\sigma_x = 1500, \quad \sigma_y = -1000,$$
$$\sigma_z = 1000, \quad \tau_{xy} = -300,$$
$$\sigma_{yz} = 0, \quad \tau_{xz} = 100 \text{ MPa}.$$

1.5 Show that

(*i*) $9\tau^2_{oct} = (\sigma_1 - \sigma_2)^2 + (\sigma_2 - \sigma_3)^2 + (\sigma_3 - \sigma_1)^2$

(*ii*) $9\tau^2_{oct} = (\sigma_x - \sigma_y)^2 + (\sigma_y - \sigma_z)^2 + (\sigma_z - \sigma_x)^2 + 6(\tau^2_{xy} + \tau^2_{yz} + \tau^2_{xz})$

(*iii*) $9\tau^2_{oct} = 2I_1^2 - 6I_2$.

1.6 For a given displacement field

$$\bar{u} = (x^2y + 5z^2)\hat{i} + (xy^2z + y^2)\hat{j} + x^2y^2z^2\hat{k}$$

determine the strain tensor, rotation tensor and the angle of rotation at the point (2, – 1, 2).

1.7 Show that

$$\varepsilon_x = c_1(x^2 + y^2), \ \varepsilon_y = c_1(y^2 + z^2),$$
$$\gamma_{xy} = c_2xyz, \ \varepsilon_z = \gamma_{xz} = \gamma_{yz} = 0.$$

where c_1 and c_2 are constants, is not a possible state of strain.

1.8 Given the following displacement field,

$$\bar{u} = (xyz\,\hat{i} + 3x^2yz\hat{j} + 4yz\hat{k}) \times 10^{-2}$$

and a very small segment Δs having the following direction cosines before deformation :

$$a_{sx} = 0.200,$$
$$a_{sy} = 0.800,$$
$$a_{sz} = 0.555.$$

This segment is directed away from the point (2, – 1, 3). What is the new vector $\Delta s'$ after the displacement field $\bar{u}$ has been imposed ?

1.9 The state of strain at a point is given by

$$\varepsilon_x = 0.001, \ \varepsilon_y = -0.003, \ \varepsilon_z = 0.002,$$
$$\gamma_{xy} = 0.001, \ \gamma_{yz} = 0.0005, \ \gamma_{xz} = -0.002.$$

Determine the strain invariants and the principal strains.

1.10 The state of stress at a point is given by

$$\sigma_x = 200, \quad \sigma_y = -100, \quad \sigma_z = 50,$$
$$\tau_{xy} = 40, \quad \tau_{yz} = 50, \quad \tau_{xz} = 60 \text{ MPa}.$$

If $E = 2.05 \times 10^5$ N/mm^2
$G = 0.8 \times 10^5$ N/mm^2
determine the strain components.

1.11 The state of strain at a point is defined by the given strain tensor below. For a fibre with direction $(0, -1/\sqrt{5}, -2/\sqrt{5})$, calculate (*a*) the normal strain for the fibre (*b*) the magnitude of the strain vector.

$$[\varepsilon_{ij}] = \begin{bmatrix} 200 & 183 & -25 \\ 183 & 100 & -125 \\ -25 & -125 & 150 \end{bmatrix} \times 10^{-6}.$$

1.12 At a point, the stress components are :

$$\sigma_x = 600, \quad \sigma_y = 300, \quad \sigma_z = 900$$
$$\tau_{xy} = 500, \quad \tau_{yz} = 400, \quad \tau_{zx} = -200 \text{ kPa}.$$

Show that the principal directions of stress and deviative stress coincide.

1.13 The state of stress at a point is given by

$$\sigma_x = 100, \sigma_y = 200, \sigma_z = -100, \tau_{xy} = -200, \tau_{yz} = 100$$

and $\tau_{zx} = -300$ kPa.

Determine :

(*a*) the stress invariants,

(*b*) the principal stresses, and

(*c*) the direction cosines of the principal planes.

1.14 The displacement field in a body is specified as :

$$u_x = (x^2 + 3) \times 10^{-3}$$
$$u_y = 3y^2z \times 10^{-3}$$
$$u_z = (x + 3z) \times 10^{-3}$$

Determine the strain components at a point whose co-ordinates are (1, 2, 3).

1.15 The state of stress at a point is given by

$$\begin{bmatrix} 20 & -6 & 10 \\ -6 & 10 & 8 \\ 10 & 8 & 7 \end{bmatrix} \text{MPa}.$$

Determine the principal stresses and principal directions.

1.16 The stress field in a continuous body is given by :

$$[\tau_{ij}] = \begin{bmatrix} 1 & 0 & 2y \\ 0 & 1 & 4x \\ 2y & 4x & 1 \end{bmatrix} \text{kPa}.$$

Find the stress vector at point $P(1, 2, 3)$, acting on a plane

$$x + y + z = 6.$$

1.17 The given displacement components are : $u_x = cx(y + z)^2$, $u_y = cy(x + z)^2$, $u_z = cz(x + y)^2$, where c is a constant. Find

(*a*) the components of linear strain.

(*b*) the components of rotation.

(*c*) the principal strains at a point whose co-ordinates are (1, 1, 1).

1.18 A plate whose thickness is 3 mm is stretched as shown in Fig. 1.6. Find the principal strains and the maximum shearing strain in the plate.

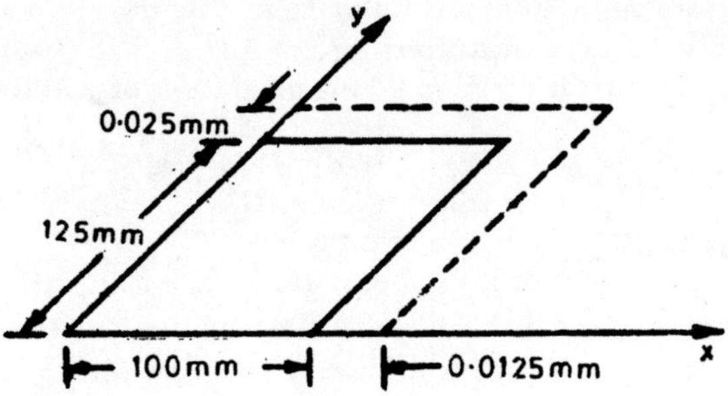

Fig. 1.6

1.19 The stress field on a body is given by

$$\sigma_x = 20x^2 + y^2 \qquad \tau_{xy} = zx$$
$$\sigma_y = 30x^3 + 200 \qquad \tau_{xz} = y^2 z$$
$$\sigma_z = 30(y^2 + z^2) \qquad \tau_{yz} = x^3 y.$$

What are the components of the body force required to ensure equilibrium ?

1.20 The state of stress at a point is given by

$$[\tau_{ij}] = 10^2 \begin{bmatrix} 10 & 5 & -10 \\ 5 & 20 & -15 \\ -10 & -15 & -10 \end{bmatrix} \text{MPa.}$$

Find the magnitude and direction of the stress vector acting on a plane whose normal has direction cosines $(1/2, 1/2, 1/\sqrt{2})$. What are the normal and tangential stresses acting on this plane ?

1.21 At a point P of a continuous body, the components of the stress tensor are :

$$[\tau_{ij}] = 10^3 \begin{bmatrix} 1 & -3 & \sqrt{2} \\ -3 & 4 & -\sqrt{2} \\ \sqrt{2} & -\sqrt{2} & 4 \end{bmatrix} \text{MPa.}$$

(*a*) Find the principal stresses and the principal directions.

(*b*) Find the octahedral normal and shearing stresses.

(*c*) What is the equation of the stress quadric ?

1.22 Obtain the principal stresses and the related direction cosines for the following state of stress :

$$\begin{bmatrix} 3 & 4 & 6 \\ 4 & 2 & 5 \\ 6 & 5 & 1 \end{bmatrix} \text{MPa.}$$

[**Ans.** 12.049, – 1.521, – 4.528 MPa, 0.618, 0.533, 0.577]

1.23 The state of strain at a point within a material is given by :

$$\begin{bmatrix} 200 & 100 & 0 \\ 100 & 300 & 400 \\ 0 & 400 & 0 \end{bmatrix} \times 10^{-6}$$

For $E = 200$ GPa and G = 80 GPa, ascertain the components of stress tensor. [Ans. 72, 99, 40, 16, 64, 0 MPa]

1.24 For the following state of strain, determine the principal strains.

$$\begin{bmatrix} 2 & 3 & 2 \\ 3 & -1 & 5 \\ 2 & 5 & -4 \end{bmatrix} \times 10^{-6}.$$

[Ans. 6.01, – 1.26, – 7.72 microns]

1.25 The state of stress at a point is given by ;

$\sigma_x = -120$, $\sigma_y = 140$, $\sigma_z = 66$, $\tau_{xy} = 45$, $\tau_{yz} = -65$ and $\tau_{xz} = 25$ MPa.

Determine the three principal stresses and the directions associated with the three principal stresses.

[Ans. 180.2, 40.1, – 143.3 MPa,
$l_1 = 0.913, m_1 = 0.8740, n_1 = -0.4773$
$l_2 = 0.2584, m_2 = 0.4422, n_2 = 0.8589$
$l_3 = 0.9598, m_3 = -0.2062, n_3 = -0.1904$]

1.26 The state of stress at a joint is given by

$\sigma_x = 120$, $\sigma_y = -55$, $\sigma_z = -85$, $\tau_{xy} = -55$, $\tau_{yz} = 33$, and $\tau_{xz} = -75$ MPa.

Determine the three principal stresses and the maximum shearing stress. [Ans. 162.5, – 114.1, – 68.4, 138.3 MPa]

2

Two-Dimensional Problems in Cartesian Coordinates

2.1 Introduction

Two dimensional problems may be either formulated as plane stress or plane strain problems. The solution of these problems may be obtained by the following methods :

1. Airy's stress function method.
2. Complex potentials method.
3. Cauchy's integral method.
4. Fourier transform method.
5. Real potential method.
6. Variational methods.
7. Numerical methods.

2.2 Airy's Stress Function Method

If there exist a body force potential V such that

$$B_x = \frac{\partial V}{\partial x}$$

$$B_y = \frac{\partial V}{\partial y}$$

and define

$$\left.\begin{aligned} \sigma_x &= \frac{\sigma^2\phi}{\partial y^2} - V \\ \sigma_y &= \frac{\partial^2\phi}{\partial x^2} - V \\ \tau_{xy} &= \frac{-\partial^2\phi}{\partial x \partial y} \end{aligned}\right\} \quad \text{...(2.1)}$$

where ϕ = Airy stress function

then the stress components given by Eq. (2.1) satisfy the equilibrium equations for two-dimensional problems and on substitution into the stress compatibility equation yields the Biharmonic equation

$$\nabla^4\phi = (1-\nu)\left(\frac{\partial B_x}{\partial x} + \frac{\partial B_y}{\partial y}\right) \qquad ...(2.2)$$

for plane stress case and

$$\nabla^4\phi = \left(\frac{1-2\nu}{1-\nu}\right)\left(\frac{\partial B_x}{\partial x} + \frac{\partial B_y}{\partial y}\right) \qquad ...(2.3)$$

for plane strain case.

If the body forces are zero, then Eqs. (2.2) and (2.3) reduce to

$$\nabla^4\phi = 0 \qquad ...(2.4)$$

where $\nabla^4 = \dfrac{\partial^4}{\partial x^4} + \dfrac{2\partial^4}{\partial x^2 \partial y^2} + \dfrac{\partial^4}{\partial y^4}$

Solutions of Eq. (2.4), which may be assumed, can be of the following form :

$$\left.\begin{aligned} \phi = {} & x, x^2, x^3, y, y^2, y^3, xy, xy, \\ & xy^2, x^3y, xy^3 \text{ ; } \cos\lambda x \cosh\lambda y, \\ & \cosh\lambda x \cos\lambda y, x\cos\lambda x\cosh\lambda y, \\ & x\cosh\lambda x\cos\lambda y, y\cos\lambda x\cosh\lambda y, \\ & y\cosh\lambda x\cos\lambda y \text{ ; } x^2-y^2, x^4-y^4, \end{aligned}\right\} \qquad ...(2.5)$$

Thus these functions may be

1. Algebraic functions
2. Polynomials
3. Trigonometric functions
4. Fourier series
5. Complex functions.

We will illustrate some of these functions in the following paragraphs.

2.2.1 Polynomials

A number of interesting solutions of Eq. (2.4) can be obtained by postulating functions in the form of polynomials of various degrees, and adjusting their coefficients to satisfy different conditions. The simplest case arise when we assume that

$$\phi = \frac{1}{2}a_2x^2 + b_2xy + \frac{1}{2}c_2y^2 \qquad ...(2.6)$$

Then, from Eq. (2.1)

$$\sigma_x = c_2,\ \sigma_y = a_2,$$

$$\tau_{xy} = -b_2 \qquad ...(2.7)$$

So that all three stress components are constant, and we have a combination of uniform normal stresses and uniform shear with no body force.

Now consider a cubic polynomial

$$\phi = \frac{1}{6} a_3 x^3 + \frac{1}{2} b_3 x^2 y + \frac{1}{2} c_3 x y^2 + \frac{1}{6} d_3 y^3 \qquad ...(2.8)$$

and, from Eq. (2.1) we see that in the absence of body forces, the corresponding stresses become,

$$\left.\begin{aligned} \sigma_x &= c_3 x - d_3 y \\ \sigma_y &= a_3 x + b_3 y \\ \tau_{xy} &= -b_3 x - c_3 y \end{aligned}\right\} \qquad ...(2.9)$$

Various cases are obtained from this set of stresses by making some of the coefficients zero. When polynomials of higher degree than the third are used, the Biharmonic Eq. (2.4) is satisfied only if certain conditions are imposed on the coefficients. For example, for the fifth degree polynomial, we have

$$\phi = \frac{1}{20} a_5 x^5 + \frac{1}{12} b_5 x^4 y + \frac{1}{6} c_5 x^3 y^2 + \frac{1}{6} d_5 x^2 y^3 + \frac{1}{12} e_5 x y^4 + \frac{1}{20} f_5 y^5 \qquad ...(2.10)$$

which satisfies Eq. (2.4) if

$$e_5 = -(2c_5 + 3a_5),$$
$$f_5 = -\frac{1}{3}(b_5 + 2d_5).$$

From Eq. (2.1) we find that in the absence of body forces the stress components are given by the expressions.

$$\left.\begin{aligned} \sigma_x &= \frac{1}{3} c_5 x^3 + d_5 x^2 y - (2c_5 + 3a_5) x y^2 \\ &\quad - \frac{1}{3}(b_5 + 2d_5) y^3 \\ \sigma_y &= a_5 x^3 + b_5 x^2 y + c_5 x y^2 + \frac{1}{3} d_5 y^3 \\ \tau_{xy} &= -\frac{1}{3} b_5 x^3 - c_5 x^2 y - d_5 x y^2 \\ &\quad + \frac{1}{3}(2c_5 + 3a_5) y^3 \end{aligned}\right\} \qquad ...(2.11)$$

2.2.1.1 Bending of a Cantilever Loaded at the End

Consider a cantilever having a narrow rectangular cross section of unit width bent by a force P applied at the end as shown in Fig. 2.1.

The upper and lower edges of the cantilever are free from load, and shearing forces, having a resultant P, are distributed along

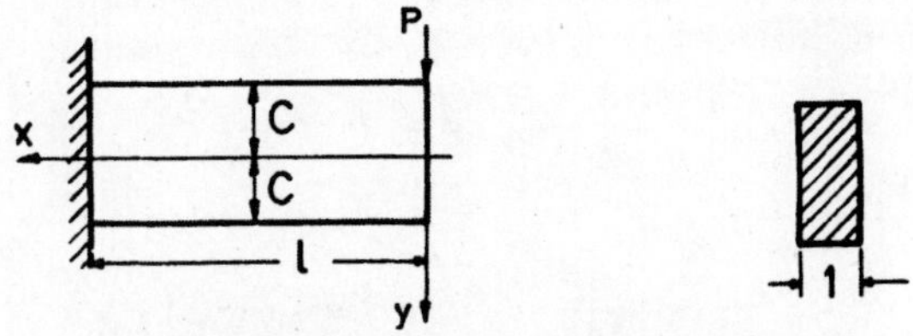

Fig. 2.1 Cantilever having end load.

the end $x = 0$. Using a polynomial of second degree with $a_2 = c_2 = 0$ and a polynomial of fourth degree with

$$a_4 = b_4 = c_4 = e_4 = 0$$

and adding both the stress functions, we get

$$\Phi = b_2xy + \frac{d_4}{6}xy^3$$

$$\therefore \quad \sigma_x = \frac{\partial^2\Phi}{\partial y^2} = d_4\,xy$$

$$\sigma_y = \frac{\partial^2\Phi}{\partial x^2} = 0$$

$$\tau_{xy} = -\frac{\partial^2\Phi}{\partial x\partial y}$$

$$= -b_2 - \frac{d_4}{2}y^2$$

The following boundary conditions must be satisfied.

(i) $(\tau_{xy})_{y=\pm c} = 0 = -b_2 - \frac{d_4}{2}c^2$

$$\therefore \quad d_4 = -\frac{2b_2}{c^2}$$

(iii) $\int_{-c}^{+c} \tau_{xy}\,dy = -P$

$$\therefore \int_{-c}^{+c}\left(-b_2 + \frac{b_2}{c^2}y^2\right)dy = -P$$

$$\therefore \quad b_2 = \frac{3}{4}\frac{P}{c}$$

Hence, we get

$$\left.\begin{aligned} \sigma_x &= -\frac{3}{2}\frac{P}{c^3}xy = -\frac{Pxy}{I} = \frac{My}{I} \\ \sigma_y &= 0 \\ \tau_{xy} &= -\frac{3P}{4c}\left(1-\frac{y^2}{c^2}\right) = -\frac{P}{2I}(c^2-y^2) \end{aligned}\right\} \quad ...(2.12)$$

where $I = \frac{2}{3}c^3$ is the moment of inertia of the beam and $M = -Px$ is the bending moment. The solution given by Eqs. (2.12) coincide exactly with the elementary solution as given by strength of materials.

Now using strain-displacement relations and applying Hooke's law, we have

$$\frac{\partial u_x}{\partial x} = \varepsilon_x = \frac{\sigma_x}{E} = -\frac{Pxy}{EI} \quad ...(2.13\ a)$$

$$\frac{\partial u_y}{\partial y} = \varepsilon_y = -\frac{\nu\sigma_a}{E} = \nu\frac{Pxy}{EI} \quad ...(2.13\ b)$$

$$\frac{\partial u_x}{\partial y} + \frac{\partial u_y}{\partial x} = \gamma_{xy} = \frac{\tau_{xy}}{G} = -\frac{P}{2IG}(c^2-y^2) \quad ...(2.14)$$

Integrating Eqs. (2.13 *a*) and (2.13 *b*), we get

$$u_x = -\frac{Px^2y}{2EI} + f_1(y) \quad ...(a)$$

$$u_y = \frac{\nu Pxy^2}{2EI} + f_2(x) \quad ...(b)$$

in which $f_1(y)$ and $f_2(x)$ are unknown functions of y and x only. Substituting these values of u_x and u_y in Eq. (2.14), we get

$$-\frac{Px^2}{2EI} + \frac{df_1(y)}{dy} + \frac{\nu Py^2}{2EI} + \frac{df_2(x)}{dx}$$

$$= -\frac{P}{2IG}(c^2-y^2)$$

$$\left[-\frac{Px^2}{2EI} + \frac{df_2(x)}{dx}\right] + \left[\frac{\nu Py^2}{2EI} - \frac{Py^2}{2IG} + \frac{df_1(y)}{dy}\right] = -\frac{Pc^2}{2IG}$$

The terms contained in the first bracket are functions of x only and in the second bracket are functions of y only and the term on the right hand side is a constant quantity. Hence

$$F_1(x) + F_2(y) = K$$

where $F_1(x) = -\dfrac{Px^2}{2EI} + \dfrac{df_2(x)}{dx}$

$$F_2(y) = \frac{\nu Py^2}{2EI} - \frac{Py^2}{2IG} + \frac{df_1(y)}{dy}$$

$$K = -\frac{Pc^2}{2IG}$$

Therefore, $F_1(x)$ and $F_2(y)$ are both constants.

Let $$-\frac{Px^2}{2EI} + \frac{df_2(x)}{dx} = K_1$$

$$\frac{\nu Py^2}{2EI} - \frac{Py^2}{2IG} + \frac{df_1(y)}{dy} = K_2$$

$$\therefore \quad \frac{df_2(x)}{dx} = K_1 + \frac{Px^2}{2EI}$$

$$\frac{df_1(y)}{dy} = K_2 - \frac{\nu Py^2}{2EI} + \frac{Py^2}{2IG}$$

Integrating, we get

$$f_2(x) = K_1 x + \frac{Px^3}{6EI} + K_2$$

$$f_4(y) = K_2 y - \frac{\nu Py^3}{6EI} + \frac{Py^3}{6IG} + K_4$$

Substituting in Eqs. (*a*) and (*b*), we get

$$u_x = -\frac{Px^2 y}{2EI} - \frac{\nu Py^3}{6EI} + \frac{Py^3}{6IG} + K_2 y + K_4$$

$$u_y = \frac{\nu Pxy^2}{2EI} + \frac{Px^3}{6EI} + K_1 x + K_3$$

The constants K_1, K_2, K_3 and K_4 are to be determined from the boundary conditions.

At the fixed end of the cantilever, $x = l$ and $y = 0$

$$u_x = u_y = 0.$$

$$\therefore \quad K_4 = 0 \text{ and } K_3 = -\frac{Pl^3}{6EI} - K_1 l$$

(*a*) Let $\left(\dfrac{\partial u_y}{\partial x}\right)_{\substack{x=l \\ y=0}} = 0$, *i.e.* a horizontal element of the axis of the beam is fixed at the fixed end.

$$\therefore \quad \frac{\nu Py^2}{2EI} + \frac{Px^2}{2EI} + K_1 = \frac{\partial u_y}{\partial x}$$

$$\therefore \quad K_1 = \frac{-Pl^2}{2EI}$$

Hence $$K_3 = -\frac{Pl^3}{6EI} + \frac{Pl^3}{2EI} = \frac{Pl^3}{3EI}$$

Also $$K_1 + K_2 = -\frac{Pc^2}{IG}$$

$$\therefore \quad K_2 = -\frac{Pc^2}{2IG} + \frac{Pl^2}{2EI}$$

$$\therefore \quad u_x = -\frac{Px^2y}{2EI} - \frac{\nu Py^3}{6EI} + \frac{Py^3}{6IG} - \frac{Pc^2y}{2IG} + \frac{Pl^2y}{2EI} \quad \text{...(2.15 a)}$$

$$u_y = \frac{\nu Pxy^2}{2EI} + \frac{Px^3}{6EI} - \frac{Pl^2x}{2EI} + \frac{Pl^3}{3EI} \quad \text{...(2.15 b)}$$

The equation of the deflection curve is

$$(u_y)_{y=0} = \frac{Px^3}{6EI} - \frac{Pl^2x}{2EI} + \frac{Pl^3}{3EI} \quad \text{...(2.16)}$$

$$(u_y)_{\substack{x=0\\y=0}} = \frac{Pl^3}{3EI} \quad \text{...(2.17)}$$

$$(u_x)_{x=l} = -\frac{\nu Py^3}{6EI} + \frac{Py^3}{6IG} - \frac{Pc^2y}{2IG} \quad \text{...(2.18)}$$

$$\left(\frac{\partial u_x}{\partial y}\right)_{x=l} = -\frac{\nu Py^2}{2EI} - \frac{Py^2}{2IG} - \frac{Pc^2}{2IG} \quad \text{...(2.19)}$$

$$\left(\frac{\partial u_x}{\partial y}\right)_{\substack{x=l\\y=0}} = -\frac{Pc^2}{2IG} = -\frac{3}{4}\frac{P}{cG} \quad \text{...(2.20)}$$

The shape of the cross-section of the cantilever after distortion is as shown in Fig. 2.2 (*a*). Owing to the shearing stress $\tau_{xy} = -\frac{3P}{4c}$ at the fixed end, an element of the cross-section rotates

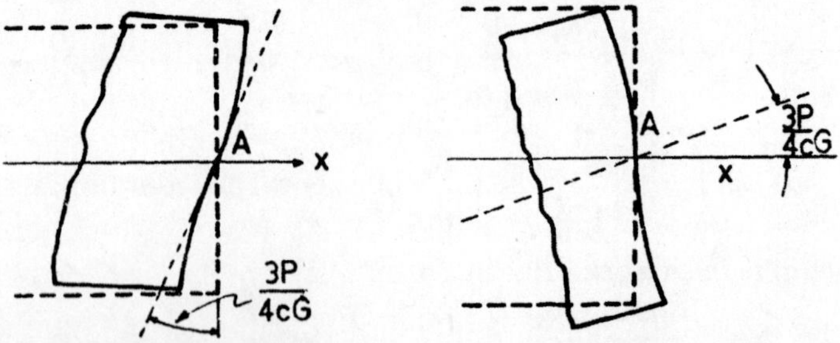

(*a*) At any location (*b*) At the fixed end

Fig. 2.2 Cantilever shape after distortion.

in the xy plane about the fixed end through an angle $\frac{3P}{4cG}$ in the clockwise direction.

(b) $\left(\frac{\partial u_x}{\partial y}\right)_{\substack{x=l\\y=0}} = 0$, *i.e.* a vertical element of the cross-section at the fixed end is fixed [Fig. 2.2 (b)].

$$\frac{\partial u_x}{\partial y} = -\frac{Px^2}{2EI} - \frac{\nu Py}{3EI} + \frac{Py^2}{2IG} + K_2$$

$$\therefore \quad K_2 = \frac{Pl^2}{2EI}$$

$$K_1 = -\frac{Pc^2}{2IG} - \frac{Pl^2}{2EI}$$

$$K_3 = -\frac{Pl^3}{6EI} + \frac{Pc^2l}{2IG} + \frac{Pl^3}{2EI}$$

$$K_3 = \frac{Pl^2}{3EI} + \frac{Pc^2l}{2IG}$$

$$\therefore \quad u_x = -\frac{Px^2y}{2EI} - \frac{\nu Py^2}{6EI} + \frac{Py^3}{6IG} + \frac{Pl^2y}{2EI} \qquad \text{...(2.21 a)}$$

$$u_y = \frac{\nu Pxy^2}{2EI} + \frac{Px^3}{6EI} - \frac{Pc^2x}{2IG} - \frac{Pl^2x}{2EI} + \frac{Pl^3}{3EI} + \frac{Pc^2l}{2IG} \qquad \text{...(2.21 b)}$$

$$(u_y)_{y=0} = \frac{Px^3}{6EI} + \frac{Pc^2}{2IG}(l-x) - \frac{Pl^2x}{2EI} + \frac{Pl^3}{3EI} \qquad \text{...(2.22)}$$

Comparing Eqs. (2.16) and (2.22) we find that in case (b) the deflection is increased by an amount

$$\frac{Pc^2}{2IG}(l-x) = \frac{3}{4}\frac{P(l-x)}{cG}$$

due to the effect of shearing force.

$$\frac{\partial u_y}{\partial x} = \frac{\nu Py^2}{2EI} + \frac{Px^2}{2EI} - \frac{Pc^2}{2IG} - \frac{Pl^2}{2EI}$$

$$\left(\frac{\partial u_y}{\partial x}\right)_{\substack{x=l\\y=0}} = -\frac{Pc^2}{2IG} = -\frac{3}{4}\frac{P}{cG} \qquad \text{...(2.23)}$$

2.2.1.2 Bending of a Beam by Uniform Load

Consider a beam of narrow rectangular cross-section of unit width, supported at the ends, be bent by a uniform distributed load of intensity q, as shown in Fig. 2.3. The conditions at the upper and lower edges of the beam are :

$$(\tau_{xy})_{y=\pm c} = 0,\ (\sigma_y)_{y=+c} = 0,\ (\sigma_y)_{y=-c} = q \qquad \text{...(a)}$$

The conditions at the ends are :

$$\left.\begin{aligned}\int_{-c}^{+c} (\tau_{xy})_{x=\pm l}\, dy &= \mp ql \\ \int_{-c}^{+c} (\sigma_x)_{x=\pm l}\, dy &= 0 \\ \int_{-c}^{+c} (\sigma_x)_{x=\pm l}\, y dy &= 0\end{aligned}\right\} \quad ...(b)$$

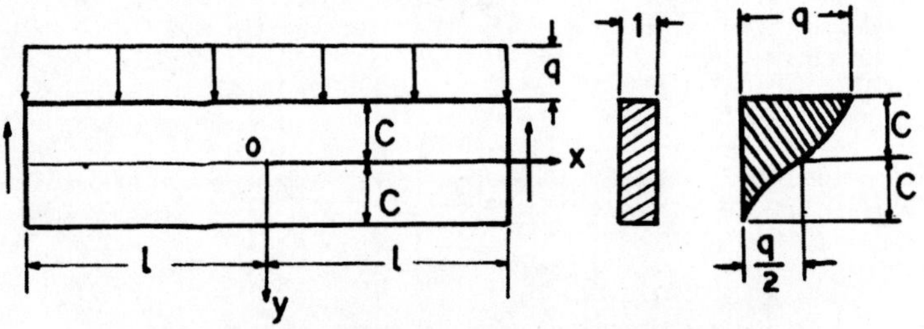

Fig. 2.3 Bending of beam by uniform load.

Consider a polynomial of second degree with

$$b_2 = c_2 = 0,$$

a polynomial of third degree with

$$a_3 = c_3 = 0,$$

and a polynomial of fifth degree with

$$a_5 = b_5 = c_5 = e_5 = 0$$

and adding, we get

$$\Phi = \frac{a_2}{2} x^2 + \frac{b_3}{2} x^2 y + \frac{d_3}{6} y^3 + \frac{d_5}{6} x^2 y^3 - \frac{2d_5}{60} y^5$$

$$\sigma_x = \frac{\partial^2 \Phi}{\partial y^2} = d_3 y + d_5 \left(x^2 y - \frac{2}{3} y^3 \right)$$

$$\sigma_y = \frac{\partial^2 \Phi}{\partial x^2} = a_2 + b_3 y + \frac{d_5}{3} y^3$$

$$\tau_{xy} = - \frac{\partial^2 \Phi}{\partial x \partial y} = - b_3 x - d_5 x y^2$$

Using the boundary conditions Eqs. (a), we get

$$- b_3 - d_5 c^2 = 0$$

$$a_2 + b_3 c + \frac{d_5}{3} c^3 = 0$$

$$a_2 - b_3 c - \frac{d_5}{3} c^3 = - q$$

which gives $a_2 = \frac{-q}{2}$, $b_3 = \frac{3q}{4c}$ and $d_5 = \frac{-3q}{4c^3}$

Now using boundary conditions Eqs. (*b*), we have

$$\int_{-c}^{+c} (\sigma_x)_{x=\pm l} \cdot y\,dy = \int_{-c}^{+c} \left[d_3 y + d_5 \left(x^2 y - \frac{2}{3} y^3 \right) \right]_{x=\pm l} y\,dy$$

$$= \int_{-c}^{+c} \left[d_3 y + d_5 \left(l^2 y - \frac{2}{3} y^3 \right) \right] y\,dy = 0$$

which gives $d_3 = -d_5 \left(l^2 - \frac{2}{5} c^2 \right)$

$$= \frac{3q}{4c^3} \left(l^2 - \frac{2}{5} c^2 \right) = \frac{3q}{4c} \left(\frac{l^2}{c^2} - \frac{2}{5} \right)$$

$$\therefore \qquad \sigma_x = \frac{3q}{4c} \left(\frac{l^2}{c^2} - \frac{2}{5} \right) y - \frac{3q}{4c^3} \left(x^2 y - \frac{2}{3} y^3 \right)$$

$$= \frac{q}{2I} (l^2 - x^2)\, y + \frac{q}{I} \left(\frac{y^3}{3} - \frac{c^2 y}{5} \right) \qquad \text{...(2.24 a)}$$

$$\sigma_y = -\frac{q}{2} + \frac{3q}{4c} y - \frac{qy^3}{4c^3}$$

$$= -\frac{q}{2I} \left(\frac{y^3}{3} - c^2 y + \frac{2}{3} c^3 \right) \qquad \text{...(2.24 b)}$$

$$\tau_{xy} = -\frac{3q}{4c} x + \frac{3q}{4c^3} xy^2$$

$$= -\frac{q}{2I} x(c^2 - y^2) \qquad \text{...(2.24 c)}$$

where $I = \frac{2}{3} c^3$

The corresponding results from strength of materials are :

$$\sigma_x = \frac{My}{I} = \frac{q}{2I} (l^2 - x^2)\, y$$

$$\sigma_y = 0 \qquad \text{...(2.25)}$$

$$\sigma_{xy} = \frac{VQ}{I} = -\frac{q}{2I} (c^2 - y^2) x$$

when $M = \frac{q}{2} (l^2 - x^2)$ = bending moment

$V = -qx$ = shear force

$Q = \frac{1}{2} (c^2 - y^2)$ = static moment w.r.t. neutral axis.

A comparison of the results obtained from the theory of elasticity and strength of materials is shown in Fig. 2.4.

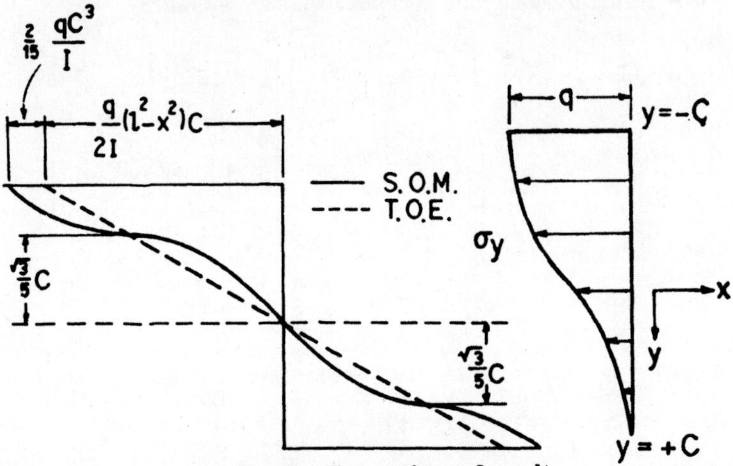

Fig. 2.4 Comparison of results.

Now
$$\varepsilon_x = \frac{\partial u_x}{\partial x} = \frac{1}{E}[\sigma_x - \nu\sigma_y]$$
$$= \frac{q}{2EI}\left\{\left[(l^2 - x^2)y + \frac{2}{3}y^3 - \frac{2}{5}c^2y\right] + \nu\left[\frac{y^3}{3} - c^2y + \frac{2}{3}c^3\right]\right\} \quad ...(c)$$

$$\varepsilon_y = \frac{\partial u_y}{\partial y} = \frac{1}{E}[\sigma_y - \nu\sigma_x]$$
$$= -\frac{q}{2EI}\left\{\left(\frac{y^3}{3} - c^2y + \frac{2}{3}c^3\right) + \nu\left[(l^2 - x^2)\,y + \frac{2}{3}y^3 - \frac{2}{5}c^2y\right]\right\} \quad ...(d)$$

$$\gamma_{xy} = \frac{\partial u_x}{\partial y} + \frac{\partial u_y}{\partial x} = \frac{\tau_{xy}}{G}$$
$$= -\frac{q}{2IG}x(c^2 - y^2) \quad ...(e)$$

Integrating Eqs. (c) and (d), we get
$$u_x = \frac{q}{2EI}\left[\left(l^2x - \frac{x^2}{3}\right)y + x\left(\frac{2}{3}y^3 - \frac{2}{5}c^2y\right) + \nu x\left(\frac{y^2}{3} - c^2y + \frac{3}{3}c^3\right)\right] + f_2(y)$$

$$u_y = -\frac{q}{2EI}\left[\left(\frac{y^4}{12} - \frac{c^2y^2}{2} + \frac{2}{3}c^3y\right)\right.$$

$$\left. + \nu\,(l^2 - x^2)\frac{y^2}{2} + \nu\left(\frac{y^4}{5} - \frac{c^2y^2}{5}\right)\right] + f_1(x)$$

Substituting in Eq. (*e*), we get

$$\frac{q}{2EI}\left[\left(l^2x - \frac{x^3}{3}\right) + x\left(2y^2 - \frac{2}{5}c^2\right) + \nu x(y^2 - c^2)\right]$$

$$+ \frac{df_2(y)}{dy} + \frac{q}{2EI} \times \nu xy^2 + \frac{df_1(x)}{dx} = -\frac{q}{2IG}x(c^2 - y^2)$$

$$\therefore \quad \frac{df_2(y)}{dy} = -\frac{df_1(x)}{dx} + \frac{qx^3}{6EI} - \frac{q}{2EI}\left[l^2 + \left(\frac{8}{5} + \nu\right)c^2\right]x = A$$

$$\frac{df_1(x)}{dx} = \frac{qx^3}{6EI} - \frac{q}{2EI}\left[l^2 + \left(\frac{8}{5} + \nu\right)c^2\right]x - A$$

Integrating we get

$$f_2(y) = Ay + B$$

$$f_1(x) = \frac{qx^4}{24EI} - \frac{q}{4EI}\left[l^2 + \left(\frac{8}{5} + \nu\right)c^2\right]x^2 - Ax + C$$

where *A*, *B*, *C* are constants.

$$u_x = \frac{q}{2EI}\left[\left(l^2x - \frac{x^3}{3}\right)y + x\left(\frac{2}{3}y^3 - \frac{2}{5}c^2y\right)\right.$$

$$\left. + \nu x\left(\frac{y^3}{3} - c^2y + \frac{2}{3}c^3\right)\right] + Ay + B$$

$$u_y = -\frac{q}{2EI}\left[\left(\frac{y^4}{12} - \frac{c^2y^2}{2} + \frac{2}{3}c^3y\right) + \nu(l^2 - x^2)\frac{y^2}{2}\right.$$

$$\left. + \nu\left(\frac{y^4}{6} - \frac{c^2y^2}{5}\right)\right] + \frac{qx^4}{24EI}$$

$$-\frac{q}{4EI}\left[l^2 + \left(\frac{8}{5} + \nu\right)c^2\right]x^2 - Ax + C$$

Now $u_x = 0$ for $x = 0$ and any value of y

$$\therefore \quad A = B = 0$$

and $\quad u_y = 0$ for $x = \pm l,\ y = 0$

$$\therefore \quad C = \frac{5}{24}\frac{ql^4}{EI}\left[1 + \frac{12}{5}\left(\frac{4}{5} + \frac{\nu}{2}\right)\frac{c^2}{l^2}\right]$$

Hence, we get

$$u_x = \frac{q}{2EI}\left[\left(l^2x - \frac{x^3}{3}\right)y + x\left(\frac{2}{3}y^3 - \frac{2}{5}c^2y\right)\right.$$

$$\left. + \nu x\left(\frac{y^2}{3} - c^2y + \frac{2}{3}c^3\right)\right] \qquad \text{...(2.26 a)}$$

$$u_y = -\frac{q}{2EI}\left[\frac{y^4}{12} - \frac{c^2y^2}{2} + \frac{2}{3}c^3y\right.$$

$$+ \nu\left\{(l^2 - x^2)\frac{y^2}{2} + \frac{y^4}{6} - \frac{c^2y^2}{5}\right\} - \frac{x^4}{12}$$

$$\left. + \frac{l^2}{2} + \left(\frac{4}{5} + \frac{\nu}{2}\right)c^2\right\}x^2\right]$$

$$+ \frac{5}{24}\frac{ql^4}{EI}\left[1 + \frac{12}{5}\left(\frac{4}{5} + \frac{\upsilon}{2}\right)\frac{c^2}{l^2}\right] \qquad \text{...(2.26 b)}$$

$$(u_y)_{max} = (u_y)_{x=y=0} = \frac{5}{24}\frac{ql^4}{EI}\left[1 + \frac{12}{5}\left(\frac{4}{5} + \frac{\nu}{2}\right)\frac{c^2}{l^2}\right] \qquad \text{...(2.27)}$$

The factor before the brackets is the deflection that is derived by the elementary analysis, assuming that cross-sections of the beam remain plane during bending. The second term in the brackets represents the correction usually called the effect of shearing force.

Example 2.1 *Prove that the following are Airy's stress functions and examine the stress distribution represented by them :*

(a) $\phi = Ax^2 + By^2$ *(b)* $\phi = Ax^3$, *and*

(c) $\phi = A(x^4 - 3x^2y^2)$.

Solution.

(a)

$$\frac{\partial\phi}{\partial x} = 2Ax, \qquad \frac{\partial\phi}{\partial y} = 2By$$

$$\frac{\partial^2\phi}{\partial x^2} = 2A, \qquad \frac{\partial^2\phi}{\partial y^2} = 2B$$

$$\frac{\partial^4\phi}{\partial x^4} = 0, \qquad \frac{\partial^4\phi}{\partial y^4} = 0$$

$$\frac{\partial^2\phi}{\partial x^2\partial y^2} = 0$$

$$\nabla^4\phi = \frac{\partial^4\phi}{\partial x^4} + \frac{2\partial^4\phi}{\partial x^2\partial y^2} + \frac{\partial^4\phi}{\partial y^4}$$

$$= 0$$

Hence, it is an Airy stress function.

$$\sigma_x = \frac{\partial^2\phi}{\partial y^2} = 2B$$

$$\sigma_y = \frac{\partial^2\phi}{\partial x^2} = 2A$$

$$\tau_{xy} = -\frac{\partial^2\phi}{\partial x\partial y} = 0$$

It represents a uniform two-dimensional stress field.

(*b*) $$\frac{\partial\phi}{\partial x} = 3Ax^2, \qquad \frac{\partial\phi}{\partial y} = 0 = \frac{\partial^2\phi}{\partial y^2} = \frac{\partial^4\phi}{\partial y^4}$$

$$\frac{\partial^2\phi}{\partial x^2} = 6Ax$$

$$\frac{\partial^3\phi}{\partial x^3} = 6A$$

$$\frac{\partial^4\phi}{\partial x^4} = 0$$

$$\nabla^4\phi = 0$$

Therefore, it is an Airy's stress function.

$$\sigma_x = 0,\ \sigma_y = 2Ax,\ \tau_{xy} = 0$$

It represents a one-dimensional linearly varying stress field.

(*c*) $$\frac{\partial\phi}{\partial x} = A(4x^3 - 6xy^2), \qquad \frac{\partial\phi}{\partial y} = A(-6x^2y)$$

$$\frac{\partial^2\phi}{\partial x^2} = A(12x^2 - 6y^2), \qquad \frac{\partial^2\phi}{\partial y^2} = -6Ax^2$$

$$\frac{\partial^3\phi}{\partial x^3} = 24Ax, \qquad \frac{\partial^4\phi}{\partial y^4} = 0$$

$$\frac{\partial^4\phi}{\partial x^4} = 24A, \qquad \frac{\partial^3\phi}{\partial x^2\partial y} - 12Ay$$

$$\frac{\partial^2\phi}{\partial x\partial y} = -12Axy, \qquad \frac{\partial^4\phi}{\partial x^2\partial y^2} = -12A$$

$$24A - 24A + 0 = 0$$

Therefore, it is an Airy's stress function.

$$\sigma_x = -6Ax^2$$

$$\sigma_y = A(12x^2 - 6y^2)$$

$$\tau_{xy} = 12Axy$$

It represents a two-dimensional complex stress field with non-linear stress distribution.

Example 2.2 *Show that the Airy's stress function* $\phi = A\left(xy^3 - \frac{3}{4}xyh^2\right)$ *represents stress distribution in a cantilever beam loaded at the free end with load P. Find the value of A if* $\tau_{xy} = 0$ *at* $y = \pm\frac{h}{2}$ *where b and h are width and depth respectively of the cantilever.*

Solution.

$$\frac{\partial\phi}{\partial x} = A\left(y^3 - \frac{3}{4}yh^2\right)$$

$$\frac{\partial^2\phi}{\partial x^2} = 0, \qquad \frac{\partial^2\phi}{\partial x\partial y} = A\left(3y^2 - \frac{3}{4}h^2\right)$$

$$\frac{\partial\phi}{\partial y} = A\left(3xy^2 - \frac{3}{4}xh^2\right)$$

$$\frac{\partial^2\phi}{\partial y^2} = A(6xy)$$

$$\sigma_x = \frac{\partial^2\phi}{\partial y^2} = 6Axy$$

$$\sigma_y = \frac{\partial^2\phi}{\partial x^2} = 0$$

$$\tau_{xy} = \frac{\partial^2\phi}{\partial x\partial y} = -A\left(3y^2 - \frac{3}{4}h^2\right)$$

At $\quad y = \pm\frac{h}{2}, \tau_{xy} = 0$

$$0 = -A\left(\frac{3h^2}{4} - \frac{3}{4}h^2\right)$$

$$= -A(0)$$

Therefore, $A \neq 0$ and the shear stress is varying parabolically, being zero at the top and bottom of cantilever.

For elementary theory of simple bending of beams,

$$\sigma_x = \frac{M_x y}{I_{xx}} = \frac{12\,Pxy}{bh^3}$$

$$\sigma_y = 0$$

$$6Axy = \frac{12\,Pxy}{bh^3}$$

$$A = \frac{2P}{bh^3}$$

At $y = 0$,

$$\tau_{xy} = \frac{2P}{bh^3} \times \frac{3}{4} h^2 = \frac{1.5P}{bh}$$

2.2.2 Direct Method for Determining Airy Polynomial Stress Functions*

In this method the Airy stress functions are expressed as doubly infinite power series to desired polynomial forms on the bases of compatibility and boundary conditions only, without resort to guessing as is necessary in the conventional superposition and "semi-inverse" methods.

We express that stress function in the form of a doubly infinite power series as :

$$\Phi = \sum_{m=0}^{\infty} \sum_{n=0}^{\infty} C_{mn} x^m y^n \qquad ...(2.28)$$

where m and n are positive integers and C_{mn} are undetermined coefficients which may be arranged in the following matrix :

$$C_{mn} = \begin{bmatrix} C_{00} & C_{01} & C_{02} \ldots\ldots C_{0n} \\ C_{10} & C_{11} & C_{12} \ldots\ldots C_{1n} \\ \ldots\ldots & \ldots\ldots & \ldots\ldots \\ \ldots\ldots & \ldots\ldots & \ldots\ldots \\ C_{m0} & C_{m1} & C_{m2} \ldots\ldots C_{mn} \end{bmatrix} \qquad ...(2.29)$$

$$\sigma_x = \frac{\partial^2 \Phi}{\partial y^2} = \sum_{m=0}^{\infty} \sum_{n=0}^{\infty} n(n-1)\, C_{mn} x^m\, y^{n-2} \qquad ...(2.30)$$

$$\sigma_y = \frac{\partial^2 \Phi}{\partial x^2} = \sum_{m=2}^{\infty} \sum_{n=0}^{\infty} m(m-1)\, C_{mn} x^m y^{n-2}\, y^n \qquad ...(2.31)$$

$$\tau_{xy} = -\frac{\partial^2 \Phi}{\partial x \partial y} = -\sum_{m=1}^{\infty} \sum_{n=1}^{\infty} mn\, C_{mn}\, x^{m-1}\, y^{n-1} \qquad ...(2.32)$$

Evidently C_{00}, C_{01} and C_{10} can be omitted from matrix (2.29) for they do not appear in the expressions for stresses.

Now the stress compatibility equation in the absence of body forces is

$$\left(\frac{\partial^2}{\partial x^2} + \frac{\partial^2}{\partial y^2} \right) (\sigma_x + \sigma_y) = 0$$

*Ching-Yuan Neou, Journal of Applied Mechanics, Vol. 24, No. 3. Sept., 1957, pp. 387—390.

or
$$\left(\frac{\partial^2}{\partial x^2}+\frac{\partial^2}{\partial y^2}\right)\left(\frac{\partial^2\Phi}{\partial y^2}+\frac{\partial^2\Phi}{\partial x^2}\right)=0$$

$$\nabla^4\Phi = 0 \qquad ...(2.33)$$

where $\nabla^4 = \dfrac{\partial^4}{\partial x^4}+\dfrac{2\partial^4}{\partial x^2\partial y^2}+\dfrac{\partial^4}{\partial y^4}$

Substituting Eqs. (2.28) in (2.33) and regrouping, we get

$$\sum_{m=2}^{\infty}\sum_{n=2}^{\infty}\left.\begin{aligned}&\{(m+2)(m+1)m(m-1)\,C_{m+2,\,n-2}\\&+2m(m-1)n(n-1)C_{mn}\\&+(n+2)(n+1)n(n-1)\,C_{m-2,\,n+2}\}\end{aligned}\right\}x^{m-2}y^{n-2}=0$$

As the condition of compatibility is independent of x and y, the coefficient for $x^{m-2}y^{n-2}$ must be zero. Thus

$$(m+2)(m+1)m(m-1)C_{m+2,\,n-2}+2m(m-1)n(n-1)C_{mn}$$
$$+(n+2)(n+1)n(n-1)\,C_{m-2,\,n+2}=0 \qquad ...(2.34)$$

Eq. (2.34) establishes the inter-relationships among any three alternate coefficients in the diagonals running from the lower left to the upper right diagonals in the matrix given by Eq. (2.29) starting with the $\{C_{40}, C_{22}, C_{04}\}$ group.

For example, C_{60}, C_{20} and C_{24} are related by

$$(6.5.4.3)\,C_{60}+2(4.3)(2.1)C_{42}+(4.3.2.1)C_{24}=0$$

Other groups in the same diagonal are

$$\{C_{42}, C_{24}, C_{06}\} \text{ and } \{C_{51}, C_{33}, C_{15}\}.$$

The whole problem is then reduced to the determination of C_{mn} is Eq. (2.28) from the prescribed forces on the boundaries and the interdependence among C_{mn} governed by Eq. (2.34). The process of eliminating unwanted terms and determining the values for C_{mn} of the remaining terms in Eq. (2.28) can best be illustrated by way of the following two examples :

2.2.2.1 Cantilever having udl and Concentrated Load at the Free End

Consider the cantilever as shown in Fig. 2.5 of rectangular cross-section having unit width

Since $\tau_{xy} = 0$ at $y = \pm c$ for $0 < x < l$ *i.e.* being independent of x, the coefficients of powers of x in Eq. (2.32) must all be zero.

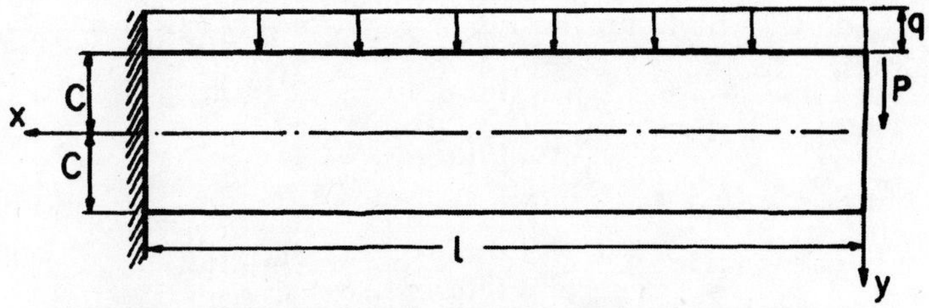

Fig. 2.5 Cantilever subjected to end load and *udl*.

$$\sum_{n=1}^{\infty} nC_{mn}c^{n-1} = 0 \text{ for } m \geq 1 \qquad ...(1)$$

$$\sum_{n=1}^{\infty} nC_{mn}(-c)^{n-1} = 0 \text{ for } m \geq 1 \qquad ...(2)$$

Addition and subtraction of Eqs. (1) and (2), multiplied by c yield

$$\sum_{n=2,4,6}^{\infty} nC_{mn}\, c^{n} = 0 \text{ for } m \geq 1 \qquad ...(3)$$

$$\sum_{n=1,3,5}^{\infty} nC_{mn}c^{n} = 0 \text{ for } n \geq 1 \qquad ...(4)$$

Similarly the boundary condition that $\sigma_y = 0$ at $y = +c$ for $0 < x < l$ requires

$$\sum_{n=0}^{\infty} C_{mn}\, c^{n} = 0 \text{ for } m \geq 2 \qquad ...(5)$$

and $$\sigma_y = -q \text{ at } y = -c \text{ for } 0 < x < l.$$

Hence Eq. (2.31) becomes

$$-q = \sum_{m=2}^{\infty} \left\{ \sum_{n=0}^{\infty} C_{mn}\,(-c)^{n} \right\} m(m-1)x^{m-2}$$

Hence $$\sum_{n=0}^{\infty} C_{mn}(-c)^{n} = 0 \text{ for } m \geq 3 \qquad ...(6)$$

$$2\sum_{n=0}^{\infty} C_{2n}(-c)^{n} = -q \qquad ...(7)$$

Addition and subtraction of Eqs. (5) for $m \geq 3$ and Eq. (6) results in

$$\sum_{n=0,2,4}^{\infty} C_{mn}\, c^n = 0 \text{ for } m \geq 3 \qquad \text{...(8)}$$

$$\sum_{n=1,3,5}^{\infty} C_{mn}\, c^n = 0 \text{ for } m \geq 3 \qquad \text{...(9)}$$

Eq. (9) for $m \geq 3$ and Eq. (8) can simultaneously be satisfied by

$$C_{mn} = 0 \text{ for } m \geq 3, n = 0, 2, 4, \ldots..$$

Similarly, Eq. (4) for $m \geq 3$ and Eq. (9) yield,

$$C_{mn} = 0 \text{ for } m \geq 3, n = 1, 3, 5, \ldots.$$

These C_{mn} can now be dropped from matrix given by Eq. (2.29), which are shown by / in matrix (10). A check on the compatibility relation (2.34) shows that $C_{14} = 0$ because $C_{50} = C_{32} = 0$. Similarly, we can put to zero all the C_{mn} below the diagonal running from C_{60} to C_{06}, which are shown by × in matrix Eq. (10).

$$\begin{bmatrix} - & - & \not{C}_{02} & C_{03} & \xcancel{C}_{04} & C_{05} & \xcancel{C}_{06} & - \\ - & C_{11} & \not{C}_{12} & C_{13} & \xcancel{C}_{14} & \xcancel{C}_{15} & \xcancel{C}_{16} & - \\ C_{20} & C_{21} & \not{C}_{22} & C_{23} & \xcancel{C}_{24} & \xcancel{C}_{25} & \xcancel{C}_{26} & - \\ \not{C}_{30} & \not{C}_{31} & \not{C}_{32} & \not{C}_{33} & \not{C}_{34} & \not{C}_{35} & \not{C}_{36} & + \\ \not{C}_{40} & \not{C}_{41} & \not{C}_{42} & \not{C}_{43} & \not{C}_{44} & \not{C}_{45} & \not{C}_{46} & + \\ \not{C}_{50} & \not{C}_{51} & \not{C}_{52} & \not{C}_{53} & \not{C}_{54} & \not{C}_{55} & \not{C}_{56} & + \\ \not{C}_{60} & \not{C}_{61} & \not{C}_{62} & \not{C}_{63} & \not{C}_{64} & \not{C}_{65} & \not{C}_{66} & + \\ + & + & + & + & + & + & + & + \end{bmatrix} \qquad \text{...(10)}$$

Now we write out Eqs. (3), (4), (5) and (7) for $m = 2$ in detail

$$2C_{22}c^2 = 0$$

$$C_{21}c + 3C_{23}c^3 = 0$$

$$C_{20} + C_{21}c + C_{22}c^2 + C_{23}c^3 = 0$$

$$C_{20} - C_{21}c + C_{22}c^2 - C_{23}c^3 = \frac{-q}{2}$$

from which we obtain

$$C_{22} = 0,\ C_{20} = -\frac{q}{4},\ C_{21} = \frac{3q}{8c},\ C_{23} = \frac{-q}{8c^3}$$

and again the compatibility relation of Eq. (2.34) shows

$$C_{04} = 0,\ C_{05} = -\frac{1}{5}C_{23} = \frac{q}{40c^3}$$

Similarly Eqs. (3) and (4) for $m = 1$ yield,

$$C_{12} = 0,\ C_{11} = -3C_{13}c^2$$

The constants $C_{22} = C_{04} = C_{12} = 0$ are shown by \ in matrix Eq. (10)

The stress function Φ now becomes

$$\Phi = C_{02}y^2 + C_{03}y^3 + \frac{q}{40c^3}y^5 - 3C_{13}c^2xy$$

$$+ C_{13}xy^3 - \frac{q}{4}x^2 + \frac{3q}{8c}x^2y - \frac{q}{8c^3}x^2y^3 \quad \text{...(11)}$$

The three unknown coefficients C_{02}, C_{03} and C_{13} can readily be determined from the boundary conditions,

$$\int_{-c}^{+c} (\sigma_x)_{x=0}\, dy = 0 \quad \text{...(12)}$$

$$\int_{-c}^{+c} (\sigma_x)_{x=0}\, y\, dy = 0 \quad \text{...(13)}$$

$$\int_{-c}^{+c} (\tau_{xy})_{x=0}\, dy = -P \quad \text{...(14)}$$

Substituting Eq. (11) in Eqs. (12) to (14) yields $C_{02} = 0$, which is shown by | in matrix (10), $C_{03} = -\frac{q}{20c}$ and $C_{13} = -\frac{P}{4c^3}$.

Hence final stress function becomes

$$\Phi = \frac{q}{40}\left[y^2\left(-2\frac{y}{c} + \frac{y^3}{c^3}\right) - 5x^2\left(2 - \frac{3y}{c} + \frac{y^3}{c^3}\right)\right]$$

$$+ \frac{Px}{4}\left(\frac{3y}{c} - \frac{y^3}{c^3}\right) \quad \text{...(2.35)}$$

$$\sigma_x = \frac{qy}{20C}\left(10\frac{y^2}{c^2} - 6 - 15\frac{x^2}{c^2}\right) - \frac{3}{2}\frac{P}{c^2}xy \quad \text{...(2.36 a)}$$

$$\sigma_y = -\frac{q}{4}\left(2 + \frac{y}{c}\right)\left(1 - \frac{y}{c}\right)^2 \quad \text{...(2.36 b)}$$

$$\tau_{xy} = -\frac{3}{4c}(qx + p)\left(1 - \frac{y^2}{c^2}\right) \quad \text{...(2.36 c)}$$

2.2.2.2 Simply Supported Rectangular Beam under a Triangular Load.

Consider the beam as shown in Fig. 2.6. The symmetry of loading w.r.t. y-axis reveals that the stress function of Eq. (2.28) should contain no terms odd in x, except for $m = 1$ and 3, because of the dependence of σ_y at the top surface on the first degree of x. Also the asymmetry of loading at $y = \pm c$ leads to the dropping of term even in y. These C_{mn} can now be dropped from matrix (2.29) at an early stage (shown by / in matrix (1).

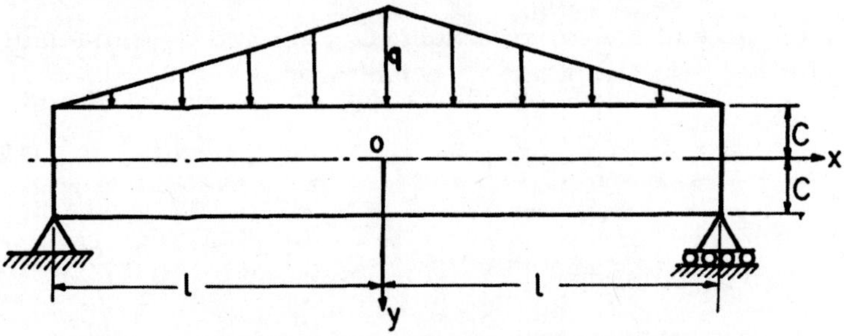

Fig. 2.6 Beam having triangular load.

$$\begin{bmatrix} - & - & \not{C}_{02} & C_{03} & \not{C}_{04} & C_{05} & \not{C}_{06} & \times \\ - & C_{11} & \not{C}_{12} & C_{13} & \not{C}_{14} & C_{15} & \not{C}_{16} & \times \\ C_{20} & C_{21} & \not{C}_{22} & C_{23} & \not{C}_{24} & \not{C}_{25} & \not{C}_{26} & \times \\ C_{30} & C_{31} & \not{C}_{32} & C_{33} & \not{C}_{34} & \not{C}_{35} & \not{C}_{36} & + \\ \not{C}_{40} & \not{C}_{41} & \not{C}_{42} & \not{C}_{43} & \not{C}_{44} & \not{C}_{45} & \not{C}_{46} & + \\ \not{C}_{50} & \not{C}_{51} & \not{C}_{52} & \not{C}_{53} & \not{C}_{54} & \not{C}_{55} & \not{C}_{56} & + \\ \not{C}_{60} & \not{C}_{61} & \not{C}_{62} & \not{C}_{63} & \not{C}_{64} & \not{C}_{65} & \not{C}_{66} & + \\ - & - & - & - & - & - & - & + \end{bmatrix} \quad \text{...(1)}$$

Hence forth, we need to consider only the right half of the beam. The boundary conditions at $y = \pm c$ yield $C_{mn} = 0$ for $m \geq 4$ (shown by/in matrix (1). The compatibility relation further demands that $C_{25} = C_{35} = 0$ and also all C_{mn} for $n > 6$ be zero (shown by $\times$ in matrix (1). For $m \leq 3$, Eqs, (3) and (4) of art 2.2.2.1 for $\tau_{xy} = 0$ at $y = \pm c$ yield

$$C_{11}c + 3C_{13}c^3 + 5C_{15}c^5 = 0 \quad \text{...(2)}$$

$$C_{21}c + 3C_{23}c^3 = 0 \quad \text{...(3)}$$

$$C_{31}c + 3C_{33}c^3 = 0 \quad \text{...(4)}$$

and Eq. (2.30) for $\sigma_x = 0$ at $y = c$ shows

$$C_{20} + C_{21}c + C_{23}c^3 = 0 \quad \text{...(5)}$$

$$C_{38} + C_{31}c + C_{33}c^3 = 0 \quad \text{...(6)}$$

while $\sigma_y = -q\left(1 - \frac{x}{c}\right)$ at $y = c$ results in

$$-q = (2C_{20} - C_{21}c - C_{23}c^3) \qquad ...(7)$$

$$\frac{q}{l} = 6(C_{30} - C_{31}c - C_{33}c^3) \qquad ...(8)$$

From Eqs. (3), (5) and (7), we find

$$C_{20} = -\frac{q}{4}, \; C_{21} = \frac{3q}{8c},$$

$$C_{23} = -\frac{q}{8c^3}$$

and from Eqs. (4), (6) and (8), we obtain

$$C_{30} = \frac{q}{12l} \cdot C_{31} = \frac{-q}{8cl}$$

$$C_{33} = \frac{q}{24c^3 l}$$

The compatibility Eq. (2.34) further shows that

$$C_{05} = -\frac{1}{5}C_{23} = \frac{q}{40c^3}$$

$$C_{15} = -\frac{3}{5}C_{33} = \frac{-q}{40c^3 l}$$

thereby from Eq. (2) we obtain

$$C_{11} = \frac{cq}{8l} - 3C_{13}c^2$$

Substituting in Eq. (2.29), we get

$$\int_{-c}^{+c} \tau_{xy}\, dy = \int_{-c}^{+c} \left(\frac{-\partial^2\Phi}{\partial x \partial y}\right)_{x=l} dy = \frac{-ql}{2} \qquad ..(10)$$

and $$\sigma_x = \left(\frac{\partial^2\Phi}{\partial y^2}\right)_{x=l} = 0 \text{ for } -c < y < +c \qquad ...(11)$$

we find $$C_{13} = \frac{q}{20cl},$$

$$C_{03} = \frac{q}{60c}\left(\frac{5l^2}{c^2} - 3\right).$$

Then from Eq. (9), we get

$$C_{11} = -\frac{qc}{40l}$$

Hence the stress function becomes

$$\Phi = \frac{q}{60c}\left(\frac{5l^2}{c^2} - 3\right)y^3 + \frac{q}{40c^3}y^5 - \frac{qc}{40l}xy + \frac{q}{20cl}xy^3$$

$$-\frac{q}{40c^3l}xy^5 - \frac{q}{4}x^2 + \frac{3q}{8c}x^2y - \frac{q}{8c^3}x^2y^3$$

$$+\frac{q}{12l}x^3 - \frac{q}{8cl}x^3y + \frac{q}{24c^3l}x^3y^3 \qquad \text{...(2.37)}$$

$$\sigma_x = \frac{q}{20}\left(\frac{y}{c}\right)\left\{-6 + 10\frac{y^2}{c^2} + 10\frac{l^2}{c^2}\right.$$

$$\left. + \frac{x}{l}\left(6 - 10\frac{y^2}{c^2} + 5\frac{x^2}{c^2} - 15\frac{lx}{c^2}\right)\right\} \qquad \text{...(2.38 } a\text{)}$$

$$\sigma_y = -\frac{q}{4}\left(1 - \frac{x}{l}\right)\left(1 - \frac{y}{c}\right)^2\left(2 + \frac{y}{c}\right) \qquad \text{...(2.38 } b\text{)}$$

$$\tau_{xy} = \frac{q}{40}\left(1 - \frac{y^2}{c^2}\right)\left\{\frac{c}{l}\left(1 - \frac{5y^2}{c^2}\right) + 15\frac{x}{c}\left(\frac{x}{l} - 2\right)\right\}$$

$$\text{...(2.38 } c\text{)}$$

2.2.3 Fourier Series

(*i*) Suppose that $f(x)$ is a periodic function of period $2l$ and that it is absolutely integrable over the interval $(-l, l)$. The following trigonometric series is called the Fourier series of $f(x)$.

$$f(x) = \frac{a_0}{2} + \sum_{m=1}^{\infty}\left(a_m \cos\frac{m\pi x}{l} + b_m \sin\frac{m\pi x}{l}\right) \qquad \text{...(2.39)}$$

The coefficients of which (the Fourier coefficients) are given by the formulae

$$a_m = \frac{1}{l}\int_{-l}^{+l} f(x)\cos\frac{m\pi x}{l}\,dx \qquad (m = 0, 1, 2, \ldots)$$

$$b_m = \frac{1}{l}\int_{-l}^{+l} f(x)\sin\frac{m\pi x}{l}\,dx \qquad (m = 1, 2, \ldots)$$

(*ii*) A function $f(x)$ defined in the interval $(0, l)$ can be expanded in a cosine series of the form

$$f(x) = \frac{a_0}{2} + \sum_{m=1}^{\infty} a_m \cos\frac{m\pi x}{l} \qquad \text{...(2.40)}$$

where $a_m = \frac{2}{l}\int_0^l f(x) \cos \frac{m\pi x}{l}\, dx$ $(m = 1, 2, ...)$

or sine series of the form

$$f(x) = \sum_{m=1}^{\infty} b_m \sin \frac{m\pi x}{l}$$

where $b_m = \frac{2}{l}\int_0^l f(x) \sin \frac{m\pi x}{l}\, dx$ $(m = 1, 2, ...)$

(*iii*) The Fourier series of a function $f(x)$ in the interval $l < x < l + 2\pi$ is given by

$$f(x) = \frac{a_0}{2} + \sum_{m=1}^{\infty} (a_m \cos mx + b_m \sin mx) \quad ...(2.41)$$

where $a_m = \frac{1}{2\pi}\int_l^{l+2\pi} f(x) \cos mx\, dx$

$b_m = \frac{1}{2\pi}\int_l^{l+2\pi} f(x) \sin mx\, dx.$

(*iv*) The Fourier series of a function $f(x)$ in the interval $l < x < l + 2\pi$ can be written in the complex form as

$$f(x) = a_0 + \sum_{m=1}^{\infty} (a_m e^{imx} + a_{-m} e^{-imx})$$

$$= \sum_{m=-\infty}^{\infty} a_m e^{imx} \quad ...(2.42)$$

where $a_m = \frac{1}{2\pi}\int_l^{l+2\pi} f(x)\, e^{-imx}\, dx$ $(m = 0, \pm 1, \pm 2, ...)$

2.2.3.1 Solution of Biharmonic Equation by Fourier Series

The Biharmonic equation is

$$\nabla^4\Phi = 0 \quad ...(1)$$

where $\nabla^4 = \frac{\partial^4}{\partial x^4} + 2\frac{\partial^4}{\partial x^2\, \partial y^2} + \frac{\partial^4}{\partial y^4}$

By taking Φ in the form

$$\Phi = \sin \frac{m\pi x}{l} \cdot f(y) \qquad ...(2)$$

Equation (1) is satisfied, in which m is an integer and $f(x)$ a function of y only. Substituting Eq. (2) in (1) and putting $\frac{m\pi}{l} = \alpha$, we get

$$\alpha^4 f(y) - 2\alpha^2 f''(y) + f^{IV}(y) = 0 \qquad ...(3)$$

The general solution of this linear differential equation with constant coefficients is

$$f(y) = C_1 \cosh \alpha y + C_2 \sinh \alpha y + C_3 y \cosh \alpha y + C_4 y \sinh \alpha y$$

The stress function then becomes

$$\Phi = \sin \alpha x \, (C_1 \cosh \alpha y + C_2 \sinh \alpha y + C_3 y \cosh \alpha y + C_4 y \sinh \alpha y) \qquad ...(4)$$

and the corresponding stress components are :

$$\sigma_x = \frac{\partial^2 \Phi}{\partial y^2} = \sin \alpha x \, [C_1 \alpha^2 \cosh \alpha y + C_2 \alpha^2 \sinh \alpha y + C_3 \alpha (2 \sinh \alpha y + \alpha y \cosh \alpha y) + C_4 \alpha \, (2 \cosh \alpha y + \alpha y \sinh \alpha y)] \qquad ...(2.43)$$

$$\sigma_y = \frac{\partial^2 \Phi}{\partial x^2} = -\alpha^2 \sin \alpha x \, [C_1 \cosh \alpha y + C_2 \sinh \alpha y + C_3 y \cosh \alpha y + C_4 y \sinh \alpha y] \qquad ...(2.44)$$

$$\tau_{xy} = \frac{-\partial^2 \Phi}{\partial x \partial y} = -\alpha \cos \alpha x \, [C_1 x \sinh \alpha y + C_2 x \cosh \alpha y + C_3 (\cosh \alpha y + \alpha y \sinh \alpha y) + C_4 (\sinh \alpha y + \alpha y \cosh \alpha y)] \qquad ...(2.45)$$

2.2.3.2 Simply Supported Beam having Sinusoidal Loading on the Upper and Lower Edges

Consider a rectangular beam as shown in Fig. 2.7 supported at the ends and subjected along the upper and lower edges to the action of continuously distributed vertical forces of intensity $A \sin \alpha x$ and $B \sin \alpha x$ respectively. Let us take

$$\alpha = \frac{4\pi}{l}.$$

The boundary conditions are

(*i*) $(\tau_{xy})_{y = \pm c} = 0$

(*ii*) $(\sigma_y)_{y = +c} = -B \sin \alpha x$ ...(1)

(*iii*) $(\sigma_y)_{y = -c} = -A \sin \alpha x$

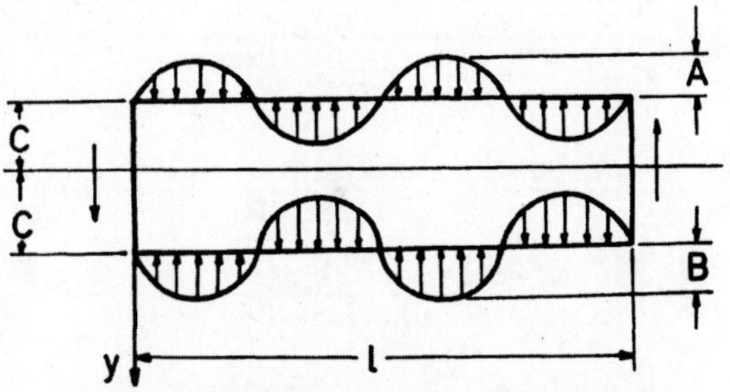

Fig. 2.7 Beam having sinusoidal loading.

Substituting in Eq. (2.45), we get

$$C_1\alpha \sin \alpha c + C_2\alpha \cosh \alpha c + C_3 (\cosh \alpha c + \alpha c \sinh \alpha c) + C_4 (\sinh \alpha c + \alpha c \cosh \alpha c) = 0$$

$$-C_1\alpha \sinh \alpha c + C_2\alpha \cosh \alpha c + C_3 (\cosh \alpha c + \alpha c \sinh \alpha c) - C_4 (\sinh \alpha c + \alpha c \cosh \alpha c) = 0$$

Solving, we find

$$\left.\begin{aligned} C_3 &= -C_2 \frac{\alpha \cosh \alpha c}{\cosh \alpha c + \alpha C \sinh \alpha c} \\ C_4 &= -C_1 \frac{\alpha \sinh \alpha c}{\sinh \alpha c + \alpha c \cosh \alpha c} \end{aligned}\right\} \quad ...(2)$$

Substituting the boundary conditions in Eq. (2.44), we get

$$\alpha^2 (C_1 \cosh \alpha c + C_2 \sin \alpha c + C_3 c \cosh \alpha c + C_4 c \sinh \alpha c) = B$$

$$\alpha^2 (C_1 \cosh \alpha c - C_2 \sinh \alpha c - C_3 c \cosh \alpha c + C_4 c \sinh \alpha c) = A$$

By adding and subtracting these equations and using Eqs. (2), we get

$$\left.\begin{aligned} C_1 &= \left(\frac{A+B}{\alpha^2}\right)\left(\frac{\sinh \alpha c + \alpha c \cosh \alpha c}{\sinh 2\alpha c + 2\alpha c}\right) \\ C_2 &= -\left(\frac{A-B}{\alpha^2}\right)\left(\frac{\cosh \alpha c + \alpha c \sinh \alpha c}{\sinh 2\alpha c - 2\alpha c}\right) \\ C_3 &= \left(\frac{A-B}{\alpha^2}\right)\left(\frac{\alpha \cosh \alpha c}{\sinh 2\alpha c - 2\alpha c}\right) \\ C_4 &= -\left(\frac{A+B}{\alpha^2}\right)\left(\frac{\alpha \sinh \alpha c}{\sinh 2\alpha c + 2\alpha c}\right) \end{aligned}\right. \quad ...(3)$$

Substituting in Eqs. (2.43) to (2.45), we get

$$\sigma_x = (A + B)\left[\frac{(\alpha c \cosh \alpha c - \sinh \alpha c) \cosh \alpha y - \alpha y \sinh \alpha y \sinh \alpha c}{\sinh 2\alpha c + 2\alpha c}\right] \times \sin \alpha x$$

$$-(A-B)\left[\frac{(\alpha c \sinh \alpha c - \cosh \alpha c)\sinh \alpha y - \alpha y \cosh \alpha y \cosh \alpha c}{\sinh 2\alpha c - 2\alpha c}\right] \times \sin \alpha x \quad \text{...(4)}$$

$$\sigma_y = -(A+B)\left[\frac{(\alpha c \cosh \alpha c + \sinh \alpha c)\cosh \alpha y - \alpha y \sinh \alpha y \sinh \alpha c}{\sinh 2\alpha c + 2\alpha c}\right] \times \sin \alpha x$$

$$+(A-B)\left[\frac{(\alpha c \sinh \alpha c + \cosh \alpha c)\sinh \alpha y - \alpha y \cosh \alpha y \cosh \alpha c}{\sinh 2\alpha c - 2\alpha c}\right] \times \sin \alpha x \quad \text{...(5)}$$

$$\tau_{xy} = -(A+B)\left[\frac{\alpha c \cosh \alpha c \sinh \alpha y - \alpha y \cosh \alpha y \sinh \alpha c}{\sinh 2\alpha c + 2\alpha c}\right] \cos \alpha x$$

$$+(A-B)\left[\frac{\alpha c \sinh \alpha c \cosh \alpha y - \alpha y \sinh \alpha y \cosh \alpha c}{\sinh 2\alpha c - 2\alpha c}\right] \cos \alpha x \quad \text{...(6)}$$

If $A = B$, then

$$(\sigma_y)_{y=0} = -2A\left[\frac{\alpha c \cosh \alpha c + \sinh \alpha c}{\sinh 2\alpha c + 2\alpha c}\right] \sin \alpha x \quad \text{...(7)}$$

For long beams $\alpha c = \frac{m\pi c}{l}$ is small, provided the number of half-waves m is not large. In that case, we have

$$\sinh \alpha c = \alpha c + \frac{(\alpha c)^3}{6} + \frac{(\alpha c)^5}{120} + \ldots\ldots$$

$$\cosh \alpha c = 1 + \frac{(\alpha c)^2}{2} + \frac{(\alpha c)^4}{24} + \ldots\ldots$$

After neglecting small quantities of higher order than $(\alpha c)^4$, we find

$$\sigma_y = -A \sin \alpha x \left[1 - \frac{(\alpha c)^4}{24}\right] \quad \text{...(2.46)}$$

Hence for small values of αc the distribution stresses over the middle plane is practically the same as on both horizontal edges ($y = \pm c$) of the beam.

$$(\sigma_y)_{y=0} = 2A\left[\frac{\alpha c \cosh \alpha c - \sinh \alpha c}{\sinh 2\alpha c + 2\alpha c}\right] \sin \alpha x \quad \text{...(2.47)}$$

$$(\tau_{xy})_{y=0} = 0 \quad \text{...(2.48)}$$

2.2.3.3 Rectangular Beam Subjected to a Band of Load of Width 2a

Consider a rectangular beam of length $2l$, unit width and depth equal to $2c$ subjected to a band of load of intensity q and width $2a$ on the upper and lower edges of the beam as shown in Fig. 2.8.

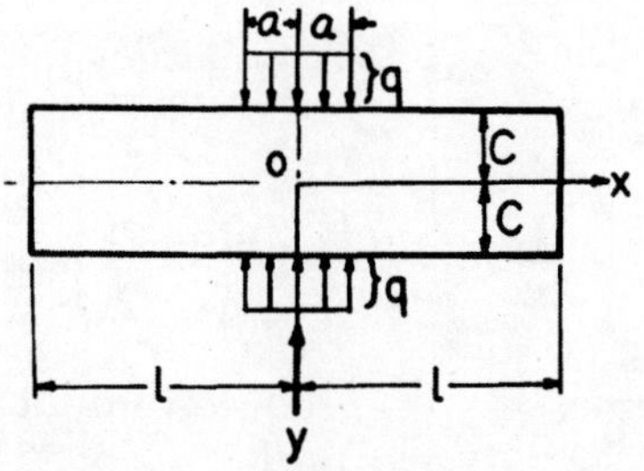

Fig. 2.8 Beam subjected to a band of load.

In the most general case, the distribution of vertical loading along the upper and lower edges of a beam can be represented by the Fourier series :

For the upper edge,

$$q_u = A_0 + \sum_{m=1}^{\infty} \left(A_m \sin \frac{m\pi x}{l} + A'_m \cos \frac{m\pi x}{l} \right)$$

For the lower edge,

$$q_l = B_0 + \sum_{m=1}^{\infty} \left(B_m \sin \frac{m\pi x}{l} + B'_m \cos \frac{m\pi x}{l} \right)$$

where $A_0 = \frac{1}{2l}\int_{-a}^{+a} q_u(x)\, dx \;;\quad A_m = \frac{1}{l}\int_{-a}^{+a} q_u(x) \sin \frac{m\pi x}{l}\, dx$

$$A'_m = \frac{1}{l}\int_{-a}^{+a} q_u(x) \cos \frac{m\pi x}{l}\, dx$$

$$B_0 = \frac{1}{2l}\int_{-a}^{+a} q_l(x)\, dx \;; \qquad B_m = \frac{1}{l}\int_{-a}^{+a} q_l(x) \sin \frac{m\pi x}{l}\, dx$$

$$B'_m = \frac{1}{l}\int_{-a}^{+a} q_l(x) \cos \frac{m\pi x}{l}\, dx.$$

Since in the present case, we have symmetrical loading, *i.e.* $q_u = q_l = q$ and the terms with $\sin\left(\frac{m\pi x}{l}\right)$ vanish. Further

$$A_0 = B_0 \text{ and } A'_m = B'_m$$

Now $\qquad A_0 = \frac{1}{2l}\int_{-a}^{+a} q\, dx = \frac{qa}{l}$

$$A'_m = \frac{1}{l}\int_{-a}^{+a} q \cos\frac{m\pi x}{l}\,dx = \frac{2q}{m\pi}\sin\left(\frac{m\pi a}{l}\right)$$

$$\therefore \qquad q = \frac{qa}{l} + \frac{2q}{\pi}\sum_{m=1}^{\infty}\frac{1}{m}\sin\left(\frac{m\pi a}{l}\right)\cos\left(\frac{m\pi x}{l}\right) \qquad \text{...(1)}$$

The first term of Eq. (1) produces a uniform compression in the y-direction equal to $\frac{qa}{l}$. The stresses produced by the trigonometric terms can be obtained by the same procedure as followed in Art. 2.2. 3.2. By inspection, we see that the solution for the present case can be obtained by exchanging sin αx for cos αx and changing the sign of τ_{xy}. On the middle plane $y = 0$, we have

$$(\sigma_y)_{y=0} = -\frac{qa}{l} - \frac{4q}{\pi}\sum_{m=1}^{\infty}\frac{\sin\left(\frac{m\pi a}{l}\right)}{m}$$

$$\times\left[\frac{\left(\frac{m\pi c}{l}\right)\cosh\left(\frac{m\pi c}{l}\right) + \sinh\left(\frac{m\pi c}{l}\right)}{\sinh\left(\frac{2m\pi c}{l}\right) + \left(\frac{2m\pi c}{l}\right)}\right]\cos\left(\frac{m\pi x}{l}\right) \qquad \text{...(2)}$$

When the dimension a is very small, *i.e.* concentrated force $P = 2qa$ and for an infinitely long strip, *i.e.*, $\frac{m\pi c}{l}$ is very small, we get

$$(\sigma_y)_{y=0} = -\frac{qa}{l} - \frac{2q}{\pi}\sum_{m=1}^{\infty}\frac{\pi a}{l}\left[1 - \frac{1}{24}\left(\frac{m\pi c}{l}\right)^4\right]\cos\left(\frac{m\pi x}{l}\right)$$

$$(\sigma_y)_{y=0} = -\frac{qa}{l} - \frac{2qa}{l}\sum_{m=1}^{\infty}\left[1 - \frac{1}{24}\left(\frac{m\pi c}{l}\right)^4\right]\cos\left[\frac{m\pi(x/c)}{(l/c)}\right] \qquad \text{...(2.49)}$$

Filon[4] evaluated Eq. (2.49) for various values of $\left(\frac{x}{c}\right)$. It was found that σ_y diminishes very rapidly with x as shown in Fig. 2.9 (*b*).

4. L.N.G. Filon, Trans. Roy. Soc. (London), Ser. A, Vol. 201, p. 67, 1903.

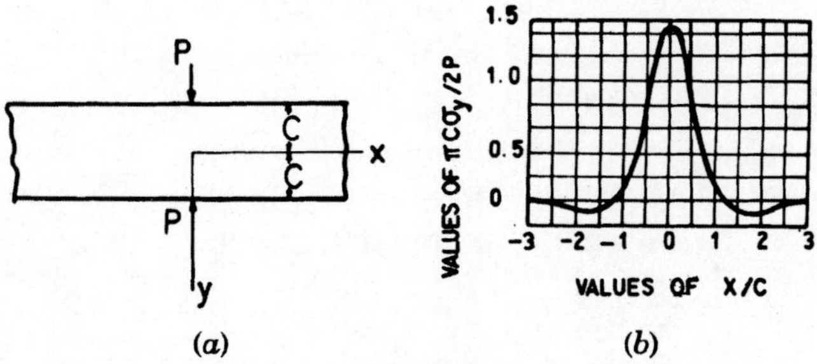

Fig. 2.9

At a value of $\frac{x}{c} = 1.35$, it becomes zero and is then replaced by tension.

2.3 Complex Potentials

Two-dimensional elasticity seems an obvious field in which the complex variable methods may be used. The early work of Kolosov and Muskhelishvili remained unknown outside Russia until recent years. The fundamental equations by complex potentials method were derived independently by Stevension[5], whose approach will be followed here.

2.3.1 Functions of a Complex Variable

If $z = x + iy$ and $\omega = \alpha + i\beta$ are two complex variables ($i = \sqrt{-1}$), and if for each value of z in some portion of the complex plane one or more values of ω are defined, then ω is said to be a function of z and we write

$$\omega = f(z)$$

If a unique value of ω exists for each value of z, then ω is called a single-valued function of z. If more than one value of ω corresponding to a given value of z exists, then ω is called a multivalued function of z.

Any function $\omega = f(z)$ can be written as

$$\omega = \alpha(x, y) + i\beta(x, y)$$

where $\alpha(x, y)$ and $\beta(x, y)$ are real functions of the real variables x and y.

The derivative of a function of a complex variable $\omega = f(z)$ is defined as

$$\frac{d\omega}{dz} = \omega' = f'(z)$$

5. A.C. Stevenson : Proc, Roy. Soc. London, Ser. A 184, 129 (1945), Phil. Mag., Ser. VII 34, 766 (1943).

$$= \underset{\Delta z \to 0}{\text{Lt}} \left[\frac{f(z + \Delta z) - f(z)}{\Delta z} \right]$$

The necessary and sufficient conditions for the existence of the derivative of a function of a complex variable defined as

$$\omega = f(z) = \alpha(x, y) + i\beta(x, y) \text{ are}$$

$$\left.\begin{aligned} \frac{\partial \alpha}{\partial x} &= \frac{\partial \beta}{\partial y} \\ \text{and} \quad \frac{\partial \alpha}{\partial y} &= -\frac{\partial \beta}{\partial x} \end{aligned}\right\} \qquad \text{...(2.50)}$$

These equations are known as the *Cauchy-Riemann* equations.

If $\omega = f(z)$ possesses a derivative at $z = z_0$ and at every point in some neighbourhood of z_0, then $f(z)$ is said to be analytic at z_0 and the point z_0 is called a regular point of the function. Analytic functions are sometimes called regular or holomorphic functions. If $f(z)$ is not analytic at z_0 but if every neighbourhood of z_0 contain points at which $f(z)$ is analytic, then z_0 is called a singular point of $f(z)$. An important property of the analytic functions is that both the real and imaginary parts satisfy Laplace equation, *i.e.*

$$\left.\begin{aligned} \frac{\partial^2 \alpha}{\partial x^2} + \frac{\partial^2 \alpha}{\partial y^2} &= 0 \\ \frac{\partial^2 \beta}{\partial x^2} + \frac{\partial^2 \beta}{\partial y^2} &= 0 \end{aligned}\right\} \qquad \text{...(2.51)}$$

A function which possesses continuous second partial derivatives and satisfies Laplace equation is called a Harmonic function.

The conjugate complex function of $z = x + iy$ is written as $\bar{z} = x - iy$. Two harmonic functions α and β so related that $\alpha + i\beta$ is an analytic function are called Conjugate Harmonic Functions.

Another important property of the analytic functions is that if $\omega = \alpha + i\beta$ is an analytic function then the curves of the family of $\alpha(x, y) = C_1$ are the orthogonal trajectories of the curves of the family $\beta(x, y) = C_2$ and *vice-versa*.

2.3.2 Derivation of Potentials

The variables (x, y) are replaced by the complex conjugate variables

$$z = x + iy$$

$$\bar{z} = x - iy$$

from which, we have $\dfrac{\partial}{\partial x} = \dfrac{\partial}{\partial z} + \dfrac{\partial}{\partial \bar{z}}$,

$$\frac{\partial}{\partial y} = i\left(\frac{\partial}{\partial z} - \frac{\partial}{\partial \bar{z}}\right) \qquad ...(1)$$

The body forces are represented in terms of a potential function $V(x, y) = U(z, \bar{z})$ such that

$$B_x = -\frac{\partial V}{\partial x},$$

$$B_y = -\frac{\partial V}{\partial y}$$

so that the equations of equilibrium can be written as

$$\frac{\partial}{\partial x}(\sigma_x - V) + \frac{\partial \tau_{xy}}{\partial y} = 0 \qquad ...(2)$$

$$\frac{\partial}{\partial y}(\sigma_y - V) + \frac{\partial \tau_{xy}}{\partial x} = 0 \qquad ...(3)$$

Multiplying Eq. (3) by i and adding to Eq. (2), we have by changing the variables to $(z, \bar{z})$ the complex equation.

$$\frac{\partial}{\partial z}(\sigma_x - \sigma_y + 2i\,\tau_{xy}) + \frac{\partial}{\partial \bar{z}}(\sigma_x + \sigma_y - 2U) = 0 \qquad ...(4)$$

which is obviously satisfied by a function $F(z, \bar{z})$, where

$$\frac{\partial F}{\partial z} = \sigma_x + \sigma_y - 2U \qquad ...(5)$$

$$-\frac{\partial F}{\partial \bar{z}} = \sigma_x - \sigma_y + 2i\,\tau_{xy} \qquad ...(6)$$

The complex displacement D is defined by the equation

$$D = u_x + iu_y$$

Using the relation (1) we have the equation

$$2\frac{\partial D}{\partial \bar{z}} = \frac{\partial u_x}{\partial x} - \frac{\partial u_y}{\partial y} + i\left(\frac{\partial u_y}{\partial x} + \frac{\partial u_x}{\partial y}\right)$$

$$= \varepsilon_x - \varepsilon_y + i\,\gamma_{xy}$$

Now

$$\varepsilon_x = \frac{1}{E}[\sigma_x - \nu\sigma_y]$$

$$\varepsilon_y = \frac{1}{E}[\sigma_y - \nu\sigma_x]$$

$$\gamma_{xy} = \frac{2(1+\nu)}{E}\tau_{xy}$$

$$\therefore\ \varepsilon_x - \varepsilon_y + i\,\gamma_{xy} = \frac{1}{E}[\sigma_x - \nu\sigma_y - \sigma_y + \nu\sigma_x + 2i(1+\nu)\,\tau_{xy}]$$

$$= \frac{1+\nu}{E}[\sigma_x - \sigma_y + 2i\tau_{xy}]$$

$$\therefore \qquad 2\frac{\partial D}{\partial \bar{z}} = \frac{1+v}{E}[\sigma_x - \sigma_y + 2i\,\tau_{xy}]$$

or

$$4G\frac{\partial D}{\partial \bar{z}} = \sigma_x - \sigma_y + 2i\,\tau_{xy} \qquad ...(7)$$

Using Eq. (6), we get

$$4G\frac{\partial D}{\partial \bar{z}} = -\frac{\partial F}{\partial \bar{z}}$$

Integrating, we get

$$4GD = f(z) - F(z, \bar{z}) \qquad ...(8)$$

where $f(z)$ is a function of z alone. Differentiating D w.r.t. z and using Eq. (1) we find that

$$2\frac{\partial D}{\partial z} = \frac{\partial u_x}{\partial x} + \frac{\partial u_y}{\partial y} + i\left(\frac{\partial u_y}{\partial x} - \frac{\partial u_x}{\partial y}\right)$$

Now for the case of plane strain, we have

$$\sigma_x = 2G\frac{\partial u_x}{\partial x} + \lambda\left(\frac{\partial u_x}{\partial x} + \frac{\partial u_y}{\partial y}\right)$$

$$\sigma_y = 2G\frac{\partial u_y}{\partial y} + \lambda\left(\frac{\partial u_x}{\partial x} + \frac{\partial u_y}{\partial y}\right)$$

$$\therefore \qquad \sigma_x + \sigma_y = (2\lambda + 2G)\left(\frac{\partial u_x}{\partial x} + \frac{\partial u_y}{\partial y}\right)$$

$$\therefore \qquad \frac{\partial u_x}{\partial x} + \frac{\partial u_y}{\partial y} = \frac{\sigma_x + \sigma_y}{2(\lambda + G)}$$

Hence $\quad 2\dfrac{\partial D}{\partial z} = \dfrac{\sigma_x + \sigma_y}{2(\lambda + G)} + i\left(\dfrac{\partial u_y}{\partial x} - \dfrac{\partial u_x}{\partial y}\right)$

$$4(\lambda + G)\frac{\partial D}{\partial z} = \sigma_x + \sigma_y + 2i(\lambda + G)\left(\frac{\partial u_y}{\partial x} - \frac{\partial u_x}{\partial y}\right) \qquad ...(9)$$

Differentiating Eq. (8) w.r.t. z and substituting for $\dfrac{\partial F}{\partial z}$ from Eq. (5) we, obtain

$$4G\frac{\partial D}{\partial z} = f'(z) - (\sigma_x + \sigma_y) + 2U \qquad ...(10)$$

Eliminating $\dfrac{\partial D}{\partial z}$ between Eqs. (9) and (10), we get

$$\sigma_x + \sigma_y - \frac{U}{1-\nu} - \frac{iG}{1-\nu}\left(\frac{\partial u_y}{\partial x} + \frac{\partial u_x}{\partial y}\right) = 4\varphi'(z) \qquad ...(11)$$

where $\varphi(z) = \dfrac{f(z)}{8(1-\nu)}$ ...(12)

The real part of Eq. (11) is

$$\sigma_x + \sigma_y - \frac{U}{1-\nu} = 2[\varphi'(z) + \overline{\varphi}'(\overline{z})]$$

so that from Eq. (5)

$$\frac{\partial F}{\partial z} = 2[\varphi'(z) + \overline{\varphi}'(\overline{z})] - \left(\frac{1-2\nu}{1-\nu}\right)\frac{\partial W}{\partial z} \qquad \text{...(13)}$$

where $\dfrac{\partial}{\partial z} W(z, \overline{z}) = U$

Then, integrating w.r.t. z, we get

$$F(z, \overline{z}) = 2[\varphi(z) + z\,\overline{\varphi}'(\overline{z})] - \left(\frac{1-2\nu}{1-\nu}\right) W(z, \overline{z}) + 2\psi^T(\overline{z}) \qquad \text{...(14)}$$

where $\overline{\psi}(\overline{z})$ is a function of $\overline{z}$ alone. From Eqs. (8), (12) and (14) the complex displacement is given by

$$2G(u_x + i\,u_y) = \chi\,\varphi(z) - z\,\overline{\varphi}'(\overline{z}) - \overline{\psi}(\overline{z}) + \tfrac{1}{2}\kappa\,W(z, \overline{z}) \qquad \text{...(2.52)}$$

where $\chi = 3 - 4\nu$, $\kappa = \dfrac{1-2\nu}{1-\nu}$

while from Eqs. (5), (6) and (14), we have

$$\sigma_x + \sigma_y = 2[\varphi'(z) = \varphi^T(\overline{z})] + (2-\kappa)\frac{\partial W}{\partial z} \qquad \text{...(2.53)}$$

$$\sigma_x - \sigma_y + 2i\tau_{xy} = -2[z\,\overline{\varphi}''(\overline{z})]\,\overline{\psi}'(\overline{z})] + \kappa\frac{\partial W}{\partial z} \qquad \text{...(2.54)}$$

Thus for a given body force function $W(z, \overline{z})$ the displacement and stress configurations are given completely in terms of the complex potential $\varphi(z)$, $\psi(z)$ through Eqs. (2.52), (2.53) and (2.54). In order to obtain the results for plane stress case, use a modified Poisson's ratio $\nu' = \left(\dfrac{\nu}{1+\nu}\right)$ instead of ν in the above equations.

Further, the body force potential $V(x, y)$ should satisfy the Biharmonic equation $\nabla^4 V = 0$.

2.3.3 Relation to Airy's Stress Function

When the body forces are zero, *i.e.* $V(x, y) = 0$, then

$$\sigma_x + \sigma_y = 4\frac{\partial^2 \Phi}{\partial z \partial \overline{z}}$$

$$\sigma_x - \sigma_y + 2i\tau_{xy} = -4\frac{\partial^2\Phi}{\partial \bar{z}^2}$$

where Φ is the stress function. Then from Eqs. (2.53) and (2.54) we have

$$\left.\begin{aligned} 2\frac{\partial^2\Phi}{\partial z\partial \bar{z}} &= \varphi'(z) + \bar{\varphi}'(\bar{z}) \\ 2\frac{\partial^2\Phi}{\partial \bar{z}^2} &= z\,\bar{\Phi}''(\bar{z}) + \bar{\psi}'(\bar{z}) \end{aligned}\right\} \qquad ...(a)$$

The stress function Φ satisfies the Biharmonic equation

$$\nabla^4\Phi = \frac{\partial^4\Psi}{\partial z^2\,\partial \bar{z}^2} = 0$$

Integrating Eq. (a), we have

$$2\frac{\partial\Phi}{\partial \bar{z}} = \varphi(z) + z\,\bar{\varphi}'(\bar{z}) + f_1(\bar{z})$$

$$2\frac{\partial\Phi}{\partial \bar{z}} = z\,\bar{\varphi}'(z) + \bar{\psi}(\bar{z}) + f_2(z)$$

from which $f_1(\bar{z}) = \bar{\psi}(\bar{z})$

$f_2(z) = \varphi(z)$

and hence $2\dfrac{\partial\Phi}{\partial \bar{z}} = \varphi(z) + z\,\bar{\varphi}'(\bar{z}) + \bar{\psi}(\bar{z})$

and
$$2\Phi = \bar{z}\,\varphi(z) + z\,\bar{\varphi}(\bar{z}) + \int(\bar{z})d\,\bar{z} + g(z) \qquad ...(b)$$

where $g(z)$ is a function of z alone. Since all its second derivatives are real Φ must be real part from a possible imaginary constant.

We must have then in Eq. (b)

$$\bar{g}(\bar{z}) = \int \bar{\psi}(\bar{z})\,d\bar{z} + C$$

where C is an arbitrary complex constant. Putting $C = 0$, we have

$$2\Phi = x\varphi_1(x, y) + y\varphi_2(x, y) + g_1(x, y) \qquad ...(2.55)$$

where $\varphi = \varphi_1 + i\varphi_2$ and $g = g_1 + ig_2$

and since φ and g are analytic functions, $\varphi_1, \varphi_2, g_1, g_2$ are harmonic functions.

2.3.4 Muskhelishvili Method

In the absence of body forces, stress compatibility equation is

$$\nabla^2(\sigma_x + \sigma_y) = 0$$

or
$$\nabla^4\Phi = 0 \qquad ...(1)$$

where Φ = Airy stress functions and

$$\sigma_x = \frac{\partial^2 \Phi}{\partial y^2}, \qquad \sigma_y = \frac{\partial^2 \Phi}{\partial x^2},$$

$$\tau_{xy} = -\frac{\partial^2 \Phi}{\partial x \partial y}$$

But

$$\left.\begin{aligned} \sigma_x &= 2G\frac{\partial u_x}{\partial x} + \lambda\left(\frac{\partial u_x}{\partial x} + \frac{\partial u_y}{\partial y}\right) \\ \sigma_y &= 2G\frac{\partial u_y}{\partial y} + \lambda\left(\frac{\partial u_x}{\partial x} + \frac{\partial u_y}{\partial y}\right) \end{aligned}\right\} \quad ...(2)$$

$$\tau_{xy} = G\left(\frac{\partial u_x}{\partial x} + \frac{\partial u_y}{\partial y}\right)$$

$$\therefore \quad 2G\frac{\partial u_x}{\partial x} = \sigma_x - \lambda\left(\frac{\partial u_x}{\partial x} + \frac{\partial u_y}{\partial y}\right)$$

Assuming plane strain conditions, we have

$$\varepsilon_x = \frac{1+\nu}{E}[\sigma_x(1-\nu) - \nu\sigma_y]$$

$$\varepsilon_y = \frac{1+\nu}{E}(\sigma_y(1-\nu) - \nu\sigma_x]$$

$$\therefore \quad \varepsilon_x + \varepsilon_y = \frac{(1+\nu)(1-2\nu)}{E}(\sigma_x + \sigma_y) = \frac{\sigma_x + \sigma_y}{2(\lambda + G)}$$

$$\therefore \quad 2G\frac{\partial u_x}{\partial x} = \frac{\partial^2 \Phi}{\partial y^2} - \frac{\lambda}{2(\lambda + G)}\left(\frac{\partial^2 \Phi}{\partial y^2} + \frac{\partial^2 \Phi}{\partial x^2}\right)$$

or

$$2G\frac{\partial u_x}{\partial x} = \frac{\partial^2 \Phi}{\partial y^2} - \frac{\lambda}{2(\lambda + G)}\nabla^2 \Phi$$

Similarly

$$2G\frac{\partial u_y}{\partial y} = \frac{\partial^2 \Phi}{\partial x^2} - \frac{\lambda}{2(\lambda + G)}\nabla^2 \Phi \quad ...(3)$$

Let $\nabla^2 \Phi = P$ and replacing $\frac{\partial^2 \Phi}{\partial y^2}$ by $P - \frac{\partial^2 \Phi}{\partial x^2}$

and $\frac{\partial^2 \Phi}{\partial x^2}$ by $P - \frac{\partial^2 \Phi}{\partial y^2}$, we obtain from Eqs. (3)

$$2G\frac{\partial u_x}{\partial x} = P - \frac{\partial^2 \Phi}{\partial x^2} - \frac{\lambda}{2(\lambda + G)}P$$

$$\left.\begin{aligned} &= -\frac{\partial^2 \Phi}{\partial x^2} + \frac{\lambda + 2G}{2(\lambda + G)} P \\ \text{and} \qquad 2G\frac{\partial u_y}{\partial y} &= -\frac{\partial^2 \Phi}{\partial y^2} + \frac{\lambda + 2G}{2(\lambda + G)} P \end{aligned}\right\} \qquad \text{...(4)}$$

The function P is harmonic, because

$$\nabla^2 P = \nabla^4 \Phi = 0 \qquad \text{...(5)}$$

Let Q be the harmonic function, conjugate to P, *i.e.* the function satisfying the Cauchy-Riemann conditions

$$\frac{\partial P}{\partial x} = \frac{\partial Q}{\partial y}, \; \frac{\partial P}{\partial y} = -\frac{\partial Q}{\partial x}$$

Then the expression

$$f(z) = P(x, y) + i\, Q(x, y) \qquad \text{...(6)}$$

will represent a function of the complex variable $z = x + iy$, holomorphic in the region S occupied by the body. The integral of this function with respect to z is another analytic function, $4\Phi\,(z)$, say. Then, writing p and q for the real and imaginary parts of $\varphi(z)$, we have

$$\varphi(z) = p + iq = \frac{1}{4}\int f(z)\, dz. \qquad \text{...(7)}$$

so that

$$\begin{aligned} \varphi'(z) &= \frac{1}{4} f(z) = \frac{\partial p}{\partial x} + i\frac{\partial q}{\partial x} \\ &= \tfrac{1}{4}(P + iQ). \end{aligned}$$

Equating real and imaginary parts, we find

$$\frac{\partial p}{\partial x} = \frac{1}{4} P, \; \frac{\partial q}{\partial x} = \frac{1}{4} Q \qquad \text{...(8)}$$

From Cauchy-Riemann conditions, we have

$$\frac{\partial p}{\partial x} = \frac{\partial q}{\partial y}, \; \frac{\partial p}{\partial y} = -\frac{\partial q}{\partial x}$$

$$\therefore \qquad \frac{\partial p}{\partial x} = \frac{\partial q}{\partial y} = \frac{1}{4} P, \; \frac{\partial p}{\partial y} = -\frac{\partial q}{\partial x} = -\frac{1}{4} Q.$$

Thus

$$P = 4\frac{\partial p}{\partial x} = 4\frac{\partial q}{\partial y} \qquad \text{...(9)}$$

Hence Eqs. (4) become

$$\left.\begin{aligned} 2G\frac{\partial u_x}{\partial x} &= -\frac{\partial^2 \Phi}{\partial x^2} + \frac{2(\lambda + 2G)}{(\lambda + G)}\frac{\partial p}{\partial x} \\ 2G\frac{\partial u_y}{\partial y} &= -\frac{\partial^2 \Phi}{\partial y^2} + \frac{2(\lambda + 2G)}{(\lambda + G)}\frac{\partial q}{\partial y} \end{aligned}\right\} \qquad \text{...(10)}$$

Integrating Eqs. (10), we get

$$\left.\begin{aligned} 2Gu_x &= -\frac{\partial \Phi}{\partial x} + \frac{2(\lambda + 2G)}{(\lambda + G)} p + f_1(y) \\ 2Gu_y &= -\frac{\partial \Phi}{\partial x} + \frac{2(\lambda + 2G)}{(\lambda + G)} p + f_2(x) \end{aligned}\right\} \quad \text{...(11)}$$

where $f_1(y)$ and $f_2(x)$ are arbitrary functions of y and x only respectively. Omitting these functions, which only give rigid body displacements, we get

$$\left.\begin{aligned} 2Gu_x &= -\frac{\partial \Phi}{\partial x} + \frac{2(\lambda + 2G)}{(\lambda + G)} p \\ 2Gu_y &= -\frac{\partial \Phi}{\partial y} + \frac{2(\lambda + 2G)}{(\lambda + G)} q \end{aligned}\right\} \quad \text{...(12)}$$

Now we have seen that every biharmonic function $\Phi(x, y)$ of the two variables x, y may be represented by the help of two functions of the complex variable.

$$z = x + iy, \quad \varphi(z) = p + iq$$

Recalling that $P = \nabla^2\Phi$, Eq. (9) enables us to show that $\Phi - xp - yq$ is a harmonic function, because

$$\begin{aligned} \nabla^2(\Phi - xp - yq) &= \nabla^2\Phi - 2\frac{\partial p}{\partial x} - 2\frac{\partial q}{\partial y} \\ &= \nabla^2\Phi - \frac{P}{2} - \frac{P}{2} \\ &= \nabla^2\Phi - P = 0 \end{aligned}$$

Thus for any stress function Φ, we have

$$\Phi - xp - yq = p_1$$

where p_1 is some harmonic function. Consequently

$$\Phi = xp + yq + p_1 \quad \text{...(13)}$$

Let us now introduce the function q_1, which is the conjugate harmonic to p_1, and write

$$\chi(z) = p_1 + iq_1$$

Then it is easily verified that the real part of

$$(x - iy)(p + iq) + (p_1 + iq_1)$$

is identical with the right-hand side of Eq. (13). Thus any stress function is expressible in the form

$$\Phi = R_e\,[\,\bar{z}\,\varphi(z) + \chi(z)] \quad \text{...(14)}$$

where R_e means "real part of" and $\bar{z} = x - iy$. $\varphi(z)$ and $\chi(z)$ are suitably chosen analytic functions.

$$\overline{\varphi}\,(\bar{z}) = p - iq$$

$\overline{\chi}(\overline{z}) = p_1 - iq_1$

Now $\quad 2R_e f(z) = f(z) + \overline{f}(\overline{z})$

$$\therefore \quad 2\Phi = \overline{z}\,\varphi(z) + z\,\overline{\varphi}(\overline{z}) + \chi(z) + \overline{\chi}(\overline{z}) \qquad \text{...(15)}$$

Differentiating, we get

$$\frac{2\partial\Phi}{\partial x} = \overline{z}\,\varphi'(z) + \varphi(z) + z\,\overline{\varphi}'(\overline{z}) + \overline{\varphi}(\overline{z}) + \chi'(z) + \overline{\chi}'(\overline{z})$$

$$\frac{2\partial\Phi}{\partial y} = i\,[\,\overline{z}\,\varphi'(z) - \varphi(z) - z\,\overline{\varphi}'(\overline{z}) + \overline{\varphi}(\overline{z}) + \overline{\chi}'(z) - \overline{\chi}'(\overline{z})]$$

$$\therefore \quad \frac{\partial\Phi}{\partial x} + i\frac{\partial\Phi}{\partial y} = \varphi(z) + z\,\overline{\varphi}'(\overline{z}) + \overline{\chi}'(\overline{z}) \qquad \text{...(16)}$$

$$= \varphi(z) + z\,\overline{\varphi}'(\overline{z}) + \overline{\psi}(\overline{z})$$

where $\psi(z) = d\chi/dz = \chi'(z)$

From Eqs. (12), we can write

$$2G\,(u_x + i\,u_y) = -\left(\frac{\partial\Phi}{\partial x} + i\frac{\partial\Phi}{\partial y}\right) + \frac{2(\lambda + 2G)}{(\lambda + G)}(p + iq)$$

Now using Eqs. (7) and (16), we get

$$2G\,(u_x + i\,u_y) + -[\varphi(z) + z\overline{\varphi}'(\overline{z}) + \overline{\psi}(\overline{z})] + \frac{2(\lambda + 2G)}{(\lambda + G)}\varphi(z)$$

$$= \left(\frac{\lambda + 2G}{\lambda + G}\right)\varphi(z) - z\overline{\varphi}'(\overline{z}) + \overline{\psi}(\overline{z})$$

$$= (3 - 4\upsilon)\varphi(z) - z\overline{\varphi}'(\overline{z}) + \overline{\psi}(\overline{z}) \qquad \text{...(2.56)}$$

Eq. (2.56) determines u_x and u_y for plane strain when the complex potentials $\varphi(z)$, $\psi(z)$ are given. For plane stress $(3 - 4\nu)$ is replaced by $\left(\frac{3 - \nu}{1 + \nu}\right)$ on the right hand side of Eq. (2.56) and Eq. (12) for plane stress case can be written as

$$\left.\begin{aligned} 2Gu_x &= -\frac{\partial\Phi}{\partial x} + \left(\frac{4}{1 + \nu}\right)p \\ 2Gu_y &= -\frac{\partial\Phi}{\partial y} + \left(\frac{4}{1 + \nu}\right)q \end{aligned}\right\} \qquad \text{...(2.57)}$$

Hence, we can write in gengral

$$2G(u_x + i\,u_y) = -\left(\frac{\partial\Phi}{\partial x} + i\frac{\partial\Phi}{\partial y}\right) + \frac{2(\lambda + 2G)}{(\lambda + G)}\varphi(z)$$

$$= \chi\varphi(z) - z\overline{\varphi}'(\overline{z}) - \overline{\psi}(\overline{z}) \qquad \text{...(2.58)}$$

where $\chi = 3 - 4\nu$ for plane strain

$$= \frac{3-\nu}{1+\nu} \text{ for plane stress}$$

Now differentiating Eq. (16) w.r.t. x, we have

$$\frac{\partial^2\Phi}{\partial x^2} + i\frac{\partial^2\Phi}{\partial x\,\partial y} = \varphi'(z)\, z\bar{\varphi}''(\bar{z}) + \bar{\varphi}'(\bar{z}) + \bar{\psi}'(\bar{z}) \qquad ...(17)$$

Differentiating Eq. (16) w.r.t. y and multiplying by i, we get

$$i\frac{\partial^2\Phi}{\partial x\partial y} - \frac{\partial^2\Phi}{\partial y^2} = -\varphi'(x) + z\bar{\varphi}''(\bar{z}) - \bar{\varphi}'(\bar{z}) + \bar{\psi}'(\bar{z}) \qquad ...(18)$$

By subtracting and adding Eqs. (17) and (18), we get

$$\frac{\partial^2\Phi}{\partial x^2} + \frac{\partial^2\Phi}{\partial y^2} = 2\varphi'(z) + 2\,\bar{\varphi}'(\bar{z})$$

$$\sigma_x + \sigma_y = 4R_e\varphi'(z) \qquad ...(2.59)$$

$$\frac{\partial^2\Phi}{\partial x^2} + \frac{\partial^2\Phi}{\partial y^2} + 2i\frac{\partial^2\Phi}{\partial x\partial y} = 2z\bar{\varphi}''(\bar{z}) + 2\bar{\psi}'(\bar{z})$$

$$\sigma_y - \sigma_x - 2i\tau_{xy} = 2[z\bar{\varphi}''(\bar{z}) + \bar{\psi}'(\bar{z})] \qquad ...(2.60)$$

Changing i to $-i$ on both sides of Eq. (2.60), we get

$$\sigma_y - \sigma_x + 2i\,\tau_{xy} = 2[\bar{z}\,\varphi''(z) + \psi'(z)] \qquad ...(2.61)$$

On separation of real and imaginary parts the right-hand side of Eq. (2.61) or (2.60), gives $\sigma_x - \sigma_y$ and $2\tau_{xy}$. The two Eqs. (2.59) and (2.60) determine the stress components in terms of the complex potentials $\varphi(z)$ and $\psi(z)$. Thus by choosing definite functions for $\varphi(z)$ and $\psi(z)$, we find a possible state of stress from Eqs. (2.59) and (2.61), and the displacements corresponding to this state of stress are obtained from Eq. (2.58).

2.3.5 Resultant Force and Moment on a Curve

Consider some arc AB in the plane XOY. Let its positive direction be from A to B and draw the normal n and tangent t to an elementary distance ds on the arc AB. Then

$$X_n = \sigma_x \cos(n, x) + \tau_{xy}\cos(n, y)$$

$$= \frac{\partial^2\Phi}{\partial y^2}\cdot\cos(n, x) + \left(\frac{-\partial^2\Phi}{\partial x\partial y}\right)\cos(n, y)$$

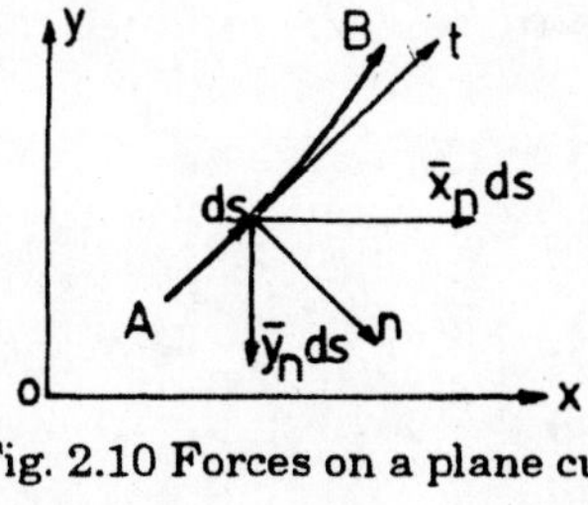

Fig. 2.10 Forces on a plane curve.

$$Y_n = \tau_{xy} \cos(n, x) + \sigma_y \cos(n, y)$$

$$= -\frac{\partial^2 \Phi}{\partial x \partial y} . \cos(n, x) + \frac{\partial^2 \Phi}{\partial x^2} \cos(n, y)$$

Now $\cos(n, x) = \cos(t, y) = \dfrac{dy}{ds}$

$$\cos(n, y) = -\cos(t, x) = -\frac{dx}{ds}$$

$$\therefore \qquad X_n = \frac{\partial^2 \Phi}{\partial y^2} \cdot \frac{dy}{ds} + \frac{\partial^2 \Phi}{\partial x \partial y} \cdot \frac{dx}{ds}$$

$$Y_n = \frac{-\partial^2 \Phi}{\partial x \partial y} \cdot \frac{dy}{ds} - \frac{\partial^2 \Phi}{\partial x^2} \cdot \frac{dx}{ds}$$

$$X_n = \frac{\partial}{\partial y}\left(\frac{\partial \Phi}{\partial y}\right)\frac{dy}{ds} + \frac{\partial}{\partial x}\left(\frac{\partial \Phi}{\partial y}\right)\frac{dx}{ds}$$

$$= \frac{d}{ds}\left(\frac{\partial \Phi}{\partial y}\right) \qquad \ldots(1)$$

$$Y_n = -\left[\frac{\partial}{\partial y}\left(\frac{\partial \Phi}{\partial y}\right)\frac{dy}{ds} + \frac{\partial}{\partial x}\left(\frac{\partial \Phi}{\partial y}\right)\frac{dx}{ds}\right]$$

$$= -\frac{d}{ds}\left(\frac{\partial \Phi}{\partial y}\right) \qquad \ldots(2)$$

$$X_n + iY_n = \frac{d}{ds}\left(\frac{\partial \Phi}{\partial y} - i\frac{\partial \Phi}{\partial x}\right) = -i\frac{d}{ds}\left(\frac{\partial \Phi}{\partial x} + \frac{\partial \Phi}{\partial y}\right)$$

If (F_x, F_y) be the resultant force, then

$$F_x + i\,F_y = \int_A^B (X_n + iY_n)\, ds = \int_A^B -id\left(\frac{\partial \Phi}{\partial x} + i\frac{\partial \Phi}{\partial y}\right)$$

$$= -i\left[\frac{\partial \Phi}{\partial y} + i\frac{\partial \Phi}{\partial x}\right]_A^B \qquad \ldots(3)$$

Using Eq. (16) of Art. 2.3.4, we get

$$F_x + iF_y = -i\left[\varphi(z) + z\overline{\varphi}'(\overline{z}) + \overline{\psi}(\overline{z})\right]_A^B \qquad \ldots(2.62)$$

Resultant moment about the origin of co-ordinate system is

$$M = \int_A^B (xY_n - yX_n)\, ds$$

$$= \int_A^B \left[-x\frac{d}{ds}\left(\frac{\partial \Phi}{\partial x}\right) - y\frac{d}{ds}\left(\frac{\partial \Phi}{\partial y}\right)\right] ds$$

$$= -\int_A^B \left[xd\left(\frac{\partial \Phi}{\partial x}\right) + yd\left(\frac{\partial \Phi}{\partial y}\right)\right]$$

$$= -\left\{\left[x\frac{\partial \Phi}{\partial y} + y\frac{\partial \Phi}{\partial y}\right]_A^B - \int_A^B \left(\frac{\partial \Phi}{\partial x}dx + \frac{\partial \Phi}{\partial y}dy\right)\right\}$$

$$= -\left[x\frac{\partial \Phi}{\partial y} + y\frac{\partial \Phi}{\partial y}\right]_A^B + \Big[\ \Phi\ \Big]_A^B$$

But $x\dfrac{\partial \Phi}{\partial x} + y\dfrac{\partial \Phi}{\partial y} = R_e\left[z\left(\dfrac{\partial \Phi}{\partial x} - i\dfrac{\partial \Phi}{\partial y}\right)\right]$

and $\dfrac{\partial \Phi}{\partial x} - i\dfrac{\partial \Phi}{\partial y} = \bar{z}\,\varphi'(z) + \bar{\varphi}(\bar{z}) + \psi(z)$

Further from Eq. (14) of Art. 2.3.4, we have

$$\Phi = R_e[\,\bar{z}\varphi(z) - \chi(z)]$$

$$\therefore \qquad M = -R_e\Big[\, z(\bar{z}\varphi'(z) + \bar{\varphi}(\bar{z}) + \psi(z)\,\Big]_A^B$$

$$+ R_e\Big[\,\bar{z}\,\varphi(z) + \chi(z)\,\Big]_A^B$$

$$= R_e\Big[-z\bar{z}\varphi'(z) - z\,\bar{\varphi}(\bar{z}) - z\psi(z) + \bar{z}\,\varphi(z) + \chi(z)\Big]_A^B$$

$$= R_e\Big[-z\bar{z}\,\varphi'(z) + \chi(z) - z\psi(z)\Big]_A^B \qquad \text{...(2.63)}$$

Because $R_e\,[\,\bar{z}\varphi(z) - z\bar{\varphi}(\bar{z})] = 0$

If the arc is a closed contour AA, when

$$F_x + iF_y = -i\Big[\,\varphi(z) + \psi\,(\bar{z})\,\Big]_A^A \qquad \text{...(2.64)}$$

$$M = R_e\Big[\,\psi(z) - z\,\psi(z)\,\Big]_A^A \qquad \text{...(2.65)}$$

For stresses to be single valued

$$\left.\begin{aligned} F_x + iF_y &= 0 \\ M &= 0 \end{aligned}\right\} \qquad \text{...(2.66)}$$

Suppose the region R occupied by the body is bounded by the closed contours C_1, C_2, ..., C_m, C_{m+1}, where C_{m+1} contains the remainder, and z_1, z_2, ..., z_m are any fixed points within the respective contours, while F_{xk} and F_{yk} are the components of the stress resultant over C_k.

Then it can be shown that[6]

$$\left.\begin{aligned} \varphi(z) &= -\frac{1}{2\pi(1+\chi)} \sum_{k=1}^{m} (F_{xk} - iF_{yk}) \log (z - z_k) + \varphi^*(z) \\ \psi(z) &= -\frac{\chi}{2\pi(1+\chi)} \sum_{k=1}^{m} (F_{xk} - iF_{yk}) \log (z - z_k) + \psi^*(z) \end{aligned}\right\} \quad \text{...(2.67)}$$

where $\varphi^*(z)$ and $\psi^*(z)$ are functions holomorphic in R.

When C_{m+1} approches infinity so that R becomes an infinite region these results still hold good. Denote by C_{R0} a circle $|z| = R_0$ containing all the finite C_k. Then for all $|z| \geq R_0$ we can write for the infinite region

$$\left.\begin{aligned} \varphi(z) &= -\frac{F_x + iF_y}{2\pi(1+\chi)} \log z + \varphi^{**}(z) \\ \psi(z) &= \frac{\chi(F_x - iF_y)}{2\pi(1+\chi)} \log z + \psi^{**}(z) \end{aligned}\right\} \quad \text{...(2.68)}$$

where $F_x = \sum_{k=1}^{m} F_{xk}$, $\quad F_y = \sum_{k=1}^{m} F_{yk}$

and $\varphi^{**}(z)$, $\psi^{**}(z)$ are functions holomorphic in $|z| \geq R_0$ except possibly at infinity. Introducing these expressions into Eqs. (2.53) and (2.54) and imposing the condition that the stress components should be bounded throughout the infinite region R it is found that

$$\left.\begin{aligned} \varphi^{**}(z) &= \Gamma(z) + \varphi_1(z) \\ \psi^{**}(z) &= \Gamma'(z) + \psi_1(z) \end{aligned}\right\} \quad \text{...(2.69)}$$

where Γ, Γ' are complex constants and $\varphi_1(z)$, $\psi_1(z)$ are functions holomorphic in $|z| \geq R_0$ including the point at infinity, thus having expansions of the form

$$a_0 + \frac{a_1}{z} + \frac{a_2}{z^2} + \ldots\ldots$$

for sufficiently large $|z|$.

6. N.I. Muskhelishvili : Some Basic Problems of the Mathematical Theory of Elasticity. P. Noordhoff Ltd. 1963, p. 124.

The constants, a_0, a_1, ... in these series and also $I_m(\Gamma)$ may be assumed zero without affecting the state of stress. Also, using Eqs. (2.53) and (2.54) with $W = 0$, it is readily shown that

$$R_e(\Gamma) = \frac{1}{4}(\sigma_1 + \sigma_2), \quad \Gamma' = -\frac{1}{2}(\sigma_1 - \sigma_2)\, e^{-2i\alpha} \qquad ...(2.70)$$

where σ_1, σ_2 are the principal stresses at infinity and α is the angle between the direction of σ_1 and ox. Using Eqs. (2.68) and (2.69) with Eq. (2.70), the complex displacement is given for large $|z|$, in the absence of body forces, by

$$2G(u_x + iu_y) = -\frac{\chi(F_x + iF_y)}{2\pi(1+\chi)} \log(z\bar{z}) + (\chi\Gamma - \bar{\Gamma})z - \Gamma'\bar{z} + ... \qquad ...(2.71)$$

where the further terms are bounded as $|z| \to \infty$.

Clearly the displacements are not bounded at infinity unless

$$F_x = F_y = 0, \qquad \Gamma = \Gamma' = 0$$

Such conditions imply that the resultant of the stresses acting on all the C_k is zero and that the stresses at infinity are zero, while φ and ψ would have expansions, for sufficiently large $|z|$,

$$\varphi(z) = z_0 + \frac{a_1}{z} + \frac{a_2}{z^2} + ...$$

$$\psi(z) = b_0 + \frac{b_1}{z} + \frac{b_2}{z^2} + ...$$

2.4 Cauchy Integral Method

The interior of a simple closed curve is called a simply connected domain. A domain which is not simply connected is said to be multiply connected. When a domain is simply connected the complex potential method may be readily applied. When a domain is multiply connected then it is not easy to find the complex potentials directly from the boundary conditions as is possible for a simply connected domain. For multiply connected domains Cauchy integral method may be used for regions which can be mapped on to a circle and on a half plane. We shall state, without proof, some results on Cauchy integrals which will be required subsequently.

(*i*) **Cauchy Integral Theorem.** If a function $f(z)$ is single-valued and analytic throughout a simply connected region R, then the integral of $f(z)$ around a closed path C which lies entirely inside R is zero, *i.e.*

$$\oint_C f(z)dz = 0 \qquad ...(2.72)$$

An important corollary of the Cauchy integral theorem is that if a function $f(z)$ is a single-valued and analytic function in a simply connected region R, then the value of the line integral $\int_{z_1}^{z_2} f(z)\, dz$ between two points z_1 and z_2 is independent of the path of integration, provided that the path lies wholly inside the region of analyticity R.

A multiply connected region can be cut such that the resulting region becomes simply connected. Thus

$$\int_{C_1} f(z)dz + \int_{C_2} f(z)dz + \int_{C_3} f(z)dz + \ldots\ldots = 0 \qquad \text{...(2.73)}$$

where C_2, C_3 .... are traversed in the same sense and C_1 is traversed in the opposite sense.

(*iii*) **Cauchy's Integral Formula.** Let $f(z)$ be analytic and single-valued in a simply connected region R. Then for any point z_0 in R and any simple closed path C in R which encloses z_0

$$\int_C \frac{f(z)\, dz}{z - z_0} = 2\pi i f(z_0) \qquad \text{...(2.74)}$$

The integration being taken in the counter-clockwise sense and z_0 does not lie on C.

For a multiply connected domain, cross-cuts may be made to convert the domain to a simply connected one. For example for a doubly connected domain, we have

$$\int_{C_1} \frac{f(z)\, dz}{z - z_0} - \int_{C_2} \frac{f(z)\, dz}{z - z_0} = 2\pi i f(z_0) \qquad \text{...(2.75)}$$

where both integrals have been taken in the counterclockwise sense.

The derivatives of Cauchy's integral are in general

$$f^{(n)}(z_0) = \frac{n\,!}{2\pi i} \int \frac{f(z)}{(z - z_0)^{n+1}}\, dz \qquad \text{...(2.76)}$$

(*iii*) A necessary and sufficient condition for $f(z)$ to be the boundary value to some function holomorphic in R^+ is

$$\frac{1}{2\pi i} \int_C \frac{f(z)\, dz}{z - z_0} = 0, \quad \text{all } z_0 \text{ in } R^- \qquad \text{...(2.77)}$$

(*iv*) A necessary and sufficient condition for $f(z)$ to be the boundary value of some function holomorphic in R^-, including infinity, is

$$\frac{1}{2\pi i}\int_C \frac{f(z)\,dz}{z - z_0} = a, \text{ all } z_0 \text{ in } R^+ \qquad \text{...(2.78)}$$

where a is a constant equal to $f(\infty)$.

When C is a unit circle, denoted by γ, the properties of complex conjugate functions can be used with these theorems to obtain the following results :

(*a*) A necessary and sufficient condition for the function $f(\sigma)$ continuous on γ, to be the boundary value of some function holomorphic outside γ is

$$\frac{1}{2\pi i}\int_\gamma \frac{\overline{f(\sigma)}\,d\sigma}{\sigma - z} = 0, \text{ all } |z| > 1 \qquad \text{...(2.79)}$$

(*b*) A necessary and sufficient condition for the function $f(\sigma)$, continuous on γ, to be the boundary value of some function holomorphic inside γ is

$$\frac{1}{2\pi i}\int_\gamma \frac{\overline{f(\sigma)}\,d\sigma}{\sigma - z} = \bar{a}, \text{ all } |z| < 1 \qquad \text{...(2.80)}$$

where $\bar{a}$ is a constant equal to $\bar{f}(0)$.

(*v*) A necessary and sufficient condition for $f(z)$ to be the boundary value of a function holomorphic in R^+ and continuous in $R^+ + C$, including infinity, is

$$\frac{1}{2\pi i}\int_C \frac{f(z)\,dz}{z - z_0} = -\frac{1}{2}a, \text{ all } z_0 \text{ in } R^- \qquad \text{...(2.81)}$$

(*vi*) A necessary and sufficient condition for $f(z)$ to be the boundary value of a function holomorphic in R^- and continuous in $R^- + C$ including infinity, is

$$\frac{1}{2\pi i}\int_C \frac{f(z)\,dz}{z - z_0} = \frac{1}{2}a, \text{ all } z_0 \text{ in } R^+ \qquad \text{...(2.82)}$$

Under the same conditions we have also.

(*a*) A necessary and sufficient condition for $f(z)$ to be the boundary value of a function holomorphic in R^- is

$$\frac{1}{2\pi i}\int_C \frac{\overline{f(z)}\,dz}{z - z_0} = -\frac{1}{2}\bar{a}, \text{ all } z_0 \text{ in } R^- \qquad \text{...(2.83)}$$

(*b*) A necessary and sufficient condition for $f(z)$ to be the boundary value of a function holomorphic in R^+ is

$$\frac{1}{2\pi i}\int_C \frac{\overline{f(z)}\,dz}{z - z_0} = -\frac{1}{2}\bar{a}, \text{ all } z_0 \text{ in } R^+ \qquad \text{...(2.84)}$$

Holder condition. Let $f(z)$ be some, in general complex, function of the point z of C, this means that

$$f(z) = f_1(z) + if_2(z)$$

where $f_1(z)$ and $f_2(z)$ are real functions of z on C.

The function $f(z)$ will be said to satisfy on C the Holder condition, or just the H condition, if for every pair of points z_1, z_2 of C the following inequality holds true :

$$| f(z_1) - f(z_2) | \leq A | t_1 - t_2 | \mu \qquad ...(2.85)$$

where A and μ are positive constants and $0 < \mu \leq 1$; A is called the Holder constant and μ the Holder index.

If this condition is satisfied in the neighbourhood of a point z_0 other than an end of C, and if $f^+(z_0)$ and $f^-(z_0)$, *i.e.* the values of $f(z)$ from the left and from the right exist, then they are given by Plemelj formulae

$$f^+(z_0) = \frac{1}{2} f(z_0) + \frac{1}{2\pi i}\int_C \frac{f(z)\, dz}{z - z_0}$$

$$f^-(z_0) = -\frac{1}{2} f(z_0) + \frac{1}{2\pi i}\int_C \frac{f(z)\, dz}{z - z_0} \qquad ...(2.86)$$

Principal value of Cauchy integral. If the point z_0 lie on C and if $f(z_0) \neq 0$, the integrand becomes infinite like $| z - z_0 |^{-1}$ as $z \to z_0$. Hence the integral has no meaning in the ordinary sense. However, for certain conditions referring to $f(z)$, the Cauchy integral may be given a definite interpretation. In fact, assume that z_0 is not an end of C and separate from C a sufficiently small arc z_1z_2 which contains z_0 in such a way that

$$| z_1 - z_0 | = | z_2 - z_0 |$$

Denote the arc z_1z_2 by C_1 and the remaining part of C by $C - C_1$ and consider the integral

$$\frac{1}{2\pi i}\int_{C - C_1} \frac{f(z)\, dz}{z - z_0} \qquad ...(2.87)$$

This integral is defined completely in the ordinary sense and its value tends to a definite limit, which is called the principal value of Cauchy integral.

Harnack's Theorem. Let C be a simple contour and let R^+, R^- be the finite and infinite parts into which the plane is divided by C. Let $f(z)$ be a real and continuous function on C. Then, if

$$\frac{1}{2\pi i}\int_C \frac{f(z)\, dz}{z - z_0} = 0 \quad \text{for all } z_0 \text{ in } R^+ \qquad ...(2.88)$$

$f(z) = 0$ everywhere on C. Also, if

$$\frac{1}{2\pi i}\int_C \frac{f(z)\,dz}{z - z_0} = 0 \text{ for all } z_0 \text{ in } R^- \qquad ...(2.89)$$

$f(z) =$ constant oı. C.

Let $f_1(z)$, $f_2(z)$ be two real continuous functions, given on C. Then, if

$$\frac{1}{2\pi i}\int_C \frac{f_1(z)\,dz}{z - z_0} = \frac{1}{2\pi i}\int \frac{f_2(z)\,dz}{z - z_0} \text{ for all } z_0 \text{ in } R^+$$

$f_1(z) = f_2(z)$ on C ; also, if

$$\frac{1}{2\pi i}\int_C \frac{f_1(z)\,dz}{z - z_0} = \frac{1}{2\pi i}\int_C \frac{f_2(z)\,dz}{z - z_0} \text{ for all } z_0 \text{ in } R^{-1}$$

$f_2(z) = f_1(z) +$ constant on C.

The first fundamental problem (Dirichlet problem). It consists of determining a function, harmonic in a region, when its boundary values are given.

The second fundamental problem (Neumann problem). It consists of determining a function, harmonic in a given region, when the boundary values of its normal derivatives are given.

(*vii*) (*a*) **First fundamental problem for a circle**

Let the radius of the circle be unity and its centre at the origin. Let us denote the circumference of the circle by γ and its points by $\sigma = e^{i\theta}$, other points of the plane be denoted by ζ. Let the unknown harmonic function be P and its conjugate complex function Q so that

$$F(\zeta) = P + iQ$$

where $F(\zeta)$ must be holomorphic inside γ. Then

$$F(\zeta) = \frac{1}{2\pi i}\int_\gamma f(\theta)\,\frac{\sigma + \zeta}{\sigma - \zeta}\,\frac{d\sigma}{\sigma} + i\beta_0 \qquad ...(2.90)$$

where $\alpha_0 - i\beta_0 = \dfrac{1}{2\pi i}\displaystyle\int_\gamma \frac{\overline{F(\sigma)}\,d\sigma}{\sigma - \zeta}$

and α_0 and β_0 are real constants.

$$\alpha_0 = \frac{1}{2\pi}\int_0^{2\pi} f(\theta)d\theta \text{ and } \beta_0 \text{ is arbitrary}$$

Thus $$P = R_e\,F(\zeta) = R_e\,\frac{1}{2\pi i}\int_\gamma f(\theta)\,\frac{\sigma + \zeta}{\sigma - \zeta}\cdot\frac{d\sigma}{\sigma}$$

Putting $\sigma = e^{i\theta}, \zeta = \rho e^{i\psi}, d\sigma = i e^{i\theta} d\theta.$

$$P = \frac{1}{2\pi}\int_0^{2\pi} \frac{(1-\rho^2) f(\theta)\, d\theta}{1 - 2\rho\cos(\theta-\psi) + \rho^2} \qquad \text{...(2.91)}$$

(b) Second fundamental problem for a circle

Let the derivative $F'(\zeta)$ takes the definite boundary value $F'(\sigma)$. Then

$$F(\zeta) = \frac{1}{\pi i}\int \frac{d\zeta}{\zeta}\int_\gamma \frac{f(\theta)\, d\sigma}{\sigma - \zeta} + \text{constant} \qquad \text{...(2.92)}$$

$$P = R_e F(\zeta)$$

$$= -\frac{1}{\pi}\int_0^{2\pi} f(\theta) \log r\, d\theta + \text{constant} \qquad \text{...(2.93)}$$

where $r = |\sigma - \zeta|$ and const. is an arbitrary real constant. Further

$$\zeta F'(\zeta) = \frac{1}{\pi i}\int_\gamma \frac{f(\theta)\, d\theta}{\sigma - \zeta} \qquad \text{...(2.94)}$$

(c) First and second fundamental problems of the half-plane

Let $f(z)$ be a real continuous function, given on the real axis C, and let it be required to find $P(x, y)$, harmonic in the upper half plane R^+ and taking the boundary value $P^+ = f(z)$ on C including the point at infinity, so that for $z \to \infty$ (in $R^+ + C$) $P \to a$, where a is the real constant

$$a = f(\infty)$$

then putting $F(z) = P + iQ$ holomorphic in R^+

$$F(z_0) = \frac{1}{\pi i}\int_C \frac{f(z)\, dz}{z - z_0} + ib \qquad \text{...(2.95)}$$

where $a + ib = F(\infty)$ and the quantity b remains arbitrary.

2.4.1 Solution of the first fundamental problem

Suppose that R is a finite or infinite region of the z-plane, bounded by the simple contour C and mapped on to the unit circle γ, in the ζ-plane by the function

$$z = \omega(\zeta) \qquad \text{...(1)}$$

It is assumed that for R finite $\zeta = 0$ corresponds to $z = 0$ and for R infinite $\zeta = 0$ corresponds to $z = \infty$.

In the first boundary value problem stress components X_n, Y_n are given everywhere on C, and from them we define the function of the arc co-ordinate, S,

$$f(s) = f_1(s) + i f_2(s)$$

$$= i\int_0^s (X_n + iY_n)ds + \text{const.} \qquad ...(2)$$

where the const. and the choice of the point $s = 0$ and C are arbitrary. Then from Eq. (2.62) with a suitable choice of constant,

$$\varphi(z) + z\,\overline{\varphi'}\,(\bar{z}) + \overline{\psi(\bar{z})} = f \text{ on } C \qquad ...(3)$$

Under the transformation (1) becomes by noting that

$$\varphi(z) = \varphi[\omega(\zeta)] = \varphi_0(\zeta)$$

$$\psi(\bar{z}) = \bar{\psi}\,[\omega(\bar{\zeta}) = \bar{\psi}_0(\bar{\zeta})$$

$$\varphi\,'(z) = \frac{d}{dz}\varphi(z) = \frac{d}{d\zeta}\varphi\,[\omega(\zeta)]\frac{d\zeta}{dz}$$

$$= \frac{d}{d\zeta}\varphi_0(\zeta)\frac{d\zeta}{dz} = \varphi'_0\zeta = \frac{d\zeta}{dz}$$

and
$$\frac{d\zeta}{dz} = \frac{1}{dz/d\zeta} = \frac{1}{\omega'(\zeta)}$$

$$\therefore \qquad z\bar{\varphi}'(\bar{z}) = \frac{\omega(\zeta)\,.\,\bar{\varphi}'_0(\bar{\zeta})}{\bar{\omega}'(\bar{\zeta})}$$

$$\therefore \qquad \phi_0(\zeta) + \frac{\omega(\zeta)}{\bar{\omega}'(\bar{\zeta})}\bar{\varphi}'_0(\bar{\zeta}) + \bar{\psi}_0\,(\bar{\zeta}) = f \text{ on } |\,\zeta\,| = 1 \qquad ...(4)$$

which may be written

$$\varphi_0(\sigma) + \bar{\psi}_0(\bar{\sigma}) + \frac{\omega(\sigma)}{\bar{\omega}\,(\bar{\sigma})}\bar{\Phi}_0'(\bar{\sigma}) = f(\sigma) \qquad ...(5)$$

where $\sigma = e^{i}$ is any point on γ and f is a known function of σ and θ.

Condition (5) is the foundations of the methods of Muskhelishvili for the determination of complex potentials from given boundary conditions.

Now by Cauchy's theorem, $\psi_0(\zeta) = \dfrac{1}{2\pi i}\displaystyle\int_\gamma \frac{\psi_0(\sigma)\,d\sigma}{\sigma - \zeta}$

while from Eq. (2.80)

$$\frac{1}{2\pi i}\int_\gamma \frac{\overline{\varphi_0(\bar{\sigma})}}{\sigma - \zeta}\,d\sigma = \bar{\varphi}_0(0) = 0$$

Then from the complex conjugate of Eq. (5)

$$\psi_0(\zeta) = \frac{1}{2\pi i}\int_\gamma \frac{\bar{f}\,d\sigma}{\sigma - \zeta} - \frac{1}{2\pi i}\int_\gamma \frac{\bar{\omega}(\bar{\sigma})\varphi_0'\,(\sigma)\,d\sigma}{\omega'\,(\sigma)(\sigma - \zeta)} \qquad ...(2.96)$$

Now from Eq. (5)

$$\overline{\psi_0}\,(\bar{\sigma}) = f(\sigma) - \varphi_0(\sigma) - \frac{\omega\,(\sigma)}{\omega'(\bar{\sigma})}\,\overline{\varphi_0}'\,(\bar{\sigma})$$

and from Eq. (2.80)

$$\frac{1}{2\pi i}\int_r \frac{\psi_0(\bar{\sigma})}{\sigma - \zeta}\,d\sigma = \overline{\psi}_0\,(0) = \bar{a}$$

$$\therefore \quad \frac{1}{2\pi i}\int_\gamma \frac{f d\sigma}{\sigma - \zeta} - \varphi_0(\zeta) - \frac{1}{2\pi i}\int_\gamma \frac{\omega(\sigma)\,\overline{\varphi}_0'\,(\bar{\sigma})\,d\sigma}{\overline{\omega}'(\bar{\sigma})\,(\sigma - \zeta)} = \bar{a} \quad ..(2.97)$$

for all $|\,\zeta\,| < 1$.

Eqs. (2.96) and (2.97) are sufficient to determine $\varphi_0(\zeta)$ and $\psi_0(\zeta)$ and thus to solve the problem. The constant $\bar{a}$ may be determined from the condition $\varphi_0\,(0) = 0$.

When R is infinite, the stress resultants (F_x, F_y) over C and the stresses at infinity are not all zero, the function $\varphi(z)$ and $\Psi(z)$ have from Eqs. (2.68) and (2.69), the forms

$$\left.\begin{aligned} \varphi(z) &= -\frac{F_x + i\,F_y}{2\pi\,(1+\chi)}\log z + \Gamma z + \varphi_1(z) \\ \psi(z) &= \frac{\chi(F_x + iF_y)}{2\pi\,(1+\chi)}\log\ z + \Gamma' z + \psi_1(z) \end{aligned}\right\} \quad ...(2.98)$$

where φ_1, (z) and $\psi_1(z)$ are holomorphic in R, including infinity. From its definition $\omega(\zeta)$ must be of the form.

$$\omega(\zeta) = \frac{C}{\zeta} + \text{a holomorphic function}$$

$$\therefore \quad \left.\begin{aligned} \varphi_0(\zeta) &= \frac{(F_x + i\,F_y)}{2\pi(1+\chi)}\log\zeta + \frac{\Gamma c}{\zeta} + \varphi_{1,0}\,(\zeta) \\ \psi_0(\zeta) &= -\frac{\chi\,(F_x - i\,F_y)}{2\pi(1+\chi)}\log\zeta + \frac{\Gamma' c}{\zeta} + \psi_{1.0}\,(\zeta) \end{aligned}\right\} \quad ...(2.99)$$

where $\varphi_{1,0}(\zeta)$ and $\psi_{1,0}(\zeta)$ are functions holomorphic inside and continuous up to γ.

Substituting Eqs. (2.99) in Eqs. (2.5), it is seen that $\varphi_{1,0}(\zeta)$ and $\psi_{1,0}(\zeta)$ must satisfy the same equations as $\varphi_0(\zeta)$ and $\psi_0(\zeta)$ did previously if $f(\sigma)$ is replaced everywhere by $f_0(\sigma)$ where

$$\begin{aligned} f_0(\sigma) = f(\sigma) &- \frac{F_x + i\,F_y}{2\pi}\log\sigma - \frac{\Gamma c}{\sigma} \\ &- \frac{\omega(\sigma)}{\overline{\omega}'\,(\sigma)}\left[\frac{F_x - i\,F_y}{2\pi\,(1+\chi)}\sigma - \Gamma\,\bar{c}\sigma^2\right] - \bar{\Gamma} \end{aligned} \quad ...(2.100)$$

where $\log \sigma = i\theta$. It is easily seen that $f_0(\sigma)$ will be a single valued, continuous function on γ the derivative of which w.r.t. θ satisfies the Holder condition, provided the given functions F_x and F_y satisfy that condition.

2.4.2 Solution of the Second Fundamental Problem

For the second fundamental problem the displacements along the boundary C of the region R are given as,

$$u_x = g_1(s), \quad u_y = g_2(s) \qquad \text{...(1)}$$

In the absence of body forces, from Eq. (2.58), we have

$$2G(u_x + i\, u_y) = \chi\varphi(z) - z\overline{\varphi}'(\bar{z}) - \overline{\psi}(\bar{z})$$

Transforming with the help of Eq. (1) of Art. 2.4.1, we get

$$\chi\varphi_0(\sigma) - \frac{\omega(\sigma)}{\overline{\omega}'(\bar{\sigma})}\overline{\varphi}_0'(\bar{\sigma}) - \overline{\psi}(\bar{\sigma})$$

$$= 2G(g_1 + i\, g_2) = 2G\, g \qquad \text{...(2)}$$

By methods similar to those used for the first boundary value problem as given in Art. 2.4.1, the following equations for $\varphi_0(\zeta)$ and $\psi_0(\zeta)$ are obtained on the assumption that

$$F_x = F_y = \Gamma = \Gamma' = 0$$

$$\chi\varphi_0(\zeta) - \frac{1}{2\pi i}\int_\gamma \frac{\omega(\sigma)\,\overline{\varphi}_0'(\bar{\sigma})}{\overline{\omega}'(\bar{\sigma})(\sigma-\zeta)}\,d\sigma - \frac{G}{\pi i}\int_\gamma \frac{g\,d\sigma}{\sigma-\zeta} = \bar{a} \qquad \text{...(2.101)}$$

$$\psi_0(\zeta) = -\frac{1}{\pi i}\int_\gamma \frac{\bar{g}\,d\sigma}{\sigma-\zeta} - \frac{1}{2\pi i}\int_\gamma \frac{\overline{\omega}(\bar{\sigma})\,\varphi_0'(\sigma)\,d\sigma}{\omega'(\sigma)(\sigma-\zeta)}$$

$$\text{...(2.102)}$$

for all $|\zeta| < 1$. When R is infinite and F_x, F_y, Γ, Γ' are not all zero the problem may be reduced to the preceding case in the same way as before, so that, in place of Eq. (2), we have an equation in $\varphi_{1.0}(\sigma)$, $\overline{\varphi}'_{1,0}(\bar{\sigma})$, $\overline{\psi}_{1,0}(\bar{\sigma})$ and g_0 appropriately defined.

The functional equation for $\varphi(\zeta)$ for the case of the mixed fundamental problem may be constructed in an analogous manner. However, in this case the equation becomes more complicated and is not attempted here.

2.4.3 Circular Disc under Concentrated Forces applied to its Boundary

Let R be the radius of the circle S with circumference C. Thus

$$Z = \omega(\zeta) = R\zeta \qquad \text{...(1)}$$

Let γ denote the unit circle $|\zeta| = 1$, $\sigma = e^{i\theta}$ a point on the circle.

The boundary condition becomes

$$\varphi(\sigma) + \sigma\varphi'(\bar{\sigma}) \mp \bar{\psi}(\sigma) = f_1 + if_2 = f \quad ...(2)$$

or
$$\varphi(\sigma) + \sigma\varphi'(\sigma) + \psi(\sigma) + f_1 - if_2 = \bar{f} \quad ...(3)$$

$$\psi(\sigma) = f - \bar{\varphi}(\bar{\sigma}) - \bar{\sigma}\varphi'(\sigma) \quad ...(3\,a)$$

For $\Psi(\sigma)$ to be the boundary-value of some function $\psi(\zeta)$, holomorphic inside γ and vanishing for $\zeta = 0$, we have

$$\frac{1}{2\pi i}\int_\gamma \frac{f d\sigma}{\sigma - \zeta} - \frac{1}{2\pi i}\int_\gamma \frac{\varphi(\sigma)\, d\sigma}{\sigma - \zeta} - \frac{1}{2\pi i}\int_\gamma \frac{\sigma\bar{\varphi}'(\bar{\sigma})\, d\sigma}{\sigma - \zeta} = 0$$

which gives

$$\varphi(\zeta) + \frac{1}{2\pi i}\int_\gamma \frac{\sigma\bar{\varphi}'(\bar{\sigma})\, d\sigma}{\sigma - \zeta} = \frac{1}{2\pi i}\int_\gamma \frac{f d\sigma}{\sigma - \zeta} \quad ...(4)$$

Let $\varphi(\zeta) = a_0 + a_1\zeta + a_2\zeta^2 + ...$ ...(5)

$\therefore$ $\varphi'(\zeta) = a_1 + 2a_2\zeta + ...$

$$\therefore \quad \frac{1}{2\pi i}\int_\gamma \frac{\sigma\bar{\varphi}'(\bar{\sigma})\, d\sigma}{\sigma - \zeta} = \bar{a}_1\zeta + 2\,\bar{a}_2$$

Because
$$\frac{1}{2\pi i}\int_\gamma \frac{\sigma\bar{\varphi}'(\bar{\sigma})\, d\sigma}{\sigma - \zeta} = \bar{a}_0\,\zeta^k + \bar{a}_1\,\zeta^{k-1} + ... + \bar{a}_k$$

[$k = 0, 1, 2 ....$]

Hence Eq. (4) becomes

$$\varphi(\zeta) + \bar{a}_1\,\zeta + 2\bar{a}_2 = \frac{1}{2\pi i}\int_\gamma \frac{f d\sigma}{\sigma - \zeta} \quad ...(6)$$

Substituting,

$$\varphi(\zeta) = a_1\zeta + a_2\zeta^2 + ...$$

in Eq. (6), we get by comparing coefficients of ζ and ζ^2

$$a_1 + \bar{a}_1 = \frac{1}{2\pi i}\int_\gamma f\frac{d\sigma}{\sigma^2} \quad ...(7)$$

$$a_2 = \frac{1}{2\pi i}\int_\gamma f\frac{d\sigma}{\sigma^2} \quad ...(8)$$

Now $f = f_1 + i\,f_2, \sigma = e^{i\theta}$

$$\therefore \quad a_1 + \bar{a}_1 = \frac{1}{2\pi}\int_0^{2\pi} (f_1 + i\,f_2)e^{-j\theta}\, d\theta$$

This condition can only be fulfilled when the right hand side is real, *i.e.* when

$$\int_0^{2\pi} (-f_1 \sin\theta + f_2 \cos\theta)\, d\theta = 0 \quad ...(9)$$

which expresses the necessity for the vanishing of the resultant moment of the external forces.

Let $I_m(a) = 0$. Hence from Eq. (7), we get

$$a_1 = \bar{a}_1 = \frac{1}{4\pi i}\int \frac{f d\sigma}{\sigma^2} \quad ...(10)$$

$$\therefore \quad \varphi(\zeta) = \frac{1}{2\pi i}\int_\gamma \frac{f d\sigma}{\sigma - \zeta} - \bar{a}_1 \zeta - 2\bar{a}_2 \quad ...(11)$$

Having found $\varphi(\zeta)$, the function $\Psi(\zeta)$ may be determined from Eq. (3 a).

Now $$\frac{1}{2\pi i}\int_\gamma \frac{\bar{\varphi}(\bar{\sigma})\, d\sigma}{\sigma - \zeta} = \bar{\varphi}\,(\bar{0})$$

$$\frac{1}{2\pi i}\int_\gamma \frac{\bar{\sigma}\varphi'(\sigma)\, d\sigma}{\sigma - \zeta} = \frac{1}{2\pi i}\int_\gamma \frac{\varphi'(\sigma)}{\sigma} \cdot \frac{d\sigma}{\sigma - \zeta}$$

$$= \frac{\varphi'(\zeta)}{\zeta} - \frac{a_1}{\zeta}$$

$$\therefore \quad \Psi(\zeta) = \frac{1}{2\pi i}\int_\gamma \frac{\bar{f}\, d\sigma}{\sigma - \zeta} - \frac{\varphi'(\zeta)}{\zeta} + \frac{a_1}{\zeta} - \bar{\varphi}(\bar{0}) \quad ...(12)$$

Omitting constant terms in Eqs. (11) and (12) which do not influence the stress distribution, we get

$$\varphi(\zeta) = \frac{1}{2\pi i}\int_\gamma \frac{f d\sigma}{\sigma - \zeta} - \bar{a}_1\zeta \quad ...(13)$$

$$\Psi(\zeta) = \frac{1}{2\pi i}\int_\gamma \frac{\bar{f} d\sigma}{\sigma - \zeta} - \frac{\varphi'(\zeta)}{\zeta} + \frac{a_1}{\zeta} \quad ...(14)$$

Let the concentrated forces acting on the disc be

$$(F_{x1}, F_{y1}), (F_{x2}, F_{y2}), \ldots, (F_{xn}, F_{yn})$$

at the points

$$Z_1 = R\, e^{i\alpha_1}, Z_2 = R\, e^{i\alpha_2}, \ldots, Z_n = R\, e^{i\alpha_n}$$

$$(0 \le \alpha_1 < \alpha_2 < \ldots < \alpha_n < 2\pi)$$

The points

$$\sigma_1 = e^{i\alpha_1}, \ldots, \sigma_n = e^{i\alpha_n}$$

correspond to these points in the ζ-plane.

Now $$\frac{1}{2\pi i}\int_{\gamma}\frac{f d\sigma}{\sigma-\zeta}=-\frac{1}{2\pi}\cdot\{(F_{x1}+i\,F_{y1})\log(\sigma_1-\zeta)+(F_{x2}+i\,F_{y2})\log(\sigma_2-\zeta)+\ldots..+(F_{xn}+i\,F_{yn})\log(\sigma_n-\zeta)\}$$

and $$\frac{1}{2\pi i}\int_{\gamma}\frac{\bar{f} d\sigma}{\sigma-\zeta}=\frac{1}{2\pi}\{(F_{x1}-i\,F_{y1})\log(\sigma_1-\zeta)+\ldots+(F_{xn}-i\,F_{yn})\log(\sigma_n-\zeta)\}$$

Since $a_1=\bar{a}_1=\varphi'(0)$.

Hence by using Eq. (13), we get by putting $\zeta=0$

$$2a_1=\frac{1}{2\pi}\sum_{k=1}^{n}\frac{F_{xk}+i\,F_{yk}}{\sigma_k}=\frac{1}{2\pi}\sum_{k=1}^{n}(F_{xk}+i\,F_{yk})\overline{\sigma_k}$$

For a_1 to be real

$$\sum_{k=1}^{n}(x_k F_{yk}-y_k F_{xk})=0$$

where $x_k+i\,y_k=z_k=\sigma_k R$, *i.e.* $M=0$.

$$\therefore\quad \varphi(\zeta)=-\frac{1}{2\pi}\sum_{k=1}^{n}(F_{xk}+i\,F_{yk})\log(\sigma_k-\zeta)-\frac{\zeta}{4\pi}\sum_{k=1}^{n}(F_{xk}+i\,F_{yk})\bar{\sigma}_k \qquad \ldots(15)$$

$$\Psi(\zeta)=\frac{1}{2\pi}\sum_{k=1}^{n}(F_{xk}-i\,F_{yk})\log(\sigma_k-\zeta)-\frac{1}{2\pi}\sum_{k=1}^{n}\frac{(F_{xk}+i\,F_{yk})\,\bar{\sigma}_k}{\sigma_k-\zeta} \qquad \ldots(16)$$

Consider, for example, the case when two equal and opposite forces (P, o) and $(-P, o)$, parallel the ox axis, act on the disc at the points $Z_1=R\,e^{i\alpha}$, $Z_2=R\,e^{i(\pi-\alpha)}=-R\,e^{-i\alpha}$ as shown in Fig. 2.11.

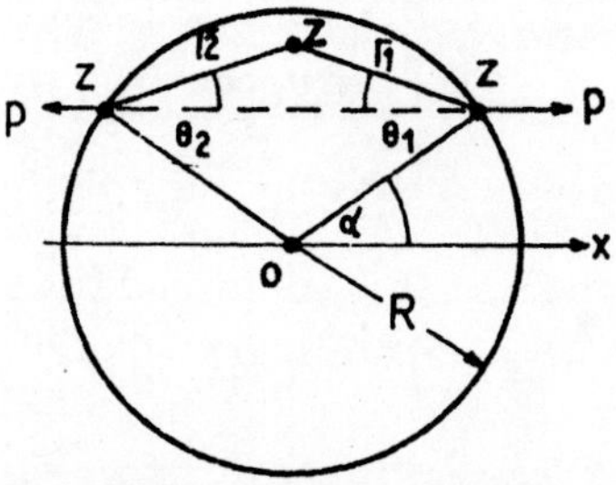

Fig. 2.11 Circular disc subjected to two forces

Then from Eqs. (15) and (16), we get

$$\varphi_1(z) = -\frac{P}{2\pi}\left\{ \log(z_1 - z) - \log(z_2 - z) + \frac{\bar{z}_1 - \bar{z}_2}{2R^2} z \right\}$$

$$\Psi_1(z) = \frac{P}{2\pi}\left\{ \log(z_1 - z) - \log(z_2 - z) - \frac{\bar{z}_1}{z_1 - z} + \frac{\bar{z}_2}{z_2 - z} \right\}$$

$$\varphi_1'(z) = \frac{P}{2\pi}\left\{ \frac{1}{z_1 - z} - \frac{1}{z_2 - z} - \frac{\bar{z}_1 - \bar{z}_2}{2R^2} \right\} = \phi_1(z)$$

$$\Psi_1'(z) = -\frac{P}{2\pi}\left\{ \frac{1}{z_1 - z} - \frac{1}{z_2 - z} + \frac{\bar{z}_1}{(z_1 - z)^2} - \frac{\bar{z}^2}{(z_2 - z)^2} \right\} = \Psi(z)$$

The stresses are given by the formula

$$\sigma_x + \sigma_y = 4Re\,[\phi_1(z)]$$

$$\sigma_y - \sigma_x + 2i\tau_{xy} = 2[\bar{z}\,\phi_1'(z) + \Psi_1(z)]$$

Also $\quad z_1 = R\,e^{i\alpha},\ z_2 = -R\,e^{-i\alpha}$

$$z_1 - z = r_1\,e^{-i\theta_1},\ z_2 - z = -r_2\,e^{i\theta_2}$$

$$\sigma_x + \sigma_y = \frac{2P}{\pi}\left\{ \frac{\cos\theta_1}{r_1} + \frac{\cos\theta_2}{r_2} - \frac{\cos\alpha}{R} \right\}$$

$$\sigma_x - \sigma_y = \frac{P}{\pi}\left\{ \frac{\cos 3\theta_1 + \cos\theta_1}{r_1} + \frac{\cos 3\theta_1 + \cos\theta_2}{r_2} \right\}$$

$$2\tau_{xy} = -\frac{P}{\pi}\left\{ \frac{\sin 3\theta_1 + \sin\theta_1}{r_1} - \frac{\sin 3\theta_2 + \sin\theta_2}{r_2} \right\}$$

Hence
$$\sigma_x = \frac{2P}{\pi}\left\{ \frac{\cos^3\theta_1}{r_1} + \frac{\cos^3\theta_2}{r_2} \right\} - \frac{p}{\pi R}\cos\alpha \qquad ...(2.103\ a)$$

$$\sigma_y = \frac{2P}{\pi}\left\{ \frac{\sin^2\theta_1\cos\theta_1}{r_1} - \frac{\sin^2\theta_1\cos\theta_2}{r_2} \right\} - \frac{p}{\pi R}\cos\alpha \qquad ...(2.103\ b)$$

$$\tau_{xy} = -\frac{2P}{\pi}\left\{\frac{\sin\theta_1\cos^2\theta_1}{r_1} - \frac{\sin\theta_2\cos^2\theta_2}{r_2}\right\} \qquad \text{...(2.103 c)}$$

The displacements can be obtained from

$$2G(u_x + i\,u_y) = \chi\varphi_1(z) - \bar{z}\varphi_1'(\bar{z}) - \overline{\Psi}_1(\bar{z})$$

which gives

$$2G(u_x + i\,u_y) = \frac{P}{2\pi}\left\{\chi\log\frac{z_2 - z}{z_1 - z} + \log\frac{\bar{z}_2 - \bar{z}}{\bar{z}_1 - \bar{z}}\right.$$

$$\left. + \frac{z_1 - z}{\bar{z}_1 - \bar{z}} - \frac{z_2 - z}{\bar{z}_2 - \bar{z}} - \frac{(\chi - 1)\cos\alpha}{R}z\right\}$$

$$\therefore \qquad u_x = \frac{P}{4\pi G}\left\{\frac{2(\lambda + 2G)}{\lambda + G}\log\frac{r_2}{r_1} + \cos 2\theta_1 - \cos 2\theta_2\right.$$

$$\left. - \frac{2G\cos\alpha}{\lambda + G}\cdot\frac{x}{R}\right\} \qquad \text{...(2.104 a)}$$

$$u_y = \frac{P}{4\pi G}\left\{\frac{2G}{\lambda + G}(\theta_1 + \theta_2) - \sin 2\theta_1 - \sin 2\theta_2\right.$$

$$\left. - \frac{2G\cos\alpha}{\lambda + G}\cdot\frac{x}{R}\right\} \qquad \text{...(2.104 b)}$$

2.5 Fourier Transform Methods

When the solid is infinite, semi-infinite, or is an infinite strip with parallel boundaries, the Fourier transform[7] method provides an effective method of solution. First of all, we will state some of the important results of Fourier transform.

(*i*) The integral transform of a function $f(x)$ is

$$I_f(x) = \int_0^\infty f(x)\,\kappa(\alpha, x)\,dx$$

where the function $\kappa\,(\alpha, x)$ is the Kernel.

If $\quad \kappa(\alpha, x) = x^{\alpha - 1}$ leading to

$$F(\alpha) = \int_0^\infty f(x)\,x^{\alpha - 1}\,dx$$

then this is known as the Mellin transform of function $f(x)$. Mellin transform may be used to solve problems on infinite wedges.

7. I.N. Sneddon ; Fourier Transform, McGraw-Hill Book Company Inc. (1951).

(*ii*) If the solution of the integral equation

$$I_f(\alpha) = \int_0^\infty f(x)\,\kappa(\alpha, x)\,dx$$

is given by

$$f(x) = \int_0^\infty I_f(\alpha)\,\kappa(\alpha, x)\,dx$$

then the function $\kappa(\alpha, x)$ is called a Fourier Kernel.

A necessary condition for the function $\kappa(\alpha, x)$ to be a Fourier Kernel is that the Mellin transform $\kappa(s)$ of the function $\kappa(x)$ should satisfy the functional equation

$$\kappa(s)\,\kappa(1 - s) = 1$$

where $\kappa(s) = \int_0^\infty \kappa(x)\,x^{s-1}\,dx$

(*iii*) If $F(\alpha)$ is the Fourier transform of $f(x)$, that is, if

$$F(\alpha) = \frac{1}{\sqrt{2\pi}}\int_{-\infty}^{\infty} f(x)\,e^{i\alpha x}\,dx$$

then $f(x)$ is given in terms of $F(\alpha)$ by the relation

$$f(x) = \frac{1}{\sqrt{2\pi}}\int_{-\infty}^{\infty} F(\alpha)\,e^{-i\alpha x}\,d\alpha$$

(*iv*) If $F_c(\alpha)$ is the Fourier cosine transform of $f(x)$, that is, if

$$F_c(\alpha) = \sqrt{\frac{2}{\pi}}\int_{-\infty}^{\infty} f(x)\cos(\alpha x)\,dx$$

then $f(x)$ is given by

$$f(x) = \sqrt{\frac{2}{\pi}}\int_0^\infty F_c(\alpha)\cos(\alpha x)\,d\alpha$$

(*v*) If $F_s(\alpha)$ is the Fourier sine transform of $f(x)$, that is, if

$$F_s(\alpha) = \sqrt{\frac{2}{\pi}}\int_0^\infty f(x)\sin(\alpha x)\,dx$$

then $f(x)$ is given by

$$f(x) = \sqrt{\frac{2}{\pi}}\int_0^\infty F_s(\alpha)\sin(\alpha x)\,d\alpha$$

(*vi*) Let $f(x, y)$ be a function of the two independent variables x and y ; then, regarded as a function of x, $f(x, y)$ has the Fourier transform

$$\bar{f}(\xi, y) = \frac{1}{\sqrt{2\pi}} \int_{-\infty}^{\infty} f(x, y)\, e^{i\xi x}\, dx$$

and this function, regarded as a function of y, has the Fourier transform

$$F(\xi, \eta) = \frac{1}{\sqrt{2\pi}} \int_{-\infty}^{\infty} \bar{f}(\xi, y)\, e^{i\eta y}\, dy$$

Combining, we get

$$F(\xi, \eta) = \frac{1}{2\pi} \int_{-\infty}^{\infty} \int_{-\infty}^{\infty} f(x, y)\, e^{i(\xi x + \eta y)}\, dx\, dy$$

and we say that $F(\xi, \eta)$ is the two-dimensional Fourier transform of the function $f(x, y)$. The inverse Fourier transform will be given by

$$f(x, y) = \frac{1}{2\pi} \int_{-\infty}^{\infty} \int_{-\infty}^{\infty} F(\xi, \eta) e^{-i(\xi x + \eta y)}\, d\xi d\eta$$

(*vii*) Convolution of Faltung theorem.

The function $fg = \frac{1}{\sqrt{2\pi}} \int_{-\infty}^{\infty} g(\eta)\, f(x - \eta)\, d\eta$

is called the convolution or Faltung of functions f and g over the interval $(-\infty, \infty)$.

This can also be written as

$$fg = \int_{-\infty}^{\infty} F(t)\, G(t)\, e^{-itx}\, dt$$

where F and G are the Fourier transforms of f and g respectively.

Similarly for double Fourier transforms, we have

$$\int_{-\infty}^{\infty} \int_{-\infty}^{\infty} F(\xi, \eta)\, G(\xi, \eta) e^{-i(\xi x + \eta y)}\, d\xi d\eta$$

$$= \int_{-\infty}^{\infty} \int_{-\infty}^{\infty} f(u, v)\, g(x - u, y - v)\, dv du$$

Hence in general, if

$$F(\xi_1, \xi_2, ..., \xi_n) = \left(\frac{1}{2\pi}\right)^{1/2n} \int_{-\infty}^{\infty} ... \int_{-\infty}^{\infty} f(x_1, x_2, ..., x_n)$$

$$\times e^{i(\xi_1 x_1 + \xi_2 x_2 + ... + \xi_n x_n)} dx_1 dx_2 ... dx_n$$

Then

$$f(x_1, x_2, ..., x_n) = \left(\frac{1}{2\pi}\right)^{1/2n} \int_{-\infty}^{\infty} ... \int_{-\infty}^{\infty} n(\xi_1, \xi_2, ..., \xi_n)$$

$$\times e^{-i(\xi_1 x_1 + \xi_2 x_2 + ... + \xi_n x_n)} d\xi_1 d\xi_2 ... d\xi_n$$

and the Faltum theorem becomes

$$\int_{-\infty}^{\infty} ... \int_{-\infty}^{\infty} F(\xi_1, \xi_2, ..., \xi_n) G(\xi_1, \xi_2, ..., \xi_n)$$

$$\times e^{-i(\xi_1 x_1 + ... + \xi_n x_n)} d\xi_1 ... d\xi_n$$

$$= \int_{-\infty}^{\infty} ... \int_{-\infty}^{\infty} F(u_1, ... u_n) g(x_1 - u_1, ..., x_n - u_n)\, du_1 ... du_n.$$

2.5.1 Infinite Solid Body with Body Forces

Suppose that the elastic solid is of infinite extent and that it is deformed by the action of known body forces $B_x(x, y)$ and $B_y(x, y)$. We shall further suppose that as $|x|$ or $|y| \to \infty$, all the components of the displacement and of the stress tensor tend to zero. Let us introduce the two-dimensional Fourier transform

$$\left(\bar{\sigma}_x, \bar{\sigma}_y, \bar{\tau}_{xy}, \bar{B}_x, \bar{B}_y\right) = \frac{1}{2\pi} \int_{-\infty}^{\infty} \int_{-\infty}^{\infty} (\sigma_x, \sigma_y, \tau_{xy}, B_x, B_y)$$

$$\times e^{i(\xi x + \eta y)} dx dy$$

Multiplying the equations of equilibrium

$$\frac{\partial \sigma_x}{\partial x} + \frac{\partial \tau_{xy}}{\partial y} + B_x = 0$$

$$\frac{\partial \tau_{xy}}{\partial x} + \frac{\partial \sigma_y}{\partial y} + B_y = 0$$

by $\quad e^{i(\xi x + \eta y)}$

and integrating over the whole plane by parts, we find the relations

$$\xi \bar{\sigma}_x + \eta \bar{\tau}_{xy} + iB_x = 0 \qquad ...(1)$$

$$\xi \bar{\tau}_{xy} + \eta \bar{\sigma}_y + iB_y = 0 \qquad ...(2)$$

Compatibility equation in terms of stress components for plane strain case is

$$\frac{\partial^2}{\partial y^2}[(1-\nu)\sigma_x - \nu\sigma_y] + \frac{\partial^2}{\partial x^2}[(1-\nu)\sigma_y - \nu\sigma_x]$$

$$= \frac{\partial^2 \tau_{xy}}{\partial x \partial y}$$

and for plane stress

$$\frac{\partial^2}{\partial y^2}[(\sigma_x - \nu\sigma_y] + \frac{\partial^2}{\partial x^2}[\sigma_y - \nu\sigma_x] = 2(1+\nu)\frac{\partial^2 \tau_{xy}}{\partial x \partial y}$$

Similarly by multiplying the compatibility equation for plane strain by $e^{i(\xi x + \eta y)}$ and integrating over the whole plane, we get

$$\bar{\sigma}_x + [(1-\nu)\eta^2 - \nu\xi^2] + \bar{\sigma}_y[(1-\nu)\xi^2 - \nu\eta^2) = 2\xi\eta\tau_{xy} \quad \text{...(3)}$$

Solving Eqs. (1) to (3), we find that

$$\bar{\sigma}_x + \bar{\sigma}_y = -\frac{1}{1-\nu}\cdot\frac{i\xi\bar{B}_x + i\eta\bar{B}_y}{\xi^2+\eta^2} \quad \text{...(4)}$$

$$\bar{\sigma}_x - \bar{\sigma}_y = -\left(\frac{1-2\nu}{1-\nu}\right)\cdot\frac{(\xi^2-\eta^2)}{(\xi^2+\eta^2)^2}\times(i\xi\bar{B}_x + i\eta\bar{B}_y) \quad \text{...(5)}$$

$$-4\,\frac{\xi\eta(i\xi\bar{B}_x - i\eta\bar{B}_y)}{(\xi^2+\eta^2)^2}$$

$$\bar{\tau}_{xy} = \frac{(\xi^2-\eta^2)}{(\xi^2-\eta^2)^2}(i\xi\,\bar{B}_y - i\eta\bar{B}_x) - \frac{(1-2\nu)}{(1-\nu)}$$

$$\times\frac{\xi\eta(i\eta\bar{B}_x + i\xi\bar{B}_y)}{(\xi^2+\eta^2)^2} \quad \text{...(6)}$$

where the functions $\bar{B}_x(\xi, \eta)$, $\bar{B}_y(\xi, \eta)$ are known from the body force distribution.

Using the inversion formula,

$$(\sigma_x, \sigma_y\, \tau_{xy}, B_x, B_y) = \frac{1}{2\pi}\int_{-\infty}^{\infty}\int_{-\infty}^{\infty}(\bar{\sigma}_x, \bar{\sigma}_y, \bar{\tau}_{xy}, \bar{B}_x, \bar{B}_y)$$

$$\times\, e^{-i(\xi x + \eta y)}\, d\xi d\eta$$

Eq. (4) becomes

$$\sigma_x + \sigma_y = -\frac{1}{2\pi(1-\nu)}\int_{-\infty}^{\infty}\int_{-\infty}^{\infty}\frac{(i\xi\bar{B}_x - i\eta\bar{B}_y)}{(\xi^2+\eta^2)^2}$$

$$\times\, e^{-(\xi x + \eta y)}\, d\xi d\eta$$

Now $$\int_{-\infty}^{\infty}\int_{-\infty}^{\infty} \frac{i\xi}{\xi^2+\eta^2} e^{-i(\xi x+\eta y)}\, d\xi d\eta$$

$$= 2\int_0^{\infty} \xi \sin(\xi x) d\xi \int_0^{\infty} \frac{\cos(\eta y)\, d\eta}{\xi^2+\eta^2}$$

and Fourier transform of $\frac{x}{x^2+y^2}$ is $\frac{i\xi}{\xi^2+\eta^2}$.

Hence by Faltung theorem

$$\int_{-\infty}^{\infty}\int_{-\infty}^{\infty} F(\xi,\eta)G(\xi,\eta)\, e^{-i(\xi x+\eta y)}\, d\xi d\eta$$

$$= \int_{-\infty}^{\infty}\int_{-\infty}^{\infty} f(\alpha,\beta).\, g(x-\alpha, y-\beta)\, d\alpha d\beta$$

We get

$$\int_{-\infty}^{\infty}\int_{-\infty}^{\infty} \frac{i\xi\, \overline{B}_x(\xi,\eta)}{\xi^2+\eta^2} e^{-i(\xi x+\eta y)}\, d\xi d\eta$$

$$= \int_{-\infty}^{\infty}\int_{-\infty}^{\infty} \frac{(x-\alpha)\, B_x(\alpha,\beta)\, d\alpha d\beta}{(x-\alpha)^2+(y-\beta)^2}$$

Similarly $$\int_{-\infty}^{\infty}\int_{-\infty}^{\infty} \frac{i\eta\, \overline{B}_y(\xi,\eta)}{\xi^2+\eta^2} e^{-i(\xi x+\eta y)}\, d\xi d\eta$$

$$= \int_{-\infty}^{\infty}\int_{-\infty}^{\infty} \frac{(y-\beta)\, B_y(\alpha,\beta)\, d\alpha d\beta}{(x-\alpha)^2+(y-\beta)^2}$$

so that $$\sigma_x + \sigma_y = -\frac{1}{2\pi(1-\nu)}$$

$$\int_{-\infty}^{\infty}\int_{-\infty}^{\infty} \frac{(x-\alpha)\, B_x(\alpha,\beta) + (y-\beta)\, B_y(\alpha,\beta)}{(x-\alpha)^2+(y-\beta)^2}\, d\alpha d\beta \qquad ...(2.7)$$

By a similar process we can show that

$$\sigma_x - \sigma_y = -\frac{(1-2\nu)}{\pi(1-\nu)}\int_{-\infty}^{\infty}\int_{-\infty}^{\infty} \frac{(x-\alpha)(y-\beta)}{[(x-\alpha)^2+(y-\beta)^2]^2}$$
$$\times [(y-\beta)\, B_x(\alpha,\beta) - (x-\alpha)\, B_y(\alpha,\beta)]\, d\alpha d\beta$$
$$-\frac{1}{\pi}\int_{-\infty}^{\infty}\int_{-\infty}^{\infty} \frac{[(x-\alpha)^2-(y-\beta)^2]}{[(x-\alpha)^2+(y-\beta)^2]} [(x-\alpha)B_x(\alpha,\beta)$$
$$+ (y-\beta)\, B_y(\alpha,\beta)]\, d\alpha d\beta \qquad ...(2.8)$$

$$\tau_{xy} = -\frac{(1-2\nu)}{4\pi(1-\nu)}\int_{-\infty}^{\infty}\int_{-\infty}^{\infty}\frac{[(y-\beta)^2-(x-\alpha)^2]}{[(x-\alpha)^2+(y-\beta)^2]^2}$$

$$\times [(y-\beta)\,B_x(\alpha,\beta)-(x-\alpha)\,B_y(\alpha,\beta)d\alpha d\beta$$

$$-\frac{1}{\pi}\int_{-\infty}^{\infty}\int_{-\infty}^{\infty}\frac{(x-\alpha)(y-\beta)[(x-\alpha)\,B_x(\alpha,\beta)-(y-\beta)\,B_y(\alpha,\beta)]}{[(x-\alpha)^2+(y-\beta)^2]^2}\,d\alpha d\beta$$

...(2.9)

A particularly simple case is that of a point force of magnitude F acting at $x = y = 0$ in the negative direction of x. The force is represented by

$$B_x(\alpha,\beta) = k\delta(\alpha)\delta(\beta),\; B_y(\alpha,\beta) = 0$$

where $\delta(\alpha)$ is the Dirac delta function, with the properties

$$\delta(\alpha) = 0,\; \alpha \neq 0,$$

$$\int_{-\infty}^{\infty} f(\alpha)\delta(\alpha)\,d\alpha = f(0)$$

and the constant k is to be found from the condition

$$\iint B_x(\alpha,\beta)\,d\alpha d\beta = -F$$

$$\therefore \quad \iint k\,\delta(\alpha)\delta(\beta)\,d\alpha d\beta = -F$$

$$\therefore \quad k = -F$$

We then have from Eqs. (2.8) to (2.9) the stress components

$$\left.\begin{aligned}\sigma_x &= \frac{Fx}{4\pi(1-\nu)r^2}\left[(1-2\nu)+\frac{2x^2}{r^2}\right]\\ \sigma_y &= \frac{Fx}{4\pi(1-\nu)r^2}\left[(1-2\nu)+\frac{2x^2}{r^2}\right]\\ \tau_{xy} &= \frac{Fy}{4\pi(1-\nu)r^2}\left[(1-2\nu)+\frac{2x^2}{r^2}\right]\end{aligned}\right\} \quad \text{...(2.105)}$$

where $r^2 = x^2 + y^2$.

2.5.2 Solution of the Two-Dimensional Biharmonic Equation

In the absence of body forces, the solution of two-dimensional equilibrium equations reduces to

$$\nabla^2.\,\nabla^2\Phi = \nabla^4\Phi = 0 \quad \text{...(1)}$$

where Φ = Airy stress function.

Now the Fourier transform of a function $\frac{d^r f}{dx^r}$ is

$$\frac{1}{\sqrt{2\pi}}\int_{-\infty}^{\infty}\frac{d^r f}{dx^r}\,e^{i\alpha x}\,dx = F^{(r)}(\alpha), \text{ (say)}$$

If we assume that $\frac{d^{r-1} f}{dx^{r-1}} \to 0$ as $|x| \to \infty$, then integrating the function on the left hand side, we get

$$F^{(r)}(\alpha) = -i\alpha\, F^{(r-1)}$$

$\therefore$ By a repetition of this rule and by the assumption that

$$\lim_{|x|\to\infty}\left(\frac{d^s f}{dx^s}\right) = 0,\ s = 1, 2, \ldots, r-1$$

we have $$F^{(r)} = (-i\sigma)^r F \qquad \ldots(2)$$

where F is the Fourier transform of $f(x)$.

Now Fourier transform of $\nabla^2\Phi$ is

$$\int_{-\infty}^{\infty}\nabla^2\Phi . e^{i\delta y}\,dy$$

By using Eq. (2), we get

$$\int_{-\infty}^{\infty}\nabla^2\Phi . e^{i\xi y}\,dy = \int_{-\infty}^{\infty}\left(\frac{\partial^2\Phi}{\partial x^2}+\frac{\partial^2\Phi}{\partial y^2}\right)e^{i\xi y}\,dy$$

$$= \left(\frac{d^2}{dx^2}-\xi^2\right)\int_{-\infty}^{\infty}\Phi . e^{i\xi y}\,dv$$

Repeating the process, we get

$$\int_{-\infty}^{\infty}\nabla^4\Phi\ e^{i\xi y}\,dy = \left(\frac{d^2}{dx^2}-\xi^2\right)^2\int_{-\infty}^{\infty}\Phi . e^{i\xi y}\,dy \qquad \ldots(3)$$

Let $$G = \int_{-\infty}^{\infty}\Phi . e^{i\xi y}\,dy$$

Hence Eq. (3) becomes,

$$\left(\frac{d^2}{dx^2}-\xi\right)^2 G = 0 \qquad \ldots(4)$$

General solution of Eq. (4) may be written in the form

$$G = (A + Bx)e^{-|\xi|x} + (C + Dx)\,e^{|\xi|x} \qquad \ldots(5)$$

where A, B, C, D are constants, which are in general, functions of ξ and are determined from the boundary conditions.

Using Fourier inverse transform, we get

$$\Phi(x, y) = \frac{1}{2\pi}\int_{-\infty}^{\infty} F(x, \xi)\, e^{-i\xi y}\, d\xi \qquad \text{...(6)}$$

Now
$$\sigma_x = \frac{\partial^2 \Phi}{\partial y^2}$$

$$\sigma_y = \frac{\partial^2 \Phi}{\partial x^2}$$

and
$$\tau_{xy} = -\frac{\partial^2 \Phi}{\partial x \partial y}$$

Multiplying by $e^{i\xi y}$ and integrating w.r.t. y over the range $(-\infty, \infty)$ we obtain

$$\int_{-\infty}^{\infty} \sigma_x \,.\, e^{i\xi y}\, dy = \int_{-\infty}^{\infty} \frac{\partial^2 \Phi}{\partial y^2} e^{i\xi y}\, dy$$

$$= -\xi^2 G \qquad \text{...(7)}$$

$$\int_{-\infty}^{\infty} \sigma_y \,.\, e^{i\xi y}\, dy = \int_{-\infty}^{\infty} \frac{\partial^2 \Phi}{\partial x^2} e^{i\xi y}\, dy$$

$$= \frac{d^2 G}{dx^2} \qquad \text{...(8)}$$

$$\int_{-\infty}^{\infty} \tau_{xy} \,.\, e^{i\xi y}\, dy = -\int_{-\infty}^{\infty} \frac{\partial^2 \Phi}{\partial x \partial y} e^{i\xi y}\, dy$$

$$= i\xi \frac{dG}{dx} \qquad \text{...(9)}$$

Therefore, by Fourier inverse transform, we get

$$\sigma_x = -\frac{1}{2\pi}\int_{-\infty}^{\infty} \xi^2 G\, e^{-i\xi y}\, dy \qquad \text{...(2.106 } a)$$

$$\sigma_y = \frac{1}{2\pi}\int_{-\infty}^{\infty} \frac{d^2 G}{dx^2} e^{-i\xi y}\, dy \qquad \text{...(2.106 } b)$$

$$\tau_y = \frac{1}{2\pi}\int_{-\infty}^{\infty} i\xi \frac{dG}{dx} e^{-i\xi y}\, dy \qquad \text{....(2.106 } c)$$

To determine the components of displacement vector, we have for plane strain case.

$$\frac{E}{1+\nu}\frac{\partial u_y}{\partial y} = (1-\nu)\,\sigma_y - \nu\sigma_x$$

Multiplying by $e^{i\xi y}$ and integrating from $-\infty$ to ∞, we have

$$\frac{E}{1+\nu}\int_{-\infty}^{\infty}\frac{duy}{dy}e^{i\xi y}\,dy = (1-\nu)\int_{-\infty}^{\infty}\sigma_y e^{i\xi y}\,dy - \nu\int_{-\infty}^{\infty}\sigma_x e^{i\xi y}\,dy$$

By using Eq. (2) and Eqs. (7) to (9), we get

$$\frac{-\xi E}{1-\nu}\int_{-\infty}^{\infty}u_y e^{i\xi y}\,dy = (1-\nu)\frac{d^2G}{dx^2} + \nu\xi^2 G$$

Hence by inversion formula, we get

$$u_y = \frac{1+\nu}{2\pi E}\int_{-\infty}^{\infty}\left[(1-\nu)\frac{d^2G}{dx^2} + \nu\xi^2 G\right]\times i\,e^{-i\xi y}\,.\,\frac{d\xi}{\xi} \qquad ...(2.107)$$

Similarly from

$$\frac{E}{2(1+\nu)}\left(\frac{\partial u_x}{\partial y} + \frac{\partial u_y}{\partial x}\right) = \tau_{xy}$$

we get

$$\frac{E}{2(1+\nu)}\left[\int_{-\infty}^{\infty}\frac{du_x}{dy}e^{i\xi y}\,dy + \int_{-\infty}^{\infty}\frac{du_y}{dx}e^{i\xi y}\,dx\right] = \int_{-\infty}^{\infty}\tau_{xy}e^{i\xi y}\,dy$$

$$\frac{-i\xi E}{2(1-\nu)}\int_{-\infty}^{\infty}u_x e^{i\xi y}\,dy + \frac{E}{2(1+\nu)}\frac{d}{dx}\int_{-\infty}^{\infty}u_y e^{i\xi y}\,dy = i\xi\,.\,\frac{dG}{dx}$$

$$\frac{-i\xi E}{2(1+\nu)}\int_{-\infty}^{\infty}u_x e^{i\xi y}\,dy + \frac{E}{2(1+\nu)}\left[-\frac{(1+\nu)}{i\xi E}\left\{(1-\nu)\frac{d^3G}{dx^3} + \nu\xi^2\frac{dG}{dx}\right\}\right]$$

$$= i\xi\frac{dG}{dx}$$

$$\frac{-i\xi E}{2(1+\nu)}\int_{-\infty}^{\infty}u_x e^{i\xi y}\,dy = i\xi\frac{dG}{dx} + \frac{1}{2i\xi}\left[(1-\nu)\frac{d^3G}{dx^3} + \nu\xi^2\frac{dG}{dx}\right]$$

$$= i\xi\frac{dG}{dx} + \frac{1}{2\xi}\left[-i(1-\nu)\frac{d^3G}{dx^3} - i\nu\xi^2\frac{dG}{dx}\right]$$

$$= i\xi\frac{dG}{dx} - \frac{i(1-\nu)}{2\xi}\frac{\partial^3G}{dx^3} - \frac{i\nu\xi}{2}\frac{dG}{dx}$$

$$= \left(1 - \frac{\nu}{2}\right) i\xi \frac{dG}{dx} - \frac{i(1-\nu)}{2\xi} \frac{d^3G}{dx^3}$$

$$= \left(\frac{2-\nu}{2}\right) i\xi \frac{dG}{dx} - \frac{i(1-\nu)}{2\xi} \frac{d^3G}{dx^3}$$

$$\int_{-\infty}^{\infty} u_x e^{i\xi y}\, dy = \frac{1+\nu}{E}\left[-(2-\nu)\frac{dG}{dx} + \frac{1-\nu}{\sigma^2}\frac{d^3G}{dx^3}\right]$$

$$= \frac{1+\nu}{E}\left[(1-\nu)\frac{d^3G}{dx^3} - (2-\nu)\xi^2 \frac{dG}{dx}\right]\frac{1}{\xi^2}$$

Hence by inversion formula, we get

$$u_x = \frac{1+\nu}{2\pi E}\int_{-\infty}^{\infty}\left[(1-\nu)\frac{d^3G}{dx^3} - (2-\nu)\,\xi^2\frac{dG}{dx}\right]\frac{e^{-i\xi y}}{\xi^2}\, d\xi \qquad \text{...(2.108)}$$

2.5.3 Semi-infinite Elastic Medium

Consider a semi-infinite elastic solid subjected to the application of an external pressure to its surface. The elastic medium is assumed to be bounded by a plane of infinite extent $x = 0$. The x-axis is taken normal to this plane to point into the medium. The surface pressure p is prescribed along the surface such that

$$\sigma_x = -p(y),\ \tau_{xy} = 0 \text{ on } x = 0.$$

As $x \to \infty$, $\sigma_x, \sigma_y, \tau_{xy} \to 0$.

Substituting these conditions in Eq. (5) of Art. 2.5.2 we find $C = D = 0$ and from Eqs. (2.106 a) and (2.106 c), we get

$$-\xi^2 A = -\int_{-\infty}^{\infty} p(y)\, e^{i\xi y}\, dy = -\bar{p}(\xi)$$

$$0 = B - |\,\xi\,|\, A$$

$$\therefore \qquad A = \frac{\bar{p}(\xi)}{\xi^2},\ B = \frac{\bar{p}(\xi)}{|\,\xi\,|}$$

$$\therefore \qquad G = \frac{\bar{p}(\xi)}{\xi^2}\left[1 + |\,\xi\,|\, x\right] e^{-|\,\xi\,|\, x} \qquad \text{...(1)}$$

$\therefore$ Eqs. (2.106) becomes

$$\sigma_x = -\frac{1}{2\pi}\int_{-\infty}^{\infty} \bar{p}(\xi)\, e^{-|\,\xi\,|x - i\xi y}\,.\,(1 + |\,\xi\,|\, x)\, d\xi \qquad \text{...(2.109 } a\text{)}$$

$$\sigma_y = -\frac{1}{2\pi}\int_{-\infty}^{\infty} \bar{p}(\xi)\, e^{-|\xi|x - i\xi y} . (1 - |\xi|x)\, d\xi \qquad \text{...(2.109 b)}$$

$$\tau_{xy} = -\frac{i}{2\pi}\int_{-\infty}^{\infty} x\xi\bar{p}(\xi)\, e^{-|\xi|x - i\xi y} . d\xi \qquad \text{...(2.109 c)}$$

$$\sigma_z = \nu(\sigma_x + \sigma_y) = -\frac{\nu}{\pi}\int_{-\infty}^{\infty} \bar{p}(\xi)\, e^{-|\xi|x - i\xi y} . d\xi \qquad \text{...(2.109 d)}$$

The displacement components from Eqs. (2.107) and (2.108) are

$$u_x = \frac{1+\nu}{2\pi E}\int_{-\infty}^{\infty} |\xi|\, \bar{p}(\xi)\, e^{-|\xi|x - i\xi y}\, [2(1-\nu) + |\xi|x]\, d\xi \qquad \text{...(2.110)}$$

$$u_y = -\frac{1+\nu}{2\pi E}\int_{-\infty}^{\infty} |\xi|\, \bar{p}(\xi)\, e^{-|\xi|x - i\xi y} \times [(1-2\nu) - |\xi|x]\, d\xi \qquad \text{...(2.111)}$$

when $\bar{p}(\xi)$ is an even function of ξ, then, we get

$$\left.\begin{aligned} \sigma_x &= -\frac{2}{\pi}\int_{-\infty}^{\infty} \bar{p}(\xi)\,(1+\xi x)\, e^{-\xi x}\cos(\xi y)\, d\xi \\ \sigma_y &= -\frac{2}{\pi}\int_{0}^{\infty} \bar{p}(\xi)(1-\xi x)\, e^{-\xi x}\cos(\xi y)\, d\xi \\ \tau_{xy} &= -\frac{2x}{\pi}\int_{0}^{\infty} \xi\bar{p}(\xi)\, e^{-\xi x}\sin(\xi y)\, d\xi \end{aligned}\right\} \qquad \text{...(2.112)}$$

$$\left.\begin{aligned} u_x &= \frac{2(1+\nu)}{\pi E}\int_{0}^{\infty} \bar{p}(\xi)\, e^{-\xi x}\, [2(1-\nu) + \xi x]\, \frac{\cos \xi y}{\xi}\, d\xi \\ u_y &= -\frac{2(1+\nu)}{\pi E}\int_{0}^{\infty} \bar{p}(\xi)\, e^{-\xi x}\, [1-2\nu - \xi x]\, \frac{\sin \xi y}{\xi}\, d\xi \end{aligned}\right\} \qquad \text{...(2.113)}$$

where $\overline{p}(\xi) = \int_0^\infty p(y) \cos(\xi y)\, dy$.

If $\overline{p}(\xi)$ is an odd function of ξ, then

$$\left.\begin{aligned}\sigma_x &= -\frac{2}{\pi}\int_0^\infty \overline{p}(\xi)(1+\xi x)\, e^{-\xi x} \sin(\xi y)\, d\xi\\ \sigma_y &= -\frac{2}{\pi}\int_0^\infty \overline{p}(\xi)(1-\xi x)\, e^{-\xi x} \sin(\xi y)\, d\xi\\ \tau_{xy} &= \frac{2x}{\pi}\int_0^\infty \xi\,\overline{p}(\xi)\, e^{-\xi x} \cos(\xi y)\, d\xi\end{aligned}\right\} \quad ...(2.114)$$

$$\left.\begin{aligned}u_x &= \frac{2(1+\nu)}{\pi E}\int_0^\infty \overline{p}(\xi)[2(1-\nu)+\xi x]\, e^{-\xi x} \frac{\sin \xi y}{\xi}\, d\xi\\ u_y &= \frac{2(1+\nu)}{\pi E}\int_0^\infty \overline{p}(\xi)[2(1-\nu)-\xi x]\, e^{-\xi x} \frac{\cos \xi y}{\xi}\, d\xi\end{aligned}\right\} \quad ...(2.115)$$

where $\overline{p}(\xi) = \int_0^\infty p(y) \sin(\xi y)\, dy$

Let $\qquad p = p_0$ for $-a \le y \le a$

Then $\qquad \overline{p}(\xi) = p_\theta \int_0^a \cos(\xi y)\, dy$

$$p_0 = \frac{\sin(\xi a)}{\xi} \quad ...(2.116)$$

Then from Eqs. (2.112), we get

$$\left.\begin{aligned}\sigma_z &= -\frac{2p_0}{\pi}\int_0^\infty \frac{1+\xi x}{\xi}\, e^{-\xi x} \sin(\xi a) \cos(\xi y)\, d\xi\\ &= \frac{p_0}{2\pi}[2(\theta_1-\theta_2) - \sin 2\theta_1 + \sin 2\theta_2]\\ \sigma_y &= -\frac{2p_0}{\pi}\int_0^\infty \frac{1-\xi x}{\xi}\, e^{-\xi x} \sin(\xi a) \cos(\xi y)\, d\xi\\ &= -\frac{p_0}{2\pi}[2(\theta_1-\theta_2) + \sin 2\theta_1 - \sin 2\theta_2]\end{aligned}\right\} \quad ...(2.117)$$

$$\tau_{xy} = -\frac{2p_0 x}{\pi}\int_0^\infty e^{-\xi x} \sin\ (\xi a) \sin\ (\xi y)\ d\xi$$

$$= \frac{p_0}{2\pi}\left[\cos 2\theta_2 - \cos 2\theta_1\right]$$

where $\theta_1 = \tan^{-1}\left(\frac{y-a}{x}\right)$, $\theta_2 = \tan^{-1}\left(\frac{y+a}{x}\right)$.

If a point force of magnitude P acts at the origin of coordinates of the boundary, then $p_0 = \frac{P}{2a}$. Hence

$$\bar{p}(\xi) = \lim_{a \to 0} \frac{P}{2a}\frac{\sin(\xi a)}{\xi} = \frac{P}{2}$$

and Eqs. (2.112) give

$$\left.\begin{aligned}
\sigma_x &= -\frac{P}{\pi}\int_0^\infty (1+\xi x)\, e^{-\xi x} \cos\ (\xi y)\ d\xi \\
&= -\frac{2Px^3}{\pi(x^2+y^2)^2} \\
\sigma_y &= -\frac{P}{\pi}\int_0^\infty (1-\xi x)\, e^{-\xi x} \cos\ (\xi y)\ d\xi \\
&= -\frac{2Pxy^2}{\pi(x^2+y^2)^2} \\
\tau_{xy} &= -\frac{Px}{\pi}\int_0^\infty \xi\, e^{-\xi x} \sin\ (\xi y)\ d\xi \\
&= -\frac{2Px^2y}{\pi(x^2+y^2)^2}
\end{aligned}\right\} \qquad \text{...(2.118)}$$

2.6 Real Potential Methods

This method is readily applied to problems involving semi-infinite bodies or infinite strips. This method is in fact based upon the decomposition of the Airy stress function into potential functions of real variables. This method was suggested by Payne in 1955.

The Airy stress function $\Phi(x, y)$ satisfying the biharmonic equation $\nabla^4\Phi = 0$ can be represented by

$$\Phi = \varphi_1{}' - x\frac{\partial \varphi_2{}'}{\partial x} \qquad \text{...(2.119)}$$

where $\varphi_1{}'$ and $\varphi_2{}'$ are any harmonic functions.

In the absence of body forces

$$\left.\begin{aligned}\sigma_x &= \frac{\partial^2\Phi}{\partial y^2} = \frac{\partial^2\varphi_1'}{\partial y^2} - x\frac{\partial^3\varphi_2'}{\partial x\,\partial y^2}\\ &= -\frac{\partial^2\varphi_1'}{\partial x^2} + x\frac{\partial^3\varphi_2'}{\partial x^3}\\ \sigma_y &= \frac{\partial^2\Phi}{\partial x^2} = \frac{\partial^2\varphi_1'}{\partial x^2} - \frac{2\partial^2\,\varphi_2}{\partial x^2} - x\frac{\partial^3\varphi_2'}{\partial x^3}\\ \tau_{xy} &= -\frac{\partial^2\Phi}{\partial x\,\partial y} = -\frac{\partial}{\partial y}\left\{\frac{\partial\varphi_1'}{\partial x} - \frac{\partial\varphi_2'}{\partial x} - x\frac{\partial^2\varphi_2'}{\partial x^2}\right\}\end{aligned}\right\} \quad \text{...(2.120)}$$

Using these expressions in the stress-strain relations for plane strain and integrating, we get

$$\left.\begin{aligned}2Gu_x &= -\frac{\partial\varphi_1}{\partial x} + x\frac{\partial^2\varphi_2}{\partial x^2} - (1-2\nu)\frac{\partial\varphi_2'}{\partial x}\\ 2Gu_y &= -\frac{\partial\varphi_1'}{\partial y} + 2(1-\nu)\frac{\partial\varphi_2'}{\partial y} + x\frac{\partial^2\varphi_2'}{\partial x\partial y}\end{aligned}\right\} \quad \text{...(2.121)}$$

Now we introduce the harmonic conjugate functions, ψ_1' and ψ_2' of φ_1' and φ_2' respectively, so that

$$\frac{\partial\varphi_i'}{\partial x} = \frac{\partial\psi_i'}{\partial y},$$

$$\frac{\partial\varphi_i'}{\partial x} = -\frac{\partial\psi_i'}{\partial y}, \quad i = 1, 2$$

and put

$$\varphi_i = \frac{\partial\varphi_i'}{\partial x},$$

$$\psi_i = \frac{\partial\psi_i'}{\partial x}, \quad i = 1, 2$$

Eqs. (2.120) and (2.121) reduce to

$$\left.\begin{aligned}\sigma_x &= -\frac{\partial\varphi_1}{\partial x} + x\frac{\partial^2\,\varphi_2}{\partial x^2}\\ \sigma_y &= \frac{\partial\varphi_1}{\partial x} - 2\frac{\partial\varphi_2}{\partial x} - x\frac{\partial^2\varphi_2}{\partial x^2}\\ \tau_{xy} &= -\frac{\partial}{\partial y}\left(\varphi_1 - \varphi_2 - x\frac{\partial\varphi_2}{\partial x}\right)\\ 2Gu_x &= -\varphi_1 + x\frac{\partial\varphi_2}{\partial x} - (1-2\nu)\varphi_2\\ 2Gu_y &= \psi_1 - 2(1-\nu)\psi_2 + x\frac{\partial\varphi_2}{\partial x}\end{aligned}\right\} \quad \text{(2.122)}$$

Now, on $x = 0$

$$\sigma_x = 0, \text{ if } \varphi_1 = 0$$

$$\sigma_y = 0, \text{ if } \varphi_1 = 2\varphi_2$$

$$\tau_{xy} = 0, \text{ if } \varphi_1 = \varphi_2$$

$$u_x = 0, \text{ if } \varphi_1 = -(1-2\nu)\varphi_2$$

$$u_y = 0, \text{ if } \varphi_1 = 2(1-\nu)\varphi_2$$

If $\varphi_1 = \varphi_2 = \varphi$ then $\psi_1 = \psi_2 = \psi$ say, and Eqs. (2.122) become

$$\left.\begin{aligned} \sigma_x &= -\frac{\partial\varphi}{\partial x} + x\frac{\partial^2\varphi}{\partial x^2} \\ \sigma_y &= -\frac{\partial\varphi}{\partial x} - x\frac{\partial^2\varphi}{\partial x^2} \\ \tau_{xy} &= x\frac{\partial^2\varphi}{\partial x\,\partial y} \\ 2Gu_x &= -2(1-\nu)\varphi + x\frac{\partial\varphi}{\partial x} \\ 2Gu_y &= -(1-2\nu)\psi + x\frac{\partial\varphi}{\partial y} \end{aligned}\right\} \qquad \text{...(2.123)}$$

2.6.1 Homogeneous Boundary Value Problems

Fourier transform-methods are used to solve the first and second boundary value problems. In the second boundary value problem we must assume that the displacement vanishes at infinity, while the displacement on the boundary is given by

$$Gu_x = -(1-\nu)f(y),\ x = 0$$

We put $$\bar{\varphi} = \frac{1}{\sqrt{2\pi}}\int_0^\infty \varphi e^{i\alpha y}\,dy \qquad \text{...(2.124)}$$

and then $\bar{\varphi}$ satisfies the transformed Laplace's equation

$$\frac{d^2\bar{\varphi}}{dx^2} - \alpha^2\bar{\varphi} = 0.$$

Using the fourth of Eqs. (2.123) with the boundary conditions, we find the solution

$$\bar{\varphi} = \bar{f}(\alpha)\,e^{-|\alpha|x} \qquad \text{...(2.125)}$$

The two potential functions can then be written in the forms.

(*a*) **f(y) even** :

$$\left.\begin{aligned}\varphi &= \frac{2}{\pi}\int_0^\infty e^{-\alpha x}\cos\alpha y\,d\alpha\int_0^\infty f(y')\cos\alpha y'\,dy'\\ \psi &= -\frac{2}{\pi}\int_0^\infty e^{-\alpha x}\sin\alpha y\,d\alpha\int_0^\infty f(y')\cos\alpha y'\,dy'\end{aligned}\right\}\quad\text{...(2.126)}$$

(*b*) **f(y) odd**

$$\left.\begin{aligned}\varphi &= \frac{2}{\pi}\int_0^\infty e^{-\alpha x}\sin\alpha y\,d\alpha\int_0^\infty f(y')\sin\alpha y'\,dy'\\ \psi &= \frac{2}{\pi}\int_0^\infty e^{-\alpha x}\cos\alpha y\,d\alpha\int_0^\infty f(y')\sin\alpha y'\,dy'\end{aligned}\right\}\quad\text{...(2.127)}$$

To solve the first boundary value problem we first note that $\frac{\partial\varphi}{\partial x}$ also satisfies Laplace equation. In addition to σ_x vanishing at infinity, we have the condition

$$\sigma_x = -g(y),\ x = 0.$$

Using these conditions, we get

(*a*) **g(y) even**

$$\left.\begin{aligned}\frac{\partial\varphi}{\partial x} &= \frac{2}{\pi}\int_0^\infty e^{-\alpha x}\cos\alpha y\,d\alpha\int_0^\infty g(y')\cos\alpha y'\,dy'\\ \frac{\partial\psi}{\partial x} &= -\frac{2}{\pi}\int_0^\infty e^{-\alpha x}\sin\alpha y\,d\alpha\int_0^\infty g(y')\cos\alpha y'\,dy'\end{aligned}\right\}\quad\text{...(2.128)}$$

(*b*) **g(y) odd**

$$\left.\begin{aligned}\frac{\partial\varphi}{\partial x} &= \frac{2}{\pi}\int_0^\infty e^{-\alpha x}\sin\alpha y\,d\alpha\int_0^\infty g(y')\sin\alpha y'\,dy'\\ \frac{\partial\psi}{\partial x} &= \frac{2}{\pi}\int_0^\infty e^{-\alpha x}\cos\alpha y\,d\alpha\int_0^\infty g(y')\sin\alpha y'\,dy'\end{aligned}\right\}\quad\text{...(2.129)}$$

EXERCISES

2.1 A cantilever beam of rectangular cross-section 40 mm wide and 60 mm thick is 800 mm in length. It carries a load of 500 N at the free end. Determine the stresses in the cantilever at mid-length.

2.2 A rectangular beam 80 mm wide and 100 mm thick is of 600 mm in length. It carries a uniformly distributed load of intensity 10 N/mm throughout its length. Plot the variation of stresses in the beam at mid-span. Also compare the results as obtained from elementary strength of materials.

2.3 In the cantilever problem of Fig. 2.1, the support conditions at $x = l$ are given as :

at $x = l, y = 0$: $u_x = u_y = 0$

at $x = l, y = \pm C$: $u_x = 0$

Show that the deflection is now

$$(u_y)_{x=y=0} = \frac{Pl^3}{3El}\left[1 + \frac{1}{2}(4 + 5\nu)\frac{C^2}{l^2}\right]$$

2.4 If the beam in problem 2.2 is loaded by its own weight instead of the load 10 N/mm on the upper edge, then find expressions for the displacement components u_x and u_y.

2.5 Show that

$$(A\,e^{\alpha y} + B\,e^{-\alpha y} + Cy\,e^{\alpha y} + Dy\,e^{-\alpha y})\sin \alpha x$$

is a stress function.

Derive series expressions for the stresses in a semi-infinite plate, $y > 0$, with normal pressure on the straight edge ($y = 0$) having the distribution

$$\sum_{m=1}^{\infty} b_m \sin\frac{m\pi x}{l}$$

2.6 Show that if V is a plane harmonic function, *i.e.* it satisfies the Laplace equation $\nabla^2 V = 0$, then the functions xV, yV, $(x^2 + y^2)V$ satisfy the biharmonic equation and so can be used as stress functions.

2.7 Determine the stress fields that arise from the following stress functions :

(*i*) $\Phi = Cy^2$

(*ii*) $\Phi = Ax^2 = Bxy + Cy^2$

(*iii*) $\Phi = Ax^3 + Bx^2y + Cxy^2 + Dy^3$.

2.8 The stress field for the cantilever beam shown in Fig. 2.1 considering bending and transverse shear effects are :

$$\sigma_x = -\frac{Pxy}{I},\ \sigma_y = 0,\ \tau_{xy} = -\frac{P}{2I}(C^2 - y^2).$$

Verify that this stress field satisfies the equilibrium equations.

2.9 Prove that if a harmonic function may be expressed as the real or imaginary part of a complex function then show that the stress field must have the form

$$\sigma_x + \sigma_y = 2\{\varphi'(z) + \overline{\varphi'(z)}\}$$

$$= 4Re\,\varphi'(z)$$

$$\sigma_x - \sigma_y + 2i\tau_{xy} = -2\{z\,\overline{\varphi''(z)} + \overline{\psi'(z)}\}.$$

Also determine the form of the complex displacement D in terms of $\varphi(z)$ and $\psi(z)$.

2.10 Find the stress and displacement fields corresponding to the complex potentials

$$\varphi(z) = \alpha + iAz,\ \psi(z) = \beta$$

where α and β are complex constants and A is real. Interpret this displacement field.

2.11 Show that the complex potentials

$$\varphi(z) = \alpha z, \quad \psi(z) = \beta z$$

Correspond to a uniform stress field. Choose α and β so that

$$\sigma_x = X, \quad \tau_{xy} = S, \quad \sigma_y = Y.$$

2.12 Determine the stress and displacement fields in an infinite medium due to equal and opposite point forces acting at different points along their common line of action.

2.13 Consider the effect of a point force $X + iY$ acting at an internal point Z_0 of the half plane S^+ when the boundary $y = 0$ is unstressed. Determine the normal displacement on $y = 0$.

2.14 Using Fourier integral method, determine the solution of biharmonic equation in Cartesian co-ordinates.

2.15 A semi-infinite elastic medium is subjected to a normal pressure of intensity p distributed over a circular area of radius a at $x = 0$. Determine the stress distribution by using Fourier integral method.

2.16 Show that the stress function

$$\phi = C\left[(x^2 + y^2)\tan^{-1}\frac{y}{x} - xy\right]$$

provides the solution to the problem of the semi-infinite elastic medium acted upon by a uniform pressure q on one side of the origin (Fig. 2.12).

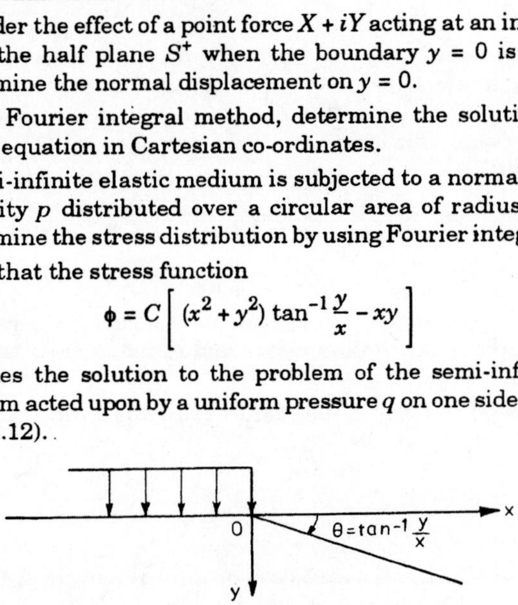

Fig. 2.12

2.17 Investigate the plane stress problem represented by the Airy's stress function

$$\phi = \frac{3F}{4h}\left(xy - \frac{xy^3}{3h^2}\right) + \frac{P}{2}y^2$$

where h is half depth of the beam and F is the concentrated load.

2.18 A cantilever beam loaded at its free end has a stress function

$$\phi = Axy + B\frac{xy^3}{6}$$

Determine an expression for the vertical deflection curve.

2.19 Using Airy's stress function, $\phi = B\left(y^3 + \frac{3}{4}yh^2\right) + F\frac{y^2}{8}$, determine the stress distribution at $0 < x < l$ for $y = \pm h$, where 'l' is the span of the beam, '$2h$' the depth of the beam and F the load.

2.20 Using the stress function $\phi = \frac{3P}{4c}\left(xy - \frac{xy^3}{3c^2}\right)$, determine the deflection at the free end of a cantilever, with the conditions $0 < x < l$ and $y = \pm c$.

3

Two-Dimensional Problems in Polar Co-ordinates

3.1 Basic Equations

(a) Plane Stress

(i) Equilibrium Equations

$$\left.\begin{aligned}\frac{\partial \sigma_r}{\partial r} + \frac{1}{r}\frac{\partial \tau_{r\theta}}{\partial \theta} + \frac{\sigma_r - \sigma_\theta}{r} + B_r = 0 \\ \frac{1}{r}\frac{\partial \sigma_\theta}{\partial \theta} + \frac{\partial \tau_{r\theta}}{\partial r} + \frac{2\tau_{r\theta}}{r} + B_\theta = 0\end{aligned}\right\} \quad \text{...(3.1)}$$

(ii) Strain-displacement relations

$$\varepsilon_r = \frac{\partial u_r}{\partial r}$$

$$\varepsilon_\theta = \frac{1}{r}\frac{\partial u_\theta}{\partial \theta} + \frac{u_r}{r}, \quad \varepsilon_z = \frac{\partial u_z}{\partial z} \quad \text{...(3.2)}$$

$$\gamma_{r\theta} = \frac{1}{r}\frac{\partial u_r}{\partial \theta} + \frac{\partial u_\theta}{\partial r} - \frac{u_\theta}{r}$$

(iii) Stress-strain relations

$$\varepsilon_r = \frac{1}{E}[\sigma_r - \nu\sigma_\theta]$$

$$\varepsilon_\theta = \frac{1}{E}[\sigma_\theta - \nu\sigma_r] \quad \text{...(3.3)}$$

$$\varepsilon_z = -\frac{\nu}{E}[\sigma_r + \sigma_\theta]$$

$$\gamma_{r\theta} = \frac{2(1+\nu)}{E}\tau_{r\theta}$$

(iv) Stress-displacement relations

$$\sigma_r = 2G\frac{\partial u}{\partial r} + \lambda\left[\frac{\partial u_r}{\partial r} + \frac{1}{r}\frac{\partial u_\theta}{\partial \theta} + \frac{u_r}{r} + \frac{\partial u_z}{\partial z}\right]$$

$$\sigma_\theta = 2G\left(\frac{1}{r}\frac{\partial u_\theta}{\partial \theta} + \frac{u_r}{r}\right) + \lambda\left[\frac{\partial u_r}{\partial r} + \frac{1}{r}\frac{\partial u_\theta}{\partial \theta} + \frac{u_r}{r} + \frac{\partial u_z}{\partial z}\right] \quad ...(3.4)$$

$$\tau_{r\theta} = G\,\gamma_{r\theta} = G\left[\frac{1}{r}\frac{\partial u_r}{\partial \theta} + \frac{\partial u_\theta}{\partial r} - \frac{u_\theta}{r}\right]$$

(*v*) *Strain-compatibility equation*

$$\frac{1}{r^2}\frac{\partial}{\partial r}\left(r\frac{\partial \gamma_{r\theta}}{\partial \theta}\right) = \frac{1}{r^2}\frac{\partial}{\partial r}\left(r^2\frac{\partial \varepsilon_\theta}{\partial r}\right) + \left(\frac{1}{r^2}\frac{\partial^2}{\partial \theta^2} - \frac{1}{r}\frac{\partial}{\partial r}\right)\varepsilon_r \quad ...(3.5)$$

(*iv*) ***Equilibrium Equations in terms of displacements***

$$\nabla^2 u_r + \left(\frac{3}{1-2\nu}\right)\frac{\partial \varepsilon_m}{\partial r} - \frac{u_r}{r^2} - \frac{2}{r^2}\frac{\partial u_\theta}{\partial \theta} + \frac{B_r}{G} = 0 \quad ...(3.6)$$

$$\nabla^2 u_\theta + \left(\frac{3}{1-2\nu}\right)\frac{\partial \varepsilon_m}{r\partial \theta} - \frac{u_\theta}{r^2} + \frac{2}{r^2}\frac{\partial u_r}{\partial \theta} + \frac{B_\theta}{G} = 0$$

where $\varepsilon_m = \frac{1}{3}(\varepsilon_r + \varepsilon_\theta + \varepsilon_z)$

$$\nabla^2 = \frac{\partial^2}{\partial r^2} + \frac{1}{r}\frac{\partial}{\partial r} + \frac{1}{r^2}\frac{\partial^2}{\partial \theta^2}$$

For axial symmetry, all derivatives w.r.t. θ are zero.

(*b*) Plane strain

(*i*) *Equilibrium Equations*

$$\frac{\partial \sigma_r}{\partial r} + \frac{1}{r}\frac{\partial \tau_{r\theta}}{\partial \theta} + \frac{\sigma_r - \sigma_\theta}{r} + B_r = 0$$

$$\frac{1}{r}\frac{\partial \sigma_\theta}{\partial \theta} + \frac{\partial \tau_{r\theta}}{\partial r} + \frac{2\tau_{r\theta}}{r} + B_\theta = 0 \quad ...(3.7)$$

$$\frac{\partial \sigma_z}{\partial z} = 0$$

(*ii*) *Strain-displacement relations*

$$\varepsilon_r = \frac{\partial u_r}{\partial r}$$

$$\varepsilon_\theta = \frac{1}{r}\frac{\partial u_\theta}{\partial \theta} + \frac{ur}{r} \quad ...(3.8)$$

$$\gamma_{r\theta} = \frac{1}{r}\frac{\partial u_r}{\partial \theta} + \frac{\partial u_\theta}{\partial r} - \frac{u_\theta}{r}$$

(*iii*) *Stress-strain relations*

$$\varepsilon_z = \frac{1}{E}\left[\sigma_z - \nu(\sigma_r + \sigma_\theta)\right] = 0$$

$$\therefore \qquad \sigma_z = \nu(\sigma_r + \sigma_\theta)$$

$$\varepsilon_r = \frac{1}{E}[\sigma_r - \nu(\sigma_\theta + \sigma_z)]$$

$$= \frac{1}{E}[\sigma_r - \nu\sigma_\theta - \nu^2(\sigma_r + \sigma_\theta)]$$

$$= \frac{1}{E}[(1 - \nu^2)\sigma_r - \nu(1 + \nu)\sigma_\theta]$$

$$= \frac{1 + \nu}{E}[1 - \nu)\sigma_r - \nu\sigma_\theta] \qquad ...(3.9)$$

$$\varepsilon_\theta = \frac{1 + \nu}{E}[(1 - \nu)\sigma_\theta - \nu\sigma_r]$$

$$\gamma_{r\theta} = \frac{2(1 + \nu)}{E}\tau_{r\theta}$$

(iv) Stress-displacement relations

$$\sigma_r = 2G\frac{\partial u_r}{\partial_r} + \lambda\left[\frac{\partial u_r}{\partial r} + \frac{1}{r}\frac{\partial u_\theta}{\partial \theta} + \frac{u_r}{r}\right]$$

$$\sigma_\theta = 2G\left(\frac{1}{r}\frac{\partial u_\theta}{\partial r} + \frac{u_r}{r}\right) + \lambda\left[\frac{\partial u_r}{\partial r} + \frac{1}{r}\frac{\partial u_\theta}{\partial \theta} + \frac{u_r}{r}\right]$$

$$\tau_{r\theta} = G\left[\frac{1}{r}\frac{\partial u_r}{\partial \theta} + \frac{\partial u_\theta}{\partial_r} - \frac{u_\theta}{r}\right] \qquad ...(3.10)$$

(v) Strain-compatibility Equation

$$\frac{1}{r^2}\frac{\partial}{\partial r}\left(r\frac{\partial \gamma_{r\theta}}{\partial \theta}\right) = \frac{1}{r^2}\frac{\partial}{\partial r}\left(r^2\frac{\partial \varepsilon_\theta}{\partial r}\right) + \left(\frac{1}{r^2}\frac{\partial^2}{\partial \theta^2} - \frac{1}{r}\frac{\partial}{\partial r}\right)\varepsilon_r \qquad ...(3.11)$$

(vi) Equilibrium Equations in terms of displacements

$$\nabla^2 u_r\left(\frac{3}{1 - 2\nu}\right)\frac{\partial \varepsilon_m}{\partial r} - \frac{u_r}{r^2} - \frac{2}{r^2}\frac{\partial u_\theta}{\partial_\theta} + \frac{B_r}{G} = 0 \qquad ...(3.12)$$

$$\nabla^2 u_\theta + \left(\frac{3}{1 - 2\nu}\right)\frac{\partial \varepsilon_m}{r\partial \theta} - \frac{u_\theta}{r^2} + \frac{2}{r^2}\frac{\partial u_\theta}{\partial \theta} + \frac{B_\theta}{G} = 0$$

where $\varepsilon_m = \frac{1}{3}(\varepsilon_r + \varepsilon_\theta)$

$$\nabla^2 = \frac{\partial^2}{\partial r^2} + \frac{1}{r}\frac{\partial}{\partial r} + \frac{\partial^2}{\partial \theta^2}$$

For axial symmetry, all derivatives w.r.t. θ are zero.

3.2 Biharmonic Equation

The relationship between the Cartesian and polar co-ordinates is

$$x = r\cos\theta,\quad y = r\sin\theta$$

$$r^2 = x^2 + y^2,\ \tan\theta = \frac{y}{x}$$

Taking the derivatives, we have

$$\frac{\partial r}{\partial x} = \frac{x}{r} = \cos\theta,\quad \frac{\partial r}{\partial y} = \frac{y}{r} = \sin\theta$$

$$\frac{\partial \theta}{\partial x} = -\frac{y}{r^2} = -\frac{\sin\theta}{r},\quad \frac{\partial \theta}{\partial y} = \frac{x}{r^2} = \frac{\cos\theta}{r}$$

For any function $\Phi(x, y)$ in polar co-ordinates $\Phi\,(r\cos\theta, r\sin\theta)$, we have

$$\frac{\partial \Phi}{\partial x} = \frac{\partial \Phi}{\partial r}\cdot\frac{\partial r}{\partial x} + \frac{\partial \Phi}{\partial \theta}\cdot\frac{\partial \theta}{\partial x}$$

$$= \frac{\partial \Phi}{\partial r}\,.\cos\theta - \frac{\partial \Phi}{\partial \theta}\frac{\sin\theta}{r}$$

$$\frac{\partial \Phi}{\partial y} = \frac{\partial \Phi}{\partial r}\cdot\frac{\partial r}{\partial y} + \frac{\partial \Phi}{\partial \theta}\cdot\frac{\partial \theta}{\partial y}$$

$$= \frac{\partial \Phi}{\partial r}\,.\sin\theta + \frac{\partial \Phi}{\partial \theta}\cdot\frac{\cos\theta}{r}$$

$$\frac{\partial^2 \Phi}{\partial x^2} = \left(\cos\theta\frac{\partial}{\partial r} - \frac{\sin\theta}{r}\frac{\partial}{\partial \theta}\right)\left(\cos\theta\frac{\partial \Phi}{\partial r} - \frac{\sin\theta}{r}\frac{\partial \Phi}{\partial \theta}\right)$$

$$= \frac{\partial^2 \Phi}{\partial r^2}\cos^2\theta - \frac{2}{r}\frac{\partial^2 \Phi}{\partial r\partial \theta}\sin\theta\cos\theta$$

$$+ \frac{1}{r}\frac{\partial \Phi}{\partial r}\sin^2\theta + \frac{2}{r^2}\frac{\partial \Phi}{\partial \theta}\sin\theta\cos\theta$$

$$+ \frac{1}{r^2}\frac{\partial^2 \Phi}{\partial \theta^2}\sin^2\theta$$

$$\frac{\partial^2 \Phi}{\partial y^2} = \left(\sin\theta\frac{\partial}{\partial r} + \frac{\cos\theta}{r}\frac{\partial}{\partial \theta}\right)\left(\sin\theta\frac{\partial \Phi}{\partial r} + \frac{\cos\theta}{r}\frac{\partial \Phi}{\partial \theta}\right)$$

$$= \frac{\partial^2 \Phi}{\partial r^2}\sin^2\theta + \frac{2}{r}\frac{\partial^2 \Phi}{\partial r\partial \theta}\sin\theta\cos\theta$$

$$+ \frac{1}{r}\frac{\partial \Phi}{\partial r}\cos^2\theta - \frac{2}{r^2}\frac{\partial \Phi}{\partial \theta}\sin\theta\cos\theta$$

$$+ \frac{1}{r^2}\frac{\partial^2 \Phi}{\partial \theta^2}\cos^2\theta$$

Now $$\nabla^4\Phi = \left(\frac{\partial^2}{\partial x^2} + \frac{\partial^2}{\partial y^2}\right)\left(\frac{\partial^2\Phi}{\partial x^2} + \frac{\partial^2\Phi}{\partial y^2}\right)$$

Also from above derivatives, we have

$$\frac{\partial^2\Phi}{\partial x^2} + \frac{\partial^2\Phi}{\partial y^2} = \frac{\partial^2\Phi}{\partial r^2} + \frac{1}{r}\frac{\partial\Phi}{\partial r} + \frac{1}{r^2}\frac{\partial^2\Phi}{\partial\theta^2}$$

$$\therefore \quad \nabla^2\Phi = \left(\frac{\partial^2}{\partial r^2} + \frac{1}{r}\frac{\partial}{\partial r} + \frac{1}{r^2}\frac{\partial^2}{\partial\theta^2}\right)\Phi$$

Hence

$$\nabla^4\Phi = \left(\frac{\partial^2}{\partial r^2} + \frac{1}{r}\frac{\partial}{\partial r} + \frac{1}{r^2}\frac{\partial^2}{\partial\theta^2}\right)\left(\frac{\partial^2\Phi}{\partial r^2} + \frac{1}{r}\frac{\partial\Phi}{\partial r} + \frac{1}{r^2}\frac{\partial^2\Phi}{\partial\theta^2}\right) = 0 \qquad ...(3.13)$$

If we put

$$\left.\begin{aligned} \sigma_r &= \frac{1}{r}\frac{\partial\Phi}{\partial r} + \frac{1}{r^2}\frac{\partial^2\Phi}{\partial\theta^2} \\ \sigma_\theta &= \frac{\partial^2\Phi}{\partial r^2} \\ \tau_{r\theta} &= \frac{1}{r^2}\frac{\partial\Phi}{\partial\theta} - \frac{1}{r}\frac{\partial^2\Phi}{\partial r\partial\theta} = -\frac{\partial}{\partial r}\left(\frac{1}{r}\frac{\partial\Phi}{\partial\theta}\right) \end{aligned}\right\} \qquad ...(3.14)$$

then the equilibrium equations are satisfied in the absence of body forces.

If $\quad B_\theta = B_z = 0$ and

$$B_r = \sqrt{B_x^{\,2} + B_y^{\,2}}$$

$$= \sqrt{\left(\frac{\partial V}{\partial x}\right)^2 + \left(\frac{\partial V}{\partial y}\right)^2}$$

$$\frac{\partial V}{\partial x} = \frac{\partial V}{\partial r}\cdot\frac{\partial r}{\partial x} = \frac{\partial V}{\partial r}\cdot\cos\theta$$

$$\frac{\partial V}{\partial y} = \frac{\partial V}{\partial r}\cdot\frac{\partial r}{\partial y} = \frac{\partial V}{\partial r}\cdot\sin\theta$$

$$\therefore \quad B_r = \sqrt{\left(\frac{\partial V}{\partial r}\right)^2\cos^2\theta + \left(\frac{\partial V}{\partial r}\right)^2\sin^2\theta}$$

$$= \frac{\partial V}{\partial r}$$

The equilibrium equations in plane strain will be

$$\frac{\partial \sigma_r}{\partial r} + \frac{1}{r}\frac{\partial \tau_{r\theta}}{\partial \theta} + \frac{\sigma_r - \sigma_\theta}{r} + \frac{\partial V}{\partial r} = 0$$

$$\frac{1}{r}\frac{\partial \sigma\theta}{\partial \theta} + \frac{\partial \tau_{r\theta}}{\partial r} + \frac{2\tau_{r\theta}}{r} = 0$$

$$\frac{\partial \sigma_z}{\partial z} = 0$$

Then

$$\left.\begin{aligned} \sigma_r &= \frac{1}{r}\frac{\partial \Phi}{\partial r} + \frac{1}{r^2}\frac{\partial^2 \Phi}{\partial \theta^2} \\ \sigma_\theta &= \frac{\partial^2 \Phi}{\partial r^2} + r\frac{\partial V}{\partial r} \\ \tau_{r\theta} &= -\frac{\partial}{\partial r}\left(\frac{1}{r}\frac{\partial \Phi}{\partial \theta}\right) \end{aligned}\right\} \quad \text{...(3.15)}$$

and compatibility equation becomes.

$$\nabla^4\Phi + \left(\frac{1-2\nu}{1-\nu}\right)\nabla^2 V = 0$$

$$\left(\frac{\partial^2}{\partial r^2} + \frac{1}{r}\frac{\partial}{\partial r} + \frac{1}{r^2}\frac{\partial^2}{\partial \theta^2}\right)\left(\frac{\partial^2 \Phi}{\partial r^2} + \frac{1}{r}\frac{\partial \Phi}{\partial r} + \frac{1}{r^2}\frac{\partial^2 \Phi}{\partial \theta^2}\right) + \left(\frac{1-2\nu}{1-\nu}\right)\left(\frac{\partial^2 V}{\partial r^2} + \frac{1}{r}\frac{\partial V}{\partial r}\right) = 0 \quad \text{...(3.16)}$$

Similarly for plane stress, we have

$$\nabla^4\Phi + (1-\nu)\nabla^2 V = 0$$

$$\left(\frac{\partial^2}{\partial r^2} + \frac{1}{r}\frac{\partial}{\partial r} + \frac{1}{r^2}\frac{\partial^2}{\partial \theta^2}\right)\left(\frac{\partial^2 \Phi}{\partial r^2} + \frac{1}{r}\frac{\partial \Phi}{\partial r} + \frac{1}{r^2}\frac{\partial^2 \Phi}{\partial \theta^2}\right) + (1-\nu)\left(\frac{\partial^2 V}{\partial r^2} + \frac{1}{r}\frac{\partial V}{\partial r}\right) = 0 \quad \text{...(3.17)}$$

3.3 Solution of Biharmonic Equation for Axial-Symmetry

For axial symmetry, the compatibility equation in the absence of body forces is :

$$\left.\begin{aligned} &\left(\frac{d^2}{dr^2} + \frac{1}{r}\frac{d}{dr}\right)\left(\frac{d^2\Phi}{dr^2} + \frac{1}{r}\frac{d\Phi}{dr}\right) = 0 \\ &\frac{d^4\Phi}{dr^4} + \frac{2}{r}\frac{d^3\Phi}{dr^3} - \frac{1}{r^2}\frac{d^2\Phi}{dr^2} + \frac{1}{r^3}\frac{d\Phi}{dr} = 0 \end{aligned}\right\} \quad \text{...(3.18)}$$

This is a fourth order differential equation. The general solution of Eq. (3.18) can be written as

$$\Phi = A \log r + Br^2 \log r + Cr^2 + D \qquad ...(3.19)$$

where A, B, C and D are constants.

The corresponding stress components are

$$\left.\begin{aligned} \sigma_r &= \frac{1}{r}\frac{\partial \Phi}{\partial r} = \frac{A}{r^2} + B(1 + 2\log r) + 2C \\ \sigma_\theta &= \frac{\partial^2 \Phi}{\partial r^2} = -\frac{A}{r^2} + B(3 + 2\log r) + 2C \\ \tau_{r\theta} &= 0 \end{aligned}\right\} \qquad ...(3.20)$$

3.4 General Solution of Biharmonic Equation

The solutions of biharmonic equations are

$$\begin{aligned} \Phi = {} & r^2, \ln r, r^2 \ln r, \theta, r^2\theta, \theta \ln r, \\ & r^2 \theta \ln r, \cos\theta \,.\, r \ln r, r \ln r \times \theta \cos\theta, \\ & r\theta \cos\theta, r^\lambda \cos\lambda\theta, r^{\lambda+2} \,.\, \cos\lambda\theta, \qquad ...(3.21) \\ & \cos(\lambda \ln r) \cosh\lambda\theta, r^2 \cos(\lambda \ln r) \cosh\lambda\theta \end{aligned}$$

where λ is any real number.

The most general solution of $\nabla^4\Phi = 0$ is

$$\begin{aligned} \Phi = {} & A_0 \ln r + B_0 r^2 + C_0 r^2 \ln r + D_0 r^2 \theta \\ & + \frac{A_1}{2} r\theta \sin\theta + (B_1 r^{-1} + C_1 r^3 + D_1 r \ln r) \cos\theta \\ & + \sum_{n-2}^{\infty} (A_n r^n + B_n r^{-n} + C_n r^{n+2} + D_n r^{-n+2}) \cos n\theta \\ & + A'_0 \theta - \frac{A_1'}{2} r\theta \cos\theta + (B_1' r^{-1} + C_1' r^3 + D_1' r \ln r) \sin\theta \\ & + \sum_{n-2}^{\infty} (A_n' r^n + B_n' r^{-n} + C_n' r^{n+2} + D_n' r^{-n+2}) \sin n\theta \end{aligned} \qquad ...(3.22)$$

where A's, B's, C's and D's are constants.

3.5 Bending of a Curved Bar

3.5.1 Exact Analysis

Consider a curved bar with a constant narrow rectangular cross-section and a circular axis bent in the plane of curvature by couples M applied at the ends as shown in Fig. 3.1.

Denoting by a and b the inner and the outer radii of the boundary and taking the width of the rectangular cross-section as

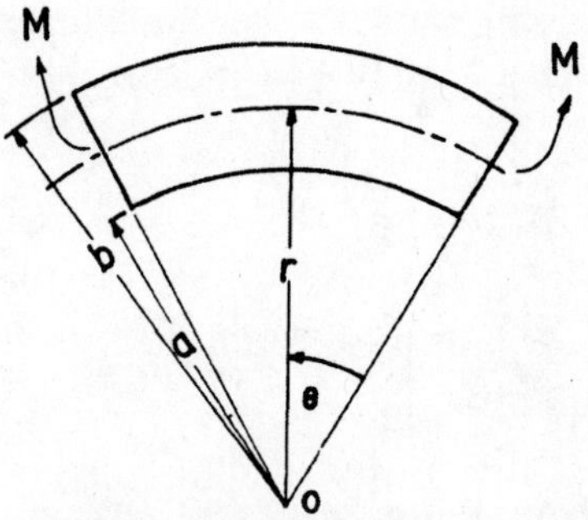

Fig. 3.1 Bending of a curved bar.

unity, the boundary conditions are

$$\left.\begin{aligned} &(i)\ \sigma_r = 0 \quad \text{at} \quad r = a \quad \text{and} \quad r = b \\ &(ii) \int_a^b \sigma_\theta\, dr = 0, \quad \int_a^b \sigma_\theta \,.\, r dr = -M \\ &(iii)\ \tau_{r\theta} = 0 \text{ at the boundary} \end{aligned}\right\} \quad \text{...(1)}$$

Here the components of stress are independent of θ, hence

$$\nabla^4 \Phi = \left(\frac{d^2}{dr^2} + \frac{1}{r}\frac{d}{dr}\right)\left(\frac{d^2\Phi}{dr^2} + \frac{1}{r}\frac{\partial \Phi}{dr}\right) = 0 \quad \text{...(2)}$$

and

$$\left.\begin{aligned} \sigma_r &= \frac{1}{r}\frac{d\Phi}{dr} \\ \sigma_\theta &= \frac{d^2\Phi}{dr^2} \\ \tau_{r\theta} &= 0 \end{aligned}\right\} \quad \text{...(3)}$$

The general solution of Eq. (2) is

$$\Phi = A \log r + Br^2 \log r + Cr^2 + D \quad \text{...(4)}$$

where A, B, C and D are constants.

Using Eqs. (3), we get

$$\left.\begin{aligned} \sigma_r &= \frac{A}{r^2} + B(1 + 2 \log r) + 2C \\ \sigma_\theta &= -\frac{A}{r^2} + B(3 + 2 \log r) + 2C \\ \tau_{r\theta} &= 0 \end{aligned}\right\} \quad \text{...(5)}$$

Using boundary conditions in Eqs. (1), we get

$$\frac{A}{a^2} + B(1 + 2 \log a) + 2C = 0 \quad \text{...(6)}$$

$$\frac{A}{b^2} + B(1 + 2\log b) + 2C = 0 \qquad \text{...(7)}$$

$$\int_a^b \frac{d^2\Phi}{dr^2}\,.\,r\,dr = \int_a^b r.\,d\left(\frac{d\Phi}{dr}\right)$$

$$= \left| r\frac{d\Phi}{dr} \right|_a^b - \int_a^b \frac{d\Phi}{dr}\,.\,dr$$

$$= 0 - \left| \Phi \right|_a^b = -M$$

$$A\log\frac{b}{a} + B(b^2\log b - a^2\log a) + C(b^2 - a^2) = M \qquad \text{...(8)}$$

Solving Eqs. (6), (7) and (8), we get

$$A = \frac{4M}{K}a^2b^2\log\frac{b}{a}$$

$$B = -\frac{2M}{K}(b^2 - a^2)$$

$$C = \frac{M}{K}[b^2 - a^2 + 2(b^2\log b - a^2\log a)]$$

where $K = (b^2 - a^2) - 4a^2b^2\left(\log\frac{b}{a}\right)^2$

Substituting in Eqs. (5), we get

$$\left.\begin{aligned} \sigma_r &= -\frac{4M}{K}\left[\frac{a^2b^2}{r^2}\log\frac{b}{a} + b^2\log\frac{r}{b} + a^2\log\frac{a}{r}\right] \\ \sigma_\theta &= -\frac{4M}{K}\left[-\frac{a^2b^2}{r^2}\log\frac{b}{a} + b^2\log\frac{r}{b} + a^2\log\frac{a}{r} + b^2 - a^2\right] \\ \tau_{r\theta} &= 0 \end{aligned}\right\} \qquad \text{...(3.23)}$$

Let $(\sigma_\theta)_{\substack{max\\min}} = \alpha\frac{M}{a^2}$

$$\therefore \qquad \alpha = \frac{(\sigma_\theta)_{\substack{max\\min}}\,a^2}{M} \qquad \text{...(3.24)}$$

Table 3.1 gives the values of α calculated by the two elementary methods and by the exact formula given by Eq. (3.24).

Table 3.1 Values of ∝

$\frac{b}{a}$	According to S.O.M.				According to T.O.E.	
	Linear stress distribution		Hyperbolic stress distribution			
	$\sigma_{\theta max}$	$\sigma_{\theta min}$	$\sigma_{\theta max}$	$\sigma_{\theta min}$	$\sigma_{\theta max}$	$\sigma_{\theta min}$
1.3	66.67	− 66.67	72.98	− 61.27	73.05	− 61.35
2	6.00	− 6.00	7.725	− 4.863	7.755	− 4.917
3	1.50	− 1.50	2.285	− 1.095	2.292	− 1.130

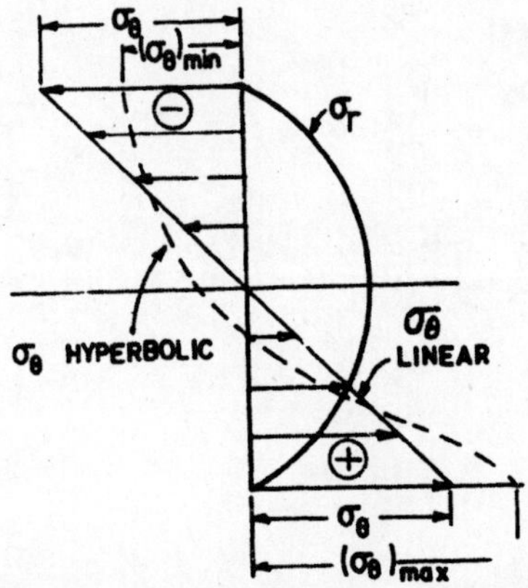

Fig. 3.2 Distribution of stress in a curved bar under bending.

The stress distribution has been plotted in Fig. 3.2.

The displacements may be calculated as :

$$\left.\begin{aligned} \varepsilon_r &= \frac{\partial u_r}{\partial r} = \frac{1}{E}[\sigma_r - \nu\sigma_\theta] \\ \varepsilon_\theta &= \frac{u_r}{r} - \frac{1}{r}\frac{\partial u_\theta}{\partial \theta} = \frac{1}{E}[\sigma_\theta - \nu\sigma_r] \\ \gamma_{r\theta} &= \frac{1}{r}\frac{\partial u_r}{\partial \theta} + \frac{\partial u\theta}{\partial r} - \frac{u_\theta}{r} = \frac{\tau_{r\theta}}{G} \end{aligned}\right\} \quad \text{...(9)}$$

$$\therefore \quad \frac{\partial u_r}{\partial r} = \frac{1}{E}\left[\frac{1+\nu}{r^2}A + 2(1-\nu)B\log r + (1-3\nu)B + 2(1-\nu)C\right]$$

Integrating, we get

$$u_r = \frac{1}{E}\left[-\left(\frac{1+\nu}{r}\right)A + 2(1-\nu)Br\log r - (1+\nu)Br + 2(1-\nu)Cr \right] + f(\theta) \qquad ...(10)$$

Now $$\frac{\partial u_\theta}{\partial \theta} = \varepsilon_\theta - u_r$$

$$= \frac{1}{E}[\sigma_\theta - \nu\sigma_r] - u_r$$

$$= \frac{4Br}{E} - f(\theta)$$

On integrating, we get

$$u_\theta = \frac{4Br\theta}{E} - \int_0^\theta f(\theta)d\theta + F(r) \qquad ...(11)$$

$$\therefore \qquad \gamma_{r\theta} = \frac{df(\theta)}{d\theta} + \int_0^\theta f(\theta)\,d\theta - F(r) + r\frac{dF(r)}{dr} = 0 \qquad ...(12)$$

which yields

$$\frac{df(\theta)}{d\theta} + \int_0^\theta f(\theta)\,d\theta = 0 \qquad ...(13)$$

$$-F(r) + r\frac{dF(r)}{dr} = 0 \qquad ...(14)$$

Differentiating Eq. (13) w.r.t. θ, we get

$$\frac{d^2 f(\theta)}{d\theta^2} + f(\theta) = 0$$

whose general solution is

$$f(\theta) = H\sin\theta + J\cos\theta \qquad ...(15)$$

where H and J are constants.

Eq. (14) may be written as

$$\frac{dF(r)}{F(r)} = \frac{dr}{r}$$

Integrating, we get

$$\ln F(r) = \ln r + \ln N$$

$$\therefore \qquad F(r) = Nr \qquad ...(16)$$

where N is another arbitrary constant.

Hence Eqs. (10) and (11) become

$$u_r = \frac{1}{E}\left[-\left(\frac{1+\nu}{E}\right)A + 2(1-\nu)Br\log r - (1+\nu)Br + 2(1-\nu)Cr\right] + H\sin\theta + J\cos\theta \quad ...(3.25)$$

$$u_\theta = \frac{4Br\theta}{E} - \int_0^\theta (H\sin\theta + J\cos\theta)d\theta + rN \quad ...(3.26)$$

If $\quad u_r = u_\theta = \frac{du_\theta}{dr} = 0$

for $\quad r = r_0 = \frac{a+b}{2}$

and $\quad \theta = 0, \text{ then } N = H = 0$

Hence $\quad u_\theta = \frac{4Br\theta}{E} - H\cos\theta - J\sin\theta + rN$

becomes $\quad u_\theta = \frac{4Br\theta}{E} - J\sin\theta \quad ...(3.27)$

This means that the displacement of any cross-section consists of a translatory displacement $-J\sin\theta$, the same for all points in the cross-section, and of rotation of the cross-section by the angle $\frac{4B\theta}{E}$ about the centre of curvature. Hence the cross-sections remain plane in pure bending as is usually assumed in the elementary theory of bending of curved bars.

3.5.2 Winkler's Theory

The following basic assumptions are made :

1. All cross-sections possess a vertical axis of symmetry lying in the plane of the centroidal axis passing through C (Fig. 3.3 a).

2. The beam is subjected to end couples M. The bending moment vector is everywhere normal to the plane of symmetry of the beam.

3. Sections originally plane and perpendicular to the centroidal beam axis remain so subsequent to bending.

The deformation of arbitrary fibre gh is

$$\varepsilon_c R d\theta + y \cdot \Delta d\theta$$

where ε_c denotes the strain of the centroidal fibre.

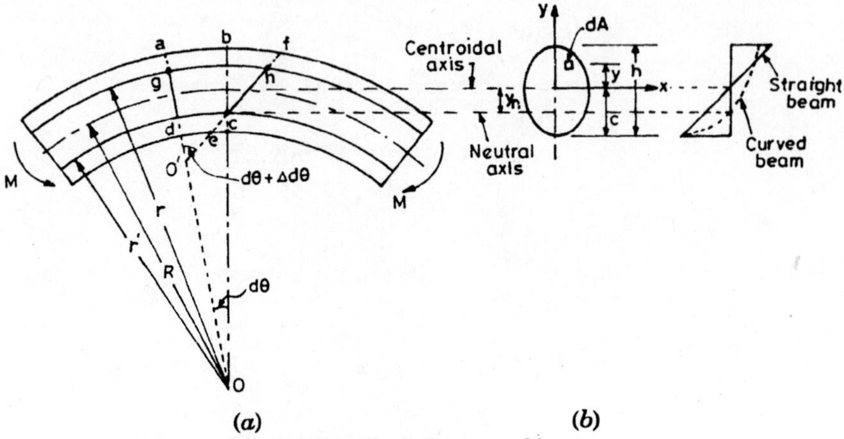

Fig. 3.3 Bending of a curved beam.

Original length of $gh = (R + y)\, d\theta$

Tangential strain of gh, $\varepsilon_\theta = \dfrac{\varepsilon_c R d\theta + y.\, \Delta\, d\theta}{(R + y)\, d\theta}$

Tangential stress acting on area dA is,

$$\sigma_\theta = \left[\frac{\varepsilon_c R + y.(\Delta d\theta / d\theta)}{R + y} \right] E$$

Denoting the angular strain $\Delta d\theta/d\theta$ by λ and adding and subtracting $\varepsilon_c y$ in the numerator, we have

$$\sigma_\theta = \left[\varepsilon_c + (\lambda - \varepsilon_c) \frac{y}{R + y} \right] E \qquad \text{...(1)}$$

For the beam cross-section to satisfy the conditions of equilibrium, $F_z = 0$ and $M_x = 0$.

$$\int \sigma_\theta\, dA = 0 \text{ and } \int \sigma_\theta\, y dA = M \qquad \text{...(2)}$$

Inserting the value of σ_θ from Eq. (1) in (2), we have

$$\varepsilon_c \int dA = -(\lambda - \varepsilon_c) \int \frac{y}{R + y}\, dA$$

$$M = \left[\varepsilon_c \int y dA\ (\lambda - \varepsilon_c) \int \frac{y^2}{R + y}\, dA \right] E \qquad \text{...(3)}$$

Now $\quad \int dA = A$ and $\int y dA = 0$

Let $\quad m = -\dfrac{1}{A} \displaystyle\int \frac{y}{R + y}\, dA \qquad \text{...(3.28)}$

It follows that

$$\int \frac{y^2}{R+y}\, dA = \int \left(y - \frac{Ry}{R+y} \right) dA = -R \int \frac{y}{R+y}\, dA = mAR \qquad ...(4)$$

Eq. (3) can be written as :

$$\varepsilon_c = (\lambda - \varepsilon_c)\, m \text{ and } M = E(\lambda - \varepsilon_c)\, mAR$$

or

$$\varepsilon_c = \frac{M}{EAR}, \ \lambda = \frac{1}{AE}\left(\frac{M}{R} + \frac{M}{mR} \right) \qquad ...(5)$$

Substituting Eq. (5) in (1), we get

$$\sigma_\theta = \frac{M}{AR}\left[1 + \frac{y}{m(R+y)} \right] \qquad ...(3.29)$$

The variation of stress over the cross-section is thus hyperbolic.

At the neutral axis, $\sigma_\theta = 0$

$$1 + \frac{y_n}{m(R+y_n)} = 0$$

$$y_n = -\left(\frac{mR}{m+1} \right) \qquad ...(3.30)$$

Total stress, $\sigma = \dfrac{P}{A} + \dfrac{M}{AR}\left[1 + \dfrac{y}{m(R+y)} \right]$ ...(3.31)

The linear and hyperbolic stress distributions are approximately the same for $R/C > 20$. Table 3.2 gives values of m for various cross-sections.

Table 3.2 Values of m

Cross-section	*Formula for m*
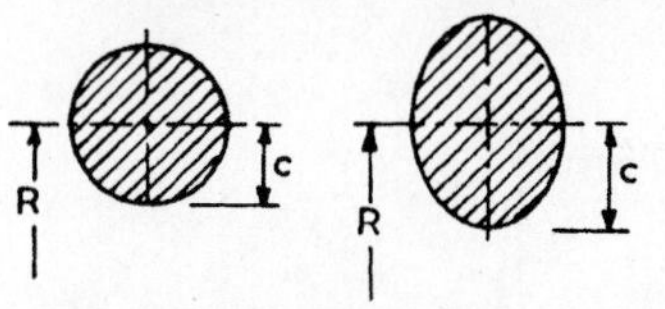	$-1 + 2\left(\frac{R}{C}\right)^2 - 2\left(\frac{R}{C}\right)\sqrt{\left(\frac{R}{C}\right)^2 - 1}$
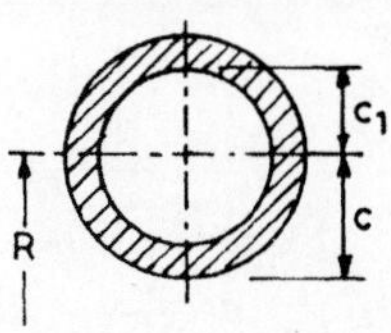	$-1 + \left(\frac{2R}{C^2 - C_1^2}\right)\left[\sqrt{R^2 - C_1^2} - \sqrt{R^2 - C^2}\right]$

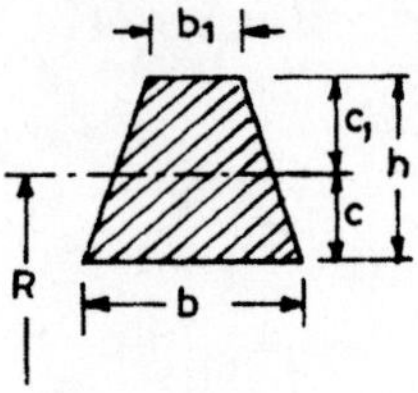	$-1+\frac{R}{Ah}\left[\left\{b_1h+(R+C)(b-b_1)\right\}\ln\left(\frac{R+C_1}{R-C}\right)-(b-b_1)\,h\right]$ **For rectangular section :** $C=C_1,\ b=b_1$ **For triangular section :** $b_1=0$ $C=\frac{h}{3}\left(\frac{b+2b_1}{b+b_1}\right)$
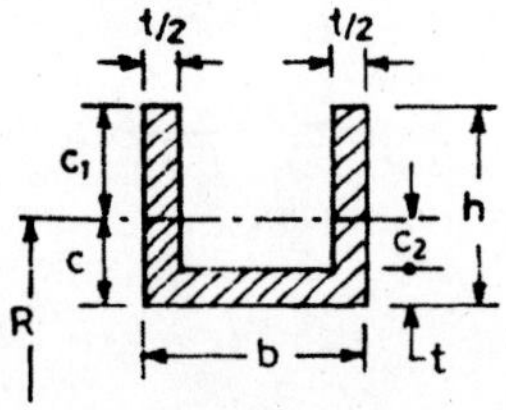	$-1+\frac{R}{A}[t\ln(R+C_1)+(b-t)$ $\times\ln(R-C_2)-b\ln(R-C)]$
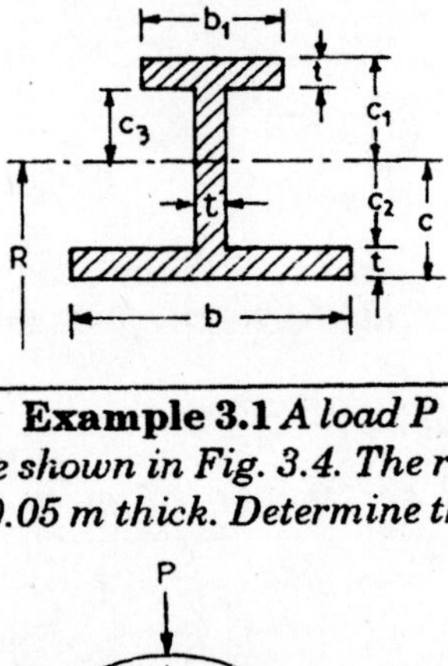	$-1+\frac{R}{A}[b_1\ln(R+C_1)$ $+(t-b_1\ln(R+C_3)$ $+(b-t)\ln(R-C_2)-b\ln(R-C)]$

Example 3.1 *A load P = 70 kN is applied to the circular steel frame shown in Fig. 3.4. The rectangular cross-section is 0.1 m wide and 0.05 m thick. Determine the tangential stress at points A and B.*

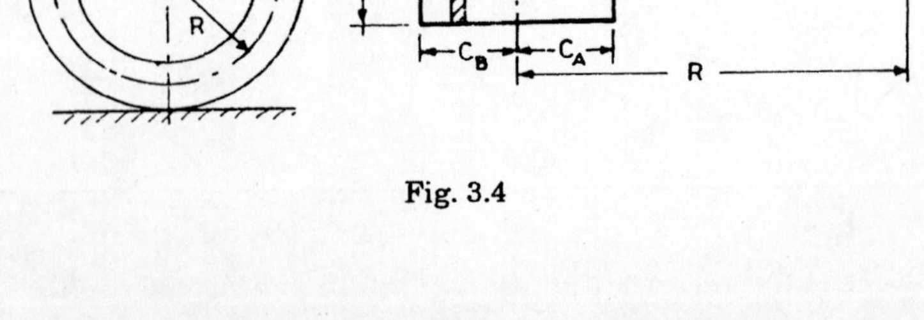

Fig. 3.4

Solution. Here $C_A = C_B = C = \dfrac{h}{2} = 0.05$ m ; $t = 0.05$ m,

$R = 0.1 + 0.05 = 0.15$ m

$$m = -\frac{1}{2tC}\int_{-C}^{+C} \frac{y}{R+y}\, t dy = -\frac{1}{2C}\int_{-C}^{C} \frac{y}{R+y}\, dy$$

$$= -\frac{1}{2C}\int_{-C}^{C}\left(1 - \frac{R}{y+R}\right) dy$$

$$= -1 + \frac{R}{2C}\ln\left(\frac{R+C}{R-C}\right)$$

$$m = -1 + \frac{0.15}{2 \times 0.05}\ln\left(\frac{0.15+0.05}{0.15-0.05}\right)$$

$$= -1 + 1.5 \ln 2 = 0.0397$$

$$\sigma_A = -\frac{P}{A} + \frac{M}{AR}\left[1 + \frac{-C_A}{m(R - C_A)}\right]$$

$$M = PR$$

$$\sigma_A = -\frac{PC_A}{mA(R - C_A)}$$

$$= -\frac{7000 \times 0.05}{0.0397 \times 0.1 \times 0.05 \times (0.15 - 0.05)}$$

$$= -176 \text{ MPa}$$

$$\sigma_B = -\frac{P}{A} + \frac{M}{AR}\left[1 + \frac{C_B}{m(R + C_B)}\right] = \frac{PC_B}{mA\,(R + C_B)}$$

$$= \frac{7000 \times 0.05}{0.0397 \times 0.1 \times 0.05 \times (0.15 + 0.05)}$$

$$= 88 \text{ MPa.}$$

Example 3.2 *A steel ring of 0.35 m mean diameter and of uniform rectangular section 0.06 m wide and 0.012 m thick is shown in Fig. 3.5. A rigid bar is fitted across diameter AB, and a tensile force P applied to the ring as shown. Assuming an allowable stress of 140 MPa, determine the maximum tensile force that can be carried by the ring.*

Solution. Let $2F$ be the thrust induced in the bar. The moment at any section $m - n$ is then,

$$M_\theta = -\frac{F}{R}\sin\theta + M_B + \frac{PR}{2}(1 - \cos\theta)$$

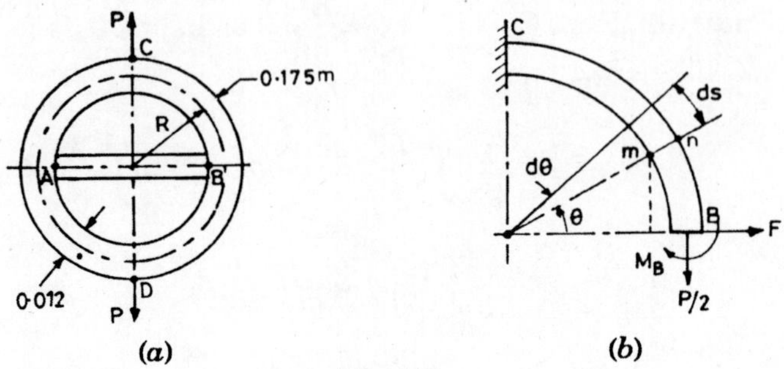

Fig. 3.5

The relative slope between B and C before and after deformation remains unchanged. Therefore the relative angular rotation between B and C is zero. Thus

$$EI\theta = 0 \int_B^C M_\theta \, dx = R \int_0^{\pi/2} M_\theta \, d\theta$$

where $ds = dx = Rd\theta$.

$$R \int_0^{\pi/2} \left[-FR \sin\theta + M_B + \frac{PR}{2}(1 - \cos\theta) \right] d\theta = 0$$

$$\frac{1}{2}\pi M_B + \frac{1}{2} PR \left(\frac{\pi}{2} - 1 \right) - FR = 0 \qquad \text{...(1)}$$

Also deflection at B is zero.

$$\delta_B = \frac{\partial U}{\partial F} = \frac{1}{EI} \int_0^{\pi/2} M_\theta \frac{\partial M_\theta}{\partial F} (Rd\theta) = 0$$

$$= \int_0^{\pi/2} \left[-FR \sin\theta + M_B + \frac{PR}{2}(1 - \cos\theta) \right] \times (-R \sin\theta) \, d\theta = 0$$

or
$$-\frac{1}{4}\pi FR + \frac{1}{4} PR + M_B = 0 \qquad \text{...(2)}$$

Solving Eqs. (1) and (2), we get

$$M_B = 0.1131\, PR$$
$$FR = 0.4625\, PR$$

For $\theta = 90°$, M_θ becomes

$$M_C = -FR + M_B + \frac{1}{2} PR = 0.1506\, PR$$

Thus $M_C > M_B$. Since $\frac{R}{C} = \frac{0.175}{0.006} = 29$, the simple bending formula gives quite accurate results.

$$\sigma_{A,B} = \frac{P}{2A} + \frac{M_B C}{I}$$

$$= 694P + 13809P = 14503P$$

$$\sigma_{C,D} = \frac{M_C C}{I} = 18387P$$

Hence $\sigma_C > \sigma_B$

$$140 \times 10^6 = 18387P$$

$$P = 7641 \text{ N}.$$

3.6. Thick Cylinder

Consider a long thick cylinder of inner radius a and outer radius b, both small compared to its length so that plane sections normal to the centre line remain plane and a state of plane strain exists in the cylinder. The cylinder is subjected to internal pressure p_1 and external pressure p_2 as shown in Fig. 3.6.

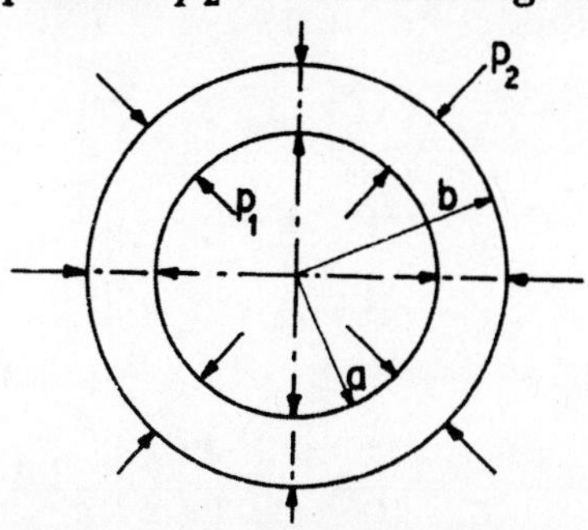

Fig. 3.6 Thick cylinder under pressure.

The cylinder is symmetrical about the Z-axis.

Therefore, the compatibility equation becomes

$$\nabla^4\Phi = \left(\frac{d^2}{dr^2} + \frac{1}{r}\frac{d}{dr}\right)\left(\frac{d^2\Phi}{dr^2} + \frac{1}{r}\frac{d\Phi}{dr}\right) = 0$$

Its general solution is

$$\Phi = A \log r + Br^2 \log r + Cr^2 + D$$

The stress components become

$$\left.\begin{aligned}\sigma_r &= \frac{1}{r}\frac{d\Phi}{dr} = \frac{A}{r^2} + B(1 + 2\log r) + 2C \\ \sigma_\theta &= \frac{d^2\Phi}{dr^2} = -\frac{A}{r^2} + B(3 + 2\log r) + 2C\end{aligned}\right\} \quad ...(1)$$

$$\tau_{r\theta} = 0$$

Since $\quad u_\theta = 0$

and $\quad u_r = f(r)$ only $\quad ...(2)$

Also $\quad \varepsilon_z = \frac{1}{E}[\sigma_z - \nu(\sigma_\theta + \sigma_r)] = 0$

$\therefore \quad \sigma_z = \nu(\sigma_\theta + \sigma_r) \quad ...(3)$

As a result of this, the stress-strain relations become :

$$\left.\begin{aligned}\varepsilon_r &= \frac{1+\nu}{E}[(1-\nu)\,\sigma_r - \nu\sigma_\theta] = \frac{du_r}{dr} \\ \varepsilon_\theta &= \frac{1+\nu}{E}[(1-\nu)\,\sigma_\theta - \nu\sigma_r] = \frac{u_r}{r}\end{aligned}\right\} \quad ...(4)$$

Substituting Eqs. (1) in (4), we get

$$\frac{du_r}{dr} = \frac{1+\nu}{E}\left[(1-\nu)\left\{\frac{A}{r^2} + B(1 + 2\log r) + 2C\right\} - \nu\left\{-\frac{A}{r^2} + B(3 + 2\log r) + 2C\right\}\right] \quad ...(5)$$

$$u_r = \frac{(1+\nu)r}{E}\left[(1-\nu)\left\{-\frac{A}{r^2} + B(3 + 2\log r) + 2C\right\} - \nu\left\{\frac{A}{r^2} + B(1 + 2\log r) + 2C\right\}\right] \quad ...(6)$$

Differentiating Eq. (6) w.r.t. r, we get

$$\begin{aligned}\frac{du_r}{dr} = \frac{1+\nu}{E}\Bigg[(1-\nu)\left\{-\frac{A}{r^2} + B(3 + 2\log r) + 2C\right\} &- \nu\left\{\frac{A}{r^2} + B(1 + 2\log r) + 2C\right\}\Bigg] \\ &+ \frac{(1+\nu)r}{E}\left[(1-\nu)\left\{\frac{2A}{r^3} + \frac{2B}{r}\right\} - \nu\left\{\frac{2A}{r^3} + \frac{2B}{r}\right\}\right]\end{aligned}$$

$$= \frac{1+\nu}{E}\left[(1-\nu)\left\{\frac{A}{r^2} + B(5 + 2\log r) + 2C\right) - \nu\left\{-\frac{A}{r^2} + B(3 + 2\log r) + 2C\right\}\right] \quad ...(7)$$

Comparing Eqs. (5) and (7), we get

$$4(1-\nu)B = 0$$

$$\therefore \quad B = 0$$

$$\therefore \quad \left.\begin{aligned} \sigma_r &= \frac{A}{r^2} + 2C \\ \sigma_\theta &= \frac{-A}{r^2} + 2C \end{aligned}\right\} \quad ...(8)$$

The boundary conditions are

$$(\sigma_r)_{r=a} = \frac{A}{a^2} + 2C = -p_1$$

$$(\sigma_r)_{r=b} = \frac{A}{b^2} + 2C = -p_2$$

Solving, we get

$$A = \frac{a^2b^2(p_2 - p_1)}{b^2 - a^2}$$

$$2C = \frac{p_1a^2 - p_2b^2}{b^2 - a^2}$$

Substituting in Eqs. (8), we get

$$\left.\begin{aligned} \sigma_r &= \frac{a^2b^2(p_2 - p_1)}{(b^2 - a^2)} \cdot \frac{1}{r^2} + \frac{p_1a^2 - p_2b^2}{b^2 - a^2} \\ \sigma_\theta &= \frac{-a^2b^2(p_2 - p_1)}{(b^2 - a^2)} \cdot \frac{1}{r^2} + \frac{p_1a^2 - p_2b^2}{b^2 - a^2} \end{aligned}\right\} \quad ...(3.32)$$

$$\sigma_z = \frac{2\nu(p_1a^2 - p_2b^2)}{(b^2 - a^2)} = \text{constant}$$

If the cylinder is free at the ends then one must superimpose upon the normal stress σ_z an equal and opposite stress $-\sigma_z$ so that

$$\varepsilon_z = -\frac{\sigma_z}{E} = -\frac{2\nu}{E}\left[\frac{p_1a^2 - p_2b^2}{b^2 - a^2}\right] \quad ...(3.33)$$

(*i*) If $p_1 \neq 0$ and $p_2 = 0$, then Eqs. (3.32) become

$$\left.\begin{aligned} \sigma_r &= \frac{p_1 a^2}{(b^2 - a^2)}\left[1 - \frac{b^2}{r^2}\right] \\ \sigma_\theta &= \frac{p_1 a^2}{(b^2 - a^2)}\left[\frac{1 + b^2}{r^2}\right] \end{aligned}\right\} \quad \text{...(3.34)}$$

At $\qquad r = a$

$$(\sigma_\theta)_{max} = (\sigma_\theta)_{r=a} = \frac{p_1 a^2}{(b^2 - a^2)}\left[1 + \frac{b^2}{a^2}\right] \quad \text{...(3.35)}$$

Also from Eq. (6), we get

$$u_r = \frac{(1 + \nu)r}{E}\left[(1 - \nu)\sigma_\theta - \nu\sigma_r\right]$$

$$u_r = \frac{(1 + \nu)p_1 a^2}{Er} \quad \text{...(3.36 a)}$$

$$\therefore \qquad (u_r)_{max} = (u_r)_{r=a} = \frac{(1 + \nu)p_1 a}{E} \quad \text{...(3.36 b)}$$

(*ii*) If $p_1 = 0$, $p_2 \neq 0$, then

$$\left.\begin{aligned} \sigma_r &= \frac{p_2 b^2}{(b^2 - a^2)}\left[-1 + \frac{a^2}{r^2}\right] \\ \sigma_\theta &= \frac{-p_2 b^2}{(b^2 - a^2)}\left[1 + \frac{a^2}{r^2}\right] \end{aligned}\right\} \quad \text{...(3.37)}$$

$$\sigma_z = \frac{-2\nu p_2 b^2}{(b^2 - a^2)} \quad \text{...(3.38)}$$

(*iii*) When $r_2 = \infty$, then from Eqs. (3.34), we get

$$-\sigma_r = \sigma_\theta = \frac{p_1 a^2}{r^2} \quad \text{...(3.39)}$$

Eq. (3.39) gives us the stresses around a circular hole of radius 'a' inside an infinite elastic medium and subjected to pressure p_1. In this particular case $\varepsilon_z = \sigma_z = 0$ and it is a case of plane strain and plain stress at the same time.

(*iv*) In order to determine the stresses in a shrink-fitted cylinder having shrink fit allowance equal to δ, we have from Eq. (6).

$$u_r = \frac{r}{E}(1 + \nu)\left[(1 - \nu)\,\sigma_\theta - \nu\sigma_r\right]$$

Let $\qquad c$ = junction radius

$$\therefore \quad u_r = \left(\frac{1-\nu}{E}\right)\left(\frac{p_2 b^2 - p_1 a^2}{b^2 - a^2}\right) r + \left(\frac{1+\nu}{E}\right)(p_2 - p_1)\frac{b^2 a^2}{b^2 - a^2} \cdot \frac{1}{r}$$

If $p_1 \neq 0$ and $p_2 = 0$, then

$$u_{r1} = \frac{-p_1 a}{E}\left[\frac{b^2 + a^2}{b^2 - a^2} + \nu\right]$$

and if $p_1 = 0$ and $p_2 \neq 0$, then

$$u_{r2} = \frac{p_2 b}{E}\left[\frac{b^2 + a^2}{b^2 - a^2} - \nu\right]$$

$$\delta = \frac{pc}{E}\left[\frac{b^2 + c^2}{b^2 - c^2} + \nu\right] + \frac{pc}{E}\left[\frac{c^2 + a^2}{c^2 - a^2} - \nu\right]$$

$$p = \frac{\delta E}{c}\left[\frac{(c^2 - a^2)(b^2 - c^2)}{2c^2(b^2 - a^2)}\right] \qquad \text{...(3.40)}$$

For $a = 0$

$$p = \frac{\delta E}{c}\left(\frac{b^2 - c^2}{2b^2}\right)$$

The stresses become for $a \le r \le c$

$$\sigma_r = \frac{pc^2}{c^2 - a^2}\left(1 - \frac{a^2}{r^2}\right) \qquad \text{...(3.41)}$$

$$\sigma_\theta = \frac{pc^2}{c^2 - a^2}\left(1 + \frac{a^2}{r^2}\right) \qquad \text{...(3.42)}$$

and for $c \le r \le b$, we have

$$\sigma_r = \frac{pc^2}{b^2 - c^2}\left(\frac{b^2}{r^2} - 1\right) \qquad \text{...(3.43)}$$

$$\sigma_\theta = \frac{-pc^2}{b^2 - c^2}\left(\frac{b^2}{r^2} + 1\right)$$

Example 3.3 *A steel gun barrel is subjected to an internal pressure of 70 MPa. The internal diameter of the barrel is 75 mm and external diameter of 225 mm. A steel band 25 mm thick and internal diameter 0.075 mm smaller than the external diameter of the gum barrel is shrunk on to the gun barrel. Calculate*

(a) the shrinkage pressure on the gun barrel,

(*b*) *maximum stress in the steel band, and*

(*c*) *minimum temperature to which the band must be heated to make the assembly.*

For steel E = 200 GPa, ν = 0.3 and coefficient of thermal expansion = 10 × 10⁻⁶/°C.

Solution. $r_1 = 37.5$ mm, $r_3 = 112.5$ mm, $r_2 = 137.5$ mm.

Let p_s = shrinkage pressure

Shrinkage stresses :

Gun barrel

At the outside surface :

$$\sigma_r = -p_s$$

$$\sigma_\theta = -p_s\left(\frac{r_3^2 + r_1^2}{r_3^2 - r_1^2}\right), \frac{r_3}{r_1} = \frac{112.5}{37.5} = 3$$

$$= -p_s\left(\frac{10}{8}\right) = -1.125\, p_s$$

$$\varepsilon_\theta = \frac{1}{\varepsilon}\left[\sigma_\theta - \nu\sigma_r\right] = -\frac{p_s}{E}(1.125 - 0.3) = \frac{-0.825\, p_s}{E}$$

Steel band

At the inside surface :

$$\sigma_r = -p_s$$

$$\sigma_\theta = p_s\left(\frac{r_2^2 + r_3^2}{r_2^2 - r_3^2}\right), \frac{r_2}{r_3} = \frac{137.5}{112.5} = 1.222$$

$$= p_s\left(\frac{2.4938}{0.4938}\right) = 5\, p_s$$

$$\varepsilon_\theta = \frac{p_s}{E}[5 + 0.3] = \frac{5.3\, p_s}{E}$$

Total strain $= (5.3 + 0.825)\dfrac{p_s}{E}$

$$= 6.125\frac{p_s}{E}$$

$$= \frac{0.075}{225}$$

$$p_s = \frac{0.075}{225} \times \frac{1}{6.125} \times 200 \times 10^3$$

$$= 10.884 \text{ MPa}$$

(*b*) **Internal pressure :**

Gun barrel :

At the inside surface :

$$\sigma_\theta = \frac{p_i\,(r_2^{\,2} + r_1^{\,2})}{(r_2^{\,2} - r_1^{\,2})}, \; \frac{r_2}{r_1} = \frac{137.5}{37.5} = 3.667$$

$$= \frac{70 \times 13.44}{11.44} = 82.23 \text{ MPa}$$

At the outside surface :

$$\sigma_\theta = p_i \left(\frac{r_1}{r_3}\right)^3 \left(\frac{r_2^{\,2} + r_3^{\,2}}{r_2^{\,2} - r_1^{\,2}}\right)$$

$$= 70\left(\frac{37.5}{112.5}\right)^2 \left[\frac{(137.5)^2 + (112.5)^2}{(137.5)^2 - (37.5)^2}\right]$$

$$= 70 \times 0.111 \times \frac{31562.5}{17500} = 14.03 \text{ MPa}$$

Maximum stress in steel band

$$= 5 \times 14.884 + 14.03 = 68.45 \text{ MPa}$$

$$c = \frac{\delta d}{d} = \alpha\, \Delta\theta$$

$$\Delta\theta = \frac{0.075}{225 \times 10 \times 10^{-6}} = 33.3°\text{C}$$

3.7 Rotating Disc and Cylinder

Consider an element of the disc as shown in Fig. 3.7. Let ω be the angular speed, ρ the mass density and the disc be of unit thickness. Centrifugal force acting per unit mass on the disc is

$$F_r = \rho\omega^2 r$$

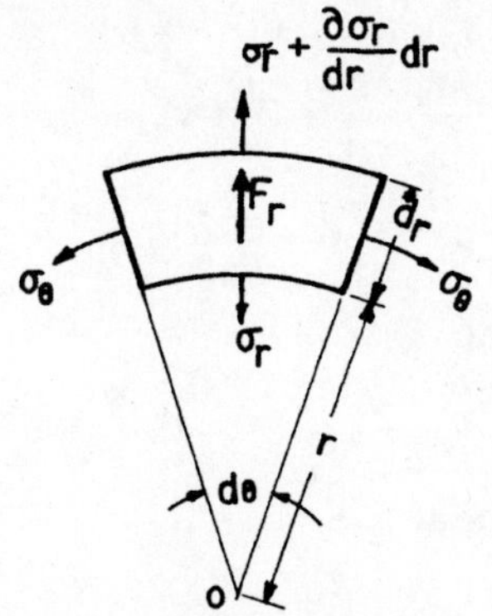

Fig. 3.7 Rotating disc.

Considering the equilibrium of forces in the radial direction, we get

$$\left(\sigma_r + \frac{d\sigma_r}{dr}\,dr\right)(r + dr)\,d\theta - \sigma_r\,.\,rd\theta$$

$$-2\,\sigma_\theta\,.\,\sigma_r \sin\frac{d\theta}{2} + F_r\,.\,rdr\,d\theta = 0$$

Simplifying, we get

$$\frac{d\sigma_r}{dr} + \frac{\sigma_r - \sigma_\theta}{r} + \rho\omega^2 r = 0$$

$$\frac{d}{dr}(r\sigma_r) - \sigma_\theta + \rho\omega^2 r^2 = 0 \qquad ...(1)$$

Eq. (1) is satisfied by a stress function $\Phi(r)$ such that

$$\sigma_r = \frac{\Phi}{r},\ \sigma_\theta = \frac{d\Phi}{dr} + \rho\omega^2 r^2 \qquad ...(2)$$

Now $$\varepsilon_r = \frac{du_r}{dr},\ \varepsilon_\theta = \frac{u_r}{r},\ \gamma_{r\theta} = 0 \qquad ...(3)$$

∴ $$\varepsilon_r = \frac{d}{dr}(r\varepsilon_\theta)$$

or $$r\frac{d\varepsilon_\theta}{dr} + \varepsilon_\theta - \varepsilon_r = 0 \qquad ...(4)$$

Also from Hooke's law, we have

$$\left.\begin{aligned}\varepsilon_r &= \frac{1}{E}[\sigma_r - \nu\sigma_\theta] \\ \sigma_\theta &= \frac{1}{E}[\sigma_\theta - \nu\sigma_r]\end{aligned}\right\} \qquad ...(5)$$

Combining Eqs. (2), (4) and (5), we get

$$\frac{d^2\Phi}{dr^2} + \frac{1}{r}\frac{d\Phi}{dr} - \frac{\Phi}{r^2} + (3 + \nu)\,\rho\omega^2 r = 0$$

or $$\frac{d}{dr}\left[\frac{1}{r}\frac{d}{dr}(r\Phi)\right] = -(3 + \nu)\,\rho\omega^2 r \qquad ...(6)$$

Integrating, we get

$$\Phi(r) = -\frac{(3 + \nu)}{8}\rho\omega^2 r^2 + C_1\frac{2}{r} + \frac{C_2}{r} \qquad ...(7)$$

where C_1 and C_2 are constants.

Using Eqs. (2), we get

$$\left.\begin{aligned}\sigma_r &= -\frac{(3+\nu)}{8}\rho\omega^2 r^2 + \frac{C_1}{2} + \frac{C_2}{r^2}\\ \sigma_\theta &= -\frac{(1+3\nu)}{8}\rho\omega^2 r^2 + \frac{C_1}{2} - \frac{C_2}{r^2}\end{aligned}\right\} \quad ...(8)$$

(*i*) *Thin solid disc of radius b.*

From Eqs. (8) we see that for stresses to be finite at $r = 0$,

$C_2 = 0$. At $r = b$, $\sigma_r = 0$.

$$\therefore \quad C_1 = \frac{(3+\nu)}{4}\rho\omega^2 b^2$$

$$\therefore \quad \left.\begin{aligned}\sigma_r &= \frac{(3+\nu)}{8}[b^2 - r^2]\\ \sigma_\theta &= \frac{\rho\omega^2}{8}[(3+\nu)b^2 - (1+3\nu)r^2]\end{aligned}\right\} \quad ...(9)$$

$$(\sigma_r)_{r=0} = (\sigma_\theta)_{r=0} = \left(\frac{3+\nu}{8}\right)\rho\omega^2 b^2$$

and are the maximum stresses in disc.

(*ii*) *Thin hollow disc of external radius b and internal radius a.*

The boundary conditions are :

$$(\sigma_r)_{r=a} = -\left(\frac{3+\nu}{8}\right)\rho\omega^2 a^2 + \frac{C_1}{2} + \frac{C_2}{a^2} = 0$$

$$(\sigma_r)_{r=b} = -\left(\frac{3+\nu}{8}\right)\rho\omega^2 b^2 + \frac{C_1}{2} + \frac{C_2}{b^2} = 0$$

$$\therefore \quad \frac{C_1}{2} = \left(\frac{3+\nu}{8}\right)\rho\omega^2 (a^2 + b^2)$$

$$C_2 = -\left(\frac{3+\nu}{8}\right)\rho\omega^2 a^2 b^2$$

Hence, we get from Eq. (8)

$$\left.\begin{aligned}\sigma_r &= \left(\frac{3+\nu}{8}\right)\rho\omega^2\left(a^2 + b^2 - \frac{a^2b^2}{r^2} - r^2\right)\\ \sigma_\theta &= \left(\frac{3+\nu}{8}\right)\rho\omega^2\left(a^2 + b^2 + \frac{a^2b^2}{r^2} - \frac{1+3\nu}{3+\nu}r^2\right)\end{aligned}\right\} \quad ...(3.45)$$

$$(\sigma_\theta)_{max} = (\sigma_\theta)_{r=a} = \left(\frac{3+\nu}{4}\right)\rho\omega^2 b^2\left[1 + \left(\frac{1-\nu}{3+\nu}\right)\frac{a^2}{b^2}\right] \quad ...(3.46)$$

If $\frac{a}{b} \to 0$, then

$$(\sigma_\theta)_{max} = \left(\frac{3+\nu}{8}\right) \rho\omega^2 b^2 \qquad ...(3.47)$$

$$\therefore \qquad \frac{(\sigma_\theta)_{max} \text{ for } \frac{a}{b} \to 0}{(\sigma_\theta)_{max} \text{ for } a = 0} = 2$$

(*iii*) *Solid rotating cylinder*

$$\left.\begin{aligned} \varepsilon_r &= \left(\frac{1+\nu}{E}\right)[(1-\nu)\,\sigma_r - \nu\sigma_\theta] \\ \varepsilon_\theta &= \left(\frac{1+\nu}{E}\right)[(1-\nu\,\sigma_\theta) - \nu\sigma_r] \\ \gamma_{r\theta} &= \frac{2(1+\nu)}{E}\tau_{r\theta} \end{aligned}\right\} \qquad ...(10)$$

Combining Eqs. (2), (3) and (13), we get

$$\frac{\partial^2\Phi}{\partial r^2} + \frac{1}{r}\frac{d\Phi}{dr} - \frac{\Phi}{r^2} + \left(\frac{3-2\nu}{1-\nu}\right)\rho\omega^2 r = 0$$

$$\frac{d}{dr}\left[\frac{1}{r}\frac{d}{dr}(r\Phi)\right] = -\left(\frac{3-2\nu}{1-\nu}\right)\rho\omega^2 r$$

Integrating, we get

$$\Phi = -\frac{1}{8}\left(\frac{3-2\nu}{1-\nu}\right)\rho\omega^2 r^3 + \frac{C_1}{2}r + \frac{C_2}{r}$$

$$\therefore \qquad \left.\begin{aligned} \sigma_r &= \frac{1}{8}\left(\frac{3-2\nu}{1-\nu}\right)\rho\omega^2 r^2 + \frac{C_1}{2} + \frac{C_2}{r^2} \\ \sigma_\theta &= -\frac{1}{8}\left(\frac{1+2\nu}{1-\nu}\right)\rho\omega^2 r^2 + \frac{C_1}{2} - \frac{C_2}{r^2} \end{aligned}\right\} \qquad ...(11)$$

For a solid cylinder $C_2 = 0$ for stresses to remain finite at $r = 0$. Therefore, for a solid cylinder of radius b, we get from Eq. (14),

$$(\sigma_r)_{r=b} = -\frac{1}{8}\left(\frac{3-2\nu}{1-\nu}\right)\rho\omega^2 b^2 + \frac{C_1}{2} = 0$$

$$\therefore \qquad \frac{C_1}{2} = \frac{1}{8}\left(\frac{3-2\nu}{1-\nu}\right)\rho\omega^2 b^2$$

$$\therefore \qquad \left.\begin{aligned} \sigma_r &= \frac{1}{8}\left(\frac{3-2\nu}{1-\nu}\right)\rho\omega^2 (b^2 - r^2) \\ \sigma_\theta &= \frac{1}{8}\left(\frac{\rho\omega^2}{1-\nu}\right)[(3-2\nu)\,b^2 - (1+2\nu)r^2] \end{aligned}\right\} \qquad ...(3.48)$$

$$(\sigma_r)_{r=0} = (\sigma_\theta)_{r=0} = \frac{1}{8}\left(\frac{3-2\nu}{1-\nu}\right)\rho\omega^2 b^2 \qquad \text{...(3.49)}$$

(iv) Hollow rotating cylinder

For a hollow cylinder of inner radius a and outer radius b, we have, proceeding as above

$$\left.\begin{aligned} \sigma_r &= \frac{1}{8}\left(\frac{3-2\nu}{1-\nu}\right)\rho\omega^2\left[b^2 + a^2 - \frac{a^2b^2}{r^2} - r^2\right] \\ \sigma_\theta &= \frac{1}{8}\left(\frac{3-2\nu}{1-\nu}\right)\rho\omega^2\left[b^2 + a^2 + \frac{a^2b^2}{r^2} - \left(\frac{1+2\nu}{3-2\nu}\right)r^2\right] \end{aligned}\right\} \qquad \text{...(3.50)}$$

$$(\sigma_\theta)_{r=a} = (\sigma_\theta)_{max} = \frac{1}{4}\left(\frac{3-2\nu}{1-\nu}\right)\rho\omega^2 b^2\left[\left(\frac{1-2\nu}{3-2\nu}\right)\frac{a^2}{b^2}\right] \qquad \text{...(3.51)}$$

If $\frac{a}{b} \to 0$, then

$$(\sigma_\theta)_{max} = \frac{(3-2\nu)}{4(1-\nu)} \cdot \rho\omega^2 b^2 \qquad \text{...(3.52)}$$

3.7.1 Shrink-fitted Rotating Disc and Shaft

Consider a disc of inside radius a and outside radius b shrunk on to a shaft of radius $a + h$, the interference h being small in comparison with a. Both disc and shaft are assumed to be at the same temperature and rotate with an angular speed ω. If p is the shrinkage pressure between the disc and the shaft, then the stresses in the disc are :

$$\sigma_r = \left(\frac{3+\nu}{8}\right)\rho\omega^2\left[a^2 + b^2 - \frac{a^2b^2}{r^2} - r^2\right] + \frac{pa^2}{b^2 - a^2}\left[1 - \frac{b^2}{r^2}\right]$$

$$\sigma_\theta = \left(\frac{3+\nu}{8}\right)\rho\omega^2\left[a^2 + b^2 + \frac{a^2b^2}{r^2} - \left(\frac{1+3\nu}{3+\nu}\right)r^2\right] + \frac{pa^2}{b^2 - a^2}\left[1 + \frac{b^2}{r^2}\right].$$

The radial displacement of the disc, measured on the inside radius is

$$Eu_r = \frac{\rho\omega^2}{4}[(3+\nu)\,ab^2 + (1-\nu)\,a^3] + \frac{p}{b^2-a^2}[(1+\nu)ab^2 + (1-\nu)\,a^3]$$

For the shaft

$$(\sigma_r)_{shaft} = \frac{(3-2\nu)}{8(1-\nu)}\rho\omega^2(a^2-r^2) - p$$

$$(\sigma_\theta)_{shaft} = \frac{(3-2\nu)}{8(1-\nu)}\rho\omega^2\left[a^2 - \left(\frac{1+2\nu}{3-2\nu}\right)r^2\right] - p$$

$$(\sigma_z)_{shaft} = \frac{\nu}{4(1-\nu)}\rho\omega^2(a^2-2r^2)$$

The radial displacement at $r = a$ is

$$(Eu_r)_{shaft} = \frac{1-\nu}{4}\rho\omega^2 a^3 + (1-\nu)pa$$

(*i*) *At stand still*

$$Eh = Eu_r - (Eu_r)_{shaft} = \frac{2ab^2}{b^2-a^2}\cdot p$$

$$\therefore \quad \left.\begin{aligned} \sigma_r &= \frac{Eha}{2b^2}\left(1-\frac{b^2}{r^2}\right) \\ \sigma_\theta &= \frac{Eha}{2b^2}\left(1+\frac{b^2}{r^2}\right) \end{aligned}\right\} \qquad ...(3.53)$$

$$(\sigma_r)_{shaft} = (\sigma_\theta)_{shaft} = -\frac{Eh}{2a}\left(1-\frac{a^2}{b^2}\right) \qquad ...(3.54)$$

(*ii*) In order to determine the speed at which the shrink fit will loosen up, we have $p = 0$

$$Eh = \left(\frac{3+4}{\nu}\right)ab^2\rho w^2$$

$$\omega = \sqrt{\frac{4Eh}{(3+\nu)\rho\, ab^2}} \qquad ...(3.55)$$

$$\left.\begin{aligned} (\sigma_r)_{shaft} &= \frac{Eha}{2b^2}\cdot\left(\frac{3-2\nu}{3+\nu}\right)\left(1-\frac{r^2}{a^2}\right)\left(\frac{1}{1-\nu}\right) \\ (\sigma_\theta)_{shaft} &= \frac{Eha}{2b^2}\left(\frac{3-2\nu}{3+\nu}\right)\left[1-\left(\frac{1+2\nu}{3-2\nu}\right)\frac{r^2}{a^2}\right]\left(\frac{1}{1-\nu}\right) \end{aligned}\right\} \qquad ...(3.56)$$

$$\left.\begin{aligned}\sigma_r &= \left(\frac{3+\nu}{8}\right)\rho\omega^2\left[a^2+b^2-\frac{a^2b^2}{r^2}-r^2\right]\\ \sigma_\theta &= \left(\frac{3+\nu}{8}\right)\rho\omega^2\left[a^2+b^2+\frac{a^2b^2}{r^2}-\left(\frac{1+3\nu}{3+\nu}\right)r^2\right]\end{aligned}\right\}\quad ...(3.57)$$

Example 3.4 *A steel rotor of an electric machine having outer diameter 1500 mm and inner diameter 200 mm, is provided with 200 mm deep trapezoidal slots axially around its outer periphery for accommodating the windings as shown in Fig. 3.8. Determine the maximum stresses induced in the rotor, if it rotates at 1500 rev / min. Assume that the weight of the windings in the slot is the same as that of the material removed. The rotor density is 7800 kg / m² and $\nu = 0.3$.*

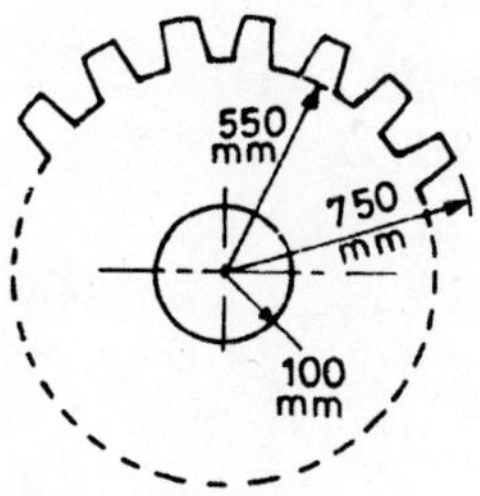

Fig. 3.8

Solution. $\omega = \dfrac{2\pi \times 1500}{60} = 157.08$ red/s

Centrifugal force at $r = 550$ mm

$$F = \int_{550}^{750} 2\,\pi r\,dr \times t \times \rho\omega^2 r$$

$$= 2\,\pi\,\rho t\,\omega^2 \int_{550}^{750} r^2\,dr$$

$$= \frac{2\pi\,\rho t\omega^2}{3}\left|\,r^3\,\right|_{550}^{750}$$

$$= \frac{2\pi \times 7800 \times t \times (157.08)^2}{3}(750^3 - 550^3) \times 10^{-9}$$

$$= 102987829.7 \times t\ N$$

Pressure, $p_2 = \dfrac{F}{2\pi r_2 t} = \dfrac{102987829.7 \times t}{2\pi \times 550 \times 10^{-3} \times t} = 29.8 \text{ MPa}$

At $\quad r = r_1$

$$\sigma_\theta = \frac{+2p_2}{1-(r_1/r_2)^2}$$

$$= \frac{+2 \times 29.8}{1-(100/550)^2} = +61.64 \text{ MPa}$$

Due to rotation, at $r = r_1$

$$(\sigma_\theta)_{max} = \left(\frac{3+\nu}{4}\right)\rho\omega^2\left[r_2^2 + \left(\frac{1-\nu}{3+\nu}\right)r_1^2\right]$$

$$= \left(\frac{3.3}{4}\right) \times 7800 \times (157.08)^2$$

$$\times \left[550^2 + \frac{0.7}{3.3} \times 100^2\right] \times 10^{-6}$$

$$= 48.37 \text{ MPa}$$

Resultant stress at $r = r_1$

$$= 61.64 + 48.37$$

$$= 110.01 \text{ MPa.}$$

Example 3.5 *A cast iron shaft 500 mm diameter is rotating at 3000 rev/min. Its density is 7470 kg/m³ and Poisson's ratio 0.28. It is constrained at its ends so that it cannot expand or contract axially. Calculate the total longitudinal thrust over its entire cross-section due to centrifugal stresses.*

Solution. $\varepsilon_z = \dfrac{1}{E}[\sigma_z - \nu(\sigma_r + \sigma_\theta)] = 0$

$$\sigma_z = \nu(\sigma_r + \sigma_\theta)$$

For a solid shaft

$$\sigma_r = \frac{1}{8}\left(\frac{3-2\nu}{1-\nu}\right)\rho\omega^2(r_2^2 - r^2)$$

$$\sigma_\theta = \frac{1}{8}\left(\frac{3-2\nu}{1-\nu}\right)\rho\omega^2\left[r_2^2 - \left(\frac{1+2\nu}{3-2\nu}\right)r^2\right]$$

$$\sigma_z = \frac{\nu}{8}\left(\frac{3-2\nu}{1-\nu}\right)\rho\omega^2\left[2r_2^2 - \left(\frac{4}{3-2\nu}\right)r^2\right]$$

$$= \frac{\nu}{4}\left(\frac{3-2\nu}{1-\nu}\right)\rho\omega^2\left[r_2^2 - \left(\frac{2}{3-2\nu}\right)r^2\right]$$

Restraining force $= \displaystyle\int_0^{r_2} \sigma_z \, 2\pi r dr$

$$= \frac{\pi\nu}{2}\left(\frac{3-2\nu}{1-\nu}\right)\rho\omega^2\int_0^{r_2}\left[r_2^2-\left(\frac{2}{3-2\nu}\right)r^2\right]r\,dr$$

$$= \frac{\pi\nu}{2}\left(\frac{3-2\nu}{1-\nu}\right)\rho\omega^2\left|\frac{r_2^2 r^2}{2}-\left(\frac{2}{3-2\nu}\right)\frac{r^4}{4}\right|_0^{r_2}$$

$$= \frac{\pi\nu}{2}\left(\frac{3-2\nu}{1-\nu}\right)\rho\omega^2\left[\frac{r_2^4}{2}-\left(\frac{2}{3-2\nu}\right)\frac{r_2^4}{4}\right]$$

$$= \frac{\pi\nu}{4}\left(\frac{3-2\nu}{1-\nu}\right)\rho\omega^2 r_2^4\left[1-\frac{1}{3-2\nu}\right]$$

$$= \frac{\pi\nu}{2}\rho\omega^2 r_2^4$$

$$= \frac{\pi}{2}\times 0.28\times 7470\times\left(\frac{2\pi\times 3000}{60}\right)^2\times(250)^4\times 10^{-12}$$

$$= 1266.65 \text{ kN}$$

Example 3.6 *A steel turbine rotor of 750 mm outer diameter, 150 mm inner diameter and 50 mm thickness, has 100 blades 150 mm long, each weighing 4 N. It is shrink-fitted on a rigid shaft. Calculate the initial shrinkage allowance on the inner diameter of the rotor so that it just loosens on the shaft at 3000 rev/min. Take E = 200 GPa, ν = 0.3. The density of shaft and rotor is 7500 kg/m³.*

Solution. $\omega = \dfrac{2\pi\times 3000}{60} = 314.16$ rad/s

Centrifugal stress at $r = r_1$,

$$\sigma_\theta = \left(\frac{3+\nu}{4}\right)\rho\omega^2\left[r_2^2+\left(\frac{1-\nu}{3+\nu}\right)r_1^2\right]$$

$$= \left(\frac{3.3}{4}\right)\times 7500\times(314.16)^2\left[375^2+\frac{0.7}{3.3}\times 75^2\right]\times 10^{-6}$$

$$= 86.6 \text{ MPa}$$

$$\sigma_r = 0$$

Pressure on the outer surface of the rotor due to centrifugal force exerted by the blades,

$$p_2 = -\frac{W}{g}\times n\times\left(r_2+\frac{h}{2}\right)\times\omega^2\times\frac{1}{2\pi r_2 t}$$

$$= -\frac{4}{9.81}\times 100\times\left(375+\frac{150}{2}\right)\times 10^{-3}\times(314.16^2)$$

$$\times\frac{1}{2\pi\times 375\times 50\times 10^{-6}}$$

$$= -18.617 \text{ MPa}$$

Pressure stresses :

At $\quad r = r_1, \sigma_r = 0$

$$\sigma_\theta = +\frac{2p_2 r_2^2}{r_2^2 - r_1^2}$$

$$= +\frac{2 \times 18.617 \times 375^2}{(375^2 - 75^2)}$$

$$= 38.78 \text{ MPa}$$

Total stresses at $r = r_1$:

$$\sigma_r = 0$$

$$\sigma_\theta = 86.6 + 38.78$$

$$= 125.38 \text{ MPa}$$

$$\frac{\delta r_1}{r_1} = \frac{\sigma_\theta}{E} \qquad \text{(assuming shaft to be rigid)}$$

$$\delta r_1 = \frac{75 \times 125.38 \times 10^6}{200 \times 10^9} = 0.047 \text{ mm}$$

3.8 Stress-concentration due to a Circular Hole in a Stressed Plate (Kirsch Problem)

Consider first an infinite plate referred to a system of orthogonal co-ordinates oxy and assume that the plate is subjected to a uniform tensile stress of intensity σ_0 in the x-direction as shown in Fig. 3.9. If there is no hole in the plate, the state of stress will be given by

$$(\sigma_z)_1 = \sigma_0, (\sigma_y)_1 = 0, (\tau_{xy})_1 = 0 \qquad ...(1)$$

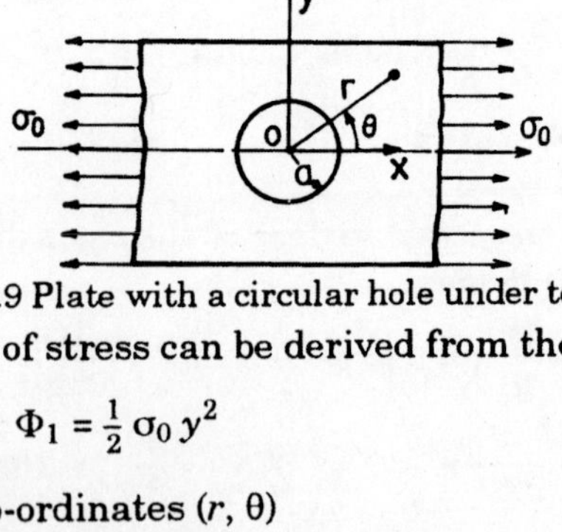

Fig. 3.9 Plate with a circular hole under tension.

This state of stress can be derived from the stress function

$$\Phi_1 = \frac{1}{2}\sigma_0 y^2 \qquad ...(2)$$

In polar co-ordinates (r, θ)

$$y = r \sin \theta$$

$$\therefore \qquad \Phi_1 = \frac{1}{2}\sigma_0 r^2 \sin^2\theta = \frac{\sigma_0}{4} r^2(1 - \cos 2\theta) \qquad \text{...(3)}$$

$$\left.\begin{aligned}(\sigma_r)_1 &= \frac{1}{r}\frac{\partial \Phi_1}{\partial r} + \frac{1}{r}\frac{\partial^2 \Phi_1}{\partial \theta^2} \\ &= \frac{\sigma_0}{2}(1 + \cos 2\theta) \\ (\sigma_\theta)_1 &= \frac{\partial^2 \Phi_1}{\partial r^2} = \frac{\sigma_0}{2}(1 - \cos 2\theta) \\ (\tau_{r\theta})_1 &= -\frac{\partial}{\partial r}\left(\frac{1}{r}\frac{\partial \Phi_1}{\partial \theta}\right) \\ &= -\frac{1}{2}\sigma_0 \sin 2\theta \end{aligned}\right\} \qquad \text{...(4)}$$

Now suppose that a hole of radius a in the centre at the origin 0 is drilled through the plate. A new system of stresses σ_r, σ_θ and $\tau_{r\theta}$ will be produced inside the plate. These stresses must satisfy the following boundary conditions :

For $\qquad r = a, \ \ \sigma_r = 0, \ \ \tau_{r\theta} = 0$

For $\qquad r = \infty, \ \ \sigma_r = (\sigma_r)_1, \ \ \sigma_\theta = (\sigma_\theta)_1, \ \tau_{r\theta} = (\tau_{r\theta})_1 \qquad$ ...(5)

These stresses can be derived from the biharmonic equation

$$\nabla^4\Phi = \left(\frac{\partial^2}{\partial r^2} + \frac{1}{r}\frac{\partial}{\partial r} + \frac{1}{r^2}\frac{\partial^2}{\partial \theta^2}\right)\left(\frac{\partial^2 \Phi}{\partial r^2} + \frac{1}{r}\frac{\partial \Phi}{\partial r} + \frac{1}{r^2}\frac{\partial^2 \Phi}{\partial \theta^2}\right)$$

$$= 0 \qquad \text{...(6)}$$

Assume this unknown stress function Φ to be of the same form as the stress function given by Eq. (3), *i.e.* composed of a function $f_1(r)$ plus a function $f_2(r)$ multiplied by $\cos 2\theta$, thus

$$\Phi(r, \theta) = f_1(r) + f_2(r)\cos 2\theta \qquad \text{...(7)}$$

Substituting Eq. (7) in Eq. (6), we get

$$\left(\frac{d^2}{dr^2} + \frac{1}{r}\frac{d}{dr}\right)\left(\frac{d^2 f_1}{dr^2} + \frac{1}{r}\frac{df_1}{dr}\right) + \left(\frac{d^2}{dr^2} + \frac{1}{r}\frac{d}{dr} - \frac{4}{r^2}\right)$$

$$\times \left(\frac{d^2 f_2}{dr^2} + \frac{1}{r}\frac{df_2}{dr} - \frac{4f_2}{r^2}\right)\cos 2\theta = 0$$

or

$$\left.\begin{aligned}\left(\frac{d^2}{dr^2} + \frac{1}{r}\frac{d}{dr}\right)\left(\frac{d^2 f_1}{dr^2} + \frac{1}{r}\frac{df_1}{dr}\right) &= 0 \\ \left(\frac{d^2}{dr^2} + \frac{1}{r}\frac{d}{dr} - \frac{4}{r^2}\right)\left(\frac{d^2 f_2}{dr^2} + \frac{1}{r}\frac{df_2}{dr} - \frac{4f_2}{r^2}\right) &= 0\end{aligned}\right\} \qquad \text{...(8)}$$

The general solution of these equations is

$$f_1(r) = C_1r^2 \log r + C_2r^2 + C_3 \log r + C_4$$

$$f_2(r) = C_5r^2 + C_6r^4 + \frac{C_7}{r^2} + C_8$$

and the function $\Phi(r, \theta)$ becomes

$$\Phi = (C_1r^2 \log r + C_2r^2 + C_3 \log r + C_4) + \left(C_5r^2 + C_6r^4 + \frac{C_7}{r^2} + C_8 \right) \cos 2\theta$$

This gives the following system of stresses

$$\sigma_r = C_1 (1 + 2 \log r) + 2C_2 + \frac{C_3}{r^2} - \left(2C_5 + \frac{6C_7}{r^4} + \frac{4C_8}{r^2} \right) \cos 2\theta$$

$$\sigma_\theta = C_1(3 + 2 \log r) + 2C_2 - \frac{C_3}{r^2} + \left(2C_5 + 12C_6r^2 + \frac{6C_7}{r^4} \right) \cos 2\theta$$

$$\tau_{r\theta} = \left(2C_5 + 6C_6r^2 - \frac{6C_7}{r^4} - \frac{2C_8}{r^2} \right) \sin 2\theta$$

where C's are constants.

Since for $r \to \infty$, σ_r, σ_θ $\tau_{r\theta}$ must be finite.

$$\therefore \quad C_1 = C_6 = 0$$

The remaining constants are determined from boundary conditions given by Eqs. (5)

$$2C_2 + \frac{C_3}{a^2} = 0$$

$$2C_5 + \frac{6C_7}{a^4} + \frac{2C_8}{a^2} = 0$$

$$2C_5 - \frac{6C_7}{a^4} - \frac{2C_8}{a^2} = 0$$

$$\therefore \quad 2C_5 = -\frac{\sigma_0}{2}, \; 2C_2 = \frac{\sigma_0}{2}$$

$$\therefore \quad C_2 = \frac{\sigma_0}{4}, \; C_3 = -\frac{a^2\sigma_0}{2}, \; C_5 = -\frac{\sigma_0}{4},$$

$$C_7 = -\frac{a^4\sigma_0}{4}, \; C_8 = \frac{a^2\sigma_0}{2}$$

Therefore, the stresses become

$$\left.\begin{aligned}
\sigma_r &= \frac{\sigma_0}{2}\left(1-\frac{a^2}{r^2}\right)+\frac{\sigma_0}{2}\left(1+\frac{3a^4}{r^4}-\frac{4a^2}{r^2}\right)\cos 2\theta \\
\sigma_\theta &= \frac{\sigma_0}{2}\left(1+\frac{a^2}{r^2}\right)-\frac{\sigma_0}{2}\left(1+\frac{3a^4}{r^4}\right)\cos 2\theta \\
\tau_{r\theta} &= -\frac{\sigma_0}{2}\left(1-\frac{3a^4}{r^4}+\frac{2a^2}{r^2}\right)\sin 2\theta
\end{aligned}\right\} \quad \text{...(9)}$$

$$(\sigma_\theta)_{max} = 3\sigma_0 \text{ for } \theta = \frac{\pi}{2}, \frac{3\pi}{2}$$

and $$r = a$$

$$(\sigma_\theta)_{min} = -\sigma_0 \text{ for } \theta = 0, \pi \text{ and } r = a$$

$$\left(\sigma_\theta\right)_{\theta=\frac{\pi}{2},\frac{3\pi}{2}} = \frac{\sigma_0}{2}\left(2+\frac{a^2}{r^2}+\frac{3a^4}{r^4}\right)$$

$$\simeq \sigma_0 \text{ for } r > 10a$$

When the plate with a circular hole is subjected to uniform shearing stress of intensity σ_0, then the state of stress in the plate can be determined by adding the states of stress as shown in Fig. 3.10.

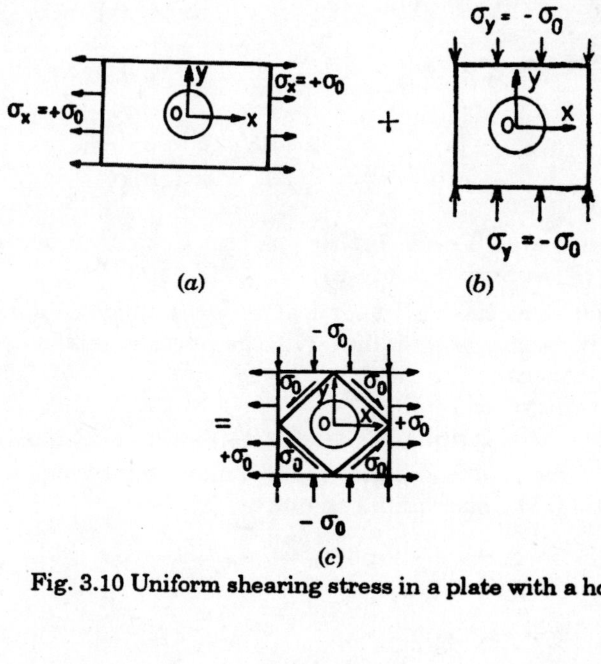

(c)

Fig. 3.10 Uniform shearing stress in a plate with a hole.

The state of stress due to $\sigma_x = +\sigma_0$ is given by Eq. (9) and that due to $\sigma_y = -\sigma_0$ is obtained by introducing $-\sigma_0$ instead of σ_0 and $\theta - \frac{\pi}{2}$ instead of θ. Hence

$$\left.\begin{aligned}
\sigma_r &= -\frac{\sigma_0}{2}\left(1 - \frac{a^2}{r^2}\right) - \frac{\sigma_0}{2}\left(1 + \frac{3a^4}{r^4} - \frac{4a^2}{r^2}\right) \\
&\qquad \times \cos 2\left(\theta - \frac{\pi}{2}\right) \\
\sigma_\theta &= \frac{-\sigma_0}{2}\left(1 + \frac{a^2}{r^2}\right) + \frac{\sigma_0}{2}\left(1 + \frac{3a^4}{r^4}\right)\cos 2\left(\theta - \frac{\pi}{2}\right) \\
\tau_{r\theta} &= \frac{\sigma_0}{2}\left(1 - \frac{3a^4}{r^4} + \frac{2a^2}{r^2}\right)\sin 2\left(\theta - \frac{\pi}{2}\right)
\end{aligned}\right\} \quad \text{...(10)}$$

Superimposing Eqs. (9) and (10), we get

$$\left.\begin{aligned}
\sigma_r &= \sigma_0\left(1 + \frac{3a^4}{r^4} - \frac{4a^2}{r^2}\right)\cos 2\theta \\
\sigma_\theta &= -\sigma_0\left(1 + \frac{3a^4}{r^4}\right)\cos 2\theta \\
\tau_{r\theta} &= -\sigma_0\left(1 - \frac{3a^4}{r^4} + \frac{2a^2}{r^2}\right)\sin 2\theta
\end{aligned}\right\} \quad \text{...(3.58)}$$

$$\left.\begin{aligned}
(\sigma_\theta)_{min} &= \left(\sigma_\theta\right)_{\theta = 0,\, \pi} = -4\sigma_0 \\
(\sigma_\theta)_{max} &= \left(\sigma_\theta\right)_{\theta = \frac{\pi}{2},\, \frac{3\pi}{2}} = +4\sigma_0
\end{aligned}\right\} \quad \text{...(3.59)}$$

3.9 Concentrated Load Acting on the Free Surface of a Plate (Flemant's Problem)

Consider a concentrated vertical force P acting on a horizontal straight boundary of an infinitely large plate as shown in Fig. 3.11. The thickness of the plate is taken as unity, so that P is the load per unit thickness.

Any element of the plate at a distance r from the point of application of the load is subjected to a simple compression in the radial direction. The stress components are

$$\sigma_r = -\frac{2P}{\pi r}\cos\theta, \quad \sigma_\theta = 0, \quad \tau_{r\theta} = 0 \qquad \text{...(1)}$$

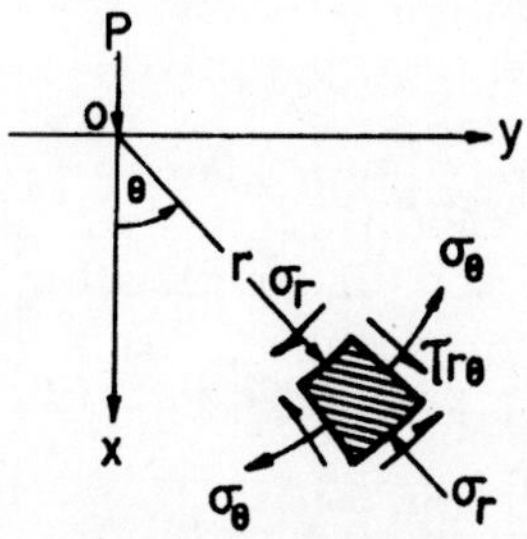

Fig. 3.11 Concentrated load on the free surace of plate.

which have been obtained from the stress function

$$\Phi = -\frac{P}{\pi} r\theta \sin \theta$$

The stress distribution given by Eqs. (1) satisfy the equilibrium and boundary conditions.

Now

$$\left.\begin{aligned} \varepsilon_r &= \frac{1}{E}[\sigma_r - \nu\sigma_\theta] \\ \varepsilon_\theta &= \frac{1}{E}[\sigma_\theta - \nu\sigma_r] \\ \gamma_{r\theta} &= \frac{\tau_{r\theta}}{G} \end{aligned}\right\} \quad \text{...(2)}$$

Also

$$\left.\begin{aligned} \varepsilon_\theta &= \frac{\partial u_r}{\partial_r} \\ \varepsilon_\theta &= \frac{u_r}{r} + \frac{1}{r}\frac{\partial u_\theta}{\partial \theta} \\ \gamma_{r\theta} &= \frac{1}{r}\frac{\partial u_r}{\partial \theta} + \frac{\partial u_\theta}{\partial r} - \frac{u_\theta}{r} \end{aligned}\right\} \quad \text{...(3)}$$

$\therefore$

$$\left.\begin{aligned} \frac{\partial u_r}{\partial r} &= -\frac{2P}{\pi E}\frac{\cos\theta}{r} \\ \frac{u_r}{r} + \frac{1}{r}\frac{\partial u_\theta}{\partial \theta} &= \frac{2\nu P}{\pi E}\frac{\cos\theta}{r} \\ \frac{1}{r}\frac{\partial u_r}{\partial \theta} + \frac{\partial u_\theta}{\partial r} - \frac{u_\theta}{r} &= 0 \end{aligned}\right\} \quad \text{...(4)}$$

Integrating the first of Eqs. (4), we get

$$u_r = -\frac{2P}{\pi E}\cos\theta . \log r + f(\theta) \quad \text{...(5)}$$

$$\therefore \quad \frac{\partial u_\theta}{\partial \theta} = \frac{2\nu P}{\pi E}\cos\theta + \frac{2P}{\pi E}\cos\theta \log r - f(\theta)$$

Integrating, we get

$$u_\theta = \frac{2\nu P}{\pi E}\sin\theta + \frac{2P}{\pi E}\sin\theta.\log r - \int f(\theta)\,d\theta + F(r) \qquad \text{...(6)}$$

Also from Eqs. (5) and (6), we get after differentiating,

$$\left.\begin{aligned} \frac{\partial u_r}{\partial\theta} &= \frac{2P}{\pi E}\sin\theta.\log r + \frac{d}{d\theta}f(\theta) \\ \frac{\partial u_\theta}{\partial r} &= \frac{2P}{\pi E}\frac{\sin\theta}{r} + \frac{d}{dr}F(r) \end{aligned}\right\} \qquad \text{...(7)}$$

Substituting Eqs. (6) and (7) into last of Eq. (4), we get

$$\frac{2P}{\pi E}\sin\theta\,(1-\nu) + \frac{d}{d\theta}f(\theta) + \int f(\theta)\,d\theta + r\frac{d}{dr}F(r) - F(r) = 0$$

This yields

$$\frac{2P}{\pi E}(1-\nu)\sin\theta + \frac{d}{d\theta}f(\theta) + \int f(\theta)\,d\theta = 0 \qquad \text{...(8)}$$

$$r\frac{d}{dr}F(r) - F(r) = 0 \qquad \text{...(9)}$$

Integrating Eqs. (8) and (9), we get

$$f(\theta) = -\frac{(1-\nu)P}{\pi E}\theta\sin\theta + A\sin\theta + B\cos\theta \qquad \text{...(10)}$$

$$F(r) = Cr \qquad \text{...(11)}$$

where A, B and C are constants.

Eqs. (5) and (6) become

$$\left.\begin{aligned} u_r &= -\frac{2P}{\pi E}\cos\theta\,.\,\log r - \frac{(1-\nu)P}{\pi E}\theta\sin\theta + A\sin\theta \\ &\qquad + B\cos\theta \\ u_\theta &= -\left(\frac{1-\nu}{\pi E}\right)P\theta\cos\theta + \frac{(1+\nu)}{\pi E}P\sin\theta + \frac{2P}{\pi E} \\ &\qquad \times \log r\sin\theta + A\cos\theta - B\sin\theta + Cr \end{aligned}\right\} \qquad \text{...(12)}$$

Since $u_\theta = 0$ for $\theta = 0$ and for any value of r,

$\therefore \quad A + Cr = 0 \quad$ or $\quad A = C = 0$

If we let $\quad u_r' = B\cos\theta$

$\qquad\qquad u_\theta' = -B\sin\theta$

Then $\quad \sqrt{u_r'^2 + u_\theta'^2} = B$

where B = rigid displacement of every point of the plate in the x-direction

$= 0$

Hence, we get

$$\left.\begin{aligned} u_r &= \frac{-P}{\pi E}[2\cos\theta\log r + (1-\nu)\,\theta\sin\theta] \\ u_\theta &= \frac{P}{\pi E}[-(1-\nu)\theta\cos\theta + (1+\nu)\sin\theta \\ &\qquad + 2\sin\theta\log r] \end{aligned}\right\} \quad ...(13)$$

$$(u_r)_{\theta=\pm\pi/2} = -\frac{(1-\nu).P}{2E} \quad ...(14)$$

$$\left.\begin{aligned} (u_\theta)_{\theta=+\pi/2} &= \frac{P}{\pi E}[(1+\nu) + 2\log r] \\ (u_\theta)_{\theta=-\pi/2} &= -\frac{P}{\pi E}[(1+\nu) + 2\log r] \end{aligned}\right\} \quad ...(15)$$

In Cartesian co-ordinates,

$$\left.\begin{aligned} \sigma_x &= \sigma_r\cos^2\theta = -\frac{2P}{\pi r}\cos^3\theta \\ \sigma_y &= \sigma_r\sin^2\theta = -\frac{2P}{\pi r}\sin^2\theta\cos\theta \\ \tau_{xy} &= \sigma_r\sin\theta\cos\theta = -\frac{2P}{\pi r}\sin\theta\cos^2\theta \end{aligned}\right\} \quad ...(3.60)$$

Now $\cos\theta = \frac{x}{r}$, $\sin\theta = \frac{y}{r}$, $r = \sqrt{x^2+y^2}$

$$\therefore \quad \left.\begin{aligned} \sigma_x &= -\frac{2Px^3}{\pi(x^2+y^2)^2} \\ \sigma_y &= -\frac{2Pxy^2}{\pi(x^2+y^2)^2} \\ \tau_{xy} &= -\frac{2Px^2y}{\pi(x^2+y^2)^2} \end{aligned}\right\} \quad ...(3.61)$$

3.10 Uniformly Distributed Load Applied on the Free Surface of a Large Plate

Consider a uniformly distributed load of intensity q_0 applied on the free surface of an infinitely large plate as shown in Fig. 3.12.

Let us take an elementary length dy of the load so that

$$dy = \frac{r\,d\theta}{\cos\theta}$$

and

$$dP = q_0 dy = \frac{q_0\, r\,d\theta}{\cos\theta}$$

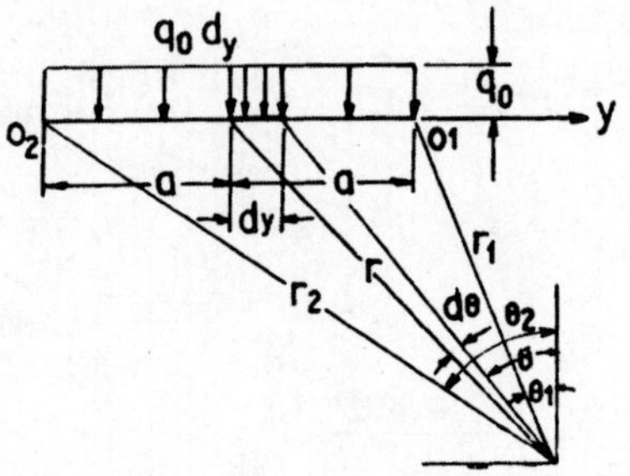

Fig. 3.12 udl on the free surface of a plate.

Using the results as obtained in Art. 3.9, we can write

$$d\sigma_x = -\frac{2q_0}{\pi}\cos^2\theta\,d\theta$$

$$d\sigma_y = -\frac{2q_0}{\pi}\sin^2\theta\,d\theta$$

$$d\tau_{xy} = -\frac{q_0}{\pi}\sin 2\theta\,d\theta$$

$$\therefore \quad \sigma_x = -\frac{2q_0}{\pi}\int_{\theta_1}^{\theta_2}\cos^2\theta\,d\theta$$

$$= -\frac{q_0}{2\pi}[2(\theta_2-\theta_1)+(\sin 2\theta_2-\sin 2\theta_1)] \quad \text{...(3.62 } a)$$

$$\sigma_y = -\frac{2q_0}{\pi}\int_{\theta_1}^{\theta_2}\sin^2\theta\,d\theta$$

$$= -\frac{q_0}{2\pi}[2(\theta_2-\theta_1)-(\sin 2\theta_2-\sin 2\theta_1] \quad \text{...(3.62 } b)$$

$$\tau_{xy} = -\frac{q_0}{\pi}\int_{\theta_1}^{\theta_2}\sin 2\theta\,d\theta$$

$$= -\frac{q_0}{2\pi}[\cos 2\theta_1-\cos 2\theta_2] \quad \text{...(3.62 } c)$$

3.11 Stresses due to a Couple on a Large Plate

Let a couple $M = Pa$ act on the surface of an infinitely large plate as shown in Fig. 3.13.

The stress function for the case when the tensile force P is at the point O_1, at a distance a from the origin can be obtained from

$$\Phi = -\frac{P}{\pi}r\theta\sin\theta$$

regarded for the moment as a function of x and y instead of r and θ, by writing $(y + a)$ instead of y and also $-P$ instead of P.

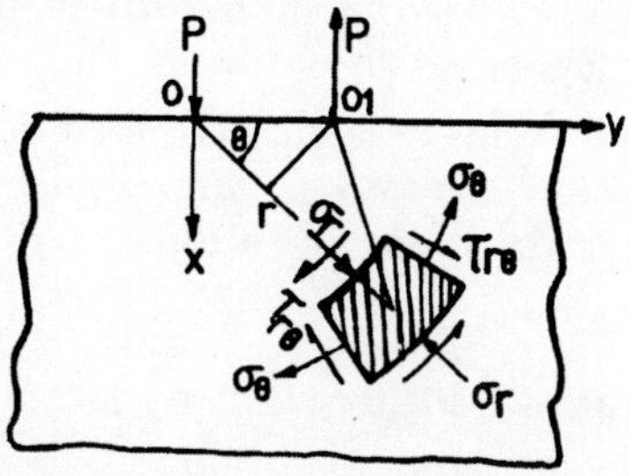

Fig. 3.13 Couple on a large plate.

This and the original stress function Φ can be combined to obtain the stress function for the two equal and opposite forces applied at O and O_1, in the form

$$-\Phi(x, y + a) + \Phi(x, y)$$

when a is very small, this approaches the value

$$\Phi_1 = -a\frac{\partial \Phi}{\partial y}$$

$$= -\frac{Pa}{\pi} \cdot \frac{\partial}{\partial y}(r\theta \sin \theta)$$

Now $$\frac{\partial \Phi}{\partial y} = \frac{\partial \Phi}{\partial r}\sin \theta + \frac{\partial \Phi}{\partial \theta}\frac{\cos \theta}{r}$$

$\therefore$ $$\Phi_1 = \frac{Pa}{\pi}(\theta + \sin \theta \cos \theta)$$

$$= \frac{M}{\pi}(\theta + \sin \theta \cos \theta)$$

$\therefore$ $$\sigma_r = \frac{1}{r}\frac{\partial \Phi_1}{\partial r} + \frac{1}{r^2}\frac{\partial^2 \Phi_1}{\partial \theta^2}$$

$$= -\frac{2M}{\pi r^2}\sin 2\theta \qquad \text{...(3.63 a)}$$

$$\left.\begin{aligned} \sigma_\theta &= \frac{\partial^2 \Phi_1}{\partial r^2} = 0 \\ \tau_{r\theta} &= -\frac{\partial}{\partial r}\left(\frac{1}{\partial r}\frac{\partial \Phi_1}{\partial \theta}\right) = \frac{2M\cos^2 \theta}{\pi r^2} \end{aligned}\right\} \qquad \text{...(3.63 b)}$$

3.12 Saint Venant's Principle

This principle states that if the forces acting on a small area be replaced by a statically equivalent system on the same area, the new system will only produce changes in the immediate vicinity of the applied loads, but that it will not affect the stresses at distances which are large compared with the linear dimensions of the area on which the boundary forces act.

Consider the case of a concentrated load P acting on the edge of a semi-infinite plate. We remove a semi-circular groove of radius r_0 under the load from the plate and subject it to radial forces only defined by

$$q = -\frac{q_0}{r_0} \cos \theta$$

as shown in Fig. 3.14 (*a*).

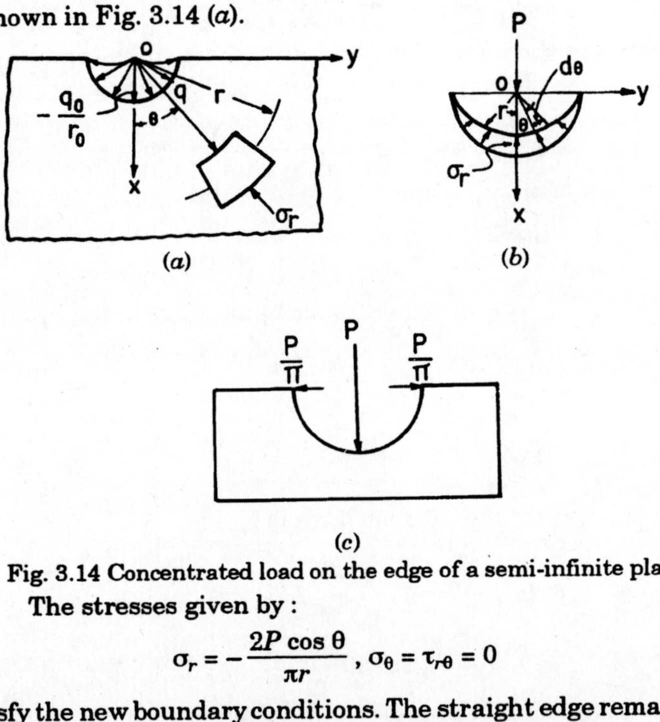

Fig. 3.14 Concentrated load on the edge of a semi-infinite plate.

The stresses given by :

$$\sigma_r = -\frac{2P \cos \theta}{\pi r}, \quad \sigma_\theta = \tau_{r\theta} = 0$$

satisfy the new boundary conditions. The straight edge remains free since σ_θ and $\tau_{r\theta}$ vanish everywhere. On the semi-circular groove the stress σ_r becomes equal to the assumed pressure q and all stresses vanish at infinity.

For statically equivalent system, the horizontal thrust [Fig. 3.14 (*b*)]

$$F_y = \int_0^{\pi/2} \sin \theta . \sigma_r . dA = \frac{P}{\pi}\int_0^{\pi/2} \sin 2\theta \, d\theta = \frac{P}{\pi}$$

Therefore, a concentrated load acting on the straight edge of thin, large plate may be replaced by an equal force on a semi-circular groove of small radius accompanied by two thrusts, numerically equal to $\frac{P}{\pi}$, and acting at right angles to the force P as shown in Fig. 3.14 (*c*).

3.12.1 Concentrated Load acting on the Vertex of a Wedge along the Line of Symmetry

First method. Consider a wedge loaded as shown in Fig. 3.15 (*a*). Let the wedge be of unit thickness.

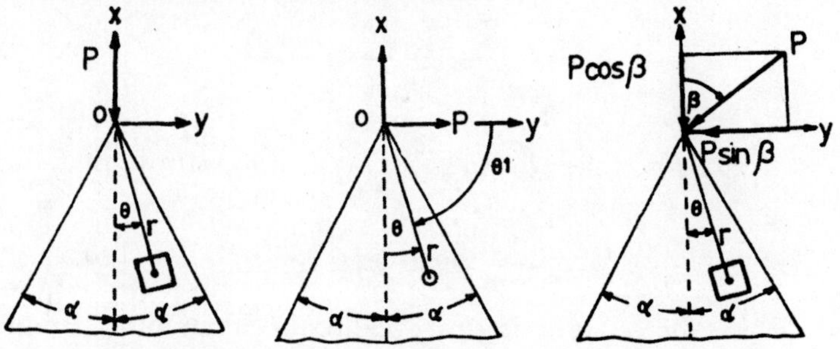

(*a*) (*b*) (*c*)

Fig. 3.15 Concentrated load on the vertex of a wedge.

The stress function

$$\Phi(r, \theta) = CPr\,\theta \sin\theta \quad ...(1)$$

satisfies the compatibility equation

$$\nabla^4\Phi = \left(\frac{\partial^2}{\partial r^2} + \frac{1}{r}\frac{\partial}{\partial r} + \frac{1}{r^2}\frac{\partial^2}{\partial\theta^2}\right) \times \left(\frac{\partial^2\Phi}{\partial r^2} + \frac{1}{r}\frac{\partial\Phi}{\partial r} + \frac{1}{r^2}\frac{\partial^2\Phi}{\partial\theta^2}\right) = 0$$

where C = Constant to be determined from boundary conditions.

$$\left.\begin{aligned} \sigma_r &= \frac{1}{r}\frac{\partial\Phi}{\partial r} + \frac{1}{r^2}\frac{\partial^2\Phi}{\partial\theta^2} = 2CP\frac{\cos\theta}{r} \\ \sigma_\theta &= \frac{\partial^2\Phi}{\partial r^2} = 0 \\ \tau_{r\theta} &= \frac{1}{r^2}\frac{\partial\Phi}{\partial\theta} - \frac{1}{r^2}\frac{\partial^2\Phi}{\partial r\partial\theta} = 0 \end{aligned}\right\} \quad ...(2)$$

The boundary conditions to be satisfied are,

$$\left.\begin{aligned} &\text{for} && \theta = \pm\alpha,\ \sigma_\theta = \tau_{r\theta} = 0 \\ &\text{and} && 2\int_0^\alpha \sigma_r \,.\, r\cos\theta\, d\theta = -P \end{aligned}\right\} \quad ...(3)$$

The first two boundary conditions are identically satisfied and from third of Eqs. (3), we get

$$-P = 2\int_0^{\alpha} 2CP\frac{\cos\theta}{r} \,.\, r\cos\theta \, d\theta$$

$$= 4CP\int_0^{\alpha} \cos^2\theta \, d\theta$$

$$-P = CP(2\alpha + \sin 2\alpha)$$

$$\therefore \quad C = \frac{-1}{(2\alpha + \sin 2\alpha)}$$

$$\therefore \quad \left.\begin{aligned} \sigma_r &= -\frac{2P}{(2\alpha + \sin 2\alpha)} \cdot \frac{\cos\theta}{r} \\ \sigma_\theta &= \tau_{r\theta} = 0 \end{aligned}\right\} \quad \text{...(4)}$$

For $\alpha = \pi/2$, the wedge becomes a semi infinite plate and we get

$$\sigma_r = -\frac{2P}{\pi r}\cos\theta, \; \sigma_\theta = \tau_{r\theta} = 0 \quad \text{...(5)}$$

If the force P is acting perpendicular to the line of symmetry, then [Fig. 3.15 (*b*)]

$$\Phi = CPr\,\theta_1 \sin\theta_1 \text{ where } \theta_1 = -(90° - \theta)$$

$$\left.\begin{aligned} \sigma_r &= \frac{2CP}{r}\cos\theta_1 \\ \sigma_\theta &= \tau_{r\theta} = 0 \end{aligned}\right\} \quad \text{...(6)}$$

$$P = \int_{90° - \alpha}^{90° + \alpha} \sigma_r \,.\, r\cos\theta_1 \, d\theta_1$$

$$= 2CP\int_{90° - \alpha}^{90° + \alpha} \cos^2\theta_1 \, d\theta_1$$

$$= (2\alpha - \sin 2\alpha)\, CP$$

$$\therefore \quad C = \frac{1}{2\alpha - \sin 2\alpha}$$

$$\sigma_r = \left(\frac{2P}{2\alpha - \sin 2\alpha}\right)\frac{\cos\theta_1}{r}$$

$$\left.\begin{aligned} \sigma_r &= \left(\frac{2P}{2\alpha - \sin 2\alpha}\right)\frac{\sin\theta}{r} \\ \sigma_\theta &= \tau_{r\theta} = 0 \end{aligned}\right\} \quad \text{...(7)}$$

For $\alpha = \pi/2$

$$\left.\begin{aligned}\sigma_r &= \frac{2P}{\pi} \cdot \frac{\sin\theta}{r} \\ \sigma_\theta &= \tau_{r\theta} = 0\end{aligned}\right\} \qquad \text{...(3.64)}$$

If the force P is inclined at an angle β to the axis of symmetry, then we can resolve this force into horizontal and vertical components as shown in Fig. 3.15 (*c*). Hence the solution can be obtained by adding Eqs. (4) and (7) by putting $P = P\cos\beta$ in Eq. (4) and $P = P\sin\beta$ in Eq. (7). Thus, we get

$$\sigma_r = \frac{-2P\cos\beta}{2\alpha + \sin 2\alpha} \cdot \frac{\cos\theta}{r} + \frac{2P\sin\beta}{2\alpha - \sin 2\alpha}\frac{\sin\theta}{r} \qquad \text{...(3.65)}$$

$$\sigma_\theta = \tau_{r\theta} = 0$$

If a moment M is acting on the vertex of the wedge, then the stress function

$$\Phi = -\frac{M(\sin 2\theta - 2\theta\cos 2\alpha)}{2(\sin 2\alpha - 2\alpha\cos 2\alpha)}$$

satisfies the compatibility equation

$$\nabla^4\Phi = 0$$

and the stress components become

$$\left.\begin{aligned}\sigma_r &= \frac{M}{2(\sin 2\alpha - 2\alpha\cos 2\alpha)} \cdot \frac{4}{r^2}\sin 2\theta \\ \sigma_\theta &= 0 \\ \tau_{r\theta} &= \frac{M}{2(\sin 2\alpha - 2\alpha\cos 2\alpha)} \cdot \frac{2}{r^2}(\cos 2\theta - \cos 2\alpha)\end{aligned}\right\} \qquad \text{...(3.66)}$$

For $\alpha = \pi/2$

$$\left.\begin{aligned}\sigma_r &= \frac{2M\sin 2\theta}{\pi r^2} \\ \sigma_\theta &= 0 \\ \tau_{r\theta} &= \frac{2M\cos^2 2\theta}{\pi r^2}\end{aligned}\right\} \qquad \text{...(3.67)}$$

Second method. By removing a groove of radius r_0 from the apex of the wedge, the boundary forces on the groove are (Fig. 3.16).

$$q = -q_0\frac{\cos\theta}{r_0} \quad \text{where } q_0 = \text{constant}$$

The state of stress can be obtained from

$$\Phi = Cr\theta\sin\theta$$

giving

$$\sigma_r = 2C\frac{\cos\theta}{r}, \ \sigma_\theta = \tau_{r\theta} = 0$$

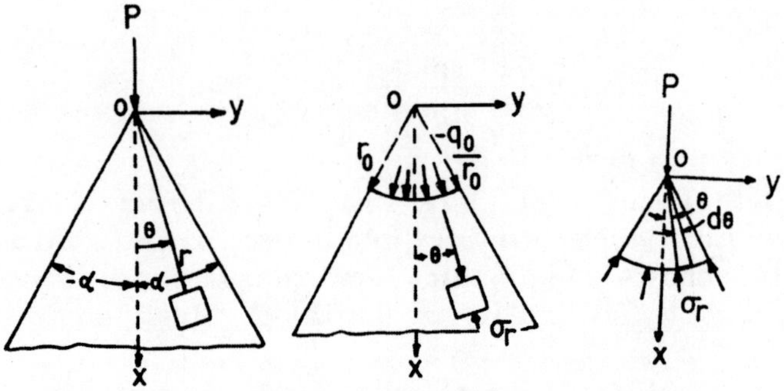

(a) (b) (c)

Fig. 3.16 Concentrated load on the vertex of a wedge.

These stresses satisfy all boundary conditions in the wedge. They vanish at infinity, leave the straight edges free and on the groove of radius r_0, they reduce to the assumed boundary forces q if C is taken equal to $\frac{-q_0}{2}$. The statical equivalent of the forces on the groove is a concentrated load parallel to the plane of symmetry of the wedge and acting at the apex.

The constant C can be determined from

$$\Sigma F_x = 0$$

$$\therefore \quad P - \int_{-\alpha}^{\alpha} \sigma_r \, . \cos\theta \, . \, dA = 0$$

$$P = \int_{-\alpha}^{\alpha} 2C \frac{\cos\theta}{r} \, . \cos\theta \, . \, r d\theta$$

$$= C\,(2\alpha + \sin 2\alpha)$$

$$\therefore \quad C = \frac{P}{2\alpha + \sin 2\alpha}$$

$$\therefore \quad \sigma_r = \frac{-P\cos\theta}{r(2\alpha + \sin 2\alpha)}, \; \sigma_\theta = \tau_{r\theta} = 0$$

If the concentrated load is acting perpendicular to the axis of symmetry then,

$$q = -q_0 \frac{\cos\theta_1}{r_0} \quad \text{where } \theta_1 = \theta - 90°$$

$$\therefore \quad \sigma_r = 2C \frac{\cos\theta_1}{r}, \; \sigma_\theta = \tau_{r\theta} = 0$$

$$C = \frac{P}{2\alpha - \sin 2\alpha}$$

$$\therefore \quad \sigma_r = -\frac{2P\cos\theta_1}{r(2\alpha - \sin 2\alpha)} = \frac{2P\sin\theta}{r(2\alpha - \sin 2\alpha)}$$

$$\sigma_\theta = \tau_{r\theta} = 0$$

In Cartesian co-orninates,

$$\sigma_x = \sigma_r \cos^2\theta$$

$$\sigma_y = \sigma_r \sin^2\theta$$

$$\tau_{xy} = \sigma_r \sin\theta\cos\theta$$

$$\therefore \quad \left.\begin{aligned} \sigma_x &= -\frac{2P\,x^2y}{(2\alpha - \sin 2\alpha)\,(x^2+y^2)^2} \\ \sigma_y &= -\frac{2Py^3}{(2\alpha - \sin 2\alpha)\,(x^2+y^2)^2} \\ \tau_{xy} &= \frac{2P\,xy^2}{(2\alpha - \sin 2\alpha)\,(x^2+y^2)^2} \end{aligned}\right\} \qquad \ldots(3.68)$$

3.12.2 Concentrated Load Acting on a Rectangular Beam

Consider a rectangular beam simply supported at the ends and carrying a concentrated load P at the mid-span as shown in Fig. 3.17. According to Euler-Bernouli theory of bending of beams, we have

$$\sigma_x = 0,\ \sigma_y = \frac{M(x-c)}{I},\ \tau_{xy} = \frac{VA\bar{y}}{It}$$

or in terms of P and l, and denoting these stresses by dashes, we get

$$\sigma'_x = 0,\ \sigma_y' = \frac{3P}{4tc^3}\,(l-y)\,(x-c)$$

$$\tau'_{xy} = \frac{-3P}{8tc^3}\,(2c-x)x$$

In this theory the effect of concentrated loads is neglected. The concentrated load introduces a radial set stresses emanating from the point of application of central region of the beam as if it were a part of a semi-infinite plate. These components of stress we denote by double dashes.

$$\sigma''_x = -\frac{2P}{\pi t}\cdot\frac{x^2}{(x^2+y^2)^2}$$

$$\sigma''_y = -\frac{2P}{\pi t}\,\frac{xy^2}{(x^2+y^2)^2}$$

$$\sigma''_{xy} = -\frac{2P}{\pi t}\,\frac{x^2y}{(x^2+y^2)^2}$$

Combining these stresses, we get

$$\left.\begin{aligned}\sigma_x &= \sigma_x' + \sigma_x'' = -\frac{2P}{t}\cdot\frac{x^2}{\pi(x^2+y^2)^2}\\ \sigma_y &= \sigma_y' + \sigma''_y = \frac{2P}{t}\left[\frac{3}{8c^2}(l-y)(x-c) - \frac{1}{\pi}\frac{xy^2}{(x^2+y^2)^2}\right]\\ \tau_{xy} &= \tau'_{xy} + \tau''_{xy} = -\frac{2P}{t}\left[\frac{3}{16c^3}(2cx-x^2) + \frac{1}{\pi}\frac{x^2y}{(x^2+y^2)^2}\right]\end{aligned}\right\} \quad ...(3.69)$$

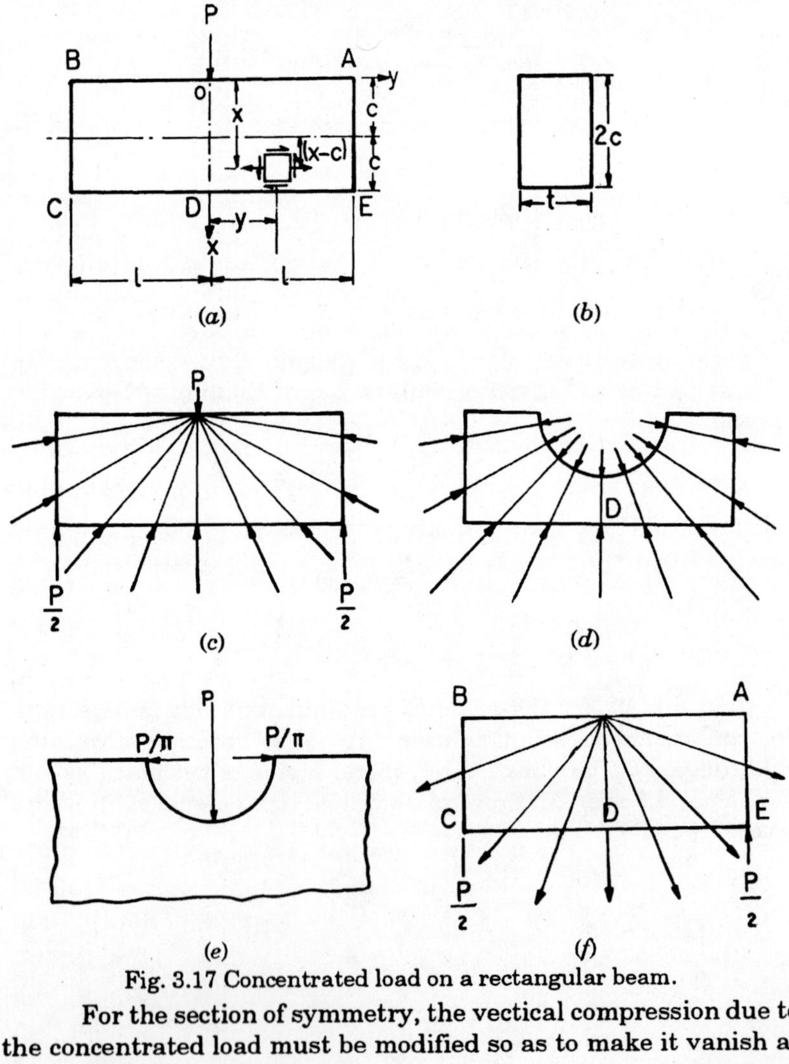

Fig. 3.17 Concentrated load on a rectangular beam.

For the section of symmetry, the vectical compression due to the concentrated load must be modified so as to make it vanish at point D of the lower fibre CE. The effect of the radial tension is, therefore, to reduce the vertical stresses. This correction is a linear function of x, so that

$$(\sigma_x)_{y=0} = -\frac{2P}{\pi t x} + kx$$

Since $$\left(\sigma_x\right)_{\substack{x=2c\\ y=0}} = 0$$

$$\therefore \qquad k = \frac{P}{2\pi t c^2}$$

Hence along the section of symmetry,

$$\sigma_x = -\frac{2P}{\pi t x} + \frac{Px}{2\pi t c^2} = \frac{2P}{\pi t}\left(\frac{x}{4c^2} - \frac{1}{x}\right) \qquad ...(3.70)$$

Now $$M = \frac{Pl}{2} - \frac{Pc}{\pi} = P\left(\frac{l}{2} - \frac{c}{\pi}\right)$$

Along the section of symmetry,

$$\sigma_y' = P\left(\frac{l}{2} - \frac{c}{\pi}\right)\left(\frac{x-c}{I}\right)$$

$$= \frac{3P}{2tc^3}\left(\frac{l}{2} - \frac{c}{\pi}\right)(x-c)$$

Since $\frac{P}{\pi}$ is an eccentric load, it produces additional tensile stress $= \frac{P}{2\pi c t}$.

$$\therefore \qquad (\sigma_x)_{y=0} = \frac{3}{2}\frac{P}{tc^2}\left(\frac{l}{2} - \frac{c}{\pi}\right)(x-c) + \frac{P}{2\pi c t} \qquad ...(3.71)$$

This is called the *Wilson-Stokes* solution.

3.12.3 Stresses in a Circular Disc under Diametral Compression

(*i*) Consider a circular disc of diameter D subjected to equal diametral compressive loads P. To determine the stresses in the disc, the following procedure will be followed :

(*a*) **Radial Forces** [Fig. 3.18 (*a*)]

Consider a disc of diameter D subjected to uniform radial tension of intensity q. The stress function for this case is

$$\Phi = cr^2 \qquad ...(1)$$

which satisfies the biharmonic equation,

$$\sigma_r = 2c,\ \sigma_\theta = 2c,\ \tau_{r\theta} = 0$$

Boundary condition, $\sigma_r = q$ at $r = \dfrac{D}{2}$ gives,

$$c = \frac{q}{2}$$

Hence $\quad \sigma_r = q,\ \sigma_\theta = q,\ \tau_{r\theta} = 0 \quad$ ...(2)

and $\quad \sigma_x = \sigma_y = q,\ \tau_{xy} = 0$

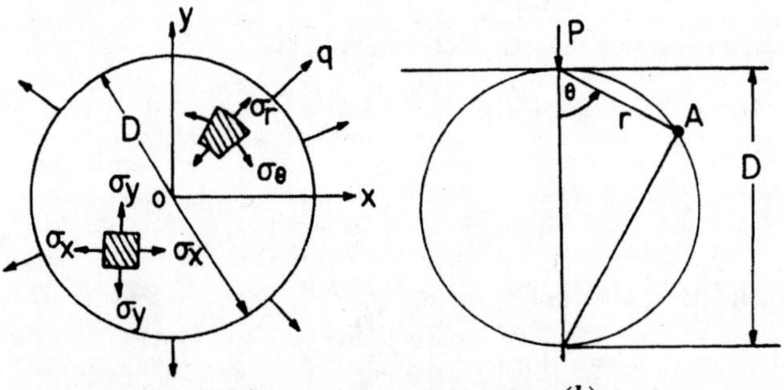

Fig. 3.18 Circular disc under diametral compression.

(b) Concentrated Diametral Load

Consider now the Flemant solution for the concentrated load of intensity P acting on the surface of a plate as shown in Fig. 3.18 (b) giving,

$$\sigma_r = -\frac{2P}{\pi}\frac{\cos\theta}{r}$$

$$\sigma_\theta = \tau_{r\theta} = 0$$

On the circle of diameter D

$$\frac{r}{\cos\theta} = D$$

$$\sigma_r = -\frac{2P}{\pi D}$$

which is constant at each point of the circle. To isolate the disc of diameter D inside the plate, a stress q_1 must be applied at each point of the surface of disc to equilibrate the stress σ_r [Fig. 3.19 (b)].

$$q_1 . \overline{AB} = \sigma_r . \overline{AC}$$

Since $OA \perp AB$ and $O_1A \perp AC$, $\angle CAB = \theta$ and

$$q_1 = \sigma_r . \frac{\overline{AC}}{\overline{AB}} = \sigma_r \cos\theta = -\frac{2P}{\pi D}\cos\theta$$

Consider now the cases of two diametrically applied forces acting on the disc. At each point A on the surface of the disc, two stresses

$$q' = -\frac{2P}{\pi}\frac{\cos\theta_1}{D} \quad \text{and} \quad q'' = -\frac{2P}{\pi}\frac{\cos\theta_2}{D}$$

must be applied as shown in Fig. 3.19 (*c*).

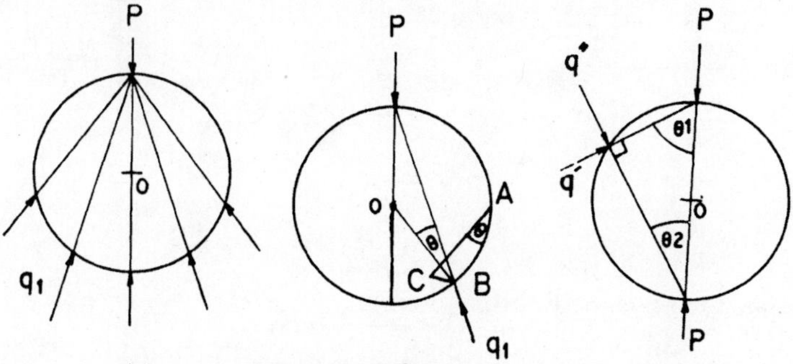

Fig. 3.19

$$\theta_2 = 90° - \theta_1$$

$$\therefore \qquad \cos\theta_2 = \sin\theta_1$$

$$\therefore \qquad q'' = -\frac{2P}{\pi}\frac{\sin\theta_1}{D}$$

$$q = -\sqrt{q'^2 + q''^2}$$

$$= -\frac{2P}{\pi D}\sqrt{\cos\theta_1{}^2 + \sin\theta_1{}^2} = -\frac{2P}{\pi D}$$

which is a constant radial compressive stress of magnitude $q = -\dfrac{2P}{\pi D}$. Thus the disc is subjected to two concentrated forces P and the distributed force q.

The state of stress in the disc is the resultant of the following states of stress (Fig. 3.20) :

$$\sigma'_r = -\frac{2P}{\pi}\frac{\cos\theta_1}{r_1}, \; \sigma'_\theta = \tau'_{r\theta} = 0 \qquad \text{...(3)}$$

$$\sigma''_r = -\frac{2P}{\pi}\frac{\cos\theta_2}{r_2}, \; \sigma''_\theta = \tau''_{r\theta} = 0 \qquad \text{...(4)}$$

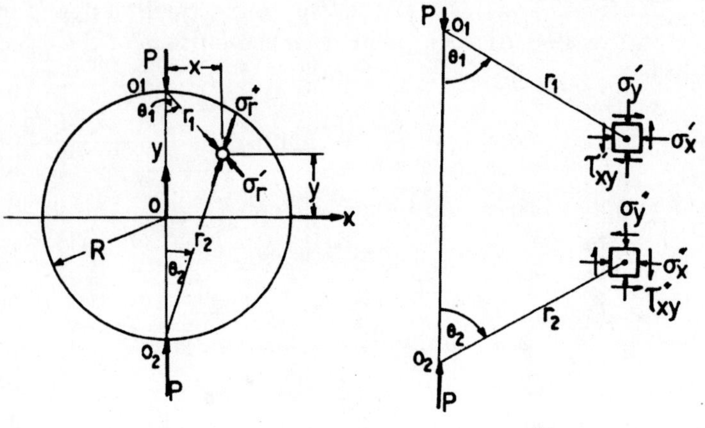

(a) (b)

Fig. 3.20

or in Cartesian co-ordinates

$$\left.\begin{aligned}\sigma'_x &= -\frac{2P}{\pi}\frac{\cos\theta_1\sin^2\theta_1}{r_1}\\ \sigma'_y &= -\frac{2P}{\pi}\frac{\cos^3\theta_1}{r_1}\\ \tau'_{xy} &= \frac{2P}{\pi}\frac{\cos^2\theta_1\sin\theta_1}{r_1}\end{aligned}\right\} \quad \text{...(5)}$$

and

$$\left.\begin{aligned}\sigma''_x &= -\frac{2P}{\pi}\frac{\cos\theta_2\sin^2\theta_2}{r_2}\\ \sigma''_y &= -\frac{2P}{\pi}\frac{\cos^3\theta_2}{r_2}\\ \tau''_{xy} &= -\frac{2P}{\pi}\frac{\cos^2\theta_2\sin\theta_2}{r_2}\end{aligned}\right\} \quad \text{...(6)}$$

If an additional uniform tension $\frac{2P}{\pi D}$ is applied to the circular boundary, then the disc will be free of externally distributed forces and subjected only to the concentrated forces P. For this

$$\sigma''_x = \sigma''_y = \frac{2P}{\pi D}, \tau''_{xy} = 0 \quad \text{...(7)}$$

By adding Eqs. (5) to (7), we get

$$\left.\begin{aligned}\sigma_x &= -\frac{2P}{\pi}\left[\frac{\cos\theta_1\sin^2\theta_1}{r_1}+\frac{\cos\theta_2\sin^2\theta_2}{r_2}-\frac{1}{D}\right]\\ \sigma_y &= -\frac{2P}{\pi}\left[\frac{\cos^3\theta_1}{r_1}+\frac{\cos^3\theta_2}{r_2}-\frac{1}{D}\right]\\ \tau_{xy} &= \frac{2P}{\pi}\left[\frac{\cos^2\theta_1\sin\theta_1}{r_1}-\frac{\cos^2\theta_2\sin\theta_2}{r_2}\right]\end{aligned}\right\}\quad ...(8)$$

Now $\sin\theta_1 = \dfrac{x}{r_1}$, $\cos\theta_1 = \dfrac{R-y}{r_1}$

$\sin\theta_2 = \dfrac{x}{r_2}$, $\cos\theta_2 = \dfrac{R+y}{r_2}$

where $r_1^2 = x^2 + (R-y)^3$

$r_2^2 = x^2 + (R+y)^2$, $R = D/2$

$$\therefore\quad \left.\begin{aligned}\sigma_x &= -\frac{2P}{\pi}\left[\frac{(R-y)\,x^2}{r_1^4}+\frac{(R+y)x^2}{r_2^4}-\frac{1}{D}\right]\\ \sigma_y &= -\frac{2P}{\pi}\left[\frac{(R-y)^3}{r_1^4}+\frac{(R+y)^3}{r_2^4}-\frac{1}{D}\right]\\ \tau_{xy} &= \frac{2P}{\pi}\left[\frac{(R-y)^2x}{r_1^4}-\frac{(R+y)^2x}{r_2^4}\right]\end{aligned}\right\}\quad ...(9)$$

On the x-axis, $y = 0$,

$$r_1 = r_2 = \sqrt{x^2+R^2}$$

$$\therefore\quad \left.\begin{aligned}\sigma_x &= \frac{2P}{\pi D}\left[\frac{D^2-4x^2}{D^2+4x^2}\right]^2\\ \sigma_y &= -\frac{2P}{\pi D}\left[\frac{4D^2}{(D^2+4x^2)^2}-1\right]\end{aligned}\right\}\quad ...(3.72)$$

$$\tau_{xy} = 0$$

For $\quad x = \pm\dfrac{D}{2}$, $\sigma_x = \sigma_y = 0$

At the centre of the disc

$$\left.\begin{aligned}(\sigma_x)_{x=y=0} &= \frac{2P}{\pi D}\\ (\sigma_y)_{x=y=0} &= -\frac{6P}{\pi D}\end{aligned}\right\}\quad ...(3.73)$$

On the y-axis, $x = 0, r_1 = R - y, r_2 = R + y$

$$\therefore \quad \left.\begin{aligned} \sigma_x &= \frac{2P}{\pi D} \\ \sigma_y &= -\frac{2P}{\pi}\left[\frac{2}{D-2y}+\frac{2}{D+2y}-\frac{1}{D}\right] \\ \tau_{xy} &= 0 \end{aligned}\right\} \quad \ldots(3.74)$$

For $\quad y = \pm\frac{D}{2}, \sigma_y \to \infty$

at $\quad x = y = 0, \sigma_y = -\frac{6P}{\pi D}$

(*ii*) Now consider the circular disc loaded along a chord O_1O_2 as shown in Fig. 3.21 (*a*).

$$\sigma_r = -\frac{2P\cos\theta}{\pi r}, \sigma_\theta = 0, \tau_{r\theta} = 0$$

At any point A on the circumference of the circle, there are two radial stresses σ'_r and σ''_r acting along the radii vector r_1 and r_2 respectively.

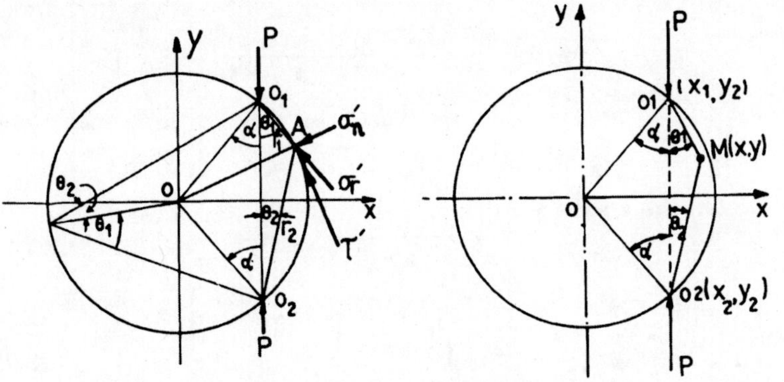

(*a*) (*b*)

Fig. 3.21

$$\sigma''_n = \sigma'_r \sin^2\theta_2, \tau' = \frac{\sigma'_r}{2}\sin 2\theta_2$$

$$\sigma''_n = \sigma''_r \sin^2\theta_1, \tau'' = \frac{\sigma''_r}{2}\sin 2\theta_1$$

where $\quad \sigma'_r = -\frac{2P\cos\theta_1}{\pi r_1}, \sigma''_r = -\frac{2P\cos\theta_2}{\pi r_2}$

Now $\quad r_1 = D\sin\theta_2, r_2 = D\sin\theta_1$

$$\sigma_n = \sigma'_n = \sigma''_n$$

$$= -\frac{2P}{\pi D}\left[\cos\theta_1 \sin\theta_2 + \cos\theta_2 \sin\theta_1\right]$$

$$= -\frac{2P}{\pi D}\sin(\theta_1 + \theta_2) \qquad \text{...(3.75)}$$

On the circular boundary

$$\theta_1 + \theta_2 = \frac{\pi}{2} \pm \alpha$$

depending on whether the point A is on the arc O_1O_2 which is less than a semi-circle or on the arc O_1O_2 which is greater than a semi-circle.

Since $\quad \sin\left(\frac{\pi}{2} + \alpha\right) = \sin\left(\frac{\pi}{2} - \alpha\right)$

$$\therefore \qquad \sigma_n = -\frac{2P}{\pi D}\sin\left(\frac{\pi}{2} + \alpha\right) \qquad \text{...(3.76)}$$

Now $\quad \tau = \tau' - \tau''$

$$= -\frac{2P}{\pi D}(\cos\theta_1 \cos\theta_2 - \cos\theta_2 \cos\theta_1) = 0$$

The stresses at any point in the disc are thus obtained by combining two radial stresses σ'_r and σ''_r with an isotropic tension of magnitude $\frac{2P}{\pi D}\sin\left(\frac{\pi}{2} + \alpha\right)$

In Cartesian co-ordinates

$$\left.\begin{aligned}
\sigma_x &= \frac{-2P}{\pi}\left[\frac{\cos\theta_1 \cos^2\theta_1}{r_1} + \frac{\cos\theta_2 \sin^2\theta_2}{r_2} - \frac{\sin\left(\frac{\pi}{2} + \alpha\right)}{D}\right] \\
\sigma_y &= \frac{-2P}{\pi}\left[\frac{\cos^3\theta_1}{r_1} + \frac{\cos^3\theta_2}{r_2} - \frac{\sin\left(\frac{\pi}{2} + \alpha\right)}{D}\right] \\
\tau_{xy} &= \frac{2P}{\pi}\left[\frac{\cos^2\theta_1 \sin\theta_1}{r_1} - \frac{\cos^2\theta_2 \sin\theta_2}{r_2}\right]
\end{aligned}\right\} \qquad \text{...(3.77)}$$

Let (x_1, y_1), (x_2, y_2) be the co-ordinates of points O_1 and O_2 respectively and let (x, y) be the co-ordinates of any point M in the disc as shown in Fig. 3.21 (*b*). Then

$$x_1 = x_2, \quad y_1 = -y_2$$

$$\sin\theta_1 = \frac{x - x_1}{r_1}$$

$$\cos\theta_1 = \frac{y_1 - y}{r_1}$$

$$\sin\theta_2 = \frac{x - x_1}{r_2}$$

$$\cos\theta_2 = \frac{y - y_2}{r_2} = \frac{y + y_1}{r_2}$$

$$\sigma_x = \frac{-2P}{\pi}\left[\frac{(y_1 - y)(x - x_1)^2}{r_1^4} + \frac{(y_1 + y)(x - x_1)^2}{r_2^4} - \frac{\sin\left(\frac{\pi}{2} + \alpha\right)}{D}\right]$$

$$\sigma_y = \frac{-2P}{\pi}\left[\frac{(y_1 - y)^3}{r_1^4} + \frac{(y_1 + y)^3}{r_2^4} - \frac{\sin\left(\frac{\pi}{2} + \alpha\right)}{D}\right]$$

$$\tau_{xy} = \frac{2P}{\pi}\left[\frac{(y_1 - y)^2(x - x_1)}{r_1^4} - \frac{(y_1 + y)^2(x - x_1)}{r_2^4}\right]$$

where $r_1^2 = (x - x_1)^2 + (y_1 - y)^2$

$$r_2^2 = (x - x_1)^2 + (y - y_2)^2$$

$$= (x - x_1)^2 + (y + y_1)^2$$

Along the x-axis, $r_1 = r_2 = r$, $y = 0$

$$\left.\begin{aligned} \sigma_x &= \frac{-2P}{\pi}\left[\frac{2y_1(x - x_1)^2}{r^4} - \frac{\sin\left(\frac{\pi}{2} + \alpha\right)}{D}\right] \\ \sigma_y &= \frac{-2P}{\pi}\left[\frac{2y_1^3}{r^4} - \frac{\sin\left(\frac{\pi}{2} + \alpha\right)}{D}\right] \\ \tau_{xy} &= 0 \end{aligned}\right\} \quad \text{...(3.78)}$$

Thus for any number of concentrated loaded on the boundary,

$$\sigma_r = -\sum_{i=1}^{n} \frac{2P_i}{\pi D}\frac{\cos\theta_i}{r_i}$$

and adding to it an isotropic tension given by :

$$\sum_{i=1}^{n} \frac{2P_i}{\pi D} \sin(\theta_1 + \theta_2)$$

we can get the stresses at any point in the disc.

3.13 Bending of a Curved Bar by a Force at the End

Consider a bar of narrow rectangular cross-section and with a circular axis which is constrained at the lower end bent by a force P applied at the upper end in the radial direction as shown in Fig. 3.22.

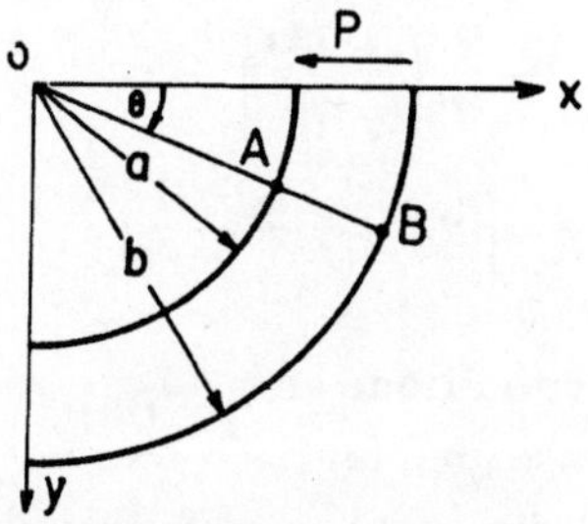

Fig. 3.22 Bending of a curved bar by end force.

The bending moment at any cross-section AB is proportional to $\sin\theta$ and the normal stress σ_θ, according to the elementary theory of bending of curved bars is proportional to the bending moment. Assuming that the stress function Φ, satisfying the biharmonic equation

$$\nabla^4\Phi = \left(\frac{\partial^2}{\partial r^2} + \frac{1}{r}\frac{\partial}{\partial r} + \frac{1}{r^2}\frac{\partial^2}{\partial \theta^2}\right)\left(\frac{\partial^2\Phi}{\partial r^2} + \frac{1}{r}\frac{\partial\Phi}{\partial r} + \frac{1}{r^2}\frac{\partial^2\Phi}{\partial \theta^2}\right)$$

$$= 0 \qquad \text{...(1)}$$

should be proportional to $\sin\theta$. Taking

$$\Phi = f(r)\sin\theta \qquad \text{...(2)}$$

and substituting in Eq. (1), we find that $f(r)$ must satisfy the following ordinary differential equation.

$$\left(\frac{d^2}{dr^2} + \frac{1}{r}\frac{d}{dr} - \frac{1}{r^2}\right)\left(\frac{d^2f}{dr^2} + \frac{1}{r}\frac{df}{dr} - \frac{f}{r^2}\right) = 0 \qquad \text{...(3)}$$

The general solution of Eq. (3) may be written as :

$$f(r) = Ar^3 + \frac{B}{r} + Cr + Dr\log r \qquad \text{...(4)}$$

where A, B, C and D are constants, which are determined from the boundary conditions..Hence Eq. (2) becomes,

$$\Phi = \left(Ar^3 + \frac{B}{r} + Cr + Dr \log r \right) \sin \theta \qquad ...(5)$$

The stress components become

$$\left.\begin{aligned} \sigma_r &= \frac{1}{r}\frac{\partial \Phi}{\partial r} + \frac{1}{r^2}\frac{\partial^2 \Phi}{\partial \theta^2} \\ &= \left(2Ar - \frac{2B}{r^3} + \frac{D}{r} \right) \sin \theta \\ \sigma_\theta &= \frac{\partial^2 \Phi}{\partial r^2} = \left(6Ar - \frac{2B}{r^3} + \frac{D}{r} \right) \sin \theta \\ \tau_{r\theta} &= -\frac{\partial}{\partial r}\left(\frac{1}{r}\frac{\partial \Phi}{\partial \theta} \right) \\ &= -\left(2Ar - \frac{2B}{r^3} + \frac{D}{r} \right) \cos \theta \end{aligned}\right\} \qquad ...(6)$$

The boundary conditions are

$$\sigma_r = \tau_{r\theta} = 0 \text{ for } r = a \text{ and } r = b$$

Substituting in Eqs. (6), we get

$$\left.\begin{aligned} 2Aa - \frac{2B}{a^3} + \frac{D}{a} &= 0 \\ 2Ab - \frac{2B}{b^3} + \frac{D}{b} &= 0 \end{aligned}\right\} \qquad ...(7)$$

Taking the width of the bar as unity, we obtain for $\theta = 0$

$$\int_a^b \tau_{r\theta}\, dr = -\int_a^b \frac{\partial}{\partial r}\left(\frac{1}{r}\frac{\partial \Phi}{\partial \theta} \right) dr = \left| \frac{1}{r}\frac{\partial \Phi}{\partial \theta} \right|_b^a$$

$$= \left| Ar^2 + \frac{B}{r^2} + C + D \log r \right|_a^b = P$$

$$-A(b^2 - a^2) + B\left(\frac{b^2 - a^2}{a^2 b^2} \right) - D \log \frac{b}{a} = P \qquad ...(8)$$

Solving Eqs. (7) and (8), we get

$$A = \frac{P}{2K},\ B = -\frac{Pa^2b^2}{2K}$$

$$D = -\frac{P}{K}(a^2 + b^2) \qquad ...(9)$$

where $K = a^2 - b^2 + (a^2 - b^2) \log \dfrac{b}{a}$.

Substituting the values of constants in Eq. (6), we get for the upper end where $\theta = 0$,

$$\sigma_r = \sigma_\theta = 0$$

$$\tau_{r\theta} = -\frac{P}{K}\left[r + \frac{a^2b^2}{r^3} - \frac{1}{r}(a^2 + b^2) \right] \qquad ...(3.79)$$

For the lower end, $\theta = \frac{\pi}{2}$

$$\left.\begin{aligned} \sigma_r &= \frac{P}{K}\left[r + \frac{a^2b^2}{r^3} - \frac{a^2 + b^2}{r} \right] \\ \sigma_\theta &= \frac{P}{K}\left[3r - \frac{a^2b^2}{r^3} - \frac{a^2 + b^2}{r} \right] \\ \tau_{r\theta} &= 0 \end{aligned}\right\} \qquad ...(3.80)$$

To determine the displacements, we have

$$\left.\begin{aligned} \frac{\partial u_r}{\partial r} &= \frac{\sin\theta}{E}\left[2Ar(1 - 3\nu) - \frac{2B}{r^3}(1 + \nu) + \frac{D}{r}(1 - \nu) \right] \\ \frac{\partial u_\theta}{\partial \theta} &= r\,\varepsilon_\theta - u_r \\ \gamma_{r\theta} &= \frac{1}{r}\frac{\partial u_r}{\partial \theta} + \frac{\partial u_\theta}{\partial r} - \frac{u_\theta}{r} \end{aligned}\right\} \qquad ...(10)$$

Integrating first of Eq. (10), we get

$$u_r = \frac{\sin\theta}{E}\left[Ar^2(1 - 3\nu) + \frac{B}{r^2}(1 + \nu) + D(1 - \nu)\log r \right] + f(\theta) \qquad ...(11)$$

where $f(\theta)$ is a function of θ only. Substituting Eq. (11) in second of Eqs. (10), we find

$$u_\theta = -\frac{\cos\theta}{E}\left[Ar^2(5 + \nu) + \frac{B}{r^2}(1 + \nu) - D\log r(1 - \nu) + D(1 - \nu) \right] - \int f(\theta)\,d\theta + F(r) \qquad ...(12)$$

where $F(r)$ is a function of r only. Now substituting Eqs. (11) and (12) in third of Eqs. (10) we get

$$\int f(\theta)\,d\theta + f'(\theta) + r F'(r) - F(r) = -\frac{4D\cos\theta}{E}$$

$$\therefore \quad \int f(\theta)\, d\theta + f'(\theta) = -\frac{4D\cos\theta}{E}$$

$$rF'(r) - F(r) = 0$$

The solution of these equations are

$$\left.\begin{aligned} f(\theta) &= -\frac{2D}{E}\theta\cos\theta + H\sin\theta + J\cos\theta \\ F(r) &= Nr \end{aligned}\right\} \quad ...(13)$$

where H, J and N are arbitrary constants. Eqs. (11) and (12) become

$$u_r = -\frac{2D}{E}\theta\sin\theta + \frac{\sin\theta}{E}\left[D(1-\nu)\log r + A(1-3\nu)r^2 + \frac{B(1+\nu)}{r^2}\right] + H\sin\theta + J\cos\theta \quad ...(14)$$

$$u_\theta = \frac{2D}{E}\theta\sin\theta - \frac{\cos\theta}{E}\left[A(5+\nu)r^2 + \frac{B(1+\nu)}{r^2} - D(1-\nu)\log r\right] + \frac{D(1+\nu)}{E}\cos\theta + H\cos\theta - J\sin\theta + Nr \quad ...(15)$$

At the upper end, $\theta = 0$

$$\therefore \quad (u_r)_{\theta=0} = J$$

For $\theta = \frac{\pi}{2}$, we have $u_\theta = 0$, $\frac{\partial u_\theta}{\partial r} = 0$

Hence from Eq. (15), we get

$$\therefore \quad N = 0,\ J = \frac{\pi D}{E} \quad ...(16)$$

$$\therefore \quad (u_r)_{\theta=0} = \frac{\pi D}{E} = \frac{\pi P(a^2+b^2)}{E\left[(a^2-b^2) + (a^2+b^2)\log\frac{b}{a}\right]} \quad ...(3.81)$$

Let $b \to a$ and $h = (b - a)$ is small as compared to a, then

$$\log\frac{b}{a} = \log\left(1 + \frac{h}{a}\right) = \frac{h}{a} - \frac{h^2}{2a^2} + \frac{1}{3}\frac{h^3}{a^3} + \ldots..$$

$$\therefore \quad (u_r)_{\theta=0} = -\frac{3\pi a^3 P}{Eh^2} \quad ...(3.82)$$

3.14 Integral Transform Method

In polar co-ordinates, the equilibrium equations in the absence of body forces are :

$$\left.\begin{aligned}\frac{\partial\sigma_r}{\partial r}+\frac{1}{r}\frac{\partial\tau_{r\theta}}{\partial\theta}&=0\\\frac{\partial\tau_{r\theta}}{\partial r}+\frac{1}{r}\frac{\partial\sigma_\theta}{\partial\theta}&=0\end{aligned}\right\}\qquad ...(3.83)$$

If we express the components of stress in terms of an Airy stress function Φ by the relations.

$$\left.\begin{aligned}\sigma_r&=\frac{1}{r}\frac{\partial\Phi}{\partial r}+\frac{1}{r^2}\frac{\partial^2\Phi}{\partial\theta^2}\\\tau_\theta&=\frac{\partial^2\Phi}{\partial r^2}\\\tau_{r\theta}&=-\frac{\partial}{\partial r}\left(\frac{1}{r}\frac{\partial\Phi}{\partial\theta}\right)\end{aligned}\right\}\qquad ...(3.84)$$

The biharmonic equation is

$$\left(\frac{\partial^2}{\partial r^2}+\frac{1}{r}\frac{\partial}{\partial r}+\frac{\partial^2}{\partial\theta^2}\right)^2\Phi=0 \qquad ...(3.85)$$

To solve this equation, we introduce the Mellin transform

$$\overline{\Phi}=\int_0^\infty r^{s-1}\,\Phi\,dr \qquad ...(3.86)$$

Now $$\int_0^\infty \frac{\partial^n\Phi}{dr^n}\,r^{s+n-1}\,dr=(-1)^n\,\frac{\Gamma(s+n)}{\Gamma(s)}\,\overline{\Phi} \qquad ...(3.87)$$

where Γ stands for the gamma function.

Eq. (3.85) thus can be written as :

$$\left(\frac{d^2}{d\theta^2}+s^2\right)\left[\frac{d^2}{d\theta^2}+(s+2)^2\right]\overline{\Phi}=0$$

whose general solution is

$$\overline{\Phi}=A\sin(s\theta)+B\cos(s\theta)+C\sin[(s+2)\theta]+D\cos[(s+2)\theta] \qquad ...(3.88)$$

The arbitrary constants A, B, C and D may be determined from the prescribed boundary conditions.

If we multiply throughout Eqs. (3.89) be r^{s+1} and integrate w.r.t. r from 0 to ∞, we find, on making use of Eq. (3.87), that

$$\left.\begin{aligned} \int_0^\infty (r^2\,\sigma_r)\, r^{s-1}\,dr &= \left(\frac{d^2}{d\theta^2} - s\right)\overline{\Phi} \\ \text{and} \qquad \int_0^\infty (r^2\,\sigma_\theta)\, r^{s-1}\,dr &= s(s+1)\,\overline{\Phi} \\ \int_0^\infty (r^2\,\tau_{r\theta})\, r^{s-1}\,dr &= (s+1)\frac{d\overline{\Phi}}{d\theta} \end{aligned}\right\} \qquad ...(3.89)$$

Once the constants A, B, C, D have been determined from these equations, the function $\overline{\Phi}$ may be written down from Eq. (3.88) and the components of stress determined by inverting Eqs. (3.89) by means of the result that

$$\left.\begin{aligned} \text{if} \qquad G(s) &= \int_0^\infty \xi^{s-1}\, g(\xi)\,d\xi \\ \text{then} \qquad g(\xi) &= \frac{1}{2\pi i}\int_{C-i\infty}^{C+i\infty} G(s)\xi^{-s}\,ds \end{aligned}\right\} \qquad ...(3.90)$$

where $C > 0$.

Thus, we get

$$\left.\begin{aligned} \sigma_r &= \frac{1}{2\pi i}\int_{C-i\infty}^{C+i\infty} \left(\frac{d^2\overline{\Phi}}{d\theta^2} - s\,\overline{\Phi}\right) r^{-s-2}\,ds \\ \sigma_\theta &= \frac{1}{2\pi i}\int_{C-i\infty}^{C+i\infty} s(s+1)\,\overline{\Phi}\, r^{-s-2}\,ds \\ \tau_{r\theta} &= \frac{1}{2\pi i}\int_{C-i\infty}^{C+i\infty} (s+1)\frac{d\overline{\Phi}}{d\theta}\, r^{-s-2}\,ds \end{aligned}\right\} \qquad ...(3.91)$$

3.14.1 Distribution of Stress in an Infinite Wedge

Consider an infinite wedge which is bounded by the lines $\theta = \pm \alpha$ as shown in Fig. 3.23.

Suppose that the face $\theta = \alpha$ of the wedge is subjected to the surface stresses

$$\sigma_\theta = f_1(r), \quad \tau_{r\theta} = g_1(r) \qquad ...(3.92)$$

and the face $\theta = -\alpha$ to the stresses

$$\sigma_\theta = f_2(r), \quad \tau_{r\theta} = g_2(r) \qquad ...(3.93)$$

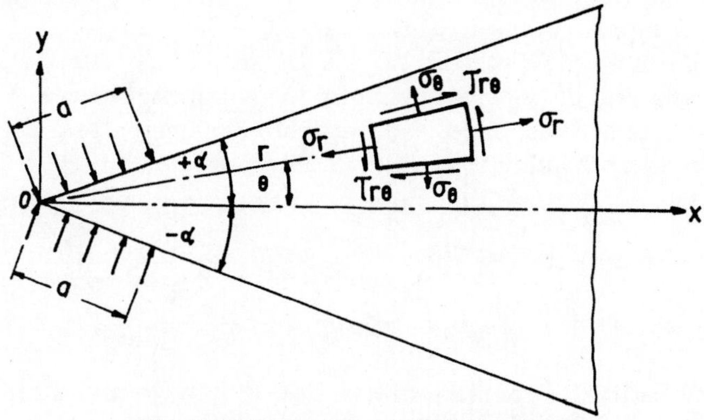

Fig. 3.23 Stress in an infinite wedge.

Substituting these conditions into Eqs. (3.89), we get

$$\left.\begin{array}{l} s(s+1)\,\bar{\Phi} = \bar{f}_1(s),\ (s+1)\dfrac{d\bar{\Phi}}{d\theta} = \bar{g}_1(s),\ \theta = \alpha \\[2ex] s(s+1)\,\bar{\Phi} = \bar{f}_2(s),\ (s+1)\dfrac{d\bar{\Phi}}{d\theta} = \bar{g}_2(s),\ \theta = -\alpha \end{array}\right\} \quad ...(3.94)$$

where $\bar{f}_i, \bar{g}_i$ $(i = 1, 2)$ denote the Mellin transforms of $r^2 f_i(r)$ and $r^2 g_i(r)$, so that

$$(\bar{f}_i, \bar{g}_i) = \int_0^\infty (f_i, g_i)\, r^{s+1}\, dr,\ i = 1, 2 \qquad ...(3.95)$$

Substituting Eq. (3.88) in Eqs. (3.90), we get four simultaneous equations for the four constants A, B, C and D whose solution may be written in the form,

$$2s(s+1)\,G(\alpha, s)A = s(\bar{g}_1 + \bar{g}_2)\sin[(s+2)\alpha] - (s+2)(\bar{f}_1 - \bar{f}_2)\cos[(s+2)\alpha]$$

$$2(s+1)\,G(\alpha, s)\,C = -(\bar{g}_1 + \bar{g}_2)\sin(s\alpha) + (\bar{f}_1 - \bar{f}_2)\cos(s\alpha)$$

$$2s(s+1)\,H(\alpha, s)B = s(\bar{g}_1 - \bar{g}_2)\cos[(s+2)\alpha] + (s+2)(\bar{f}_1 + \bar{f}_2)\sin[(s+2)\alpha]$$

$$2(s+1)\,H(\alpha, s)\,D = -(\bar{g}_1 - \bar{g}_2)\cos(s\alpha) - (\bar{f}_1 + \bar{f}_2)\sin(s\alpha)$$

...(3.96)

where $G(\alpha, s) = (s+1)\sin 2\alpha - \sin[2(s+1)\alpha]$

$$H(\alpha, s) = (s+1)\sin 2\alpha + \sin[2(s+1)\alpha] \qquad ...(3.97)$$

The stress components can then be determined by means of Eqs. (3.91) by knowing $\bar{\Phi}$ from Eq. (3.88).

As a particular case, let the faces of the wedge be subjected to a uniform normal pressure p_0 for a distance 'a' measured from the vertex, the rest of the faces being free from normal stress. Also we assume that both the faces are free from shearing stress. In these circumstances it follows from Eq. (3.95) that

$$\bar{g}_1(s) = \bar{g}_2(s) = 0$$

$$\bar{f}_1(s) = \bar{f}_2(s)$$

$$= -p_0 \int_0^a r^{s+1}\, dr = -p_0 \frac{a^{s+2}}{s+2}$$

Substituting these results in Eqs. (3.96), we find that

$$A = C = 0$$

and B and D are given by

$$s(s+1) \quad H(\alpha, s)B = -p_0\, a^{s+2} \sin[(s+2)\alpha]$$

$$(s+2)(s+1)\, H(\alpha, s)\, D = p_0\, a^{s+2} \sin(s\alpha) \qquad \ldots(3.98)$$

Thus stress components from Eqs. (3.91) can be obtained by knowing $\bar{\Phi}$ from Eq. (3.88) and using (3.98)

$$\sigma_\theta = -\frac{p_0}{2\pi i} \int_{c-i\infty}^{c+i\infty} \left(\frac{a}{r}\right)^{s+2} \Big\{ \sin[(s+2)\alpha] \cos(s\theta) - \left(\frac{s}{s+2}\right) \sin(s\alpha) \cos[(s+2)\theta] \Big\} \frac{ds}{H(\alpha, s)} \qquad \ldots(3.99\, a)$$

$$\sigma_r = -\frac{p_0}{2\pi i} \int_{c-i\infty}^{c+i\infty} \left(\frac{a}{r}\right)^{s+2} \times \Big\{ \left(\frac{s+4}{s+2}\right) \sin(s\alpha) \cos[(s+2)\theta] - \sin[(s+2)\alpha] \cos(s\alpha) \Big\} \frac{ds}{H(\alpha, s)} \qquad \ldots(3.99\, b)$$

$$\tau_{r\theta} = \frac{p_0}{2\pi i} \int_{C-i\infty}^{C+i\infty} \left(\frac{a}{r}\right)^{s+2} \Big\{ \sin[(s+2)\alpha] \sin(s\theta) - \sin(s\alpha) \sin[(s+2)\theta] \Big\} \frac{ds}{H(\alpha, s)} \qquad \ldots(3.99\, c)$$

For α to lie between 0 and $\frac{\pi}{2}$, the only zero of the function $H(\alpha, s)$ is at $s = -1$. The contour integrals in the Eqs. (3.99) can, therefore, be replaced by integrals from $-\infty$ to 0 and from 0 to ∞ minus πi times the residue at the point $s = -1$. Thus

$$\frac{\pi r}{2p_0 a}(\sigma_\theta - \sigma_r) = \frac{\pi \sin \alpha \cos \theta}{2\alpha + \sin 2\alpha} - \int_0^\infty P(u) \sin\left[u \log\left(\frac{a}{r}\right)\right] du \qquad \text{...(3.100 a)}$$

$$\frac{\pi r}{2p_0 a}(\sigma_\theta + \sigma_r) = -\frac{\pi \sin \alpha \cos \theta}{2\alpha + \sin 2\alpha} + \int_0^\infty \left\{(P - uQ) \times \sin\left[u \log\left(\frac{a}{r}\right)\right] - (Q + uP) \cos\left[u \log\left(\frac{a}{r}\right)\right]\right\} \frac{du}{1 + u^2} \qquad \text{...(3.100 b)}$$

$$\frac{\pi r}{p_0 a}\tau_{r\theta} = \int_0^\infty R(u) \cos\left[u \log\left(\frac{a}{r}\right)\right] du \qquad \text{...(3.100 c)}$$

where the functions P, Q and R are given by

$$(u \sin 2\alpha + \sinh 2\alpha u)P(u) = \sin(\alpha - \theta)\cosh[(\alpha + \theta)u] + \sin(\alpha + \theta)\cosh[(\alpha + \theta)u]$$

$$(u \sin 2\alpha + \sinh 2\alpha u)\,Q(u) = \cos(\alpha - \theta)\sinh[(\alpha + \theta)u] + \cos(\alpha + \theta)\sinh[(\alpha - \theta)u]$$

$$(u \sin 2\alpha + \sinh 2\alpha u)\,R(u) = \sin(\alpha - \theta)\sinh[(\alpha + \theta)u] - \sin(\alpha + \theta)\sinh[(\alpha + \theta)u] \qquad \text{...(3.101)}$$

For $\alpha = \frac{\pi}{2}$, the wedge becomes a semi-infinite solid, and the integrals become,

$$\int_0^\infty \frac{\cosh(px)}{\sinh(\frac{1}{2}\pi x)} \sin(mx)\,dx = \frac{\sin(2m)}{\cos(2p) + \cosh(2m)}$$

$$\int_0^\infty \frac{\sinh(px)}{\sinh(\frac{1}{2}\pi x)} \cos(mx)\,dx = \frac{\sin(2p)}{\cos(2p) + \cosh(2m)}$$

provided that $4p^2 \leq \pi^2$ ...(3.102)

Hence the stresses become

$$\left.\begin{aligned}
\sigma_\theta - \sigma_r &= \frac{4ar\cos\theta}{\pi}\left(\frac{r^2 + a^2\cos 2\theta}{r^4 + 2a^2r^2\cos 2\theta + a^4}\right) \\
\sigma_\theta + \sigma_r &= -\frac{2}{\pi}\left[\pi - \tan^{-1}\left(\frac{2ar\cos\theta}{a^2 - r^2}\right)\right], \; r < a \\
&= -\frac{2}{\pi}\tan^{-1}\left(\frac{2ar\cos\theta}{r^2 - a^2}\right), \; r > a \\
\tau_{r\theta} &= \frac{2ar\cos\theta}{\pi}\left(\frac{a^2\sin\theta}{r^4 + 2a^2r^2\cos 2\theta + a^4}\right)
\end{aligned}\right\} \quad ...(3.103)$$

3.15 Analysis of Plane Stress with Varying Thickness[8]

Let us consider an element of plate whose thickness varies as :

$$h = f(r).$$

The forces acting on the plate element are shown in Fig. 3.24. Equations of equilibrium become

$$\left.\begin{aligned}
\frac{\partial}{\partial r}(rh - \sigma_r) - h\sigma_\theta + h\frac{\partial \tau_{r\theta}}{\partial \theta} &= 0 \\
\frac{\partial}{\partial r}(r^2 h\,\tau_{r\theta}) + rh\frac{\partial \sigma_\theta}{\partial \theta} &= 0
\end{aligned}\right\} \quad ...(1)$$

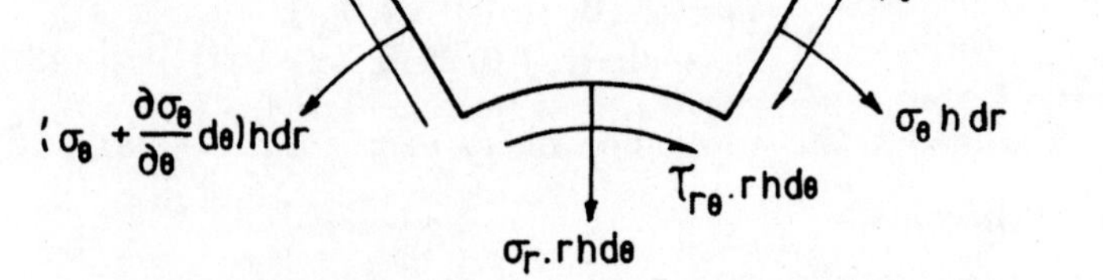

Fig. 3.24 Plate with varying thickness.

8. H.D. Conway : Journal of Applied Mechanics, Vol. 26, No. 3, 1959, pp. 437—39.

Eqs. (1) are satisfied if

$$\left.\begin{aligned} h\sigma_r &= \frac{1}{r}\frac{\partial \Phi}{\partial r} + \frac{1}{r^2}\frac{\partial^2 \Phi}{\partial \theta^2} \\ h\sigma_\theta &= \frac{\partial^2 \Phi}{\partial r^2} \\ h\tau_{r\theta} &= -\frac{\partial}{\partial r}\left(\frac{1}{r}\frac{\partial \Phi}{\partial \theta}\right) \end{aligned}\right\} \qquad \text{...(2)}$$

Now

$$\left.\begin{aligned} \varepsilon_r &= \frac{\partial u_r}{\partial r} \\ \varepsilon_\theta &= \frac{u_r}{r} + \frac{1}{r}\frac{\partial u_\theta}{\partial r} \\ \gamma_{r\theta} &= \frac{1}{r}\frac{\partial u_r}{\partial \theta} + \frac{\partial u_\theta}{\partial r} - \frac{u_\theta}{r} \end{aligned}\right\} \qquad \text{...(3)}$$

Strain-compatibility equation is

$$\frac{\partial}{\partial r}\left(r\frac{\partial \gamma_{r\theta}}{\partial \theta} - r^2\frac{\partial \varepsilon_\theta}{\partial r}\right) + r\frac{\partial \varepsilon_r}{\partial r} - \frac{\partial^2 \varepsilon_\theta}{\partial \theta^2} = 0 \qquad \text{...(4)}$$

and stress-strain relations are

$$\left.\begin{aligned} E\varepsilon_r &= \sigma_r - \nu\sigma_\theta \\ E\varepsilon_\theta &= \sigma_\theta - \nu\sigma_r \\ G\gamma_{r\theta} &= \tau_{r\theta} \end{aligned}\right\} \qquad \text{...(5)}$$

Substituting Eqs. (3) in (4) by using Eqs. (5), we get

$$\left[r^2\frac{\partial^2}{\partial r^2} - (2+\nu)r\frac{\partial}{\partial r} - \nu\frac{\partial^2}{\partial \theta^2}\right]\left(\frac{1}{\lambda} - \frac{\partial^2 \Phi}{\partial r^2}\right)$$

$$-\left[\nu r^2\frac{\partial^2}{\partial r^2} + (2\nu+1)\,r\frac{\partial}{\partial r} - \frac{\partial^2}{\partial \theta^2}\right]\left(\frac{1}{hr}\frac{\partial \Phi}{\partial r}\right.$$

$$\left.+\frac{1}{hr^2}\frac{\partial^2 \Phi}{\partial \theta^2}\right) + 2(1+\nu)\frac{\partial}{\partial r}\left[r\frac{\partial}{\partial \theta} - \left\{\frac{1}{h}\frac{\partial}{\partial r}\right.\right.$$

$$\left.\left.\times\left(\frac{1}{r}\frac{\partial \Phi}{\partial \theta}\right)\right\}\right] = 0 \qquad \text{...(6)}$$

If h is constant, then Eq. (6) reduces to

$$\nabla^2 . \nabla^2 \Phi = \left(\frac{\partial^2}{\partial r^2} + \frac{1}{r}\frac{\partial}{\partial r} + \frac{1}{r^2}\frac{\partial^2}{\partial \theta^2}\right)^2 \Phi = 0 \qquad \text{...(7)}$$

The general solution of Eq. (7) is

$$\Phi = A_0 \log r + B_0 r^2 + C_0 r^2 \log r + D_0 r^2 \theta + A_0' \theta$$
$$+ \frac{A_1}{2} r\theta \sin \theta + (B_1 r^3 + A_1' r^{-1} + B_1' r \log r) \cos \theta$$
$$- \frac{C_1}{2} r\theta \cos \theta + (D_1 r^3 + C_1' r^{-1} + D_1' r \log r) \sin \theta$$
$$+ \sum_{n=2}^{\infty} (A_n r_n + B_n r^{n+2} + A_n' r^{-n} + B_n' r^{-n+2}) \cos n\theta$$
$$+ \sum_{n=2}^{\infty} (C_n r^n + D_n r^{n+2} + C_n' r^{-n} + D_n' r^{-n+2}) \sin n\theta \quad ...(8)$$

Let $\qquad h = h_0 r^{-m} \qquad ...(9)$

where h_0 and m are arbitrary constants.

For $r = 0$, $h = \infty$ if m is positive and

$h = 0$ if m is negative

Substituting Eq. (9) in Eq. (6), we write

$$\Phi = R_0 + \sum_{n=2}^{\infty} R_n \sin n\theta + \sum_{n=2}^{\infty} S_n \cos n\theta \quad ...(10)$$

The following Euler equation is then obtained

$$r^4 R_n^{IV} + 2(m+1) r^3 R'''_n - (1 + \nu m + 2n^2 - m - m^2) r^2 R''_n - (m-1)$$
$$(1 + \nu m + 2n^2) r R'_n - n^2 (4 + \nu m - 3m - \nu m^2 - n^2) R_n = 0 \quad ...(11)$$

Putting $r = e^z$, we get

$$\frac{d^4 R_n}{dz^4} + 2(m-2) \frac{\partial^3 R_n}{dz^3} + (4 - 5m - \nu m - 2n^2 + m^2) \frac{d^2 R_n}{dz^2} - (m-2)$$
$$\times (m + \nu m + 2n^2) \frac{dR_n}{dz} - n^2 (4 + \nu m - 3m - \nu m^2 - n^2) R_n = 0 \quad ...(12)$$

A similar equation is obtained for S_n.

Hence the characteristic equation to be solved is

$$\lambda^4 + 2(m-2)\lambda^3 + (4 - 5m - \upsilon m - 2n^2 + m^2)\lambda^2 - (m-2)$$
$$\times (m + \nu m + 2n^2)\lambda - n^2(4 + \nu m - 3m - \nu m^2 - n^2) = 0 \quad ...(13)$$

To find the roots, we tentatively write,

$$(\lambda + \alpha + \beta_1)(\lambda + \alpha - \beta_1)(\lambda + \alpha + \beta_2)(\lambda + \alpha - \beta_2) = 0$$

where $\alpha = \dfrac{m-2}{2}$.

After multiplication, we get

$$\lambda^4 + 2(m-2)\lambda^3 + \left[\frac{3(m-2)^2}{2} - \beta_1^2 - \beta_2^2\right]\lambda^2 + (m-2)$$
$$\times\left[\frac{(m-2)^2}{2} - \beta_1^2 - \beta_2^2\right]\lambda + \left[\frac{(m-2)^2}{2} - \beta_1^2\right]$$
$$\times\left[\frac{(m-2)^2}{4} - \beta_2^2\right] = 0 \qquad ...(14)$$

Comparing the coefficients of λ^2 and λ in Eqs. (13) and (14), we get

$$\beta_1^2 + \beta_2^2 = \frac{(m-2)^2}{2} + m(1+\nu) + 2n^2 \qquad ...(15)$$

Also comparing terms independent of λ, we get

$$\left[\frac{(m-2)^2}{4} - \beta_1^2\right]\left[\frac{(m-2)^2}{4} - \beta_2^2\right]$$
$$= -n^2(4 + \nu m - 3m - \nu m^2 - n^2) \qquad ...(16)$$

$$\therefore \quad (\beta_1^2 - \beta_2^2)^2 = m^2(1+\nu)^2 - 4n^2 m(2 - 2\nu + \nu m) + 16n^2 \qquad ...(17)$$

and with $\alpha = \frac{1}{2}(m-2)$, the roots of Eq. (13) are obtained.

For $m = 2$ and $h = h_0 r^{-2}$, we get

$$\lambda^4 - 2(n^2 + 1 + \nu)\lambda^2 + n^2(n^2 + 2 + 2\nu) = 0 \qquad ...(18)$$

The roots of this equations are

$$\lambda = \pm n, \pm (n^2 + 2 + 2\nu)^{1/2} \qquad ...(19)$$

For constant thickness, $m = 0$, then

$$\lambda = \pm n, 2 \pm n$$

For axi-symmetrical case, $n = 0$ and

$$\lambda = 0, 2 - m, -\frac{(m-2)}{2} \pm \tfrac{1}{2}[(m-2)^2 + 4m(1+\nu)]^{1/2} \qquad ...(21)$$

The equation for R_n is finally obtained as

$$R_n = A_n r^{\lambda_1} + B_n r^{\lambda_2} + C_n r^{\lambda_3} + D_n r^{\lambda_4} \qquad ...(22)$$

with a corresponding equation for S_n. For $m = 2$,

$$R_0 = A_0 + B_0 \log r + C_0 r^{\lambda_3} + D_0 r^{\lambda_4} \qquad ...(23)$$

For constant thickness case, the Michell solution is

$$\Phi = E[r\theta \sin\theta + F(r)\cos\theta] \qquad ...(24)$$

Substituting in Eq. (6) with $h = h_0 r^{r-m}$, we get

$$r^4F^{IV} + 2(m+1)r^2 F''' - (3 + \nu m - m - m^2)r^2F''$$
$$+ (3 + \nu m - 3m - \nu m^2)rF' - (3 + \nu m - 3m - \nu m^2)F$$
$$= (2\nu m - 2m - 2\nu m^2)r$$

The particular solution of Eq. (25) is,

$$F(r) = \frac{2m(\nu - 1 - \nu m)\, r \log r}{3 + m + \nu m - m^2} \quad \text{...(26)}$$

The complementary solution of Eq. (25) is accounted for in Eq. (10) and the comparison solution to Eq. (24) is obtained in a similar manner as :

$$\Phi = G[r\theta \cos\theta + F(r)\sin\theta] \quad \text{...(27)}$$

Finally, assume with $h = h_o r^{-m}$ a solution of Eq. (6) in the form

$$\Phi = \theta H(r) \quad \text{...(28)}$$

and obtain

$$H(r) = Ar^{\lambda_1} + Br^{\lambda_2} + Cr^{\lambda_3} + Dr^{\lambda_4} \quad \text{...(3.104)}$$

where λ's are given by Eq. (21).

3.15.1 Stresses in a Tapering Beam

Consider a tapering beam of inner radius a and outer radius $b = 2a$ subjected to bending moment M as shown in Fig. 3.25.

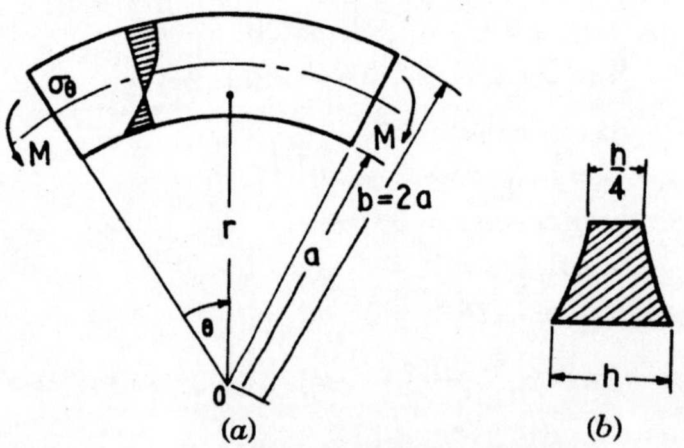

Fig. 3.25 Tapering beam.

For $m = 2, h = h_0 r^{-2}$

$$\Phi = B \log r + Cr^{\lambda} + Dr^{-\lambda},$$
$$\lambda = (2 + 2\nu)^{1/2} \quad \text{...(1)}$$
$$\sigma_r = (B + C\lambda r^{\lambda} - D\lambda r^{-\lambda})/h_0$$
$$\sigma_\theta = [-B + C\lambda(\lambda - 1)r^{\lambda} + D\lambda(\lambda + 1)r^{-\lambda}]/h_0 \quad \text{...(2)}$$

$$\tau_{r\theta} = 0$$

Since $\sigma_r = 0$ at $r = a$ and $r = b$

$$B + C\lambda a^{\lambda} - D\lambda a^{-\lambda} = 0$$

$$B + C\lambda b^{\lambda} - D\lambda b^{-\lambda} = 0 \qquad \text{...(3)}$$

For zero normal force on any radial cross-section,

$$\int_a^b \sigma_\theta \, . \, h dr = \int_a^b \frac{\partial^2 \Phi}{\partial r^2} dr = \left| \frac{\partial \Phi}{\partial r} \right|_a^b$$

$$\left| Br^{-1} + C\lambda r^{\lambda - 1} - D\lambda \, r^{-\lambda - 1} \right|_a^b = 0 \qquad \text{...(4)}$$

Similarly

$$\int_a^b \sigma_\theta \, . \, r \, h \, dr = \int_a^b r \frac{\partial^2 \Phi}{\partial r^2} dr = \left| r \frac{\partial \Phi}{\partial r} \right|_a^b - | \, \Phi \, |_a^b$$

$$= - B \log \frac{b}{a} - C(b^{\lambda} - a^{\lambda}) - D(b^{-\lambda} - a^{-\lambda}) = M \qquad \text{...(5)}$$

The constants B, C and D can be determined from Eqs. (3) and (5), then

$$\sigma_r = \frac{M\lambda}{h_0 N} \left[\left(\frac{b}{a} \right)^{\lambda} - \left(\frac{a}{b} \right)^{\lambda} + \left(\frac{r}{a} \right)^{\lambda} \left\{ \left(\frac{a}{b} \right)^{\lambda} - 1 \right\} - \left(\frac{r}{a} \right)^{-\lambda} \left\{ \left(\frac{b}{a} \right)^{\lambda} - 1 \right\} \right] \qquad \text{...(3.105)}$$

$$\sigma_\theta = \frac{M\lambda}{h_0 N} \left[- \left(\frac{b}{a} \right)^{\lambda} - \left(\frac{a}{b} \right)^{\lambda} + (\lambda - 1) \left(\frac{r}{a} \right)^{\lambda} \left\{ \left(\frac{a}{b} \right)^{\lambda} - 1 \right\} + (\lambda + 1) \left(\frac{r}{a} \right)^{-\lambda} \left\{ \left(\frac{b}{a} \right)^{\lambda} - 1 \right\} \right] \qquad \text{...(3.106)}$$

where $N = 2 \left(\frac{a}{b} \right)^{\lambda} + 2 \left(\frac{b}{a} \right)^{\lambda} - 4 + \lambda \left[\left(\frac{a}{b} \right)^{\lambda} - \left(\frac{b}{a} \right)^{\lambda} \right] \log \frac{b}{a}$

$$(\sigma_\theta)_{r=a} = -(\sigma_\theta)_{r=b} = \frac{-M\lambda^2\left[\left(\frac{b}{a}\right)^{\lambda/2} - \left(\frac{a}{b}\right)^{\lambda/2}\right]}{h_o\left[\left\{\left(\frac{b}{a}\right)^{\lambda/2} + \left(\frac{a}{b}\right)^{\lambda/2}\right\}\lambda \log\frac{b}{a} - 2 \times \left\{\left(\frac{b}{a}\right)^{\lambda/2} - \left(\frac{a}{b}\right)^{\lambda/2}\right\}\right]} \quad ...(3.107)$$

3.15.2 Rotating Disc of Variable Thickness

Consider a disc of variable thickness as shown in Fig. 3.26 in which the thickness is a function of the distance r from the axis. Let σ_r and σ_θ denote the mean radial and tangential stresses at r, and h the variable thickness, the equation of equilibrium of such an element is :

$$\frac{d}{dr}(hr\,\sigma_r) - h\sigma_\theta + \rho\omega^2 hr^2 = 0 \quad ...(1)$$

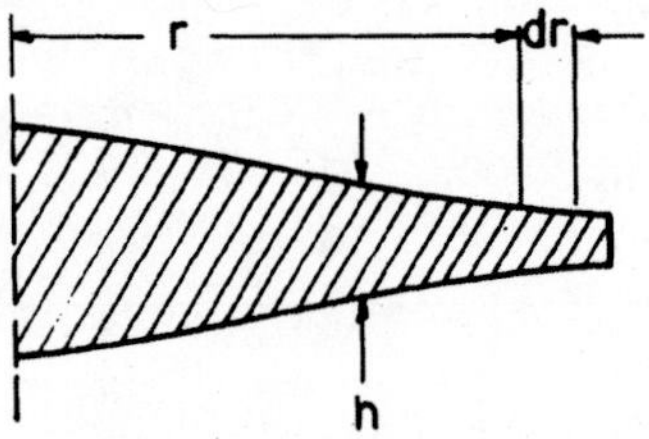

Fig. 3.26 Variable thickness rotating disc.

This equation is satisfied if we introduce a stress function Φ such that

$$hr\sigma_r = \Phi,\ h\sigma_\theta = \frac{d\Phi}{dr} + \rho\omega^2 hr^2 \quad ...(2)$$

In the case of rotational symmetry,

$$u_r = u_r(r) \text{ and } u_\theta = 0$$

Hence $\varepsilon_r = \dfrac{du_r}{dr},\ \varepsilon_\theta = \dfrac{u_r}{r}$

Hence $\dfrac{d}{dr}(r\varepsilon_\theta) - \varepsilon_r = 0$ or $r\dfrac{d\varepsilon_\theta}{dr} + \varepsilon_\theta - \varepsilon_r = 0$

Using Hooke's law and Eqs. (2), we get from Eq. (3), the compatibility equation as :

$$r^2\frac{d^2\Phi}{dr^2} + \left(1 - \frac{r}{h}\frac{dh}{dr}\right)r\frac{d\Phi}{dr} + \left(\nu\frac{r}{h}\frac{dh}{dr} - 1\right)\Phi = -(3+\nu)\rho\omega^2 hr^2 \quad ...(4)$$

from which Φ can be found when h is given.

If the thickness of the disc varies according to the equation

$$h = Cr^{-\beta}$$

where C is a constant and β is any number, Eq. (4) reduces to

$$r^2 \frac{d^2\Phi}{dr^2} + (1 + \beta)\, r \frac{d\Phi}{dr} - (1 + \nu\beta)\Phi$$

$$= -(3 + \nu)\rho\omega^2 C r^{3-\beta}$$

Let $\xi = \log r$, then above equation becomes

$$\frac{d^2\Phi}{d\xi^2} + \beta \frac{d\Phi}{\partial\xi} - (1 + \nu\beta)\,\phi$$

$$= -(3 + \nu)\,\rho\omega^2 C e^{(3-\beta)\xi}$$

The general solution of this equation is,

$$\Phi = C_1 r^{q_1} + C_2 r^{q_2} - \frac{3 + \nu}{8 - (3 + \nu)\beta} C\rho\omega^2 r^{3-\beta}$$

q_1, q_2 are the roots of the equation

$$q^2 + \beta q - (1 + \nu\beta) = 0 \qquad \text{...(5)}$$

The corresponding stress components are

$$\sigma_r = \frac{\Phi}{hr} = \frac{C_1}{C} r^{(q_1 + \beta - 1)} + \frac{C_2}{C} r^{(q_2 + \beta - 1)}$$

$$- \frac{3 + \nu}{3 - (3 + \nu)\beta} \rho\omega^2 r^2$$

$$\sigma_\theta = \frac{1}{h}\frac{\partial\Phi}{\partial r} + \rho\omega^2 r^2 = \frac{C_1}{C} q_1 r^{(q_1 + \beta - 1)}$$

$$+ \frac{C_2}{C} q_2 r^{(q_2 + \beta - 1)} - \frac{1 + 3\nu}{8 - (3 + \nu)\beta} \rho\omega^2 r^2$$

From Eq. (5), we find

$$q = -\frac{\beta}{2} \pm \sqrt{\left(\frac{\beta}{2}\right)^2 + (1 + \nu\beta)}$$

If q_2 is the smaller root and β is positive, than

$$(q_2 + \beta) = \frac{\beta}{2} - \sqrt{\left(\frac{\beta}{2}\right)^2 + (1 + \nu\beta)}$$

Since $\sqrt{\left(\frac{\beta}{2}\right)^2 + (1 + \nu\beta)} > \frac{\beta}{2}$; $(q_2 + \beta)$ is always a negative quantity.

For a solid disc, if β is positive, we must have $C_2 = 0$ because otherwise σ_r and σ_θ would be infinite at the centre of the disc. If the surface of the disc at $r = b$ is free from tractions then at $r = b$, $\sigma_r = 0$ and we get

$$\frac{C_1}{C} = \frac{3+\nu}{8-(3+\nu)\beta}\,\rho\omega^2 e^{(3-q_1-\beta)}$$

Hence

$$\left.\begin{aligned}\sigma_r &= \frac{3+\nu}{8-(3+\nu)\beta}\,\rho\omega^2 b^2\left[\left(\frac{r}{b}\right)^{(q_1+\beta-1)} - \left(\frac{r}{b}\right)^2\right]\\ \sigma_r &= \frac{3+\nu}{8-(3+\nu)\beta}\,\rho\omega^2 b^2\left[q_1\left(\frac{r}{b}\right)^{(q_1+\beta-1)}\right.\\ &\qquad\left. - \left(\frac{1+3\nu}{3+\nu}\right)\left(\frac{r}{b}\right)^2\right]\end{aligned}\right\} \quad \text{...(6)}$$

For a solid disc with uniform stress, the thickness h should be so proportioned that $\sigma_r = \sigma_\theta = \sigma$ at every point in the disc. Substituting into Hooke's law, we find

$$\varepsilon_r = \varepsilon_\theta$$

The compatibility Eq. (3) then becomes

$$\frac{d\varepsilon_\theta}{dr} = 0 \quad \text{or} \quad \varepsilon_\theta = \text{const.}$$

Therefore, σ_r and ε_θ are also constant throughout the disc. Equation of equilibrium (1) reduces to

$$\sigma\frac{d(hr)}{dr} - \sigma h + \rho\omega^2 hr^2 = 0.$$

Differentiating and combining we find

$$\frac{1}{h}\frac{dh}{dr} = -\frac{\rho w^2}{\sigma}r$$

Integrating, we get $\log h = \dfrac{\rho\omega^2}{2\sigma}r^2 + C_3$

or

$$h = e^{\left(-\frac{\rho\omega^2 r^2}{2\sigma} + C_3\right)}$$

$$= e^{C_2} \,.\, e^{-\frac{\rho\omega^2 r^2}{2\sigma}}$$

$$= C_4 \,.\, e^{-\frac{\rho\omega^2 r^2}{2\sigma}} \quad \text{...(3.108)}$$

where C_3 and C_4 are constants.

EXERCISES

3.1 A curved bar of inner radius 200 mm and outer radius 250 mm is subjected to a bending moment of 8 kN-mm. Assuming unit thickness for the bar, determine the radial and hoop stresses in the bar according to the theory of elasticity.

3.2 A thick cylinder of internal radius 100 mm and external radius 150 mm is subjected to an internal pressure of 6 N/mm^2. Determine the variation of radial and hoop stresses in the cylinder wall.

3.3 An aluminium cylinder ($E_{al} = 0.7 \times 10^5$ N/mm^2, $\nu_{al} = 0.33$) with an outer radius of 150 mm and inner radius of 100 mm, is to be press-fitted over a steel cylinder ($E_s = 2 \times 10^5$ N/mm^2, $\nu_s = 0.29$) with an outer radius of 100.25 mm and inner radius 50 mm. Determine the interference pressure p.

3.4 A disc with outer radius 150 mm and inner radius 25 mm is press fitted on to a shaft of radius 25.1 mm. Both members are made of steel with $E = 2.1 \times 10^5$ N/mm^2, $\nu = 0.29$, and with weight densities of 7470 kg/m^3. Determine

(*a*) the stress distribution in the disc at 5000 r.p.m., and

(*b*) the speed for which the interference pressure goes to zero.

3.5 A compound tube is made by shrinking one tube on another, the final dimensions being internal diameter, 80 mm, external diameter, 160 mm ; diameter at the junction of the tubes, 120 mm. If the radial pressure at the common 60 mm radius is 15 N/mm^2, find the greatest hoop tension and hoop pressure in the compound cylinder. What difference must there be in the external diameter of the inner tube and the internal diameter of the outer tube before shrinking on, and what is the least difference of temperature necessary to allow the outer one passing over the inner one ? Take the coefficient of expansion as 11.2×10^{-6} per degree C, and $E = 2 \times 10^5$ N/mm^2.

3.6 A circular disc of uniform thickness has external radius of 250 mm and internal radius of 50 mm. It is rotated at 750 rpm. Determine the maximum circumferential, radial and shear stresses in the disc. Density of material is 7470 kg/m^3 and $\nu = 0.28$.

3.7 Determine the maximum stress in a long cylinder 75 mm inside diameter and 250 mm outside diameter rotating at 3000 rpm. Take $\nu = 0.32$ and density of material 6700 kg/m^3.

3.8 A steam turbine disc is to be designed so that between radii of 200 mm and 400 mm the radial and circumferential stresses are to be constant and both equal to 65 N/mm^2, when running at 3000 rpm. If the axial thickness is 10 mm at the outer edge of this zone, what should it be at the inner edge ? Assume $\rho = 7470$ kg/m^3.

3.9 (*a*) Plot the variation of stress distribution in a plate of infinite dimensions with a circular hole under uniaxial uniform tension.

(*b*) Derive the equations used for stresses.

3.10 Determine the stress in a semi-infinite plate due to normal load w per unit length, acting on its edge, as shown in Fig. 3.27. Try the

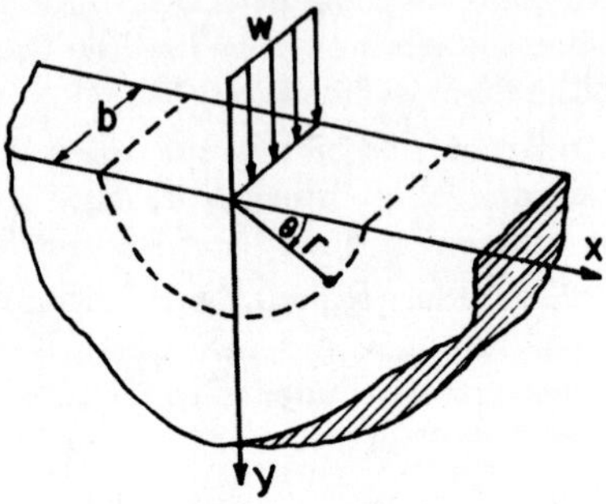

Fig. 3.27

stress function.

$$\Phi = A\theta + br^2\theta + Cr\theta \sin\theta + Dr\,\theta\cos\theta$$

where A, B, C and D are constants.

3.11 Find the stress function of the type

$$\Phi = Ar^3 \cos 3\theta + Br^3 \cos\theta$$

which satisfies the conditions

$$\sigma_\theta = 0, \tau_{r\theta} = Sr \text{ on } \theta = \alpha$$

$$\sigma_\theta = 0, \tau_{r\theta} = -Sr \text{ on } \theta = -\alpha$$

S being a constant. Sketch the loading for positive S.

3.12 Find the stress function of the type

$$C_1r^4 \cos 4\theta + C_2r^4 \cos 2\theta$$

which satisfies the conditions

$$\sigma_\theta = 0, \tau_{r\theta} = Sr^2 \text{ on } \theta = \alpha, \sigma_\theta = 0, \tau_{r\theta} = -Sr^2 \text{ on } \theta = -\alpha.$$

S being a constant. Sketch the loading.

3.13 Determine the value of the constant C in the stress function.,

$$\Phi = Cr^2 (\cos 2\theta - \cos 2\alpha)$$

required to satisfy the conditions

$$\sigma_\theta = 0, \tau_{r\theta} = S \text{ on } \theta = \alpha\,;\ \sigma_\theta = 0, \tau_{r\theta} = -S \text{ on } \theta = -\alpha$$

corresponding to uniform shear loading on each edge of a wedge, directed away from the wedge vertex. Verify that no concentrated force or couple acts on the vertex.

3.14 A circular disc of 8 cm diameter and 5 mm thick is subjected to diametral compression. If the applied load is 800 N, determine the stress distribution in the disc at the centre.

3.15 A curved bar of inner radius 250 mm and outer radius 300 mm is subjected to a concentrated force of 500 N applied at the end of the bar. Determine the stresses in the bar at a radius of 270 mm and

$\theta = 45°$.

3.16 A tapering beam of inner radius 300 mm and outer radius 450 mm is subjected to a bending moment of 12 N-m. If the thickness of the beam at 300 mm radius is 50 mm and at 450 m radius is 25 mm., determine the radial and hoop stresses in the beam at the inner and outer radii.

3.17 Obtain the solutions of following problems in polar coordinates by means of the complex potentials indicated. Evaluate the stress and displacement components.

(*a*) The ring under internal pressure p_i, external pressure p_0

$$\psi(z) = A(z),\ \chi(z) = B \log z.$$

(*b*) The curved bar of inner radius a and outer radius b subjected to end force P. $\psi(z) = Az^2 + B \log z$, $\chi(z) = Cz \log z + (D/z)$

(c) The plate under tension with a circular hole

$$\psi(z) = Az + (B/z),\ \chi(z) = C \log z + Dz^2 + (F/z^2)$$

(*d*) The force at a point of an infinite plate

$$\psi(z) = A \log z,\ \chi(z) = Bz \log z.$$

3.18 Find the ratio of thickness to internal diameter for a tube subjected to internal pressure when the pressure is equal in magnitude to 3/4 of the maximum hoop stress. If the internal diameter of the tube is 100 mm, determine the increase in the external diameter when the internal pressure is 84 MPa and the tube is prevented from changing length. (E = 200 GPa. ν = 0.3).

3.19 A steel bar of 50 mm diameter is pressed into a steel sleeve so that when assembled, the magnitude of radial stress between the two is 14 MPa, and that of the circumferential stress at the inside of the sleeve is 21 MPa. Assuming a close fit and neglecting friction, determine the change of radial stress when the bar is subjected to an axial compressive load of 75 kN. (ν = 0.3).

3.20 A short steel rod of 50 mm diameter is subjected to an axial compressive load of 300 kN. It is surrounded by a sleeve 15 mm thick, slightly shorter than the rod so that the load is carried only by the rod. Assuming a close fit before the load is applied and neglecting friction, find the pressure between the sleeve and the rod, and the maximum tensile stress in the sleeve (ν = 0.3).

3.21 The external diameter of a steel hub is 250 mm and the internal diameter increases 0.125 mm when shrunk on to a solid steel shaft of 125 mm diameter. Find the reduction in diameter of the shaft, the radial pressure between the hub and the shaft, and the circumferential stress at the inner surface of the hub. (E = 200 GPa, ν = 0.3)

3.22 A steel hollow sphere, whose inside diameter is 125 mm, is subjected to an internal pressure of 35 MPa. Determine the thickness of the material if the magnitude of the maximum stress is not to exceed 70 MPa.

3.23 Determine the greatest value of the radial and circumferential stresses for a thin disc rotating at an angular velocity of 200 rad/s, the inner and outer radii of the disc are 150 mm and 300 mm respectively, and the density of its material is 7500 kg/m^3 (E = 200 GPa, ν = 0.3).

3.24 A solid steel shaft 900 mm in diameter is rotating at 250 rpm. If the shaft cannot deform longitudinally, calculate the total longitudinal thrust over a cross-section due to rotational stresses. (ρ = 7470 kg/m^3, E = 200 GPa, ν = 0.3).

3.25 The Airy's stress function for the stress distribution under a knife edge on the boundary of an infinite solid is given by $\phi = \frac{P}{\pi} r\theta \sin\theta$, where P is the uniformly distributed load per unit length of the edge. Determine the stresses.

3.26 Determine the stress components from the following stress function :

$$\phi = Cr^2(\alpha - \theta) + r^2 \sin\theta \cos\theta - r^2 \cos^2\theta \tan\alpha$$

and calculate their values for α = 20° and 30°.

3.27 Verify that the stress function $\phi = Cr\,\theta \sin\theta$ satisfy the biharmonic equation. Then find the expressions for stresses.

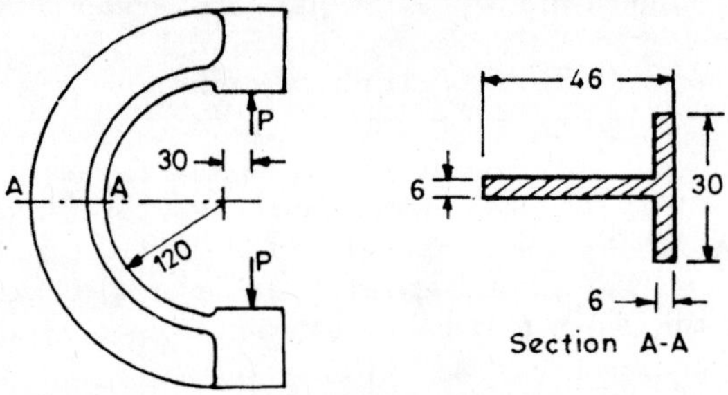

Fig. 3.28 Section A-A.

3.28 The allowable stress in tension and compression for the clamp body shown in Fig. 3.28 is 80 MPa. Calculate the maximum permissible load the member can resist. [**Ans.** 1517 N]

3.29 Calculate the tangential stress at points A and B of the U-frame shown in Fig. 3.29.

3.30 A flat 0.5 m outer diameter, 0.1 m inner diameter, and 0.075 m thick steel disc is shrunk onto a steel shaft. The assembly is to run at speeds upto 6900 rpm. Determine (*a*) the shrinkage allowance, (*b*) the maximum stress when not rotating, and (*c*) the maximum stress when rotating. E = 200 GPa, ν = 0.3 and ρ = 7800 N . s^2/m^4, g = 9.8 m/s^2. [**Ans.** 5.25 × 10^{-5} m, 109.2 MPa, 211.78 MPa]

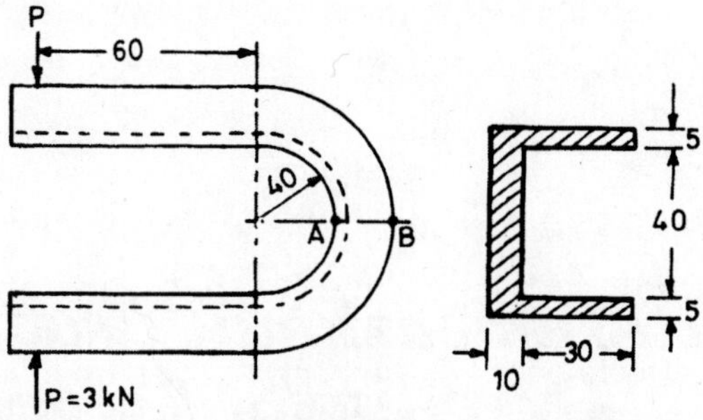

Fig. 3.29

3.31 A flywheel of 0.5 m diameter and 0.1 m inner diameter is pressed onto a solid shaft. The maximum tangential stress induced in the flywheel is 35 MPa. The length of the flywheel parallel to the shaft axis is 0.05 m. Assuming a coefficient of static friction of 0.2 at the common surface, find the maximum torque that may be transmitted by the flywheel without slippage. [**Ans.** 5072.67 N.m]

3.32 A flat annular steel disc of 0.8 m outer diameter and 0.15 m inner diameter is to be shrunk around a solid steel shaft. The shrinkage allowance is 1 part per 1000. For $\nu = 0.3$, $E = 210$ GPa and $\rho = 7.8$ kN . s^2/m^4, determine, neglecting shaft expansion :

(*a*) the maximum stress in the system at stand-still, and

(*b*) the rpm at which the shrink fit will loosen as a result of rotation.

[**Ans.** 108.68 MPa, 4300 rpm]

3.33 A steel turbine disc with outer radius 0.5 m, inner radius 0.0625 m and tip thickness 0.05 m rotates at 5000 rpm carrying blades weighing a total of 540 N. The centre of gravity of each blade lies on a circle of 0.575 m radius. Assuming zero pressure at the bore, determine (*a*) the maximum stress for a disc of constant thickness. $\rho = 7.8$ kN . s^2/m^4, $g = 9.81$ m/s^2. [**Ans.** 554.58 MPa]

4

Two-Dimensional Problems in Curvilinear Co-ordinates

4.1 Introduction

In many applications of the two-dimensional theory of elasticity it is necessary to have expressions for the stress and displacement in an orthogonal curvilinear co-ordinate system. Elliptical and Bipolar co-ordinates are the most important examples of curvilinear co-ordinates which is of much use in solving two-dimensional problems having curvilinear boundaries. In order to understand these co-ordinate systems it is essential to have a knowledge of the conformal transformation.

4.2 Conformal Transformation

Let z and ζ be two complex variables such that

$$z = \omega(\zeta) \qquad \ldots(4.1)$$

where $\omega(\zeta)$ is a single-valued analytic function in some region R in the ζ-plane. Equation (4.1) relates every point ζ of R to some definite point z in the z-plane. These latter points will cover in the z-plane some region S. Conversely, let it be assumed that each point z of S, by equation (4.1), corresponds to some definite point of R. It will then be said that Eq. (4.1) determines an invertible single-valued conformal transformation or conformal mapping of the region S into the region R or *vice-versa*. Now we give some of the properties of conformal transformation without proof.

(*i*) If in region R two linear elements be taken which extend from some ζ and form between them an angle α, the corresponding elements in region S will form the same angle α and the sense of the angle will be maintained.

(*ii*) The mapping defined by an analytic function $f(z)$ is conformal except at points where the derivative $f'(z)$ is zero.

(*iii*) A harmonic function $\phi(x, y)$ remains harmonic under a change of the variables arising from one-to-one conformal transformation.

(*iv*) Let R be a finite or infinite region in the ζ-plane, bounded by a simple contour γ, and let $\omega(\zeta)$ be a function, holomorphic in R and continuous upto the contour. Further, let the points defined by $z = \omega(\zeta)$, describe in the z-plane some simple contour C, when ζ describes γ. Then $z = \omega(\zeta)$ gives the conformal transformation of the region S, contained inside C, on the region R and *vice-versa*.

(*v*) Let R be a finite or infinite region, bounded by several simple contours $\gamma_1, \gamma_2, \ldots\ldots \gamma_k$. Let $\omega(\zeta)$ be a function, holomorphic in R and continuous upto the boundary, and let the point z, defined by $z = \omega(\zeta)$, describe in the z-plane the simple contours $C_1, C_2, \ldots C_k$, bounding some region S, when ζ describes the contours $\gamma_1, \gamma_2, \ldots, \gamma_k$. Under these conditions $z = \omega(\zeta)$ represents the conformal transformation of S on to R and *vice versa*.

Stretch Ratio. If the length of an element $PQ = ds$ in the z plane changes to $P'Q' = ds'$ in the ζ-plane, then the ratio of ds' to ds is defined as the stretch ratio and is denoted by J.

$$J = \frac{ds'}{ds}$$

Now
$$ds = \sqrt{dx^2 + dy^2}$$
$$ds' = \sqrt{d\xi^2 + d\eta^2}$$

where $z = x + iy$, $\qquad \zeta = \xi + i\eta$

$$d\xi = \frac{\partial \xi}{\partial x} dx + \frac{\partial \xi}{\partial y} dy, \; d\eta = \frac{\partial \eta}{\partial x} dx + \frac{\partial \eta}{\partial y} dy$$

$$\delta\xi^2 + d\eta^2 = \left[\left(\frac{\partial \xi}{\partial x}\right)^2 + \left(\frac{\partial \eta}{\partial x}\right)^2\right](dx^2 + dy^2)$$

$$\therefore \qquad ds'^2 = \left[\left(\frac{\partial \xi}{\partial x}\right)^2 + \left(\frac{\partial \eta}{\partial x}\right)^2\right] ds^2$$

$$\therefore \qquad J = \sqrt{\left(\frac{\partial \xi}{\partial x}\right)^2 + \left(\frac{\partial \eta}{\partial x}\right)^2}$$

4.3 Curvilinear Co-ordinates

A curvilinear co-ordinate system in two-dimensions consists of a two-fold family of surfaces whose equations in terms of the rectangular co-ordinates are given by equations of the form

$$\left.\begin{aligned} \xi &= f_1(x, y) \\ \eta &= f_2(x, y) \end{aligned}\right\} \qquad \ldots(4.2)$$

where f_1 and f_2 are functions of given form (x, y) are rectangular coordinates in a plane which may be referred to as the xy-plane.

Similarly (ξ, η) are rectangular co-ordinates in the $\xi\eta$ plane. O and O' are the origins of co-ordinates in the xy and $\xi\eta$ planes respectively as shown in Fig. 4.1. If P is the point (x, y), then the point Q in the $\xi\eta$ plane, whose co-ordinates (ξ, η) are given in terms of x, y by the equations of transformation Eqs. (4.2) is said to correspond to P.

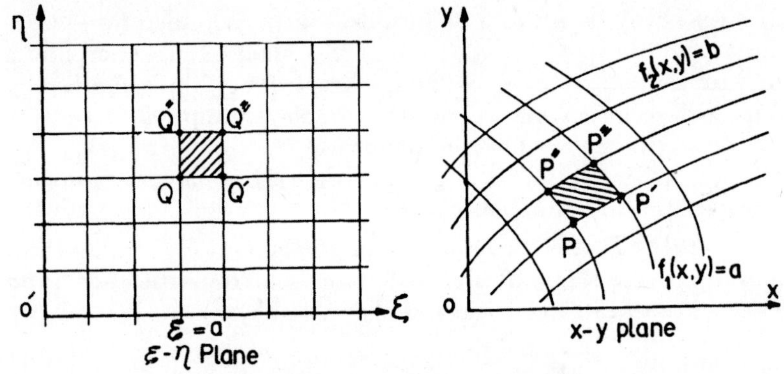

(a) (b)

Fig. 4.1 Curvilinear coordinates.

If P describes any locus in the xy plane, then Q describes a corresponding locus in the $\xi\eta$ plane. In particular, a rectangular network in the $\xi\eta$ plane, corresponding to the set of lines

$$\xi = a, \eta = b \qquad \text{...(4.3)}$$

where a and b are constants, will correspond in the xy plane to the set of curves

$$f_1(x, y) = a, f_2(x, y) = b \qquad \text{...(4.4)}$$

forming a curvilinear network. The Eqs. (4.2) can be solved back for (x, y) in terms of (ξ, η) giving

$$\left.\begin{aligned} x &= F_1(\xi, \eta) \\ y &= F_2(\xi, \eta) \end{aligned}\right\} \qquad \text{...(4.5)}$$

so that a rectangular network in the xy plane will correspond to a curvilinear network in the $\xi\eta$-plane.

Now consider a rectangle $QQ'Q''Q'''$ bounded by straight lines ξ, $\xi + \delta\xi$, η, $\eta + \delta\eta$. This becomes in the xy-plane a small 'curved' parallelogram $PP'P''P'''$, of sides $PP' = \delta S_\xi$, $PP'' = \delta S_\eta$ and of angle $P'PP'' = \alpha$, thus δS_ξ is the arc of the curve η = const. corresponding to an increase $\delta\xi$ of the parameter. The ratio

$$\frac{1}{h_1} = \frac{\delta S_\xi}{\delta \xi_1}, \frac{1}{h_2} = \frac{\delta S_\eta}{\delta \eta} \qquad \text{...(4.6)}$$

are the stretch ratios of the transformation, we have

$$(\delta S_\xi)^2 = \left(\frac{\partial x}{\partial \xi}\right)^2 (\delta\xi)^2 + \left(\frac{\partial y}{\partial \xi}\right)^2 (\delta\xi)^2$$

$$\therefore \qquad \frac{1}{h_1{}^2} = \left(\frac{\partial x}{\partial \xi}\right)^2 + \left(\frac{\partial y}{\partial \xi}\right)^2 \qquad \text{...(4.7)}$$

Similarly,

$$\frac{1}{h_2{}^2} = \left(\frac{\partial x}{\partial \eta}\right)^2 + \left(\frac{\partial y}{\partial \eta}\right)^2 \qquad \text{...(4.8)}$$

Now $$\delta\xi = \frac{\partial \xi}{\partial x}\delta x + \frac{\partial \xi}{\partial y}\delta y$$

$$0 = \frac{\partial \eta}{\partial x}\delta x + \frac{\partial \xi}{\partial y}\delta y$$

$$\therefore \qquad \frac{\delta x}{\frac{\partial \xi}{\partial y}} = \frac{\delta y}{-\frac{\partial \eta}{\partial x}}$$

$$= \frac{(\delta x)^2 + (\delta y)^2}{\sqrt{\left(\frac{\partial \eta}{\partial x}\right)^2 + \left(\frac{\partial \eta}{\partial y}\right)^2}}$$

$$= \frac{\partial S_\xi}{\sqrt{\left(\frac{\partial \eta}{\partial x}\right)^2 + \left(\frac{\partial \eta}{\partial y}\right)^2}}$$

$$\therefore \qquad h_1 = \frac{\delta\xi}{\delta S_\xi} = \frac{\frac{\partial \xi}{\partial x}\cdot\frac{\partial \eta}{\partial y} - \frac{\partial \xi}{\partial y}\cdot\frac{\partial \eta}{\partial x}}{\sqrt{\left(\frac{\partial \eta}{\partial x}\right)^2 + \left(\frac{\partial \eta}{\partial y}\right)^2}} \qquad \text{...(4.9)}$$

$$h_2 = \frac{\delta\eta}{\delta S_\eta} = \frac{\frac{\partial \xi}{\partial x}\cdot\frac{\partial \eta}{\partial y} - \frac{\partial \xi}{\partial y}\cdot\frac{\partial \eta}{\partial x}}{\sqrt{\left(\frac{\partial \xi}{\partial x}\right)^2 + \left(\frac{\partial \xi}{\partial y}\right)^2}} \qquad \text{...(4.10)}$$

To find angle α, we have

$$\left(\frac{dy}{dx}\right)_{\eta = \text{const}} = -\frac{\partial \eta}{\partial x} \Big/ \frac{\partial \eta}{\partial y}$$

$$\left(\frac{dy}{dx}\right)_{\xi = \text{const}} = -\frac{\partial \xi}{\partial x} \Big/ \frac{\partial \xi}{\partial y}$$

$$\tan \alpha = \frac{-\dfrac{\left(\dfrac{\partial \xi}{\partial x}\right)}{\left(\dfrac{\partial \xi}{\partial y}\right)} + \dfrac{\left(\dfrac{\partial \eta}{\partial x}\right)}{\left(\dfrac{\partial \eta}{\partial y}\right)}}{1 + \dfrac{\left(\dfrac{\partial \xi}{\partial x} \cdot \dfrac{\partial \eta}{\partial x}\right)}{\left(\dfrac{\partial \xi}{\partial y} \cdot \dfrac{\partial \eta}{\partial y}\right)}}$$

$$\tan \alpha = -\frac{\dfrac{\partial \xi}{\partial x} \cdot \dfrac{\partial \eta}{\partial y} - \dfrac{\partial \eta}{\partial x} \cdot \dfrac{\partial \xi}{\partial y}}{\dfrac{\partial \xi}{\partial x} \cdot \dfrac{\partial \eta}{\partial x} + \dfrac{\partial \xi}{\partial y} \cdot \dfrac{\partial \eta}{\partial y}} \qquad \text{...(4.11)}$$

4.4 Stress Components in Curvilinear Co-ordinates

For the conformal transformation $z = \omega(\zeta)$ where $\zeta = \xi + \eta$, first consider the rectangular co-ordinate axes Pn, Ps formed by the normal, n, to the curve η = const. at the point $P(x, y)$ and the normal, s, to the curve ξ = const. at the same point, taken in the directions ξ increasing, η increasing respectively. If n makes an angle α with the x-axis as shown in Fig. 4.2, the stress components σ_n, σ_s, τ_{ns} in the (n, s) co-ordinate system are obtained by the transformation

$$\sigma_n = \sigma_x \cos^2 \alpha + \sigma_y \sin^2 \alpha + 2\tau_{xy} \sin \alpha \cos \alpha$$

$$\sigma_s = \sigma_x \sin^2 \alpha + \sigma_y \cos^2 \alpha - 2\tau_{xy} \sin \alpha \cos \alpha$$

$$\tau_{ns} = -\sigma_x \sin \alpha \cos \alpha, + \sigma_y \sin \alpha \cos \alpha + \tau_{xy} (\cos^2 \alpha - \sin^2 \alpha)$$

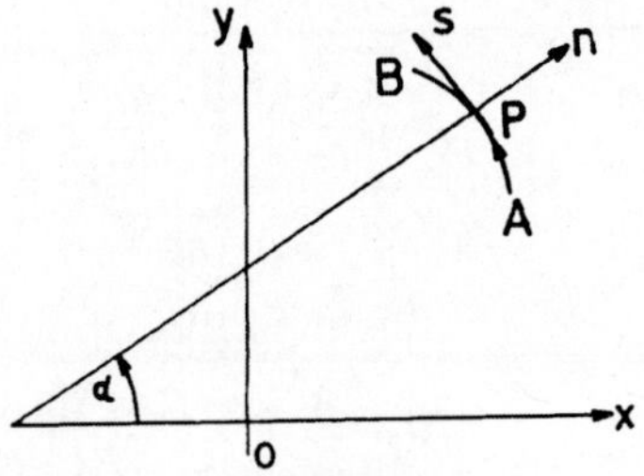

Fig. 4.2

It follows that

$$\left.\begin{aligned} \sigma_n + \sigma_s &= \sigma_x + \sigma_y \\ \sigma_n - \sigma_s + 2i\tau_{ns} &= e^{-2i\alpha} (\sigma_x - \sigma_y + 2i\tau_{xy}) \\ \sigma_n - \sigma_s - 2i\tau_{ns} &= e^{2i\alpha} (\sigma_x - \sigma_y - 2i\tau_{xy}) \end{aligned}\right\} \qquad \text{...(4.12)}$$

Now if an increment dz_1 is taken along the curve η = const. at the point P

$$dz_1 = J\, e^{i\alpha}\, d\xi$$

where J is real, and since $z = \omega(\zeta)$ is analytic at P, therefore,

$$J\, e^{i\alpha} = \omega'(\zeta)$$

Hence $$J^2 = |\,\omega'(\zeta)\,|^2$$

$$= \omega'(\zeta)\,\overline{\omega}'(\overline{\zeta}) \qquad \text{...(4.13)}$$

and $$e^{-2i\alpha} = \left[\frac{\overline{\omega}'(\overline{\zeta})}{|\,\omega(\zeta)\,|}\right]^2 = \frac{\overline{\omega}'(\overline{\zeta})}{\omega'(\zeta)} \qquad \text{...(4.14)}$$

Then the stress components in the (ξ, η) co-ordinate system become, for any point $P(x, y)$:

$$\left.\begin{aligned} \sigma_\xi + \sigma_\eta &= \sigma_x + \sigma_y \\ \sigma_\xi - \sigma_\eta + 2i\tau_{\xi\eta} &= \frac{\overline{\omega'(\zeta)}}{\omega'(\zeta)}(\sigma_x - \sigma_y + 2i\tau_{xy}) \\ \sigma_\xi - \sigma_\eta - 2i\tau_{\xi\eta} &= \frac{\omega'(\zeta)}{\overline{\omega}'(\overline{\zeta})}(\sigma_x - \sigma_y - 2i\tau_{xy}) \end{aligned}\right\} \qquad \text{...(4.15)}$$

If u_ξ, u_η denote the components of displacement at $P(x, y)$ resolved along the normals n, s respectively, then

$$u_\xi + iu_\eta = e^{-i\alpha}(u_x + iu_y) \qquad \text{...(4.16)}$$

where $e^{-i\alpha}$ is given by Eq. (4.14).

Now in the absence of body forces

$$2G(u_x + iu_y) = \chi\ \varphi(z) - z\,\overline{\varphi}'(\overline{z}) - \overline{\psi}(\overline{z})$$

where $\chi = 3 - 4\nu$

Putting $\varphi(z) = \varphi_0(\zeta)$, $\psi(z) = \psi_0(\zeta)$ we get

$$2G(u_\xi + iu_\eta) = \frac{\overline{\omega}'(\zeta)}{|\,\omega'(\zeta)\,|}\left[x\varphi_0(\zeta) - \frac{\omega(\zeta)}{\overline{\omega}'(\overline{\zeta})}\overline{\varphi}'_0(\overline{\zeta}) - \overline{\psi}_0(\xi)\right] \qquad \text{...(4.17)}$$

Also $\sigma_x + \sigma_y = 2[\varphi'(z) + \overline{\varphi}'(\overline{z})] = 4R_e\,\varphi'(z)$

$$\sigma_x - \sigma_y + 2i\tau_{xy} = -2[z\,\overline{\varphi}''(\overline{z}) + \overline{\psi}'(\overline{z})]$$

$$\sigma_x - \sigma_y - 2i\tau_{xy} = -2[\overline{z}\,\varphi''(z) + \psi'(z)]$$

$\therefore$ Eq. (4.15) can be written as

$$\sigma_\xi + \sigma_\eta = 2\{\Phi_0(\zeta) + \overline{\Phi}_0(\overline{\zeta})]$$

$$\sigma_\xi - \sigma_\eta + 2i\tau_{\xi\eta} = -\frac{2}{\omega'(\zeta)}\,\omega(\zeta)\,\overline{\Phi}'_0(\overline{\zeta}) + \overline{\omega}'(\overline{\zeta})\,\overline{\Psi}_0(\overline{\zeta})]$$

$$\sigma_\xi - \sigma_\eta - 2i\tau_{\xi\eta} = -\frac{2}{\overline{\omega'(\zeta)}}\left[\overline{\omega(\zeta)}\,\Phi_0'(\zeta) + \omega'(\zeta)\Psi_0(\zeta)\right]$$

where $\varphi'(z) = \Phi(z) + \Phi_0(\zeta)$ ...(4.18)

$$\psi'(z) = \Psi(z) = \Psi_0(\zeta)$$

and
$$\varphi'(z) = \frac{d\varphi}{dz} = \frac{d\varphi}{d\zeta}\cdot\frac{d\zeta}{dz} = \frac{\varphi'(\zeta)}{\omega'(\zeta)}$$

$$\psi'(z) = \frac{\partial\psi}{dz} = \frac{d\psi}{d\zeta}\cdot\frac{d\zeta}{dz} = \frac{\psi'(\zeta)}{\omega'(\zeta)}$$

When mapping into circular regions, Muskhelishvili put $\zeta = \rho e^{i\theta}$ and considers the orthogonal curvilinear co-ordinate system (ρ, θ). If, at P, an increment dz_1 is taken along the curve θ = const., in the direction of ρ increasing, the corresponding increment $d\zeta$ will be in the radial direction. Then if β is the angle between the tangent to the curve θ = const. and Ox,

$$dz_1 = e^{i\beta}\,|\,dz_1\,|$$

at P, while
$$d\zeta = e^{i\beta}\,|\,d\zeta\,|$$

It follows that $e^{i\beta} = \dfrac{dz_1}{|\,dz_1\,|} = \dfrac{\omega'(\zeta)d\zeta}{|\,\omega'(\zeta)\,d\zeta\,|}$

$$\left.\begin{aligned} e^{i\beta} &= e^{i\theta}\frac{\omega'(\zeta)}{|\,\omega'(\zeta)\,|} = \frac{\zeta\omega'(\zeta)}{\rho\,|\,\omega'(\zeta)\,|} \\ \text{and}\quad e^{-i\beta} &= \frac{\bar{\zeta}}{\rho}\cdot\frac{\overline{\omega'}\,(\bar{\zeta})}{|\,\omega'(\zeta)\,|} \end{aligned}\right\} \quad ...(4.19)$$

Comparing Eqs. (4.14) and (4.19), we find

$$\beta = \alpha + \theta$$

Hence replacing $e^{i\alpha}$ by $e^{i\beta}$, we get from Eq. (4.17)

$$2G(u_\rho + iu_\theta) = \frac{\bar{\zeta}\,\overline{\omega'}\,(\bar{\zeta})}{\rho\,|\,\omega'(\zeta)\,|}\left[\chi\varphi_0(\zeta) - \frac{\omega(\zeta)}{\overline{\omega'}\,(\bar{\zeta})}\,\overline{\varphi'}_0(\bar{\zeta}) - \overline{\psi'}_0(\bar{\zeta})\right] \quad ...(4.20)$$

and Eqs. (4.18) give

$$\sigma_\rho + \sigma_\theta = 2\left[\Phi_0(\zeta) + \overline{\Phi}_0(\bar{\zeta})\right] \quad ...(4.21)$$

$$\sigma_\rho - \sigma_\theta + 2i\,\tau_{\rho\theta} = \frac{2\zeta^2}{\rho^2\,\overline{\omega'}\,(\bar{\zeta})}\left[\bar{\omega}\,(\bar{\zeta})\,\Phi_0'(\zeta) + \omega'(\zeta)\Psi_0(\zeta)\right] \quad ...(4.22)$$

when $z = \omega(\zeta) = \zeta$, then

$$2G(u_\rho + iu_\theta) = e^{-i\theta}\left[\chi\varphi(z) - z\,\overline{\varphi'}\,(\bar{z}) - \overline{\psi}(\bar{z})\right] \quad ...(4.23)$$

$$\sigma_\rho + \sigma_\theta = 2[\Phi(z) + \overline{\Phi}\,(\bar{z})] \quad ...(4.24)$$

$$\sigma_\rho - \sigma_\theta + 2i\tau_{\rho\theta} = 2\, e^{2i\theta}\, [\bar{z}\, \Phi'(z) + \Psi(z)] \qquad ...(4.25)$$

where $z = \rho e^{i\theta}$, $\psi(z) = \Psi'(z)$, $\Phi(z) = \varphi'(z)$

4.5 Polar Co-ordinates

Consider the analytic function

$$x + iy = e^{f(z)}$$

$$\therefore \quad f(z) = \log (x + iy)$$

$$= \log (r \cos \theta + i\, r \sin \theta)$$

$$= \log (r\, e^{i\theta})$$

$$= \log r + i\theta$$

$$= \log (x^2 + y^2)^{1/2} + i \tan^{-1} \frac{y}{x}$$

$$= \xi(x, y) + i^{\eta}(x, y)$$

$$\therefore \quad \xi(x, y) = \log r = \log (x^2 + y^2)^{1/2}$$

$$\eta(x, y) = \theta = \tan^{-1} \frac{y}{x}$$

Hence equations of transformation are

$$\xi = \log r \qquad ...(4.26)$$

$$\eta = \theta \qquad ...(4.27)$$

Eq. (4.26) transforms concentric circles with centres at the origin into straight lines parallel to the η-axis and Eq. (4.27) transforms radial lines into lines parallel to the η-axis as shown in Fig. 4.3. The circle of unit radius is transformed into the η-axis and that

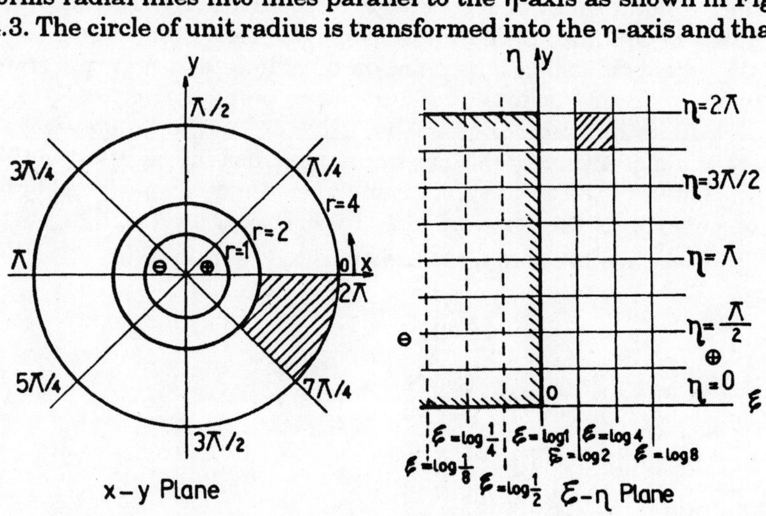

(a) (b)

Fig. 4.3

the x-axis is transformed into ξ-axis. All circles lying inside the unit circle give negative values. The entire xy-plane is transformed into a strip of width equal to 2π in η-direction and of infinite length in the ξ-direction.

4.6 Elliptic Co-ordinates

Co-ordinates which are constant over elliptic boundaries are obtained from the transformation

$$z = c \cosh \zeta \qquad \text{...(4.28)}$$

whence equating real and imaginary parts, we get

$$x = c \cosh \xi \cos \eta, y = c \sinh \xi \sin \eta \qquad \text{...(4.29)}$$

where $z = x + iy$ and $\zeta = \xi + i\eta$.

The curves ξ = constant are given by

$$\frac{x^2}{c^2 \cosh^2 \xi} + \frac{y^2}{c^2 \sinh^2 \xi} = \cos^2 \eta + \sin^2 \eta = 1 \quad \text{...(4.30)}$$

which are a set of confocal ellipses whose foci are $(\pm c, 0)$. The curves η = constant are given by

$$\frac{x^2}{c^2 \cos^2 \eta} - \frac{y^2}{c^2 \sin^2 \eta} = \cosh^2 \xi - \sinh^2 \xi = 1 \quad \text{...(4.31)}$$

which are a set of hyperbolas confocal with and orthogonal to the previous ellipses.

The whole of the xy-plane does not transform into the whole of the $\xi\eta$-plane. ξ may be restricted to be positive and the same value of ξ may be assigned to the whole of one ellipse. η then corresponds to the eccentric angle of the point on the ellipse and may run from 0 to 2π or from $-\pi$ to $+\pi$. Thus the xy-plane corresponds to a semi-infinite rectangle $0 < \xi < +\infty$, $0 < \eta < 2\pi$ in the $\xi\eta$-plane, and there is a discontinuity in η whenever we cross the segment of the x-axis between the two foci, a given value of η being assigned only to one quarter of the corresponding hyperbola as shown in Fig. 4.4.

Since x, y are conjugate functions of ξ, η, therefore,

$$\frac{\partial x}{\partial \xi} = \frac{\partial y}{\partial \eta}, \frac{\partial x}{\partial \eta} = \frac{-\partial y}{\partial \xi}$$

$$\therefore \quad \frac{1}{h_1^2} = \frac{1}{h_2^2} = \left(\frac{\partial x}{\partial \xi}\right)^2 + \left(\frac{\partial x}{\partial \eta}\right)^2$$

$$= J^2 \text{ (say)} \qquad \text{...(4.32)}$$

and also

$$J^2 = \left(\frac{\partial x}{\partial \xi}\right)^2 + \left(\frac{\partial y}{\partial \xi}\right)^2$$

$$= \left(\frac{\partial x}{\partial \xi} + i \frac{\partial y}{\partial \xi} \right) \left(\frac{\partial x}{\partial \xi} - i \frac{\partial y}{\partial \xi} \right)$$

$$= f'(\xi + i\eta)\, f'(\xi - i\eta)$$

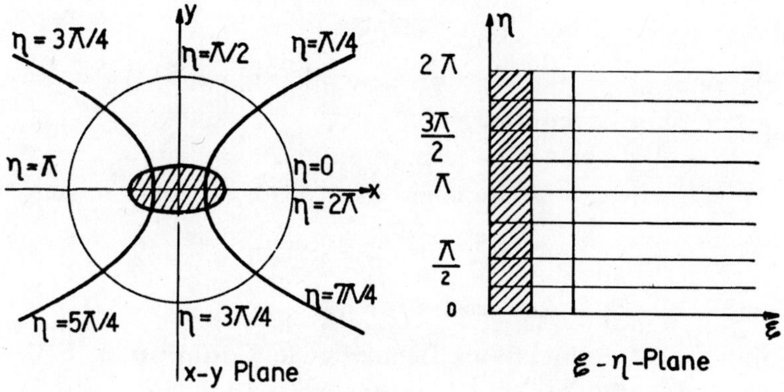

(a) (b)

Fig. 4.4

If $(x + dx, y + dy)$ corresponds to $(\xi + d\xi, \eta)$, then

$$dx = ds_\xi \cos \alpha, \; dy = ds_\xi \sin \alpha$$

where $\dfrac{ds_\xi}{d\xi} = J$

Thus
$$\frac{\partial x}{\partial \xi} = J \cos \alpha, \; \frac{\partial y}{\partial \eta} = J \sin \alpha$$

$$\frac{\partial x}{\partial \eta} = -J \sin \alpha, \; \frac{\partial y}{\partial \eta} = J \cos \alpha$$

$$Je^{i\alpha} = \frac{\partial x}{\partial \xi} + i \frac{\partial y}{\partial \xi} = f'(\xi + i\eta)$$

$$Je^{-i\alpha} = f'(\xi - i\eta)$$

Taking logs, we get

$$\log J + ix = \log f'(\xi + i\eta)$$

so that $\log J$ and α are real and imaginary parts of a function of $\xi + i\eta$, that is of $x + iy$, and so are conjugate functions in either set of variables.

$$\frac{1}{J}\frac{\partial J}{\partial \xi} = \frac{\partial \alpha}{\partial \eta}, \; \frac{1}{J}\frac{\partial J}{\partial \eta} = -\frac{\partial \alpha}{\partial \xi}$$

If ρ_1 is the radius of carvature of the curve

η = const. and ρ_2 that of the curve ξ = const., then

$$\frac{1}{\rho_1} = \frac{\partial \alpha}{\partial S_\xi} = \frac{1}{J}\frac{\partial \alpha}{\partial \xi} = -\frac{1}{J^2}\frac{\partial J}{\partial \eta}$$

$$\frac{1}{\rho_2} = \frac{\partial \alpha}{\partial S_\eta} = \frac{1}{J}\frac{\partial \alpha}{\partial \eta} = \frac{1}{J^2}\frac{\partial J}{\partial \xi} \qquad ...(4.33)$$

To find J and α in elliptic co-ordinates, we have

$$Je^{i\alpha} = \frac{dz}{d\zeta} = c \sinh \zeta$$

$$= c(\sinh \xi \cos \eta + i \sin \eta \cosh \xi)$$

Thus $J \cos \alpha = c \sinh \xi \cos \eta$

$$J \sin \alpha = c \cosh \xi \sin \eta$$

$$J^2 = c^2(\sinh^2 \xi \cos^2 \eta + \cosh^2 \xi \sin^2 \eta)$$

$$= \frac{c^2}{2}(\cosh 2\xi - \cos 2\eta) \qquad ...(4.34)$$

and $\tan \alpha = \cosh \xi \tan \eta \qquad ...(4.35)$

4.6.1 Solution of Biharmonic Equation in Elliptical Co-ordinates

Since $\nabla^2\phi = 0$ transforms into

$$\frac{\partial^2\phi}{\partial\xi^2} + \frac{\partial^2\phi}{\partial\eta^2} = 0, \text{ where } \phi \text{ is any function}$$

we have typical values of ϕ of the form $e^{\pm n\xi \pm ni\eta}$, *i.e.* of the form $e^{\pm n\xi} \cos n\eta$ and $e^{\pm n\xi} \sin n\eta$, n being any constant.

Now $\Phi = x\phi$ and $\Phi = y\phi$

are solutions of the biharmonic equation

$$\nabla^4\Phi = 0.$$

Hence

$$2e^{\pm n\xi} \cosh \xi \,.\, \cos \eta \,.\, \cos n\eta$$

and

$$2e^{\pm n\xi} \sinh \xi \,.\, \sin \eta \,.\, \sin n\eta$$

are solutions for the stress function. These give

$$e^{\pm n\xi} \cosh \xi[\cos (n-1)\eta + \cos (n+1)\eta]$$

$$e^{\pm n\xi} \sinh \xi\, [\cos(n-1)\eta - \cos (n+1)\eta]$$

Whence by addition and subtraction, we obtain the typical solutions.

$$e^{\pm n\xi + \xi} \cos (n-1)\, \eta + e^{\pm n\xi - \xi} \,.\, \cos (n+1)\eta$$

and

$$e^{\pm n\xi - \xi} \,.\, \cos (n-1)\, \eta + e^{\pm n\xi + \xi} \,.\, \cos (n+1)\eta$$

The solution corresponding to the lower signs in the first of these and that corresponding to the upper signs in the second are of the form

$$e^{\pm m\xi} \,.\, \cos mx, \text{ (} m \text{ is any constant).}$$

and are solutions of $\nabla^2\Phi = 0$ as well as $\nabla^4\Phi = 0$. The solutions characteristics of $\nabla^4\Phi = 0$ are, therefore,

$$\Phi = e^{(n+1)\xi}.\cos(n-1)\eta + e^{(n-1)\xi}\cos(n+1)\eta \quad ...(4.36)$$

and $$\Phi = e^{-(n+1)\xi}.\cos(n-1)\eta + e^{-(n-1)\xi}\cos(n+1)\eta \quad ...(4.37)$$

Similarly

$$\Phi = e^{(n+1)\xi}\sin(n-1)\eta + e^{(n-1)\xi}\sin(n+1)\eta \quad ...(4.38)$$

and $$\Phi = e^{-(n+1)\xi}\sin(n-1)\eta + e^{-(n-1)\xi}\sin(n+1)\eta \quad ...(4.39)$$

With these we shall require to combine the four harmonic solutions,

$$\Phi = e^{\pm n\xi}\cos n\eta$$

and $$\Phi = e^{\pm n\xi}\sin n\eta \quad ...(4.40)$$

giving altogether eight solutions, which may refer to as the solutions of order n.

4.6.2 Stresses in Curvilinear Coordinates

Now $$\sigma_{\xi x} = \sigma_x\cos\alpha + \tau_{xy}\sin\alpha$$

$$= \left(\cos\alpha.\frac{\partial}{\partial y} - \sin\alpha.\frac{\partial}{\partial x}\right)\frac{\partial\Phi}{\partial y}$$

$$= \left\{\cos\left(\frac{\pi}{2}+\alpha\right)\frac{\partial}{\partial x} + \sin\left(\frac{\pi}{2}+\alpha\right)\frac{\partial}{\partial y}\right\}\frac{\partial\Phi}{\partial y}$$

$$= \frac{\partial}{\partial S_\eta}\left(\frac{\partial\Phi}{\partial y}\right)$$

$$= \frac{\partial}{\partial y}\left(\frac{\partial\Phi}{\partial S_\eta}\right) - \frac{\partial\cos\alpha}{\partial y}.\frac{\partial\Phi}{\partial y} + \frac{\partial\sin\alpha}{\partial y}.\frac{\partial\Phi}{\partial x}$$

$$= \frac{\partial}{\partial y}\left(\frac{\partial\Phi}{\partial S_\eta}\right) + \left(\sin\alpha\frac{\partial\Phi}{\partial y} + \cos\alpha.\frac{\partial\Phi}{\partial x}\right)\frac{\partial\Phi}{\partial y}$$

$$= \frac{\partial}{\partial y}\left(\frac{\partial\Phi}{\partial S_\eta}\right) + \left(\frac{\partial\Phi}{\partial S_\xi}\right)\frac{\partial\alpha}{\partial y}$$

Similarly

$$\sigma_{\xi y} = \tau_{xy}\cos\alpha + \sigma_y\sin\alpha$$

$$= -\frac{\partial}{\partial S_\eta}\left(\frac{\partial\Phi}{\partial x}\right)$$

$$= -\frac{\partial}{\partial x}\left(\frac{\partial\Phi}{\partial S_\eta}\right) - \left(\frac{\partial\Phi}{\partial S_\xi}\right)\frac{\partial\alpha}{\partial x}$$

$$\sigma_{\eta x} = \sigma_x \cos\left(\frac{\pi}{2} + \alpha\right) + \tau_{xy} \sin\left(\frac{\pi}{2} + \alpha\right) = -\frac{\partial}{\partial S_\xi}\left(\frac{\partial \Phi}{\partial y}\right)$$

$$= -\frac{\partial}{\partial y}\left(\frac{\partial \Phi}{\partial S_\xi}\right) + \left(\frac{\partial \Phi}{\partial S_\eta}\right)\frac{\partial \alpha}{\partial y}$$

$$\sigma_{\eta y} = \frac{\partial}{\partial S_\xi}\left(\frac{\partial \Phi}{\partial x}\right)$$

$$= \frac{\partial}{\partial x}\left(\frac{\partial \Phi}{\partial S_\xi}\right) - \left(\frac{\partial \Phi}{\partial S_\eta}\right)\frac{\partial \alpha}{\partial x}$$

$$\therefore \quad \sigma_\xi = \sigma_{\xi x} \cos\alpha + \sigma_{\xi y} \sin\alpha$$

$$= \left(-\sin\alpha \frac{\partial}{\partial x} + \cos\alpha \frac{\partial}{\partial y}\right)\frac{\partial \Phi}{\partial S_\eta}$$

$$+ \frac{\partial \Phi}{\partial S_\xi}\left(-\sin\alpha \frac{\partial}{\partial x} + \cos\alpha \frac{\partial}{\partial y}\right)\alpha$$

$$= \frac{\partial^2 \Phi}{\partial S^2_\eta} + \frac{\partial \Phi}{\partial S_\xi}\frac{\partial \alpha}{\partial S_\eta}$$

$$= \frac{\partial^2 \Phi}{\partial S^2_\eta} + \frac{1}{\rho_2} + \frac{\partial \Phi}{\partial S_\xi} \qquad \text{...(4.41)}$$

$$\tau_{\xi\eta} = \tau_{\xi x} \cos\left(\alpha + \frac{\pi}{2}\right) + \sigma_{\xi y} \sin\left(\alpha + \frac{\pi}{2}\right)$$

$$= -\frac{\partial}{\partial S_\xi}\left(\frac{\partial \Phi}{\partial S_\eta}\right) - \frac{\partial \Phi}{\partial S_\xi} \cdot \frac{\partial \alpha}{\partial S_\xi}$$

$$= -\frac{\partial}{\partial S_\xi}\left(\frac{\partial \Phi}{\partial S_\eta}\right) - \frac{1}{\rho_1}\frac{\partial \Phi}{\partial S_\xi} \qquad \text{...(4.42)}$$

or

$$\tau_{\xi\eta} = \sigma_{\eta x} \cos\alpha + \sigma_{\eta y} \sin\alpha$$

$$= -\frac{\partial}{\partial S_\eta}\left(\frac{\partial \Phi}{\partial S_\xi}\right) + \frac{\partial \Phi}{\partial S_\eta} \cdot \frac{\partial \alpha}{\partial S_\eta}$$

$$= -\frac{\partial}{\partial S_\eta}\left(\frac{\partial \Phi}{\partial S_\xi}\right) + \frac{1}{\rho_2}\frac{\partial \Phi}{\partial S_\eta} \qquad \text{...(4.43)}$$

$$\sigma_\eta = -\sigma_{\eta x} \sin\alpha + \sigma_{\eta y} \sin\alpha$$

$$= -\frac{\partial}{\partial S_\eta}\left(\frac{\partial \Phi}{\partial S_\xi}\right) - \left(\frac{\partial \Phi}{\partial S_\eta}\right) \cdot \frac{\partial \alpha}{\partial S_\xi}$$

$$= \frac{\partial^2 \Phi}{\partial S^2_\xi} - \frac{1}{\rho_1}\frac{\partial \Phi}{\partial S_\eta} \qquad \text{...(4.44)}$$

These stresses can be written as

$$\sigma_\xi = \frac{1}{J}\cdot\frac{\partial}{\partial\eta}\frac{\partial\Phi}{\partial\eta} + \frac{1}{J}\frac{\partial\alpha}{\partial\eta}\cdot\frac{1}{J}\frac{\partial\Phi}{\partial\xi}$$

$$= \frac{1}{J^2}\frac{\partial^2\Phi}{\partial\eta^2} + \frac{1}{J^3}\left(\frac{\partial\Phi}{\partial\xi}\cdot\frac{\partial J}{\partial\xi} - \frac{\partial\Phi}{\partial\eta}\cdot\frac{\partial J}{\partial\eta}\right) \qquad \text{...(4.45)}$$

$$\sigma_\eta = \frac{1}{J}\frac{\partial}{\partial\xi}\cdot\frac{1}{J}\cdot\frac{\partial\Phi}{\partial\xi} - \frac{1}{J}\cdot\frac{\partial\alpha}{\partial\xi}\cdot\frac{1}{J}\frac{\partial\Phi}{\partial\eta}$$

$$= \frac{1}{J^2}\frac{\partial^2\Phi}{\partial\xi^2} + \frac{1}{J^3}\left(\frac{\partial\Phi}{\partial\eta}\cdot\frac{\partial J}{\partial\eta} - \frac{\partial\Phi}{\partial\xi}\cdot\frac{\partial J}{\partial\xi}\right) \qquad \text{...(4.46)}$$

$$\tau_{\xi\eta} = -\frac{1}{J}\cdot\frac{\partial}{\partial\eta}\cdot\frac{1}{J}\cdot\frac{\partial\Phi}{\partial\xi} + \frac{1}{J}\frac{\partial\alpha}{\partial\eta}\cdot\frac{1}{J}\frac{\partial\Phi}{\partial\eta}$$

$$= -\frac{1}{J}\cdot\frac{\partial}{\partial\xi}\cdot\frac{1}{J}\cdot\frac{\partial\Phi}{\partial\eta} - \frac{1}{J}\frac{\partial\alpha}{\partial\xi}\cdot\frac{1}{J}\frac{\partial\Phi}{\partial\xi}$$

$$= -\frac{1}{J^2}\frac{\partial^2\Phi}{\partial\xi\partial\eta} + \frac{1}{J^3}\left(\frac{\partial\Phi}{\partial\xi}\cdot\frac{\partial J}{\partial\eta} + \frac{\partial\Phi}{\partial\eta}\cdot\frac{\partial J}{\partial\xi}\right) \qquad \text{...(4.47)}$$

4.6.3 Stresses in Elliptic Coordinates.

Now

$$J^4\sigma_\xi = J^2\frac{\partial^2\Phi}{\partial\eta^2} + \frac{1}{2}\left(\frac{\partial\Phi}{\partial\xi}\cdot\frac{\partial J^2}{\partial\xi} - \frac{\partial\Phi}{\partial\eta}\cdot\frac{\partial J^2}{\partial\eta}\right)$$

or

$$(2J^2)^2\sigma_\xi = (2J^2)\,.\,2\frac{\partial^2\Phi}{\partial\eta^2} + \frac{\partial\Phi}{\partial\xi}\cdot\frac{\partial(2J^2)}{\partial\xi} - \frac{\partial\Phi}{\partial\xi}\cdot\frac{\partial(2J^2)}{\partial\eta}$$

Similarly

$$(2J^2)^2\,\sigma_\eta = (2J^2).\,2\frac{\partial^2\Phi}{\partial\xi^2} + \frac{\partial\Phi}{\partial\xi}\cdot\frac{\partial(2J^2)}{\partial\eta} - \frac{\partial\Phi}{\partial\xi}\cdot\frac{\partial(2J^2)}{\partial\xi}$$

$$(2J^2)\,\tau_{\xi\eta}^2 = -(2J^2)\,.\,2\frac{\partial^2\Phi}{\partial\xi\partial\eta} + \frac{\partial\Phi}{\partial\xi}\cdot\frac{\partial(2J^2)}{\partial\eta} + \frac{\partial\Phi}{\partial\eta}\cdot\frac{\partial(2J^2)}{\partial\xi}$$

Also

$$\Phi = e^{(n+1)\xi}\cos(n-1)\,\eta + e^{(n-1)\xi}\cos(n+1)\,\eta$$

$$(2J^2)^2\,\sigma_\xi = c^2(\cosh 2\xi - \cos 2\eta)\,\{-2(n-1)^2 e^{(n+1)}.\cos(n-1)\,\eta$$
$$-2(n+1)^2 e^{(n-1)\xi}\cos(n+1)\eta\}$$
$$+2c^2\sinh 2\xi\{(n+1)\,e^{(n+1)\xi}\cos(n-1)\eta$$
$$+(n-1)e^{(n-1)\xi}\cos(n+1)\,\eta\}$$
$$+2c^2\sin 2\eta\{(n-1)\,e^{(n+1)\xi}\sin(n-1)\eta$$
$$+(n+1)\,e^{(n-1)\xi}\sin(n+1)\eta\}$$

$$+ c^2 \cos (n-1)\eta \{-2(n-1)^2 e^{(n+1)\xi} \cosh 2\xi$$
$$+ 2(n+1)\, e^{(n+1)\xi} \sinh 2\xi + (n+1)^2\, e^{(n-1)\xi}$$
$$+ (n+1)\, e^{(n-1)\xi}\}$$
$$+ c^2 \cos (n+1)\, \eta\{-(n+1)^2\, e^{(n-1)\xi} \cosh 2\xi +$$
$$+ 2(n-1)\, e^{(n-1)\xi} \sinh 2\xi$$
$$+ (n-1)^2\, e^{(n+1)\xi} - (n-1)\, e^{(n+1)\xi}\}$$
$$+ c^2 \cos (n+3)\eta\{(n+1)^2 e^{(n-1)\xi} - (n+1)e^{(n-1)\xi}\}$$
$$+ c^2 \cos (n+3)\eta\{(n-1)^2 e^{(n+1)\xi} + (n-1)e^{(n+1)\xi}\}$$
$$= c^2 \cos (n-1)\, \eta\{4ne^{(n-1)\xi} - n(n-3)e^{(n+1)\xi}\}$$
$$+ c^2 \cos (n+1)\eta\{-4ne^{(n+1)\xi} - n(n+3)e^{(n+1)\xi}\}$$
$$+ c^2 \cos (n+3)\eta\{n(n+1)\}e^{(n-1)\xi}$$
$$+ c^2 \cos (n-3)\eta\{n(n-1)\}e^{(n+1)\xi}$$

Similarly

$$\frac{c^2}{n} (\cosh 2\xi - \cos 2\eta)^2 \sigma_\xi$$
$$= \cos (n+3)\eta \,.\, (n+1)e^{(n-1)\xi}$$
$$- \cos (n+1)\eta \,.\, \{4e^{(n-1)\xi} + (n+3)e^{(n-3)\xi}\}$$
$$+ \cos (n-1)\eta \,.\, \{4e^{(n-1)\xi} - (n-3)e^{(n+3)\xi}\}$$
$$+ \cos (n-3)\eta \,.\, (n-1)\, e^{(n+1)\xi} \qquad \text{...(4.48)}$$

$$\frac{c^2}{n} (\cosh 2\xi - \cos 2\eta)^2 \sigma_\eta$$
$$= -\cos (n+3)\eta \,.\, (n-3)e^{(n-1)\xi}$$
$$- \cos (n+1)\eta \,.\, \{4e^{(n+1)\xi} - (n-1)e^{(n-3)\xi}\}$$
$$+ \cos (n-1)\eta \,.\, \{4e^{(n-1)\xi} + (n+1)e^{(n+3)\xi}\}$$
$$- \cos(n-3)\eta \,.\, (n+3)e^{(n+1)\xi} \qquad \text{...(4.49)}$$

$$\frac{c^2}{n} (\cosh 2\xi - \cos 2\eta)^2 . \tau_{\xi\eta}$$
$$= -\sin (n+3)\eta \,.\, (n-1)\, e^{(n-1)\xi}$$
$$+ \sin (n+1)\eta \,.\, (n+1)e^{(n-3)\xi}$$
$$+ \sin (n-1)\eta \,.\, (n-1)\, e^{(n+3)\xi}$$
$$- \sin (n-3)\eta \,.\, (n+1)e^{(n+1)\xi} \qquad \text{...(4.50)}$$

Taking $\Phi = e^{n\xi} \cos n\eta$, we obtain

$$\frac{c^2}{n} (\cosh 2\xi - \cos 2\eta)^2 \sigma_\xi = \cos (n+2)\eta \quad (n-1)\, e^{n\xi}$$

$$- \cos n\eta[(n-1)\, e^{(n+2)\xi} + (n+1)\, e^{(n-2)\xi}]$$

$$+ \cos (n-2)\eta \,.\, (n+1)\, e^{n\xi} \qquad \text{...(4.51)}$$

$$\sigma_\eta = -\sigma_\xi \qquad \text{...(4.52)}$$

$$\frac{c^2}{n} (\cosh 2\xi - \cos 2\eta)^2 \tau_{\xi\eta} = -\sin (n+2)\eta\, (n-1)\, e^{n\xi}$$

$$+ \sin n\eta[(n-1)e^{(n+2)\xi} + (n+1)\, e^{(n-2)\xi}]$$

$$- \sin (n-2)\eta \,.\, (n+1)e^{n\xi} \qquad \text{...(4.53)}$$

4.6.4 Infinite Plate with an Elliptic Hole under All-round Tension

For this problem $\xi = \alpha$ will be taken to be the boundary of the hole. At infinity the stress function Φ must reduce to :

$$\Phi = \frac{T}{2} (x^2 + y^2) \qquad \text{...(1)}$$

where T is the given all round tension. Expressing this in terms of ξ, η, we have

$$\left.\begin{aligned} \Phi &= \tfrac{1}{4}\, C^2 T (\cosh 2\xi + \cos 2\eta) \\ &= \tfrac{1}{8}\, C^2\, T\, (e^{2\xi} + 2 \cos 2\eta) \end{aligned}\right\} \qquad \text{...(2)}$$

approximately, when ξ is large.

For $n = 1$, $n = -1$ and $n = 2$, we have

$$\left.\begin{aligned} \Phi_1 &= e^{2\xi} + \cos 2\eta \\ \Phi_2 &= e^{-2\xi} + \cos 2\eta \\ \Phi_3 &= e^{-2\xi} \cos 2\eta \\ \Phi_4 &= \xi \end{aligned}\right\} \qquad \text{...(3)}$$

$e^{2\xi} \cos 2\eta$ being rejected, as it leads to an unsuitable term when

$$\xi = +\infty$$

The corresponding stresses are given by

$$C^2 (\cosh 2\xi - \cos 2\eta)^2 (\sigma_\xi)_1 = 2 \cos 4\eta - 8 \cos 2\eta \cosh 2\xi$$

$$+ 4 + 2\, e^{4\xi}$$

$$C^2 (\cosh 2\xi - \cos 2\eta)^2 (\sigma_\eta)_1 = 2 \cos 4\eta$$

$$- 8 \cos 2\eta\, e^{2\xi} + 4 + 2e^{4\xi}$$

$$C^2 (\cosh 2\xi - \cos 2\eta)^2 (\tau_{\xi\eta})_1 = 4 \sin 2\eta \cosh 2\xi$$

$$C^2 (\cosh 2\xi - \cos 2\eta)^2 (\sigma_\xi)_2 = 2 \cos 4\eta - 8 \cos 2\eta \cosh 2\xi + 4 + 2e^{-4\xi}$$

$$C^2 (\cosh 2\xi - \cos 2\eta)^2 (\sigma_\eta)_2 = 2 \cos 4\eta - 8 \cos 2\eta\, e^{-2\xi} + 4 + 2\, e^{-4\xi}$$

$$C^2 (\cosh 2\xi - \cos 2\eta)^2 (\tau_{\xi\eta})_2 = -4 \sin 2\eta \cosh 2\xi$$

$$\frac{C^2}{2} (\cosh 2\xi - \cos 2\eta)^2 (\sigma_\xi)_3 = \cos 4\eta e^{-2\xi} - \cos 2\eta\, (e^{-4\xi} + 3) + 3e^{-2\xi}$$

$$\frac{C^2}{2} (\cosh 2\xi - \cos 2\eta)^2 (\sigma_\eta)_3 = -\cos 4\eta\, e^{-2\xi} - 3\, e^{-2\xi} + \cos 2\eta\, (e^{-4\xi} + 3)$$

$$\frac{C^2}{2} (\cosh 2\xi - \cos 2\eta)^2 (\sigma_{\xi\eta})_3 = \sin 4\eta\, e^{-2\xi} - \sin 2\eta\, (e^{-4\xi} + 3)$$

$$C^2 (\cosh 2\xi - \cos 2\eta)^2 (\sigma_\xi)_4 = 2 \sinh 2\xi$$

$$C^2 (\cosh 2\xi - \cos 2\eta)^2 (\sigma_\eta)_4 = -2 \sinh 2\xi$$

$$C^2 (\cosh 2\xi - \cos 2\eta)^2 (\tau_{\xi\eta})_4 = 2 \sin 2\eta$$

We have to combine these stresses so as to make σ_ξ, $\tau_{\xi\eta}$ both vanish over $\xi = a$. Φ_3 is the only solution which introduces a term in $\sin 4\eta$ in $\tau_{\xi\eta}$ and this term cannot vanish unless the coefficient of Φ_3 also vanishes. We may, therefore, omit Φ_3 altogether from the solution. Assume then

$$\Phi = A\Phi_1 + B\Phi_2 + C\Phi_4 \qquad \text{...(4)}$$

The conditions for $\sigma_\xi = \tau_{\xi\eta} = 0$ for $\xi = a$ give $2A + 2B = 0$

$$-8A \cosh 2a - 8B \cosh 2a = 0$$

$$A(4 + 2\, e^{4a}) + B(4 + 2\, e^{-4a}) + 2C \sinh 2a = 0$$

$$4A \cosh 2a - 4B \cosh 2a + 2C = 0$$

$$\therefore \quad B = -A$$

$$4A \sinh 4a + 2C \sinh 2a = 0$$

$$8A \cosh 2a + 2C = 0$$

$$\therefore \quad C = -4A \cosh 2a$$

$$\therefore \quad \Phi = A\{\Phi_1 - \Phi_2 - 4\phi_4 \cosh 2a\}$$

$$= 2A\, \{\sinh 2\xi - 2\xi \cosh 2a\} \qquad \text{...(5)}$$

If $A = \frac{1}{8} C^2T$, then the above expression leads to the same solution at infinity, for the stresses due to cos 2η and ξ respectively vanish when ξ is large. The complete expression for the stresses become.

$$(\cosh 2\xi - \cos 2\eta)^2 \sigma_\xi = T \sinh 2\xi (\cosh 2\xi - \cosh 2a)$$

$$(\cosh 2\xi - \cos 2\eta)^2 \sigma_\eta = T \sinh 2\xi (\cosh 2\xi + \cosh 2a - 2 \cos 2\eta) \quad ...(6)$$

$$(\cosh 2\xi - \cos 2\eta)^2 \tau_{\xi\eta} = T \sin 2\eta (\cosh 2\xi - \cosh 2a)$$

$$\left(\sigma_\eta\right)_{\xi = a} = \frac{2T \sinh 2a (\cosh 2a - \cos 2\eta)}{(\cosh 2a - \cos 2\eta)^2}$$

$$= \frac{2T \sinh 2a}{\cosh 2a - \cos 2\eta} \quad ...(4.54)$$

At $\eta = 0$

$$\left(\sigma_\eta\right)_{\substack{\xi = a \\ \eta = 0}} = \frac{2T \sinh 2a}{\cosh 2a - 1} = 2T \left(\frac{C \cosh a}{C \sinh a}\right) \quad ...(4.55)$$

At $\eta = 90°$

$$\left(\sigma_\eta\right)_{\substack{\xi = a \\ \eta = 90°}} = \frac{2T \sinh 2a}{\cosh 2a + 1} = 2T \left(\frac{C \sinh a}{C \cosh a}\right) \quad ...(4.56)$$

where $C \cosh a$ and $C \sinh a$ are the semi-axes of the elliptical hole.

4.6.5 Plate with an Elliptic Hole under Simple Tension

Suppose first of all that at a distance from the hole there is a tension T parallel to the y-axis. Thus at infinity

$$\Phi = \frac{1}{2} T x^2 = T \frac{C^2}{8} e^{2\xi} \cos^2 \eta$$

$$= \frac{TC^2}{16} e^{2\xi} + \frac{TC^2}{16} . e^{2\xi} \cos 2\eta$$

In this case, in addition to the functions already considered in art. 4.6.4, we introduce

$$\Phi_5 = e^{2\xi} \cos 2\eta$$

with the corresponding stress components,

$$C^2 (\cosh 2\xi - \cos 2\eta)^2 (\sigma_\xi)_5$$

$$= 2 \cos 4\eta . e^{2\xi} - \cos 2\eta (2 e^{4\xi} + 6) + 6 e^{2\xi}$$

$$C^2 (\cosh 2\xi - \cos 2\eta)^2 (\sigma_\eta)_5$$

$$= - 2 \cos 4\eta . e^{2\xi} + \cos 2\eta (2 e^{4\xi} + 6) - 6 e^{2\xi}$$

$$C^2 (\cosh 2\xi - \cos 2\eta)^2 (\tau_{\xi\eta})_5$$
$$= -2 \sin 4\eta . e^{2\xi} + \sin 2\eta \, (2 e^{4\xi} + 6)$$

Now we assume

$$\Phi = A\Phi_1 + B\Phi_2 + C\Phi_3 + D\Phi_4 + F\Phi_5$$

where A, B, C, D and F are constants.

For $\xi = a$, $\sigma_\xi = \tau_{\xi\eta} = 0$. Hence

$$A + B + C e^{-2a} + F e^{2a} = 0$$
$$4A \cosh 2a + 4B \cosh 2a + C (3 + e^{-4a}) + F (3 + e^{4a}) = 0$$
$$A (2 + e^{4a}) + B (2 + e^{-4a}) + 3C e^{-2a} + 3F e^{2a} + D \sinh 2a = 0$$
$$C e^{-2a} - F e^{2a} = 0$$
$$2A \cosh 2a - 2B \cosh 2a - (3 + e^{-4a}) + F(3 + e^{4a}) + D = 0$$

Solving, we get

$$A = \frac{1}{16} TC^2, B = -\frac{T}{16} C^2 (1 + e^{2a})$$
$$C = \frac{1}{16} TC^2 e^{4a}, D = -\frac{T}{4} C^2 (\cosh 2a + 1)$$
$$F = \frac{1}{16} TC^2$$

Thus
$$\Phi = \frac{1}{16} TC^2 [\Phi_1 - (1 + 2e^{2a}) \Phi_2 + e^{4a} \Phi_3$$
$$- 4 (\cosh 2a + 1) \Phi_4 + \Phi_5]$$

$$\left(\sigma_\eta\right)_{\xi = a} = T \left[\frac{e^{2a} \cos 2\eta + \sinh 2a - 1}{\cosh 2a - \cos 2\eta} \right]$$
$$= Te^{2a} \left[\frac{\sinh 2a \, (1 + e^{-2a})}{\cosh 2a - \cos 2\eta} - 1 \right] \quad ...(4.57)$$

$$\left(\sigma_\eta\right)_{\substack{\xi = a \\ \eta = 0}} = T \left[\frac{\cosh 2a - 1 + 2 \sinh 2a}{\cosh 2a - 1} \right]$$
$$= T(1 + 2 \coth a) \quad ...(4.58)$$

$$\left(\sigma_\eta\right)_{\substack{\xi = a \\ \eta = \pi/2}} = -T \quad ...(4.59)$$

When the tension T is parallel to the x-axis, then we would get

$$\left(\sigma_\eta\right)_{\xi = a} = T \left[\frac{\sinh 2a + 1 - e^{2a} \cos 2\eta}{\cosh 2a - \cos 2\eta} \right]$$

$$= T\left[e^{2a} + \frac{(1 - e^{2a})\sinh 2a}{\cosh 2a - \cos 2\eta}\right] \quad ...(4.60)$$

$$\left(\sigma_\eta\right)_{\substack{\xi = a \\ \eta = \pi/2}} = T\,(1 + 2\tanh a) \quad ...(4.61)$$

$$\left(\sigma_\eta\right)_{\substack{\xi = a \\ \eta = 0}} = -T \quad ...(4.62)$$

4.7 Bipolar-Coordinates

Consider the analytic function

$$\zeta = \log\left(\frac{z + ia}{z - ia}\right) = \log\left[\frac{x + i(y + a)}{x + i(y - a)}\right] \quad ...(1)$$

where $\zeta = \xi + i\eta$.

The quantity $z \pm ia$ is represented by the line joining the point $\mp ia$ to the point z in the xy plane, in the sense that its projections on the axes give the real and imaginary parts.

Let $z + ia = r_1 e^{i\theta_1}$ and $z - ia = r_2 e^{i\theta_2}$.

Then Eq. (1) becomes

$$\xi + i\eta = \log\left[\frac{r_1}{r_2} . e^{i(\theta_1 - \theta_2)}\right]$$

$$= \log\frac{r_1}{r_2} + i(\theta_1 - \theta_2) \quad ...(4.63)$$

$$\therefore \qquad \xi = \log\frac{r_1}{r_2}, \; \eta = \theta_1 - \theta_2$$

Conversely we can write

$$x = \frac{a \sin\eta}{\cosh\xi - \cos\eta}$$

$$y = \frac{a \sinh\xi}{\cosh\xi - \cos\eta} \quad ...(4.64)$$

It may be seen from Fig. 4.5 that $\theta_1 - \theta_2$ is the angle between the two lines joining the poles, $-ia$, ia to the typical point z, when this point lies to the right of the y-axis, and is minus this angle when the point lies to the left. It follows that a curve η = constant is an arc of a circle passing through the poles. A curve ξ = constant will be a curve for which $\frac{r_1}{r_2}$ = const. Such a curve is also a circle. It surrounds the pole ia if $\frac{r_1}{r_2}$ exceeds unity, *i.e.* if ξ is positive and surrounds the other pole $-ia$ if ξ is negative.

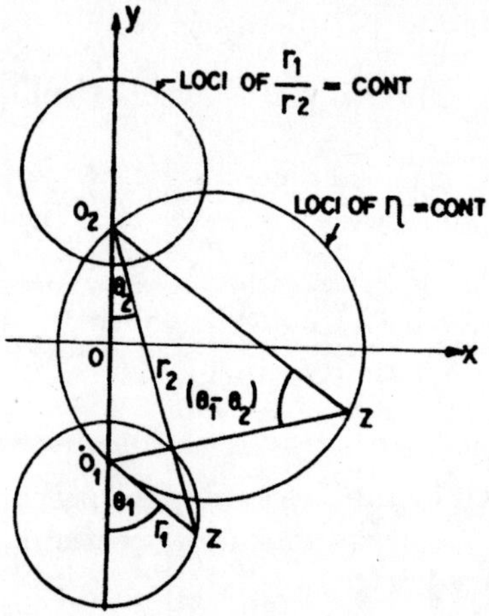

Fig. 4.5 Bipolar coordinates.

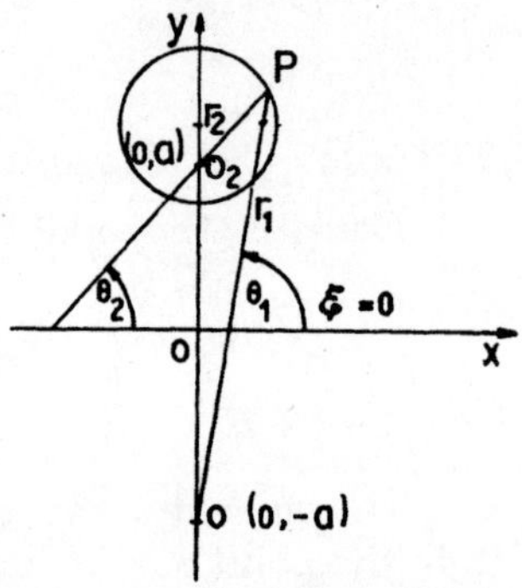

Fig. 4.6

The co-ordinate η changes from π to $-\pi$ on crossing the segment of the y-axis joining the poles. Now

$$z = ia \coth \tfrac{1}{2} \zeta$$

Differentiating, we get

$$J\, e^{i\alpha} = \frac{dz}{d\zeta} = -\tfrac{1}{2}\, ia \operatorname{cosech}^2 \tfrac{1}{2} \zeta$$

and
$$e^{2i\alpha} = \frac{dz/d\zeta}{d\bar{z}/d\bar{\zeta}} = -\sinh^2 \tfrac{1}{2} \bar{\zeta} \operatorname{cosech}^2 \tfrac{1}{2} \zeta$$

Problems involving two non-concentric circular boundaries are very much suitable to be solved by using bipolar co-ordinates.

4.7.1 Stresses in Bipolar Co-ordinates

Consider the co-ordinate system as shown in Fig. 4.6 such that

$$\xi + i\eta = \log\left[\frac{x + i(y + a)}{x + i(y - a)}\right]$$

Solving for x and y, we have

$$x = \frac{a \sin \eta}{\cosh \xi - \cos \eta}$$

$$y = \frac{a \sinh \xi}{\cosh \xi - \cos \eta}$$

Elements of arc measured along the normals to the curve ξ, η = constant are respectively $\frac{\delta\xi}{h}, \frac{\delta\eta}{h}$

where $\frac{1}{h^2} = \left(\frac{\partial x}{\partial \xi}\right)^2 + \left(\frac{\partial y}{\partial \xi}\right)^2$

or $\quad h = \frac{1}{a}(\cosh \xi - \cos \eta)$.

If u_ξ and u_η are the displacements normal to the curves ξ and η, the strains are ;

$$\left.\begin{aligned} \varepsilon_\xi &= h\left(\frac{\partial u_\xi}{\partial \xi}\right) - u_\eta\left(\frac{\partial h}{\partial \eta}\right) \\ \varepsilon_\eta &= h\left(\frac{\partial u_\eta}{\partial \eta}\right) - u_\xi\left(\frac{\partial h}{\partial \xi}\right) \\ \gamma_{\xi\eta} &= \frac{\partial}{\partial \xi}(hu_\eta) + \frac{\partial}{\partial \eta}(hu_\xi) \end{aligned}\right\} \quad \text{...(4.65)}$$

and stresses are

$$\left.\begin{aligned} \sigma_\xi &= \lambda(\varepsilon_\xi + \varepsilon_\eta) + 2G\,\varepsilon_\xi \\ \sigma_\eta &= \lambda(\varepsilon_\xi + \varepsilon_\eta) + 2G\,\varepsilon_\eta \\ \tau_{\xi\eta} &= G\gamma_{\xi\eta} \end{aligned}\right\} \quad \text{...(4.66)}$$

Now in rectangular co-ordinates,

$$\sigma_x = \frac{\partial^2\Phi}{\partial y^2},\ \sigma_y = \frac{\partial^2\Phi}{\partial x^2},\ \tau_{xy} = \frac{-\partial^2\Phi}{\partial x \partial y}$$

where Φ is Airy's stress function.

In curvilinear coordinates

$$\left.\begin{aligned} \sigma_\xi &= h\frac{\partial}{\partial \eta}\left(h\frac{\partial \Phi}{\partial \eta}\right) - h\frac{\partial h}{\partial \xi}\cdot\frac{\partial \Phi}{\partial \xi} \\ \sigma_\eta &= h\frac{\partial}{\partial \xi}\left(h\frac{\partial \Phi}{\partial \xi}\right) - h\frac{\partial h}{\partial \eta}\cdot\frac{\partial \Phi}{\partial \eta} \\ \tau_{\xi\eta} &= -h\frac{\partial^2}{\partial \xi \partial \eta}(h\Phi) + h\Phi\frac{\partial^2 h}{\partial \xi \partial \eta} \end{aligned}\right\} \quad \text{...(4.67)}$$

Using $h\Phi$ instead of Φ, we have

$$a\sigma_\xi = \left[(\cosh \xi - \cos \eta) \frac{\partial^2}{\partial \eta^2} - \sinh \xi \frac{\partial}{\partial \xi} - \sin \eta \frac{\partial}{\partial \eta} + \cosh \eta \right] (h\Phi)$$

$$a\sigma_\eta = \left[(\cosh \xi - \cos \eta) \frac{\partial^2}{\partial \xi^2} - \sinh \xi \frac{\partial}{\partial \xi} - \sin \eta \frac{\partial}{\partial \eta} + \cosh \eta \right] (h\Phi)$$

$$a\tau_{\xi\eta} = -(\cosh \xi - \cos \eta) \frac{\partial^2}{\partial \xi \partial \eta} (h\Phi)$$

Thus $a(\sigma_\xi - \sigma_\eta) = (\cosh \xi - \cos \eta) \left(\frac{\partial^2}{\partial \eta^2} - \frac{\partial^2}{\partial \xi^2} + 1 \right) (h\Phi)$

If $(h\Phi)$ and its second differential are finite at infinity ($\xi = \eta = 0$), then

$$a\sigma_\xi = h\Phi = a\sigma_\eta \text{ and } \tau_{\xi\eta} = 0$$

and
$$\nabla^2 = h^2 \left(\frac{\partial^2}{\partial \xi^2} + \frac{\partial^2}{\partial \eta^2} \right)$$

$\therefore$
$$a\nabla^2\Phi = \left[(\cosh \xi - \cos \eta) \left(\frac{\partial^2}{\partial \xi^2} + \frac{\partial^2}{\partial \eta^2} \right) - 2 \sinh \xi \frac{\partial}{\partial \xi} - 2 \sin \eta \frac{\partial}{\partial \eta} + \cosh \xi + \cos \eta \right] (h\Phi)$$

Then biharmonic equation $\nabla^4\Phi = 0$ becomes

$$\left(\frac{\partial^4}{\partial \xi^4} + 2 \frac{\partial^4}{\partial \xi^2 \partial \eta^2} + \frac{\partial^4}{\partial \eta^4} - 2 \frac{\partial^2}{\partial \xi^2} + 2 \frac{\partial^2}{\partial \eta^2} + 1 \right) (h\Phi) = 0 \quad \text{...(4.68)}$$

For the displacements to be single-valued, we have

$$\frac{\partial}{\partial \xi} \left[\frac{\partial \Phi}{\partial \xi} - 2(\lambda + G) \frac{u_\xi}{h} \right] + \frac{\partial}{\partial \eta} \left[\frac{\partial \Phi}{\partial \eta} - 2(\lambda + G) \frac{u_\eta}{h} \right] = 0 \quad \text{...(4.69)}$$

$$\frac{\partial}{\partial \xi} \left[h^2 \frac{\partial \Phi}{\partial \xi} + 2Gh\, u_\xi \right] - \frac{\partial}{\partial \eta} \left[h^2 \frac{\partial \Phi}{\partial \eta} + 2Gh\, u_\eta \right] = 0$$

$$\frac{\partial}{\partial \xi} \left[h^2 \frac{\partial \Phi}{\partial \eta} + 2Gh\, u_\eta \right] - \frac{\partial}{\partial \eta} \left[h^2 \frac{\partial \Phi}{\partial \xi} + 2Gh\, u_\xi \right] = 0$$

If we define

$$\frac{\partial P}{\partial \xi} = h^2 \frac{\partial \Phi}{\partial \eta} + 2Gh\, u_\eta \qquad ...(4.70)$$

$$\frac{\partial P}{\partial \eta} = h^2 \frac{\partial \Phi}{\partial \xi} + 2Gh\, u_\xi \qquad ...(4.71)$$

$$\nabla^2 P = 0$$

To satisfy Eq. (4.69), substitute u_ξ, u_η in terms of P, we find

$$h^2 \frac{\partial}{\partial \xi}\left(\frac{1}{h^2}\frac{\partial P}{\partial \eta}\right) + h^2 \frac{\partial}{\partial \eta}\left(\frac{1}{h^2}\frac{\partial P}{\partial \xi}\right) = \left(\frac{\lambda + 2G}{\lambda + G}\right)\nabla^2 \Phi$$

On rearranging, we get

$$\frac{\partial}{\partial \xi}\left[\left(\frac{\lambda + 2G}{\lambda + G}\right)\frac{\partial \Phi}{\partial \xi} - \frac{1}{h^2}\frac{\partial P}{\partial \eta}\right] + \frac{\partial}{\partial \eta}\left[\left(\frac{\lambda + 2G}{\lambda + G}\right)\frac{\partial \Phi}{\partial \eta} - \frac{1}{h^2}\frac{\partial P}{\partial \xi}\right] = 0$$

It follows that a function Q exists such that

$$h^2 \frac{\partial Q}{\partial \xi} = \frac{\partial P}{\partial \xi} - \left(\frac{\lambda + 2G}{\lambda + G}\right) h^2 \frac{\partial \Phi}{\partial \eta} \qquad ...(4.72)$$

$$h^2 \frac{\partial Q}{\partial \eta} = -\frac{\partial P}{\partial \eta} + \left(\frac{\lambda + 2G}{\lambda + G}\right) h^2 \frac{\partial \Phi}{\partial \xi} \qquad ...(4.73)$$

Eliminating P by differentiating w.r.t. η and ξ respectively Eqs. (4.72) and (4.73) and adding, we have

$$\frac{\partial^2 (hQ)}{\partial \xi\, \partial \eta} - Q\frac{\partial^2 h}{\partial \xi\, \partial \eta} = \frac{\lambda + 2G}{2(\lambda + G)}\left[\frac{1}{h}\frac{\partial}{\partial \xi}\left(h^2 \frac{\partial \Phi}{\partial \xi}\right) - \frac{1}{h}\frac{\partial}{\partial \eta}\left(h^2 \frac{\partial \Phi}{\partial \eta}\right)\right]$$

which becomes

$$\frac{\partial^2 (hQ)}{\partial \xi\, \partial \eta} = \frac{\lambda + 2G}{2(\lambda + G)}\left[\frac{\partial^2 (h\Phi)}{\partial \xi^2} + \frac{\partial^2 (h\Phi)}{\partial \eta^2} - h\Phi\right] \qquad ...(4.74)$$

To satisfy $\nabla^2 P = 0$, differentiate Eqs. (4.72) and (4.73) w.r.t. ξ and η respectively and subtract,

$$\frac{\partial}{\partial \xi}\left(h^2 \frac{\partial Q}{\partial \xi}\right) - \frac{\partial}{\partial \eta}\left(h^2 \frac{\partial Q}{\partial \eta}\right) = -\left(\frac{\lambda + 2G}{\lambda + G}\right)\left[\frac{\partial}{\partial \xi}\left(h^2 \frac{\partial \Phi}{\partial \eta}\right) + \frac{\partial}{\partial \eta}\left(h^2 \frac{\partial \Phi}{\partial \xi}\right)\right]$$

$$\frac{\partial^2}{\partial \xi^2}(hQ) - \frac{\partial^2}{\partial \eta^2}(hQ) - hQ = -\frac{2(\lambda + 2G)}{\lambda + G}\frac{\partial^2 (hQ)}{\partial \xi\, \partial \eta} \quad ...(4.75)$$

Eliminating Q, we have

$$\left[\frac{\partial^2}{\partial \xi^2} - \frac{\partial^2}{\partial \eta^2} - 1\right]^2 (h\Phi) = -4\,\frac{\partial^4 (h\Phi)}{\partial \xi^2 \partial \eta^2}$$

$\therefore$ (hQ) is also a solution of $\nabla^4 Q = 0$.

From Eq. (4.74), we get on integration

$$hQ = \left[\frac{\lambda + 2G}{2(\lambda + G)}\right]\iint\left[\frac{\partial^2 (h\Phi)}{\partial \xi^2} + \frac{\partial^2 (h\Phi)}{\partial \eta^2} - h\Phi\right]\partial \xi\, \partial \eta \quad ..(4.76)$$

and from Eqs. (4.70), (4.71) and (4.72), (4.73)

$$2Gu_\xi = \left(\frac{G}{\lambda + G}\right) h\,\frac{\partial \Phi}{\partial \xi} - h\,\frac{\partial Q}{\partial \eta} \quad ...(4.77)$$

$$2Gu_\eta = \left(\frac{G}{\lambda + G}\right) h\,\frac{\partial \Phi}{\partial \eta} - h\,\frac{\partial Q}{\partial \xi} \quad ...(4.78)$$

$$hQ = aA(\cosh \xi + \cos \eta) + B(\cosh \xi - \cos \eta) + Ca \sinh \xi + Da \sin \eta \quad ...(4.79)$$

$$hQ = Ar^2 + aB + Cy + Dx \quad ...(4.80)$$

where r is the distance from the origin.

Solution of Eq. (4.68) can be obtained from

$$(h\Phi) = \eta[F \cos \eta + H \sin \eta + K \cosh \xi + M \sinh \xi] \quad ...(4.81)$$

and the general expression can be written as :

$$\begin{aligned}(h\Phi) = {} & [F \cos \eta + H \sin \eta + K \cosh \xi + M \sinh \xi]\eta \\ & + (A_1 \cosh 2\xi + B_1 + C_1 \sinh 2\xi + D_1) \cos \eta \\ & + (B_0 \cosh \xi + D_0 \sinh \xi)\,\xi + (A'_1 \cosh 2\xi \\ & + C'_1 \sinh 2\xi + D_1') \sin \eta \\ & + \sum_{n=2}^{\infty} [\{A_n \cosh (n+1)\xi + B_n \cosh (n-1)\xi \\ & + C_n \sinh (n+1)\xi + D_n \sinh (n-1)\xi\} \cos n\eta \\ & + \{A'_1 \cosh (n+1)\xi + B'_n \cosh (n-1)\xi \\ & + C'_n \sinh (n+1)\xi + D'_n \sinh (n-1)\xi\}] \sin n\eta]\end{aligned} \quad ...(4.82)$$

where F, H, K, M and A's, B's, C's, and D's are arbitrary constants.

For displacements to be single-valued

$$F + K = 0, \qquad B_0 + D_1 = 0$$

$$H - (\lambda + 2G)D_0 = 0, \quad GM + (\lambda + 2G)\, D'_1 = 0.$$

These terms correspond to the resultant of the forces and couple applied over the boundaries.

Then we have

$$h\Phi = (\cosh \xi - \cos \eta) \log (\cosh \xi - \cos \eta) \quad \text{...(4.83)}$$

This can be expanded in the Fourier series as :

$$\left.\begin{aligned} \phi_n(\xi) &= A_n \cosh (n+1)\,\xi + B_n \cosh (n-1)\xi \\ &\quad + C_n \sinh (n+1)\,\xi + D_n \sinh (n-1)\,\xi \\ \psi_n(\xi) &= A'_n \cosh (n+1)\,\xi + B'_n \cosh (n-1)\xi \\ &\quad + C'_n \sinh (n+1)\,\xi + D'_n \sinh (n-1)\,\xi \end{aligned}\right\} \quad \text{...(4.84)}$$

If $n \geq 2$ and

$$\left.\begin{aligned} \phi_1(\xi) &= A_1 \cosh 2\xi + B_1 + C_1 \sinh 2\xi \\ \psi_1(\xi) &= A'_1 \cosh 2\xi + C'_1 \sinh 2\xi \end{aligned}\right\} \quad \text{...(4.85)}$$

$$(h\Phi) = [B_0\xi + \log (\cosh \xi - \cos \eta)] (\cosh \xi - \cos \eta)$$

$$+ \sum_{n=1}^{\infty} [\phi_n(\xi) \cos n\eta + \psi_n(\xi) \sin n\eta] \quad \text{...(4.86)}$$

When a boundary is free from stress, the boundary conditions become

$$\frac{\partial}{\partial \xi} (h\Phi) = \text{const} = \rho \text{ (say)} \quad \text{...(4.87)}$$

and $$(\cosh \xi - \cos \eta) \frac{\partial^2}{\partial \eta^2} (h\Phi) - \sin \eta \frac{\partial}{\partial \xi} (h\Phi) + \cosh \xi (h\Phi)$$

$$= \rho \sinh \xi$$

The solution is

$$h\Phi = \rho \tanh \xi + \sigma(\cosh \xi \cos \eta - 1) + \tau \sin \eta \quad \text{...(4.88)}$$

Eqs. (4.80) and (4.86) are necessary and sufficient conditions for a boundary ξ = constant to be free from stress, ρ, σ, τ are Michell's constants of the boundary.

4.7.2 Stresses in a Semi-infinite Plate with a Circular Hole under Tension

Consider a circular boundary defined by $\xi = \xi_0$ so that r is its radius and d the distance of its centre from its straight edge. Then

$$r = a \operatorname{cosech} \xi_0, \quad d = a \coth \xi_0$$

$$\frac{d}{r} = \cosh \xi_0$$

Far away from the hole, $\Phi = \frac{1}{2} T y^2$, where T is the tension, so that if $\xi > 0$,

$$(h\Phi_0) = \frac{1}{2} a\, T \sinh^2 \xi/(\cosh \xi - \cos \eta)$$

$$= \frac{1}{2} a\, T \sinh \xi \left(1 + \xi \sum_{n=1}^{\infty} e^{-n\xi} \cos n\eta \right) \quad ...(4.89)$$

We have to add to this a stress function which gives no stress at infinity and no stress over $\xi = 0$ and is such that the complete stress function gives no stress over $\xi = \xi_0$.

If $\phi_n(0) = 0$ in Eq. (4.89), then σ_ξ and $\tau_{\xi\eta}$ becomes zero over $\xi = 0$ and $\phi_n(0) = 0$ for $n \geq 1$ and hence from Eqs. (4.84) and (4.85), we get

$$A_n + B_n = 0$$

$$(n + 1)C_n + (n + 1) D_n = 0$$

On the circular boundary $\xi = \xi_0$:

$$\left(\sigma_\eta\right)_{\xi = \xi_0} = 2T (\cosh \xi_0 - \cos \eta)$$

$$\times \left[\sinh \xi_0 \operatorname{sech} 2\,\xi_0 + \sum_{n=2}^{\infty} M_n \cos n\eta \right] \quad ...(4.90)$$

where $M_n = \dfrac{n(n-1)\sinh (n+1)\xi_0 - n(n+1)\sinh (n-1)\xi_0}{2[\sinh^2 n\xi_0 - n^2 \sinh^2 \xi_0]}$

When $\xi = 0$, then from Eq. (4.90), we get

$$\left(\sigma_\eta\right)_{\xi = 0} = T\left[1 + (1 - \cos \eta) \sum_{n=1}^{\infty} P_n \cos n\eta \right] \quad ...(4.91)$$

where $P_n = 4n\, A_n$.

Let $\quad M_n = 2n(n \sinh \xi_0 - \cosh \xi_0)e^{-n\xi_0} + N_n$

then we get

$$2(\cosh \xi_0, \cos \eta) \sum_{n=1}^{\infty} n(n \sinh \xi_0 - \cosh \xi_0)\, e^{-n\xi_0} \cos n\eta$$

$$= 1 - \frac{2 \sinh^2 \xi_0 \sin^2 \eta}{(\cosh \xi_0 - \cos \eta)^2}$$

Substituting in Eq. (4.90), we get

$$\sigma_\eta = 2T\left[1 - \frac{2\sinh^2\xi_0 \sin^2\eta}{(\cosh\xi_0 - \cos\eta)^2}\right] + 2T(\cosh\xi_0 - \cos\eta)$$

$$\times\left[\sinh\xi_0 \operatorname{sech} 2\xi_0 + 2e^{-2\xi_0}\cos\eta + \sum_{n=2}^{\infty} N_n \cos n\eta\right] \quad ...(4.92)$$

If θ is the angle between the radius to the point ξ_0, η and the perpendicular to the straight edge, then

$$\sin\theta = \frac{\sinh\xi_0 \sin\eta}{\cosh\xi_0 - \cos\eta}$$

If ξ_0 is large, then Eq. (4.92) reduces to

$$\sigma_\eta = T(1 + 2\cos 2\theta) \quad ...(4.93)$$

For $\theta = 0°$, $\sigma_\eta = 3T$ and for $\theta = 90°$, $\sigma_\eta = -T$

Values of N_n for several values of ξ_0 are given in Table 1.

Table 4.1 Values of N_n

ξ_0	0.6	0.8	1.0	1.2	1.4	1.6	1.8	2.0	2.2	2.4
N_2	1.4640	0.7716	0.4139	0.2240	0.1219	0.0665	0.0364	0.0199	0.0109	0.0060
N_3	0.7457	0.2647	0.0914	0.0306	0.0100	0.0032	0.0010	0.0033	0.0001	
N_4	0.3238	0.0713	0.0148	0.0029	0.0005	0.0010				
N_5	0.1232	0.0162	0.0019	0.0002						
N_6	0.0421	0.0032	0.0002							
N_7	0.0131	0.0005								
N_8	0.0038	0.0001								
N_9	0.0010									
N_{10}	0.0003									
N_{11}	0.0001									

4.7.3 Stresses due to Diametral Forces on a Circular Disc with an Eccentric Hole[9]

Let us introduce the curvilinear co-ordinate system defined by

$$\xi + i\eta = \log\left[\frac{x + i(y + a)}{x + i(y - a)}\right] \quad ...(1)$$

9. A.M. Sen Gupta : Journal of Applied Mechanics, Vol. 22, No. 2, June 1955, pp. 263—266.

Solving for x and y, we have

$$x = \frac{a \sin \eta}{\cosh \xi - \cos \eta}, \; y = \frac{a \sinh \xi}{\cosh \xi - \cos \eta} \qquad \ldots(2)$$

From the relation

$$\frac{1}{h^2} = \left(\frac{\partial x}{\partial \xi}\right)^2 + \left(\frac{\partial y}{\partial \xi}\right)^2, \text{ we get}$$

$$h = (\cosh \xi - \cos \eta)/a \qquad \ldots(3)$$

The stresses in terms of stress function Φ are given by

$$a\sigma_\xi = \left[(\cosh \xi - \cos \eta)\frac{\partial^2}{\partial \eta^2} - \sinh \xi \frac{\partial}{\partial \xi} - \sin \eta \frac{\partial}{\partial \eta} + \cosh \xi\right](h\Phi)$$

$$a\sigma_\eta = \left[(\cosh \xi - \cos \eta)\frac{\partial^2}{\partial \xi^2} - \sinh \xi \frac{\partial}{\partial \xi} - \sin \eta \frac{\partial}{\partial \eta} + \cosh \eta\right](h\Phi)$$

$$a\tau_{\xi\eta} = -(\cosh \xi - \cos \eta)\frac{\partial^2 (h\Phi)}{\partial \xi \partial \eta} \qquad \ldots(4)$$

so that, we have

$$a(\sigma_\xi - \sigma_\eta) = (\cosh \xi - \cos \eta)\left(\frac{\partial^2}{\partial \eta^2} - \frac{\partial^2}{\partial \xi^2} + 1\right)(h\Phi) \qquad \ldots(5)$$

The differential equation satisfied by $(h\Phi)$ is

$$\left(\frac{\partial^4}{\partial \xi^4} + 2\frac{\partial^4}{\partial \xi^2 \partial \eta^2} + \frac{\partial^4}{\partial \eta^4} - \frac{\partial^2}{\partial \xi^2} + 2\frac{\partial^2}{\partial \eta^2} + 1\right)(h\Phi) = 0 \qquad \ldots(6)$$

The necessary and sufficient conditions for the boundary ξ = constant to be free from stress are that on the boundary

$$\frac{\partial}{\partial \xi}(h\Phi) = \rho \qquad \ldots(7)$$

and $h\Phi = \rho \tanh \xi + \sigma(\cosh \xi \cos \eta - 1) + \tau \sin \eta$ ...(8)

where ρ, σ and τ are Michell's three constants of the boundary.

Let us consider a circular disc of unit thickness (Fig. 4.7) the xy-plane of Cartesian co-ordinates being the middle plane of the disc. The disc contains an eccentric circular hole. The axis of y is taken to be the line passing through the centres of the disc and the hole and coinciding with the diameter of the disc at the extremities A and B of which two equal and opposite forces each of intensity P per unit

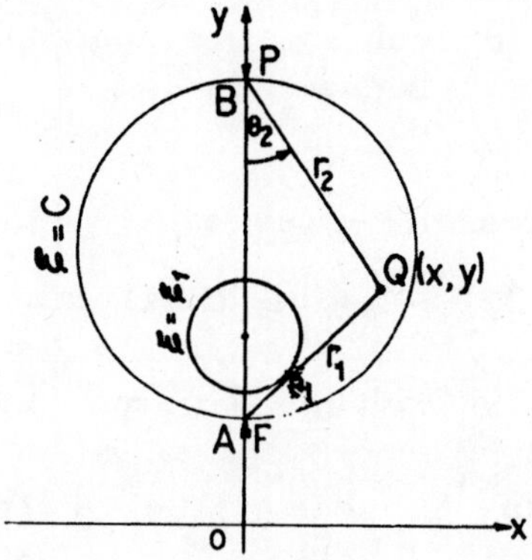

Fig. 4.7

thickness are applied. Let the outer boundary of the disc be defined by $\xi = C$, $C > 0$ and the hole by $\xi = \xi_1$, $\xi_1 > C$. We assume that each of the forces P produces a simple radial stress distribution, so that the stress function Φ_0 required to make the boundary of the outer circle free from stress at all points other than A and B is given by :

$$\Phi_0 = -\frac{P}{2\pi}\left[2x(\theta_1 + \theta_2) - \frac{r_1^2}{a \operatorname{cosech} c}\right] \qquad ...(9)$$

where (r_1, θ_1), (r_2, θ_2) are the co-ordinates of any point $Q(x, y)$ on the disc.

Noting that for $\xi > C > 0$

$$\theta_1 = \sum_{n=1}^{\infty} \frac{e^{-n\xi}}{n}\left[1 + (-1)^{n-1} e^{nc}\right] \sin n\xi$$

$$\theta_2 = \sum_{n=1}^{\infty} \frac{e^{-n\xi}}{n}\left[e^{nc} - 1\right] \sin n\,\eta \qquad ...(10)$$

$$r_1^2 = a^2 \operatorname{sech}^2 \frac{c}{2}\left[\frac{\cosh(\xi - c) + \cos\eta}{\cosh\xi - \cos\eta}\right].$$

We have for $\alpha = (\xi - c) > 0$

$$h\Phi_0 = \frac{P}{\pi}\left[-e^{-\alpha} + \sum_{n=2,4}^{\infty}\left\{\frac{e^{-(n-1)\alpha}}{n-1} - \frac{e^{-(n+1)\alpha}}{n+1}\right\}\cos n\,\eta \right.$$

$$\left. + \tanh\frac{c}{2}(\cosh\alpha + \cos\eta)\right] \qquad ...(11)$$

For an auxiliary stress function Φ_1, we have to assume a function which will produce no stress on the boundary $\xi = c$ and which is such that the complete stress function $\Phi = \Phi_0 + \Phi_1$ will produce no stress on the boundary $\xi = \xi_1$.

We assume

$$h\Phi_1 = \frac{P}{\pi}[B_0\{\alpha(\cosh\xi - \cos\eta)\cosh c + \sinh\alpha(\cosh\xi\cos\eta - \cosh^2 c)\} + A_1(\cosh 2\alpha - 1)\cos\eta + \sum_{n=2}^{\infty}\{A_n[\cosh(n+1)\alpha - \cosh(n-1)\alpha] + B_n[(n-1)\sinh(n+1)\alpha - (n+1)\sinh(n-1)\alpha\}] \times \cos n\eta] \quad ...(12)$$

Since on the boundary $\xi = c$ or $\alpha = 0$, $h\Phi_1 = 0$ and $\frac{\partial}{\partial\xi} - h\Phi_1 = 0$, the stresses on the boundary $\alpha = 0$ are uneffected by the introduction of Φ_1, each of the Michell's constants being zero in this case. Further, it can be seen that the displacements produced by $h\Phi_1$ are single-valued.

The complete stress function $\Phi = \Phi_0 + \Phi_1$ given by Eqs. (11) and (12) must satisfy the conditions (7) and (8) for the boundary $\xi = \xi_1$ or $\alpha = \alpha_1$.

From the boundary conditions the values of the constants B_0, A_1, A_n and B_n are determined as :

$$\left.\begin{aligned} B_0 &= -\frac{\cosh c\cosh\alpha_1\operatorname{cosech}\alpha_1}{\sinh\xi_1\sinh\alpha_1 + \sinh c\sinh(\xi_1 + c)\cosh\alpha_1} \\ A_1 &= \frac{1}{2}\,\frac{\cosh c\sinh\xi_1\operatorname{cosech}\alpha_1}{\sinh\xi_1\sinh\alpha_1 + \sinh c\sinh(\xi_1 + c)\cosh\alpha_1} \end{aligned}\right\} \quad ...(13)$$

For n even ≥ 2

$$A_n[\sinh^2 n\alpha_1 - n^2\sinh^2\alpha_1] = -n\sinh^2\alpha_1$$

$$B_n[\sinh^2 n\alpha_1 - n^2\sinh^2\alpha_1] = \frac{1}{n^2 - 1}[n^2\sinh^2\alpha_1 - n\sinh\alpha_1\cosh\alpha_1 + e^{-n\alpha_1}\sinh n\alpha_1] \quad ...(14)$$

and for n odd ≥ 3, $\quad A_n = B_n = 0 \quad ...(15)$

On the hole $\xi = \xi_1$, $\sigma_\xi = 0$

so that σ_η can be easily calculated from Eq. (5). We have

$$\left(\sigma_\eta\right)_{\xi=\xi_1} = \frac{P}{\pi a} \times$$

$$\left\{\frac{\cosh^2 c[1 - \coth\alpha_1 \sinh 2\xi_1 - 2\operatorname{cosech}\alpha_1 \sinh c\cos\eta + \cos 2\eta]}{\sinh\xi_1 + \sinh c\sinh(\xi_1 + c)\cosh\alpha_1}\right\}$$

$$+ \frac{4P}{\pi a}(\cosh\xi_1 - \cos\eta)\sum_{n=2,4\ldots}^{\infty} R_n \cos n\eta \qquad \ldots(16)$$

where $\left[R_n + \frac{n}{n-1}e^{-(n-1)\alpha_1}\right](\sinh^2 n\alpha_1 - n^2\sinh^2\alpha_1)$

$$= \frac{n}{n-1}\left(n^2\sinh^2\alpha_1 + n\sinh\alpha_1\cosh\alpha_1 + e^{-n\alpha_1}\sinh n\alpha_1\right)$$

$$\times \sinh(n-1)\alpha_1 - n^2\sinh^2\alpha_1\cosh(n-1)\alpha_1$$

The series in Eq. (16) converges slowly for small values of α_1. To make it rapidly converge, we put

$$R_n = 2ne^{-n\alpha_1}\sinh\alpha_1 + Q^n$$

where $Q_n[\sinh^2 n\alpha_1 - n^2\sinh^2\alpha_1] = e^{-n\alpha_1}[2n^3\sinh^3\alpha_1$

$$+ ne^{-n\alpha_1}\sinh\alpha_1\sinh n\alpha_1]$$

Now noting that

$$\sum_{n=2,4,\ldots}^{\infty} ne^{-n\alpha_1}\cos n\eta = \frac{\cosh 2\alpha_1\cos 2\eta - 1}{(\cosh 2\alpha_1 - \cos 2\eta)^2} \qquad \ldots(17)$$

We have $\left(\sigma_\eta\right)_{\xi=\xi_1} = \frac{P}{\pi a} \times$

$$\left\{\frac{\cosh^2 c[1 - \coth\alpha_1\sinh 2\xi_1 - 2\operatorname{cosech}\alpha_1\sinh c\cos\eta + \cos 2\eta}{\sinh\xi_1\sinh\alpha_1 + \sinh c\sinh(\xi_1 + c)\cosh\alpha_1}\right\}$$

$$+ \frac{8P}{\pi a}(\cosh\xi_1 - \cos\eta)\sinh\alpha_1\left[\frac{\cosh 2\alpha_1\cos 2\eta - 1}{(\cosh 2\alpha_1 - \cos 2\eta)^2}\right]$$

$$+ \frac{4P}{\pi a}(\cosh\xi_1 - \cos\eta)\sum_{n=2,4,\ldots}^{\infty} Q_n\cos n\eta \qquad \ldots(18)$$

And for the outer boundary $\xi = c$

$$\left(\sigma_\eta\right)_{\xi=c} = \frac{P}{\pi a}(\cosh c - \cos\eta)$$

$$\times\left\{\left[\frac{(1 + \cosh 2c)\cos\eta - \cosh c\coth\alpha_1\sinh 2c}{\sinh\xi_1\sinh\alpha_1 + \sinh c\sinh(\xi_1 + c)\cosh\alpha_1}\right]\right\}$$

$$+ \sum_{n=2,4,\ldots}^{\infty} N_n \cos n\eta \Bigg\} \qquad \ldots(19)$$

where $N_n = 4n\, A_n$.

The stresses are symmetrical about the y-axis. Hence the values of $\sigma\eta$ on the boundaries $\xi = \xi_1$ and $\xi = c$ respectively, are given for values of η between 0 and π when $\xi_1 = 2$ and $c = 0.6$, *i.e.* when

$$\frac{\sinh \alpha_1}{\sinh c} = 5.697 = \frac{\text{dia of disc}}{\text{dia of hole}}.$$

It is found that the maximum tensions are at the extremities of the diameter of the hole passing through A and B, and it is more marked at the point where the distance between the two boundaries is least.

Table 4.2 Numerical Values of Coefficients N_n

α_1	0.8	1.0	1.2	1.4	1.6	1.8	2.0
$-N_2$	5.0750	2.8960	1.7560	1.1030	0.7081	0.4617	0.3042
$-N_4$	0.3671	0.1223	0.0398	0.0127	0.0039	0.0012	0.0003
$-N_6$	0.0309	0.0049	0.0007	0.0001	0.0001		
$-N_8$	0.0022	0.0001					
$-N_{10}$	0.0001						

Table 4.3 Numerical Values of Q_n

α_1	0.8	1.0	1.2	1.4	1.6	1.8	2.0
Q_2	0.9770	0.4810	0.2475	0.1299	0.0693	0.0374	0.0203
Q_4	0.0271	0.0053	0.0009	0.0002	0.0000		
Q_6	0.0007	0.0000					

Table 4.4 Values of σ_η on Boundary $\xi = \xi_1$ and $\xi = c$ ($\xi_1 = 2$, $c = 0.6$)

η, *deg.*	$\frac{\pi a}{P}\left(\sigma_\eta\right)_{\xi=\xi_1}$	$\frac{\pi a}{P}\left(\sigma_\eta\right)_{\xi=c}$
0	4.6608	– 0.1985
15	3.4704	– 0.2019
30	0.4327	– 0.1662
45	– 3.3331	0.00506
60	– 6.8547	0.3563
75	– 9.5040	0.8046
90	– 10.8410	1.1440
105	– 10.4630	1.1230
120	– 8.0770	– 0.5890
135	– 3.5670	– 0.4485
150	2.3510	– 1.6880
165	6.8690	– 2.7020
180	10.0170	– 3.0980

Putting $c = 0$, we have the case of a semi-infinite plate with an unstressed circular hole subjected to a concentrated force P normal to its straight boundary. Hence from Eq. (18), we get

$$\left(\sigma_\eta\right)_{\substack{\xi=\xi_1\\ c=0}} = \frac{4P}{\pi a}\frac{\sinh\xi_1(\cosh 2\xi_1\cos 2\eta - 1)}{(\cosh 2\xi_1 - \cos 2\eta)(\cosh\xi_1 + \cos\eta)}$$

$$-\frac{2P}{\pi a}\left(\frac{\cosh 2\xi_1 - \cos 2\eta}{\cosh 2\xi_1 - 1}\right)$$

$$+\frac{4P}{\pi a}(\cos\xi_1 - \cos\eta)\sum_{n=2,4,\ldots}^{\infty} M_n \cos n\eta \quad \ldots(4.94)$$

where $M_n[\sinh^2 n\xi_1 - n^2\sinh^2\xi_1] = e^{-n\xi_1}[2n^3\sinh^3\xi_1$

$$+ ne^{-n\xi_1}\sinh\xi_1\sinh n\xi_1]$$

and for $\xi = 0$, *i.e.* for the straight boundary, we have from Eq. (19)

$$\left(\sigma_\eta\right)_{\substack{\xi=0\\ c=0}} = \frac{P}{\pi a}(1-\cos\eta)(2\,\mathrm{cosech}^2\xi_1 + \sum_{n=2,4,\ldots}^{\infty} P_n\cos n\eta)$$

$$\ldots(4.95)$$

where $P_n(\sin n\xi_1 - n^2\sinh^2\xi_1) = -4n^2\sinh^2\xi_1$.

4.7.4 Solution of an Elliptic Hole in Uniformly Stressed Plate by Complex Potentials

In elliptic co-ordinates ξ, η, we have

$$\omega(\zeta) = z = c \cosh \zeta, \zeta = \xi + i\eta \qquad ...(1)$$

which gives $\quad x = c \cosh \xi \cos \eta, y = c \sinh \xi \sin \eta$

and $\quad \dfrac{dz}{d\zeta} = c \sinh \zeta = \omega'(\zeta)$

For $\quad z = \omega(\zeta)$

$$\omega'(\zeta) = J\, e^{2i\alpha}, \overline{\omega}'(\overline{\zeta}) = J\, e^{-i\alpha}$$

so that $$e^{2i\alpha} = \frac{\omega'(\zeta)}{\overline{\omega}'(\overline{\zeta})} = \frac{\sinh \zeta}{\sinh \overline{\zeta}} \qquad ...(2)$$

The co-ordinate ξ is constant and equal to ξ_0 on an ellipse of semi axes $a = c \cosh \xi_0$, $b = c \sinh \xi_0$. A point on the ellipse goes once around the ellipse as η goes from zero to 2π (for positive x-axis). In this respect η resembles the angle θ of polar co-ordinates. Continuity of displacement and stress components requires that they be periodic in η with period 2π so that they have the same values for $\eta = 2\pi$ and $\eta = 0$.

Consider now an infinite plate in a state of uniform all-round tension T disturbed by an elliptical hole of semi-axes a and b, which is free from stress. Thus

$$\sigma_x = \sigma_y = T \text{ as } \xi \to \infty \qquad ...(3)$$

$$\sigma_\xi = \tau_{\xi\eta} = 0 \quad \text{as } \xi = \xi_0 \qquad ...(4)$$

Now from Eq. (4.18), we have

$$\sigma_\xi + \sigma_\eta = 2\{\Phi_0(\zeta) + \overline{\Phi}_0(\overline{\zeta})]$$

$$\sigma_\xi - \sigma_\eta + 2i\tau_{\xi\eta} = -\frac{2}{\omega'(\zeta)}[\omega(\zeta)\overline{\Phi'}_0(\zeta) + \overline{\omega'}(\overline{\zeta})\Psi'_0(\overline{\zeta})]$$

$$\sigma_\xi - \sigma_\eta - 2i\tau_{\xi\eta} = -\frac{2}{\overline{\omega'}(\overline{\zeta})}[\overline{\omega}(\overline{\zeta})\Phi'_0(\zeta) + \omega'(\zeta)\Psi_0(\zeta)]$$

where $\varphi'(z) = \Phi_0(\zeta)$, $\Psi'(z) = \Psi_0(\zeta)$.

Therefore, in the absence of body forces.

$$\left.\begin{aligned} \Phi_0(\zeta) + \overline{\Phi}_0(\overline{\zeta}) &= T \\ \overline{\omega}(\zeta)\Phi_0'(\zeta) + \omega'(\zeta)\Psi_0(\zeta) &= 0 \end{aligned}\right\} \qquad ...(5)$$

as $|z| \to \infty$. Stevenson found that complex potentials of the forms

$$\varphi(z) = AC \sinh \zeta, \qquad \psi(z) = BC \operatorname{cosech} \zeta \qquad ...(6)$$

A, B real, were sufficient to describe the stress distribution. They give rise to stresses and moment which are zero over a circuit containing the hole.

$$\left.\begin{aligned}\therefore \quad \Phi_0(\zeta) &= \varphi'(z)\\ &= AC\cosh\zeta\,.\,\frac{d\zeta}{dz}\\ &= \frac{AC\cosh\zeta}{C\sinh\zeta} = A\coth\zeta\\ \Psi_0(\zeta) &= \psi'(z)\\ &= -BC\,\text{cosech}\,\zeta\coth\zeta\,.\,\frac{d\zeta}{dz}\\ &= -B\,\text{cosech}^2\zeta\coth\zeta\end{aligned}\right\} \qquad \ldots(7)$$

Substituting in Eqs. (5), we get

$$A\coth\zeta + A\coth\bar{\zeta} = T$$

$$C\cosh\bar{\zeta}\left(-\frac{A\,\text{cosech}^2\zeta}{C\sinh\zeta}\right) + C\sinh\zeta\,(-B\,\text{cosech}^2\zeta\coth\zeta) = 0$$

As $\xi \to \infty$, $\coth\zeta = 1 = \coth\bar{\zeta}$

Hence $\qquad T = 2A \qquad \ldots(8)$

$$\begin{aligned}\text{Now}\quad \sigma_\xi - i\tau_{\xi\eta}\,\tau_{\xi\eta} &= \Phi_0(\zeta) + \overline{\Phi}_0(\bar{\zeta})\\ &\quad - \frac{2}{\omega'(\zeta)}\,[\overline{\omega}(\bar{\zeta})\,\Phi_0'(\zeta) + \omega'(\zeta)\Psi_0(\zeta)]\\ &= \coth\zeta\,[A + A\coth\zeta\coth\bar{\zeta}\\ &\quad + B\,\text{cosech}\,\zeta\,.\,\text{cosech}\,\bar{\zeta}]\end{aligned}$$

This expression vanishes on the boundary

$$\zeta + \bar{\zeta} = 2\xi_0, \text{ if}$$

$$A[\sinh\zeta\sinh(2\xi_0 - \zeta) + \cosh\zeta\cosh(2\xi_0 - \zeta)] = -B$$

i.e. $\quad$ if $B = -A\cosh 2\xi_0$

$$= -\tfrac{1}{2}T\cosh 2\xi_0 \qquad \ldots(9)$$

$$\left.\begin{aligned}\text{Hence}\quad \varphi(z) &= \tfrac{1}{2}TC\sinh\zeta\\ \psi(z) &= -\tfrac{1}{2}TC\cosh 2\xi_0\,\text{cosech}\,\zeta\end{aligned}\right\} \qquad \ldots(10)$$

$$\therefore\ \sigma_\xi + \sigma_\eta = 4Re\varphi'(z) = 2TRe\coth\zeta$$

$$\begin{aligned}\text{But}\quad \coth\zeta &= \coth(\xi + i\eta)\\ &= \frac{\cosh(\xi + i\eta)}{\sinh(\xi + i\eta)} = \frac{\sinh 2\xi - i\sin 2\eta}{\cosh 2\xi - \cos 2\eta}\end{aligned}$$

$$\therefore \qquad \sigma_\xi + \sigma_\eta = \frac{2T\sinh 2\xi}{\cosh 2\xi - \cos 2\eta}$$

On the hole boundary $\sigma_\xi = 0$.

$$\therefore \quad \left(\sigma_\eta\right)_{\substack{\xi = \xi_0 \\ \eta = 0, \pi}} = \frac{2T \sinh 2\xi_0}{\cosh 2\xi_0 - \cos 2\eta}$$

$$\left(\sigma_\eta\right)_{\xi = \xi_0} = \frac{2T \sinh 2\xi_0}{\cosh 2\xi_0 - 1} = \left(\sigma_\eta\right)_{\max}$$

Now $\qquad c^2 = a^2 - b^2$

$$\sinh 2\xi_0 = \frac{2ab}{c^2}$$

$$\cosh 2\xi_0 = \frac{a^2 + b^2}{c^2}.$$

$$\therefore \quad \left(\sigma_\eta\right)_{max} = 2T\frac{a}{b} \qquad \text{...(4.96)}$$

$$\left(\sigma_\eta\right)_{min} = \left(\sigma_\eta\right)_{\substack{\xi = \xi_0 \\ \eta = \frac{\pi}{2}, \frac{3\pi}{2}}}$$

$$= \frac{2T \sinh 2\xi_0}{\cosh 2\xi_0 + 1}$$

$$= 2T \, . \frac{b}{a} \qquad \text{...(4.97)}$$

4.7.5 Solution of Elliptic Hole in a Plate under Simple Tension by Complex Potentials Method

Consider the infinite plate in a state of simple tensile stress T in a direction at an angle β below the positive x-axis as shown in Fig. 4.8, disturbed by an elliptic hole, with its major axis along the x-axis. Let ox', oy' be Cartesian axes obtained by rotating ox through the angle β so as to bring it parallel to the tension T. Then

$$\sigma_x + \sigma_y = \sigma_x' + \sigma_y'$$

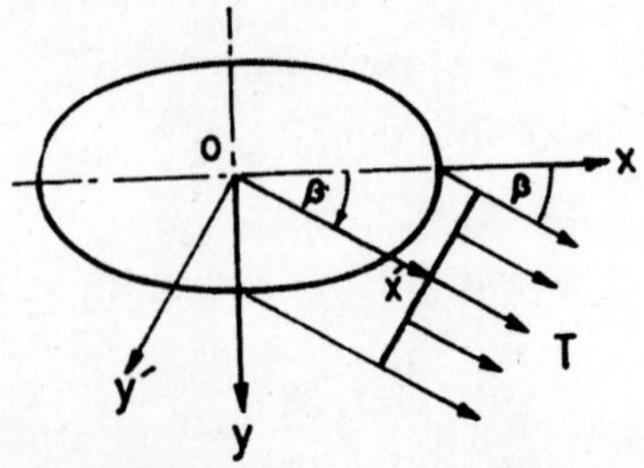

Fig. 4.8 Elliptic hole in a plate.

$$\sigma_y' - \sigma_x' + 2i\tau_{x'y'} = e^{2i\beta}(\sigma_y - \sigma_x + 2i\tau_{xy})$$

Since at infinity $\sigma_x' = T$

$$\sigma_y' = \tau_{x'y'} = 0.$$

$$\therefore \quad \sigma_x + \sigma_y = T$$

$$\sigma_y - \sigma_x + 2i\tau_{xy} = -T\,e^{-2i\beta}$$

$$\therefore \quad 4Re\,\varphi'(z) = T$$

$$2[\bar{z}\,\varphi''(z) + \psi'(z)] = -T\,e^{-2i\beta} \qquad ...(1)$$

at infinity.

At the boundary of the hole, $\xi = \xi_0$, we must have

$$\sigma_\xi = \tau_{\xi\eta} = 0$$

All these boundary conditions can be satisfied by taking

$$4\varphi(z) = AC\cosh\zeta + BC\sinh\zeta$$

$$4\chi(z) = Cc^2\zeta + Dc^2\cosh 2\zeta + Fc^2\sinh 2\zeta$$

where $\psi(z) = \chi'(z)$ and A, B, C, D, F are constants to be determined. Constants A and C are real and B, D, F are complex.

Let
$$B = B_1 + iB_2$$
$$D = D_1 + iD_2$$
$$F = F_1 + iF_2$$

Substituting in Eq. (1), we get

$$A + B_1 = T$$

$$2(D + F) = -T\,e^{-2i\beta} \qquad ...(2)$$

Also
$$4(\sigma_\xi - i\,\tau_{\xi\eta}) = \text{cosech}\,\bar{\zeta}\,[(2A + B\coth\zeta)\sinh\bar{\zeta} + (\bar{B} + B\,\text{cosech}^2\zeta)\cosh\bar{\zeta} + (C + 2F)\,\text{cosech}\,\zeta\coth\zeta - 4D\sinh\zeta - 4F\cosh\zeta]$$

At the boundary of the hole,

$$\xi = \xi_0$$

and
$$\bar{\zeta} = 2\xi_0 - \zeta$$

The expression in the square brackets becomes ;

$$(2A\sinh 2\xi_0 - 2iB_2\cosh 2\xi_0 - 4F)\cosh\zeta - (2A\cosh 2\xi_0 - 2iB_2\sinh 2\xi_0 + 4D)\sinh\zeta + (C + 2F + B\cosh 2\xi_0)\coth\zeta\,\text{cosech}\,\zeta$$

For
$$\sigma_\xi = \tau_{\xi\eta} = 0$$

at the hole, we must have

$$\left.\begin{aligned} 2A \sinh 2\xi_0 - 2i B_2 \cosh 2\xi_0 - 4F = 0 \\ 2A \cosh 2\xi_0 - 2iB_2 \sinh 2\xi_0 + 4D = 0 \\ C + 2F + B \cosh 2\xi_0 = 0 \end{aligned}\right\} \quad ...(3)$$

Solving, we get

$$A = Te^{2\xi_0} \cos 2\beta$$

$$B = T(1 - e^{2\xi_0 + 2i\beta})$$

$$C = -T(\cosh 2\xi_0 - \cos 2\beta)$$

$$D = -\tfrac{1}{2}Te^{2\xi_0} \cosh 2(\xi_0 + i\beta)$$

$$F = \tfrac{1}{2}Te^{2\xi_0} \sinh 2(\xi_0 + i\beta)$$

$$\therefore \quad 4\varphi(z) = Tc\left[e^{2\xi_0} \cos 2\beta \coth \zeta + (1 - e^{2\xi_0 + 2i\beta}) \sinh \zeta\right]$$

$$4\chi(z) = -Tc^2\left[(\cosh 2\xi_0 - \cos 2\beta)\zeta + \frac{1}{2} e^{2\xi_0} \cosh 2(\zeta - \xi_0 - i\beta)\right]$$

At the hole

$$\left(\sigma_\eta\right)_{\xi = \xi_0} = T\frac{\sinh 2\xi_0 + \cos \beta - e^{2\xi_0} \cos 2(\beta - \eta)}{\cos 2\xi_0 - \cos 2\eta} \quad ...(4.98)$$

When $\beta = \dfrac{\pi}{2}$, then

$$\left(\sigma_\eta\right)_{\xi = \xi_0} = T e^{2\xi_0}\left[\frac{\sinh 2\xi_0 (1 + e^{-2\xi_0})}{\cosh 2\xi_0 - \cos 2\eta} - 1\right] \quad ...(4.99)$$

$$\left(\sigma_\eta\right)_{\substack{\xi = \xi_0 \\ \eta = 0, \pi}} = T e^{2\xi_0}\left[\frac{\sinh 2\xi_0 (1 + e^{-2\xi_0})}{\cosh 2\xi_0 - 1} - 1\right]$$

$$= T\left(1 + 2\frac{a}{b}\right) \quad ...(4.100)$$

EXERCISES

4.1 Use the series method to solve the problem of an unstressed elliptical hole in an infinite plate under a uniform tension T at infinity acting at an angle α to the major axis of the hole. If the transformation is written in the form

$$z = R\left(\zeta + \frac{m}{\zeta}\right)$$

Confirm that $\varphi(\zeta) = \dfrac{RT}{4}\zeta + \dfrac{RT}{4}(2e^{2i\alpha} - m)\dfrac{1}{\zeta}$

and $$\psi(\zeta) = -\frac{1}{2}RT\,e^{-2i\alpha}\zeta.$$

4.2 Use the series method to show that the complex potentials for an epitrochoidal disc deformed under a uniform pressure p on its boundary are $\varphi(\zeta) = -\frac{1}{2}p\,m(\zeta)$, $\psi(\zeta) = 0$ where the disc is defined by the transformation

$$z = m(\zeta) = R(\zeta + m\zeta^n),\ (|\zeta| < 1).$$

4.3 An infinite plate contains an elliptical hole, with major and minor axes a and b. If the plate is under all round tension and the hole is unstressed, determine the hoop stress around the hole and show that the stress concentration factors at the ends of the major and minor axes are $\dfrac{2a}{b}$ and $\dfrac{2b}{a}$ respectively.

4.4 An infinite plate containing an equilateral triangular hole may be approximately mapped onto the region $|\zeta| \geq 1$ by

$$z = R\left\{\zeta + \frac{1}{3\zeta^2} + \frac{1}{45\,\zeta^5} + \ldots\right\}$$

If the hole is unstressed and a tension T acts at infinity at an angle α to the real axis, determine the hoop stress at the point corresponding to $\zeta = 1$. As α varies show that this stress has a maximum of $3T$ for the one term approximation, $11T$ for the two term approximation and $17.9\,T$ for the three term approximation.

4.5 An infinite plate with an elliptic hole has at infinity, uniform stress

$$\sigma_x = T_1,\ \sigma_y = T_2,\ \tau_{xy} = 0.$$

(*a*) Find an expression for the stress at the hole.

(*b*) Show that if $\dfrac{T_2}{T_1} = \dfrac{b}{a}$, the stress at the hole is the same all around the hole.

(*c*) Show that when the stress at infinity is a pure shear at 45° to the axes of the ellipse, the greatest stress on the hole occurs at the ends of the major axis, and corresponds to a stress-concentration factor $2\left(1 + \dfrac{a}{b}\right)$.

5

Torsion of Prismatic Bars

5.1 Introduction

The elementary theory of torsion of prismatic bars is restricted to bars of circular cross-section only. This simple theory was first given by Coulomb in 1784 and results in the equation :

$$\frac{T}{J} = \frac{\tau}{r} = G\theta \qquad ...(5.1)$$

where T is the applied torque, J the polar moment of inertia, τ the shear stress on a transverse section at a distance r from the axis of the bar, G the shear modulus and θ the angle of twist over a unit length of the bar. Eq. (5.1) is based upon the assumption that initially plane sections of the bar remain plane after application of the torque. And the magnitude of the shear stress at a point varies directly as its distance from the axis of the bar. In this chapter, we will study the theory of torsion of prismatic bars of non-circular cross-sections, as put forward by St. Venant in 1855.

5.2 St. Venant's Theory

Consider a portion of a prismatic bar subjected to twist T. oz is the axis of the bar parallel to the generator and the mutually perpendicular axes ox and oy lie in the plane of the cross-section as shown in Fig. 5.1.

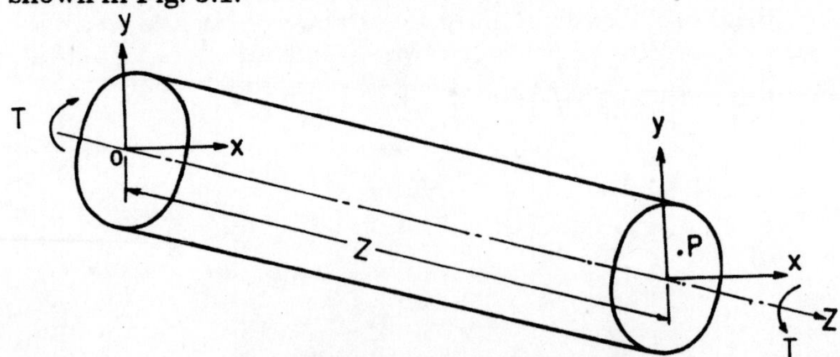

Fig. 5.1 Prismatic bar under torsion.

Body forces due to gravitational weight, etc. are neglected. According to St. Venant's theory the deformation of the twisted bar consist of (*i*) rotations of the cross-sections of the bar as in the case of a circular shaft, (*ii*) warping of the cross-sections which is the same for all cross-sections.

Let the initial co-ordinates of a point P be (x, y, z) and on twisting positively, P changes to P' $(x + u_x, y + u_y, z + u_z)$, where u_x, u_y, u_z are small displacements, as shown in Fig. 5.2. The angle of rotation of a cross-section at distance z from the origin is $z\theta$, where θ is the angle of twist per unit length. For elastic deformation θ is small and constant along the length of the bar. Then from Fig. 5.2, we have

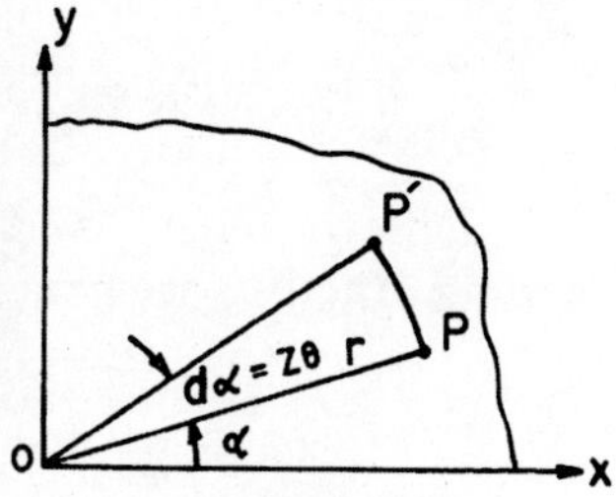

Fig. 5.2

$$x = r\cos\alpha,\ y = r\sin\alpha$$

$$\therefore \quad \delta x = -r\sin\alpha\,.\,\delta\alpha,\ \delta y = r\cos\alpha\,.\,\delta\alpha$$

Hence $\quad u_x = \delta x = -y\delta\alpha,\ u_y = x\delta\alpha$

Thus $\quad u_x = -y.z\theta,\ u_y = x.z\theta \qquad \text{...(5.2)}$

P' is taken to have moved out of the original plane of P by amount u_z ; this amount varies from point to point, and for a given point is also a function of θ, and is expressed by

$$u_z = \theta\,\varphi(x, y) \qquad \text{...(5.3)}$$

where $\varphi(x, y)$ is called the warping or torsion function.

Since the warping function is independent of z, therefore, warping takes place freely at all sections including the ends of the bar. Now

$$\left.\begin{aligned}
\varepsilon_x &= \frac{\partial u_x}{\partial x} = 0\ ;\ \varepsilon_y = \frac{\partial u_y}{\partial y} = 0\ ;\ \varepsilon_z = \frac{\partial u_z}{\partial z} = 0\\
\gamma_{xy} &= \frac{\partial u_x}{\partial y} + \frac{\partial u_y}{\partial x} = -z\theta + z\theta = 0\\
\gamma_{yz} &= \frac{\partial u_y}{\partial z} + \frac{\partial u_z}{\partial y} = x\theta + \theta\frac{\partial\varphi}{\partial y}\\
&= \theta\left(x + \frac{\partial\varphi}{\partial y}\right)\\
\gamma_{zx} &= \frac{\partial u_z}{\partial x} + \frac{\partial u_z}{\partial x} = \theta\frac{\partial\varphi}{\partial x} - y\theta\\
&= \theta\left(-y + \frac{\partial\varphi}{\partial x}\right)
\end{aligned}\right\} \qquad \text{...(5.4)}$$

Using Hooke's law, we get

$$\left.\begin{aligned} \sigma_x = \sigma_y = \sigma_z = \tau_{xy} = 0 \\ \tau_{yz} = G\gamma_{yz} = G\theta\left(x + \frac{\partial\varphi}{\partial y}\right) \\ \tau_{zx} = G\gamma_{zx} = G\theta\left(-y + \frac{\partial\varphi}{\partial x}\right) \end{aligned}\right\} \qquad \text{...(5.5)}$$

Differentiating and subtracting Eqs. (5.5) to eliminate the warping function, we get

$$\frac{\partial\tau_{zx}}{\partial y} - \frac{\partial\tau_{yz}}{\partial x} = -2G\theta \qquad \text{...(5.6)}$$

Now the only equilibrium equation by neglecting body force is :

$$\frac{\partial\tau_{yz}}{\partial y} + \frac{\partial\tau_{zx}}{\partial x} = 0 \qquad \text{...(5.7)}$$

Substituting Eqs. (5.5) in Eq. (5.7), we get

$$\frac{\partial^2\varphi}{\partial x^2} + \frac{\partial^2\varphi}{\partial y^2} = \nabla^2\varphi = 0 \qquad \text{...(5.8)}$$

Hence the warping function must satisfy the Laplace equation.

Let
$$\left.\begin{aligned} \tau_{zx} = \frac{\partial\phi}{\partial y} \\ \tau_{yz} = -\frac{\partial\phi}{\partial x} \end{aligned}\right\} \qquad \text{...(5.9)}$$

where ϕ is a function of x and y, called the stress function. Eqs. (5.9) satisfy Eq. (5.7) and Eq. (5.6) becomes

$$\frac{\partial^2\phi}{\partial x^2} + \frac{\partial^2\phi}{\partial y^2} = -2G\theta$$

or

$$\nabla^2\phi = -2G\theta \qquad \text{...(5.10)}$$

Eq. (5.10) is the well-known Poisson's equation for torsion of prismatic bars of non-circular cross-sections.

Suppose that the boundary of the section is defined by points (x, y) such that $x = x(s)$ and $y = y(s)$. Fig. 5.3 shows a part of a section of the bar perpendicular to the axis, where s is increasing and x is decreasing. Consider the triangular element ABC at the boundary. The shear stresses over the triangular area dA are τ_{zx} and τ_{zy}. Considering the equilibrium of forces in a direction normal to the curved surface of the bar on which there is no normal stress, we get

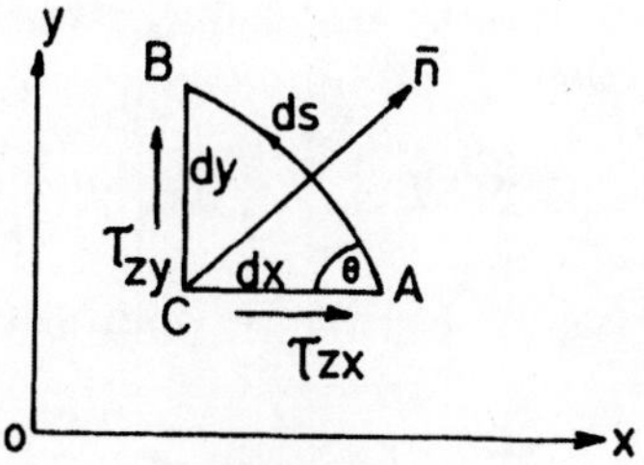

Fig. 5.3

$$\tau_{zx} \,.\, dA \,.\, \sin\theta + \tau_{zy} \,.\, dA \cos\theta = 0$$

$$\therefore \qquad \frac{\tau_{zy}}{\tau_{zx}} = -\frac{\sin\theta}{\cos\theta}$$

Now $\quad \sin\theta = \dfrac{dy}{ds}$

$$\cos\theta = -\frac{dx}{ds}$$

$$\therefore \qquad \frac{\tau_{zy}}{\tau_{zx}} = \frac{dy/ds}{dx/ds} \qquad ...(5.11)$$

Substituting Eqs. (5.9) in Eq. (5.11), we get

$$\frac{-\dfrac{\partial\phi}{\partial x}}{\dfrac{\partial\phi}{\partial y}} = \frac{dy/ds}{dx/ds}$$

$$\therefore \qquad \frac{\partial\phi}{\partial x}\cdot\frac{dx}{ds} + \frac{\partial\phi}{\partial y}\cdot\frac{dy}{ds} = 0$$

or

$$\frac{d\phi}{ds} = 0 \qquad ...(5.12)$$

Hence ϕ = a constant along the boundary of the cross-section. In the case of singly-connected sections, *e.g.* for solid bars, this constant can be chosen arbitrarily, and we shall take it zero.

The torque T, required to give a twist θ is obviously

$$T = \int (x\tau_{yz} - y\tau_{zx})\, dA$$

Using Eq. (5.9), we have

$$T = -\iint\left(x\frac{\partial\phi}{\partial x} + y\frac{\partial\phi}{\partial y}\right)dxdy$$

$$= -\int dy \int x\frac{\partial\phi}{\partial x}\,dx - \int dx \int y\frac{\partial\phi}{\partial y}\,dy$$

Integrating by parts, we get

$$T = -\int dy \left[x\phi - \int \phi dx \right]_{x_1}^{x_2} - \int dx \left[y\phi - \int \phi dy \right]_{y_3}^{y_4} \qquad \ldots(5.13)$$

where (x_1, y_1), (x_2, y_1), (x_3, y_3), (x_3, y_4) are points on the boundary as shown in Fig. 5.4.

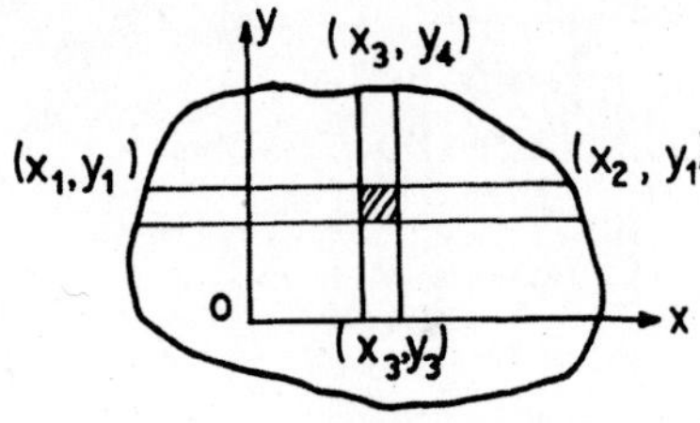

Fig. 5.4

Thus
$$T = -\int dy \left[x_2\phi_2 - x_1\phi_1 - \int_{x_1}^{x_2} \phi dx \right] - \int dx \left[y_4\phi_4 - y_3\phi_3 - \int_{y_3}^{y_4} \phi dy \right]$$

$$= 2\iint_A \phi \, dxdy \qquad \ldots(5.14)$$

Since $\phi_1 = \phi_2 = \phi_3 = \phi_4 = 0$ on the boundary.

Thus, if a function $\phi(x, y)$ can be found which is zero round a boundary, and if on calculating $\nabla^2\phi$ it is found to be a constant, the equation $\nabla^2\phi + 2G\theta = 0$ can be satisfied. Hence the shear stress distribution through the section and the torque can be found. The magnitude of the resultant shear stress at a point is

$$\tau = \sqrt{\left(\frac{\partial\phi}{\partial x}\right)^2 + \left(\frac{\partial\phi}{\partial y}\right)^2} = \text{grad } \phi \qquad \ldots(5.15)$$

5.2.1 Torsion of Elliptical Cross-section Bar

The equation for an ellipse having major and minor axes $2a$ and $2b$ respectively is :

$$\frac{x^2}{a^2} + \frac{y^2}{b^2} = 1 \qquad \ldots(1)$$

The torsion equation $\nabla^2\phi = -2G\theta$ and the boundary condition ϕ = constant = 0 is satisfied by assuming

$$\phi = m\left(\frac{x^2}{a^2} + \frac{y^2}{b^2} - 1\right) \qquad \ldots(2)$$

where m is a constant whose value will be determined. Substituting Eq. (2) in torsion equation

$$\nabla^2\phi = -2G\theta, \text{ we get}$$

$$\frac{\partial^2\phi}{\partial x^2} = \frac{2m}{a^2}, \frac{\partial^2\phi}{\partial y^2} = \frac{2m}{b^2}$$

$$\therefore \quad 2m\left(\frac{1}{a^2} + \frac{1}{b^2}\right) = -2\,G\theta$$

$$\therefore \quad m = -\left(\frac{a^2b^2}{a^2+b^2}\right) G\theta$$

$$\therefore \quad \phi = -\left(\frac{a^2b^2}{a^2+b^2}\right)\left(\frac{x^2}{a^2} + \frac{y^2}{b^2} - 1\right) G\theta \qquad \text{...(3)}$$

The stresses become :

$$\left.\begin{aligned} \therefore \quad \tau_{yz} &= -\frac{\partial\phi}{\partial x} = \left(\frac{2b^2}{a^2+b^2}\right) G\theta\, x \\ \tau_{zx} &= \frac{\partial\phi}{\partial y} = -\left(\frac{2a^2}{a^2+b^2}\right) G\theta\, y \\ \text{Thus} \quad \tau &= \sqrt{\tau_{yz}^2 + \tau_{zx}^2} \\ &= \frac{\pm 2\,G\theta}{\sqrt{a^2+b^2}} \sqrt{a^4y^2 + b^4x^2} \end{aligned}\right\} \qquad \text{...(4)}$$

$$\text{Torque, } T = 2\iint \phi\, dx\, dy$$

$$= -2\iint\left(\frac{a^2b^2}{a^2+b^2}\right)\left(\frac{x^2}{a^2} + \frac{y^2}{b^2} - 1\right) G\theta \,.\, dxdy$$

$$= -2\left(\frac{a^2b^2}{a^2+b^2}\right) G\theta \,.\iint\left(\frac{x^2}{a^2} + \frac{y^2}{b^2} - 1\right) dx\, dy$$

$$= -2\left(\frac{a^2b^2}{a^2+b^2}\right) G\theta \,.\left[\frac{1}{a^2}\int_0^b\left(\int_0^x x^2\, dx\right) dy\right.$$

$$\left. + \frac{1}{b^2}\int_0^a\left(\int_0^y y^2 dy\right) dx - \int_0^b\left(\int_0^x dx\right) dy\right]$$

$$= -2\left(\frac{a^2b^2}{a^2+b^2}\right) G\theta\left[\frac{1}{a^2}\cdot\frac{\pi a^2 b}{4} + \frac{1}{b^2}\cdot\frac{\pi a b^2}{4} - \pi ab\right]$$

$$= \frac{\pi a^3 b^3}{(a^2 + b^2)} G\theta. \qquad ...(5.16)$$

Since
$$\iint x^2 \, dx \, dy = I_y = \frac{\pi a^3 b}{4}$$

$$\iint y^2 \, dx \, dy = I_x = \frac{\pi ab^3}{4}$$

$$\iint dx \, dy = \pi ab$$

Using Eq. (5), Eq. (3) can be written as :

$$\phi = -\frac{T}{\pi ab}\left(\frac{x^2}{a^2} + \frac{y^2}{b^2} - 1\right) \qquad ...(5)$$

and the stress components become

$$\tau_{zx} = -\frac{2Ty}{\pi ab^3},$$

$$\tau_{yz} = \frac{2Tx}{\pi a^3 b} \qquad ...(6)$$

The greatest stresses occur at the boundary of the bar, given by,

$$\tau_{max}\Big|_{y=b} = \pm \frac{2G\theta}{(a^2 + b^2)} \sqrt{a^4 y^2 + b^4 x^2}$$

$$= \pm \frac{2G\theta}{(a^2 + b^2)} \sqrt{a^2 y^2 (a^2 - b^2) - a^2 b^4}$$

$$= \pm \frac{2G\theta \, a^2 b}{(a^2 + b^2)} = \frac{2T}{\pi ab^2} \qquad ...(5.17)$$

The angle of twist θ is given by

$$\theta = \frac{T \, . \, (a^2 + b^2)}{\pi a^3 b^3 G} \qquad ...(5.18)$$

Torsional rigidity,

$$\frac{T}{\theta} = D = \frac{\pi a^3 b^3 G}{(a^2 + b^2)} \qquad ...(5.19)$$

To investigate the warping which occurs, we have from Eqs. (5.5) and (4).

$$\tau_{zx} = G\theta\left(\frac{\partial \varphi}{\partial x} - y\right)$$

$$= -\frac{2a^2}{(a^2 + b^2)} G\theta y$$

$$\therefore \quad \frac{\partial \varphi}{\partial x} - y = -\frac{2a^2 y}{a^2 + b^2}$$

$$\frac{\partial\varphi}{\partial x} = -\left(\frac{a^2 - b^2}{a^2 + b^2}\right) y$$

Integrating, we get

$$\varphi = -\left(\frac{a^2 - b^2}{a^2 + b^2}\right) yx + C$$

where C is an arbitrary constant.

At $x = y = 0$,

$\varphi = 0$.

Hence $C = 0$

$$\therefore \quad \varphi = -\left(\frac{a^2 - b^2}{a^2 + b^2}\right) xy \quad \text{...(5.20)}$$

which are rectangular hyperbolas, whose asymptotes are the axes of the ellipse as shown in Fig. 5.5.

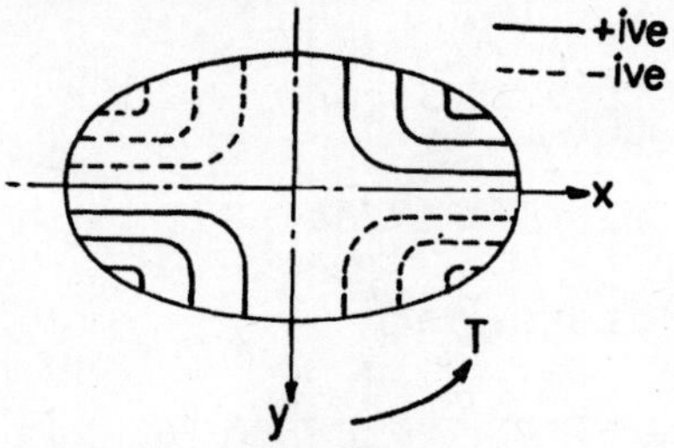

Fig. 5.5

The displacement components are given by

$$u_x = -\theta\, zy$$

$$u_y = \theta\, zx$$

and

$$u_z = \theta\, \varphi(x, y)$$

$$= \frac{T(a^2 + b^2)}{\pi\, a^3 b^3\, G} \cdot \frac{(b^2 - a^2)}{(a^2 + b^2)} xy$$

$$= \frac{T(b^2 - a^2)\, xy}{\pi\, a^3 b^3\, G} \quad \text{...(5.21)}$$

5.2.2 Torsion of Equilateral Triangle Cross-section Bar

Let the triangle be of side length $2a$ as shown in Fig. 5.6. If the stress function is chosen to be

$$\phi = m(\sqrt{3}\, x - a)\, (\sqrt{3}\, x - 3y + 2a)(\sqrt{3}\, x + 3y + 2a)$$

it will be found to be zero on its perimeter :

$$\frac{\partial^2\phi}{\partial x^2} = 18\, m\, (\sqrt{3}x + a)$$

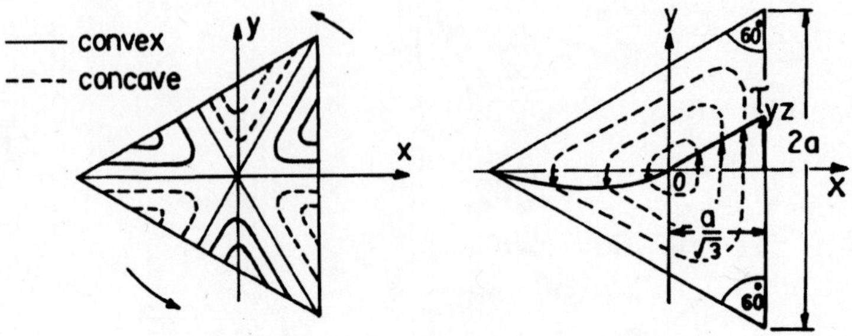

Fig. 5.6 Equilateral triangle bar under torsion.

$$\frac{\partial^2 \phi}{\partial y^2} = 18\, m\ (a - \sqrt{3}x)$$

The torsion equation, $\nabla^2 \phi = -2G\theta$

is satisfied when $m = -\dfrac{G\theta}{18a}$.

$$\therefore \qquad \tau_{zx} = \frac{\partial \phi}{\partial y} = \frac{G\theta y}{a}\,(x\sqrt{3} - a)$$

$$\tau_{yz} = -\frac{\partial \phi}{\partial x} = \frac{\sqrt{3}\ G\theta}{2a}\left(x^2 + \frac{2a}{\sqrt{3}}x - y^2\right)$$

when $y = 0,\ \tau_{zx} = 0$

and
$$\tau_{yz} = \frac{\sqrt{3}\ G\theta}{2a}\left(x^2 + \frac{2a}{\sqrt{3}}x\right) \qquad ...(5.22)$$

τ_{yz} is maximum when $x = \dfrac{a}{\sqrt{3}}$

and then
$$(\tau_{yz})_{max} = \frac{\sqrt{3}}{2}\,.\,aG\theta \qquad ...(5.23)$$

and
$$\tau_{yz} = 0 \ \text{ when } \ x = -\frac{2a}{\sqrt{3}}$$

The distribution of shear stress τ_{yz} has been shown in Fig. 5.6 at $y = 0$.

Torque
$$T = 2\iint \phi\, dx\, dy$$
$$= G\theta\, a^4 \frac{\sqrt{3}}{5} \qquad ...(5.24)$$

Example 5.1 *A square shaft rotating at 250 rpm, transmits torque to a crane which is designed to lift maximum load of 150 kN at a speed of 10 m/min. If the efficiency of crane gearing is 65%, estimate the size of the shaft for the maximum permissible shear* *stress of 35 MPa. Also calculate the angle of twist of the shaft for a length of 3 m. Take G = 100 GPa.*

Solution. Power consumed by the crane

$$= \frac{150 \times 10^3 \times 10}{60 \times 10^3} = 25 \text{ kW}$$

Power delivered by the shaft

$$= \frac{25}{0.65} = 38.46 \text{ kW}$$

$$\frac{2\pi NT}{60 \times 10^3} = 38.46$$

$$T = \frac{38.46 \times 60 \times 10^3}{2\pi \times 250} = 1469 \text{ N.m.}$$

For a square shaft,

$$\tau_{max} = \frac{T}{k_1 a^3}, \text{ where } k_1 = 0.208$$

$$a^3 = \frac{1469}{0.208 \times 35 \times 10^6} = 201.786 \times 10^{-6}$$

$$a = 5.8654 \times 10^{-2} \text{ m or } 58.654 \text{ mm}$$

$$\theta = \frac{Tl}{k_2 G a^4}, \text{ where } k_2 = 0.141$$

$$= \frac{1469 \times 3}{0.141 \times 100 \times 10^9 \times (58.654)^4 \times 10^{-12}} \times \frac{180}{\pi}$$

$$= 1.513°.$$

Example 5.2. *A 300 mm steel I-beam (Fig. 5.7) with flanges and web 12.5 mm thick is subjected to a torque of 4 kN.m. Find the maximum shear stress and angle of twist per unit length. G = 100 GPa.*

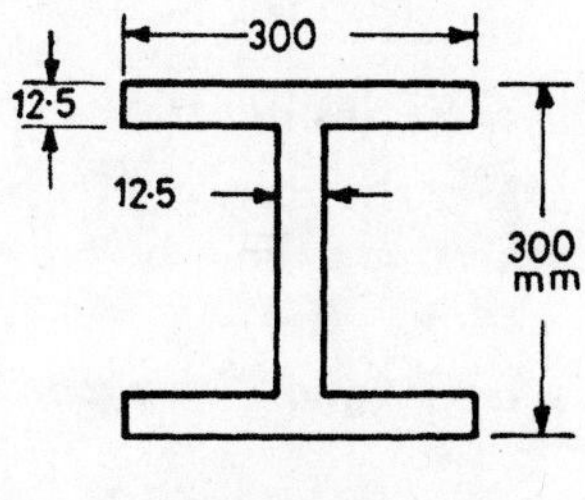

Fig. 5.7

Solution. $\tau_{max} = \dfrac{T}{\Sigma k_1 db^2}$

d	b	d/b	k_1	k_2
300	12.5	24	0.333	0.333
250	12.5	20	0.333	0.333

$$\Sigma k_1 db^2 = 0.333\,[2 \times 300 \times (12.5)^2 + 200 \times (12.5)^2] \times 10^{-9}$$
$$= 41.67 \times 10^{-6}\ \text{m}^3$$
$$\tau_{max} = \frac{4 \times 10^3}{41.67 \times 10^{-6}} = 96\ \text{MPa}$$
$$\theta = \frac{Tl}{G\,\Sigma k_2 db^3}$$
$$\Sigma k_2 db^3 = 0.333\,[2 \times 300 \times (12.5)^3 + 250 \times (12.5)^3] \times 10^{-12}$$
$$= 520.83 \times 10^{-9}\ \text{m}^4$$
$$\theta = \frac{4 \times 10^3 \times 1}{100 \times 10^9 \times 520.83 \times 10^{-9}} \times \frac{180}{\pi}$$
$$= 4.4°.$$

Example 5.3. *An elliptical shaft of semi axes a = 0.05 m, b = 0.025 m, and G = 80 GPa is subjected to a twisting moment of 1200 π N.m. Determine the maximum shearing stress and the angle of twist per unit length.*

Solution. $\tau_{max} = \dfrac{2T}{\pi ab^2}$

$$= \frac{2 \times 1200\,\pi}{\pi \times 0.05 \times (0.025)^2} = 76.8\ \text{MPa}$$
$$\theta = \frac{(a^2 + b^2)\,T}{\pi \times a^3 b^3\,G}$$
$$= \frac{(0.05^2 + 0.025^2) \times 1200\,\pi}{\pi \times 0.05^3 \times 0.025^3 \times 80 \times 10^9}$$
$$= 0.024\ \text{rad/m}.$$

5.3. Torsion of Hollow Cross-sections

Now let us consider the torsion of a hollow cylinder of cross-sectional area A containing a hole of cross-sectional area A_1 as shown in Fig. 5.8.

The expression for torque T becomes :

$$T = -\int dy \left[x\phi - \int \phi dx \right]_{x'_3}^{x_3} - \int dy \left[x\phi - \int \phi dx \right]_{x'_4}^{x_4}$$

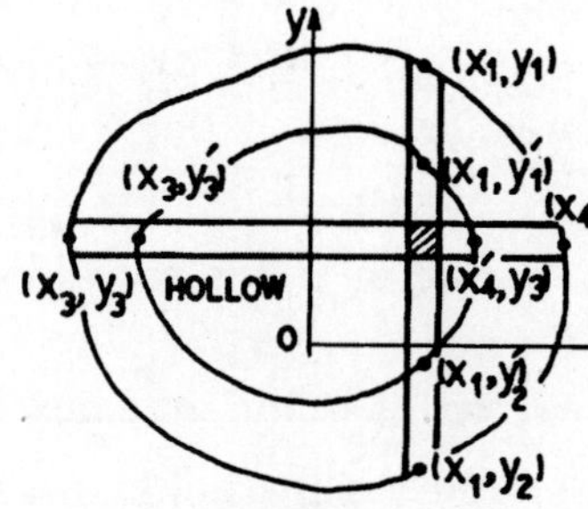

Fig. 5.8

$$-\int dx\left[y\phi-\int\phi dy\right]_{y_1'}^{y_1}-\int dx\left[y\phi-\int\phi dy\right]_{y_2'}^{y_2}$$

$$=-\int(x_3\phi_3-x_3'\phi_3')\,dy+\iint_{x_3'}^{x_3}\phi\,dxdy$$

$$-\int(x_4\phi_4-x_4'\phi_4')\,dy+\iint_{x_4'}^{x_4}\phi\,dx\,dy$$

$$-\int(y_1\phi_1-y_1'\phi_1')\,dx+\int_{y_1'}^{y_1}\phi\,dxdy$$

$$-\int(y_2\phi_2-y_2'\phi_2')\,dx+\iint_{y_2'}^{y_2}\phi\,dxdy$$

But $\phi_1=\phi_2=\phi_3=\phi_4=0$

and $\phi_1'=\phi_2'=\phi_3'=\phi_4'=\phi_1$ say, thus

$$T=2\phi_1 I_1+I_A \qquad \text{...(5.25)}$$

where I_A refers to the annulus over which $\iint\phi\,dx\,dy$ is taken and I_1 is the area of the 'hollow', ϕ_1 being a constant value of the stress function along the internal boundary as shown in Fig. 5.9.

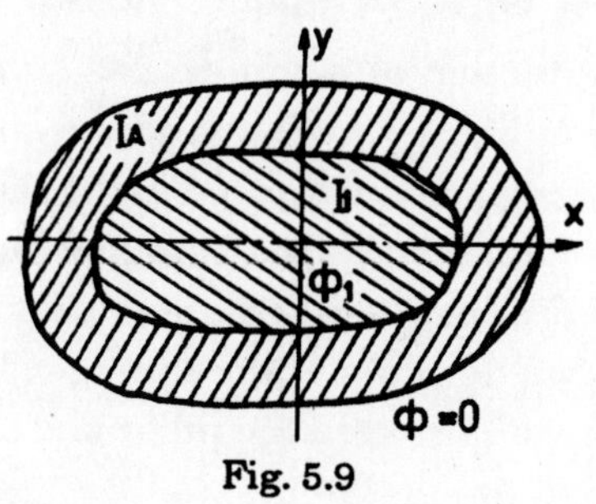

Fig. 5.9

Eq. (5.25) may be written as

$$T = 2(I_A + \phi_1 I_1 + I_H) - 2I_H$$
$$= 2(I_0 - I_H) \qquad ...(5.26)$$

where $I_0 = \iint \phi \, dx \, dy$

is taken over the whole section as if it were solid and

$$I_H = \iint \phi \, dx \, dy$$

is taken over the 'hollow', with $\phi = 0$ on the inner boundary.

5.3.1 Torsion of Hollow Elliptical Cross-section Bar

Let the major and minor axes of the inner ellipse be half of that of the outer ellipse.

i.e. $k = \frac{1}{2}$, then applying eq. (5.26), we get

$$T = 2\,[1 - k^4]\,\frac{\pi a^3 b^3}{(a^2 + b^2)}\,G\theta$$

$$= \frac{15\pi}{16} \cdot \frac{a^3 b^3}{(a^2 + b^2)}\,G\theta \qquad ...(5.27)$$

$$\theta = \frac{T}{(1 - k^4)} \cdot \frac{(a^2 + b^2)}{\pi a^3 b^3 G}$$

$$= \frac{16T}{15} \cdot \frac{(a^2 + b^2)}{\pi a^3 b^3 G} \qquad ...(5.28)$$

$$\phi = -\frac{T}{\pi ab(1 - k^4)}\left(\frac{x^2}{a^2} + \frac{y^2}{b^2} - 1\right)$$

$$= -\frac{16T}{15\pi\, ab}\left(\frac{x^2}{a^2} + \frac{y^2}{b^2} - 1\right) \qquad ...(5.29)$$

$$\tau_{max} = \frac{2T}{\pi ab^2} \cdot \frac{1}{(1 - k^4)}$$

$$= \frac{32T}{15\pi\, ab^2} \qquad ...(5.30)$$

5.4 Torsion of Thin-Walled Tubes

Consider the torsion of a thin-walled tube of non-uniform thickness as shown in Fig. 5.10. We consider a small portion of the tube as a free body diagram as indicated in the figure. Owing to the narrow width of the boundary, we shall assume that the shear stress is tangent to the midline between the inner and outer boundaries at a section and has a value equal to the average shear stress across the thickness of the wall between the inner and outer boundaries.

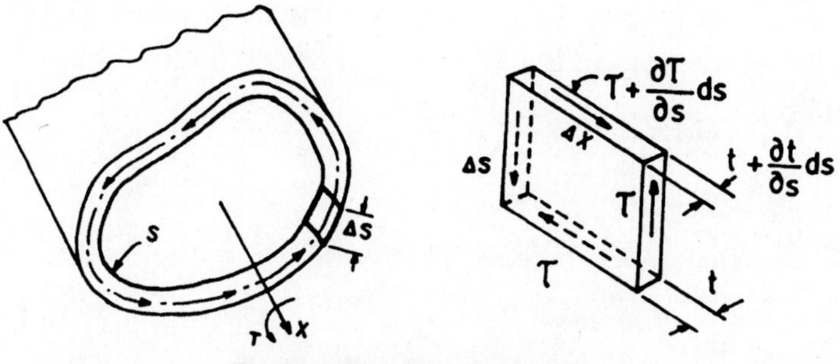

Fig. 5.10 Torsion of thin tubes.

Applying the equations of equilibrium in the x-direction, we get

$$-\tau t\,\Delta x+\left(\tau+\frac{\partial \tau}{\partial s}ds\right)\left(t+\frac{\partial t}{\partial s}ds\right)\Delta x=0$$

where s is measured along the mid-line between the inner and outer boundaries. Simplifying, we get

$$\left(\tau\frac{\partial t}{\partial s}+t\frac{\partial \tau}{\partial s}\right)\Delta s+\frac{\partial \tau}{\partial s}\cdot\frac{\partial t}{\partial s}(\Delta s)^2=0$$

Neglecting last term, being small, we get

$$\frac{d}{ds}(t\tau)=0$$

$\therefore$ $\qquad t\tau=\text{constant}$

Let $\qquad q=t\tau=\text{shear flow}$ ...(5.31)

Thus we say that the shear flow is constant for a thin-walled tube under simple torsion.

Now torque
$$\left.\begin{aligned}T&=\int r\,.\,(t\tau)\,ds\\&=t\tau\int r\,ds\end{aligned}\right\}\qquad ...(5.32)$$

where r is measured from some convenient point O in the section to the element ds. With reference to Fig. 5.11 we find that

$$\int r ds=2A$$

where A is the total plane area enclosed by the mid-line s. Thus

$$T=2t\tau A$$

$$\therefore\qquad \tau=\frac{T}{2tA}\qquad ...(5.33)$$

To determine the deformation of the thin walled tube under simple torsion, let u_s be the displacement component in the s direction, then

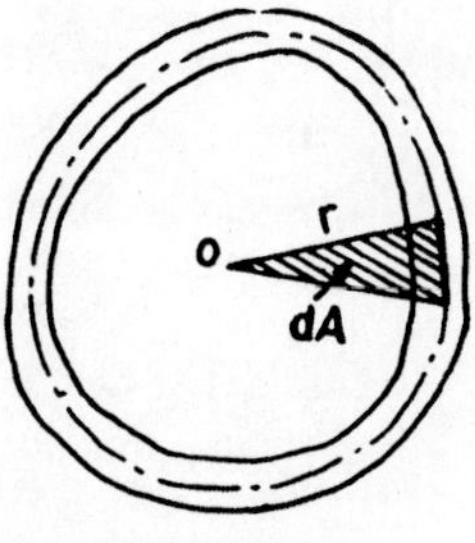

Fig. 5.11

$$\gamma_{xs} = \frac{\partial u_s}{\partial x} + \frac{\partial u_x}{\partial s} \qquad \text{...(1)}$$

Using Hooke's law,

$$\gamma_{xs} = \frac{\tau}{G} \qquad \text{...(2)}$$

Comparing Eqs. (1) and (2), we get

$$\frac{\partial u_s}{\partial x} + \frac{\partial u_x}{ds} = \frac{\tau}{G} \qquad \text{...(3)}$$

$$\frac{\partial u_x}{\partial s} = \frac{\tau}{G} - \frac{\partial u_s}{\partial x}$$

Integrating from s_1 to s_2, we get

$$(u_x)_2 - (u_x)_1 = \int_{s_1}^{s_2} \left(\frac{\tau}{G} - \frac{\partial u_s}{\partial x} \right) ds \qquad \text{...(4)}$$

If $s_1 = s_2$, *i.e.* we make a complete circuit around the section, we require that $(u_x)_2 = (u_x)_1$.

Hence Eq. (3) becomes.

$$\oint \left(\frac{\tau}{G} - \frac{\partial u_s}{\partial x} \right) ds = 0$$

or

$$\oint \frac{\partial u_s}{\partial x} ds = \oint \frac{\tau}{G} ds \qquad \text{...(5)}$$

If we now assume that there exists an axis of rotation 0—0 along the cylinder about which, if we do not include warping displacement, there is a simple rotation of the elements by an amount α. Then observing Fig. 5.12, it can be seen that for small deformations we can express u_s in the following way.

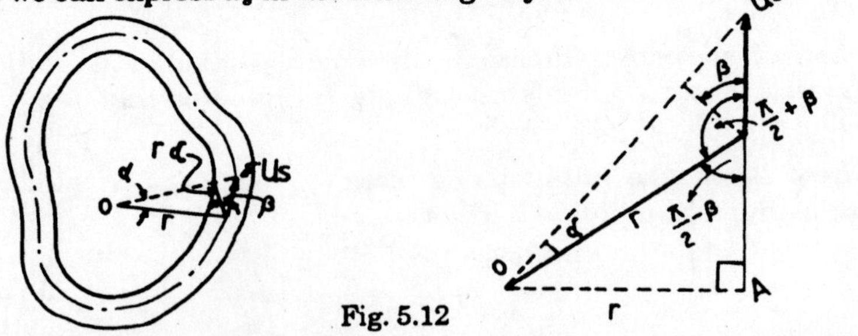

Fig. 5.12

$$u_s = r\alpha \cos\beta \qquad \text{...(5.28)}$$

Let us draw $OA \perp$ to the line of action of u_s. Then

$$r_\perp = r \sin\left(\frac{\pi}{2} - \beta \right) = r \cos\beta$$

Hence Eq. (5) becomes

$$\oint \frac{\partial (\alpha r \cos \beta)}{\partial x} ds = \frac{1}{G} \oint \tau \, ds$$

Now $r \cos \beta = | r \times \bar{s} |$

where $\bar{s}$ is the unit vector along the centre-line of the tube wall. Hence, we have

$$\oint \frac{\partial(| r \times \bar{s} | \alpha)}{\partial x} ds = \frac{1}{G} \oint \tau \, ds$$

$$\oint | r \times \bar{s} | \frac{\partial \alpha}{\partial x} ds = \frac{1}{G} \oint \tau \, ds$$

Now $\frac{\partial \alpha}{\partial x}$, the rate of twist, is not a function of s. Hence are find that,

$$\frac{d\alpha}{dx} \oint | r \times \bar{s} | \, ds = \frac{1}{G} \oint \tau ds \quad ...(6)$$

Now $\oint | r \times \bar{s} | \, ds = \oint | r \times ds | = | \oint r \times ds |$

$$= 2A$$

Where A is the area of the plane cross-section of the cylinder inside the mid line between the inner and outer walls. Hence, we get

$$\frac{d\alpha}{dx} = \frac{1}{2AG} \oint \tau ds \quad ...(5.34)$$

This is known as *Bredt's formula*. It gives us the rate of twist in terms of the shear-stress distribution. Substituting the value of τ from Eq. (1) in (5.34), we get

$$\frac{d\alpha}{dx} = \frac{1}{2AG} \oint \frac{T}{2At} ds$$

$$= \frac{T}{4A^2 G} \oint \frac{ds}{t} \quad ...(5.35)$$

If t is constant, then $\frac{d\alpha}{dx} = \frac{TS}{4A^2 Gt}$ ...(5.36)

where S is the perimeter length of the area A.

From exact analysis, $\frac{d\alpha}{dx} = \frac{T}{GJ}$ ...(5.37)

For example for a thin circular tube of inner radius r_i and outer radius r_o,

$$\left(\frac{d\alpha}{dx}\right)_{Bredt} = \frac{T \cdot [2\pi (r_i + r_o)/2]}{4\left[\pi\left(\frac{r_o + r_i)}{2}\right)^2\right]^2 G(r_o - r_i)}$$

$$= \frac{4T}{\pi G(r_o + r_i)^3 (r_o - r_i)} \quad ...(5.38)$$

$$\left(\frac{d\alpha}{dx}\right)_{exact} = \frac{T}{G \cdot \frac{\pi}{2} (r_o^4 - r_i^4)} \quad ...(5.39)$$

$$\therefore \quad \frac{\left(\frac{d\alpha}{dx}\right)_{Bredt}}{\left(\frac{d\alpha}{dx}\right)_{exact}} = \frac{2(r_o^2 + r_i^2)}{(r_o + r_i)^2}$$

$$= \frac{2\left[1 + \left(\frac{r_i}{r_o}\right)^2\right]}{\left(1 + \frac{r_i}{r_o}\right)^2} \quad ...(5.40)$$

when $\frac{r_i}{r_0} \to 1$,

then the ratio of rate of twists approaches unity. Thus for very thin tubes, Bredt's formula gives highly accurate results.

Example 5.4. *A hollow aluminium section is designed as shown in Fig. 5.13 for a maximum shear stress of 35 MPa. Find maximum permissible twisting moment for this section and the angle of twist under this moment per metre length. G = 28 GPa.*

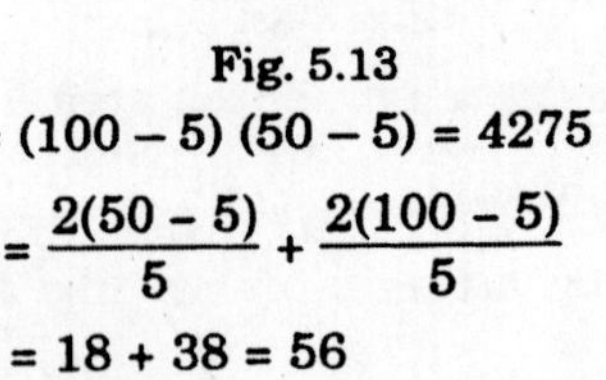

Fig. 5.13

Solution. $A = (100 - 5)(50 - 5) = 4275 \text{ mm}^2$

$$\int \frac{ds}{t} = \frac{2(50-5)}{5} + \frac{2(100-5)}{5}$$

$$= 18 + 38 = 56$$

$$T = 2At\tau$$

$$= 2 \times 4275 \times 5 \times 35 \times 10^6 \times 10^{-9}$$

$$= 1496.25 \text{ N.m.}$$

$$\theta = \frac{Tl}{4A^2G}\int\frac{ds}{t}$$

$$= \frac{1496.25 \times 1}{4 \times (4275)^2 \times 10^{-12} \times 28 \times 10^9} \times 56 \times \frac{180}{\pi}$$

$$= 2.345°.$$

Example 5.5. *A multiply connected hollow steel tube (Fig. 5.14) resists a torque of 12 kN.m. The wall thicknesses are $t_1 = t_2 = t_3 = 6$ mm and $t_4 = t_5 = 3$ mm. Determine the maximum shearing stresses and the angle to twist per unit length. Take G = 80 GPa.*

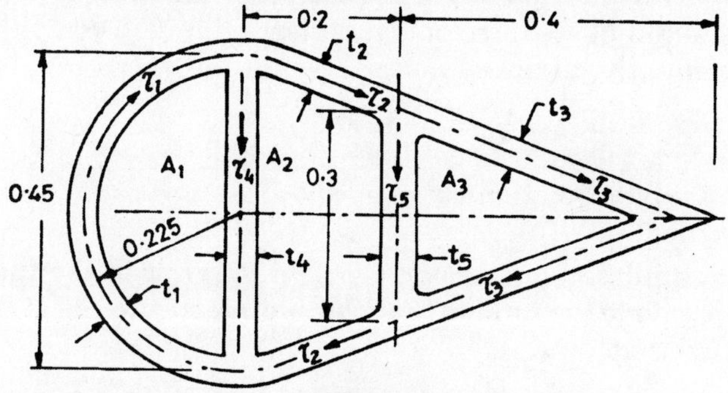

Fig. 5.14

Solution. Consideration of shear flow yields :

$$\tau_1 t_1 = \tau_2 t_2 + \tau_4 t_4 \quad ...(1)$$

$$\tau_2 t_2 = \tau_3 t_3 + \tau_5 t_5 \quad ...(2)$$

$$T = 2A_1 t_1 \tau_1 + 2A_2 t_2 \tau_2 + 2A_3 t_3 \tau_3 = 12000 \text{ N.m} \quad ...(3)$$

$$\tau_1 s_1 + \tau_4 s_4 = 2G\theta A_1 \quad ...(4)$$

$$2\tau_2 s_2 + \tau_5 s_5 - \tau_4 s_4 = 2G\theta A_2 \quad ...(5)$$

$$2\tau_3 s_3 - \tau_5 s_5 = 2G\theta A_3 \quad ...(6)$$

where $s_1 = \pi \times 0.225 = 0.7069$ m

$$s_2 = \sqrt{(0.2)^2 + (0.075)^2} = 0.2136 \text{ m}$$

$$s_3 = \sqrt{(0.4)^2 + (0.15)^2} = 0.4272 \text{ m}$$

$$s_4 = 0.45 \text{ m}, s_5 = 0.3 \text{ m}$$

$$A_1 = \frac{\pi \times (0.45)^2}{2} = 0.079522 \text{ m}^2$$

$$A_2 = \frac{1}{2}(0.45 + 0.3) \times 0.2 = 0.075 \text{ m}^2$$

$$A_3 = \frac{1}{2} \times 0.3 \times 0.4 = 0.06 \text{ m}^2$$

Solving Eqs. (1) to (6), we get

$$\tau_1 = 4.902 \text{ MPa}, \tau_2 = 5.088 \text{ MPa}$$
$$\tau_3 = 3.809 \text{ MPa}, \tau_4 = -0.373 \text{ MPa}$$
$$\tau_5 = 2.558 \text{ MPa}$$
$$\theta = 0.000\,2591 \text{ rad/m.}$$

5.5 Solution of Torsion Problems by Complex Torsion Function

Since the torsion function $\varphi(x, y)$ is harmonic in the region R representing the cross-section of the bar, therefore, we can construct the analytic function $\varphi + i\psi$, of the complex variable $x + iy$, where $\psi(x, y)$ is the conjugate harmonic function, related to $\varphi(x, y)$ by the Cauchy-Riemann equations

$$\frac{\partial \varphi}{\partial x} = \frac{\partial \psi}{\partial y}, \frac{\partial \varphi}{\partial y} = -\frac{\partial \psi}{\partial x} \qquad \text{...(1)}$$

The analytic function $\varphi(x, y) + i\psi(x, y)$ is known as the complex torsion function.

To formulate the torsion problem in terms of the function $\psi(x, y)$ conjugate to the torsion function, we substitute Eq. (1) in Eq. (5.5) and find that

$$\tau_{yz} = G\theta\left(x - \frac{\partial \psi}{\partial x}\right), \tau_{xz} = G\theta\left(-y + \frac{\partial \psi}{\partial y}\right) \qquad \text{...(2)}$$

The boundary condition Eq. (5.11) becomes

$$\frac{x - \dfrac{\partial \psi}{\partial x}}{-y + \dfrac{\partial \psi}{\partial y}} = \frac{dy/ds}{dx/ds}$$

$$\left(x - \frac{\partial \psi}{\partial x}\right)\frac{dx}{ds} = \left(-y + \frac{\partial \psi}{\partial y}\right)\frac{dy}{ds}$$

$$x\frac{dx}{ds} - \frac{\partial \psi}{\partial x}\frac{dx}{ds} = -y\frac{dy}{ds} + \frac{\partial \psi}{\partial y}\cdot\frac{dy}{ds}$$

$$\frac{\partial \psi}{\partial x}\cdot\frac{dx}{ds} + \frac{\partial \psi}{\partial y}\cdot\frac{dy}{ds} = x\frac{dx}{ds} + y\frac{dy}{ds}$$

$$\frac{d\psi}{ds} = \frac{d}{ds}\left[\frac{1}{2}\left(x^2 + y^2\right)\right]$$

$$\therefore \quad \frac{d}{ds}\left\{\psi - \frac{1}{2}\left(x^2 + y^2\right)\right\} = 0 \qquad \text{...(3)}$$

The torsion problem is therefore solved if we can find a function which satisfies

$$\left.\begin{array}{l}(i)\ \nabla^2\psi = 0 \text{ in } R \\ (ii)\ \psi - \frac{1}{2}(x^2 + y^2) = \text{constant on } C.\end{array}\right\} \quad ...(4)$$

Torque T becomes

$$T = \iint (x\tau_{yz} - y\tau_{zx})\, dx\, dy$$

$$= G\theta \iint \left(x^2 - x\frac{\partial\psi}{\partial x} + y^2 - y\frac{\partial\psi}{\partial y} \right) dxdy$$

$$= G\theta \iint \left(x^2 + y^2 - x\frac{\partial\psi}{\partial x} - y\frac{\partial\psi}{\partial y} \right) dxdy \quad ...(5)$$

Torsional rigidity

$$D = G\iint_R \left(x^2 + y^2 - x\frac{\partial\psi}{\partial x} - y\frac{\partial\psi}{\partial y} \right) dxdy \quad ...(6)$$

$$= G\iint_R \left[\left(x - \frac{\partial\psi}{\partial x} \right)^2 + \left(y - \frac{\partial\psi}{\partial y} \right)^2 \right] dxdy$$

...(7)

The boundary value problem (4) is a *Dirichlet* problem so that its solution is unique.

We shall now formulate the torsion problem in terms of the function Ψ, introduced by Prandtl, which is defined as follows :

$$\Psi(x, y) = \psi(x, y) - \frac{1}{2}(x^2 + y^2) \quad ...(8)$$

We have

$$\frac{d\Psi}{\partial x} = \frac{\partial\psi}{\partial x} - x,\ \frac{\partial\Psi}{\partial y} = \frac{\partial\psi}{\partial y} - y$$

Hence Eqs. (2) become

$$\tau_{yz} = -G\theta\frac{\partial\Psi}{\partial x},\ \tau_{zx} = G\theta\frac{\partial\Psi}{\partial y} \quad ...(5.41)$$

Since the stress components τ_{yz} and τ_{zx} are obtained from the function $\Psi(x, y)$ by differentiation, the latter is called the stress function.

Substituting Eqs. (5.41) in (5.6), we get

$$\frac{\partial^2\Psi}{\partial x^2} + \frac{\partial^2\Psi}{\partial y^2} = -2 \text{ in } R \quad ...(5.42)$$

and Eqs. (4) and (8) give

$$\Psi = \text{constant on } C. \quad ...(5.43)$$

The torsional rigidity from Eq. (7) becomes

$$\left.\begin{aligned} D &= -G\int\!\!\int_R \left\{ x\frac{\partial \Psi}{\partial x} + y\frac{\partial \Psi}{\partial y} \right\} dxdy \\ &= 2G\int\!\!\int_R \Psi\, dxdy + G\oint \Psi(ydx - xdy) \\ &= G\int\!\!\int_R \left\{ \left(\frac{\partial \Psi}{\partial x}\right)^2 + \left(\frac{\partial \Psi}{\partial y}\right)^2 \right\} dxdy \end{aligned}\right\} \qquad \ldots(9)$$

By choosing $\Psi = 0$ on C.

$$D = 2G\int\!\!\int_R \Psi\, dx\, dy \qquad \ldots(5.44)$$

Now consider a family of curves, in the plane of the cross-section of the bar, obtained by setting

$$\psi(x, y) = \text{constant} \qquad \ldots(5.45)$$

The slope $\frac{dy}{dx}$ of the tangent line to any curve of the family defined by Eq. (5.45) is determined from the formula

$$\frac{\partial \Psi}{\partial x} + \frac{\partial \Psi}{\partial y}\frac{dy}{dx} = 0$$

Substituting Eqs. (5.41), we get

$$\frac{dy}{dx} = \frac{\tau_{yz}}{\tau_{zx}} \qquad \ldots(5.46)$$

In other words, at each point of the curve (5.45) the stress vector is directed along the tangent to the curve. For this reason the curves (5.45) are called lines of shearing stress.

5.5.1 Solution for Certain Particular Cases by Complex Torsion Function

(a) Elliptic cylinder.

The stress function $\Psi(x, y)$ is to be constant on

$$\frac{x^2}{a^2} + \frac{y^2}{b^2} = 1$$

This suggests that we try

$$\Psi = A\left(\frac{x^2}{a^2} + \frac{y^2}{b^2}\right)$$

This function satisfies the equation

$$\nabla^2\Psi = -2$$

if, $$-2 = 2A\left(\frac{1}{a^2}+\frac{1}{b^2}\right)$$

$\therefore$ $$\Psi = -\frac{(b^2x^2 + a^2y^2)}{(a^2+b^2)}$$

Function ψ obtained thus satisfies the conditions

(*i*) $\nabla^2\Psi = -2$ in R

(*ii*) Ψ = constant on C

where C is the boundary

$$\frac{x^2}{a^2}+\frac{y^2}{b^2} = 1$$

Hence $$\psi = \Psi + \frac{1}{2}(x^2+y^2)$$

$$= \frac{(a^2-b^2)(x^2-y^2)}{2(a^2+b^2)}$$

The conjugate (torsion) function is

$$\varphi = -\frac{(a^2-b^2)xy}{(a^2+b^2)}$$

Hence the complex torsion function is

$$\varphi + i\psi = \frac{(a^2-b^2)}{2(a^2+b^2)}[-2xy + i(x^2-y^2)]$$

$$= \frac{(a^2+b^2)\, iz^2}{2(a^2+b^2)}$$

where $z = x + iy$

$$\left.\begin{aligned} \tau_{zx} &= G\theta\,.\frac{\partial\Psi}{\partial y} = -\frac{2G\theta\, a^2y^2}{(a^2+b^2)} \\ \tau_{yz} &= -G\theta\frac{\partial\Psi}{\partial x} = \frac{2G\theta\, b^2x}{(a^2+b^2)} \end{aligned}\right\} \quad \text{...(5.47)}$$

Torque T is,

$$T = \int_R\int (x\,\tau_{yz} - y\,\tau_{zx})\,dxdy$$

$$= \frac{2G\theta}{(a^2+b^2)}\left[b^2\int_R\int x^2\,dxdy + a^2\int_R\int y^2\,dxdy\right]$$

$$= \frac{2G\theta}{(a^2+b^2)}\left[b^2\,.\,\frac{\pi a^2 b}{4} + a^2\,.\,\frac{\pi a b^3}{4}\right]$$

$$= \frac{\pi\, G\theta\, a^3b^3}{(a^2+b^2)} \quad \text{...(5.48)}$$

∴ Torsional rigidity

$$D = \frac{\pi G a^3 b^3}{(a^2 + b^2)} \qquad \ldots(5.49)$$

(b) *Equilateral triangle*

Suppose the triangle is bounded by the lines

$$y = c \pm mx, y = 0.$$

Then the function

$$\Psi = Ay\{(y-c)^2 + m^2x^2\}$$

vanishes along the boundry. For this function

$$\nabla^2\Psi = \{(6 - 2m^2)y - 4c\}A$$

so that in order that $\nabla^2\Psi = -2$, we must choose

$$m^2 = 3, A = \frac{1}{2c}$$

$$\therefore \qquad y = c \pm \sqrt{3}x$$

and

$$\Psi = \frac{y}{2c}\{(y-c)^2 - 3x^2\}$$

$$\tau_{zx} = G\theta\frac{\partial\Psi}{\partial y} = G\theta\left[\frac{1}{2c}\{(y-c)^2 - 3x^2\} + \frac{y}{2c}\{2(y-c)\}\right]$$

$$= \frac{G\theta}{2c}[(y-c)^2 - 3x^2 + 2y(y-c)]$$

$$= \frac{G\theta}{2c}[3y^2 + c^2 - 4yc - 3x^2] \qquad \ldots(5.50)$$

$$\tau_{yz} = -G\theta\frac{\partial\Psi}{\partial x} = -G\theta\cdot\frac{y}{2c}(-6x)$$

$$= \frac{3G\theta\, xy}{c} \qquad \ldots(5.51)$$

Torsional rigidity

$$D = \frac{G}{c}\iint[-3x^2y + y(y-c)^2]\,dxdy$$

$$= \frac{\sqrt{3}}{45}Gc^4 \qquad \ldots(5.52)$$

And $\varphi(x, y) = (3xy^2 - x^3 - 2cxy + c^2x)/2c$

$$\psi(x, y) = (y^3 - 3x^2y + c^2y + cx^2 - cy^2)/2c$$

(c) *Rectangle*

Consider the rectangle bounded by the lines $x = \pm a$, $y = \pm b$. The torsion problem will be solved if we can find a harmonic function $\psi(x, y)$ which takes the form $\frac{1}{2}(x^2 + y^2)$ on the boundary of the square, *i.e.* we must have

$$\left.\begin{aligned}\psi(\pm a, y) &= \frac{1}{2}(a^2 + y^2)\\ \psi(x, \pm b) &= \frac{1}{2}(x^2 + b^2)\end{aligned}\right\} \quad \text{...(1)}$$

If we write

$$\psi(x, y) = a^2 + \frac{1}{2}(y^2 - x^2) + f(x, y) \quad \text{...(2)}$$

then it follows that $f(x, y)$ must be a harmonic function satisfying the boundary conditions

$$f(\pm a, y) = 0 \quad \text{...(3)}$$

$$f(x \pm b) = x^2 - a^2 \quad \text{...(4)}$$

The harmonic function

$$f(x, y) = \sum_{n=0}^{\infty} C_n \cosh(k_n y) \cos(k_n x) \quad \text{...(5)}$$

will satisfy the condition (3) provided that

$$k_n = \frac{\pi}{a}\left(n + \frac{1}{2}\right) \quad \text{...(6)}$$

and the condition (4) provided that the constants C_n are chosen, so that

$$x^2 - a^2 = \sum_{n=0}^{\infty} C_n \cosh(k_n b) \cos(k_n x)$$

We must, therefore, have

$$C_n = \frac{1}{a \cosh(k_n b)} \int_{-a}^{a} (x^2 - a^2) \cos(k_n x)\, dx$$

$$= -\frac{4}{a k_n^3} \frac{\sin(k_n a)}{\cosh(k_n b)}$$

$$\therefore \quad \psi(x, y) = a^2 + \frac{1}{2}(y^2 - x^2)$$

$$-\frac{32a^2}{\pi^3} \sum_{n=0}^{\infty} \frac{(-1)^n}{(2n+1)^3} \cdot \frac{\cosh(k_n y)}{\cosh(k_n b)} \cdot \cos(k_n x)$$

$$\frac{\partial \psi}{\partial x} = -x \times \frac{32a^2}{\pi^3} \sum_{n=0}^{\infty} \frac{(-1)^n}{(2n+1)^3} \cdot \frac{\cosh(k_n y)}{\cosh(k_n b)} \sin(k_n x)\, k_n$$

$$= -x + \frac{6a}{\pi^2} \sum_{n=0}^{\infty} \frac{(-1)^n}{(2n+1)^2} \frac{\cosh(k_n y)}{\cosh(k_n b)} \cdot \sin(k_n x)$$

$$\frac{\partial\psi}{\partial y} = y - \frac{16a}{\pi^2} \sum_{n=0}^{\infty} \frac{(-1)^n}{(2n+1)^2} \frac{\sinh(k_n y)}{\cosh(k_n b)} \cdot \cos(k_n x)$$

$$\therefore \quad \tau_{zx} = -\frac{16aG\theta}{\pi^2} \sum_{n=0}^{\infty} \frac{(-1)^n}{(2n+1)^2} \frac{\cosh(k_n y)}{\cosh(k_n b)} \cdot \sin(k_n x) \qquad \text{...(5.53)}$$

$$\tau_{yz} = G\theta \left[2x - \frac{16a}{\pi^2} \sum_{n=0}^{\infty} \frac{(-1)^n}{(2n+1)^2} \frac{\cosh(k_n y)}{\cosh(k_n b)} \cdot \sin(k_n x) \right] \qquad \text{...(5.54)}$$

$$\tau_{yz}\bigg|_{\substack{x=a\\y=0}} = 2G\theta a \left[1 - \frac{8}{\pi^2} \sum_{n=0}^{\infty} \frac{1}{(2n+1)^2} \operatorname{sech}(k_n \cdot b) \right] \qquad \text{...(5.55)}$$

For a square beam $a = b$, and

$$\tau_{max} = 2G\theta\, a \left[1 - \frac{8}{\pi^2} \left\{ \operatorname{sech} \frac{\pi}{2} + \sum_{n=1}^{\infty} \frac{1}{(2n+1)^2} \operatorname{sech}(2n+1)\frac{\pi}{2} \right\} \right]$$

$$\simeq 2G\theta\, a \left(1 - \frac{8}{\pi^2} \operatorname{sech} \frac{\pi b}{2a} \right) \qquad \text{...(5.56)}$$

The torque,

$$T = \int_{-b}^{b} \int_{-a}^{a} (x\, \tau_{yz} - y\, \tau_{zx})\, dx\, dy$$

$$= \frac{8G\theta\, ba^3}{3} + \frac{256G\theta\, a^3 b}{\pi^4} \sum_{n=0}^{\infty} \frac{1}{(2n+1)^4} - \frac{1024}{\pi^5} G\theta\, a^4 \sum_{n=0}^{\infty} \frac{\tanh(k_n b)}{(2n+1)^5}$$

Now $\sum_{n=0}^{\infty} \frac{1}{(2n+1)^4} = \frac{\pi^4}{96}$

$$\therefore \quad T = \frac{16}{3} G\theta\, a^3 b - \frac{1024}{\pi^5} G\theta\, a^4 \sum_{n=0}^{\infty} \frac{\tanh(k_n b)}{(2n+1)^5}$$

$$T \simeq \frac{16G\theta\, a^3 b}{3} - \frac{1024\, G\theta\, a^4}{\pi^5} \tanh \frac{\pi b}{2a} \qquad ...(5.57)$$

and $$\varphi(x, y) = xy - \frac{32a^2}{\pi^3} \sum_{n=0}^{\infty} \frac{(-1)^n}{(2n+1)^3} \frac{\sinh (k_n y)}{\cosh (k_n b)} \sin k_n x$$

$$\therefore \quad u_z = \theta\varphi(x, y)$$

The contour lines of the surface $\varphi(x, y)$ = constant for the case $a = b$ are shown in Fig. 5.15.

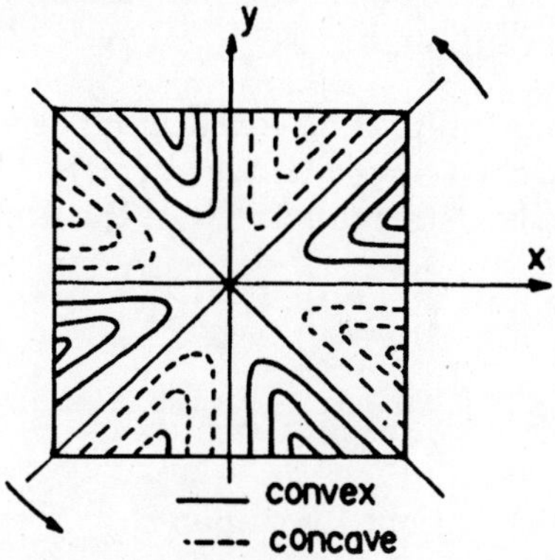

Fig. 5.15

5.6 Torsion of Hollow Bars

Let it be assumed that a bar has several longitudinal hollows so that the boundary of the cross-section of the bar is made up of several simple closed contours. Denote the exterior contour by C_0 and let $C_1, C_2, \ldots, C_n$ be the simple closed contours lying entirely within the contour C_0 as shown in Fig. 5.16.

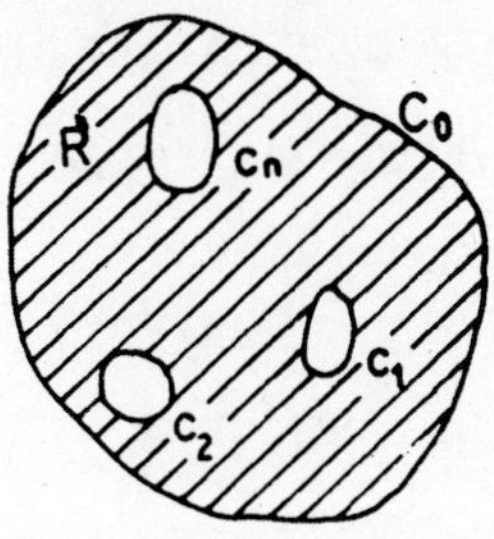

Fig. 5.16

The contours $C_1, C_2, \ldots, C_n$ correspond to the hollows of the bar. By reasoning similar to that employed in the derivation of Eq. (4) of Art. 5.5, we can show that the torsion function φ must satisfy

$\nabla^2\varphi = 0$ in R and that on each of the contours $C_0, C_1, C_2, \ldots\ldots, C_n$

$$\frac{d\psi}{dn} = y\cos\alpha - x\sin\alpha \qquad \ldots(1)$$

The boundary conditions can be expressed in terms of the conjugate function ψ as :

$$\frac{\partial\psi}{ds} = \frac{d}{ds}\left(\frac{x^2+y^2}{2}\right) \text{ on } C_i,\ (i = 0, 1, 2, \ldots, n)$$

and the integration along each contour yields,

$$\psi = \frac{1}{2}(x^2+y^2) + k_i \text{ on } C_i \qquad \ldots(2)$$

where the k_i are the integration constants. The value of one of these constants, say k_0, can be specified arbitrarily, but the remaining n constants k_i must be determined so that the function

$$\varphi(x, y) = \int\left(\frac{\partial\varphi}{\partial x}dx + \frac{\partial\varphi}{\partial y}dy\right)$$

$$= \int\left(\frac{\partial\psi}{\partial y}dx - \frac{\partial\psi}{\partial x}dy\right) \qquad \ldots(3)$$

is single-valued throughout the region R.

Since ψ is a harmonic function, $\varphi(x, y)$ will be single-valued in R, if and only if,

$$\int_{C_i}\left(\frac{\partial\psi}{\partial y}dx - \frac{\partial\psi}{dx}dy\right) = 0,\ (i = 1, 2, \ldots, n) \qquad \ldots(4)$$

Thus the constants k_i $(i = 1, 2, \ldots., n)$ in Eq. (2) must be chosen so that the solution of the Dirichlet problem

$$\left.\begin{aligned}\nabla^2\psi &= 0 \text{ in } R\\ \psi &= \tfrac{1}{2}(x^2+y^2) + k_i \text{ on } C_i,\ (i = 0, 1, 2, \ldots, n)\end{aligned}\right\} \qquad \ldots(5)$$

satisfies the set of n conditions (4).

In terms of Prandtl stress function

$$\Psi = \psi(x, y) - \tfrac{1}{2}(x^2+y^2)$$

the system (5) leads to the new system

$$\left.\begin{aligned}\nabla^2\Psi &= -2 \text{ in } R\\ \Psi &= k_i \text{ on } C_i,\ (i = 0, \ldots, 2, \ldots., n)\end{aligned}\right\} \qquad \ldots(6)$$

and $$\frac{\partial\psi}{\partial x} = \frac{\partial\Psi}{\partial x} + x,\ \frac{\partial\psi}{\partial y} = \frac{\partial\Psi}{\partial y} + y$$

Accordingly, the set of conditions (4) becomes

$$\int_{C_i} \left(\frac{\partial \Psi}{\partial y} dx - \frac{\partial \Psi}{\partial x} dy \right) + \int_{C_i} (ydx - xdy) = 0$$

Now $\int_{C_i} (ydx - xdy) = 2A_i$

where A_i is the area enclosed by C_i and

$$\int_{C_i} \left(\frac{\partial \Psi}{\partial y} dx - \frac{\partial \Psi}{\partial x} dy \right) = \int_{C_i} \left(\frac{\partial \Psi}{\partial y} \cdot \frac{dx}{ds} - \frac{\partial \Psi}{\partial x} \cdot \frac{dy}{ds} \right) ds$$

$$= \int_{C_i} \frac{d\Psi}{dn} ds$$

$$\therefore \quad \oint_{C_i} \frac{d\Psi}{dn} ds = -2A_i, \ (i = 1, 2, ..., n) \qquad ...(5.58)$$

Torsional rigidity,

$$D = G \iint_R (x\tau_{yz} - y\tau_{zx})\, dxdy$$

$$= G \iint_R \left(x^2 + y^2 + x \frac{\partial \varphi}{\partial y} - y \frac{\partial \varphi}{\partial x} \right) dxdy$$

$$= G \iint_R - \left(x \frac{\partial \Psi}{\partial x} + y \frac{\partial \Psi}{\partial y} \right) dxdy$$

where $\qquad \dfrac{\partial \varphi}{\partial x} = y + \dfrac{\partial \Psi}{\partial y}$

and $\qquad \dfrac{\partial \varphi}{\partial y} = -\left(x + \dfrac{\partial \Psi}{\partial x} \right)$

This equation can be written as :

$$D = G \iint_R \left[2\Psi - \frac{\partial}{\partial x}(x\Psi) - \frac{\partial}{\partial y}(y\Psi) \right] dxdy$$

$$= 2G \iint_R \Psi\, dxdy + G \int_C \Psi(ydx - xdy)$$

If we assume $k_0 = 0$, then, we get

$$D = 2G \iint_R \Psi\, dxdy + G \sum_{i=1}^{n} k_i \int_{C_i} (ydx - xdy)$$

But $\quad \oint (ydx - xdy) = 2 \iint_{A_i} dxdy = 2A_i$

$$\therefore \qquad D = 2G \iint_R \Psi \, dxdy + \sum_{i=1}^{n} 2G\, k_i A_i \qquad ...(5.59)$$

and

$$T = 2G\theta \left[\iint_R \Psi \, dxdy + \sum_{i=1}^{n} k_i A_i \right] \qquad ...(5.60)$$

5.7 Solution of the Torsion Problem By Conformal Mapping

Let $\quad F[z] = \varphi(x, y) + i\psi(x, y)$

where $z = x + iy$.

The function $F(z)$ is analytic throughout the region R representing the cross-section of the bar. If the region R is simply connected, we can map it conformally on a unit circle in the ζ-plane. Let

$$z = \omega(\zeta) \qquad ...(1)$$

be the function that maps the region R on the unit circle $|\zeta| \le 1$. The function $F(z)$ can be expressed in terms of the variable ζ, so that

$$\varphi + i\psi = F[\omega(\zeta)]$$
$$= f(\zeta) \qquad ...(2)$$

where $f(\zeta)$ is analytic in the interior of the circle $|\zeta| = 1$.

The function ψ satisfies on the boundary C of the region R the condition

$$\psi = \frac{1}{2}(x^2 + y^2) + \text{const.}$$
$$= \frac{1}{2} z\bar{z} + \text{const.}$$

Hence the imaginary part of the function $f(\zeta)$ defined by Eq. (2) must satisfy the condition

$$\psi = \frac{1}{2}\omega(\zeta)\,\bar{\omega}(\bar{\zeta}) \text{ on } \gamma \qquad ...(3)$$

where γ is the boundary of the unit circle $|\zeta| = 1$.

Thus the torsion problem will be solved if we succeed in determining the real part ψ of the analytic function

$$\frac{1}{i} f(\zeta) = \psi - i\varphi \qquad ...(4)$$

which on the boundary γ of the unit circle $|\zeta| = 1$ assumes the values

$$\psi = \frac{1}{2}\omega(\zeta)\bar{\omega}(\bar{\zeta})$$

Now for a function to be holomorphic inside γ, when the boundary value of its real part is known is given by

$$\frac{1}{i} f(\zeta) = \frac{1}{\pi i} \int_\gamma \frac{1}{2} \frac{\omega(\sigma)\, \overline{\omega}(\overline{\sigma})}{\sigma - \zeta}\, d\sigma + \text{const.}$$

or

$$f(\zeta) = \frac{1}{2\pi} \int_\gamma \frac{\omega(\sigma)\, \overline{\omega}(\overline{\sigma})}{\sigma - \zeta}\, d\sigma + \text{const.} \qquad \text{...(5)}$$

Noting that, on the boundary γ the unit circle

$$|\zeta| = 1, \sigma = e^{i\theta} \text{ and } \overline{\sigma} = e^{-i\theta} = \frac{1}{\sigma},$$

hence Eq. (5) reduces to

$$f(\zeta) = \varphi + i\psi = \frac{1}{2\pi} \int_\gamma \frac{\omega(\sigma)\, \overline{\omega}\left(\frac{1}{\sigma}\right)}{\sigma - \zeta}\, d\sigma + \text{constant}$$

...(6)

If $\omega(\zeta)$ is a rational function, then $\omega(\sigma)$, $\omega(\sigma)$ will likewise be a rational function of σ. The integral on the right hand side of Eq. (6) can be evaluated by use of the residual theorem and this will lead to a rational function in ζ, so that the solution will be expressed in terms of elementary functions. The torsional rigidity D becomes

$$D = G \iint_R (x^2 + y^2)\, dxdy + G \iint_R \left(x \frac{\partial \varphi}{\partial y} - y \frac{\partial \varphi}{\partial x} \right) dxdy$$

$$= GI_0 + GD_0 \qquad \text{...(7)}$$

where I_0 is the polar moment of inertia of R about O and

$$D_0 = \iint_R \left(x \frac{\partial \varphi}{\partial y} - y \frac{\partial \varphi}{\partial x} \right) dxdy$$

$$= \iint_R \left[\frac{\partial}{\partial y} (x\varphi) - \frac{\partial}{\partial x} (y\varphi) \right] dxdy$$

Applying Green's formula, we get

$$D_0 = -\int_C \varphi(xdx + yd\varphi)$$

$$= -\int_C \varphi d(\tfrac{1}{2} r^2) \qquad \text{...(8)}$$

where $r^2 = x^2 + y^2$,

and C is the boundary of the region.

Noting that on the boundary

$$r^2 = z\bar{z} = \omega(\sigma)\,\bar{\omega}(\bar{\sigma})$$

and $$\varphi = \frac{1}{2}[f(\sigma) + \bar{f}(\bar{\sigma})],$$

one may rewrite Eq. (8) as

$$D_0 = -\frac{1}{4}\int_\gamma [f(\sigma) + \bar{f}(\bar{\sigma})]\, d\,[\omega(\sigma)\,\bar{\omega}(\bar{\sigma})] \qquad \text{...(9)}$$

The integral on the right hand side can be evaluated by using the theorem of residues.

Also $$I_0 = \iint_R (x^2 + y^2)\, dxdy$$

$$= \iint_R \left[\frac{\partial}{\partial y}(x^2y) + \frac{\partial}{\partial x}(xy^2)\right] dxdy$$

$$= -\int_C xy\,(xdx - ydy)$$

But $$x = \frac{z + \bar{z}}{2}$$

$$y = \frac{z - \bar{z}}{2i}$$

$\therefore$ $$I_0 = -\frac{1}{8i}\int_C (z^2 - \bar{z}^2)\,(zdz + \bar{z}\,d\bar{z})$$

But $\int_C z^2\, dz = 0, \int_C \bar{z}^2\, d\bar{z} = 0$

and $$\int_C z^2\,\bar{z}\, d\bar{z} = \int_C z^2 d\left(\frac{1}{2}\bar{z}^2\right)$$

$$= -\int_C \bar{z}^2\, z\, dz$$

$\therefore$ $$I_0 = -\frac{i}{4}\int_C \bar{z}^2\, zdz$$

$$= -\frac{i}{4}\int_\gamma [\bar{\omega}(\bar{\sigma})]^2\;\omega(\sigma)\, d\omega(\sigma) \qquad \text{...(10)}$$

If $\omega(\zeta)$ is a rational function, formula (10) permits calculation of I_0 in closed form by elementary means.

In the case of doubly connected regions, the torsion problem is likewise easily solved, provided the function $\omega(\zeta)$ is known which maps this region on to the circular ring. In fact, in this case the problem is reduced to the determination of a function $f(\zeta)$, holomorphic inside the ring and satisfying the following boundary conditions :

$$\left.\begin{aligned} R_e\left[\frac{1}{i} f(\zeta)\right] &= \frac{1}{2}\,\omega(\zeta)\,\overline{\omega}\,(\overline{\zeta}) + k_1 \text{ on } \gamma_1 \\ R_e\left[\frac{1}{i} f(\zeta)\right] &= \frac{1}{2}\,\omega(\zeta)\,\overline{\omega}\,(\overline{\zeta}) + k_2 \text{ on } \gamma_2 \end{aligned}\right\} \qquad ...(11)$$

where γ_1 and γ_2 are the circles, bounding the ring, and k_1, k_2 are two real constants, one of which may be fixed arbitrarily.

$$\text{Now } \tau_{zx} - i\tau_{yz} = G\theta\left(\frac{\partial \varphi}{\partial x} - i\frac{\partial \varphi}{\partial y} - y - ix\right)$$

$$= G\theta\left[\frac{\partial \varphi}{\partial x} + i\frac{\partial \psi}{\partial y} - i(x - iy)\right]$$

$$\text{But } \frac{\partial \varphi}{\partial x} + i\frac{\partial \psi}{\partial x} = \frac{dF}{dz}$$

$$\therefore \quad \tau_{zx} - i\tau_{yz} = G\theta\,[F'(z) - i\overline{z}] \qquad ...(12)$$

Since $F(z) = F[\omega(\zeta)] = f(\zeta)$, we have

$$F'(z) = f'(\zeta)\frac{d\zeta}{dz} = f'(\zeta)\cdot\frac{1}{\omega'(\zeta)}$$

Hence Eq. (12) becomes

$$\tau_{zx} - i\,\tau_{yz} = G\theta\left[\frac{f'(\zeta)}{\omega'(\zeta)} - i\,\overline{\omega}\,(\overline{\zeta})\right] \qquad ...(13)$$

If the mapping function is written in the form :

$$z = \omega(\zeta) = \sum_{n=0}^{\infty} a_n \zeta^n$$

then, we have

$$\omega(\sigma)\overline{\omega}\left(\frac{1}{\sigma}\right) = \sum_{m=0}^{\infty} a_m\,\sigma^m \sum_{n=0}^{\infty} \overline{a}_n\,\sigma^{-n} \qquad ...(14)$$

$$= \sum_{n=0}^{\infty} b_n\,\sigma^n + \sum_{n=1}^{\infty} \overline{b}_n\,\sigma^{-n}$$

$$\text{where } b_n = \sum_{r=0}^{\infty} a_{n+r}\,\overline{a}_n \qquad ...(15)$$

Substituting Eqs. (14) in (16), we find

$$f(\zeta) = \frac{1}{2\pi}\int_{\gamma}\left[\sum_{n=0}^{\infty} b_n\,\sigma^n + \sum_{n=1}^{\infty} \bar{b}_n\,\sigma^{-n}\right]\frac{d\sigma}{\sigma-\zeta}$$

or

$$f(\zeta) = \varphi + i\psi = i\sum_{n=0}^{\infty} b_n\,\zeta^n \qquad \text{...(16)}$$

Because $\dfrac{1}{2\pi i}\displaystyle\int \frac{F(\theta)\,d\sigma}{\sigma-\zeta} = \frac{1}{2\pi i}\int \sum_{n=1}^{\infty} \zeta^n\,\frac{F(\theta)\,d\sigma}{\sigma^{n+1}}$

$$= \sum_{n=0}^{\infty}(a_n - ib_n)\,\zeta^n$$

Now Eq. (10) can be written as

$$I_0 = -\frac{i}{4}\int_{\gamma}\left[\omega(\sigma)\bar{\omega}\left(\frac{1}{\sigma}\right)\right]\bar{\omega}\left(\frac{1}{\sigma}\right)d\omega(\sigma)$$

From $\omega(\sigma) = \displaystyle\sum_{n=0}^{\infty} a_n\,\sigma^n$, it follows that

$$\bar{\omega}\left(\frac{1}{\sigma}\right)d\omega(\sigma) = i\sum_{n=-\infty}^{\infty} C_n\,\sigma^n\,d\theta$$

where

$$\left.\begin{aligned} C_n &= \sum_{r=0}^{\infty}(n+r)\,a_{n+r}\,\bar{a}_r \\ C_{-n} &= \sum_{r=0}^{\infty} r\bar{a}_{n+r}\,a_r,\ n = 0, 1, 2, \ldots \end{aligned}\right\} \qquad \text{...(17)}$$

Using Eqs. (17) and (14), we get

$$I_\theta = \frac{1}{4}\int_0^{2\pi}\left(\sum_{m=0}^{\infty} b_m\,\sigma^n + \sum_{m=1}^{\infty}\bar{b}_m\,\sigma^{-m}\right)$$

$$\times\left(\sum_{n=0}^{\infty} C_n\,\sigma^n + \sum_{n=1}^{\infty} C_{-n}\,\sigma^{-n}\right)d\theta$$

Since $\sigma = e^{i\theta}$, therefore, the integral of every term involving $\sigma^n (n \neq 0)$ vanishes and we are left with

$$I_\theta = \frac{\pi}{2}\left[b_0\, C_0 + \sum_{n=1}^{\infty} (b_n\, C_{-n} + \bar{b}_n\, C_n) \right]$$

Similarly, we can get

$$D_0 = \frac{1}{4}\int_0^{2\pi} \left(\sum_{m=0}^{\infty} b_m\, \sigma^m - \sum_{m=0}^{\infty} b_m\, \sigma^{-m} \right)$$

$$\times \left(\sum_{n=1}^{\infty} n b_n\, \sigma^n - \sum_{n=1}^{\infty} n \bar{b}_n\, \sigma^{-n} \right) d\theta$$

$$= -\pi \sum_{n=1}^{\infty} n\, b_n\, \bar{b}_n$$

Combining the expressions, we get finally

$$D = \frac{\pi G}{2} [b_0 C_0 + \sum_{n=1}^{\infty} (C_n \bar{b}_n + C_{-n}\, b_n - 2n b_n\, \bar{b}_n)] \qquad \text{...(5.61)}$$

5.7.1 Applications of Conformal Mapping

(a) Epitrochoidal section

Let the section R be bounded by the epitrochoids as shown in Fig. 5.17. Then

$$z = \omega(\zeta) = b(\zeta^n) + a(\zeta) \qquad \text{...(1)}$$

where n is an integer > 1, $b > 0$, $a \geq n$.

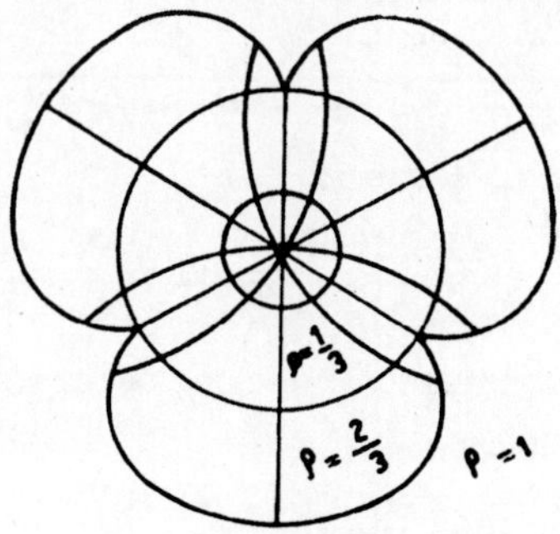

Fig. 5.17

Putting $\quad z = x + iy,\ \zeta = \rho\, e^{i\theta_1}$, we find

$$x + iy = b[\rho^n\, e^{in\theta_1} + a\, \rho\, e^{in\theta_1}]$$

$$= b\,\rho^n\,(\cos n\theta_1 + i \sin n\,\theta_1) + ba\,\rho(\cos\theta_1 + i\sin\theta_1)$$

$$\therefore \quad x = b\,\rho^n \cos n\theta_1 + ab\,\rho\cos\theta_1$$

$$y = b\,\rho^n \sin n\theta_1 + ab\,\rho\sin\theta_1$$

The circle $|\zeta| = \rho = 1$ corresponds in the z-plane to the curve C with the parametric representation

$$x = b(\cos n\theta_1 + a\cos\theta_1)$$

$$y = b(\sin n\theta_1 + a\sin\theta_1).$$

These curves are epitrochoids.

$$\text{Now } f(\zeta) = \frac{1}{2\pi}\int_\gamma b^2(\sigma^n + a\sigma)\left(\frac{1}{\sigma^n} + \frac{a}{\sigma}\right)\frac{d\sigma}{\sigma - \zeta} + \text{constant}$$

$$= \frac{ib^2}{2\pi i}\int_\gamma \left(1 + a^2 + a\,\sigma^{n-1} + \frac{a}{\sigma^{n-1}}\right)\frac{d\sigma}{\sigma - \zeta} + \text{constant}$$

$$= ib^2\,a\zeta^{n-1} \qquad \text{...(2)}$$

where the arbitrary constant has been omitted.

The stress components become

$$\tau_{zx} - i\,\tau_{yz} = G\,\theta\zeta\,\frac{iab^2\,(n-1)\,\zeta^{n-2} - ib^2\,(\zeta^{-n} + a\zeta)(n\zeta^{n-1} - a)}{\rho\,|\,\omega'(\zeta)\,|}$$

or putting $\zeta = \rho\,e^{i\theta_1}$ and separating real and imaginary parts,

$$\tau_{zx} = -\frac{G\theta\,ab^2\,(n-1)\,\rho^{n-2}\,(1-\rho^2)\sin(n-1)\,\theta_1}{|\,\omega'(\zeta)\,|} \qquad \text{...(3)}$$

$$\tau_{yz} = -\,G\theta\,b^2\,\frac{a\rho^{n-2}[n-1-(n+1)\rho^2]\cos(n-1)\theta_1 - n\rho^{2n-1} - a^2\rho}{|\,\omega'(\zeta)\,|} \qquad \text{...(4)}$$

where $|\,\omega'(\zeta)\,| = \sqrt{\omega'(\zeta)\,\overline{\omega}'(\zeta)}$

$$= b\sqrt{n^2\,\rho^{2n-2} + 2an\,\rho^{n-1}\cos(n-1)\theta + a^2}$$

on the boundary $\rho = 1$, $\tau_{zx} = 0$, and

$$\tau_{yz} = G\theta b\,\frac{n + 2a\cos(n-1)\,\theta + a^2}{\sqrt{n^2 + 2an\cos(n-1)\,\theta + a^2}} \qquad \text{...(5.62)}$$

If $n < a$ then maximum value of τ_{yz} occurs where $\cos(n-1)\theta = -1$. Hence in that case

$$(\tau_{yz})_{max} = G\theta b\left(\frac{a^2 - 2a + n}{a - n}\right) \qquad \text{...(5.63)}$$

If $a \to n$ then $\tau_{yz} \to \infty$.

Torsional rigidity D can be obtained as :

$$D = \frac{\pi G b^4}{2}(a^4 + 4a^2 + n) \qquad ...(5.64)$$

(b) Booth's Lemniscate

For this curve

$$z = w(\zeta) = \frac{k\zeta}{\zeta^2 + a} \; (a > 1, k > 0) \qquad ...(1)$$

as shown in Fig. 5.18.

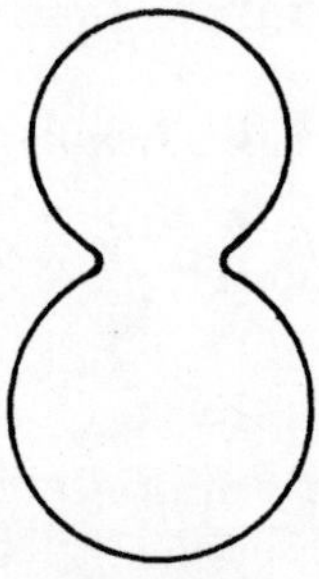

Fig. 5.18

$$f(\zeta) = \frac{1}{2\pi}\int \frac{k^2 \sigma^2 \, d\sigma}{(\sigma^2 + a)(1 + a\sigma^2)(\sigma - \zeta)} \qquad ...(2)$$

The integrand has outside γ two simple poles :

$$\sigma_1 = i\sqrt{a} \text{ and } \sigma_2 = -i\sqrt{a} \;;$$

for large $|\sigma|$, it is of order $\frac{1}{\sigma^2}$. Therefore, by the residue theorem

$$f(\zeta) = -i(R_1 + R_2)$$

where R_1, R_2 are the residues, corresponding to the point σ_1, σ_2 respectively.

$$R_1 = \left[(\sigma - i\sqrt{a})\frac{k^2\sigma^2}{(\sigma^2 + a)(1 + a\sigma^2)(\sigma - \zeta)}\right]_{\sigma = i\sqrt{a}}$$

$$= \frac{-k^2\sqrt{a}}{2i(1 - a^2)(i\sqrt{a} - \zeta)}$$

$$R_2 = \frac{-k^2\sqrt{a}}{2i(1 - a^2)(i\sqrt{a} + \zeta)}$$

$$\therefore \qquad f(\zeta) = \frac{iak^2}{(a^2 - 1)(\zeta^2 + a)} \qquad ...(3)$$

The stress components may be calculated as before.

On the boundary

$$\tau_{yz} = \frac{G\theta k\,(1 + a^2)}{(a^2 - 1)\sqrt{1 - 2a\cos 2\theta_1 + a^2}} \qquad ...(5.65)$$

The maximum value of τ_{yz} occurs for $\cos 2\theta_1 = 1$.

$$\therefore \qquad (\tau_{yz})_{max} = \frac{G\theta k\,(a^2 + 1)}{(a + 1)\,(a - 1)^2} \qquad ...(5.66)$$

and torsional rigidity

$$D = \frac{\pi\, Gk^4\,(a^4 + 1)}{2\,(a^2 - 1)^4} \qquad ...(5.67)$$

(c) The loop of Bernoulli's Lemniscate

This is an example of the type when the mapping function is not rational. Consider a map of the unit circle obtained with the aid of

$$z = \omega(\zeta) = a\sqrt{1 + \zeta} \quad (a > 0) \qquad ...(1)$$

and select that branch of the multi-valued function $\sqrt{1 + \zeta}$ which is equal to unity for $\zeta = 0$.

In other words

$$z = \omega(\zeta) = a\sqrt{r}\, e^{i\varphi/2} \left(-\frac{\pi}{2} \le \varphi \le \frac{\pi}{2} \right) \qquad ...(2)$$

as shown in Fig. 5.19 (*a*).

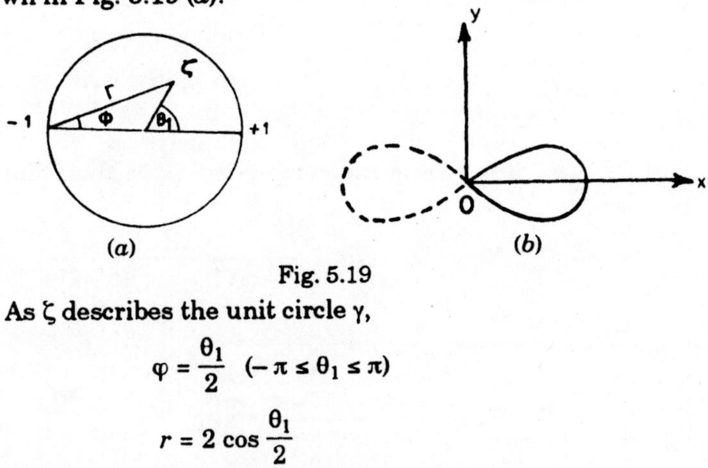

Fig. 5.19

As ζ describes the unit circle γ,

$$\varphi = \frac{\theta_1}{2} \quad (-\pi \le \theta_1 \le \pi)$$

and

$$r = 2\cos\frac{\theta_1}{2}$$

Hence

$$z = a\sqrt{2\cos\frac{\theta_1}{2}}\; e^{i\theta_1/4}$$

If R and ψ denote the modulus and argument of z, then

$$R\,e^{i\psi} = a\sqrt{2\cos\frac{\theta_1}{2}}\;e^{\frac{i\theta_1}{4}}$$

Hence $$R = a\sqrt{2\cos\frac{\theta_1}{2}}$$

$$\psi = \frac{\theta_1}{4}$$

whence $R = a\sqrt{2\cos 2\psi}$

Thus z describes one loop of Bernoulli's lemniscate as shown in Fig. 5.19 (*b*) and Eq. (1) maps the region inside this loop on to the circle $|\zeta| < 1$.

One finds for $f(\zeta)$,

$$f(\zeta) = \frac{a^2}{2\pi}\int_\gamma \sqrt{1+\sigma}\sqrt{1+\frac{1}{\sigma}}\cdot\frac{d\sigma}{\sigma-\zeta}$$

$$= \frac{a^2}{2\pi}\int_\gamma \frac{1+\sigma}{\sqrt{\sigma}}\,\frac{d\sigma}{\sigma-\zeta} \qquad \text{...(3)}$$

Since the sign of the square root must be chosen positive, we can write

$$\sqrt{\sigma} = e^{\frac{i\theta_1}{2}}$$

If we cut the negative axis as shown in Fig. 5.20, then the integrand of Eq. (3) will be single-valued in the region bounded by γ and the only singularity of the integrand is the pole at $\sigma = \zeta$. Hence

$$\frac{1}{2\pi i}\left[\int_\gamma + \int_\alpha + \int_{\gamma_1} + \int_\beta\right] = R$$

where γ_1 denotes an infinitely small circle and R is the residue at $\sigma = \zeta$

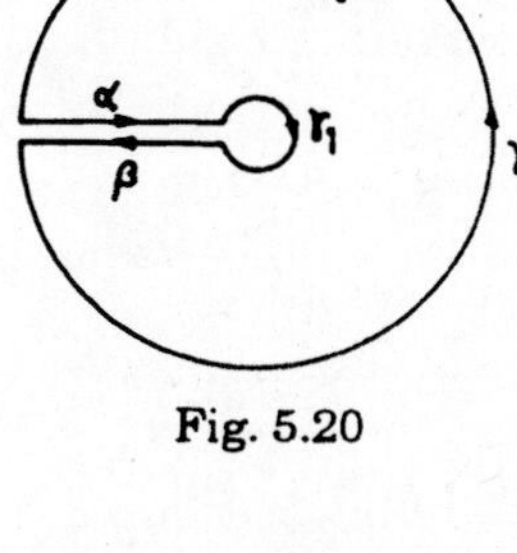

Fig. 5.20

or
$$\frac{1}{2\pi i}\left[\int_{\gamma} g(\sigma,\zeta)\,d\sigma+\int_{-1}^{0} g(\sigma,\zeta)\,d\sigma+\int_{\gamma_1} g(\sigma,\zeta)\,d\sigma + \int_{0}^{-1} g(\sigma,\zeta)\,d\sigma\right]=R$$

where $g(\sigma,\zeta)=\dfrac{1+\sigma}{\sqrt{\sigma}\,(\sigma-\zeta)}$

Now R at $\sigma=\zeta$ is obviously
$$R=\frac{1+\zeta}{\sqrt{\zeta}}$$

Thus
$$\frac{1}{2\pi i}\int_{\gamma} g(\sigma,\zeta)\,d\sigma=-\frac{1}{2\pi i}\left[\int_{-1}^{0} g(\sigma,\zeta)\,d\sigma+\int_{0}^{-1} g(\sigma,\zeta)\,d\sigma + \int_{\gamma_1} g(\sigma,\zeta)\,d\sigma\right]+R$$
$$=-\frac{1}{\pi}\int_{0}^{1}\frac{1-t}{\sqrt{t}}\cdot\frac{dt}{t+\zeta}+\frac{1+\zeta}{\sqrt{\zeta}}$$

where we have omitted the integral over the small circle γ_1, since it vanishes when the radius of γ_1 tends to zero. Omitting an arbitrary constant, we find
$$f(\zeta)=\frac{2i\,a^2}{\pi}\cdot\frac{1+\zeta}{\sqrt{\zeta}}\tan^{-1}\sqrt{\zeta}$$
$$=\frac{a^2}{\pi}\cdot\frac{1+\zeta}{\sqrt{\zeta}}\cdot\log\left[\frac{1+i\sqrt{\zeta}}{1-i\sqrt{\zeta}}\right] \qquad \text{...(4)}$$

Now $\log\left[\dfrac{1+i\sqrt{\zeta}}{1-i\sqrt{\zeta}}\right]=2i\sqrt{\zeta}\left(1-\dfrac{\zeta}{3}+\dfrac{\zeta^2}{5}-\ldots\ldots\right)$

$$\therefore \quad f(\zeta)=\frac{2i\,a^2}{\pi}(1+\zeta)\left(1-\frac{\zeta}{3}+\frac{\zeta^2}{5}-\ldots\ldots\right) \qquad \text{...(5)}$$

The stress and torsional rigidity can then be determined easily.

(*d*) Let $z=c\sec(\omega+ik),\ v=0$ ...(1)

With $z=x+iy$, $\omega=u+iv$. If we put $\zeta=e^{i\omega}$, then we get the resulting function
$$z=\omega(\zeta)=\frac{2c\,e^{k}\zeta}{\zeta+e^{2k}},\ c>0,\ k>0 \qquad \text{...(2)}$$

which maps the cross-section R of the cylinder upon the interior of the unit circle $|\zeta| \le 1$.

Let
$$z = \omega(\zeta) = \sum_{n=1}^{\infty} a_n \zeta^n = \sum_{n=0}^{\infty} a_n e^{in\omega}$$

Then
$$f(\zeta) = \varphi + i\psi$$
$$= \frac{2c^2}{\pi}\int_\gamma \frac{\sigma^2\, d\sigma}{(\sigma^2 + e^{2k})(\sigma^2 + e^{-2k})(\sigma - \zeta)} + \text{const.}$$
$$= -4c^2 i(R_1 + R_2) + \text{const.}$$
where R_1, R_2 are the residues of the integrand at $\sigma = ie^k$ and $\sigma = -ie^k$ respectively. We have
$$R_1 = \left[\frac{\sigma^2}{(\sigma + ie^k)(\sigma^2 + e^{-2k})(\sigma - \zeta)}\right]_{\sigma = ie^k}$$
$$= \frac{ie^k}{4(\zeta - ie^k)\sinh 2k}$$
$$R_2 = \frac{-ie^k}{4(\zeta + ie^k)\sinh 2k}.$$
$$f(\zeta) = c^2 \operatorname{cosech} . 2k \tan(\omega + ik) \qquad ...(3)$$
where the const. $= -ic^2 \operatorname{cosech} 2k$.

The shear stresses become
$$\tau_{zx} = -2G\theta\, c \sin u \left[\frac{\operatorname{cosech} 2k \cosh(v + k)}{\cos 2u - \cosh 2(v + k)} + \frac{\sinh(v + k)}{\cos 2u + \cosh 2(v + k)}\right] \qquad ...(5.68)$$
$$\tau_{yz} = -2G\theta\, c \cos u \left[\frac{\operatorname{cosech} 2k \sinh(v + k)}{\cos 2u - \cosh 2(v + k)} - \frac{\cosh(v + k)}{\cos 2u + \cosh 2(v + k)}\right] \qquad ...(5.69)$$
$$I_0 = 4c^2 i \int_\gamma \frac{\sigma^3(\sigma^2 - e^{2k})\, d\sigma}{(\sigma^2 + e^{2k})^3(\sigma^2 + e^{-2k})^2}$$
$$= -8\pi c^4 (R_3 + R_4)$$
where R_3, R_4 are the residues of the integrand at
$$\sigma = i e^{-k}$$
and
$$\sigma = -i e^{-k} \text{ respectively.}$$

$$R_3 = \frac{d}{d\sigma}\left[\frac{\sigma^3(\sigma^2 - e^{2k})}{(\sigma^2 + e^{2k})^3\,(\sigma + i\,e^{-k})^2}\right]_{\sigma = i\,e^{-k}}$$

$$= -\frac{\text{cosech}^4\,2k\,.\,(2 + \cosh 4k)}{16} = R_4$$

$\therefore \qquad I_o = \pi c^4(2 + \cosh 4k)\,\text{cosech}^2\,2k$

Similarly

$$D_o = -\,i\,4c^4\,\text{cosech}\,2k \int_\gamma \frac{\sigma(1 - \sigma^4)^2 d\sigma}{(\sigma^2 + e^{2k})^3\,(\sigma^2 + e^{-2k})^3}$$

$$= 8\pi c^4\,\text{cosech}\,2k(R_5 + R_6)$$

where R_5 and R_6 are the residues at $\sigma = i\,e^{-k}$ and $\sigma = -\,i\,e^{-k}$.

We find that

$$R_5 = \frac{1}{2}\frac{d^2}{d\sigma^2}\left[\frac{\sigma(1 - \sigma^4)^2}{(\sigma^2 + e^{2k})^3\,(\sigma + i\,e^{-k})^3}\right]_{\sigma = i\,e^{-k}}$$

$$= \frac{1}{8}\,\text{cosech}^3\,2k = R_6 \qquad \text{...(5.70)}$$

$$\therefore \qquad D_o = -\,2\pi c^4\,\text{cosech}^4\,2k \qquad \text{...(5.70)}$$

Hence

$$T = G\theta(I_o + D_o)$$

$$= G\theta\,\pi c^4\,(2\,\text{cosech}^2\,2k + \text{cosech}^4\,2k) \qquad \text{...(5.71)}$$

5.8 Analogous Methods

The following analogous methods have been found to be very useful in solving torsion problems.

(*i*) Membrane or soap film analogy
(*ii*) Conducting sheet analogy
(*iii*) Hydrodynamic analogy.

5.8.1 Membrane Analogy

5.8.1.1 Torsion of Solid Sections

This analogy was introduced by Prandtl in 1903 and further developments were carried out by Griffith and Taylor.

Imagine a homogeneous membrane (Fig. 5.21) supported at the edges, with the same outline as that of the cross-section of the twisted bar, subjected to uniform tension at the edges and a uniform lateral pressure. q is the pressure per unit area of the membrane and S is the uniform tension per unit length of the boundary. Considering the vertical equilibrium of the membrane and remembering that angle α is small, we get

$$q\,dx\,dy + \frac{\partial}{\partial x}\left(S\frac{\partial z}{\partial x}dy\right)dx + \frac{\partial}{\partial y}\left(S\frac{\partial z}{\partial y}dx\right)dy = 0$$

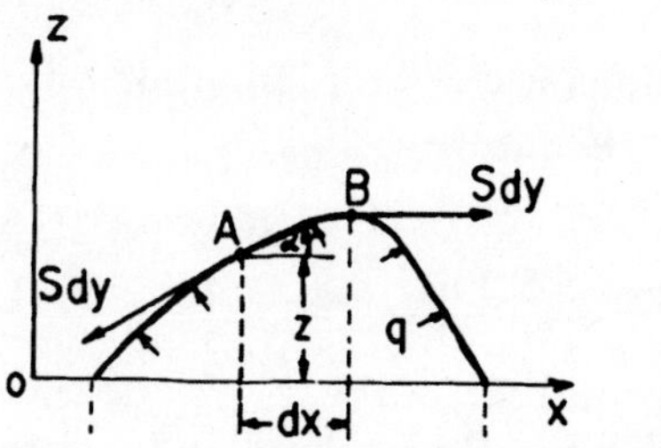

(a)

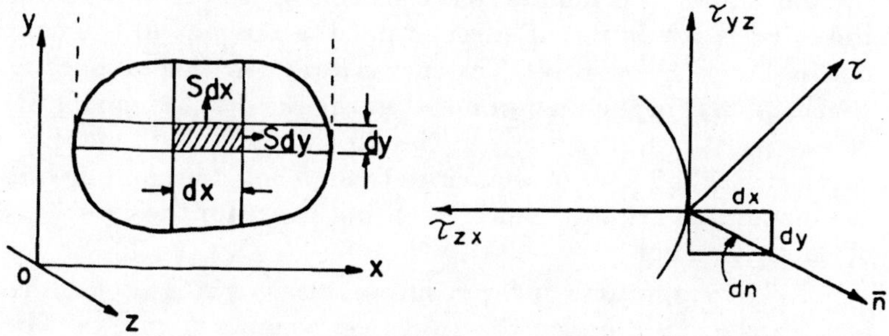

(b)

Fig. 5.21 Membrane analogy.

Simplifying, we get

$$q + S\left(\frac{\partial^2 z}{\partial x^2} + \frac{\partial^2 z}{\partial y^2}\right) = 0$$

or

$$\nabla^2 z = -\frac{q}{S} \qquad \text{...(5.72)}$$

This is Poisson's equation and is identical with the torsion equation

$$\nabla^2 \phi = -2G\theta \qquad \text{...(5.73)}$$

where ϕ is the stress function for torsion.

If $\frac{q}{S} = 2G\theta$, then the membrane equation is identical with the torsional stress function. If we arrange the membrane so that its height z remains zero at the boundary contour of the section, then the heights z of the membrane are numerically equal to the stress function ϕ.

Now consider any point B on the membrane. The deflection of the membrane along the contour line through this point is constant, and we have

$$\frac{\partial z}{\partial s} = 0$$

The corresponding equation for the stress function ϕ is

$$\frac{\partial \phi}{\partial s} = \left(\frac{\partial \phi}{\partial y} \cdot \frac{\partial y}{\partial s} + \frac{\partial \phi}{\partial x} \cdot \frac{\partial x}{\partial s} \right)$$

$$= \tau_{zx} \cdot \frac{dy}{ds} - \tau_{yz} \frac{dx}{ds} = 0$$

This expresses that the projection of the resultant shearing stress at a point B on the normal $\bar{n}$ to the contour line is zero and therefore we may conclude that the shearing stress at a point B in the twisted bar is in the direction of the tangent to the contour outline through this point. The curves drawn in the cross-section of a twisted bar, in such a manner that the resultant shearing stress at any point of the curve is in the direction of the tangent to the curve, are called lines of shearing stress. Thus, the contour lines of the membrane are the lines of shearing stress for the cross-section of the twisted bar.

The magnitude of the resultant stress τ at B is obtained by projecting on the tangent the stress components τ_{zx} and τ_{yz}. Then

$$\tau = \tau_{yz}\, a_{nx} - \tau_{zx}\, a_{ny}$$

$$= -\frac{\partial \phi}{\partial x} \cdot \frac{dx}{dn} - \frac{\partial \phi}{\partial y} \cdot \frac{dy}{dn}$$

$$= -\frac{d\phi}{dn} \qquad \text{...(5.74)}$$

Thus, the magnitude of the shearing stress at B is given by the maximum slope of the membrane at this point. It is only necessary in the expression for the slope to replace $\frac{q}{S}$ by $2G\theta$. From this it can be concluded that the maximum shear acts at the point where the contour lines are closest to each other.

Now torque

$$T = 2 \iint \phi \, dx \, dy \qquad \text{...(5.75)}$$

and the volume under the membrane

$$V = \iint z \, dx \, dy \qquad \text{...(5.76)}$$

Thus it can be concluded that double the volume bounded by the deflected membrane and the xy plane represents the torque, provided $\frac{q}{S}$ is replaced by $2G\theta$.

The apparatus developed by Griffith and Taylor (1917) is now widely used in experiments. In their method, two soap films or rubber sheets in the lid of an air-tight box are blown up side by side, one being of circular form, the other of the form to be investigated. $\frac{q}{S}$ is the same for both shapes and corresponds to the same value of $2G\theta$ for the two bars under twist ; the torques required to give these twists are in the ratio of the volumes between the membranes and the $z = 0$ plane. The volume enclosed by the circular membrane which is a paraboloidal of revolution is found, say V_1, and if the other membrane be flattened using a plate, the circular membrane will expand to enclose a new paraboloidal segment of volume V_2. Thus for the same value of $2G\theta$, the ratio of the torques in the bars is $V_1/(V_2 - V_1)$.

The inclination of the film surface to the $z = 0$ plane at any point may be determined by using an optical reflection method and is simply related to the total shear stress at the point.

The method may be examplified by blowing up a membrane on a circular cross-section of radius a and subjected to a transverse pressure q. Equating vertical forces, we find

$$\pi r^2 q = -S\frac{\partial z}{\partial r} . 2\pi r$$

or

$$\frac{\partial z}{\partial r} = -\frac{q}{2S}r$$

$$\therefore \quad z = -\frac{qr^2}{4S} + C$$

If $\quad z = 0$

when $\quad r = a$

then $\quad C = \frac{qa^2}{4S}$

$$\therefore \quad z = \frac{q}{4S}(a^2 - r^2)$$

$$\therefore \quad \phi = \frac{G\theta}{2}(a^2 - r^2)$$

The maximum shear stress at a point is given by :

$$\tau = -\frac{\partial \phi}{\partial r} = G\theta r \qquad ...(5.77)$$

$$\therefore \quad (\tau)_{r=a} = \tau_{max} = G\theta a \qquad ...(5.78)$$

The volume V between $z = 0$ and the membrane is given by

$$V = \int_A z\,dA = \int_0^a \frac{q}{4S}(a^2 - r^2).\, 2\pi r dr$$

$$= \frac{\pi a^4 q}{8S} \qquad \text{...(5.79)}$$

Hence $$T = 2V = \frac{2\pi a^4 q}{8S} = \frac{G\theta\pi a^4}{2}$$

$$= G\theta J \qquad \text{...(5.80)}$$

where $J = \dfrac{\pi a^4}{2}$ = polar moment of inertia.

$$\therefore \qquad \frac{T}{J} = \frac{\tau}{r} = G\theta$$

The value of $\dfrac{T}{\theta}$ for the circular membrane may be evaluated easily. Thus

$$\frac{\text{Volume of circular membrane}}{\text{Volume of other membrane}} = \frac{\frac{T}{\theta}\text{ of circle}}{\frac{T}{\theta}\text{ of other section}} \qquad \text{...(5.81)}$$

Hence $\dfrac{q}{S}$ need not be measured.

5.8.1.2 Torsion of a Thin Rectangular Section

Consider a thin rectangular section as shown in Fig. 5.22. If b is very much larger than t, then the bulges of the membrane across AA, BB or CC are all the same and that only near the ends DD the

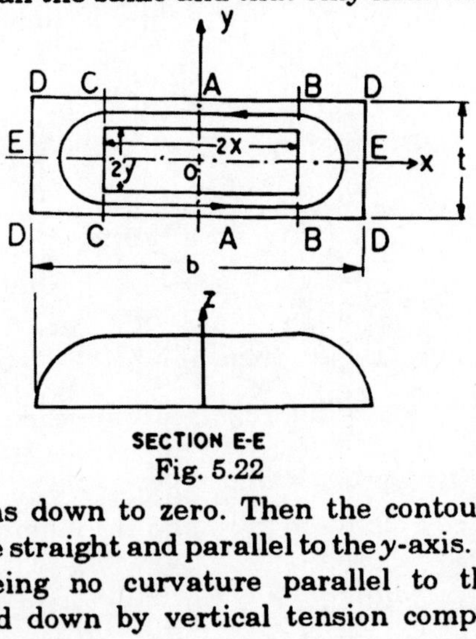

Fig. 5.22

membrane flattens down to zero. Then the contour lines in the central portion are straight and parallel to the y-axis. In this central portion, there being no curvature parallel to the y-axis the membrane is held down by vertical tension components in the

x-direction only. Cutting out a central piece of membrane of dimension $2x$ and $2y$, the equilibrium equation is :

$$4Sx\left(-\frac{\partial z}{\partial y}\right) = q \cdot 2x\, 2y$$

$$\therefore \qquad \frac{\partial z}{\partial y} = -\frac{q}{S}y$$

Thus $$z = -\frac{q}{S}\int y\,dy = -\frac{qy^2}{2S} + \text{const.}$$

Since $$z = 0 \text{ when } y = \frac{t}{2},$$

then $$z = \frac{q}{2S}\left(\frac{t^2}{4} - y^2\right)$$

and $$\phi = G\theta\left(\frac{t^2}{4} - y^2\right) \text{ which is a parabola.}$$

Now $$\left.\begin{aligned} \tau_{zx} &= \frac{\partial \phi}{\partial y} = -2\,G\theta\, y \\ \tau_{yz} &= -\frac{\partial \phi}{\partial x} = 0 \end{aligned}\right\} \qquad \text{...(5.82)}$$

$$\therefore \qquad |\,\tau\,| = |\,\sqrt{\tau_{zx}^2 + \tau_{zy}^2}\,|$$

$$= \frac{q}{S}y = 2\,G\theta\, y \qquad \text{...(5.83)}$$

$$\tau_{max} = \left(\frac{dz}{dx}\right)_{max}\Bigg|_{y = \pm\frac{t}{2}}$$

$$= \frac{q}{S}\cdot\frac{t}{2} = G\theta t \qquad \text{...(5.84)}$$

The volume under the membrane is :

$$V = \int_A z\,dA = \frac{qb}{S}\int_0^{t/2}\left(\frac{t^2}{4} - y^2\right)dy$$

$$= \frac{qb}{S}\cdot\frac{t^3}{12}$$

Torque, $$T = 2V = \frac{bt^3\,G\theta}{3} \qquad \text{...(5.85)}$$

$$\therefore \qquad \tau_{max} = \frac{3T}{bt^2} \qquad \text{...(5.86)}$$

$$\therefore \qquad \frac{\tau_{max}}{\left(\dfrac{T}{bt^2}\right)} = 3 \text{ as } \frac{b}{t} \to \infty \qquad \text{...(5.87)}$$

The values of $\tau_{max}/(T/bt^2)$ for various values of $\frac{b}{t}$ are as given below in Table 5.1.

Table 5.1 Values of $\tau_{max}/(T/bt^2)$

b/t	∞	10	5	3	2.5	2	1.5	1.0
$\frac{\tau_{max}}{\left(\frac{T}{bt^2}\right)}$	3	3.20	3.44	3.74	3.86	4.06	4.33	4.80

The maximum shear stress occurs at that point of the periphery which is closest to the centre of the section, whereas the corners has zero stresses.

On examining the nature of the shear stress components, *i.e.* τ_{zx} and τ_{yz} round the periphery of the rectangle, it is seen that one of them is always zero (because there is zero normal stress along each side or surface) and in particular, at each of the four corners, they are both zero. Thus at each corner of this rectangle, *e.g.* point *A* in Fig. 5.23 the total shear stress is zero. The stress at point *B* is maximum and depends upon the fillet radius.

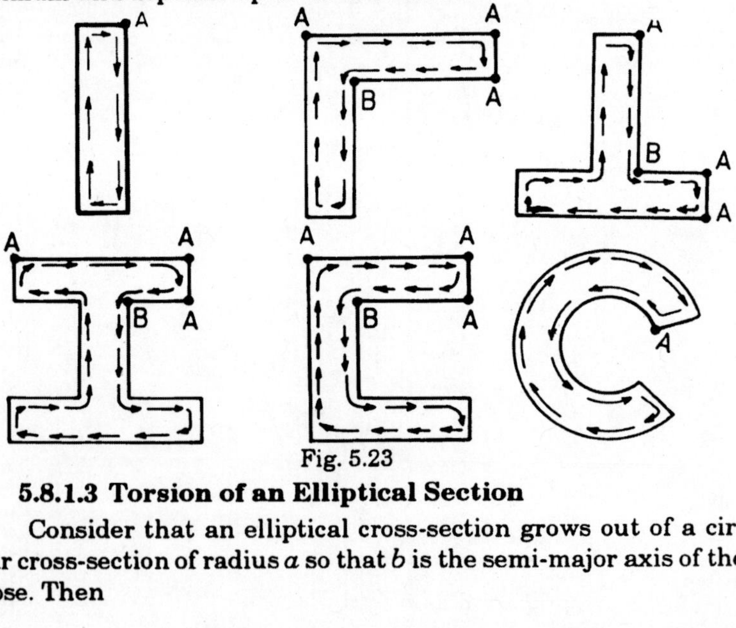

Fig. 5.23

5.8.1.3 Torsion of an Elliptical Section

Consider that an elliptical cross-section grows out of a circular cross-section of radius *a* so that *b* is the semi-major axis of the ellipse. Then

$$\left(\frac{q/S}{V}\right)_{\text{ellipse}} = \left(\frac{1+\dfrac{a^2}{b^2}}{2\,\dfrac{b}{a}}\right) \cdot \left(\frac{\dfrac{q}{S}}{V}\right)_{\text{circle}}$$

$$\left(\frac{2\,G\theta}{T/2}\right)_{\text{ellipse}} = \left(\frac{1+\dfrac{a^2}{b^2}}{\dfrac{2b}{a}}\right)\left(\frac{\dfrac{2\,G\theta}{T}}{2}\right)_{\text{circle}}$$

For the circle, $\dfrac{T}{\theta} = GJ = \dfrac{\pi}{2}\, a^4\, G.$

$\therefore$ For the ellipse,

$$D = \frac{T}{\theta} = \frac{\pi a^4 G}{2} \cdot \frac{2\,(b/a)}{1 + a^2/b^2}$$

$$= \left(\frac{\pi a^3 b^3}{a^2 + b^2}\right) G$$

$$\simeq \frac{1}{4\pi^2}\frac{GA^4}{J} \cong \frac{GA^4}{40\,J}$$

where $A = \pi\, ab$

$$J = \frac{A}{4}\,(a^2 + b^2)$$

$$\left(\frac{\text{Slope}}{V}\right)_{\text{ellipse}} = \frac{a}{b}\left(\frac{\text{Slope}}{\text{V}}\right)_{\text{circle}}$$

$$\left(\frac{\tau_{max}}{T}\right)_{\text{ellipse}} = \frac{a}{b}\left(\frac{\tau_{max}}{T}\right)_{\text{circle}}$$

$$\therefore \qquad \tau_{max} = \frac{2\,T}{\pi a^2 b}\,. \qquad \ldots(5.88)$$

These formulae are found to be very useful and correct to better than 10 per cent for sections of the type shown in Fig. 5.24. The contours indicated in Fig. 5.24 are curves of constant ϕ ; the

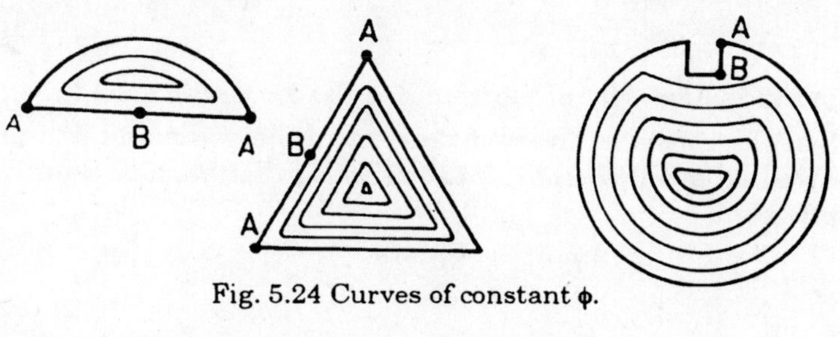

Fig. 5.24 Curves of constant ϕ.

tangent at any point on the curve gives the shear direction at that point.

5.8.1.4 Torsion of Hollow Sections. The membrane anology may be applied to hollow sections with some slight modification as shown in Fig. 5.25 (*a*). The centre portion of the membrane is replaced by a thin, weightless, stiff, flat plate identical to the

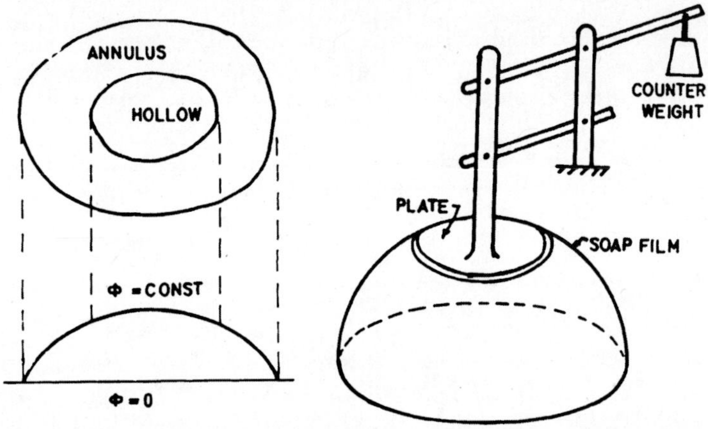

(*a*) (*b*)

Fig. 5.25

cross-section of the hollow. The plate would be in equilibrium at the same height as the periphery of the inner portion of the membrane, because the upward air push and the downward membrane pull from the annular part would be the same for the plate and for the centre membrane. A typical mechanism for effecting a displacement of the plate and rendering it, in effect, weightless, so that the air pressure does not have to support it, is shown in Fig. 5.25 (*b*).

5.8.1.5 Torsion of Thin-walled Hollow Sections

Consider the torsion of a given rectangular annulus of wall thickness t (Fig. 5.26). Let the height of the central plate be h. Therefore membrane slope is equal to h/t. Since slope $\times t = h =$ const.

$$\therefore \qquad \tau = \frac{h}{t} \qquad ...(5.89)$$

or stress is inversely proportional to the local wall thickness.

The stress is directed tangent to the walls and is assumed distributed evenly across the thickness t. For vertical equilibrium of plate let

$$A = \text{area of plate}$$

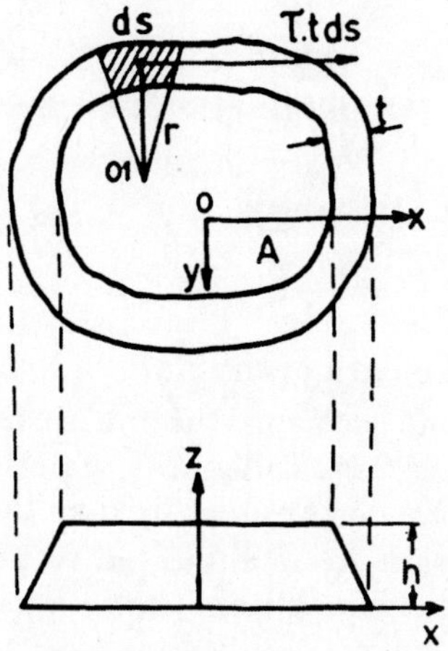

Fig. 5.26 Single cell thin-walled section.

Upward push $= q\,A$

Downward pull $= \oint S ds \,.\, \dfrac{h}{t}$

$$\therefore \qquad qA = \oint (Sds)\,\frac{h}{t}$$

Now $\qquad \dfrac{q}{S} = 2\,G\theta$ and $\dfrac{h}{t} = \tau$

$$\therefore \qquad 2\,G\theta = \frac{1}{A}\oint \tau\,.\,ds$$

Torque about any point O, is

$$T = \tau\,.\,tds\,.\,r$$

where r = normal or moment arm.

Bus rds = 2 × area of small triangle with ds as base and r as height with its apex at O_1.

$$\therefore \quad T = \int (\tau t)\,rds = \tau t\int rds = 2\,A\tau t$$

where A = sum of all little triangles of base ds and height h.

= total area of plate or entire area enclosed by the thin-walled section.

$$\therefore \qquad \tau = \frac{T}{2\,At} \qquad \text{...(5.90)}$$

Angle of twist $\theta = \dfrac{T}{2\,GA^2}\oint \dfrac{ds}{t}$

or
$$D = \frac{T}{\theta} = \frac{4\,GA^2}{\int \left(\frac{ds}{t}\right)} \qquad ...(5.91)$$

For constant wall thickness
$$D = \frac{T}{\theta} = \frac{4\,GA^2}{L} \qquad ...(5.92)$$

where L = peripheral length of the wall.

For multi-cellular sections, the membrane analogy will involve several of the weightless stiff plates, which all have to be blown up to such heights as they will naturally take by the air pressure q. The individual plates are to be subjected to pure couples only to make them horizontal. The heights of the n plates are unknown, and call them $h_1, h_2, ..., h_n$. For the outside walls the slopes will be $\frac{h}{t}$ while for the inside walls the membrane slopes will be $\frac{\Delta h}{t}$, where Δh is the difference in height of two adjoining plates. Then

$$qA_n = T \oint \frac{\Delta h}{t}\, ds$$

where A_n = area of the nth plate.

There will be n such equations which can be solved for the heights $h_1, h_2, ..., h_n$. Then

$$\frac{\Delta h}{t} = \text{slope} = \tau \qquad ...(5.93)$$

and
$$\frac{q}{S} = 2\,G\theta \qquad ...(5.94)$$

$$T = 2 \sum_n A_n\, h_n \qquad ...(5.95)$$

For example, consider the thin section as shown in Fig. 5.27.

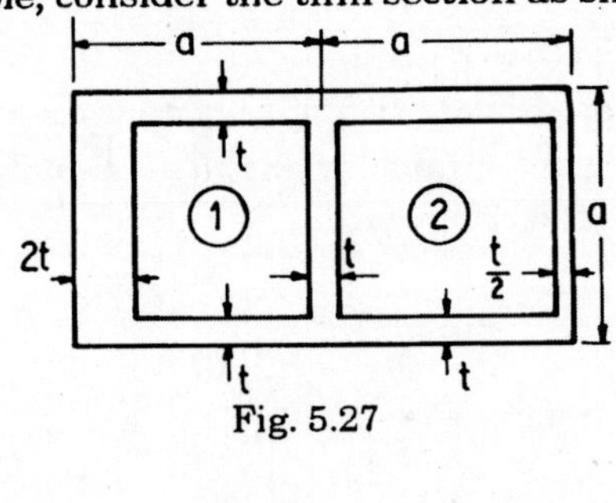

Fig. 5.27

Equilibrium equations for plates (1) and (2) gives,

$$qa^2 = Sa\left[\frac{h_1}{2t} + \frac{h_1}{t} + \frac{h_1 - h_2}{t} + \frac{h_1}{t}\right]$$

$$qa^2 = Sa\left[\frac{h_2 - h_1}{t} + \frac{h_2}{t} + \frac{h_2}{t/2} + \frac{h_2}{t}\right]$$

$$\therefore \quad \frac{qat}{S} = 3.5\,h_1 - h_2$$

$$\frac{qat}{S} = 5\,h_2 - h_1$$

$$\therefore \quad h_1 = \frac{4}{11}\frac{qat}{S}$$

$$h_2 = \frac{3}{11}\frac{qat}{S}$$

The maximum slope occurs in thin leg having thickness $\frac{t}{2}$.

Thus $$\frac{h_2}{t/2} = \frac{6qa}{11\,S}$$

Volume under membrane and plate

$$= a^2(h_1 + h_2)$$

$$= \frac{7\,qa^3\,t}{11\,S}$$

Now $$\frac{q}{S} = 2\,G\theta$$

$$\text{slope} = \tau,\ T = 2V$$

$$\therefore \quad \tau_{max} = \frac{12}{11}\,a\,G\theta$$

$$T = \frac{28}{11}\,a^3\,t\,G\theta.$$

Example 5.6. *A hollow circular torsion member has an outside diameter of 22 mm and inside diameter of 18 mm, with mean diameter D = 20 mm and t/D = 0.10. Calculate the torque and angle of twist per unit length if shearing stress at mean diameter is 70 MPa. Calculate these values if a cut is made through the wall thickness along the entire length. G = 77.5 GPa.*

Solution. $A = \frac{\pi D^2}{4} = \frac{\pi \times 400}{4} = 100\pi \text{ mm}^2$

$$T = 2At\tau = 2 \times 100\pi \times 0.1 \times 20 \times 70 \times 10^{-3} = 87.96 \text{ N.m}$$

$$\theta = \frac{\pi D\tau}{2GA}$$

$$= \frac{\pi \times 20 \times 70}{2 \times 77.5 \times 100\pi} = 0.0903 \text{ rad/m}$$

when a cut is made, the torsion member becomes a narrow rectangle of width $b = \frac{20\pi}{2} = 10\,\pi$ mm, $h = \frac{0.1 \times 20}{2} = 1$ mm

$$\theta = \frac{\tau_{max}}{2Gh}$$

$$= \frac{70 \times 10^6}{2 \times 77.5 \times 10^9 \times 1 \times 10^{-3}} = 0.04515 \text{ rad/m}$$

$$T = \frac{8\, bh^2\, \tau_{max}}{3}$$

$$= \frac{8 \times 10\pi \times 1^2 \times 10^{-9} \times 70 \times 10^6}{3} = 5.865 \text{ N.m.}$$

Example 5.7. *A hollow aluminium tube of rectangular cross-section as shown in Fig. 5.28 is subjected to a torque of 56.5 kN.m along its longitudinal axis. Determine the shearing stresses and the angle of twist. Assume G = 28 GPa.*

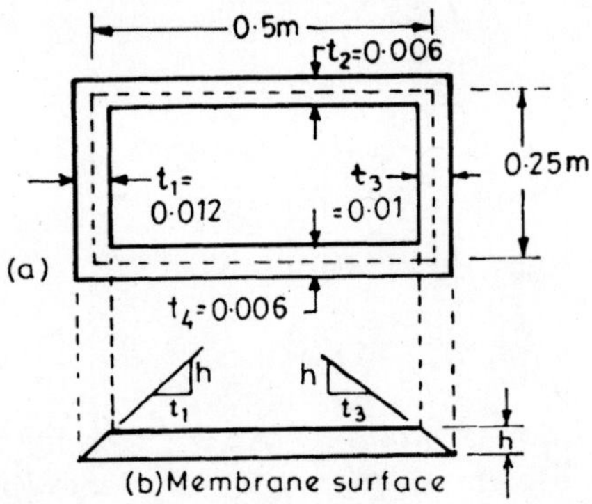

Fig. 5.28

Solution. $T = 2Ah = 2 \times 0.5 \times 0.25 \times h = 0.25\,h = 56500$

$h = 226000$ N/m

$$\tau_1 = \frac{h}{t_1} = \frac{226000}{0.012} = 18.83 \text{ MPa,}$$

$$\tau_2 = \frac{h}{t_2} = \frac{226000}{0.006} = 37.67 \text{ MPa}$$

$$\tau_3 = \frac{h}{t_3} = \frac{226000}{0.01} = 22.6 \text{ MPa}$$

$$\theta = \frac{h}{2GA} \oint \frac{ds}{t}$$

$$= \frac{226000}{2 \times 28 \times 10^9 \times 0.125} \left(\frac{0.25}{0.012} + \frac{2 \times 0.5}{0.006} + \frac{0.25}{0.01} \right)$$

$$= 0.00686 \text{ rad/m.}$$

5.8.2 Hydrodynamic Analogy

Lord Kelvin pointed out that the stress function ϕ for the torsion problems is identical with the stream function of a certain irrotational motion of ideal fluid contained in a vessel of the same cross-section as the twisted bar. Let u and v be the components of velocity of a circulating fluid at a point. Then a stream function ψ is given by

$$u = -\frac{\partial \psi}{\partial y}, \; v = \frac{\partial \psi}{\partial x}$$

and the continuity equation for incompressible fluids is

$$\frac{\partial u}{\partial x} + \frac{\partial v}{\partial y} = 0$$

Also vorticity

$$2\omega = \frac{\partial u}{\partial y} - \frac{\partial v}{\partial x}$$

$$\therefore \quad \frac{\partial^2 \psi}{\partial x^2} + \frac{\partial^2 \psi}{\partial y^2} = -2\omega$$

or

$$\nabla^2 \psi = -2\omega \qquad ...(5.96)$$

Also

$$\nabla^2 \phi = -2G\theta \qquad ...(5.97)$$

If $\omega = G\theta$ then Eqs. (5.96) and (5.97) are identical. The line ψ = constant are the stream lines and the boundary condition is that ψ = constant along the boundary. This is the same boundary condition as imposed on the stress function. The stresses are analogous to the velocities of the fluid flow. The stream lines in some of the cross-sections have been shown in Fig. 5.29. This analogy is very much suitable for determining stress concentration factors (*SCF*) around holes, notches and sharp corners.

5.8.3 Conducting Sheet Analogy

In this method an electrically conducting sheet is cut to the shape of the cross-section of the twisted bar. If constant current density i (per unit area) is fed into the surface over its whole area,

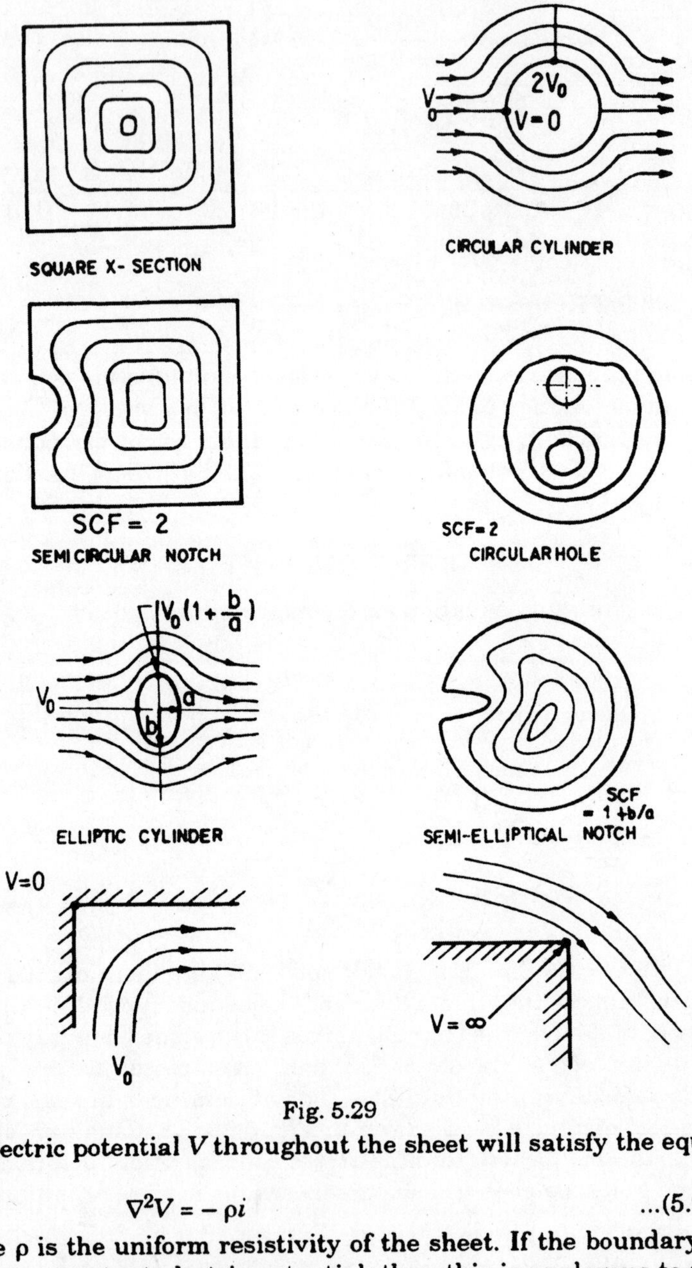

Fig. 5.29

the electric potential V throughout the sheet will satisfy the equation

$$\nabla^2 V = -\rho i \qquad \text{...(5.98)}$$

where ρ is the uniform resistivity of the sheet. If the boundary is kept at a constant electric potential, then this is analogous to the torsion problem represented by

$$\nabla^2 \phi = -2G\theta. \qquad \text{...(5.99)}$$

5.9 Torsion of Bars of Variable Diameter

Consider a bar in the form of a body of revolution twisted by couples applied at the ends as shown in Fig. 5.30. Using r, θ, z

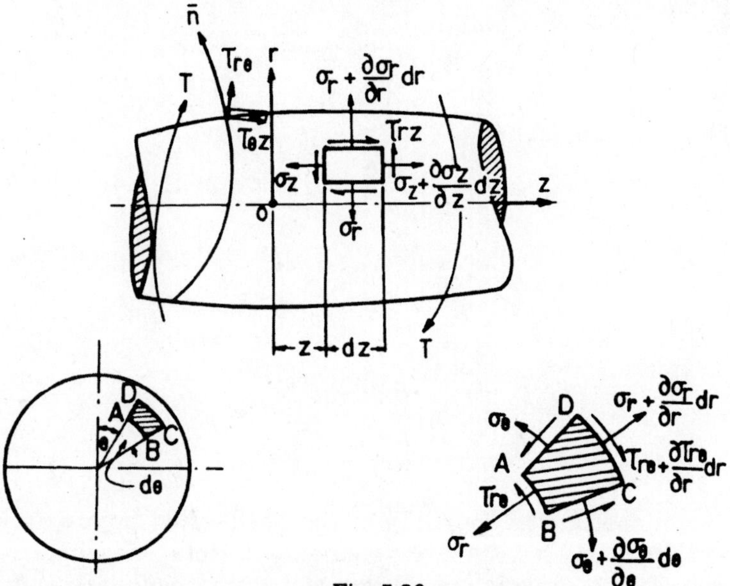

Fig. 5.30

co-ordinates, and assuming that the displacement of points in any cross-section is in the tangential direction only

i.e. $u_r = u_z = 0$

and $u_\theta \neq 0$

Further, on account of the circular symmetry, the tangential displacement is independent of θ. Hence, the strain components become,

$$\left.\begin{aligned} &\varepsilon_r = \varepsilon_\theta = \varepsilon_z = \gamma_{rz} = 0 \\ &\gamma_{r\theta} = \frac{\partial u_\theta}{\partial r} - \frac{u_\theta}{r}, \ \gamma_{\theta z} = \frac{\partial u_\theta}{\partial z} \end{aligned}\right\} \quad \ldots(1)$$

The stress components become,

$$\left.\begin{aligned} &\sigma_r = \sigma_\theta = \sigma_z = \tau_{rz} = 0 \\ &\tau_{r\theta} = G\left(\frac{\partial u_\theta}{\partial r} - \frac{u_\theta}{r}\right), \ \tau_{\theta z} = G\frac{\partial u_\theta}{\partial z} \end{aligned}\right\} \quad \ldots(2)$$

The only equilibrium equation is

$$\frac{\partial \tau_{r\theta}}{\partial r} + \frac{\partial \tau_{\theta z}}{\partial z} + \frac{2\tau_{r\theta}}{r} = 0$$

$$\frac{\partial}{\partial r}(r^2 \tau_{r\theta}) + \frac{\partial}{\partial z}(r^2\tau_{\theta z}) = 0 \qquad ...(3)$$

Eq. (1) is satisfied by assuming a stress function ϕ of r and such that

$$r^2\tau_{r\theta} = -\frac{\partial\phi}{\partial z},\ r^2\tau_{\theta z} = \frac{\partial\phi}{\partial r} \qquad ...(4)$$

From Eqs. (2) and (4), we get

$$\left.\begin{aligned} \tau_{r\theta} &= Gr\frac{\partial}{\partial r}\left(\frac{u_\theta}{r}\right) = -\frac{1}{r^2}\frac{\partial\phi}{\partial z} \\ \tau_{\theta z} &= Gr\frac{\partial}{\partial z}\left(\frac{u_\theta}{r}\right) = \frac{1}{r^2}\frac{\partial\phi}{\partial r} \end{aligned}\right\} \qquad ...(5)$$

Hence $\frac{\partial}{\partial r}\left(\frac{1}{r^3}\frac{\partial\phi}{\partial r}\right) + \frac{\partial}{\partial z}\left(\frac{1}{r^3}\frac{\partial\phi}{\partial z}\right) = 0$

or
$$\frac{\partial^2\phi}{\partial r^2} - \frac{3}{r}\frac{\partial\phi}{\partial r} + \frac{\partial^2\phi}{\partial z^2} = 0 \qquad ...(6)$$

Let the lateral boundary of the bar be free from external forces then at any point A at the boundary the total shearing stress must be in the direction of the tangent to the boundary and its projection on the normal $\bar{n}$ to the boundary must be zero. Hence

$$\tau_{r\theta}\cdot\frac{ds}{dz} - \tau_{\theta z}\cdot\frac{dr}{ds} = 0$$

where ds is the element of the boundary.

Substituting from Eq. (4) we find

$$\frac{\partial\phi}{\partial z}\cdot\frac{dz}{ds} + \frac{\partial\phi}{\partial r}\cdot\frac{dr}{ds} = 0$$

or
$$\frac{d\phi}{ds} = 0 \qquad ...(7)$$

Hence ϕ is constant along the boundary of the axial section of the shaft.

Eq. (6) together with Eq. (7) completely determine the stress function ϕ, from which we may obtain the stresses by using Eq. (4).

The twisting moment T is given by :

$$T = \int_0^{2\pi}\int_0^a \tau_{\theta z}\,.\,r^2 dr d\theta = 2\pi\int_0^a r^2\tau_{\theta z}\,dr$$

$$= 2\pi \int_0^a \frac{\partial \phi}{\partial r}\, dr = 2\pi \left| \phi \right|_0^a \qquad \text{...(5.100)}$$

where a is the outer radius of the cross-section.

To determine the twist of the bar, let

$$\psi = \frac{u_\theta}{r}$$

then Eqs. (5) become, $Gr^3 \dfrac{\partial \psi}{\partial r} = -\dfrac{\partial \phi}{\partial z}$

$$Gr^3 \frac{\partial \psi}{\partial z} = \frac{\partial \phi}{\partial r}$$

from which $\dfrac{\partial}{\partial r}\left(r^3 \dfrac{\partial \psi}{\partial r}\right) + \dfrac{\partial}{\partial z}\left(r^3 \dfrac{\partial \psi}{\partial z}\right) = 0.$

or

$$\frac{\partial^2 \psi}{\partial r^2} + \frac{3}{r}\frac{\partial \psi}{\partial r} + \frac{\partial^2 \psi}{\partial z^2} = 0 \qquad \text{...(6)}$$

A solution of Eq. (6) gives the angle of twist as a function of r and z. If we put

$$\psi = \text{const.} \qquad \text{...(7)}$$

then we obtain a surface in which all points have the same angle of twist.

5.9.1 Torsion of a Conical Shaft

Consider a conical shaft as shown in Fig. 5.31.

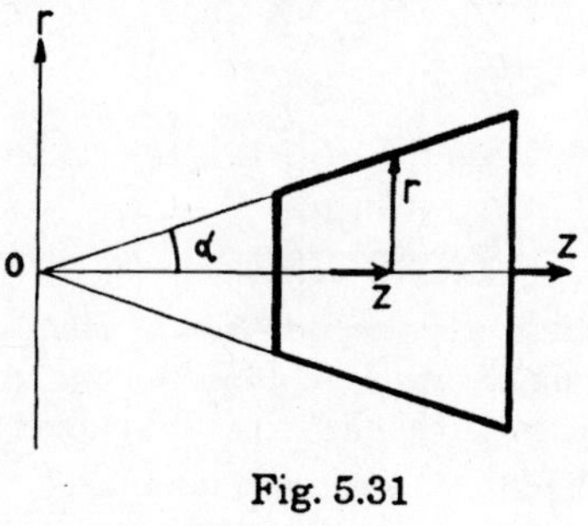

Fig. 5.31

Now $\cos \alpha = \dfrac{z}{(r^2 + z^2)^{1/2}}$

is constant on the boundary. The function

$$\phi = C\left[\frac{z}{\sqrt{r^2 + z^2}} - \frac{1}{3}\left\{\frac{z}{\sqrt{r^2 + z^2}}\right\}^3\right] \qquad \text{...(1)}$$

satisfy the condition at the boundary *i.e.* $\frac{d\phi}{ds} = 0$ where C is a constant.

$$\therefore \quad \left.\begin{aligned} \tau_{\theta z} &= \frac{1}{r^2}\frac{\partial \phi}{\partial r} = -\frac{Crz}{(r^2+z^2)^{5/2}} \\ \tau_{r\theta} &= -\frac{Cr^2}{(r^2+z^2)^{5/2}} \end{aligned}\right\} \quad \text{...(2)}$$

The constant C can be known when the twisting couple T is known. Now

$$T = 2\pi[\phi(a, z) - \phi(0, z)]$$
$$= 2\pi C[\cos\alpha - \tfrac{1}{3}\cos^3\alpha - \tfrac{2}{3}]$$

$$\therefore \quad C = \frac{-T}{2\pi[\frac{2}{3} - \cos\alpha + \frac{1}{3}\cos^2\alpha]} \quad \text{...(3)}$$

Now $$\gamma_{\theta z} = \frac{\tau_{\theta z}}{G} = \frac{\partial u_\theta}{\partial z}$$

$$\gamma_{r\theta} = \frac{\tau_{r\theta}}{G} = \frac{\partial u_\theta}{\partial r} - \frac{u_\theta}{r}$$

$$\therefore \quad \frac{\partial u_\theta}{\partial z} = \frac{-Crz}{G(r^2+z^2)^{5/2}}$$

Integrating, we get

$$u_\theta = \frac{T}{6\pi Gk}\cdot\frac{r}{(r^2+z^2)^{3/2}} + u_z \cdot r \quad \text{...(4)}$$

where $k = \cos\alpha - \frac{1}{3}\cos^3\alpha - \frac{2}{3}$

u_z = const. representing a rigid rotation of the cone about the z-axis.

EXERCISES

5.1 A shaft is of elliptical cross-section having semi-major axis 50 mm and semi-minor axis 25 mm. It is subjected to a torque of 1000 N-m. Determine the maximum shear stress developed in the shaft.

5.2 A torque T is applied to (*a*) a solid square bar, (*b*) a rectangular bar with ratio of sides as 1 : 2 (*c*) a triangular bar with equal sides and (*d*) an elliptic bar with the ratio of major axis to minor axis as 2 : 1. The areas of cross-section are equal. Find their angles of twist and maximum stresses and compare these with solid circular bar of the same area.

5.3 Compare the weights per unit length of shafts of circular and elliptic cross-sections when the shafts are subjected to the same torsional moment and angle of twist per unit length is same.

5.4 A 250 mm × 250 mm angle section with 10 mm thickness is subjected to a torque of 2500 N-m. Find the maximum shear stress induced in the section and the angle of twist per unit length. Assume $G = 0.8 \times 10^5$ N/mm^2.

5.5 A thin-walled circular cylinder of outside diameter 125 mm and 5 mm wall thickness is 500 mm long. It is subjected to a torque of 100 N-m. Estimate the shear stress and the total angle of twist. Assume $E = 2 \times 10^5$ N/mm^2 and $\nu = 0.30$.

5.6 The cross-section shown in Fig. 5.32 is subjected to a torque of 5 kN-m. Estimate the shear stress in each wall.

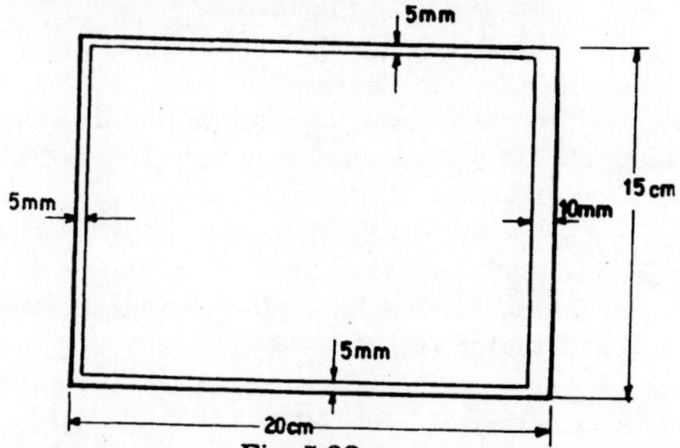

Fig. 5.32

5.7 Each of the sections shown in Fig. 5.33 is transmitting a torque of 20 N-m. Estimate the maximum shear stress in each section and the angle of twist per unit length. Take $E = 2 \times 10^5$ N/mm^2, $\nu = 0.3$.

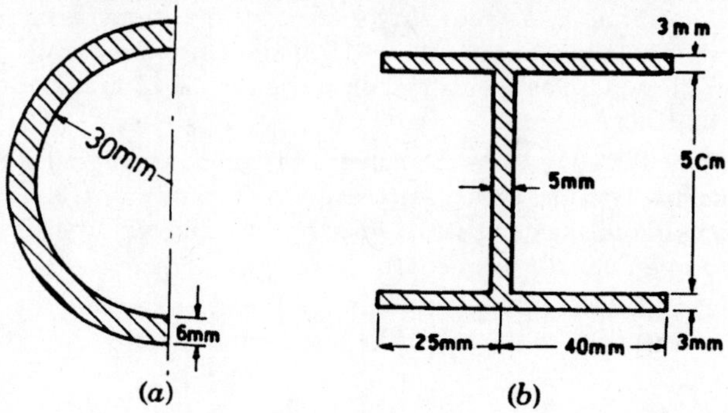

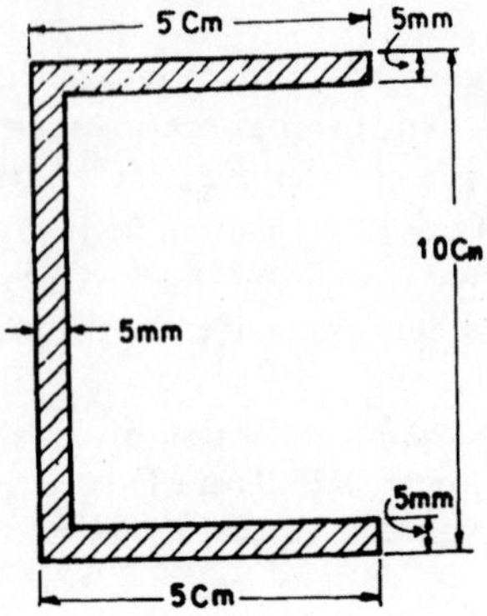

(c)

Fig. 5.33

5.8 A thin-walled rectangular shaft of 50 mm × 60 mm is having a wall thickness of 2.5 mm. The length of the shaft is 3 m and is subjected to a torque of 150 N-m at one end while the other end is rigidly fixed. Taking $G = 1.05 \times 10^5$ N/mm^2, determine the maximum shear stress developed and the angle of twist of the shaft.

5.9 Derive Bredt's formula for the rate of twist in terms of the shear-stress distribution for the torsion of thin-walled sections. Use this formula for the case of a thin-walled circular shaft and compare this result with that of the exact value.

5.10 An experiment with a thin rubber membrane under a pressure of 1.5 kN/m^2 gives a volume of 300×10^{-6} m^3. If the tensile force per unit length of the membrane is 0.2 N/mm for a material having a shear modulus of 0.7×10^5 N/mm^2, what is the rate of twist corresponding to a torque of 800 N-m predictable from the membrane analogy ?

5.11 Determine the decrease in torsional stiffness and torsional strength when a 10 mm slit is cut in a 1500 mm long tube parallel to the axis of the tube. The inside radius of the tube is 75 mm and the outside radius is 80 mm.

5.12 Show that $\phi = A(r^2 - a^2)$ solves the torsion problem for the solid or hollow circular shaft. Determine A in terms of $G\theta$. Evaluate the maximum shearing stress and the torsional rigidity in terms of the torque T for the solid shaft.

5.13 (a) Show that the expression for the stress function ϕ which corresponds to the parabolic membrane is

$$\phi = -G\theta\left(x^2 - \frac{c^2}{4}\right).$$

(b) Prove that for the triangular section of height b and width a, the torque T is given by approximately

$$T = \frac{1}{12} G\theta b a^3.$$

5.14 A closed thin-walled tube has a perimeter l and a uniform wall thickness t. An open tube is made by making a fine longitudinal cut in it. Show that when the maximum shear stress is the same in both closed and open tubes,

$$\frac{T_{open}}{T_{closed}} = \frac{lt}{6A}, \quad \frac{\theta_{open}}{\theta_{closed}} = \frac{2A}{lt}$$

where A is the area of slit.

5.15 Given the thin tubular bars with the dimensions as shown in Fig. 5.34. The bars are under the action of a torque of 10 kN.m. Find the shearing stresses in the tubes.

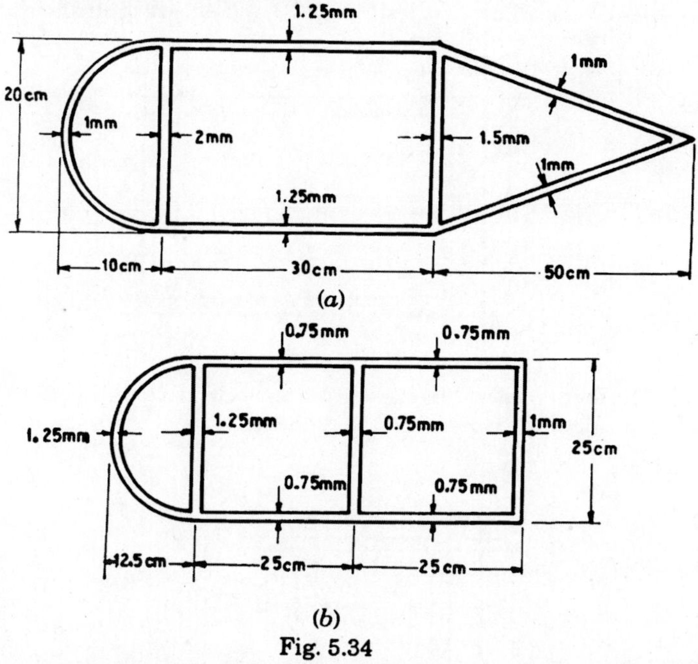

Fig. 5.34

5.16 Prove that the relative rotation of the ends of the conical shaft shown in Fig. 5.35 due to torque T is

$$\frac{T}{2\pi\left(\frac{2}{3} - \cos\alpha + \frac{1}{3}\cos^4\alpha\right)} \cdot \frac{1}{3G}\left(\frac{1}{a^3} - \frac{1}{b^3}\right)$$

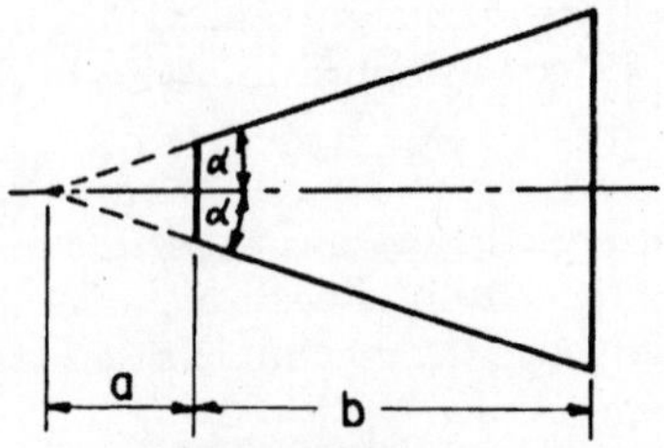

Fig. 5.35

5.17 A square shaft has 50 mm sides and has the same cross-sectional area as shafts having circular and equilateral triangular cross-sections. If each shaft is subjected to a torque of 2 kN-m, calculate the maximum shearing stress for each of the three shafts.

5.18 A hollow thin-wall torsion member of Fig. 5.36 has uniform thickness walls. Show that walls *BC, CD* and *AC* are stress free.

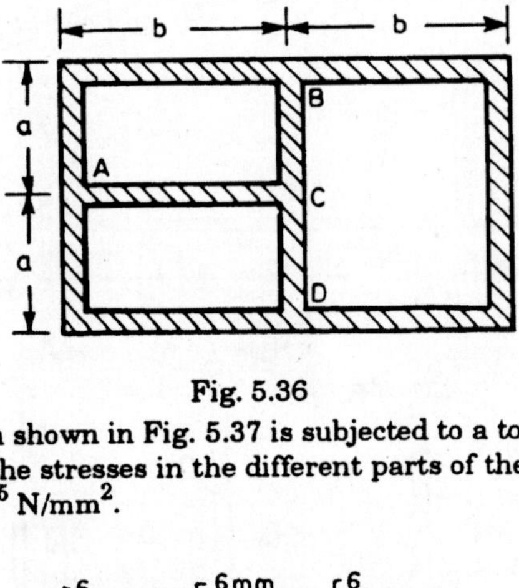

Fig. 5.36

5.19 The section shown in Fig. 5.37 is subjected to a torque of 10^5 N-m. Calculate the stresses in the different parts of the section. Take $G = 0.84 \times 10^5$ N/mm^2.

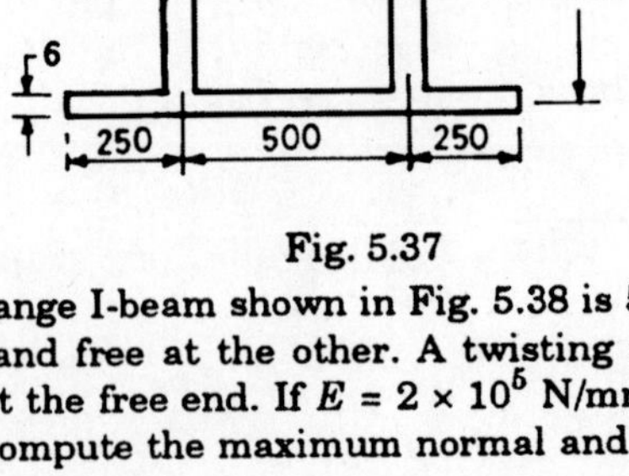

Fig. 5.37

5.20 A wide flange I-beam shown in Fig. 5.38 is 5 m long. It is fixed at one end and free at the other. A twisting moment of 3 kN-m is applied at the free end. If $E = 2 \times 10^5$ N/mm^2 and $G = 0.84 \times 10^5$ N/mm^2, compute the maximum normal and shearing stresses due to bending and the maximum shearing stress due to twist. What is the angle of twist of the beam at the free end ?

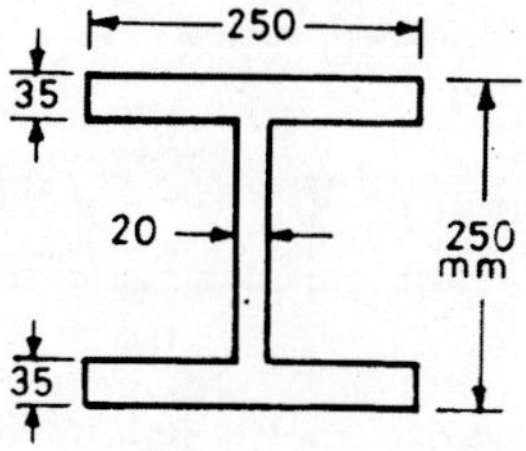

Fig. 5.38

5.21 A bar of rectangular cross-section measuring 250 mm × 200 mm is subjected to a twisting moment of 100 kN.m. Determine the maximum shear stresses and the angle of twist if G = 100 GPa and length of bar is 3 m. [**Ans.** 38.2 MPa, 0.461°]

5.22 A 100 mm × 250 mm overall depth steel channel section with flanges and web 12.5 mm thick, is used to transmit torque. Estimate the maximum permissible torque that can be transmitted if the shear stress is limited to 40 MPa. Also determine the angle of twist per metre length under this torque.

5.23 Fig. 5.39 shows a cross-section of a hollow tube used for transmitting 20 kW at 100 rpm. Calculate shear stress and angular deformation per metre length of the tube. Take G = 100 GPa.

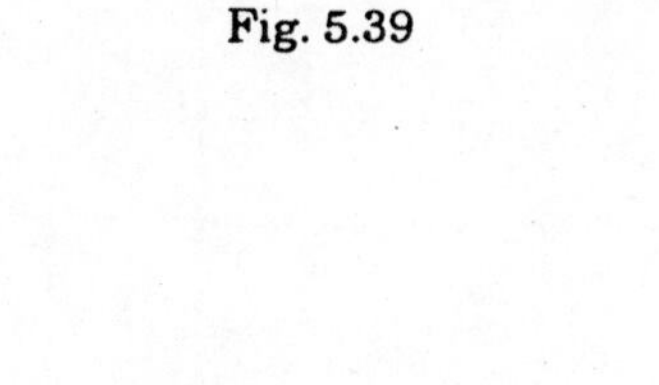

Fig. 5.39

6

Bending of Prismatic Bars

6.1 Introduction

The problem of bending of a prismatic bar of symmetrical cross-section under the action of bending moments only was attempted by many well known research workers like, Galileo, Bernoulli and Coulomb for its complete development. The well known formula for the bending of symmetrical cross-sections is given by

$$\frac{M}{I} = \frac{\sigma}{y} = \frac{E}{R} \qquad \ldots(6.1)$$

where M is the bending moment, I the moment of inertia, σ is the bending stress at a distance y from the neutral axis, E is the modulus of elasticity and R is the radius of curvature. The formula given by Eq. (6.1) is not applicable for prismatic bars of general cross-section. In this chapter, we shall study to determine stresses due to bending in prismatic bars of general cross-section.

6.2 Simple Bending

6.2.1 Bending of a Cantilever by Terminal Load

Consider the bending of a cantilever of constant cross-section of any shape by a force P applied at the end and parallel to one of the principal axes of the cross-section as shown in Fig. 6.1.

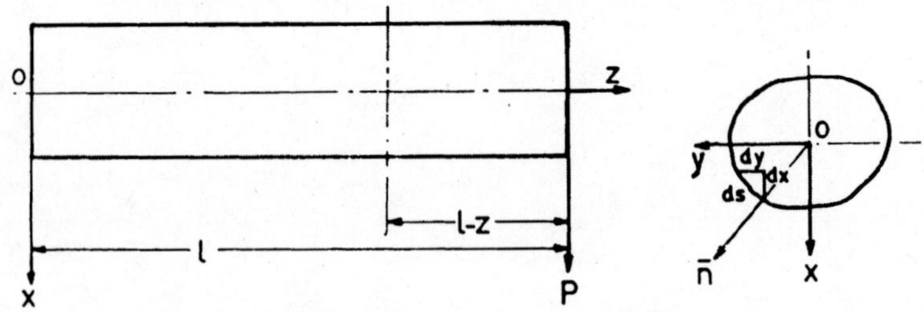

(a) (b)

Fig. 6.1

The z-axis is taken along the length of the bar through the centre while the x-axes and y-axes are any orthogonal axes intersecting at the centroid of the end $z = 0$. The lateral surface of the bar is free from external forces and the body forces are assumed to vanish.

We shall follow the semi-inverse method of St. Venant ; that is, we put

$$\sigma_x = \sigma_y = \tau_{xy} = 0 \qquad \text{...(6.2)}$$

The stresses σ_z, τ_{zx}, τ_{yz} will be so chosen that the equations of equilibrium and compatibility as well as the boundary conditions are satisfied. We assume that normal stresses over a cross-section at a distance z from the fixed end are distributed in the same manner as in the case of pure bending.

$$\sigma_z = \frac{-P(l-z)}{I} \qquad \text{...(6.3)}$$

With these assumptions, the equations of equilibrium become :

$$\frac{\partial \tau_{zx}}{\partial z} = 0, \quad \frac{\partial \tau_{yz}}{\partial z} = 0 \qquad \text{...(6.4)}$$

$$\frac{\partial \tau_{zx}}{\partial x} + \frac{\partial \tau_{yz}}{\partial y} = -\frac{Px}{I} \qquad \text{...(6.5)}$$

From Eq. (6.4), we conclude that shearing stresses do not depend on z and are same in all cross-sections of the bar.

The boundary conditions are

$$\sigma_x \cdot a_{nx} + \tau_{xy}\, a_{ny} + \tau_{xz}\, a_{nz} = 0$$

$$\tau_{xy}\, a_{nx} + \sigma_y\, a_{ny} + \tau_{yz}\, a_{nz} = 0$$

$$\tau_{zx}\, a_{nx} + \tau_{yz}\, a_{ny} + \sigma_z\, a_{nz} = 0$$

The first two equations are identically satisfied and the third gives,

$$\tau_{zx}\, a_{nx} + \tau_{yz}\, a_{ny} = 0$$

where $a_{nx} = \cos(\overline{n}, x) = \dfrac{dy}{ds}$

$$a_{ny} = \cos(\overline{n}, y) = -\frac{dx}{ds}$$

in which ds is an element of the bounding curve of the cross-section. Hence, the boundary condition becomes

$$\tau_{zx}\frac{dy}{ds} - \tau_{yz}\frac{dx}{ds} = 0 \qquad \text{...(6.6)}$$

Now the Beltrami-Michell compatibility Eq. (1.22) reduces in this case to

$$\left.\begin{aligned} \nabla^2 \tau_{yz} &= 0 \\ \nabla^2 \tau_{zx} &= -\frac{P}{I(1+\nu)} \end{aligned}\right\} \qquad \text{...(6.7)}$$

Thus the solution of the problem of bending of a prismatical cantilever of any cross-section reduces to finding, for τ_{zx} and τ_{yz}, functions of x and y that satisfy the equation of equilibrium (6.5), the boundary condition (6.6) and the compatibility Eqs. (6.7).

By taking

$$\tau_{zx} = \frac{\partial \phi}{\partial y} - \frac{Px^2}{2I} + f(y), \ \tau_{yz} = -\frac{\partial \phi}{\partial x} \qquad \text{...(6.8)}$$

the equations of equilibrium Eq. (6.5) are satisfied. Where ϕ is the stress function of x and y, and $f(y)$ is a function of y only, to be determined from the boundary conditions.

Substituting Eqs. (6.8) in (6.7), we find

$$\frac{\partial}{\partial x}\left(\frac{\partial^2 \phi}{\partial x^2} + \frac{\partial^2 \phi}{\partial y^2}\right) = 0$$

$$\frac{\partial}{\partial y}\left(\frac{\partial^2 \phi}{\partial x^2} + \frac{\partial^2 \phi}{\partial y^2}\right) = \left(\frac{\nu}{1+\nu}\right)\frac{P}{I} - \frac{d^2 f}{dy^2}$$

Integrating, we get

$$\therefore \qquad \frac{\partial^2 \phi}{\partial x^2} + \frac{\partial^2 \phi}{\partial y^2} = \left(\frac{\nu}{1+\nu}\right)\frac{Py}{I} - \frac{df}{dy} + C \qquad \text{...(6.9)}$$

where C is the constant of integration. This constant of integration can be interpreted physically. Consider the rotation of an element of area in the plane of a cross-section of the cantilever given by

$$\omega_z = \frac{1}{2}\left(\frac{\partial u_y}{\partial x} - \frac{\partial u_x}{\partial y}\right)$$

The local twist at a point (x, y) of a cross-section is defined as

$$\frac{\partial \omega_z}{\partial z} = \frac{1}{2}\left(\frac{\partial^2 u_y}{\partial z \partial x} - \frac{\partial^2 u_x}{\partial z \partial y}\right)$$

$$= \frac{1}{2}\left(\frac{\partial \gamma_{yz}}{\partial x} - \frac{\partial \gamma_{zx}}{\partial y}\right)$$

$$= \frac{1}{2G}\left(\frac{\partial \tau_{yz}}{\partial y} - \frac{\partial \tau_{zx}}{\partial y}\right)$$

$$= -\frac{1}{2G}\left(\frac{\partial^2 \phi}{\partial x^2} + \frac{\partial^2 \phi}{\partial y^2} + \frac{df}{dy}\right)$$

Substituting in Eq. (6.9), we find

$$-2G\frac{\partial \omega_z}{\partial z} = \left(\frac{\nu}{1+\nu}\right)\frac{Py}{I} + C \qquad \text{...(6.10)}$$

The mean value of the local twist over the section or at the centroid of the section is equal to the constant C. If the x axis is an axis of symmetry of the cross-section, bending by a force P in this axis will result in a symmetrical pattern of rotation ω_z of elements of the cross-section, with a mean value of zero over the whole cross-section. The mean value of $\frac{\partial \omega_z}{\partial z}$ will then also be zero and thus C will also be zero in Eq. (6.10). If the cross-section is not symmetrical then again defining bending without torsion, the value of C will be zero.

Hence Eq. (6.10) becomes

$$\frac{\partial^2 \phi}{\partial x^2} + \frac{\partial^2 \phi}{\partial y^2} = \left(\frac{\nu}{1+\nu}\right)\frac{Py}{I} - \frac{df}{dy} \qquad \text{...(6.11)}$$

Substituting Eqs. (6.11) in (6.6), we get

$$\frac{\partial \phi}{\partial y}\frac{dy}{ds} + \frac{\partial \phi}{\partial x}\frac{dx}{ds} = \frac{d\phi}{ds} = \left[\frac{Px^2}{2I} - f(y)\right]\frac{dy}{ds} \qquad \text{...(6.12)}$$

From Eq. (6.12), the value of the function ϕ along the boundary of the cross-section can be calculated if the function $f(y)$ is chosen. Eq. (6.11) together with Eq. (6.12) determines the stress function ϕ. Generally the function $f(y)$ is chosen in such a manner that the right hand side of Eq. (6.11) becomes zero. Then ϕ becomes constant along the boundary.

6.2.2 Bending of a Bar of Circular Cross-section

Let the boundary of the cross-section of the bar be given by

$$x^2 + y^2 = r^2 \qquad \text{...(1)}$$

The right hand of boundary condition (6.12) becomes zero, if

$$f(y) = \frac{P}{2I}x^2 = \frac{P(r^2 - y^2)}{2I} \qquad \text{...(2)}$$

Substituting Eq. (2) in Eq. (2.11), we find

$$\left.\begin{aligned} \frac{\partial^2 \phi}{\partial x^2} + \frac{\partial^2 \phi}{\partial y^2} &= \left(\frac{1+2\nu}{1+\nu}\right)\frac{Py}{I} \\ \text{or} \qquad \nabla^2 \phi &= \left(\frac{1+2\nu}{1+\nu}\right)\frac{Py}{I} \end{aligned}\right\} \qquad \text{...(3)}$$

The boundary condition and Eq. (3) are satisfied by taking

$$\phi = m(x^2 + y^2 - r^2)y \qquad \text{...(4)}$$

where m is a constant factor.

Eq. (3) is satisfied if we take

$$m = \frac{(1+2\nu)}{8(1+\nu)}\frac{P}{I}$$

Eq. (4) then becomes

$$\phi = \frac{(1+2\nu)}{8(1+\nu)}\frac{P}{I}(x^2+y^2-r^2)y \qquad ...(5)$$

The stress components become from Eq. (6.8) as

$$\left.\begin{aligned} \tau_{xz} &= \frac{(3+2\nu)}{8(1+\nu)}\frac{P}{I}\left(r^2-x^2-\frac{1-2\nu}{3+2\nu}y^2\right) \\ \tau_{yz} &= \frac{-(1+2\nu)}{4(1+\nu)}\frac{Pxy}{I} \end{aligned}\right\} \qquad ...(6)$$

Along the horizontal diameter of the cross-section,

$$x = 0$$

Hence, from Eq. (6), we find

$$\tau_{xz} = \frac{(3+2\nu)}{8(1+\nu)}\frac{P}{I}\left(r^2-\frac{1-2\nu}{3+2\nu}y^2\right),\ \tau_{yz} = 0 \qquad ...(6.13)$$

$$(\tau_{xz})_{y=0} = (\tau_{xz})_{max} = \frac{(3+2\nu)}{8(1+\nu)}\frac{P}{I} \qquad ...(6.14)$$

$$(\tau_{xz})_{y=\pm r} = \frac{(1+2\nu)}{4(1+\nu)}\frac{Pr^2}{I} \qquad ...(6.15)$$

From the elementary beam theory, based on the assumption that the shearing stress τ_{xz} is uniformly distributed along the horizontal diameter of the cross-section, is :

$$\tau_{xz} = \frac{4}{3}\frac{P}{A} \qquad ...(6.16)$$

where A is the area of cross-section of the bar.

6.2.3 Bending of a Prismatic Bar of Elliptic Cross-section

Let $\frac{x^2}{a^2}+\frac{y^2}{b^2} = 1$ ...(1)

be the boundary of the cross-section. The right hand side of Eq. (6.12) will vanish it we take

$$f(y) = -\frac{P}{2I}\left(\frac{a^2}{b^2}y^2-a^2\right) \qquad ...(2)$$

Substituting into Eq. (6.11), we find

$$\nabla^2\phi = \frac{Py}{I}\left(\frac{a^2}{b^2}+\frac{\nu}{1+\nu}\right) \qquad ...(3)$$

The boundary condition and Eq. (3) are satisfied by taking

$$\phi = \frac{(1+\nu)a^2 + \nu b^2}{2(1+\nu)(3a^2+b^2)} \cdot \frac{P}{I}\left(x^2 + \frac{a^2}{b^2}y^2 - a^2\right)y \quad ...(4)$$

The stress components become from Eq. (6.2) as

$$\tau_{xz} = \frac{2(1+\nu)a^2 + b^2}{(1+\nu)(3a^2+b^2)} \cdot \frac{P}{2I}\left[a^2 - x^2 - \frac{(1-2\nu)\,a^2}{2(1+\nu)a^2+b^2}y^2\right] \quad ...(6.17\,a)$$

$$\tau_{yz} = \frac{-(1+\nu)a^2 + \nu b^2}{(1+\nu)(3a^2+b^2)} \cdot \frac{Pxy}{I} \quad ...(6.17\,b)$$

$$(\tau_{xz})_{x=0} = \frac{2(1+\nu)a^2 + b^2}{(1+\nu)(3a^2+b^2)} \cdot \frac{P}{2I}\left[a^2 - \frac{(1-2\nu)\,a^2}{2(1+\nu)a^2+b^2} \cdot y^2\right] \quad ...(6.17\,c)$$

$$(\tau_{yz})_{x=0} = 0 \quad ...(6.17\,d)$$

If $b << a$, then neglecting the $\frac{b^2}{a^2}$ term, we get

$$(\tau_{xz})_{max} = \frac{Pa^2}{3I} = \frac{4P}{3A} \quad ...(6.18)$$

which coincides with the solution of elementry beam theory.

If $b >> a$, then

$$\left.\begin{aligned} (\tau_{xz})_{max} &= \left(\frac{2}{1+\nu}\right)\frac{P}{A} \\ \text{Also}\quad (\tau_{xz})_{y=\pm b} &= \left(\frac{4\nu}{1+\nu}\right)\frac{P}{A} \end{aligned}\right\} \quad ...(6.19)$$

6.2.4 Bending of a Prismatic Bar of Rectangular Cross-section

Consider a rectangular bar of width $2b$ and thickness $2a$. The equation for the boundary line is

$$(x^2 - a^2)(y^2 - b^2) = 0 \quad ...(1)$$

If we substitute

$$f(y) = \frac{Pa^2}{2I}$$

then right hand side of Eq. (6.12) becomes zero along the sides $x = \pm a$ of the rectangle. Along the vertical sides $y = \pm b$ the derivative $\frac{dy}{ds}$ is zero. Thus the right hand side of Eq. (6.12) is zero along the boundary line and we can take $\phi = 0$ at the boundary. Eq. (6.11) becomes

$$\nabla^2\phi = \left(\frac{\nu}{1+\nu}\right)\frac{Py}{I} \qquad \text{...(2)}$$

From Eq. (6.8) the shearing stresses can be resolved into the following two systems

$$\left.\begin{aligned} \tau'_{xz} &= \frac{P}{2I}(a^2 - x^2),\ \tau'_{yz} = 0 \\ \tau''_{xz} &= \frac{\partial\phi}{\partial y},\ \tau''_{yz} = -\frac{\partial\phi}{\partial x} \end{aligned}\right\} \qquad \text{...(3)}$$

The single primed stresses represent the parabolic stress distribution given by the elementary beam theory. The double primed stresses represent the necessary corrections to the elementary solution. The boundary conditions are satisfied by taking the stress function ϕ in the form of double Fourier series,

$$\phi = \sum_{m=0}^{\infty}\sum_{n=1}^{\infty} A_{2m+1,n}\cos\frac{(2m+1)\pi x}{2a}\cdot\sin\frac{n\pi y}{b} \qquad \text{...(4)}$$

Substituting Eq. (4) in Eq. (2) and calculating the Fourier coefficients, we get

$$A_{2m+1,n}\,\pi^2 ab\left[\left(\frac{2m+1}{2a}\right)^2\left(\frac{n}{b}\right)^2\right]$$

$$= -\left(\frac{\nu}{1+\nu}\right)\frac{P}{I}\int_{-a}^{+a}\int_{-b}^{+b} y\cos\frac{(2m+1)\pi x}{2a}\sin\frac{n\pi y}{b}\,dx\,dy$$

$$A_{2m+1,n} = -\left(\frac{\nu}{1+\nu}\right)\frac{P}{I}\left[\frac{8b(-1)^{m+n-1}}{\pi^4(2m+1)\,n\left\{\frac{(2m+1)^2}{(2a)^2}+\left(\frac{n}{b}\right)^2\right\}}\right]$$

Substituting in Eq. (4), we find

$$\phi = -\left(\frac{\nu}{1+\nu}\right)\frac{P}{I}\cdot\frac{8b^3}{\pi^4}$$

$$\sum_{m=0}^{\infty}\sum_{n=1}^{\infty}\frac{(-1)^{m+n-1}\cos\left[(2m+1)\frac{\pi x}{2a}\right]\sin\left(\frac{nxy}{b}\right)}{(2m+1)\,n\left\{(2m+1)^2\left(\frac{b^2}{4a^2}\right)+n^2\right\}}$$

$$(\tau''_{xz})_{x=0} = \left(\frac{\partial\phi}{\partial y}\right)_{x=0}$$

$$= -\frac{\nu}{1+\nu}\frac{P}{I}\frac{8b^2}{\pi^3}\sum_{m=0}^{\infty}\sum_{n=1}^{\infty}\frac{(-1)^{m+n-1}\cos\left(\frac{n\pi y}{b}\right)}{(2m+1)\left[(2m+1)^2\left(\frac{b^2}{4a^2}\right)+n^2\right]}$$

$$(\tau''_{xz})_{\substack{x=0\\y=0}} = -\frac{\nu}{1+\nu}\cdot\frac{P}{I}\cdot\frac{8b^2}{\pi^3}$$

$$\sum_{m=0}^{\infty}\sum_{n=1}^{\infty}\frac{(-1)^{m+n-1}}{(2m+1)\left[(2m+1)^2\left(\frac{b^2}{4a^2}\right)+n^2\right]}$$

$$(\tau''_{yz})_{\substack{x=0\\y=0}} = -\frac{\nu}{1+\nu}\cdot\frac{P}{I}\cdot\frac{8b^2}{\pi^3}$$

$$\sum_{m=0}^{\infty}\sum_{n=1}^{\infty}\frac{(-1)^{m-1}}{(2m+1)\left[(2m+1)^2\left(\frac{b^2}{4a^2}\right)+n^2\right]}$$

Now $$\sum_{n=1}^{\infty}\frac{1}{n^2}=\frac{\pi^2}{6}$$

$$\sum_{n=1}^{\infty}\frac{(-1)^n}{n^2}=-\frac{\pi^2}{12}$$

and $$\sum_{m=0}^{\infty}\frac{(-1)^m}{(2m+1)[(2m+1)^2+k^2]}=\frac{\pi^3}{32}\frac{\left[1-\operatorname{sech}\left(\frac{k\pi}{2}\right)\right]}{\frac{1}{2}\left(\frac{k\pi}{2}\right)^2}$$

$$\therefore\quad (\tau''_{xz})_{\substack{x=0\\y=0}} = -\frac{\nu}{1+\nu}\cdot\frac{3}{2}\frac{P}{A}\frac{b^2}{a^2}$$

$$\left[\frac{1}{3}+\frac{4}{\pi^2}\sum_{n=1}^{\infty}\frac{(-1)^n}{n^2\cosh\left(\frac{n\pi a}{b}\right)}\right] \quad \text{...(6.20 a)}$$

$$(\tau''_{yz})_{\substack{x=0\\y=0}} = \frac{\nu}{1+\nu}\cdot\frac{3}{2}\frac{P}{A}\frac{b^2}{a^2}\left[\frac{2}{3}-\frac{4}{\pi^2}\sum_{n=1}^{\infty}\frac{1}{n^2\cosh\left(\frac{n\pi a}{b}\right)}\right] \quad \text{...(6.20 b)}$$

where $A = 4ab$.

6.3 Unsymmetrical Bending

We now consider the pure bending of a beam where either (*i*) the beam does not have a symmetric cross-sectional area, or (*ii*) the beam has a symmetric cross-sectional area but the applied end couples are not in the plane of symmetry of the beam.

Consider a cross-section of arbitrary shape where the applied couple moment M_z is taken parallel to a principal axis z at the centroid as shown in Fig. 6.2 (*a*). If plane sections remain plane for small deformations then along some axis shown as *A-A* there will be no deformation, and at distance η normal to this axis, the stress will be given by :

$$\sigma_x = -\frac{E\eta}{R} \quad \text{...(1)}$$

where R is radius of curvature of the beam. Since there is no normal force acting on the cross-section, hence

$$\int \sigma_x dA = -\frac{E}{R}\iint \eta \, dA = 0 \quad \text{...(2)}$$

This indicates that axis *A-A* must be a centroidal axis, such as shown in Fig. 6.2 (*b*).

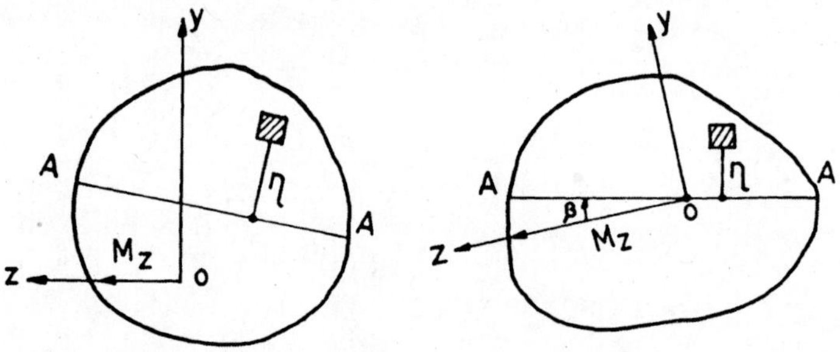

(*a*) (*b*)

Fig. 6.2

Now we take moments about the y and z axes

$$\left.\begin{aligned} \iint \sigma_x \, . \, ydA &= -\frac{E}{R}\iint \eta y dA = -M_z \\ \iint \sigma_x \, zdA &= -\frac{E}{R}\iint \eta z \, dA = 0 \end{aligned}\right\} \quad \text{...(3)}$$

Now $\quad \eta = y \cos \beta - z \sin \beta \quad$ ...(4)

$$\therefore \quad \left.\begin{aligned} M_z &= \frac{E}{R}\cos\beta \iint y^2 dA - \frac{E}{R}\sin\beta \iint yz\, dA \\ 0 &= -\frac{E}{R}\cos\beta \iint yz\, dA + \frac{E}{R}\sin\beta \iint z^2 dA \end{aligned}\right\} \quad ...(5)$$

Since the y and z-axes are principal axes, therefore, Eqs. (5) become,

$$\left.\begin{aligned} M_z &= \frac{E}{R}\cos\beta . I_{zz} \\ 0 &= \frac{E}{R}\sin\beta . I_{yy} \end{aligned}\right\} \quad ...(6)$$

where $I_{yy} = \iint z^2\, dA, \quad I_{zz} = \iint y^2 dA$

and $\quad I_{yz} = \iint yz dA = 0$

The second of Eqs. (6) can only be satisfied by taking $\beta = 0$.

Hence the neutral axis must coincide with the principal axis along which we have applied the bending moment. Hence from first of Eq. (6), we get by setting

$$\beta = 0$$

$$R = \frac{M_z}{EI_{zz}} \quad ...(6.21)$$

and Eq. (3) becomes,

$$\sigma_x = -\frac{M_z \eta}{I_{zz}} \quad ...(6.22)$$

Thus we conclude that if the applied couple M_z is parallel to a principal axis at the centroid of a cross-section, the neutral axis coincides with this principal axis and the flexure formula is valid with I_{zz} measured about this principal axis and η measured for this principal axis.

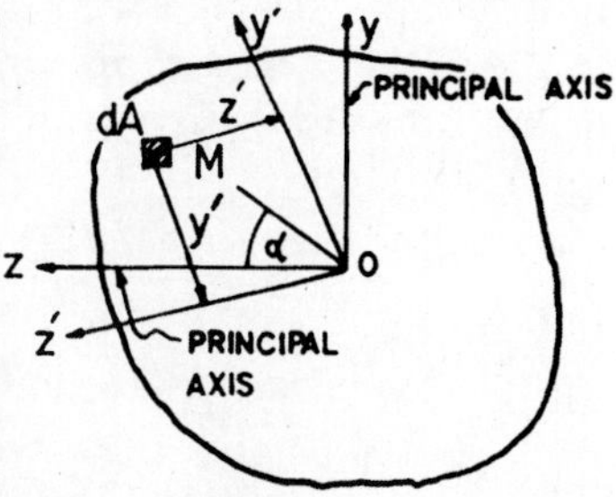

Fig. 6.3

If the applied couple is oriented at an inclination to the principal directions at the centroid at an angle α as shown in Fig. 6.3, then we can solve this problem by decomposing the couple moment into components along the principal directions. Then using the principal of superposition, we can employ the flexure formula for each component of M and the normal stress distribution can be computed as :

$$\sigma_x = \frac{(M\cos\alpha)\,y}{I_{zz}} + \frac{(M\sin\alpha)\,z}{I_{yy}} \qquad \text{...(6.23)}$$

When we have any set of axes $y'z'$ instead of principal axes yz, then since σ_x is a linear function of the co-ordinates as is obvious from Eq. (6.23), therefore, we assume

$$\sigma_x' = a + by' + cz' \qquad \text{...(7)}$$

where a, b, c are constants to be determined from the following equations :

$$\left.\begin{aligned} -M_z' &= \iint \sigma_x'\, y'dA \\ M_y' &= \iint \sigma_x'\, z'dA \\ 0 &= \iint \sigma_x'\, dA \end{aligned}\right\} \qquad \text{...(8)}$$

Substituting Eq. (6.22) in (6.23), we get

$$\left.\begin{aligned} -M_z' &= a\iint y'dA + b\iint (y')^2\, dA + c\iint y'z'dA \\ M_y' &= a\iint z'\,dA + b\iint y'z'dA + c\iint (z')^2\, dA \\ 0 &= a\iint dA = b\iint y'dA + c\iint z'\,dA \end{aligned}\right\} \qquad \text{...(9)}$$

Noting that y' and z' are centroidal axes, therefore, from third of Eqs. (6.24), we find that $a = 0$ and other equations give

$$\left.\begin{aligned} -M_z' &= b\, I_{z'z'} + c\, I_{y'z'} \\ M_y' &= b\, I_{y'z'} + c\, I_{y'y'} \end{aligned}\right\} \qquad \text{...(10)}$$

Solving for b and c, we get

$$\left.\begin{aligned} b &= -\frac{M_z'\, I_{y'y'} + M_y'\, I_{y'z'}}{I_{z'z'} I_{y'y'} - I^2_{y'z'}} \\ c &= \frac{M_y'\, I_{z'z'} + M_z'\, I_{y'z'}}{I_{z'z'}\, I_{y'y'} - I^2_{y'z'}} \end{aligned}\right\} \qquad \text{...(11)}$$

Since y' and z' are any set of axes, therefore, dropping the prime notations our result is valid for any set of axes yz at the centroid. Thus

$$\sigma_x = \frac{(M_y I_{zz} + M_z I_{yz})z - (M_z I_{yy} + M_y I_{yz})y}{I_{zz} I_{yy} - I^2_{yz}} \qquad \text{...(6.24)}$$

This is called the generalized flexure formula.

Let ϕ be the angle between the plane of the loads and the z-axis ; ϕ is measured positive counter clockwise

$$M_z = M\sin\phi,\quad M_y = -M\cos\phi,\quad \cot\phi = -\frac{M_y}{M_z} \qquad \text{...(6.25)}$$

Neutral axis :

For $\sigma_x = 0$

$$\tan\alpha = \frac{M_z I_{yz} + M_y I_z}{M_z I_y + M_y I_{yz}}$$

$$= \frac{I_{yz} - I_z \cot\phi}{I_y - I_{yz}\cot\phi} \qquad ...(6.26)$$

Now σ_x can be written as :

$$\sigma_x = \frac{M_z(y - z\tan\alpha)}{I_z - I_{yz}\tan\alpha} \qquad ...(6.27)$$

Example 6.1 *Determine the maximum stress in the cantilever beam shown in Fig. 6.4.*

Solution. $M_z = -(1000\cos 30°) \times 5$

$$= -1000 + \frac{\sqrt{3}}{2} \times 5 = -4330 \text{ N-m}$$

$$M_y = -(1000\sin 30°) \times 5$$

$$= -1000 \times \tfrac{1}{2} \times 5 = -250 \text{ N-m}$$

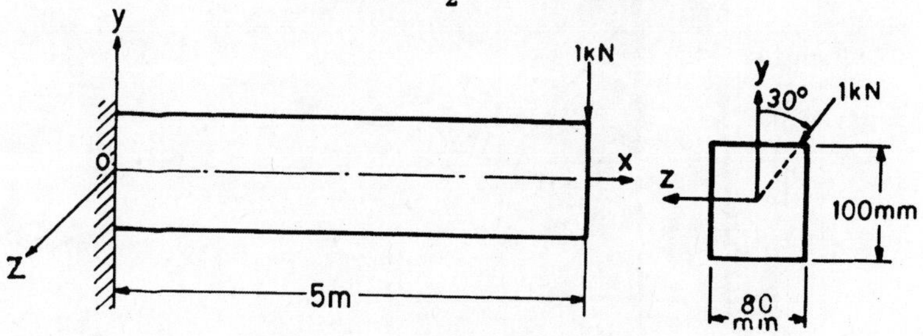

(a) (b)

Fig. 6.4

$$I_{yy} = \frac{100 \times 80^3}{12} = 42.67 \times 10^5 \text{ mm}^4$$

$$I_{zz} = \frac{80 \times 100^3}{12} = 66.67 \times 10^5 \text{ mm}^4$$

$$I_{yz} = 0$$

$$\sigma_x = \frac{M_y \,.\, z}{I_{yy}} - \frac{M_z \,.\, y}{I_{zz}}$$

$$= \frac{-2500 \times 10^3 \times z}{42.67 \times 10^5} + \frac{4330 \times 10^3 \times y}{66.67 \times 10^5}$$

$$= -58.593 \times 10^{-2} \times z + 64.95 \times 10^{-2} \times y$$

The maximum tensile stress occurs when $z = -50$ mm, $y =$ 50 mm.

$$(\sigma_x)_{max} = 58.593 \times 10^{-2} \times 40 \times 64.95 \times 10^{-2} \times 50$$
$$= 23.4372 + 32.475 = 55.9122 \text{ N/mm}^2$$

Example 6.2 *An 150 mm × 100 mm × 12 mm unequal angle bar is placed with the long leg vertical and used as a beam supported at each end, the span being 3 m. If a load of 6500 N is placed at the mid length of the bar, determine the maximum stress due to bending.*

Solution. From Fig. 6.5

$$A_1 = 150 \times 12 = 1800 \text{ mm}^2$$
$$A_2 = 88 \times 12 = 1056 \text{ mm}^2$$
$$\therefore \quad A = A_1 + A_2 = 2856 \text{ mm}^2$$

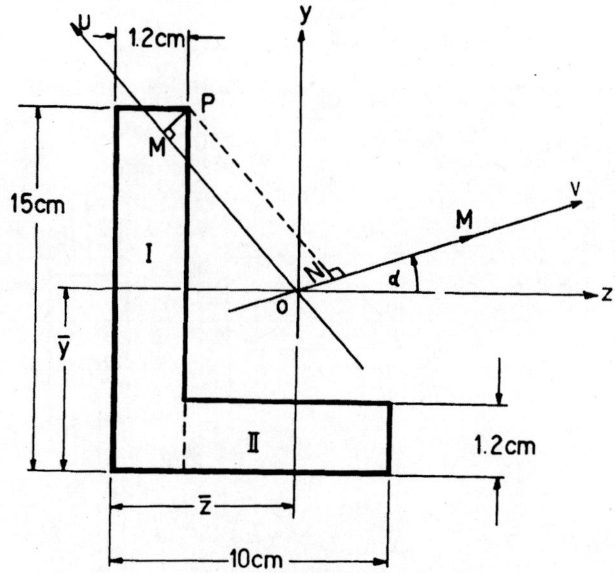

Fig. 6.5

$$\bar{z} = \frac{1800 \times 6 + 1056 \times 56}{2856} = 24.5 \text{ mm}$$

$$\bar{y} = \frac{1800 \times 75 + 1056 \times 6}{2856} = 495 \text{ mm.}$$

$$I_{zz} = \left[\frac{88 \times 12^3}{12} + 1056 \times 43.5^2\right]$$
$$+ \left[\frac{12 \times 150^3}{12} + 1800 \times 25.5^2\right]$$

$$= 656 \times 10^4 \text{ mm}^4$$

$$I_{yy} = \left[\frac{1.2 \times 88^3}{12} + 1056 \times 31.5^2\right] + \left[\frac{150 \times 12^3}{12} + 1800 \times 18.5^2\right]$$

$$= 237 \times 10^4 \text{ mm}^4$$

$$I_{yz} = 1056 \times 31.5 \times (-43.5) + 1800 \times (-18.5) \times 25.5$$

$$= -230 \times 10^4 \text{ mm}^4$$

$$M = \frac{6500 \times 3}{4} = 4875 \text{ N-m}$$

$$\tan 2\alpha = \frac{2I_{yz}}{I_z - I_y} = \frac{2 \times 230}{656 - 237} = 1.098$$

$$2\alpha = 47° \, 42'$$

$$\therefore \quad \alpha = 23° \, 51'.$$

$$M_z = M \cos\alpha = 4875 \cos(23° \, 51')$$

$$= 4875 \times 0.9145 = 4458 \text{ N-m}$$

$$M_y = M \sin\alpha = 4875 \sin(23° \, 51')$$

$$= 4875 \times 0.4043 = 1971 \text{ N-m}.$$

The stress at the point P will be maximum, whose co-ordinates are

$$v = 100.5 \sin\alpha - 12.5 \cos\alpha = 29.2 \text{ mm}$$

$$u = 100.5 \cos\alpha + 12.5 \sin\alpha = 97 \text{ mm}$$

$$I_u + I_v = I_y + I_z$$

$$I_v - I_u = (I_z - I_y) \sec 2\alpha$$

$$= (656 - 237) \times 10^4 \sec(47° \, 42')$$

$$= 419 \times 1.4859 \times 10^4 = 623 \times 10^4 \text{ mm}^4$$

$$\therefore \quad I_u = 135 \times 10^4 \text{ mm}^4, I_y = 758 \times 10^4 \text{ mm}^4$$

$$\sigma_x = \frac{M \cos\alpha}{I_v} \cdot u + \frac{M \sin\alpha}{I_u} \cdot v$$

$$= \frac{M_z}{I_y} \cdot u + \frac{M_y}{I_u} \cdot v$$

$$= \frac{4458 \times 10^3}{758 \times 10^4} \times 97 + \frac{1971 \times 10^3}{135 \times 10^4} \times 29.2$$

$$= 57 + 42.7 = 99.7 \text{ N/mm}^2$$

Example 6.3 *A 150 mm × 150 mm slender angle of 20 mm thickness is subjected to oppositely directed end couples M_z = 11 kN.m, at the centroid of the cross- section. What bending stresses exist at paints A and B on a section away from the ends (Fig. 6.6) ? Determine the orientation of the neutral axis.*

Solution.

Location of the centroid C :

$$A_1 = 130 \times 20 = 2600 \text{ mm}^2, A_2 = 150 \times 20 = 3000 \text{ mm}^2$$

$$A = A_1 + A_2 = 5600 \text{ mm}^2$$

$$5600 \times \bar{z} = 2600 \times 10 + 3000 \times 75$$

$$\bar{z} = \frac{251000}{5600} = 45 \text{ mm}$$

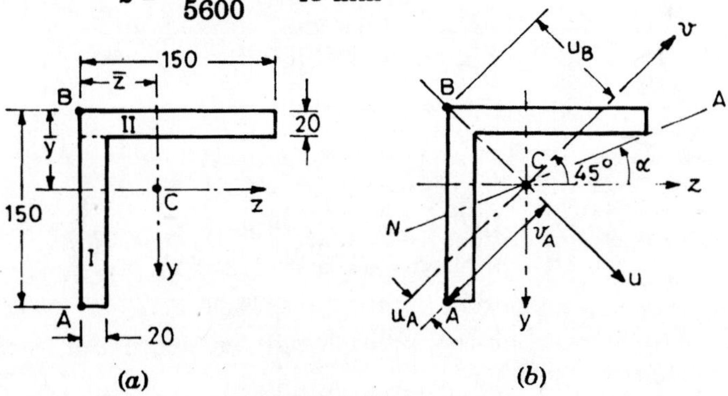

Fig. 6.6

Since the cross-section is symmetrical, $\bar{z} = \bar{y}$

Moments and product of inertia :

$$I_y = I_z = \left[\frac{20 \times 130^3}{12} + 2600 \times (85 - 45)^2 + \frac{150 \times 20^3}{12} + 3000 \times (45 - 10)^2 \right] \times 10^{-12}$$

$$= 11.596 \times 10^{-6} \text{ m}^4$$

$$I_{yz} = 10^{-12}[0 + 130 \times 20 \times 40 \times (-35) + 0 + 150 \times 20 \times (-35) \times 30]$$

$$= -6.79 \times 10^{-6} \text{ m}^4$$

Directions of principal axes and the principal moments of inertia :

$$\tan 2\theta = \frac{-2(-6.79)}{11.596 - 11.596} = \infty$$

$$\theta = 45° \text{ and } 135°.$$

$$I_u = I_v = \frac{I_y - I_z}{2} + \frac{I_y - I_z}{2}\cos 2\theta - I_{yz}\sin 2\theta$$

$$= (11.596 + 0 + 6.79) \times 10^{-6} = 18.386 \times 10^{-6}\ \text{m}^4$$

$$I_{1,2} = \frac{I_y + I_z}{2} \pm \frac{1}{2}\sqrt{(I_y - I_z)^2 + 4I_{yz}^2}$$

$$= (11.596 \pm 6.79) \times 10^{-6}$$

$$= 18.386 \times 10^{-6} \text{ and } 4.806 \times 10^{-6}\ \text{m}^4$$

Stresses :

$$\sigma_A = \frac{M_z(I_{yz}z_A - I_y y_A)}{I_y I_z - I_{yz}^2}$$

$$= \frac{11 \times 10^3\,[(-6.79)(-0.045) - (11.596)(0.105)]}{[(11.596)^2 - (-6.79)^2] \times 10^{-6}}$$

$$= -114\ \text{MPa}$$

$$\sigma_B = \frac{11 \times 10^3\,[(-6.79)(-0.045) - (11.596)(-0.045)]}{[(11.596)^2 - (-6.79)^2] \times 10^{-6}}$$

$$= 103\ \text{MPa}$$

Neutral axis :

$$(M_z I_y)y - (M_z I_{yz})z = 0$$

$$11.596y + 6.79z = 0$$

$$\tan\alpha = \frac{y}{z} = -\frac{6.79}{11.596} = -0.585547$$

$$\alpha = -30.35°$$

Example 6.4 *An angle section beam simply supported over a span of 5 m is shown in Fig. 6.7. It is subjected to a concentrated load of 4 kN at a distance of 2 m from left hand support. The plane of load makes an angle of $\frac{2\pi}{3}$ rad with the z-axis and passes through the shear centre. Determine the maximum tensile and compressive stresses in the section.*

Solution.

$$A = 80 \times 10 + 110 \times 10 = 1900\ \text{mm}^2$$

$$\bar{z} = \frac{800 \times 40 + 1100 \times 5}{1900} = 19.74\ \text{mm}$$

$$\bar{y} = \frac{800 \times 5 + 1100 \times 65}{1900} = 39.74\ \text{mm}$$

$$I_y = \frac{10 \times 80^3}{12} + 800 \times (40 - 19.74)^2 + \frac{110 \times 10^3}{12} + 1100 \times (19.74 - 5)^2$$

$$= 1.003 \times 10^6\ \text{mm}^4$$

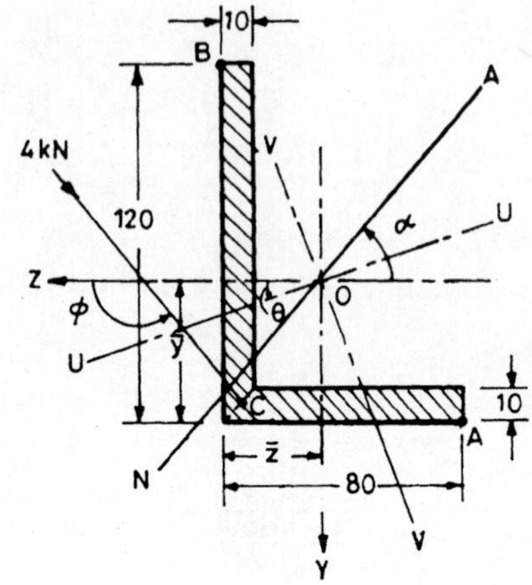

Fig. 6.7

$$I_z = \frac{80 \times 10^3}{12} + 800 \times (39.74 - 5)^2 + \frac{10 \times 110^3}{12} + 1100(65 - 39.74)^2$$

$$= 2.783 \times 10^6 \text{ mm}^4$$

$$I_{yz} = 800 \times (-34.74)(+20.26) + 1100 \times (-14.74)(65 - 39.74)$$

$$= -0.973 \times 10^6 \text{ mm}^4$$

$$\tan \alpha = \frac{I_{yz} - I_z \cot \phi}{I_y - I_{yz} \cot \phi}$$

$$= \frac{-0.973 - 2.783 \times (-0.5774)}{1.003 - (-0.973)(-0.5774)} = 1.4368$$

$$\alpha = 0.9628 \text{ rad}$$

$$M = 1.2 \times 4 = 4.8 \text{ kN.m}$$

$$M_z = M \sin \phi = 4.8 \sin \left(\frac{2\pi}{3}\right) = 4.157 \text{ kN.m}$$

$$\sigma_A = \frac{M_z(y_A - z_A \tan \alpha)}{I_z - I_{yz} \tan \alpha}$$

$$= \frac{4.157 \times 10^3 [39.74 - (-60.26)(1.4368)] \times 10^{-3}}{[2.783 \times 10^6 - (-0.973 \times 10^6)(1.4368)] \times 10^{-12}}$$

$$= 125.6 \text{ MPa}$$

$$\sigma_B = \frac{4.157 \times 10^3 [-80.26 - 19.74 \times 1.4368] \times 10^{-3}}{[2.783 \times 10^6 - (-0.973 \times 10^6)(-0.5801)] \times 10^{-12}}$$

$$= -108 \text{ MPa}$$

6.4 Shear Centre

When a beam has unsymmetrical cross-section and if the load passes through the centroid of the beam cross-section, then there is twisting in addition to bending of the beam. In order to avoid twisting of the beam and to have pure bending, the load has to be applied through some appropriate point. This point is called the shear centre. The shear centre for any transverse section of a beam is defined as the point of intersection of the bending axis and the plane of the transverse section.

For thin sections with no shear stresses acting on the lateral boundary in the form of surface tractions we know that the shear stresses at a cross-section of the beam must be tangent to the boundaries at the boundaries. And since the section is thin, we can assume with reasonable accuracy that the shear stress is uniform along lines normal to the boundaries and that the direction of the shear stress is that of the centre line of the section.

In order to determine the shear centre for a given cross-section, we assume that there is no torsion transmitted and we set forth a shear stress distribution for bending without torsion resulting from an applied shear force whose direction but not whose line of action is specified. Next we find the line of action of the resultant of this shear stress distribution. For no twisting, the applied shear force at the section must have this very same line of action. Thus we establish the proper line of action for an applied shear force in a particular direction at a section as a requirement for bending without twisting. We then do the same for an applied shear force in a second direction at right angles to the first. The intersection of these two required lines of action is called the shear centre. Since any applied shear force can be decomposed into orthogonal components and since we can superpose the stress distribution associated with each applied shear force component, we can then conclude that for bending without twisting it is necessary that the total applied shear force at each section must pass through the shear centre. It may be pointed out that shear centre is a property of the geometry of the section and not of the loading.

6.4.1 Shear Stress Distribution in Thin-walled Sections

Consider a thin-walled open cross-section, as shown in Fig. 6.8. The thickness at any section is t and s gives the position along the centreline of the section. Let τ_{xs} be the shear stress distribution in the section.

We now draw the free-body diagram of an infinitesimal portion of the beam as shown in Fig. 6.9.

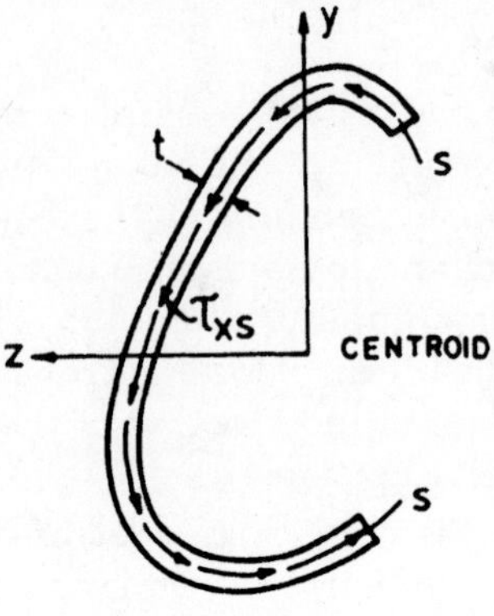

Fig. 6.8

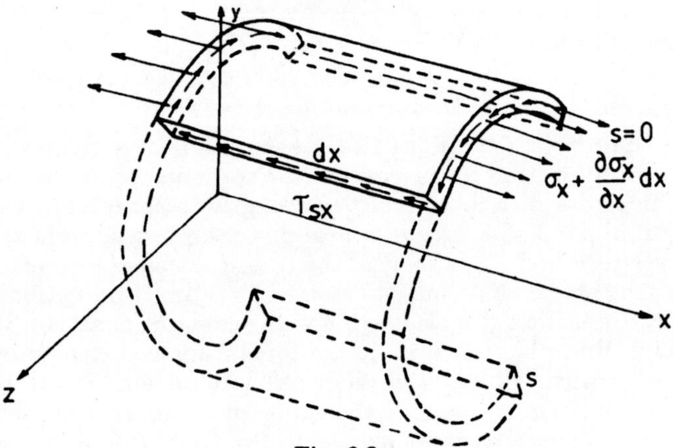

Fig. 6.9

$$-\tau_{sx}\, tdx - \int_0^s \sigma_x tds + \int_0^s \left(\sigma_x + \frac{\partial \sigma_x}{\partial x} dx \right) tds = 0$$

Cancelling terms and solving for τ_{sx}, we get

$$\tau_{sx} = \frac{1}{t} \int_0^s \frac{\partial \sigma_x}{\partial x} tds \qquad \text{...(1)}$$

From Eq. (6.24) by setting $M_y = 0$, we get

$$\sigma_x = \frac{M_z I_{yz} \cdot z - M_z I_{yy} \cdot y}{I_{yy} I_{zz} - I_{yz}^2}$$

$$\therefore \qquad \tau_{sx} = \frac{1}{t} \int_0^s \frac{\partial}{\partial x} \left[\frac{z M_z I_{yz} - y M_z I_{yy}}{I_{yy} I_{zz} - I_{yz}^2} \right] tds \qquad \text{...(2)}$$

Now $\dfrac{\partial M_z}{\partial x} = -V_y$, where V is the shear force.

$$\therefore \qquad \tau_{sx} = \frac{V_y}{t(I_{yy}I_{zz} - I_{yz}^2)}\left[I_{yy}\int_0^s ytds - I_{yz}\int_0^s ztds\right] \qquad ...(3)$$

The integrals in Eq. (3) represent first moments of that portion of the cross-section between $s = 0$ and s, where these moments are taken about the z and y axes respectively.

Now consider for the same beam a hypothetical coplanar load parallel to the xz plane and oriented normal to the x-axis. This would induce a shear force V_z and a bending moment M_y. Then

$$\tau_{sx} = \frac{V_z}{t(I_{yy}I_{zz} - I_{yz}^2)}\left[I_{zz}\int_0^s ztdz - I_{yz}\int_0^s ytds\right] \qquad ...(4)$$

For the case of loading parallel to the yx plane, the resultant of the shear stress distribution must have an equal but opposite moment about the x-axis as that of the applied shear force as shown in Fig. 6.10.

$$-(-V_y)z = \int \bar{r}d\bar{F}$$

$$= \int_0^{s_f} \bar{r} \times [\tau_{sx}(stds)] \qquad ...(5)$$

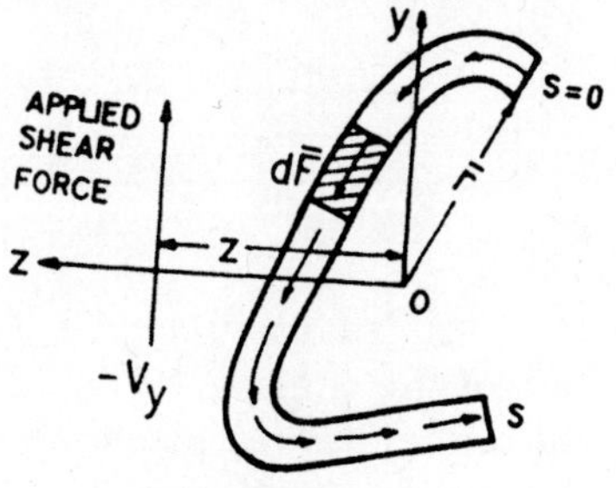

Fig. 6.10

where s_f is the extreme limit of s so as to cover the entire cross-section.

Substituting for τ_{sx} from Eq. (6.3), we find that

$$z = \int_0^{s_f}\left[\bar{r} \times s \cdot \frac{1}{t(I_{yy}I_{zz} - I^2_{yz})}\left\{I_{yy}\int_0^s ytds - I_{yz}\int_0^s ztds\right\}\right]ds$$

...(6.28)

By a similar procedure we compute y giving the line of action of V_z for the case with coplanar loading parallel to the xz plane. We get

$$y = \int_0^{s_f}\left[\bar{r} \times s\,\frac{1}{t(I_{yy}I_{zz} - I^2_{yz})} \times \left\{I_{zz}\int_0^s ztds - I_{yz}\int_0^s ytds\right\}\right]ds$$

...(6.29)

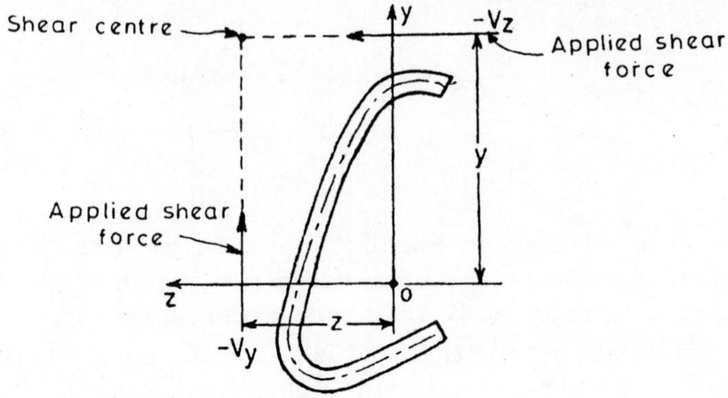

Fig. 6.11

We have shown the lines of action of V_y and V_z in Fig. 6.11. The point of intersection of these forces is the shear centre.

Example 6.5 *Determine the shear centre of the beam shown in Fig. 6.12.*

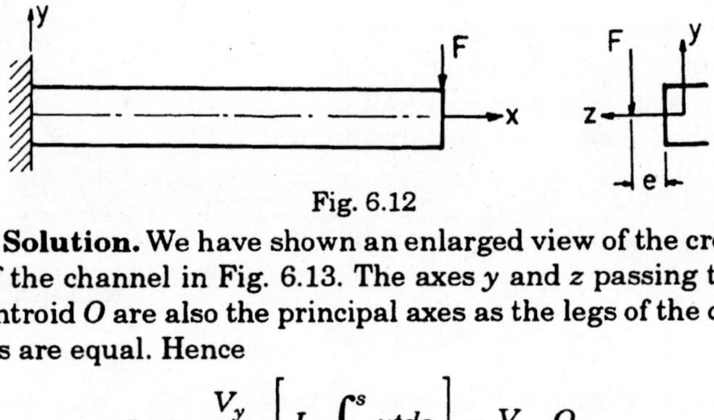

Fig. 6.12

Solution. We have shown an enlarged view of the cross-section of the channel in Fig. 6.13. The axes y and z passing through the centroid O are also the principal axes as the legs of the channel flanges are equal. Hence

$$\tau_{xy} = \frac{V_y}{l_{xy} I_{zz}} \left[I_{yy} \int_0^s ytds \right] = V_y \,.\, Q_z$$

For the upper flange, $(\tau_{xs})_{upper} = \dfrac{V_y \,.\, s\,(h - t_1)}{2I_{zz}}$

From Fig. 6.13, we have

$$I_{zz} = \frac{1}{12} t_2 (h - 2t_1)^3 + 2\left[\frac{bt_1^{\,3}}{12} + bt_1 \left(\frac{h}{2} - \frac{t_1}{2} \right)^2 \right]$$

$$Q_z = bt_1 \left(\frac{h - t_1}{2} \right) + t_2 \left(\frac{h}{2} - t_1 - y \right) \frac{\left[\frac{h}{2} - t_1 + y \right]}{2}$$

$$\therefore \quad (\tau_{xs})_{web} = \frac{V_y}{I_{zz} t_2} \left[\frac{bt_1}{2} (h - t_1) + \frac{t_2}{2} \left(\left\{ \frac{h}{2} - t_1 \right)^2 - y^2 \right\} \right]$$

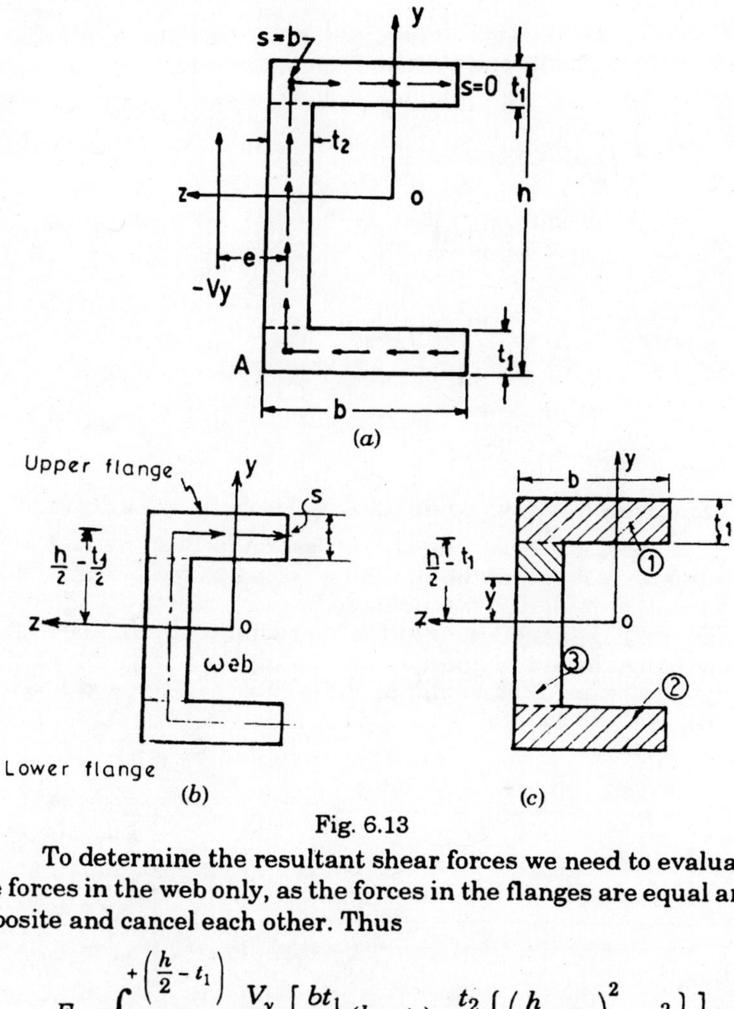

Fig. 6.13

To determine the resultant shear forces we need to evaluate the forces in the web only, as the forces in the flanges are equal and opposite and cancel each other. Thus

$$F_R = \int_{-\left(\frac{h}{2}-t_1\right)}^{+\left(\frac{h}{2}-t_1\right)} \frac{V_y}{I_{zz}t_2}\left[\frac{bt_1}{2}(h-t_1)+\frac{t_2}{2}\left\{\left(\frac{h}{2}-t_1\right)^2-y^2\right\}\right]t_2dy$$

$$= \frac{V_y}{I_{zz}}\left[bt_1(h-t_1)\left(\frac{h}{2}-t_1\right)+\frac{1}{12}t_2(h-2t_1)^3\right]$$

The bracketed term is approximately equal to I_{zz}. Hence

$$F_R = V_y$$

To get the distance e, we may measure $\bar{r}$ from A instead of O. Now $\bar{r}\times\bar{s}=0$ for the web and the lower flange and $\bar{r}\times\bar{s}=(h-t_1)$ for the upper flange, we get

$$e = \int_0^b \left[(h - t_1) \frac{1}{I_{zz}} \left\{ \int_0^s ytds \right\} \right] ds$$

Now $\int_0^s ytds = \frac{1}{2}(h - t_1)\, t_1 s$

$$\therefore \qquad e = \frac{(h - t_1)^2}{2I_{zz}} t_1 \int_0^b sds = \frac{(h - t_1)^2 t_1 b^2}{4I_{zz}}$$

Also $\quad I_{zz} \simeq \frac{bt_1h^2}{2} + \frac{1}{12} t_2 h^3$

$$\therefore \qquad e = \frac{t_1 b^2}{2bt_1 + \frac{t_2 h}{3}}$$

6.5 Solution of Bending of Bars by Harmonic Functions

Consider the bending of cantilever as solved in Art. 6.2 with the difference now that the cantilever is loaded by some distribution of forces that is statically equivalent to a single force (P_x, P_y, 0) lying in the plane $z = l$ and acting at the load point (x_o, y_o. l). The z-axis is taken along the centre line of the bar, while the x and y axes are any orthogonal axes intersecting at the centroid of the end x = 0, as shown in Fig. 6.14.

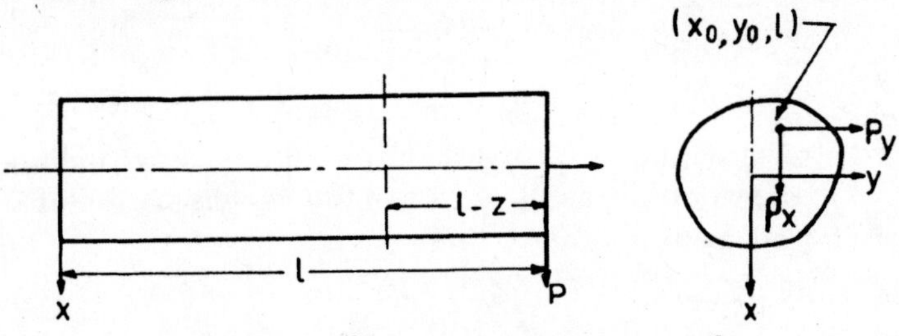

(a) (b)

Fig. 6.14

Following the same procedure as in Art. 6.2, we find

$$\sigma_x = \sigma_y = \tau_{xy} = 0 \qquad \text{...(1)}$$

$$M_y = P_x(l - z) \qquad \text{...(2)}$$

$$\left.\begin{aligned} \sigma_z &= -\frac{M_y}{I_y} x \\ \text{where} \quad I_y &= \iint_R x^2 dxdy \end{aligned}\right\} \qquad \text{...(3)}$$

We assume, $\sigma_z = -E(l-z)(K_x x + K_y y)$ ...(4)

where the constants K_x and K_y are to be determined from the conditions

$$\iint_R \tau_{zx}\, dx\, dy = P_x, \qquad \iint_R \tau_{yz}\, dx\, dy = P_y \qquad ...(5)$$

Substituting from Eqs. (1) and (4) in equilibrium equations, we get

$$\left.\begin{aligned} \frac{\partial \tau_{zx}}{\partial z} = 0, \frac{\partial \tau_{zy}}{\partial z} = 0 \\ \frac{\partial \tau_{zx}}{\partial x} + \frac{\partial \tau_{zy}}{\partial y} + E(K_x x + K_y y) = 0 \end{aligned}\right\} \qquad ...(6)$$

It follows from the first two of Eqs. (6) that the shear components τ_{zx} and τ_{zy} have the same value in all cross-sections of the beam, while the third equation can be arranged as :

$$\frac{\partial}{\partial x}\left(\tau_{zx} + \frac{1}{2} E\, K_x x^2\right) + \frac{\partial}{\partial y}\left(\tau_{zy} + \frac{1}{2} E\, K_y y^2\right) = 0$$

This equation is of the form,

$$\frac{\partial}{\partial x}\left(\frac{\partial F}{\partial y}\right) + \frac{\partial}{\partial y}\left(-\frac{\partial F}{\partial x}\right) = 0$$

Therefore, there exists a function $F(x, y)$, such that

$$\left.\begin{aligned} \tau_{xz} &= \frac{\partial F}{\partial y} - \frac{1}{2} E\, K_x x^2 \\ \tau_{zy} &= -\frac{\partial F}{\partial x} - \frac{1}{2} E\, K_y y^2 \end{aligned}\right\} \qquad ...(7)$$

The conditions to be satisfied by the function $F(x, y)$ can be determined from the Beltrami-Michell compatibility equations (1.22), which reduces in this case to

$$\nabla^2 \tau_{zx} + \frac{EK_x}{1+\nu} = 0, \quad \nabla^2 \tau_{zy} + \frac{EK_y}{1+\nu} = 0$$

Substituting Eqs. (7), we get

$$\frac{\partial}{\partial y}(\nabla^2 F) = 2G\nu K_x, \quad \frac{\partial}{\partial x}(\nabla^2 F) = -2G\nu\, K_y$$

From which it follows that,

$$\nabla^2 F(x, y) = -2G\nu K_y x + 2G\nu K_x y - 2G\theta \qquad ...(8)$$

where $-2G\theta$ is the constant of integration.

Solution of Eq. (8) is

$$F(x, y) = f(x, y) - \tfrac{1}{3}G\nu(K_y x^3 - K_x y^3) - \tfrac{1}{2}G\theta(x^2 + y^2) \qquad ...(9)$$

where $f(x, y)$ is a harmonic function.

If $g(x, y)$ is the harmonic conjugate of $f(x, y)$ then

$$\frac{\partial f}{\partial x} = \frac{\partial g}{\partial y}, \quad \frac{\partial f}{\partial y} = -\frac{\partial g}{\partial x} \qquad \text{...(10)}$$

Eq. (7) can then be written as :

$$\left.\begin{aligned} \tau_{zx} &= -\frac{\partial g}{\partial x} - G\theta y + G\nu K_x y^2 - \frac{1}{2} E K_x x^2 \\ \tau_{zy} &= -\frac{\partial g}{\partial y} + G\theta x + G\nu K_y x^2 - \frac{1}{2} E K_y y^2 \end{aligned}\right\} \qquad \text{...(11)}$$

Now the rotation of each element of area of cross-section in its own plane is

$$\omega_z = \frac{1}{2}\left(\frac{\partial u_y}{\partial x} - \frac{\partial u_x}{\partial y}\right)$$

The local twist at a point (x, y) of the cross-section is defined as :

$$\begin{aligned} \frac{\partial \omega_z}{\partial z} &= \frac{1}{2}\left(\frac{\partial^2 u_y}{\partial z \partial x} - \frac{\partial^2 u_x}{\partial z \partial y}\right) \\ &= \frac{1}{2}\left(\frac{\partial \gamma_{zy}}{\partial x} - \frac{\partial \gamma_{zx}}{\partial y}\right) \\ &= \frac{1}{2G}\left(\frac{\partial \tau_{zy}}{\partial x} - \frac{\partial \tau_{zx}}{\partial y}\right) \end{aligned}$$

Substituting the values of shear stresses from Eq. (11), one gets

$$\frac{\partial \omega_z}{\partial z} = \theta + \nu\,(K_y x - K_x y)$$

The mean value of the local twist over the section is just the constant θ. Thus the terms in Eq. (11) that involve θ represent a twist of the beam.

Now for pure torsion

$$\tau_{xz} = G\theta\left(\frac{\partial \varphi}{\partial x} - y\right), \quad \tau_{zy} = G\theta\left(\frac{\partial \varphi}{\partial y} + x\right)$$

Thus, we introduce the torsion function $\varphi\,(x, y)$, by writing

$$g(x, y) = -G\theta\varphi(x, y) - G[K_x\varphi_1(x, y) + K_y\varphi_2(x, y)] \qquad \text{...(12)}$$

where $\varphi(x, y)$, $\varphi_1(x, y)$ and $\varphi_2(x, y)$ are harmonic functions. Thus, we can write

$$\left.\begin{aligned} \tau_{zx} &= G\theta\left(\frac{\partial \varphi}{\partial x} - y\right) + GK_x\left[\frac{\partial \varphi_1}{\partial x} - x^2 - \nu(x^2 - y^2)\right] + GK_y\frac{\partial \varphi_2}{\partial x} \\ \tau_{zy} &= G\theta\left(\frac{\partial \varphi}{\partial y} + x\right) + GK_y\left[\frac{\partial \varphi_2}{\partial y} - y^2 - \nu(y^2 - x^2)\right] + GK_x\frac{\partial \varphi_1}{\partial y} \end{aligned}\right\} \qquad \text{...(6.30)}$$

The boundary conditions for the functions φ_1 and φ_2 may be derived from the relation,

$$\tau_{zx}.a_{nx} + \tau_{zy}.a_{ny} = 0 \quad ...(13)$$

Which expresses the vanishing of the external forces on the lateral surfaces of the bar. Now boundary condition on the torsion function φ is,

$$\frac{d\varphi}{dn} = y.a_{nx} - x.a_{ny} \quad ...(14)$$

Substituting Eq. (6.30) in (13) and using (14), we get

$$K_x \frac{\partial \varphi_1}{\partial n} + K_y \frac{\partial \varphi_2}{\partial n} = K_x[(1+\nu)x^2 - \nu y^2]a_{nx}$$

$$+ K_y[(1+\nu)y^2 - \nu x^2]a_{ny} \text{ on } C,$$

and this will be satisfied if the functions φ_1 and φ_2 are subject to the conditions

$$\left.\begin{aligned} \frac{d\varphi_1}{dn} &= [(1+\nu)x^2 - \nu y^2]\, a_{nx} \text{ on } C \\ \frac{d\varphi_2}{dn} &= [(1+\nu)y^2 - \nu x^2]\, a_{ny} \text{ on } C \end{aligned}\right\} \quad ...(6.31)$$

The flexure problem has thus been reduced to the determination of three functions, harmonic within the region R of the cross-section, whose normal derivatives are prescribed on the boundary C. Thus, we are led to the problem of Neumann.

In considering the torsion of a beam by couples, the solution could be made to depend upon either a problem of Neumann, *i.e.* the problem of finding a function $\varphi(x, y)$, harmonic in R and such that

$$\frac{d\varphi}{dn} = y\, a_{nx} - x\, a_{ny} \text{ on } C$$

or upon a problem of Dirichlet, with

$$\psi = \frac{1}{2}(x^2 + y^2) + \text{const. on } C.$$

The torsion functions φ and ψ are harmonic conjugates, so that $\varphi + i\psi$ is an analytic function of $x + iy$. The flexure problem may also be reduced to a problem of Dirichlet by introducing the harmonic functions ψ_1, ψ_2 conjugate to φ_1 and φ_2 respectively. Then

$$\frac{\partial \varphi_i}{\partial x} = \frac{\partial \psi_i}{\partial y}, \; \frac{\partial \varphi_i}{\partial y} = -\frac{\partial \psi_i}{\partial x}, \; \frac{\partial \varphi_i}{dn} = \frac{d\psi_i}{ds} \; (i = 1, 2)$$

and the boundary conditions (6.50) can be written as :

$$\frac{d\psi_1}{ds} = [(1+\nu)x^2 - \nu y^2]\frac{dy}{ds}$$

$$\frac{d\psi_2}{ds} = -[(1+\nu)y^2 - \nu x^2]\frac{dx}{ds}$$

or
$$\left.\begin{aligned}\psi_1 &= -\frac{1}{3}\nu y^3 + (1+\nu)\int_{(x_0,y_0)}^{(x,y)} x^2 dy + \text{const. on } C\\ \psi_2 &= \frac{1}{3}\nu x^3 - (1+\nu)\int_{(x_0,y_0)}^{(x,y)} y^2 dx + \text{const. on } C\end{aligned}\right\} \quad \ldots(15)$$

Now $P_x = \iint_R \tau_{zx}\, dx\, dy$

Substituting from Eq. (6.30), we get

$$\left.\begin{aligned}P_x &= G\theta\iint_R \frac{\partial\varphi}{\partial x}dx\,dy + GK_x\iint\frac{\partial\varphi_1}{\partial x}dx\,dy\\ &\quad + GK_y\iint_R\frac{\partial\varphi_2}{\partial x}dx\,dy\\ &\quad + GK_x[-(1+\nu)I_y + \nu I_x]\end{aligned}\right\} \quad \ldots(16)$$

where $I_x = \iint_R y^2 dx\, dy,\ \ I_y = \iint_R x^2\, dx\, dy$

Now if Φ is any harmonic function then

$$\iint_R \frac{\partial\Phi}{\partial x}dx\,dy = \int_C x\frac{\partial\Phi}{\partial n}ds.$$

With the aid of this identity and boundary conditions (14) and (6.31), Eq. (16) becomes

$$P_x = G\theta\int_C (xy\,dy + x^2dx) + GK_x\int_C[(1+\nu)x^3 - \nu xy^2]dy$$
$$+ GK_y\int_C[-(1+\nu)xy^2 + \nu x^3]\,dx + GK_x[-(1+\nu)I_y + \nu I_x]$$

Upon applying Green's theorem and putting

$$E = 2G(1+\nu)$$

the above equation becomes

$$P_x = E[K_x I_y + K_y I_{xy}] \quad \ldots(17)$$

where $I_{xy} = \iint_R xy\, dx\, dy$

Similarly from $P_y = \iint_R \tau_{zy}\, dx\, dy$

it follows that

$$P_y = E[K_y I_x + K_x I_{xy}] \qquad \text{...(18)}$$

Eqs. (17) and (18) can be solved for K_x and K_y to give

$$\left.\begin{aligned} EK_x &= \frac{I_y P_x - I_{xy} P_y}{I_x I_y - I_{xy}^2} \\ EK_y &= \frac{I_x P_y - I_{xy} P_x}{I_x I_y - I_{xy}^2} \end{aligned}\right\} \qquad \text{...(6.32)}$$

Substituting these values in Eq. (4) we check that

$$M_x = \iint_R y\sigma_z\, dx\, dy = -(l - z)\, P_y$$

$$M_y = \iint_R -x\sigma_z\, dx\, dy = (l - z)\, P_x$$

$$\sigma_z = -(l - z)\left[\frac{I_y P_x \cdot x - I_{xy} P_y \cdot x + I_x P_y \cdot y - I_{xy} P_x \cdot y}{I_x I_y - I_{xy}^2}\right]$$

and $\iint_R \sigma_z\, dx\, dy = P_x$, $\iint_R \tau_{zy}\, dx\, dy = P_y$, $\iint_R \sigma_z\, dx\, dy = 0.$

The constant θ in Eq. (6.30) is determined from the condition that the twisting moment M_z be such that

$$\iint_R (x\tau_{yz} - y\tau_{xz})\, dx\, dy = x_0 P_y - y_0 P_x \qquad \text{...(19)}$$

where $(x_0 : y_0)$ are the co-ordinates of the load point.

Thus the solution of general St. Venant flexure problem is reduced to the determination of harmonic functions φ, φ_1 and φ_2 that satisfy the boundary conditions (14) and (6.31). The boundary condition (6.31) is somewhat difficult to satisfy. However, this condition can be simplified by introducing the idea of centre of flexure.

The Eqs. (6.30) suggest that the general flexure problem can be resolved into the following simpler problems :

(*i*) A flexure problem in which the mean local twist θ is set equal to zero. The position of the load point $(\bar{x}, \bar{y}, l)$, corresponding to this stress distribution, is then determined by the condition

$$\iint_R (x\tau_{zy} - y\tau_{zx})\, dx\, dy = \bar{x} P_y - \bar{y} P_x$$

which must hold for an arbitrary choice of P_x and P_y. The load point $(\bar{x}, \bar{y}, l)$ corresponding to $\theta = 0$ is called the centre of flexure and is denoted by (x_{cf}, y_{cf}, l).

(*ii*) A torsion problem with the twist θ due to a couple of moment.

$$P_y(x_0 - x_{cf}) - P_x(y_e - y_{cf})$$

and with shear stresses determined by Eq. (6.30) with K_x and K_y set equal to zero.

We can think of the load P at the point (x_0, y_0, l) as being replaced by an equal load at the centre of flexure and by a couple producing the twist θ. The solution of the general flexure problem is then got by superposing the solutions of these two simple problems. The decomposition of the general flexure problem into problems (*i*) and (*ii*) amounts of resolving the twisting moment

$$M_z = x_0 P_y - y_0 P_x.$$

determined by Eq. (6.56) into two parts :

$$\iint_R (x\tau_{zy} - y\tau_{zx})\, dxdy = x_{cf} P_y - y_{cf} P_x \qquad \text{...(6.33)}$$

where τ_{by} and τ_{zx} are given by Eq. (6.47) with $\theta = 0$, and

$$\iint_R (x\tau_{zy} - y\tau_{zx})\, dxdy = (x_0 - x_{cf})\, P_y - (y_0 - y_{cf})\, P_x \qquad \text{...(20)}$$

where τ_{zy} and τ_{zx} are given by Eq. (6.30) with $K_x = K_y = 0$.

The position of the centre of flexure is determined from the formula (6.33) as :

$$\left.\begin{aligned} x_{cf} &= J(I_y S_2 - I_{xy} S_1) \\ y_{cf} &= J(I_{xy} S_2 - I_x S_1) \end{aligned}\right\} \qquad \text{...(6.34)}$$

where $S_1 = \iint_R \left[x\dfrac{d\varphi_1}{\partial y} - y\dfrac{\partial\varphi_1}{\partial x} + (1+\nu)\, x^2 y - \nu y^3 \right] dx\, dy$

$$S_2 = \iint_R \left[x\frac{\partial\varphi_1}{\partial y} - y\frac{\partial\varphi_2}{\partial x} - (1+\nu)\, xy^2 + \nu x^3 \right] dx\, dy$$

$$\frac{1}{J} = 2(1+\nu)(I_x I_y - I_{xy}^2)$$

If the cross section R is symmetric about the x-axis, then

$$S_1 = 0$$

and $$\frac{1}{J} = 2(1+\nu) I_x I_y$$

and

$$\left.\begin{aligned} x_{cf} &= \frac{1}{2(1+\nu)I_x}\iint_R \left[x\frac{\partial \varphi_2}{\partial y} - y\frac{\partial \varphi_2}{\partial x} \right. \\ &\qquad \left. -(1+\nu)\,xy^2 + \nu x^3 \right] dx\,dy \\ y_{cf} &= 0 \end{aligned}\right\} \qquad \text{...(21)}$$

which state that the centre of flexure lies on the axis of symmetry of the section. Accordingly, if the cross-section has two perpendicular axes of symmetry, then the centre of flexure coincides with the centroid of the section.

The solution of the simple flexure problem is given by the harmonic functions φ_1 and φ_2 which satisfy the conditions (6.31) on the boundary. We define the harmonic functions $\varphi_{11}, \varphi_{12}, \varphi_{21}, \varphi_{22}$ by the relations

$$\left.\begin{aligned} \varphi_1 &= (1+\nu)\,\varphi_{11} - \nu\,\varphi_{12} \\ \varphi_2 &= (1+\nu)\,\varphi_{22} + \nu\,\varphi_{21} \end{aligned}\right\} \qquad \text{...(22)}$$

Eqs. (6.31) now become

$$\left.\begin{aligned} (1+\nu)\frac{d\varphi_{11}}{dn} - \nu\frac{d\varphi_{12}}{dn} &= (1+\nu)\,x^2 a_{nx} - \nu y^2 a_{nx} \\ (1+\nu)\frac{d\varphi_{22}}{dn} + \nu\frac{d\varphi_{21}}{dn} &= (1+\nu)\,y^2 a_{ny} - \nu x^2 a_{ny} \end{aligned}\right\} \qquad \text{...(23)}$$

The function φ_{ij} must satisfy the conditions

$$\left.\begin{aligned} \frac{\partial \varphi_{11}}{dn} &= x^2 a_{nx} = \frac{d}{dn}\left(\frac{x^3}{3}\right) \\ \frac{\partial \varphi_{22}}{dn} &= y^2 a_{ny} = \frac{d}{dn}\left(\frac{y^3}{3}\right) \\ \frac{d\varphi_{12}}{dn} &= y^2 a_{nx} = \frac{d}{ds}\left(\frac{y^3}{3}\right) \\ \frac{d\varphi_{12}}{dn} &= -x^2 a_{ny} = \frac{d}{ds}\left(\frac{x^3}{3}\right) \end{aligned}\right\} \qquad \text{...(24)}$$

We introduce the conjugate harmonic functions ψ_{12}, ψ_{21} with

$$\frac{\partial \varphi_{12}}{\partial x} = \frac{\partial \psi_{12}}{\partial y}, \quad \frac{\partial \varphi_{12}}{\partial y} = -\frac{\partial \psi_{12}}{\partial x}$$

$$\frac{\partial \varphi_{21}}{\partial x} = \frac{\partial \psi_{21}}{\partial}, \quad \frac{\partial \varphi_{21}}{\partial y} = -\frac{\partial \psi_{21}}{\partial x}$$

Thus the last two boundary conditions in Eqs. (24) can be written as

$$\frac{\partial \psi_{12}}{ds} = \frac{d}{ds}\left(\frac{y^3}{3}\right), \quad \frac{d\psi_{21}}{ds} = \frac{d}{ds}\left(\frac{x^2}{3}\right)$$

or

$$\psi_{12} = \frac{y^3}{3} + \text{const.},$$

$$\psi_{21} = \frac{x^3}{3} + \text{const. on } C \qquad \text{...(25)}$$

The solution of the simple flexure problem in which the applied load $(P_x, P_y, 0)$ acts at the center of flexure (with $\theta = 0$) is thus given by the stresses :

$$\left.\begin{aligned}
&\sigma_x = \sigma_y = \tau_{xy} = 0 \\
&\sigma_z = -E(l - z)(K_x x + K_y y) \\
&\tau_{zx} = GK_x\left[(1+\nu)\left(\frac{\partial \varphi_{11}}{\partial x} - x^2\right) - \nu\left(\frac{\partial \psi_{12}}{\partial y} - y^2\right)\right] \\
&\qquad + GK_y\left[(1+\nu)\frac{\partial \varphi_{22}}{\partial x} + \nu\frac{\partial \psi_{21}}{\partial y}\right] \\
&\tau_{zy} = GK_y\left[(1+\nu)\left(\frac{\partial \varphi_{22}}{\partial y} - y^2\right) - \nu\left(\frac{\partial \psi_{21}}{\partial x} - x^2\right)\right] \\
&\qquad + GK_x\left[(1+\nu)\frac{\partial \varphi_{11}}{\partial y} + \nu\frac{\partial \psi_{12}}{\partial x}\right]
\end{aligned}\right\} \qquad \text{...(6.35)}$$

where

$$EK_x = \frac{I_x P_x - I_{xy} P_y}{I_x I_y - I_{xy}^2}$$

$$EK_y = \frac{I_y P_y - I_{xy} P_x}{I_x I_y - I_{xy}^2}$$

and the harmonic functions φ_{11}, φ_{22}, ψ_{12}, ψ_{21} satisfy the boundary conditions

$$\left.\begin{aligned}
&\frac{d\varphi_{11}}{dn} = x^2 a_{nx} \\
&\frac{d\varphi_{22}}{dn} = y^2 a_{ny} \\
&\psi_{21} = \frac{x^3}{3} + \text{const.}, \\
&\psi_{12} = \frac{y^3}{3} + \text{const. on } C
\end{aligned}\right\} \qquad \text{...(26)}$$

6.5.1 Bending by a Load Along a Principal Axis

Let the axes taken to be the principal axes of the section and the load $(P_x, 0, 0)$ is directed paaallel to one of these axes. Thus $I_{xy} = 0$ and Eqs. (6.32) yield

$$K_x = \frac{P_x}{EI_y}, \quad K_y = 0$$

while Eqs. (6.30) become

$$\tau_{zx} = G\theta\left(\frac{\partial\varphi}{\partial x} - y\right) + \frac{P_x}{2(1+\nu)\,I_y}\left[\frac{\partial\varphi_1}{\partial x} - (1+\nu)\,x^2 + \nu y^2\right]$$

$$\tau_{xy} = G\theta\left(\frac{\partial\varphi}{\partial y} + x\right) + \frac{P_x}{2(1+\nu)\,I_y}\cdot\frac{\partial\varphi_1}{\partial y}$$

In terms of the harmonic function $\Phi(x, y)$, the stresses can be written as

$$\left.\begin{aligned} &\sigma_x = \sigma_y = \tau_{xy} = 0 \\ &\sigma_z = -\frac{P_x}{I_y}(l - z)\,x \\ &\tau_{zx} = G\theta\left(\frac{\partial\varphi}{\partial x} - y\right) - \frac{P_x}{2(1+\nu)\,I_y}\left[\frac{\partial\Phi}{\partial x} + \frac{1}{2}\nu x^2 + \left(1 - \frac{\nu}{2}\right)y^2\right] \\ &\tau_{zy} = G\theta\left(\frac{\partial\varphi}{\partial y} + x\right) - \frac{P_x}{2(1+\nu)\,I_y}\left[\frac{\partial\Phi}{\partial_y} + (2+\nu)\,xy\right] \end{aligned}\right\} \quad \text{...(6.36)}$$

where $\varphi_1 + i\psi_1 = -(\Phi + i\Psi) + \frac{1}{3}\left(1 + \frac{\nu}{2}\right)(x + iy)^3$ ...(6.37)

From Eq. (6.37) and (6.31) it follows that the harmonic function Φ must satisfy the condition

$$\frac{d\Phi}{dn} = -\left[\frac{1}{2}\nu x^2 + \left(1 - \frac{1}{2}\nu\right)y^2\right]a_{nx} - (2+\nu)\,xy\,a_{ny} \text{ on } C \quad \text{...(6.38)}$$

The y-coordinate of the centre of flexure is found from Eq. (6.34) and (6.37) to be given by

$$y_{cf} = \frac{1}{2(1+\nu) I_y} \iint_R \left(x \frac{\partial \Phi}{\partial y} - y \frac{\partial \Phi}{\partial x} \right) dx\, dy$$

$$+ \frac{1}{2(1+\nu) I_y} \iint_R \left[\left(2 + \frac{\nu}{2} \right) x^2 y \right.$$

$$\left. - \left(1 - \frac{\nu}{2} \right) y^3 \right] dx\, dy$$

$$= \frac{1}{2(1+\nu) I_y} \iint_R \left(y \frac{\partial \varphi_1}{\partial x} - x \frac{\partial \varphi_1}{\partial y} \right) dx\, dy$$

$$- \frac{1}{2(1+\nu) I_y} \iint_R [(1+\nu) x^2 y - \nu y^3]\, dx\, dy \qquad ...(6.39)$$

Since $P_y = 0$, we cannot determine x_{cf} from Eq. (6.33). The mean twist over every section will vanish, provided the load P_x is applied at any point along the line $y = y_{cf}$.

For example for a beam of circular cross-section given by

$$x^2 + y^2 = a^2$$

and subjected to terminal load P at the centroid directed along the x-axis, the boundary conditions become

$$\Psi = -\left(\frac{3}{4} + \frac{\nu}{2} \right) a^2 r \sin \theta + \frac{1}{4} r^3 \sin 3\theta \text{ on } C$$

$$\therefore \quad \Phi = -\left(\frac{3}{4} + \frac{\nu}{2} \right) a^2 r \cos \theta + \frac{1}{4} r^3 \cos 3\theta$$

$\therefore$ From Eq. (6.36) we get the stresses as with $\theta = 0$.

$$\left.\begin{aligned} \tau_{zx} &= \frac{(3+2\nu) P}{2\pi a^4 (1+\nu)} \left(a^2 - x^2 - \frac{1-2\nu}{3+2\nu} y^2 \right) \\ \tau_{zy} &= -\frac{(1+2\nu) Pxy}{\pi a^4 (1+\nu)} \\ \sigma_z &= -\frac{4P}{\pi a^4} (l - z)\, x \end{aligned}\right\} \qquad ...(6.40)$$

These results are the same as obtained in art. 6.2.2.

6.6 Solution of Bending Problems by Soap-Film Method

Flexure problem can also be solved by the soap-film method analogous to that used in solving torsion problems. Taking

$$f(y) = \frac{\nu}{2(1+\nu)} \frac{Py^2}{I}$$

Eq. (6.9) for the stress function is

$$\nabla^2\phi = 0 \qquad ...(1)$$

This is the same equation as for an unloaded and uniformly stretched membrane. The boundary conditions Eq. (6.12) becomes

$$\frac{\partial\phi_1}{\partial s} = \left[\frac{Px^2}{2I} - \frac{\nu}{2(1+\nu)}\frac{Py^2}{I}\right]\frac{dy}{ds} \qquad ...(2)$$

Integrating along the boundary s, we get

$$\phi = \frac{P}{I}\int\frac{x^2dy}{2} - \frac{\nu}{2(1+\nu)}\frac{Py^3}{3I} + \text{const.} \qquad ...(3)$$

From which the value of ϕ for every point of the boundary can be calculated. $\int\frac{x^2dy}{2}$ vanishes when taken around the boundary. Hence ϕ, calculated from Eq. (3), is represented along the boundary by a closed curve. If now a soap film is stretched over this curve, then the surface of the film satisfies Eq. (1) and boundary condition Eq. (3). Hence the ordinates of the film represent the stress function ϕ at all points of the cross-section.

If necessary, the range of the function along the boundary can be reduced by introducing a new function ϕ_1 given by

$$\phi = \phi_1 + ax + by \qquad ...(4)$$

where a and b are arbitrary constants.

Eq. (4) also satisfies the condition $\nabla^2\phi = 0$.

From Eqs. (3) and (4), we get

$$\phi_1 = \frac{P}{I}\int\frac{x^2dy}{2} - \frac{\nu}{2(1+\nu)}\frac{Py^3}{3I} - ax - by + \text{const.}$$

The shearing stresses now become from eq. (6.8) as

$$\left.\begin{aligned} \tau_{zx} &= \frac{\partial\phi}{\partial y} - \frac{Px^2}{2I} + \frac{\nu}{2(1+\nu)}\frac{Py^2}{I} \\ \tau_{yz} &= -\frac{\partial\phi}{\partial x} \end{aligned}\right\} \qquad ...(6.41)$$

By knowing the values of the derivatives $\frac{\partial\phi}{\partial y}$ and $\frac{\partial\phi}{\partial x}$ we can determine the shear stress components. These derivatives are given by the slopes of the soap film in the y and x directions.

Example 6.5 *The cantilever beam 3 m long and shown in Fig. 6.15 has a channel section. The concentrated end load is 12 kN and lies in the plane making an angle of 60° with the z-axis. Locate points of maximum tensile and compressive stresses in the beam and determine their magnitudes.*

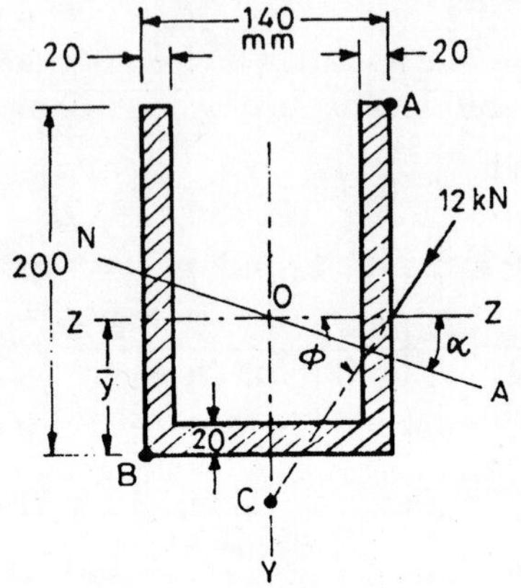

Fig. 6.15

Solution. $A = 140 \times 20 + 2 \times 180 \times 20$

$$= 2800 + 7200 = 10{,}000 \text{ mm}^2$$

$$\bar{y} = \frac{2800 \times 10 + 2 \times 180 \times 20 \times 110}{10{,}000} = 82 \text{ mm}$$

$$I_z = \frac{140 \times 20^3}{12} + 2800 \times (82 - 10)^2$$

$$+ \frac{2 \times 20 \times 180^3}{12} + 2 \times 180 \times 20 \times (110 - 82)^2$$

$$= 39.69 \times 10^6 \text{ mm}^4$$

$$I_y = \frac{20 \times 140^3}{12} + 2 \times \left[\frac{180 \times 20^3}{12} + 180 \times 20 \times 60^2 \right]$$

$$= 30.73 \times 10^6 \text{ mm}^4$$

$$I_{yz} = 0$$

$$\tan\alpha = -\frac{I_z}{I_y}\cot\phi = -\frac{39.69}{30.73}\cot 60° = -0.7457$$

$$\alpha = -0.6407 \text{ rad or } -32.65°$$

$$M = -3 \times 12 = -36 \text{ kN.m.}$$

$$M_z = M\sin\phi = -36\sin 60° = -31.18 \text{ kN.m}$$

$$\sigma_A = \frac{M_z\,(y_A - z_A\tan\alpha)}{I_z}$$

$$= \frac{-31.18 \times 10^3\,[-118 - (-70)(-0.7457)]}{39.69 \times 10^6}$$

$$= 133.7 \text{ MPa}$$

$$\sigma_B = \frac{-31.18 \times 10^3 [82 - 70(-0.7457)]}{39.69 \times 10^6}$$

$$= -105.6 \text{ MPa}$$

EXERCISES

6.1 A prismatic bar of circular cross-section of radius 25 mm is subjected to a terminal load of 5 kN. Determine the stresses in the bar at the end of the horizontal diameter. Compare the result with the elementary solution. Assume ν = 0.3.

6.2 A prismatic bar of elliptical cross-section has its semi-minor axes as 40 mm and 20 mm respectively. This bar is subjected to an end load of 2500 N. Determine the stresses in the bar at the end of major and minor axes. Assume ν = 0.28.

6.3 A prismatic bar of rectangular cross-section 50 mm × 30 mm is subjected to an end load of 4500 N. Determine the stresses at the centre of the bar and the corner. Assume ν = 0.3.

6.4 (*a*) Differentiate between symmetrical and unsymmetrical bending.

(*b*) Derive the generalised flexure formula.

6.5 A rectangular beam 120 mm × 100 mm is 3 min length and is simply supported at the ends. It carries a load of 5 kN at mid-span inclined at 45° with the vertical axis and passing through the centroid. Determine the maximum bending stress in the beam.

6.6 An unequal leg angle iron section as shown in Fig. 6.16 is used as a beam, freely supported on a span of 2 m carrying a vertical load of 2 kN at the centre. Find the maximum bending stress in the angle.

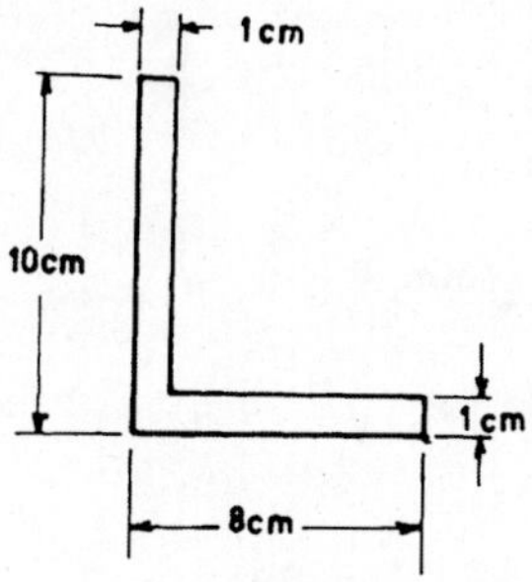

Fig. 6.16

6.7 Locate the shear centre of the cross-sections shown in Figs. 6.17, 6.18 and 6.19.

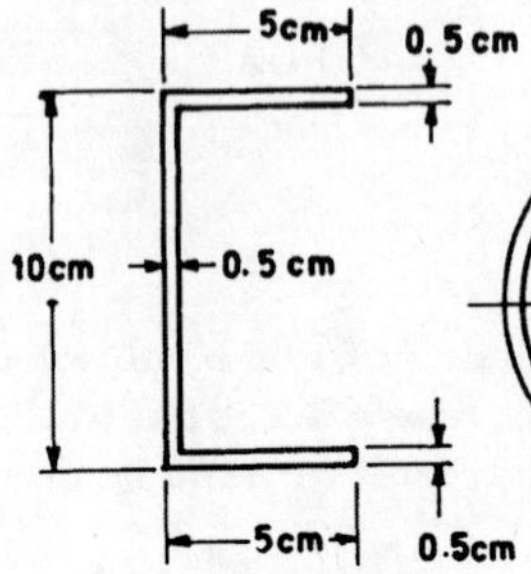

Fig. 6.17

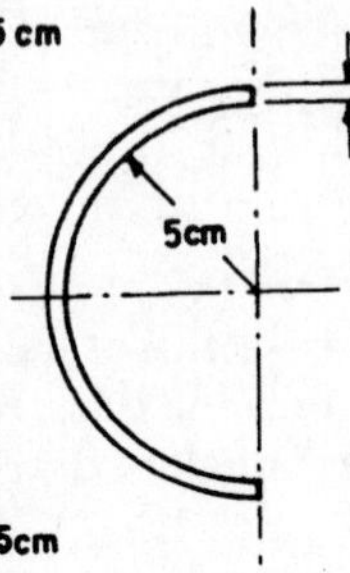

Fig. 6.18

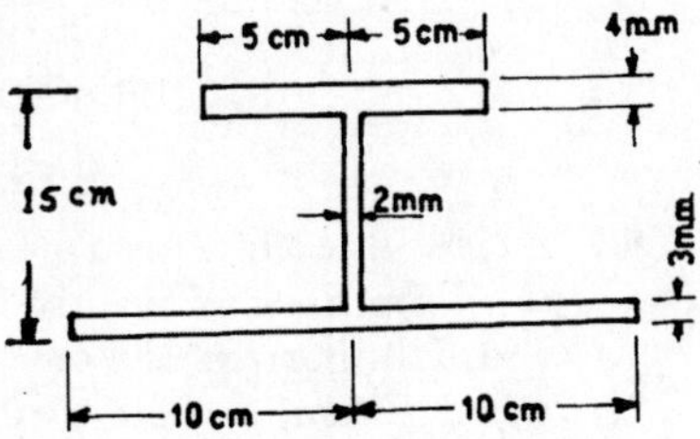

Fig. 6.19

6.8 A 4 mm thickness plate of steel of formed into the cross-section shown in Fig. 6.20. Locate the shear centre for the cross-section.

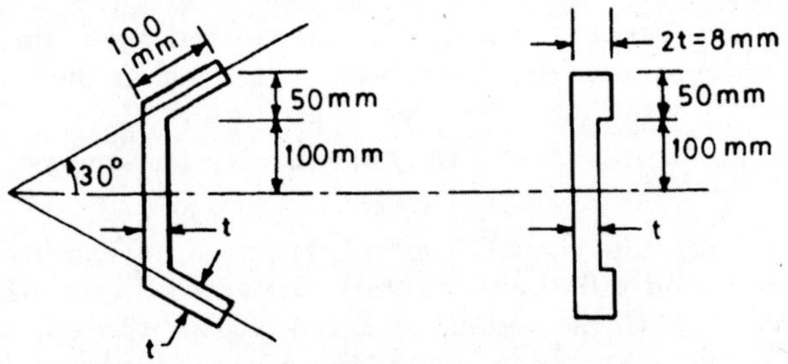

Fig. 6.20

6.9 A 3 mm thick plate of steel is formed into the cross-section shown in Fig. 6.21. Locate the shear centre for the cross-section.

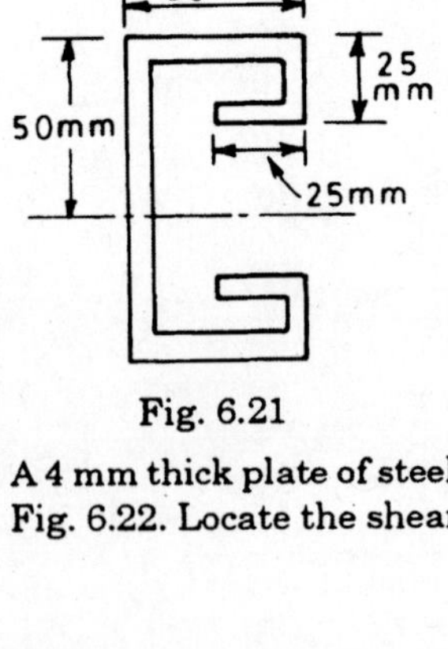

Fig. 6.21

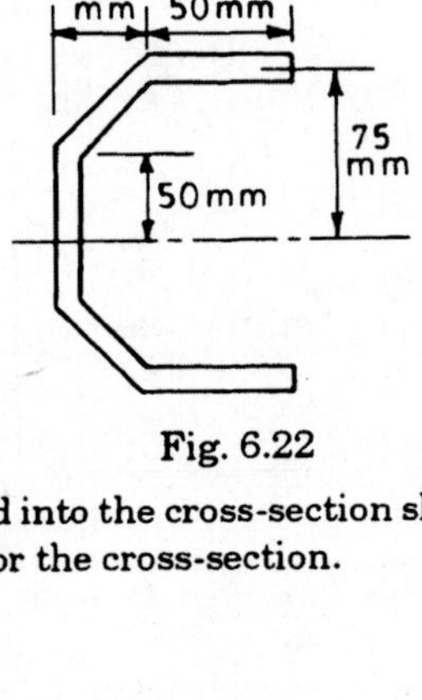

Fig. 6.22

6.10 A 4 mm thick plate of steel is formed into the cross-section shown in Fig. 6.22. Locate the shear centre for the cross-section.

6.11 A simply supported beam is loaded as shown in Fig. 6.23. Calculate the stress at A and the orientation of the neutral axis.

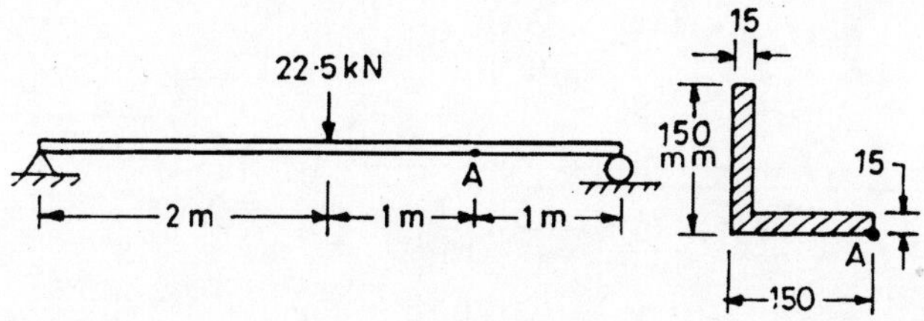

Fig. 6.23

6.12 For the simply supported beam of Fig. 6.24, determine the bending stress at points D and E.

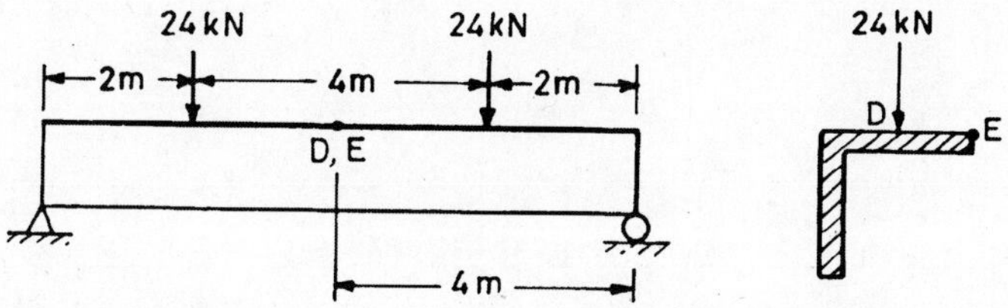

Fig. 6.24

6.13 A concentrated load P acts on a cantilever as shown in Fig. 6.25. L = 1.5 m, t = 20 mm, c = 60 mm, and b = 80 mm. If yield strength of the material is 290 MPa and factor of safety 1.2 is desire, calculate the magnitude of P for (*a*) α = 0°, and (*b*) α = 15°.

[Ans. 3.6 kN, 3.36 kN]

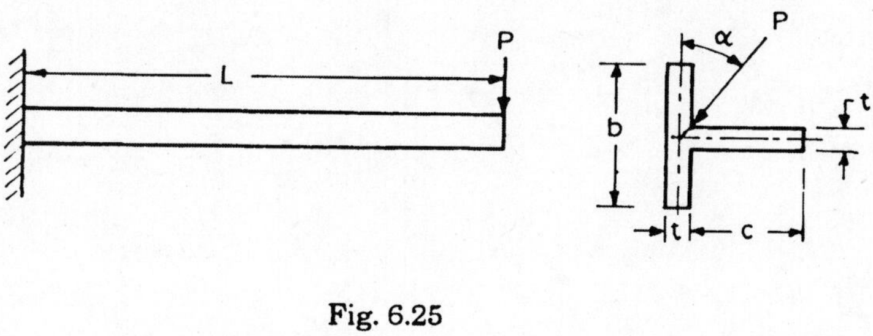

Fig. 6.25

6.14 A box beam supports the loading shown in Fig. 6.26. Determine the maximum value of P such that a flexural stress $\sigma = 7$ MPa or a shearing stress $\tau = 0.7$ MPa will not be exceeded. [**Ans.** 9320 N]

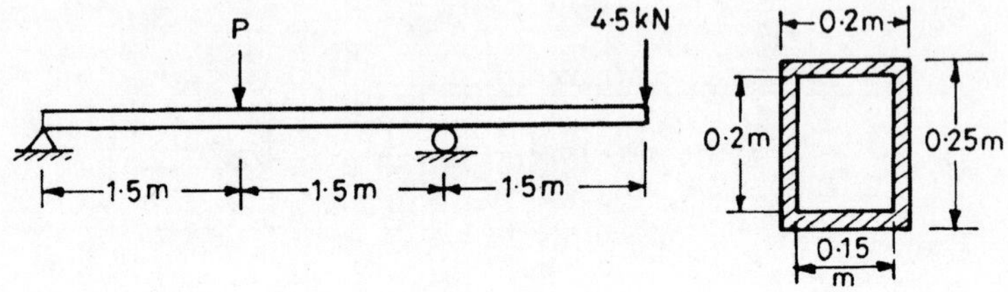

Fig. 6.26

7

Three Dimensional Problems

7.1 Introduction

In this chapter, we shall consider the solution of elasto-static problems in three-dimensions. The solution of three-dimensional elasticity problems is generally very difficult. However, we shall discuss some general methods of obtaining solutions to three dimensional problems and illustrate them by simple examples.

7.2 Stretching of a Prismatic Bar by its Own Weight

Consider a straight prismatic homogeneous bar hanging by its own weight as shown in Fig. 7.1.

The body forces are

$$B_x = B_y = 0$$

$$B_z = -\rho g \qquad ...(1)$$

where ρg is the weight per unit volume of the bar. The differential equation of equilibrium in Cartesian coordinates are satisfied if

$$\sigma_z = \rho g z,$$

$$\sigma_x = \sigma_x = \tau_{xy} = \tau_{yz} = \tau_{zx} = 0 \qquad ...(2)$$

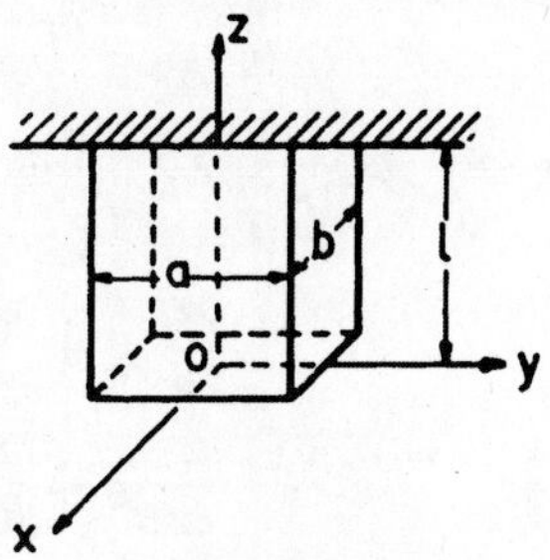

Fig. 7.1

The boundary condition at the lateral surface which is free from forces are satisfied.

The only condition we get is

$$T_z = \sigma_z a_{nz} = 0$$

Now $a_{nz} = 0$ for the surfaces of the bar parallel to the z-axis. From Eq. (2) we see that $\sigma_z = 0$ at the bottom surface of the bar, *i.e.* when $z = 0$ and at $z = l$, $\sigma_z = \rho g l$, in which l is the length of the bar.

The compatibility equations are also satisfied by Eq. (2), hence, it is the correct solution of the problem for a uniform distribution of forces at the top. The resultant of the uniform stress at the upper surface is

$$\int_A \sigma_z \, dA = \int_A \rho g l \, dA = \rho g \, abl \qquad ...(3)$$

Let us consider now the displacements. From Hooke's law, we have

$$\varepsilon_x = \frac{1}{E}(-\nu \rho g z) = \frac{\partial u_x}{\partial x} \qquad ...(4\,a)$$

$$\varepsilon_y = \frac{1}{E}(-\nu \rho g z) = \frac{\partial u_y}{\partial y} \qquad ...(4\,b)$$

$$\varepsilon_z = \frac{1}{E}(\rho g z) = \frac{\partial u_z}{\partial z} \qquad ...(4\,c)$$

$$\gamma_{xy} = \frac{\partial u_x}{\partial y} + \frac{\partial u_y}{\partial x} = 0 \qquad ...(4\,d)$$

$$\gamma_{yz} = \frac{\partial u_y}{\partial z} + \frac{\partial u_z}{\partial y} = 0 \qquad ...(4\,e)$$

$$\gamma_{zx} = \frac{\partial u_z}{\partial x} + \frac{\partial u_x}{\partial z} = 0 \qquad ...(4\,f)$$

Integrating Eq. (4 *c*), we find

$$u_z = \frac{\rho g z^2}{2E} + f(x, y) \qquad ...(5)$$

where $f(x, y)$ is an arbitrary function of x, y.

Substituting Eq. (5) in Eqs. (4 *e*) and (4 *f*), we get

$$\left.\begin{aligned} \frac{\partial u_y}{\partial z} &= -\frac{\partial f}{\partial y} \\ \frac{\partial u_x}{\partial z} &= -\frac{\partial f}{\partial x} \end{aligned}\right\} \qquad ...(6)$$

Integrating Eq. (6), we find

$$\left.\begin{aligned} u_y &= -\frac{\partial f}{\partial y} z + g(x, y) \\ u_x &= -\frac{\partial f}{\partial x} z + h(x, y) \end{aligned}\right\} \qquad ...(7)$$

where g and h are also arbitrary functions of x and y. Now substitute Eqs. (7) in Eqs. (4 *a*) and (4 *b*), we get

$$-\frac{\partial^2 f}{\partial x^2} z + \frac{\partial h}{\partial x} = -\frac{\nu \rho g z}{E} \qquad ...(8)$$

$$-\frac{\partial^2 f}{\partial y^2} z + \frac{\partial g}{\partial y} = -\frac{\nu \rho g z}{E} \qquad ...(9)$$

Since f, g and h do not depend on z, hence, we get

$$\left.\begin{aligned}\frac{\partial h}{\partial x} &= 0\\ \frac{\partial g}{\partial y} &= 0\\ \frac{\partial^2 f}{\partial x^2} &= \frac{\nu\rho g}{E}\\ \frac{\partial^2 f}{\partial y^2} &= \frac{\nu\rho g}{E}\end{aligned}\right\} \quad \text{...(10)}$$

Now substitute Eq. (7) in Eq. (4 d), we have

$$-\frac{\partial^2 f}{\partial x \partial y} z + \frac{\partial h}{\partial y} - \frac{\partial^2 f}{\partial x \partial y} z + \frac{\partial g}{\partial x} = 0$$

or

$$-2\frac{\partial^2 f}{\partial x \partial y} z + \frac{\partial h}{\partial y} + \frac{\partial g}{\partial x} = 0 \quad \text{...(11)}$$

Since $\frac{\partial h}{\partial y}$ and $\frac{\partial g}{\partial x}$ are functions of only x and y we can conclude from Eq. (11) that

$$\left.\begin{aligned}\frac{\partial h}{\partial y} + \frac{\partial g}{\partial x} &= 0\\ \frac{\partial^2 f}{\partial x \partial y} &= 0\end{aligned}\right\} \quad \text{...(12)}$$

From first two equations of (10), on integration, we get

$$\left.\begin{aligned}g &= C_1 x + C_2\\ h &= -C_1 y + C_3\end{aligned}\right\} \quad \text{...(13)}$$

Eqs. (10) and (12) indicate that f has the following form

$$f = \frac{\nu\rho g}{2E}(x^2 + y^2) + C_4 x + C_5 y + C_6 \quad \text{...(14)}$$

Hence, we get

$$\left.\begin{aligned}u_x &= -\frac{\nu\rho g\, xz}{E} - C_4 z - C_1 y + C_3\\ u_y &= -\frac{\nu\rho g\, yz}{E} - C_5 z + C_1 x + C_2\\ u_z &= \frac{\rho g z^2}{2E} + \frac{\nu\rho g}{2E}(x^2 + y^2) + C_4 x + C_5 y + C_6\end{aligned}\right\} \quad \text{...(15)}$$

The six constants C_1 to C_6 can be determined from the boundary conditions. We know that at

$$x = 0, y = 0, z = l,$$

we have $u_x = u_y = u_z = 0$

$$\therefore \quad \left.\begin{aligned} -C_4 l + C_3 &= 0 \\ -C_5 l + C_2 &= 0 \\ \frac{\rho g l^2}{2E} + C_6 &= 0 \end{aligned}\right\} \quad \text{...(16)}$$

Also at $x = 0, y = 0, z = l$, there will be no rotation. Hence at point $(0, 0, l)$

$$\left.\begin{aligned} \frac{1}{2}\left(\frac{\partial u_z}{\partial y} - \frac{\partial u_y}{\partial z}\right) &= 0 \\ \frac{1}{2}\left(\frac{\partial u_x}{\partial z} - \frac{\partial u_z}{\partial x}\right) &= 0 \\ \frac{1}{2}\left(\frac{\partial u_y}{\partial x} - \frac{\partial u_x}{\partial y}\right) &= 0 \end{aligned}\right\} \quad \text{...(17)}$$

$$\therefore \quad \left.\begin{aligned} \tfrac{1}{4}(C_5 + C_5) &= 0 \\ \tfrac{1}{2}(-C_4 - C_4) &= 0 \\ \tfrac{1}{2}(C_1 + C_1) &= 0 \end{aligned}\right\} \quad \text{...(18)}$$

$\therefore \quad C_1 = C_4 = C_5 = 0.$

Hence from Eq. (16), we get

$$C_3 = 0,\ C_2 = 0,\ C_6 = -\frac{\rho g l^2}{2E}$$

$$\therefore \quad \left.\begin{aligned} u_x &= -\frac{\nu \rho g x z}{E} \\ u_y &= -\frac{\nu \rho g y z}{E} \\ u_z &= \frac{\rho g z^2}{2E} + \frac{\nu \rho g}{2E}(x^2 + y^2) - \frac{\rho g l^2}{2E} \end{aligned}\right\} \quad \text{...(7.1)}$$

Points along the z-axis have the vertical displacements only given by

$$u_z = \frac{\rho g}{2E}(z^2 - l^2) \quad \text{...(7.2)}$$

7.3 Pure Bending of a Prismatic Bar

Consider a prismatic bar bent in one of its principal planes by two equal and opposite couples M as shown in Fig. 7.2.

Taking the origin of the co-ordinates at the centroid of the cross-section and the xz plane in the principal plane of bending, the stress components are :

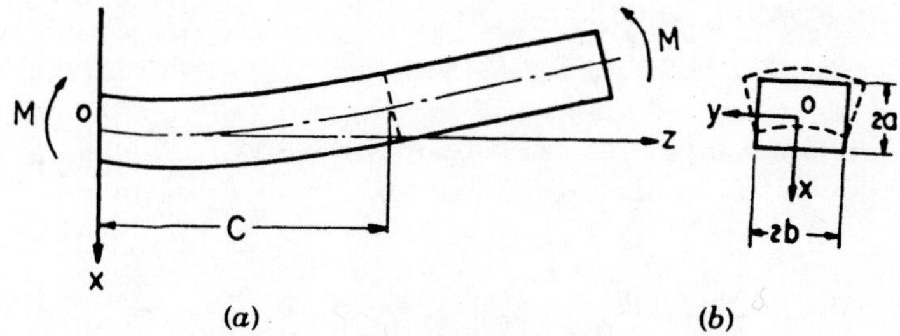

Fig. 7.2

$$\sigma_z = \frac{Ex}{R}, \sigma_x = \sigma_y = \tau_{xy} = \tau_{yz} = \tau_{zx} = 0 \quad ...(1)$$

where R is the radius of curvature of the bar after bending. The stress distribution given by Eq. (1) satisfy the equilibrium equations in the absence of body forces and also satisfy the boundary conditions. The bending moment M is given by

$$M = \iint \sigma_z x dA = \iint \frac{Ex^2}{R} dA$$

$$= \frac{E}{R}\iint x^2 dA = \frac{EI_y}{R}$$

$$\therefore \quad \frac{M}{EI_y} = \frac{1}{R} \quad ...(2)$$

Let us now consider the displacements. From Hooke's law, we have

$$\varepsilon_z = \frac{\partial u_z}{\partial z} = \frac{\sigma_z}{E} = \frac{x}{R} \quad ...(3\,a)$$

$$\varepsilon_x = \frac{\partial u_x}{\partial x} = -\nu\frac{x}{R} \quad ...(3\,b)$$

$$\varepsilon_y = \frac{\partial u_y}{\partial y} = -\nu\frac{x}{R} \quad ...(3\,c)$$

$$\gamma_{xy} = \frac{\partial u_x}{\partial y} + \frac{\partial u_y}{\partial x} = 0 \quad ...(3\,d)$$

$$\gamma_{yz} = \frac{\partial u_y}{\partial z} + \frac{\partial u_z}{\partial y} = 0 \quad ...(3\,e)$$

$$\gamma_{zx} = \frac{\partial u_z}{\partial x} + \frac{\partial u_x}{\partial z} = 0 \quad ...(3\,f)$$

Integrating Eq. (3 *a*), we get

$$u_z = \frac{xz}{R} + f(x, y)$$

Substituting in Eqs. (3 *e*) and (3 *f*), we find

$$\frac{\partial u_x}{\partial z} = -\frac{z}{R} - \frac{\partial f}{\partial x}$$

$$\frac{\partial u_y}{\partial z} = -\frac{\partial f}{\partial y}$$

which gives on integration

$$\left.\begin{aligned} u_x &= -\frac{z^2}{2R} - z\frac{\partial f}{\partial x} + g(x, y) \\ u_y &= -z\frac{\partial f}{\partial y} + h(x, y) \end{aligned}\right\} \qquad ...(4)$$

Substituting these equations in Eqs. (3 *b*) and (3 *c*), we get

$$-z\frac{\partial^2 f}{\partial x^2} + \frac{\partial g}{\partial x} = -\frac{\nu x}{R}$$

$$-z\frac{\partial^2 f}{\partial y^2} + \frac{\partial h}{\partial y} = -\frac{\nu x}{R}$$

These equations must be satisfied for any value of z, hence

$$\frac{\partial^2 f}{\partial x^2} = 0, \quad \frac{\partial^2 f}{\partial y^2} = 0 \qquad ...(5)$$

Integrating, we get

$$\left.\begin{aligned} g &= -\frac{\nu x^2}{R} + f_1(y) \\ h &= -\frac{\nu xy}{R} + f_2(x) \end{aligned}\right\} \qquad ...(6)$$

Substituting Eqs. (4) by using Eqs. (6) in Eq. (3 *d*), we find

$$-z\frac{\partial^2 f}{\partial x \partial y} + \frac{\partial f_1(y)}{\partial y} - z\frac{\partial^2 f}{\partial x \partial y} + \frac{\partial f_2(x)}{\partial x} - \frac{\nu y}{R} = 0$$

or

$$2z\frac{\partial^2 f}{\partial x \partial y} - \frac{\partial f_1(y)}{\partial y} - \frac{\partial f_2(x)}{\partial x} + \frac{\nu y}{R} = 0.$$

Since only first term in this equation depends on z, therefore, it is necessary to have

$$\frac{\partial^2 f}{\partial x \partial y} = 0, \quad \frac{\partial f_1(y)}{\partial y} + \frac{\partial f_2(x)}{\partial x} - \frac{\nu y}{R} = 0.$$

Eqs. (5), require that

$$f = C_1 x + C_2 y + C_3$$

$$f_1(y) = \frac{\nu y^2}{2R} + C_4 y + C_5$$

$$f_2(x) = -C_4 x + C_6$$

where C_1 to C_6 are arbitrary constants. Hence we find that

$$u_x = -\frac{z^2}{2R} - C_1 z - \frac{\nu x^2}{2R} + \frac{\nu y^2}{2R} + C_4 y + C_5$$

$$u_y = -C_2 z - \frac{\nu xy}{R} - C_4 x + C_6$$

$$u_z = \frac{xz}{R} + C_1 x + C_2 y + C_3.$$

The constants are to be determined from the boundary conditions. At $x = y = z = 0$, we have

$$u_x = u_y = u_z = 0$$

$$\frac{\partial u_x}{\partial z} = \frac{\partial u_y}{\partial z} = \frac{\partial u_y}{\partial x} = 0$$

These conditions are satisfied by taking all the constants equal to zero. Hence

$$\left.\begin{aligned} u_x &= -\frac{1}{2R}[z^2 + \nu(x^2 - y^2)] \\ u_y &= -\frac{\nu xy}{R} \\ u_x &= \frac{xz}{R} \end{aligned}\right\} \quad ...(7)$$

At $\quad x = y = 0$

$$u_x = -\frac{z^2}{2R},\ u_y = u_z = 0 \quad ...(8)$$

At a distance $z = c$ from 0, we have

$$z = c + u_z = c + \frac{cx}{R}$$

i.e. in pure bending, the cross-section remains plane.

At $\quad y = \pm b$

$$y = \pm b + u_y$$

$$= \pm b\left(1 - \frac{\nu x}{R}\right)$$

Therefore, the sides become inclined as shown in Fig. 7.2 (*b*) by dotted lines.

At $x = \pm a$, we have

$$x = \pm a + u_x = \pm a - \frac{1}{2R}[c^2 + \nu(a^2 - y^2)]$$

They are, therefore, bent to parabolic curves. By taking x and u_x constant in first of Equations (7), we find that the equation for the contour lines is

$$z^2 - \nu y^2 = \text{const.}$$

They are, therefore, hyperbolas with the asymptotes

$$z^2 - \nu y^2 = 0$$

which is inclined at an angle α with the z-axis given by

$$\tan^2 \alpha = \frac{1}{\nu}$$

This equation can be used to determine Poisson's ratio.

7.4 The Vector Form of Equilibrium Equations

The equilibrium equations is Cartesian co-ordinates are :

$$\frac{\partial \tau_{ij}}{\partial x_j} + B_i = 0 \qquad ...(1)$$

where $$\tau_{ij} = \lambda e \delta_{ij} + G\left(\frac{\partial u_i}{\partial x_j} + \frac{\partial u_j}{\partial x_i}\right) \qquad ...(2)$$

and dilatation $$e = \frac{\partial u_i}{\partial x_i} = \frac{\partial u_x}{\partial x} + \frac{\partial u_y}{\partial y} + \frac{\partial u_z}{\partial z}$$

Substituting Eq. (2) in (1), we find

$$\lambda \frac{\partial e}{\partial x_j} + G\frac{\partial^2 u_j}{\partial x_i \partial x_i} + G\frac{\partial^2 u_i}{\partial x_i \partial x_j} + B_i = 0.$$

Now $$\frac{\partial^2 u_i}{\partial x_i \partial x_j} = \frac{\partial}{\partial x_j}\left(\frac{\partial u_i}{\partial x_i}\right) = \frac{\partial e}{\partial x_j}$$

$$\therefore \qquad (\lambda + G)\frac{\partial e}{\partial x_j} + G\frac{\partial^2 u_j}{\partial x_i \partial x_i} + B_i = 0$$

which may be written in the vectorial form as

$$G\nabla^2 \underline{u} + (\lambda + G)\,\text{grad}\, e + \underline{B} = 0 \qquad ...(7.3)$$

where bar indicates vector quantities.

In the absence of body forces $\underline{B} = 0$, we get

$$(1 - 2\nu)\,\nabla^2 \underline{u} + \text{grad div.}\, \underline{u} = 0 \qquad ...(4)$$

Alternatively, we know that

$$\text{curl curl} = \text{grad div.} - \nabla^2$$

$\therefore$ Eq. (7.4) becomes,

$$(\lambda + 2G)\text{ grad div. } \underline{u} - G \text{ curl curl } \underline{u} + \underline{B} = 0 \qquad \text{...(7.5)}$$

7.5 Kelvin's Solution of the Equilibrium Equations

We express the displacement by means of a scalar potential φ and a vector potential $\underline{\psi}$ by means of the formula.

$$\underline{u} = \text{grad } \varphi + \text{curl } \underline{\psi} \qquad \text{...(1)}$$

We suppose that the body forces are of such a kind that they can be expressed in a like manner by means of the formula

$$\underline{B} = \text{grad } \Phi + \text{curl } \underline{\Psi} \qquad \text{...(2)}$$

Since $\quad \text{grad div. } \underline{u} = \text{grad } \nabla^2\varphi$

and $\quad \text{curl curl } \underline{u} = -\text{curl } \nabla^2\underline{\psi}$

Therefore, equations of equilibrium (7.5) become

$$\text{grad } [(\lambda + 2G)\nabla^2\varphi + \Phi] + \text{curl } [G\nabla^2\underline{\psi} + \underline{\Psi}] = 0.$$

Thus the particular solutions of the equilibrium equations can be obtained from particular solutions of the scalar equation

$$(\lambda + 2G)\, \nabla^2\varphi + \Phi = 0 \qquad \text{...(3)}$$

and of the vector equation

$$G\nabla^2\underline{\psi} + \underline{\Psi} = 0.$$

Now the body force $\underline{B}$ can be expressed by a formula of the type (2) by putting

$$\left.\begin{aligned} \Phi(r) &= -\frac{1}{4\pi}\int_V \underline{B}\,(\underline{r}')\text{ grad. } \frac{1}{|\underline{r} - \underline{r}'|}\, d^3\tau' \\ \underline{\Psi}(r) &= -\frac{1}{4\pi}\int_V \underline{B}\,(\underline{r}')\text{ grad. } \frac{1}{|\underline{r} - \underline{r}'|}\, d^3\tau' \end{aligned}\right\} \qquad \text{...(4)}$$

where the integration is taken over the volume V outside of which the body forces vanish. Using the relation

$$\text{grad. } \frac{1}{|\underline{r} - \underline{r}'|} = -\frac{\underline{r} - \underline{r}'}{|\underline{r} - \underline{r}'|^3}.$$

We have $\Phi(\underline{r}) = \dfrac{1}{4\pi}\displaystyle\int_V \frac{\underline{B}\,(\underline{r}') \cdot (\underline{r} - \underline{r}')}{|\underline{r} - \underline{r}'|^3}\, d\tau'$ ...(5)

$$\underline{\Psi}\,(\underline{r}) = \frac{1}{4\pi}\int_V \frac{\underline{B}\,(\underline{r}') \times (\underline{r} - \underline{r}')}{|\underline{r} - \underline{r}'|^3}\, d\tau' \qquad \text{...(6)}$$

Once the scalar potential Φ and the vector $\underline{\Psi}$ have been found by means of these equations it is a simple matter to obtain particular solutions of Eqs. (3) and (4). The particular solutions of these equations are :

$$\varphi(r) = \frac{1}{4\pi(\lambda + 2G)} \int \frac{\Phi(r')d\tau'}{|r - r'|} \qquad ...(7)$$

$$\underline{\psi}(r) = \frac{1}{4\pi G} \int \frac{\Psi(r')\,d\tau'}{|r - r'|} \qquad ...(8)$$

7.6 Point Force on an Infinite Solid

Consider the case of a point force of magnitude P_z acting at the origin parallel to the z-axis in an infinite elastic solid. We may take V to be the cubical volume bounded by the planes $x = \pm\frac{1}{2}\varepsilon$, $y = \pm\frac{1}{2}\varepsilon$, $z = \pm\frac{1}{2}\varepsilon$ and put

$$\underline{B} = \left(0, 0, \frac{P_z}{\varepsilon^2}\right)$$

where ε is infinitesimal. Substituting these results in Eqs. (5) and (6) of art 7.5, we obtain the expressions

$$\Phi = \frac{P_z}{4\pi}\left(\frac{z}{r^3}\right), \underline{\Psi} = \frac{P_z}{4\pi}\left(-\frac{y}{r^3}, \frac{x}{r^3}, 0\right) \qquad ...(9)$$

for the functions corresponding to a point force.

We must, therefore, find particular solutions of the equations :

$$\nabla^2\varphi + \frac{P_z}{4\pi(\lambda + 2G)}\left(\frac{z}{r^3}\right) = 0 \qquad ...(10)$$

$$\nabla^2\underline{\psi} + \frac{P_z}{4\pi G}\left(-\frac{y}{r^3}, \frac{x}{r^3}, 0\right) = 0 \qquad ...(11)$$

Now $\qquad \nabla^2\left(\frac{x}{r}\right) = -\frac{2x}{r^3}$

etc. so that we can write these equations in the form

$$\nabla^2\left[\varphi - \frac{zP_z}{8\pi(\lambda + 2G)r}\right] = 0$$

$$\nabla^2\left[\underline{\psi} + \frac{P_z}{8\pi G}\left(\frac{y}{r}, -\frac{x}{r}, 0\right)\right] = 0$$

from which it follows that

$$\varphi = \frac{P_z}{8\pi(\lambda + 2G)}\left(\frac{z}{r}\right) \qquad ...(12)$$

$$\underline{\psi} = \frac{P_z}{8\pi G}\left(-\frac{y}{r}, \frac{x}{r}, 0\right) \qquad ...(13)$$

Now from Eq. (12), we have

$$\text{grad}\,\varphi = \frac{P_z}{8\pi(\lambda + 2G)}\left(-\frac{xz}{r^3}, -\frac{yz}{r^3}, \frac{1}{r} - \frac{z^2}{r^3}\right)$$

and from Eq. (13),

$$\text{curl}\,\underline{\psi} = \frac{P_z}{8\pi G}\left(\frac{xz}{r^3}, \frac{yz}{r^3}, \frac{1}{r} + \frac{z^2}{r^3}\right)$$

so that substituting in Eq. (1) of art. 7.5, we get

$$\underline{u} = \frac{(\lambda + G)\,P_z}{8\pi G(\lambda + 2G)}\left(\frac{xz}{r^3}, \frac{yz}{r^3}, \frac{z^2}{r^3} + \frac{\lambda + 3G}{\lambda + G}\cdot\frac{1}{r}\right) \qquad \text{...(14)}$$

Eqs. (7.57) can be written as

$$\left.\begin{aligned} u_x &= \frac{P_z}{16\pi G(1-\nu)}\cdot\frac{xz}{r^3} \\ u_y &= \frac{P_z}{16\pi G(1-\nu)}\cdot\frac{yz}{r^3} \\ u_z &= \frac{P_z}{16\pi G(1-\nu)}\left[\frac{z^2}{r^3} + (3-4\nu)\frac{1}{r}\right] \end{aligned}\right\} \qquad \text{...(7.6)}$$

and the components of stress tensor become

$$\left.\begin{aligned} \sigma_x &= \frac{P_z\,.\,z}{8\pi(1-\nu)\,r^3}\left(1 - 2\nu - \frac{3x^2}{r^2}\right) \\ \sigma_y &= \frac{P_z\,.\,z}{8\pi(1-\nu)\,r^3}\left(1 - 2\nu - \frac{3y^2}{r^2}\right) \\ \sigma_z &= \frac{P_z\,.\,z}{8\pi(1-\nu)\,r^3}\left(1 - 2\nu + \frac{3z^2}{r^2}\right) \\ \tau_{xy} &= -\frac{3P_z\,xyz}{8\pi(1-\nu)\,r^5} \\ \tau_{yz} &= \frac{P_z y}{8\pi(1-\nu)\,r^3}\left(1 - 2\nu + \frac{3z^2}{r^2}\right) \\ \tau_{zx} &= -\frac{P_z x}{8\pi(1-\nu)\,r^3}\left(1 - 2\nu + \frac{3z^2}{r^2}\right) \end{aligned}\right\} \qquad \text{...(7.7)}$$

In spherical polar co-ordinates (r, θ, ϕ), we have

$$\left.\begin{aligned} u_r &= \frac{P_z \cos\theta}{4\pi\,Gr}\,.\; u_\theta = -\frac{3-4\nu}{4(1-\nu)}\cdot\frac{P_z \sin\theta}{4\pi\,Gr} \\ u_\phi &= 0 \end{aligned}\right\} \qquad \text{...(7.8)}$$

$$\left.\begin{aligned}
\sigma_r &= -\left(\frac{4-2\nu}{1-\nu}\right)\frac{P_z\cos\theta}{8\pi r^2}\\
\sigma_\theta &= \left(\frac{1-2\nu}{1-\nu}\right)\frac{P_z\cos\theta}{8\pi r^2}\\
\sigma_\phi &= \left(\frac{1-2\nu}{1-\nu}\right)\frac{P_z\cos\theta}{8\pi r^2}\\
\tau_{r\theta} &= \left(\frac{1-2\nu}{1-\nu}\right)\frac{P_z\sin\theta}{8\pi r^2}\\
\tau_{r\phi} &= 0,\ \tau_{\theta\phi} = 0.
\end{aligned}\right\} \quad \text{...(7.9)}$$

If the force P_x is directed along the x-axis, *i.e.* along the line of intersection of the equatorial plane ($\theta = \pi/2$) and the azimuthal plane from which ϕ is measured, then

$$\left.\begin{aligned}
u_r &= \frac{P_z}{4\pi Gr}\sin\theta\cos\phi\\
u_\theta &= \frac{3-4\nu}{4(1-\nu)}\cdot\frac{P_z}{4\pi Gr}\cos\theta\cos\phi
\end{aligned}\right\} \quad \text{...(7.10)}$$

$$\left.\begin{aligned}
u_\phi &= -\frac{3-4\nu}{4(1-\nu)}\frac{P_z}{8\pi Gr}\sin\phi\\
\sigma_r &= -\left(\frac{4-2\nu}{1-\nu}\right)\frac{P_z}{8\pi r^2}\sin\theta\cos\phi\\
\sigma_\theta = \sigma_\phi &= \left(\frac{1-2\nu}{1-\nu}\right)\frac{P_z}{8\pi r^2}\sin\theta\cos\phi\\
\tau_{r\theta} &= -\left(\frac{1-2\nu}{1-\nu}\right)\frac{P_z}{8\pi r^2}\cos\theta\cos\phi\\
\tau_{r\phi} &= \left(\frac{1-2\nu}{1-\nu}\right)\frac{P_z}{8\pi r^2}\sin\phi
\end{aligned}\right\} \quad \text{...(7.11)}$$

$$\tau_{\theta\phi} = 0$$

More generally, the displacement due to a force $\underline{B}$ acting at the point with position vector $\underline{a}$, is given by the equation

$$\underline{u} = \frac{\lambda+G}{8\pi G(\lambda+2G)}\left[(\underline{r}-\underline{a})\frac{(\underline{r}-\underline{a})\underline{B}}{|\underline{r}-\underline{a}|^3} + \frac{\lambda+3G}{\lambda+G}\cdot\frac{\underline{B}}{|\underline{r}-\underline{a}|}\right] \quad \text{...(7.12)}$$

when force $\underline{B}$ ($\underline{r}$) act through a volume V of finite size, then

$$\underline{u} = \frac{\lambda+G}{8\pi G(\lambda+2G)}\int_V\left[(\underline{r}-\underline{r}')\frac{(\underline{r}-\underline{r}')\cdot\underline{B}(\underline{r}')}{|\underline{r}-\underline{r}'|^3} + \frac{\lambda+3G}{\lambda+G}\cdot\frac{\underline{B}(\underline{r}')}{|\underline{r}-\underline{r}'|}\right]d\tau' \quad \text{...(7.13)}$$

7.7 The Boussinesq-Papkovich Solution

Now we shall derive a general solution of the equations of equilibrium in terms of a scalar function and a vector function. Let

$$\underline{u} = A \text{ grad } (\varphi + \underline{r} \cdot \underline{\psi}) + B\,\underline{\psi} \qquad ...(1)$$

where $\underline{r}$ denotes the position vector of a field point and A and B are constants, whose values are as yet undetermined. Making use of vector identity

$$\text{curl curl} = \text{grad div.} - \nabla^2$$

and the fact that the operator curl grad is a null operator we find that

$$\text{curl curl } \underline{u} = B\,(\text{grad div. } \underline{\psi} - \nabla^2 \underline{\psi}) \qquad ...(2)$$

Since div. grad = ∇^2

and $$\nabla^2(\underline{r} \cdot \underline{\psi}) = \underline{r} \cdot \nabla^2 \underline{\psi} + 2 \text{ div. } \underline{\psi}$$

we find that

$$\text{div. } \underline{u} = A\nabla^2\varphi + A\underline{r} \cdot \nabla^2\underline{\psi} + (2A + B) \text{ div. } \underline{\psi} \qquad ...(3)$$

Substituting from Eqs. (7.61) and (7.62) into Eq. (7.43) we obtain

$$(\lambda + 2G)\,A \text{ grad } (\nabla^2\varphi + \underline{r} \cdot \nabla^2 \underline{\psi})$$

$$+ [(\lambda + 2G)(B + 2A) - GB] \text{ grad div } \underline{\psi} + B\nabla^2\underline{\psi} + \frac{\underline{B}}{G} = 0$$

This equation is satisfied if we can choose A, B, φ, $\underline{\psi}$ so that

$$\nabla^2\varphi + \underline{r} \cdot \nabla^2\underline{\psi} = 0$$

$$(\lambda + 2G)(B + 2A) - GB = 0$$

$$B\nabla^2\underline{\psi} + \frac{\underline{B}}{G} = 0$$

The second of these equations is satisfied if we take

$$A = 1$$

and $$B = -\frac{2(\lambda + 2G)}{\lambda + G} = -4(1 - \nu)$$

Putting these in Eq. (2), we get

$$\underline{u} = \text{grad } (\varphi + \underline{r} \cdot \underline{\psi}) - 4\,(1 - \nu)\,\underline{\psi} \qquad ...(4)$$

Eq. (4) is a solution of the vector equation of equilibrium (7.5) provided that φ and $\underline{\psi}$ are solutions of the equations

$$\left.\begin{aligned} 4(1 - \nu)\,\nabla^2\varphi + \frac{1}{B}(\underline{r} \cdot \underline{B}) &= 0 \\ 4(1 - \nu)\nabla^2\,\underline{\psi} &= \frac{\underline{B}}{G} \end{aligned}\right\} \qquad ...(5)$$

In the absence of body forces the Eqs. (5) reduce to

$$\nabla^2\varphi = 0, \quad \nabla^2 \underline{\psi} = 0 \qquad \text{...(6)}$$

showing that, in this case, Eq. (4) yields a solution provided that φ and each component of $\underline{\psi}$ is a harmonic function.

In the absence of body forces if we form the divergence of both sides of Eq. (4), we find that

$$\nabla^2 (\text{div. } \underline{u}) = 0$$

showing that e = div. $\underline{u}$ is harmonic function. Similarly if we form the curl of both sides of Eq. (4), then we find

$$\nabla^2(\text{curl } \underline{u}) = 0$$

showing that each of the components of rotation

$$\omega = \frac{1}{2} \text{ curl } \underline{u}$$

is a harmonic function.

When the body force is a constant, $\underline{F}$ say, then

$$\nabla^2(r^2 \underline{F}) = 6 \underline{F}$$

$$\nabla^2(\underline{R} \cdot \underline{F}) = 6(\underline{r} \cdot \underline{F})$$

where $\underline{R}$ denotes the vector with components (x^3, y^3, z^3) and the equations of equilibrium (5) have a solution

$$\varphi = \varphi_1 - \frac{\underline{R} \cdot \underline{F}}{24(1-\nu)G}$$

$$\underline{\psi} = \underline{\psi}_1 + \frac{r^2 \underline{F}}{24(1-\nu)G} \qquad \text{...(7)}$$

where φ_1 and each of the components of $\underline{\psi}_1$ is a harmonic function.

If we put $\underline{\psi} = 0$ in Eq. (2), then we get

$$\underline{u} = A \text{ grad } \varphi \qquad \text{...(8)}$$

For such a displacement field

$$\text{div. } \underline{u} = A\nabla^2\varphi = 0 \qquad \text{...(9)}$$

so that the dilatation is zero everywhere.

The components of stress are given by

$$\left.\begin{aligned} \sigma_x &= 2GA \frac{\partial^2\varphi}{\partial x^2}, \ \sigma_y = 2GA \frac{\partial^2\varphi}{\partial y^2}, \ \sigma_z = 2GA \frac{\partial^2\varphi}{\partial z^2} \\ \sigma_{xy} &= 2GA \frac{\partial^2\varphi}{\partial x \partial y}, \ \sigma_{yz} = 2GA \frac{\partial^2\varphi}{\partial y \partial z}, \ \tau_{zx} = 2GA \frac{\partial^2\varphi}{\partial z \partial x} \end{aligned}\right\} \qquad \text{...(7.14)}$$

For example, let $\varphi = \log(z + r)$

then $$\underline{u} = A\left[\frac{x}{r(z+r)}, \frac{y}{r(z+r)}, \frac{1}{r}\right]$$

and $$\sigma_x = 2GA\left[\frac{y^2+z^2}{r^3(z+r)} - \frac{x^2}{r^2(z+r)^2}\right]$$

$$\sigma_y = 2GA\left[\frac{z^2+x^2}{r^3\,(z+r)} - \frac{y^2}{r^2(z+r)^2}\right]$$

$$\sigma_z = -\,2GA\,\frac{x}{r^3}$$

$$\tau_{xy} = -\,2GA\,\frac{xy(z+2r)}{r^3\,(z+r)^2}$$

$$\tau_{yz} = -\,2GA\,\frac{y}{r^3}$$

$$\tau_{zx} = -\,2GA\,\frac{x}{r^3}$$

7.8 The Use of Potential Functions in Elasticity

If we choose only one harmonic function which is such that the shearing stress vanishes across a plane which may be chosen to be $z = 0$. Using rectangular Cartesian co-ordinates, let us put

$$\varphi = (1-2\nu)\,\chi,$$

$$\underline{\psi} = \left(0,\,0,\,\frac{\partial\chi}{\partial z}\right)$$

in Eq. (4) of art. 7.7. Where χ is a harmonic function. Then we obtain

$$\left.\begin{aligned} u_x &= z\,\frac{\partial^2\chi}{\partial x\,\partial z} + (1-2\nu)\,\frac{\partial\chi}{\partial x} \\ u_y &= z\,\frac{\partial^2\chi}{\partial y\,\partial z} + (1-2\nu)\,\frac{\partial\chi}{\partial y} \\ u_z &= z\,\frac{\partial^2\chi}{\partial z^2} - 2(1-\nu)\,\frac{\partial\chi}{\partial z} \end{aligned}\right\} \quad \text{...(7.15)}$$

and the components of stress tensor become

$$\sigma_x = 2G\left[z\,\frac{\partial^3\chi}{\partial x^2\,\partial z} + (1-2\nu)\,\frac{\partial^2\chi}{\partial x^2} - 2\nu\,\frac{\partial^2\chi}{\partial z^2}\right] \quad \text{...(7.16}a\text{)}$$

$$\sigma_y = 2G\left[z\,\frac{\partial^3\chi}{\partial y^2\,\partial z} + (1-2\nu)\,\frac{\partial^2\chi}{\partial y^2} - 2\nu\,\frac{\partial^2\chi}{\partial z^2}\right] \quad \text{...(7.16}b\text{)}$$

$$\sigma_z = 2G\left(z\,\frac{\partial^3\chi}{\partial z^3} - \frac{\partial^2\chi}{\partial z^2}\right) \quad \text{...(7.16}c\text{)}$$

$$\tau_{xy} = 2G\left[z\,\frac{\partial^3\chi}{\partial x\,\partial y\,\partial z} + (1-2\nu)\,\frac{\partial^2\chi}{\partial x\,\partial y}\right] \quad \text{...(7.16}d\text{)}$$

$$\tau_{yz} = 2Gz \frac{\partial^3 \chi}{\partial y \partial z^2} \qquad \text{...(7.16}e\text{)}$$

$$\tau_{zx} = 2Gz \frac{\partial^3 \chi}{\partial x \partial z^2} \qquad \text{...(7.16}f\text{)}$$

As $z \to 0$, then

$$\tau_{xz} = \tau_{yz} = 0 \qquad \text{...(7.17)}$$

and

$$(\sigma_z)_{z=0} = -2G\left(\frac{\partial^2 \chi}{\partial z^2}\right)_{z=0} \qquad \text{...(7.18)}$$

$$(u_z)_{z=0} = -2(1-\nu)\left(\frac{\partial \chi}{\partial z}\right)_{z=0} \qquad \text{...(7.19)}$$

In cylindrical polar co-ordinates (r, θ, z) we obtain the following solution

$$\left.\begin{aligned} u_r &= z\frac{\partial^2 \chi}{\partial r \partial z} + (1-2\nu)\frac{\partial \chi}{\partial r} \\ u_\theta &= \frac{z}{r}\frac{\partial^2 \chi}{\partial \theta \partial z} + (1-2\nu)\frac{1}{r}\frac{\partial \chi}{\partial \theta} \\ &= z\frac{\partial^2 \chi}{\partial z^2} - 2(1-\nu)\frac{\partial \chi}{\partial z} \end{aligned}\right\} \qquad \text{...(7.20)}$$

and the components of stress tensor are given by :

$$\left.\begin{aligned} \sigma_r &= 2G\left[z\frac{\partial^3 \chi}{\partial r^2 \partial z} + \frac{\partial^2 \chi}{\partial r^2} - 2\nu\left(\frac{\partial^2 \chi}{\partial r^2} + \frac{\partial^3 \chi}{\partial z^2}\right)\right] \\ \sigma_\theta &= -2G\left[z\frac{\partial^3 \chi}{\partial r^2 \partial z} + z\frac{\partial^3 \chi}{\partial z^3} + \frac{\partial^2 \chi}{\partial z^2} - (1-2\nu)\frac{\partial^2 \chi}{\partial r^2}\right] \\ \sigma_z &= 2G\left[z\frac{\partial^3 \chi}{\partial z^3} - \frac{\partial^2 \chi}{\partial z^2}\right] \\ \tau_{r\theta} &= G\left[\frac{2z}{r}\frac{\partial^3 \chi}{\partial r \partial \theta \partial z} - \frac{z}{r^2}\cdot\frac{\partial^2 \chi}{\partial \theta \partial z} - \frac{(1-2\nu)}{r^2}\frac{\partial \chi}{\partial \theta}\right] \\ \tau_{rz} &= 2Gz\frac{\partial^3 \chi}{\partial r \partial z^2} \\ \tau_{\theta z} &= \frac{2Gz}{r}\frac{\partial^3 \chi}{\partial \theta \partial z^2} \end{aligned}\right\} \qquad \text{...(7.21)}$$

where $\nabla^2 \chi = 0$ and

$$\nabla^2 = \frac{\partial^2}{\partial r^2} + \frac{1}{r}\frac{\partial}{\partial r} + \frac{1}{r^2}\frac{\partial^2}{\partial \theta^2} + \frac{\partial^2}{\partial z^2}$$

In problems in which there is axial symmetry

$$\frac{\partial \chi}{\partial \theta} = 0$$

7.9 Betti's Method

Betti's method rests on the fact that dilatation e is a harmonic function as proved in art. 7.7. From this it follows that

$$\nabla^2(\underline{r}e) = 2 \text{ grad } e$$

so that equations of motion (7.4) can be written in the form

$$(1 - 2\nu)\, \nabla^2 \underline{u} + \tfrac{1}{2}\, \nabla^2(\underline{r}\, e) = 0$$

which is equivalent to

$$\nabla^2 \left[\underline{u} + \frac{1}{2(1-2\nu)} \underline{r}\, e \right] = 0 \qquad \text{...(7.22)}$$

Hence if e is known throughout the solid and the surface value of the vector $\underline{u}$ is prescribed, then the determination of the displacement field reduces to a problem in Potential theory. The basic problem, therefore, consists in determining e when the surface values of either the displacement or the applied forces are known. The basic tool in solving problems is Betti's reciprocal theorem which states that, if there are on body forces acting in the region bounded by a surface S, then

$$\int_S (\underline{F}, \underline{u}')\, dS = \int_S (\underline{F}' \,.\, \underline{u}) dS \qquad \text{...(7.23)}$$

where $\underline{u}$ is the displacement satisfying the equilibrium equations and $\underline{F}$ is the corresponding surface force, and also $\underline{u}'$ is a second displacement and $\underline{F}'$ is the corresponding surface force.

7.10 Formulae for the Dilatation

For convenience let us take the point in which we are interested to determine the dilatation as the origin of co-ordinates. We surround that point by a sphere Σ of infinitesimal radius, and we pass to a limit by letting this radius tend to zero. Eq. (7.22) then takes then form

$$\int_S (X_i u_i' - X_i' u_i)\, dS = \int_S (X_i' u_i - X_i u_i') dS \qquad \text{...(1)}$$

where we have written

$$\underline{F} = (X_1, X_2, X_3)$$

$$\underline{u} = (u_1, u_2, u_3) \text{ etc.}$$

If we take

$$u_i' = \frac{\partial}{\partial x_i}\left(\frac{1}{r}\right) \qquad ...(2)$$

then
$$X_i' = 2Gn_j \frac{\partial^2}{\partial x_i \partial x_j}\left(\frac{1}{r}\right) \qquad ...(3)$$

where $\underline{n} = (n_1, n_2, n_3)$ is the normal to the surface. On the sphere Σ

$$n_j = -\frac{x_j}{r} \qquad ...(4)$$

so that
$$\int_\Sigma X_i u_i' dS = -\int_\Sigma \frac{x_j}{r}\left(\lambda e\delta_{ij} + G\frac{\partial u_i}{\partial x_j} + G\frac{\partial u_j}{\partial x_i}\right) \times \frac{\partial}{\partial x_i}\left(\frac{1}{r}\right) dS$$

$$= \int_\Sigma \frac{x_i x_j}{r^4}\left(\lambda e\delta_{ij} + G\frac{\partial u_i}{\partial x_j} + G\frac{\partial u_j}{\partial x_i}\right) dS.$$

Now each integral of the type $\int_\Sigma x_1\, x_2\, dS = 0$ and each of the type $\int_\Sigma x_1^2\, dS = \frac{4}{3}\pi r^4$ so that in the limit as the radius of Σ tends to zero.

$$\int_\Sigma X_i u_i'\, dS = 4\pi\left(\lambda + \frac{2}{3}G\right) e_0$$

where e_0 denotes the value of e at the origin.

Similarly, $\int_\Sigma u_i X_i'\, dS = -4G\int_\Sigma \frac{x_i u_i}{r^4}\, dS$

Expanding u_i by Maclaurin's theorem

$$u_i = (u_i)_0 + x_j\left(\frac{\partial u_i}{\partial x_j}\right)_0 + \ldots$$

We find that $\int_\Sigma u_i X_i'\, dS = -4G\int_\Sigma \dfrac{x_i(u_i)_0 + x_i x_j\left(\dfrac{\partial u_i}{\partial x_j}\right)_0}{r^4}\, dS$

and in the limit as the radius of Σ tends to zero, we get

$$\int_\Sigma u_i X_i'\, dS = -\frac{16}{3}\pi G\left(\frac{\partial u_i}{\partial x_j}\right)_0$$

$$= -\frac{16}{3}\pi G e_0$$

Substituting these values in Eq. (7.79), we get

$$e_0 = \frac{1}{4\pi(\lambda + 2G)} \int_S (X_i' u_i - X_i u_i') dS \qquad \text{...(5)}$$

To evaluate the integral on the right hand side of Eq. (5) we should need to know the surface values of both the displacement and the applied force. To convert the result into one containing the surface values of one or the other, we consider two cases.

(*i*) *Surface displacements prescribed.* Suppose we can find a displacement vector $\underline{u}'$ which satisfies the vector equation of equilibrium at all points within S and which is identical with the vector u' at all points on S. Let the corresponding surface forces be X_i'' (i = 1, 2, 3). By the reciprocal theorem (7.23), we have

$$\int X_i'' u_i dS = \int X_i u_i'' dS \int X_i u_i' dS \qquad \text{...(6)}$$

Combining Eqs. (5) and (6), we obtain

$$e_0 = \frac{1}{4\pi(\lambda + 2G)} \int_S (X_i' - X_i'') u_i dS \qquad \text{...(7)}$$

By definition $(X_i' - X_i'')$ are the components of the surface force necessary to hold the surface fixed when there is a 'centre of compression' at the origin. Thus to determine the dilatation at any point we must therefore find a displacement which (*a*) satisfies the usual conditions of continuity and the equations of equilibrium everywhere except at the point, (*b*) in the neighbourhood of the point tends to become infinite, as if there were a centre of compression at the point, (*c*) vanishes at the surface.

(*ii*) *Surface forces prescribed.* The components X_i' form a system of surface forces which satisfy the conditions of rigid body equilibrium. Suppose that u_i'' are the components of the displacement produced in the body by the application of these surface tractions. The applying reciprocal theorem we obtain

$$\int X_i' u_i dS = \int X_i u_i'' \, dS \qquad \text{...(8)}$$

Combining Eqs. (5) and Eq. (8), we get

$$e_0 = \frac{1}{4\pi(\lambda + 2G)} \int_S X_i (u_i'' - u_i') dS \qquad \text{...(9)}$$

To find the dilatation at any point we need therefore only find the displacement field produced in the body when the surface is free from traction and there is a centre of dilatation at the point.

7.11 The Method of Integral Transforms

7.11.1 The Stresses in an Infinite Medium due to Body Forces

Equilibrium equations in Cartesian coordinates are :

$$\left.\begin{aligned}\frac{\partial \sigma_x}{\partial x}+\frac{\partial \tau_{xy}}{\partial y}+\frac{\partial \tau_{xz}}{\partial z}+B_x=0\\ \frac{\partial \tau_{xy}}{\partial x}+\frac{\partial \sigma_y}{\partial y}+\frac{\partial \tau_{yz}}{\partial z}+B_y=0\\ \frac{\partial \tau_{xz}}{\partial x}+\frac{\partial \tau_{yz}}{\partial y}+\frac{\partial \sigma_z}{\partial z}+B_z=0\end{aligned}\right\} \quad ...(1)$$

Stress-displacement relations are

$$(\sigma_x, \sigma_y, \sigma_z)=\lambda\left(\frac{\partial u_x}{\partial x}+\frac{\partial u_y}{\partial y}+\frac{\partial u_z}{\partial z}\right)+2G\left(\frac{\partial u_x}{\partial x}, \frac{\partial u_y}{\partial y}, \frac{\partial u_z}{\partial z}\right) \quad ...(2)$$

$$(\tau_{xy}, \tau_{yz}, \tau_{xz})=G\left(\frac{\partial u_x}{\partial y}+\frac{\partial u_y}{\partial x}, \frac{\partial u_y}{\partial z}+\frac{\partial u_z}{\partial y}, \frac{\partial u_x}{\partial z}+\frac{\partial u_z}{\partial x}\right) \quad ...(3)$$

To solve these equations we introduce the Fourier transform $\bar{f}$ of every quantity f occurring in them, defined by the equation

$$\bar{f}=\frac{1}{(2\pi)^{3/2}}\int_{-\infty}^{\infty}\int_{-\infty}^{\infty}\int_{-\infty}^{\infty} f e^{i(\xi x+\eta y+\zeta z)}\, dx dy dz \quad ...(4)$$

If we multiply both sides of each equation of the sets (7.88), (2) and (3) by $e^{i(\xi x+\eta y+\zeta z)}$, integrate throughout the entire space and make use of the results

$$\frac{1}{(2\pi)^{3/2}}\int_{-\infty}^{\infty}\int_{-\infty}^{\infty}\int_{-\infty}^{\infty}\frac{\partial f}{\partial x}\cdot e^{i(\xi x+\eta y+\zeta z)}\, dx dy dz=-i\xi\bar{f}$$

etc., we see that these partial differential equations are equivalent to the set of algebraic equations :

$$\left.\begin{aligned}\xi\bar{\sigma}_x+\eta\bar{\tau}_{xy}+\zeta\bar{\tau}_{xz}=-i\bar{B}_x\\ \xi\bar{\tau}_{xy}+\eta\sigma_y+\zeta\bar{\tau}_{yz}=-i\bar{B}_y\\ \xi\bar{\tau}_{xz}+\eta\bar{\tau}_{yz}+\zeta\bar{\sigma}_z=-i\bar{B}_z\end{aligned}\right\} \quad ...(5)$$

$$i(\bar{\sigma}_x, \bar{\sigma}_y, \bar{\sigma}_z)=\lambda(\xi\bar{u}_x+\eta\bar{u}_y+\zeta\bar{u}_z)\,|\,2G\,(\xi\bar{u}_x, \eta\bar{u}_y, \zeta\bar{u}_z) \quad ...(6)$$

$$i(\bar{\tau}_{xy}, \bar{\tau}_{yz}, \bar{\tau}_{xz}) = G(\eta\bar{u}_x + \xi\bar{u}_y, \zeta\bar{u}_y + \eta\bar{u}_z, \xi\bar{u}_z + \zeta\bar{u}_x) \qquad ...(7)$$

Substituting from Eqs. (6), (7) into Eqs. (5), we obtain

$$\left.\begin{aligned}
[\xi^2 + (1-\nu)(\xi^2+\eta^2+\zeta^2)]\,\bar{u}_x + \xi\eta\bar{u}_y + \xi\zeta\bar{u}_z - (1-2\nu)\frac{\bar{B}_x}{G} &= 0\\
\xi\eta\bar{u}_x + [\eta^2 + (1-2\nu)(\xi^2+\eta^2+\zeta^2)]\,\bar{u}_y + \eta\zeta\bar{u}_z - (1-2\nu)\frac{\bar{B}_y}{G} &= 0\\
\xi\zeta\bar{u}_x + \eta\zeta\bar{u}_y + [\zeta^2 + (1-2\nu)(\xi^2+\eta^2+\zeta^2)]\,\bar{u}_z - (1-2\nu)\frac{\bar{B}_z}{G} &= 0
\end{aligned}\right\} \qquad ...(8)$$

Solving these equations, we find

$$\bar{u}_x = \frac{1}{2(1-\nu)G(\xi^2+\eta^2+\zeta^2)^2}[2(1-\nu)(\xi^2+\eta^2+\zeta^2)\,\bar{B}_x - \xi\,(\xi\,\bar{B}_x + \eta\bar{B}_y + \zeta\bar{B}_z)] \qquad ...(7.24)$$

$$\bar{u}_y = \frac{1}{2(1-\nu)G(\xi^2+\eta^2+\zeta^2)^2}[2(1-\nu)(\xi^2+\eta^2+\zeta^2)\,\bar{B}_y - \eta\,(\xi\,\bar{B}_x + \eta\bar{B}_y + \zeta\bar{B}_z)] \qquad ...(7.25)$$

$$\bar{u}_z = \frac{1}{2(1-\nu)G(\xi^2+\eta^2+\zeta^2)^2}[2(1-\nu)(\xi^2+\eta^2+\zeta^2)\,\bar{B}_z - \zeta\,(\xi\,\bar{B}_x + \eta\bar{B}_y + \zeta\bar{B}_z)] \qquad ...(7.26)$$

for the Fourier transforms of the components of the displacement vector.

EXERCISES

7.1 A rectangular bar of length a, breadth b and height h is standing under its own weight. Determine the stress distribution in the bar. Show that this is an exact unique solution away from the support.

7.2 A prismatic bar of $2a \times 2b$ cross-section is bent by two equal and opposite couples. Determine the equations for the bent shape of the prismatic bar.

7.3 Explain the different methods of solving three-dimensional elasticity problems.

8

Axisymmetric Problems

8.1 Introduction

The determination of the distribution of stress in a solid of revolution which is deformed symmetrically with respect to its axis of revolution is of great practical importance. The deformation being symmetrical with respect to the axis of revolution, it is most convenient to employ cylindrical coordinates r, θ, z and to take the z-axis along the axis of symmetry. In these circumstances, the components of the displacement vector and of the stress tensor will all be independent of the angle θ, and the operator $\frac{\partial}{\partial \theta}$ will be a null operator. In this chapter, we shall study those problems having torsionless axisymmetry.

8.2 Basic Equations

(i) *Equations of equilibrium neglecting body forces*

$$\left.\begin{aligned}\frac{\partial \sigma_r}{\partial r} + \frac{\partial \tau_{rz}}{\partial z} + \frac{\sigma_r - \sigma_\theta}{r} &= 0 \\ \frac{\partial \tau_{rz}}{\partial r} + \frac{\partial \sigma_z}{\partial z} + \frac{\tau_{rz}}{r} &= 0\end{aligned}\right\} \qquad \text{...(8.1)}$$

(ii) *Strain-displacement relations*

$$\left.\begin{aligned}\varepsilon_r &= \frac{\partial u_r}{\partial r},\ \varepsilon_\theta = \frac{u_r}{r},\ \varepsilon_z = \frac{\partial u_z}{\partial z} \\ \gamma_{rz} &= \frac{\partial u_r}{\partial z} + \frac{\partial u_z}{\partial r}\end{aligned}\right\} \qquad \text{...(8.2)}$$

(iii) *Stress-strain relations*

$$\left.\begin{aligned}\varepsilon_r &= \frac{1}{E}[\sigma_r - \nu(\sigma_\theta + \sigma_z)] \\ \varepsilon_\theta &= \frac{1}{E}[\sigma_\theta - \nu(\sigma_r + \sigma_z)] \\ \varepsilon_z &= \frac{1}{E}[\sigma_z - \nu(\sigma_r + \sigma_\theta)] \\ \gamma_{rz} &= \frac{\tau_{rz}}{G}\end{aligned}\right\} \qquad \text{...(8.3)}$$

(iv) Stress-displacement relations

$$\left.\begin{aligned}
\sigma_r &= \lambda\left(\frac{\partial u_r}{\partial r}+\frac{u_r}{r}+\frac{\partial u_z}{\partial_z}\right)+2G\,.\,\frac{\partial u_r}{\partial r}\\
\sigma_\theta &= \lambda\left(\frac{\partial u_r}{\partial r}+\frac{u_r}{r}+\frac{\partial u_z}{\partial z}\right)+2G\,.\,\frac{u_r}{r}\\
\sigma_z &= \lambda\left(\frac{\partial u_r}{\partial r}+\frac{u_r}{r}+\frac{\partial u_z}{\partial z}\right)+2G\,.\,\frac{\partial u_z}{\partial z}\\
\tau_{rz} &= G\left(\frac{\partial u_r}{\partial z}+\frac{\partial u_z}{\partial r}\right)
\end{aligned}\right\}\quad ...(8.4)$$

(v) Stress-compatibility equations

$$\left.\begin{aligned}
\nabla^2\sigma_r-\frac{2}{r^2}(\sigma_r-\sigma_\theta)+\left(\frac{1}{1+\nu}\right)\frac{\partial^2}{\partial r^2}(\sigma_r+\sigma_\theta+\sigma_z)=0\\
\nabla^2\sigma_\theta+\frac{2}{r^2}(\sigma_r-\sigma_\theta)+\left(\frac{1}{1+\nu}\right)\frac{1}{r}\frac{\partial}{\partial r}(\sigma_r+\sigma_\theta+\sigma_z)=0
\end{aligned}\right\}\quad ...(8.5)$$

where $\nabla^2=\frac{\partial^2}{\partial r^2}+\frac{1}{r}\frac{\partial}{\partial r}+\frac{\partial}{\partial z}$

(vi) Stress function Φ. If we put

$$\left.\begin{aligned}
u_r &= -\frac{1+\nu}{E}\cdot\frac{\partial^2\Phi}{\partial r\partial z}\\
u_z &= \frac{1+\nu}{E}\left[(1-2\nu)\,\nabla^2\Phi+\frac{\partial^2\Phi}{\partial r^2}+\frac{1}{r}\frac{\partial\Phi}{\partial r}\right]
\end{aligned}\right\}\quad ...(8.6)$$

in Eqs. (8.4) we find that

$$\left.\begin{aligned}
\sigma_r &= \frac{\partial}{\partial z}\left[\nu\,\nabla^2\Phi-\frac{\partial^2\Phi}{\partial r^2}\right]\\
\sigma_\theta &= \frac{\partial}{\partial z}\left[\nu\,\nabla^2\Phi-\frac{1}{r}\frac{\partial\Phi}{\partial r}\right]\\
\sigma_z &= \frac{\partial}{\partial z}\left[(2-\nu)\,\nabla^2\Phi-\frac{\partial^2\Phi}{\partial z^2}\right]\\
\tau_{rz} &= \frac{\partial}{\partial r}\left[(1-\nu)\,\nabla^2\Phi-\frac{\partial^2\Phi}{\partial z^2}\right]
\end{aligned}\right\}\quad ...(8.7)$$

If we substitute these equations into Eqs. (8.1), we find they are satisfied identically and the compatibility equation $\nabla^4\Phi = 0$ is also satisfied.

Eqs. (8.6) may be written as

$$\left.\begin{aligned} 2Gu_r &= -\frac{\partial^2\Phi}{\partial r\partial z} \\ 2Gu_z &= 2(1-\nu)\nabla^2\Phi - \frac{\partial^2\Phi}{\partial z^2} \end{aligned}\right\} \qquad \text{...(8.8)}$$

If we begin with a displacement so expressed in terms of a function Φ satisfying the compatibility equation $\nabla^4\Phi = 0$, we can derive from it the strain components for Eqs. (8.2) and then the stress components from Eqs. (8.7). A particular problem is solved if we can find such a function Φ that also satisfies the boundary conditions.

Now the compatibility equation is

$$\left(\frac{\partial^2}{\partial r^2} + \frac{1}{r}\frac{\partial}{\partial r} + \frac{\partial^2}{\partial z^2}\right)\left(\frac{\partial^2\Phi}{\partial r^2} + \frac{1}{r}\frac{\partial\Phi}{\partial r} + \frac{\partial^2\Phi}{\partial z^2}\right) = \nabla^2.\nabla^2\phi = 0 \qquad \text{...(8.9)}$$

This equation can be expressed in spherical polar coordinates (R, ϕ) as

$$\left(\frac{\partial^2}{\partial R^2} + \frac{2}{R}\frac{\partial}{\partial R} + \frac{1}{R^2}\cot\phi\frac{\partial}{\partial\phi} + \frac{1}{R^2}\frac{\partial^2}{\partial\phi^2}\right)^2 \Phi = 0 \qquad \text{...(8.10)}$$

8.3 Solution by Polynomials

Now solutions of Eq. (8.10) are also the solutions of the Laplace equation

$$\left(\frac{\partial^2}{R^2} + \frac{2}{R}\frac{\partial}{\partial R} + \frac{1}{R^2}\cot\phi\frac{\partial}{\partial\phi} + \frac{1}{R^2}\frac{\partial^2}{\partial\phi^2}\right)\Phi = 0 \qquad \text{...(8.11)}$$

A particular solution of this equation can be taken in the form of a polynomial

$$\Phi_n = R^n\varphi_n \qquad \text{...(8.12)}$$

in which φ_n is a function of angle ϕ only.

Substituting Eq. (8.12) in (8.11) we find

$$\frac{1}{\sin\phi}\frac{d}{d\phi}\left(\sin\phi\frac{\partial\varphi_n}{\partial\phi}\right) + n(n+1)\,\varphi_n = 0 \qquad \text{...(8.13)}$$

Let $x = \cos\phi$, then Eq. (8.13) becomes

$$(1-x^2)\frac{d^2\varphi_n}{dx^2} - 2x\frac{d\varphi_n}{dx} + n(n+1)\,\varphi_n = 0 \qquad \text{...(8.14)}$$

This is Legendre's equation. Its two basic solutions are given by $P_n(x)$ and $Q_n(x)$ known as the Legendre functions of the first and second kinds respectively. For $n = 0, 1, 2, 3, \ldots$, the $P_n(x)$ are the Legendre polynomials.

$$P_0(x) = 1, P_1(x) = x, \quad P_2(x) = \frac{1}{2}(3x^2 - 1)$$

$$P_3(x) = \frac{1}{2}(5x^3 - 3x), \quad P_4(x) = \frac{1}{4}(35x^4 - 30x^2 + 3)$$

etc. or in general

$$P_n(x) = \frac{1}{2^n \cdot n!} \frac{d^n(x^2 - 1)^n}{dx^n}$$

Using these for φ_n is Eq. (8.12) we have corresponding solution of Eq. (8.11). Now

$$x = \cos\phi, \quad Rx = z, \quad R = \sqrt{r^2 + z^2}.$$

Thus polynomial solutions of Eq. (8.11) are

$$\left.\begin{aligned} \Phi_0 &= A_0 \\ \Phi_1 &= A_1 z \\ \Phi_2 &= A_2[z^2 - \tfrac{1}{3}(r^2 + z^2)] \\ \Phi_3 &= A_3[z^3 - \tfrac{3}{5}z(r^2 + z^2)] \\ \Phi_4 &= A_4[z^4 - \tfrac{6}{7}z^2(r^2 + z^2) + \tfrac{3}{35}(r^2 + z^2)^2] \end{aligned}\right\} \quad ...(8.15)$$

where A's are arbitrary constants.

These polynomials are also solutions of Eq. (8.10). It can be shown that if $R^n\varphi_n$ is a solution of Eq. (8.11) then $R^{n+2}\varphi_n$ is a solution of Eq. (8.10). Thus, we get from Eq. (8.9)

$$\left(\frac{\partial^2}{\partial R^2} + \frac{2}{R}\frac{\partial}{\partial R} + \frac{1}{R^2}\cot\phi\frac{\partial}{\partial\phi} + \frac{1}{R^2}\frac{\partial^2}{\partial\phi^2}\right)R^{n+2}\varphi_n = 2(2n+3)R^n\varphi_n \quad ...(8.16)$$

Repeating the operation once more, we obtain zero, since right hand side of Eq. (8.16) is a solution of Eq. (8.11). Hence R^{n+2} φ_n is a solution of Eq. (8.10). Multiplying solutions given by Eqs. (8.15) by $R^2 = r^2 + z^2$, we obtain the following solutions of biharmonic equation.

$$\left.\begin{aligned} \Phi_2 &= B_2(r^2 + z^2) \\ \Phi_3 &= B_3 z(r^2 + z^2) \\ \Phi_4 &= B_4(2z^2 - r^2)(r^2 + z^2) \\ &\text{and so on.} \end{aligned}\right\} \quad ...(8.17)$$

where B's are arbitrary constants.

Example 8.1 *The rotating disc as a three-dimensional problem.*

Solution. The differential equations of equilibrium are :

$$\left.\begin{aligned}\frac{\partial\sigma_r}{\partial r}+\frac{\partial\tau_{rz}}{\partial z}+\frac{\sigma_r-\sigma_\theta}{r}+\rho\omega^2 r=0\\ \frac{\partial\tau_{rz}}{\partial r}+\frac{\partial\sigma_z}{\partial z}+\frac{\tau_{rz}}{\partial r}=0\end{aligned}\right\}\quad ...(1)$$

where ρ is the mass per unit volume and ω is the angular speed of the disc.

The compatibility equations in terms of stress components are

$$\left.\begin{aligned}&\nabla^2\sigma_r-\frac{2}{r^2}(\sigma_r-\sigma_\theta)+\left(\frac{3}{(1+\nu)}\right)\frac{\partial^2\sigma_m}{\partial r^2}=-\left(\frac{2\rho\omega^2}{1-\nu}\right)\\ &\nabla^2\sigma_\theta+\frac{2}{r^2}(\sigma_r-\sigma_\theta)+\left(\frac{3}{(1+\nu)}\right)\frac{1}{r}\frac{\partial\sigma_m}{\partial r}=-\left(\frac{2\rho\omega^2}{1-\nu}\right)\\ &\nabla^2\sigma_z+\left(\frac{3}{1+\nu}\right)\frac{\partial^2\sigma_m}{\partial z^2}=-\left(\frac{2\nu\rho\omega^2}{1-\nu}\right)\\ &\nabla^2\tau_{zr}-\frac{\tau_{zr}}{r^2}+\left(\frac{3}{1+\nu}\right)\frac{\partial^2\sigma_m}{\partial r\partial z}=0\\ &\text{where }\ \sigma_m=\tfrac{1}{3}(\sigma_r+\sigma_\theta+\sigma_z)\end{aligned}\right\}\quad ...(2)$$

Assuming solution of Eqs. (1) in the form of polynomials, we have

$$\sigma_r=Br^2+Dz^2,\quad \sigma_\theta=Cr^2+Dz^2,\quad \sigma_z=Ar^2,\quad \tau_{rz}=0 \quad ...(3)$$

Eqs. (3) satisfy the second equilibrium equation and fourth compatibility equation. Substituting Eqs. (3) in first equilibrium equation and the first three compatibility equations, we find

$$A=\frac{\rho\omega^2(1+3\nu)}{6\nu},\quad B=-\frac{\rho\omega^2}{3},\quad C=0,$$

$$D=-\frac{\rho\omega^2(1+2\nu)(1+\nu)}{6\nu(1-\nu)}$$

Hence, we get

$$\sigma_r=-\frac{\rho\omega^2}{3}r^2-\frac{\rho\omega^2(1+2\nu)(1+\nu)}{6\nu(1-\nu)}z^2 \quad ...(4a)$$

$$\sigma_\theta=-\frac{\rho\omega^2(1+2\nu)(1+\nu)}{6\nu(1-\nu)}z^2 \quad ...(4b)$$

$$\sigma_z = \frac{\rho\omega^2(1+3\nu)}{6\nu} r^2 \qquad \text{...(4c)}$$

$$\tau_{rz} = 0 \qquad \text{...(4d)}$$

In the case of a circular disc of constant thickness, we superpose on the solution given by Eq. (4) the stress distribution derived from a stress function having the form of a polynomial of the fifth degree given by

$$\Phi = a_5(8z^5 - 40r^2z^3 + 15r^4z) + b_5(2z^5 - r^2z^3 - 3r^4z) \qquad \text{...(5)}$$

Then, from Eqs. (8.7), we find

$$\left.\begin{aligned} \sigma_r &= -a_5(180r^2 - 240z^2) + b_5[(36-54\nu)r^2 + (1+18\nu)6z^2] \\ \sigma_\theta &= a_5(-60r^2 + 240z^2) + b_5[(6+108\nu)z^2 + (12-54\nu)r^2] \\ \sigma_z &= -a_5(-240r^2 + 480z^2) + b_5[(96-108\nu)z^2 + (54\nu - 102)r^2] \\ \tau_{rz} &= 480a_5rz - b_5(96-108\nu)rz \end{aligned}\right\} \qquad \text{...(6)}$$

Adding Eqs. (4) and (6) and determining the constants a_5 and b_5 so as to make the resultant stresses $\tau_{rz} = \sigma_z = 0$, we find

$$\left.\begin{aligned} \sigma_r &= -\rho\omega^2\left[\frac{\nu(1+\nu)}{2(1-\nu)} z^2 + \frac{3+\nu}{8} r^2\right] \\ \sigma_\theta &= -\rho\omega^2\left[\frac{(1+3\nu)}{8} r^2 + \frac{\nu(1+\nu)}{2(1-\nu)} z^2\right] \end{aligned}\right\} \qquad \text{...(7)}$$

To eliminate the resultant radial compression along the boundary, *i.e.*, to make

$$\left(\int_{-c}^{c} \sigma_r \, dz\right)_{r=a} = 0$$

We superpose on Eqs. (7) a uniform radial tension of magnitude

$$\frac{\rho\omega^2}{8}(3+\nu)a^2 + \rho\omega^2 \frac{\nu(1+\nu)}{2(1-\nu)} \frac{c^2}{3}$$

where $2c$ is the thickness of the disc and a is its radius, z is measured from the middle plane of the disc.

Then final stresses become

$$\sigma_r = \rho\omega^2\left[\left(\frac{3+\nu}{8}\right)(a^2 - r^2) + \frac{\nu(1+\nu)}{6(1-\nu)}(c^2 - 3z^2)\right] \qquad \text{...(8a)}$$

$$\sigma_\theta = \rho\omega^2 \left[\left(\frac{3 + \nu}{8} \right) a^2 - \left(\frac{1 + 3\nu}{8} \right) r^2 + \frac{\nu(1 + \nu)}{6(1 - \nu)} (c^2 - 3z^2) \right] \quad \text{...(8b)}$$

$$\sigma_z = \tau_{rz} = 0 \quad \text{...(8c)}$$

Example 8.2 *Force at a point in an infinite solid.*

Solution. Consider a concentrated force acting at a point in an infinite solid. In this case we assume

$$\Phi = B(r^2 + z^2)^{1/2} \quad \text{...(1)}$$

where B is a constant to be determinec 1.

Substituting in Eqs. (8.7), we get

$$\left.\begin{aligned} \sigma_r &= BZ[(1 - 2\nu)(r^2 + z^2)^{-3/2} - 3r^2 z(r^2 + z^2)^{-5/2}] \\ \sigma_\theta &= B(1 - 2\nu)z(r^2 + z^2)^{-3/2} \\ \sigma_z &= -B[(1 - 2\nu)z(r^2 + z^2)^{-3/2} + 3z^2(r^2 + z^2)^{-5/2} \\ \tau_{rz} &= -B[(1 - 2\nu)r(r^2 + z^2)^{-3/2} + 3rz^2(r^2 + z^2)^{-5/2} \end{aligned}\right\} \quad \text{...(2)}$$

At the origin of coordinates, where the concentrated load is applied, all these stresses are singular. We therefore take the origin to be the centre of a small spherical cavity as shown in Fig. 8.1, and consider forces over the surface of the cavity as calculated from Eqs. (2).

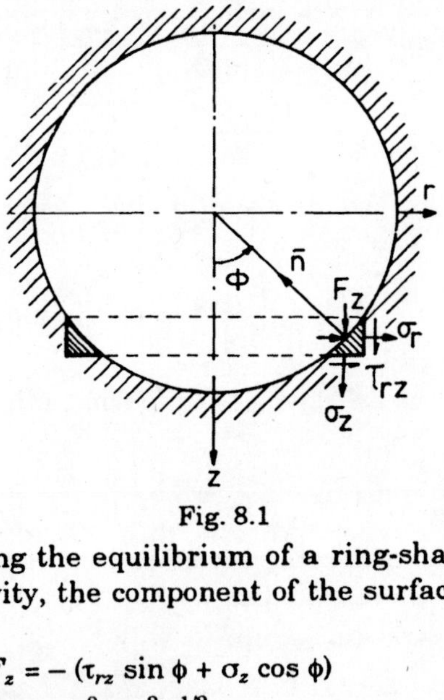

Fig. 8.1

Considering the equilibrium of a ring-shaped element, adjacent to the cavity, the component of the surface forces in the z-direction is

$$F_z = -(\tau_{rz} \sin\phi + \sigma_z \cos\phi) \quad \text{...(3)}$$

Now $\sin\phi = r(r^2 + z^2)^{-1/2}$

$$\cos\phi = z(r^2 + z^2)^{-1/2}$$

Substituting the values of stresses from Eq. (2) in Eq. (3), we find

$$F_z = B[(1-2\nu)(r^2+z^2)^{-1} + 3z^2(r^2+z^2)^{-2}]$$

The resultant of these forces over the surface of the cavity in the z-direction is

$$2\int_0^{\pi/2} F_z(r^2+z^2)^{1/2} \,.\, 2\pi r\, d\phi = 8\,B\pi(1-\nu) = P$$

where P is the magnitude of the applied force.

$$\therefore \qquad B = \frac{P}{8\pi(1-\nu)} \qquad ...(4)$$

The resultant of the forces in the z-direction is zero. Substituting the value of B in Eqs. (2), we get the stress components.

At $\qquad z = 0$

$$\sigma_r = \sigma_\theta = \sigma_z = 0$$

and $\qquad \tau_{rz} = -\dfrac{P(1-2\nu)}{8\pi(1-\nu)r^2} = -\dfrac{B(1-2\nu)}{r^2} \qquad ...(5)$

Example 8.3 *Spherical container under pressure.*

Solution. Consider a spherical container of inner radius a and outer radius b subjected to internal pressure p_1 and external pressure p_0 as shown in Fig. 8.2.

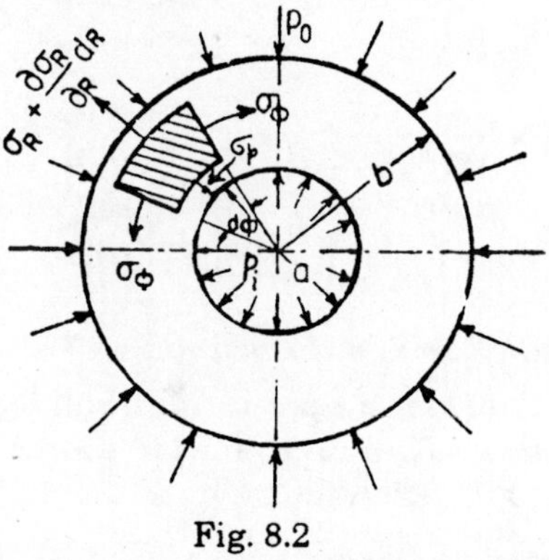

Fig. 8.2

We take a general expression for the radial normal stresses in the form

$$\sigma_R = \frac{A}{R^3} + B \qquad ...(1)$$

where A and B are constants.

At $\quad R = a, \sigma_R = -p_i$ and at

$R = b, \sigma_R = -p_o$

$$\therefore \quad \frac{A}{a^3} + B = -p_i$$

$$\frac{A}{b^3} + B = -p_o$$

$$\therefore \quad A = \frac{(p_i - p_o)\, a^3 b^3}{(a^3 - b^2)}$$

$$B = \frac{p_o b^2 - p_i a^3}{a^3 - b^3}$$

$$\therefore \quad \sigma_R = \frac{p_o b^3 (R^3 - a^3)}{R^3 (a^3 - b^3)} + \frac{p_i\, a^3 (b^3 - R^3)}{R^3 (a^3 - b^3)} \qquad \text{...(2)}$$

Consider the equilibrium of a small element of the sphere in the tangential direction, we get

$$\sigma_\phi \cdot \frac{\pi R}{2}\, dR\, (d\phi)^2 = \frac{d\sigma_R}{dR} \cdot \frac{\pi R^2}{4}\, dR (d\phi)^2 + \sigma_R \cdot \frac{\pi R}{2}\, dR (d\phi)^2$$

$$\therefore \quad \sigma_\phi = \frac{d\sigma_R}{dR} \cdot \frac{R}{2} + \sigma_R \qquad \text{...(3)}$$

Substituting Eq. (2) in Eq. (3), we get

$$\sigma_\phi = \frac{p_c b^3 (2R^3 + a^2)}{2R^3 (a^3 - b^3)} - \frac{p_i a^3 (2R^3 + b^3)}{2R^3 (a^3 - b^3)} \qquad \text{...(4)}$$

If $p_o = 0$, then

$$\sigma_\phi = \frac{p_i\, a^3 (2R^3 + b^3)}{2R^3 (b^3 - a^3)} \qquad \text{...(5)}$$

$$(\sigma_\phi)_{max} \Big|_{R = a} = \frac{p_i}{2} \left(\frac{2a^3 + b^3}{(b^3 - a^3)} \right) \qquad \text{...(6)}$$

Example 8.4 *Force on the boundary of a semi-infinite body.*

Solution. Consider a semi-infinite solid whose boundary is defined by the plane $z = 0$ and subjected to a force P acting along the z-axis as shown in Fig. 8.3.

In this case we assume

$$\sigma_R = \frac{A}{R^3},\ \sigma_\phi = \frac{d\sigma_R}{dR} \cdot \frac{R}{2} + \sigma_R = -\frac{1}{2} \frac{A}{R^3} \qquad \text{...(1)}$$

where A is a constant.

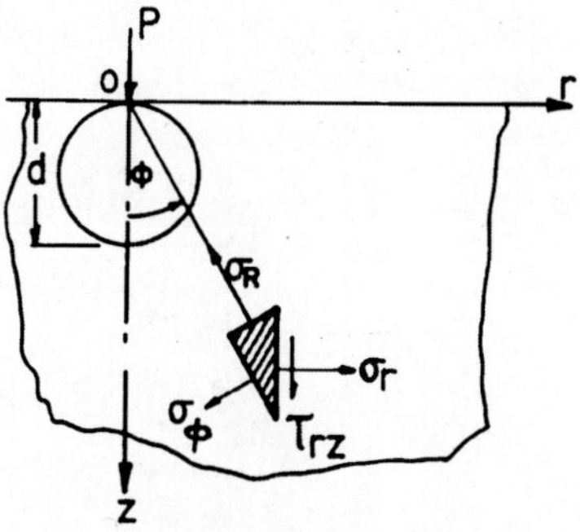

Fig. 8.3

In cylindrical coordinates the stress components become :

$$\left.\begin{aligned}\sigma_r &= \sigma_R \sin^2\phi + \sigma_\phi \cos^2\phi \\ &= A(r^2 - \tfrac{1}{2}z^2)(r^2+z^2)^{-5/2} \\ \sigma_\theta = \sigma_\phi &= -\tfrac{1}{2}A(r^2+z^2)^{-3/2} \\ \sigma_z &= \sigma_R \cos^2\phi + \sigma_\phi \sin^2\phi \\ &= A(z^2 - \tfrac{1}{2}r^2)(r^2+z^2)^{-5/2} \\ \tau_{rz} &= \tfrac{1}{2}(\sigma_R - \sigma_\phi)\sin 2\phi \\ &= \tfrac{3}{2}Arz(r^2+z^2)^{-5/2}\end{aligned}\right\} \quad \ldots(2)$$

Suppose now that the centres of pressure are uniformly distributed along the z-axis from $z = 0$ to $z = -\infty$. Then the stress components produced in an indefinitely extended solid are obtained from Eqs. (2) with a new constant A_1.

$$\sigma_r = A_1 \int_z^\infty (r^2 - \tfrac{1}{2}z^2)(r^2+z^2)^{-5/2}\,dz$$

$$= \frac{A_1}{2}\left[\frac{1}{r^2} - \frac{z}{r^2}(r^2+z^2)^{-1/2} - z(r^2+z^2)^{-3/2}\right]$$

$$\sigma_\theta = -\tfrac{1}{2}A_1 \int_z^\infty (r^2+z^2)^{-3/2}\,dz$$

$$= -\frac{A_1}{2}\left[\frac{1}{r^2} - \frac{z}{r^2}(r^2+z^2)^{-1/2}\right]$$

$$\sigma_z = A_1 \int_z^\infty (z^2 - \tfrac{1}{2}r^2)(r^2+z^2)^{-5/2}\,dz$$

$$= \frac{A_1}{2} z(r^2 + z^2)^{-3/2}$$

$$\tau_{rz} = \frac{3}{2} A_1 \int_z^{\infty} rz(r^2 + z^2)^{-5/2}\, dz \qquad ...(3)$$

$$= \frac{A_1}{2} r(r^2 + z^2)^{-3/2}$$

At $\quad z = 0, \sigma_s = \sigma_\theta = \sigma_z = 0$

and $$\tau_{rz} = \frac{A_1}{2r^2} \qquad ...(4)$$

Also due to a force at a point in an infinite solid the shear stress at $z = 0$ is given by (See example 8.2)

$$\tau_{rz} = -\frac{B(1 - 2\nu)}{r^2} \qquad ...(5)$$

τ_{rz} will be zero at $z = 0$ if

$$-B(1 - 2\nu) + \frac{A_1}{2} = 0$$

or $$A_1 = 2B(1 - 2\nu)$$

Substituting in Eqs. (3), we get

$$\sigma_r = B\left[(1 - 2\nu)\left\{\frac{1}{r^2} - \frac{z}{r^2}(r^2 + z^2)^{-1/2}\right\} - 3r^2 z(r^2 + z^2)^{-5/2}\right]$$

$$\sigma_\theta = B(1 - 2\nu)\left[-\frac{1}{r^2} + \frac{z}{r^2}(r^2 + z^2)^{-1/2} + z(r^2 + z^2)^{-3/2}\right]$$

$$\sigma_z = -3Bz^3 (r^2 + z^2)^{-5/2}$$

$$\tau_{rz} = -3Brz^2 (r^2 + z^2)^{-5/2}. \qquad ...(6)$$

The stress distribution given by Eqs. (6) satisfy the boundary conditions, since $\sigma_z = \tau_{rz} = 0$ at $z = 0$. To determine the constant B, we consider the equilibrium of the forces distributed over a hemispherical surface with centre at the origin in the z-direction.

$$F_z = -(\tau_{rz} \sin\phi + \sigma_z \cos\phi)$$
$$= 3Bz^2 (r^2 + z^2)^{-2}$$

$$\therefore \quad P = 2\pi \int_0^{\pi/2} F_z \,.\, r(r^2 + z^2)^{1/2}\, d\phi$$

$$= 6\pi B \int_0^{\pi/2} \cos^2\phi \sin\phi d\phi = 2\pi B$$

$$\therefore \qquad B = \frac{P}{2\pi}$$

Substituting in Eqs. (6), we get

$$\sigma_r = \frac{P}{2\pi}\left[(1-2\nu)\left\{\frac{1}{r^2} - \frac{z}{r^2}(r^2+z^2)^{-1/2}\right\} - 3r^2 z(r^2+z^2)^{-5/2}\right]$$

$$\sigma_\theta = \frac{P}{2\pi}(1-2\nu)\left[-\frac{1}{r^2} + \frac{z}{r^2}(r^2+z^2)^{-1/2} + z(r^2+z^2)^{-3/2}\right]$$

$$\sigma_z = -\frac{3P}{2\pi} z^2(r^2+z^2)^{-5/2}$$

$$\tau_{rz} = -\frac{3P}{2\pi} rz^2(r^2+z^2)^{-5/2} \qquad ...(7)$$

Now $\quad \dfrac{\sigma_z}{\tau_{rz}} = \dfrac{z}{r} \qquad ...(8)$

Hence the direction of the resultant stress passes through the origin O. The resultant stress is given by

$$\sigma_{Res.} = \sqrt{\sigma_z^{\,2} + \tau_{rz}^{\,2}}$$

$$= \frac{3P}{2\pi} \cdot \frac{z^2}{(r^2+z^2)^2} = \frac{3P}{2\pi}\frac{\cos^2\phi}{(r^2+z^2)} \qquad ...(9)$$

Now imagine a spherical surface of diameter d as shown in Fig. 8.3, tangent to the plane $z = 0$ at the origin. For each point of this surface,

$$r^2 + z^2 = d^2\cos^2\phi \qquad ...(10)$$

Substituting Eq. (10) in Eq. (9), we get

$$\sigma_{Res.} = \frac{3P}{2\pi d^2}$$

Hence the resultant stress on horizontal planes of the sphere is constant.

8.4 Symmetrical Deformation of a Circular Cylinder

Let us consider the deformation of a circular cylinder subjected to the action of forces applied to the lateral surface and distributed symmetrically with respect to the axis of the cylinder. Now the compatibility equation for axial symmetry is :

$$\left(\frac{\partial^2}{\partial r^2} + \frac{1}{r}\frac{\partial}{\partial r} + \frac{\partial^2}{\partial z^2}\right)\left(\frac{\partial^2\Phi}{\partial r^2} + \frac{1}{r}\frac{\partial\Phi}{\partial r} + \frac{\partial^2\Phi}{\partial z^2}\right) = 0 \qquad ...(8.18)$$

We assume a solution of the Laplace equation

$$\frac{\partial^2\Phi}{\partial r^2}+\frac{1}{r}\frac{\partial\Phi}{\partial r}+\frac{\partial^2\Phi}{\partial z^2}=0 \qquad ...(8.19)$$

in the form

$$\Phi = f(r)\sin kz \qquad ...(8.20)$$

Substituting Eq. (8.20) in (8.19), we get

$$\frac{d^2f}{dr^2}+\frac{1}{r}\frac{df}{dr}-k^2f=0 \qquad ...(8.21)$$

A solution of this equation is

$$f(r) = a_0 I_0(kr) \qquad ...(8.22)$$

where a_0 is a constant and I_0 is the modified Bessel function of the first kind of order zero.

$$\therefore \qquad \Phi_1 = a_0 I_0(kr)\sin kz \qquad ...(8.23)$$

Now consider

$$f_1(r) = r\frac{d}{dr}I_0(kr) = krI_1(kr)$$

By differentiation it can be shown that

$$\left(\frac{d^2}{dr^2}+\frac{1}{r}\frac{d}{dr}-k^2\right)\left(\frac{d^2f_1}{dr^2}+\frac{1}{r}\frac{df_1}{dr}-k^2f_1\right)=0$$

Hence a solution of Eq. (8.18) can be taken in the form

$$\Phi_2 = a_1\, krI_1(kr)\sin kz \qquad ...(8.24)$$

Combining solutions (8.23) and (8.24), we get

$$\Phi = \sin kz\,[a_0I_0(kr) + a_1kr\,I_1(kr)] \qquad ...(8.25)$$

Substituting Eq. (8.25) is Eqs. (8.7), we get

$$\left.\begin{aligned}\sigma_r &= \cos kz\,[a_0F_1(r) + a_1F_2(r)]\\ \tau_{rz} &= \sin kz\,[a_0F_3(r) + a_1F_4(r)]\end{aligned}\right\} \qquad ...(8.26)$$

where $F_1(r), \ldots\ldots, F_4(r)$ are certain functions of r containing $I_0(kr)$ and $I_1(kr)$ whose values can be calculated for any value of r.

If 'a' is the external radius of the solid cylinder, then

$$\left.\begin{aligned}\sigma_r &= \cos kz\,[a_0F_1(a) + a_1F_2(a)]\\ \tau_{rz} &= \sin kz\,[a_0F_3(a) + a_1F_4(a)]\end{aligned}\right\} \qquad ...(8.27)$$

Eqs. (8.27) can be used as boundary conditions to determine the values of the unknown constants. Let l be the length of the cylinder, then

$$k = \frac{n\pi}{l}$$

and let normal pressure given by $A_n \cos\left(\frac{n\pi z}{l}\right)$ act on the lateral surface of the cylinder. Eqs. (8.27) give

$$a_0F_1(a) + a_2F_1(a) = -A_n$$
$$a_0F_3(a) + a_1F_4(a) = 0$$

Similarly when tangential force of intensity $B_n \sin\left(\frac{n\pi z}{l}\right)$ act on the surface of the cylinder, then

$$a_0F_1(a) + a_1F_2(a) = 0$$
$$a_0F_3(a) + a_1F_4(a) = B_n.$$

By taking n = 1, 2, 3, ... and using the superposition principle, we arrive at solution of problems in which the normal pressures on the surface of the cylinder can be represented by the series

$$A_1 \cos\frac{\pi z}{l} + A_2 \cos\frac{2\pi z}{l} + A_2 \cos\frac{3\pi z}{l} + ...$$

and the shearing forces by series

$$B_1 \sin\frac{\pi z}{l} + B_2 \sin\frac{2\pi z}{l} + B_3 \sin\frac{3\pi z}{l} + ...$$

If we take

$$\Phi = f(r)\cos kz$$

then we would obtain proceeding as above.

$$\Phi = \cos kz[b_0I_0(kr) + b_1I_1(kr)] \qquad ...(8.28)$$

where b_0, b_1 are constants.

Hence, by combining solutions of Eqs. (8.25) and (8.28), we can get any axially symmetric distribution of normal and shearing forces over the surface of the cylinder.

Example 8.5 *The circular cylinder with a band of pressure.*

Solution. Consider the case of a short collar shrunk on a much longer solid shaft as shown in Fig. 8.4 (*a*). Let p be the uniform normal pressure due to the band. The solution of this problem can be obtained by superposing the effects of the two pressure distributions indicated in Fig. 8.4 (*b*). The basic problem is therefore that of pressure $\frac{p}{2}$ on the lower half of the cylindrical surface and $-\frac{p}{2}$ on the upper half.

We assume the stress function in the form

$$\Phi = [\rho I_0(kr) - kr\, I_1(kr)]\, b_1 \cos kz \qquad ...(1)$$

where we have taken $b_0 = \rho b_1$.

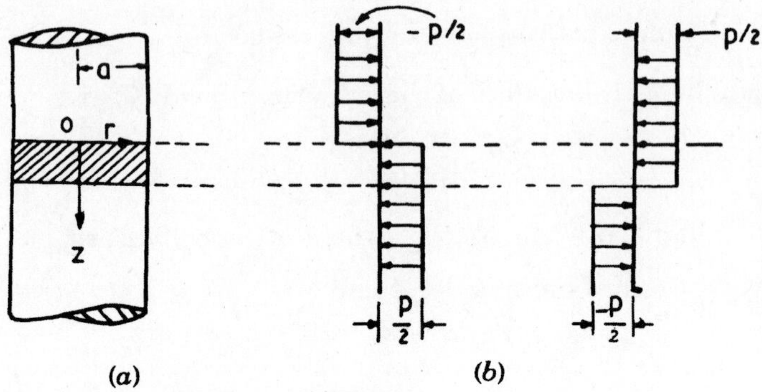

Fig. 8.4

Eq. (1) satisfies $\nabla^4\Phi = 0$

where $\nabla^2 = \left(\frac{\partial^2}{\partial r^2} + \frac{1}{r}\frac{\partial}{\partial r} + \frac{\partial^2}{\partial z^2}\right)$ for all values of k. If we consider k to take a range of values then we can allow b_1 to depend on k and an increment dk by writing

$$b_1 = f(k)\, dk$$

Putting this in Eq. (1) and adding up all such stress functions, we obtain a more general stress function given by

$$\Phi = \int_0^\infty [\rho I_0(kr) - krI_1(kr)]\, f(k) \cos kz.dk \qquad ...(2)$$

From Eq. (8.7), we find that

$$\tau_{rz} = \int_0^\infty [\rho k I_0'(kr) - k^2 r\, I_1(kr) - kI_1(kr) - 2k(1-\nu)\, I_0'(kr)]\, k^2 f(k) \cos kz.dk \qquad ...(3)$$

where primes denote differentiation w.r.t. kr.

At $\qquad r = a, \tau_{rz} = 0$

Hence, we get

$$\rho = 2(1-\nu) + ka\frac{I_0(ka)}{I_1(ka)} \qquad ...(4)$$

The remaining boundary condition is

$$\left.\begin{aligned} \sigma_r &= \frac{p}{2} \text{ for } r = a,\ z > 0 \\ a_r &= -\frac{p}{2} \text{ for } r = a,\ z < 0 \end{aligned}\right\} \qquad ...(5)$$

The value of σ_r obtained from Eq. (8.7) is

$$\sigma_r = -\int_0^\infty \left[(1 - 2\nu - \rho)\, I_0\,(kr) + \left(kr + \frac{\rho}{kr} \right) I_1\,(kr) \right] \times k^3 f(k) \sin kz.dk \qquad ...(6)$$

$$\text{Now} \int_0^\infty \frac{\sin kz}{k}\, dk = \begin{cases} \frac{\pi}{2} \text{ for } z > 0 \\ 0 \text{ for } z = 0 \\ -\frac{\pi}{2} \text{ for } z < 0 \end{cases} \qquad ...(7)$$

$$\therefore \quad \frac{p}{\pi}\int_0^\infty \frac{\sin kz}{k}\, dk = \begin{cases} \frac{p}{2} \text{ for } z > 0 \\ 0 \text{ for } z = 0 \\ -\frac{p}{2} \text{ for } z < 0 \end{cases} \qquad ...(8)$$

Equating Eqs. (6) and (8), we find

$$-\left[(1 - 2\nu - \rho)\, I_0\,(ka) + \left(ka + \frac{\rho}{ka} \right) I_1\,(ka) \right] k^3 f(k) = \frac{p}{\pi} \cdot \frac{1}{k} \qquad ...(9)$$

From Eq. (a) $f(k)$ can be determined. The stress components are then found from the stress function given by Eq. (2) by means of Eq. (8.7).

8.5 The Integral Transform Method

8.5.1 Solution of the Equations of Equilibrium

$$\text{Now} \int_0^\infty r\nabla^2 f J_0(\xi r)dr = \left(\frac{d^2}{dz^2} - \xi^2 \right) \int_0^\infty rfJ_0(\xi r)dr$$

Replacing f by $\nabla^2\Phi$, we find

$$\int_0^\infty r\nabla^4 \Phi J_0(\xi r)dr = \left(\frac{d^2}{dz^2} - \zeta^2 \right)^2 \int_0^\infty r\Phi J_0(\xi r)dr \qquad ...(8.29)$$

The compatibility equation is

$$\nabla^4\Phi = \left(\frac{\partial^2}{\partial r^2} + \frac{1}{r}\frac{\partial}{\partial r} + \frac{\partial^2}{\partial z^2} \right)\left(\frac{\partial^2 \Phi}{\partial r^2} + \frac{1}{r}\frac{\partial \Phi}{\partial r} + \frac{\partial^2 \Phi}{\partial z^2} \right) = 0$$

Multiplying both sides by $rJ_0(\xi r)$ and integrating over the whole range of r, we find

$$\left(\frac{d^2}{dz^2} - \xi^2\right)^2 G(\xi, z) = 0 \qquad \text{...(8.30)}$$

where $G(\xi, z) = \int_0^\infty r\Phi J_0(\xi r)dr$

is the zero order Hankel transform of the function $\Phi(r, z)$. Integrating Eq. (8.30), we get

$$G(\xi, z) = (A + Bz)e^{-\xi z} + (C + Dz)e^{\xi z} \qquad \text{...(8.31)}$$

where the constant A, B, C and D are determined from the boundary conditions.

Now from Eqs. (8.8), we have

$$u_r = -\left(\frac{1+\nu}{E}\right)\frac{\partial^2\Phi}{\partial r\partial z}$$

$$u_z = \frac{2(1+\nu)(1-\nu)}{E}\nabla^2\Phi - \left(\frac{1+\nu}{E}\right)\frac{\partial^2\Phi}{\partial z^2}$$

Multiplying both sides by $rJ_0'(\xi r)$ and integrating over the whole range, we get

$$\int_0^\infty ru_rJ_1(\xi r)dr = \left(\frac{1+\nu}{E}\right)\xi\frac{dG}{dz}$$

Inverting, we get

$$u_r = \left(\frac{1+\nu}{E}\right)\int_0^\infty \xi^2\frac{dG}{dz}J_1(\xi r)d\xi \qquad \text{...(8.32)}$$

and $\displaystyle\int_0^\infty ru_zJ_0(\xi r)\,dr = \frac{2(1-\nu^2)}{E}\left(\frac{d^2}{dz^2} - \xi^2\right)G - \left(\frac{1+\nu}{E}\right)\frac{d^2G}{dz^2}$

Inverting, we get

$$u_z = \frac{(1+\nu)}{E}\int_0^\infty \xi\left[\frac{d^2G}{dz^2}(1-2\nu) - 2(1-\nu)\xi^2G\right]\times J_0(\xi r)\,dr \qquad \text{...(8.33)}$$

To determine the expressions for stresses, we have

$$\sigma_z = \frac{\partial}{\partial z}\left[(2-\nu)\nabla^2\Phi - \frac{\partial^2\Phi}{\partial z^2}\right]$$

Multiplying by $r\,J_0(\xi r)$, integrating over r, we find

$$\int_0^\infty r\sigma_zJ_0(\xi r)dr = \int_0^\infty \frac{\partial}{\partial z}\left[(2-\nu)\nabla^2\Phi - \frac{\partial^2\Phi}{\partial z^2}\right]r\,J_0(\xi r)dr$$

$$= (2-\nu)\frac{\partial}{\partial z}\left[\frac{d^2}{dz^2}-\xi^2\right]G-\frac{d^3G}{dz^3}$$

$$= (2-\nu)\left[\frac{d^3G}{dz^3}-\xi^2\frac{dG}{dz}\right]-\frac{d^3G}{dz^3}.$$

Inverting, we get

$$\sigma_z = \int_0^\infty \xi\left[(1-\nu)\left(\frac{d^3G}{dz^3}\right)-(2-\nu)\,\xi^2\frac{dG}{dz}\right]\times J_0(\xi r)\,d\xi \qquad ...(8.34)$$

Now $\tau_{rz} = \frac{\partial}{\partial r}\left[(1-\nu)\,\nabla^2\Phi-\frac{\partial^2\Phi}{\partial z^2}\right]$

Multiplying by $rJ_1(\xi r)$ and integrating over r, we find

$$\int_0^\infty r\tau_{rz}J_1(\xi r)\,dr = \int_0^\infty \frac{\partial}{\partial r}\left[(1-\nu)\,\nabla^2\Phi-\frac{\partial^2\Phi}{\partial z^2}\right]r\,J_1(\xi r)\,dr$$

$$= -(1-\nu)\left[\frac{d^2}{dz^2}-\xi^2\right]G+\frac{d^2G}{dz^2}$$

$$= \nu\frac{d^2G}{dz^2}+(1-\nu)\,\xi^2G$$

Inverting, we get

$$\tau_{rz} = \int_0^\infty \xi^2\left[\nu\frac{d^2G}{dz^2}+(1-\nu)\,\xi^2G\right]J_1(\xi r)\,d\xi \qquad ...(8.35)$$

Now $\sigma_r = \frac{\partial}{\partial z}\left[\nu\,\nabla^2\Phi-\frac{\partial^2\Phi}{\partial r^2}\right]$

$$\sigma_\theta = \frac{\partial}{\partial z}\left[\nu\nabla^2\Phi-\frac{1}{r}\frac{\partial\Phi}{\partial r}\right].$$

Adding, we get

$$\sigma_r+\sigma_\theta = \frac{\partial}{\partial z}\left[2\nu\,\nabla^2\Phi-\frac{\partial^2\Phi}{\partial r^2}-\frac{1}{r}\frac{\partial\Phi}{\partial r}\right]$$

$$= \frac{\partial}{\partial z}\left[2\nu\nabla^2\Phi-\nabla^2\Phi+\frac{\partial\Phi}{\partial z}\right]$$

$$= \frac{\partial}{\partial z}\left[(2\nu-1)\,\nabla^2\Phi+\frac{\partial\Phi}{\partial z}\right]$$

Multiplying both sides by $rJ_0(\xi r)$ and integrating over r from 0 to ∞, we find

$$\int_0^\infty r(\sigma_r + \sigma_\theta)J_0(\xi r)\, dr = \int_0^\infty \frac{\partial}{\partial z}\left[(2\nu - 1)\, \nabla^2\Phi + \frac{\partial \Phi}{\partial z} \right] rJ_0(\xi r) dr$$

$$= (2\nu - 1)\left(\frac{d^2}{dz^2} - \xi^2 \right)\frac{dG}{dz} + \frac{d^3G}{dz^3}$$

$$= 2\nu \frac{d^3G}{dz^3} - (2\nu - 1)\, \xi^2 \frac{dG}{dz}$$

Inverting, we get

$$\sigma_r + \sigma_\theta = \int_0^\infty \left[2\nu\xi \frac{d^3G}{dz^3} - (2\nu - 1)\, \xi^3 \frac{dG}{dz} \right] J_0(\xi r)\, d\xi \qquad ...(8.36)$$

Equation of equilibrium (8.1) may be written as :

$$\frac{\partial}{\partial r}(r^2\sigma_r) = r(\sigma_r + \sigma_\theta) - r^2 \frac{\partial \tau_{rz}}{\partial z}$$

Substituting on the right hand side from Eqs. (8.35) and (8.36), we get

$$\frac{\partial}{\partial r}(r^2\sigma_r) = r\int_0^\infty \left[2\nu\xi \frac{d^3G}{dz^3} - (2\nu - 1)\, \xi^3 \frac{dG}{dz} \right] J_0(\xi r)\, d\xi$$

$$- r^2 \int_0^\infty \xi^2 \left[\nu \frac{d^3G}{dz^3} + (1 - \nu)\, \xi^2 \frac{dG}{dz} \right] J_1(\xi r)\, d\xi$$

Integrating w.r.t. r and making use of the integrals

$$\int_0^r rJ_0(\xi r)\, dr = \frac{r}{\xi} J_1(\xi r)$$

$$\int_0^r r^2 J_1(\xi r)\, dr = \frac{2r}{\xi^2} J_1(\xi r) - \frac{r^2}{\xi} J_0(\xi r)$$

We get

$$\sigma_r = \int_0^\infty \xi \left[2\nu \frac{d^3G}{dz^3} - (2\nu - 1)\, \xi^2 \frac{dG}{dz} \right] \frac{J_1}{r\xi}(\xi r)\, d\xi$$

$$- \int_0^\infty \left[\xi^2\nu \frac{d^3G}{dz^3} + (1 - \nu)\, \xi^4 \frac{dG}{dz} \right] \left[\frac{2J_1(\xi r)}{r\xi^2} - \frac{J_0(d\xi)}{\xi} \right] d\xi$$

$$= \int_0^\infty \left[\xi\nu \frac{d^3G}{dz^3} + (1-\nu)\,\xi^3 \frac{dG}{dz} \right] J_0(\xi r)\, d\xi$$

$$- \frac{1}{r}\int_0^\infty \xi^2 \left[\nu \frac{dG}{dz} \right] J_1(\xi r)\, d\xi \quad ...(8.37)$$

Subtracting Eq. (8.37) from (8.36), we get

$$\sigma_\theta = \frac{\nu}{r}\int_0^\infty \xi^2 \frac{dG}{dz} \,.\, J_1(\xi r)\, d\xi$$

$$+ \int_0^\infty \left[\xi\nu \frac{d^3G}{dz^3} - \nu\xi^3 \frac{dG}{dz} \right] J_0(\xi r)\, d\xi \quad ...(8.38)$$

Example 8.6 *Terezawa's solution for a semi-infinite medium.*

Solution. Consider an infinite elastic medium bounded by the infinite plane $z = 0$. Choose the z-axis normal to this plane and pointing into the medium. Let at $z = 0$

$$\sigma_z = -p(r),\ \tau_{rz} = 0 \quad ...(1)$$

Also as r and z tend to infinity, all the components of stress and displacement tend to zero.

Hence, we assume a function of the form

$$G(\xi, z) = (A + Bz)\, e^{-\xi z} \quad ...(2)$$

where $G(\xi, z) = \int_0^\infty r\,\Phi\, J_0(\xi r)\, d\xi$.

From Eqs. (8.34), (8.35) and (2), we have

$$\int_0^\infty (r)\sigma_z J_0(\xi r) dr = (1-\nu)\frac{d^3G}{dz^3} - (2-\nu)\,\xi^2 \frac{dG}{dz}$$

$$= \xi^2 e^{-\xi z}\,[\xi A + (\xi z + 1 - 2\nu)B]$$

$$\int_0^\infty r\,\tau_{rz} J_1(\xi r) dr = \nu \frac{dG}{dz^2} + (1-\nu)\,\xi^2 G$$

$$= \xi e^{-\xi z}[A\xi + (\xi z - 2\nu)B]$$

At $\qquad z = 0$, let

$$z(\xi) = -\xi\,\overline{p}(\xi) = -\xi \int_0^\infty r\,p(r)\, J_0(\xi r)\, dr$$

then on using the conditions (1), we get

$$\xi^{-3}z(\xi) = \xi A + (1 - 2\nu)B$$

$$0 = \xi A - 2\nu B$$

$$\therefore \quad B = \xi^{-3}z(\xi)$$

$$A = \xi^{-4} \,.\, 2\nu z(\xi)$$

$$\therefore \quad G(\xi, z) = e^{-\xi z}\left[\frac{2\nu z(\xi)}{\xi^4} + \frac{z}{\xi^8} \,.\, z(\xi)\right]$$

$$G(\xi, z) = \frac{z(\xi)}{\xi^4}[2\nu + \xi z]e^{-\xi z} \qquad \text{...(3)}$$

Substituting in Eqs. (8.32) and (8.33), we get

$$u_r = \left(\frac{1+\nu}{E}\right)\int_0^\infty \xi^2 \frac{dG}{dz} J_1(\xi r)d\xi$$

$$= \left(\frac{1+\nu}{E}\right)\int_0^\infty \frac{z(\xi)}{\xi}[1 - 2\nu - \xi z]\, e^{-\xi z} J_1(\xi r)d\xi$$

$$= -\left(\frac{1+\nu}{E}\right) z\int_0^\infty z(\xi)e^{-\xi z} J_1(\xi r)d\xi$$

$$+ \frac{(1-2\nu)(1+\nu)}{E}\int_0^\infty z(\xi)\, e^{-\xi z}\frac{J_1(\xi r)}{\xi}d\xi \qquad \text{...(4)}$$

$$u_z = \frac{(1+\nu)}{E}\int_0^\infty \xi\left[\frac{d^2G}{dz^2}(1-2\nu) - 2(1-\nu)\xi^2 G\right]J_0(\xi r)d\xi$$

$$= -\left(\frac{1+\nu}{E}\right) z\int_0^\infty z(\xi)\, e^{-\xi z} J_0(\xi r)\, d\xi$$

$$- \frac{2(1-\nu^2)}{E}\int_0^\infty z(\xi)e^{-\xi z} J_0(\xi r)\frac{d\xi}{\xi} \qquad \text{...(5)}$$

Also from Eqs. (8.34) and (8.35), we get

$$\sigma_z = \int_0^\infty \xi^3 e^{-\xi z}\left[\frac{2\nu z(\xi)}{\xi^3} + (\xi z + 1 - 2\nu)\frac{z(\xi)}{\xi^3}\right]J_0(\xi r)\, d\xi$$

$$= z\int_0^\infty \xi z(\xi)\, e^{-\xi z} J_0(\xi r)d\xi + \int_0^\infty z(\xi)\, e^{-\xi z} J_0(\xi r)d\xi \qquad \text{...(6)}$$

$$\tau_{rz} = \int_0^\infty \xi^3 e^{-\xi z}\left[\frac{2\nu z(\xi)}{\xi^3} + (\xi z - 2\nu)\frac{z(\xi)}{\xi^3}\right]J_1(\xi r)d\xi \qquad \text{...(7)}$$

$$= z\int_0^\infty z(\xi)\, e^{-\xi z}\, J_1(\xi r)\xi d\xi$$

and Eqs. (8.37) and (8.38) give

$$\sigma_r = \int_0^\infty \left[\xi\nu \frac{d^3G}{dz^3} + (1-\nu)\,\xi^3 \frac{dG}{dz} \right] J_0(\xi r)\, d\xi$$

$$- \frac{1}{r}\int_0^\infty \xi^2\nu \frac{dG}{dz} J_1(\xi r)\, d\xi$$

$$\sigma_r = \int_0^\infty \left[\xi\nu . \frac{z(\xi)}{\xi} . e^{-\xi z} (3 - 2\nu - \xi z) \right.$$

$$\left. + \;(1-\nu)\,\xi^3 \frac{z(\xi)}{\xi^3} . e^{-\xi z} (1 - 2\nu - \xi z) \right] \times J_0(\xi r)\, d\xi$$

$$- \frac{1}{r}\int_0^\infty \xi^2\nu \frac{z(\xi)}{\xi^3} e^{-\xi z} (1 - 2\nu - \xi z)\, J_1(\xi r)\, d\xi$$

$$= \int_0^\infty z(\xi)\, e^{-\xi z} [1 - \xi z] J_0\, (\xi r)\, d\xi$$

$$- \frac{1}{r}\int_0^\infty z(\xi)\, e^{-\xi z} [\nu - 2\nu^2 - \nu\xi z]\, J_1(\xi r) \frac{d\xi}{\xi} \qquad \text{...(8)}$$

$$\sigma_\theta = \frac{\nu}{r}\int_0^\infty \xi^2 \frac{dG}{dz} J_1(\xi r)\, d\xi$$

$$+ \int_0^\infty \left[\xi\nu \frac{d^3G}{dz^3} - \nu\xi^3 \frac{dG}{dz} \right] J_0(\xi r) d\xi$$

$$= \frac{\nu}{r}\int_0^\infty \left[\xi^2 \frac{z(\xi)}{\xi^3} e^{-\xi z} (1 - 2\nu - \xi z) \right] J_1(\xi r)\, d\xi$$

$$+ \int_0^\infty \left[\xi\nu \frac{z(\xi)}{\xi} . e^{-\xi z} (3 - 2\nu - \xi z) \right.$$

$$\left. - \nu\, \xi^3 \frac{z(\xi)}{\xi^3} . e^{-\xi z} (1 - 2\nu - \xi z) \right] \times J_0(\xi r)\, d\xi$$

$$= \frac{1}{r}\int_0^\infty z(\xi)\, e^{-\xi z}\,(\nu - 2\nu^2 - \nu\xi z) J_1(\xi r)\,\frac{d\xi}{\xi}$$

$$+ \int_0^\infty z(\xi)\, e^{-\xi z}\,[2\nu] J_0'(\xi r)\, d\xi \qquad \ldots(9)$$

$$\sigma_r + \sigma_\theta + \sigma_z = 2(1+\nu)\int_0^\infty z(\xi) J_0(\xi r) e^{-\xi z}\, d\xi \qquad \ldots(10)$$

If $\quad p(r) = \dfrac{P}{\pi a^2}$ for $0 \le r \le a$

$\quad = 0$ for $r > a$

where P is the total force applied to the circular area. Then

$$z(\xi) = -\xi \int_0^\infty r\,\frac{P}{\pi a^2} J_0(\xi r)\, dr$$

$$= -P\,\frac{J_1(\xi a)}{\pi a} \qquad \ldots(11)$$

This value of $z(\xi)$ may be inserted in Eq. (4) to Eq. (9) and displacement and stress components can be evaluated.

For a point force

$$z(\xi) = -\frac{P}{\pi}\,\underset{a \to 0}{\text{Lt}}\,\frac{J_1(\xi a)}{a} = -\frac{P\xi}{2\pi}$$

Substituting in Eq. (4) to Eq. (9), we get

$$u_r = \left(\frac{1+\nu}{E}\right) z \int_0^\infty \frac{P}{2\pi}\,\xi\, e^{-\xi z} J_1(\xi r)\, d\xi$$

$$- \frac{(1-2\nu)(1+\nu)}{E}\int_0^\infty \frac{P}{2\pi}\,\xi\, e^{-\xi z} J_1(\xi r)\,\frac{d\xi}{\xi}$$

$$= \left(\frac{1+\nu}{E}\right)\frac{Pz}{2\pi}\int_0^\infty \xi\, e^{-nz} J_1(\xi r)\, d\xi$$

$$- \frac{(1-2\nu)(1+\nu)}{E}\cdot\frac{P}{2\pi}\int_0^\infty e^{-\xi z} J_1(\xi r)\, d\xi$$

$$= \left(\frac{1+\nu}{E}\right)\frac{Pz}{2\pi}\,[r(r^2+z^2)^{-3/2}]$$

$$- \frac{(1-2\nu)(1+\nu)}{E}\cdot\frac{P}{2\pi}\left[\frac{1}{r} - \frac{z}{r(r^2+z^2)^{1/2}}\right]$$

$$= \frac{(1+\nu)}{E}\frac{P}{2\pi}\left[\frac{rz}{(r^2+z^2)^{3/2}}\right.$$

$$\left. -(1-2\nu)\left\{\frac{1}{r}-\frac{z}{(r^2+z^2)^{1/2}}\right\}\right] \quad ...(12)$$

Similarly

$$u_z = \frac{(1+\nu)}{E}\cdot\frac{P}{2\pi}\left[\frac{2(1-\nu)}{(r^2+z^2)^{1/2}}+\frac{z^2}{(r^2+z^2)^{3/2}}\right] \quad ...(13)$$

At $z = 0$

$$u_r = -\frac{(1+\nu)(1-2\nu)}{E}\cdot\frac{P}{2\pi r} \quad ...(14)$$

$$u_z = \frac{(1+\nu)(1-\nu)}{E}\frac{P}{\pi r} \quad ...(15)$$

$$\sigma_z = -z\int_0^\infty \xi^2 \cdot \frac{P}{2\pi} \cdot e^{-\xi z} J_0(\xi r)\, d\xi$$

$$-\int_0^\infty \frac{P\xi}{2\pi} e^{-\xi z} J_0(\xi r) d\xi$$

$$= -\frac{P}{2\pi}\left[z\int_0^\infty \xi^2 e^{-\xi z} J_0(\xi r)\, d\xi\right.$$

$$\left. + \int_0^\infty \xi e^{-\xi z} J_0(\xi r)\, d\xi\right]$$

$$\sigma_z = \frac{-P}{2\pi}\left[\frac{2z^3-r^2z}{(r^2+z^2)^{5/2}}+\frac{z}{(r^2+z^2)^{3/2}}\right]$$

$$= -\frac{P}{2\pi}\left[\frac{3z^3}{(r^2+z^2)^{5/2}}\right] \quad ...(16)$$

$$\tau_{rz} = -z\int_0^\infty \frac{P\xi^2}{2\pi} e^{-\xi z} J_1(\xi r)\, d\xi$$

$$= -\frac{P}{2\pi}\cdot z\int_0^\infty \xi^2 e^{-\xi z} J_1(\xi r)\, d\xi$$

$$= -\frac{P}{2\pi}\cdot z\left[\frac{3rz}{(r^2+z^2)^{5/2}}\right]$$

$$= -\frac{P}{2\pi}\left[\frac{3rz^2}{(r^2+z^2)^{5/2}}\right] \quad ...(17)$$

$$\sigma_r = -\int_0^\infty \frac{P\xi}{2\pi} e^{-\xi z} (1 - \xi z) J_0 (\xi r) (d\xi)$$

$$+ \frac{1}{r}\int_0^\infty \frac{P\xi}{2\pi} e^{-\xi z}(\nu - 2\nu^2 - \nu\xi z) J_1(\xi r) \frac{d\xi}{\xi}$$

$$= -\frac{P}{2\pi}\int_0^\infty (\xi - \xi^2 z) e^{-\xi z} J_0(\xi r)\, d\xi$$

$$+ \frac{P}{2\pi r}\int_0^\infty (\nu - 2\nu^2 - \nu\xi z) e^{-\xi z} J_1(\xi r)\, d\xi$$

$$= -\frac{P}{2\pi}\left[\frac{z}{(r^2 + z^2)^{3/2}} - \frac{2z^3 - r^2 z}{(r^2 + z^2)^{5/2}} - \left(\frac{\nu - 2\nu^2}{r}\right)\left(\frac{1}{r} - \frac{z}{r(r^2 + z^2)^{1/2}}\right) + \frac{\nu z}{r}\frac{r}{(r^2 + z^2)^{3/2}}\right]$$

$$= -\frac{P}{2\pi}\left[\frac{r^2 z - z^3 - 2z^3 + r^2 z + \nu r^2 z + \nu z^3}{(r^2 + z^2)^{5/2}} - \frac{\nu (1 - 2\nu)}{r}\left(\frac{1}{r} - \frac{z}{r(r^2 + z^2)^{1/2}}\right)\right]$$

$$= -\frac{P}{2\pi}\left[\frac{2r^2 z - z^3 + \nu r^2 z + \nu z^3}{(r^2 + z^2)^{5/2}} - \frac{\nu(1 - 2\nu)}{z^2}\left(1 - \frac{z}{(r^2 + z^2)^{1/2}}\right)\right]$$

$$= -\frac{P}{2\pi}\left[\frac{r^2 z(2 + \nu) + z^3(\nu - 1)}{(r^2 + z^2)^{5/2}} - \frac{\nu(1 - 2\nu)}{z^2}\left\{1 - \frac{z}{(r^2 + z^2)^{1/2}}\right\}\right] \quad ...(18)$$

$$\sigma_\theta = -\frac{P}{2\pi r}\int_0^\infty e^{-\xi z} (\nu - 2\nu^2 - \nu\xi z) J_1(\xi r)\, d\xi$$

$$- \frac{P}{2\pi}\int_0^\infty 2\nu\, \xi\, e^{-\xi z} J_0(\xi r)\, d\xi$$

$$= -\frac{P}{2\pi r}\left[\nu(1-2\nu)\left\{ \frac{1}{r} - \frac{z}{r(r^2+z^2)^{1/2}} \right\} - \nu z\left\{ \frac{r}{(r^2+z^2)^{3/2}} \right\}\right] - \frac{P\nu}{\pi}\left[\frac{z}{(r^2+z^2)^{3/2}} \right]$$

$$= -\frac{P}{2\pi}\left[\nu(1-2\nu)\left\{ \frac{1}{r^2} - \frac{z}{r^2(r^2+z^2)^{1/2}} \right\} - \frac{\nu z}{(r^2+z^2)^{3/2}} + \frac{2\nu z}{(r+z^2)^{3/2}} \right]$$

$$= -\frac{P}{2\pi}\left[\nu(1-2\nu)\left\{ \frac{1}{r^2} - \frac{z}{r^2(r^2+z^2)^{1/2}} \right\} + \frac{\nu z}{(r^2+z^2)^{3/2}} \right] \quad ...(19)$$

At

$$\left.\begin{aligned} z &= 0 \\ \sigma_z &= \tau_{rz} = 0 \\ \sigma_r &= \frac{P}{2\pi}\cdot\frac{\nu(1-2\nu)}{r^2} \\ \sigma_\theta &= -\frac{P}{2\pi}\cdot\frac{\nu(1-2\nu)}{r^2} \end{aligned}\right\} \quad ...(20)$$

Example 8.7 *Symmetrical deformation of a thick plate.*

Solution. Consider a thick plate of thickness $2h$ and subjected to normal pressure on the surface as shown in Fig. 8.5.

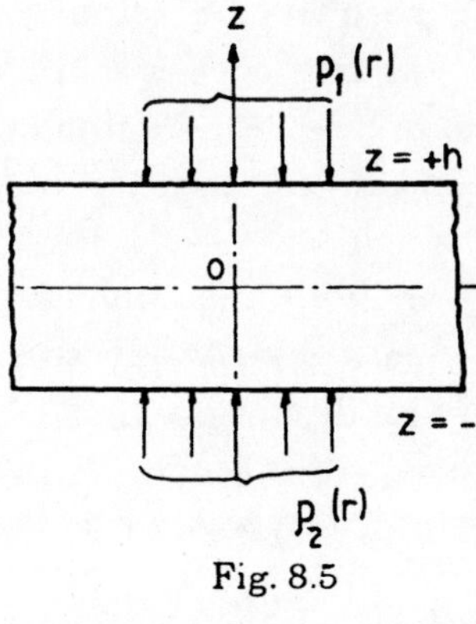

Fig. 8.5

The boundary conditions are

$$\left.\begin{aligned} \sigma_z &= -p_1(r),\ \tau_{rz} = 0 \text{ on } z = +h \\ \sigma_z &= -p_2(r),\ \tau_{rz} = 0 \text{ on } z = -h \end{aligned}\right\} \qquad \ldots(1)$$

For equilibrium of the plate

$$\int_0^\infty rp_1(r)\,dr = \int_0^\infty rp_2(r)\,dr$$

Now compatibility equation is

$$\left(\frac{d^2}{dz^2} - \xi^2\right)^2 G(\xi, z) = 0 \qquad \ldots(2)$$

where $G(\xi, z) = \int_0^\infty r\Phi\, J_0(\xi r)\,dr.$

Assuming a solution of Eq. (2) in the form

$$G(\xi, z) = (A + Bz)\cosh(\xi z) + (C + Dz)\sinh(\xi z) \qquad \ldots(3)$$

From Eq. (8.34), we have

$$\int_0^\infty r\sigma_z J_0(\xi r)\,dr = (1-\nu)\frac{d^3G}{dz^3} - (2-\nu)\,\xi^2\frac{dG}{dz}$$

$$\begin{aligned} &= (1-\nu)[A\xi^3\sinh\xi z + B\xi^2(\xi z\sinh\xi z + 3\cosh\xi z) \\ &\quad + C\xi^3\cosh\xi z + D\xi^2(3\sinh\xi z + \xi z\cosh\xi z)] \\ &\quad - (2-\nu)[A\xi^3\sinh\xi z + B\xi^2(\xi z\sinh\xi z + \cosh\xi z) \\ &\quad + C\xi^3\cosh\xi z + D\xi^2(\sinh\xi z + \xi z\cosh\xi z)] \\ &= -[A\xi^3\sinh\xi z + B\xi^2(\xi z\sinh\xi z + \cosh\xi z) \\ &\quad + C\xi^3\cosh\xi z + D\xi^2(\sinh\xi z + \xi z\cosh\xi z)] \\ &\quad + (1-\nu)[2B\xi^2\cosh\xi z + 2D\xi^2\sinh\xi z] \\ &= -[A\xi^3\sinh\xi z + B\xi^2.\{\xi z\sinh\xi z + [1 - 2(1-\nu)] \\ &\quad \times\cosh\xi z\} + C\xi^3\cosh\xi z + D\xi^2\{\xi z\cosh\xi z + (1 - 2 \\ &\qquad (1-\nu))\sinh\xi z\}] \\ &= -[A\xi^3\sinh\xi z + B\xi^2\{\xi z\sinh\xi z - (1-2\nu)\cosh\xi z\} \\ &\quad + C\xi^3\cosh\xi z + D\xi^2\{\xi z\cosh\xi z - (1-2\nu)\sinh\xi z]\} \end{aligned}$$

Using the boundary conditions, we find

$$+\int_0^\infty rp_1(r)\,J_0(\xi r)\,dr = +[A\xi^3\sinh\xi h + B\xi^2\{\xi h\sinh\xi h - (1-2\nu)$$

$$\times\cosh\xi h\} + C\xi^2\cosh\xi h + D\xi^2\{\xi h\cosh\xi h - (1-2\nu)\sinh\xi h\}]$$

$$+\int_0^\infty rp_2(r)J_0(\xi r)\,dr = +\,[-A\xi^3 \sinh \xi h + B\xi^2\{\xi h \sinh \xi h - (1-2\nu)\cosh \xi h\} + C\xi^3 \cosh \xi h + D\xi^2\{-\xi h \cosh \xi h + (1-2\nu)\sinh \xi h\}]$$

Let $\overline{p}\,(\xi) = \int_0^\infty r\,p(r)\,J_0\,(\xi r)\,dr$

$$\therefore\ \ \overline{p}_1(\xi) = A\xi^3 \sinh \xi h + B\xi^2\{\xi h \sinh \xi h - (1-2\nu)\cosh \xi h\} + C\xi^3 \cosh \xi h + D\xi^2\{\xi h \cosh \xi h - (1-2\nu)\sinh \xi h\} \quad ...(4)$$

$$\overline{p}_2(\xi) = -A\xi^3 \sinh \xi h + D\xi^2\{\xi h \sinh \xi h - (1-2\nu)\cosh \xi h\xi + C\xi^3 \cosh \xi h + D\xi^2\{-\xi h \cosh \xi h + (1-2\nu)\sinh \xi h\} \quad ...(5)$$

And from Eq. (8.35), we have

$$\int_0^\infty r\tau_{rz}\,J_1(\xi r)dr = \nu\,\frac{d^2G}{dz^2} + (1-\nu)\,\xi^2 G$$

$$= \nu[A\xi^2 \cosh \xi z + B\xi\{2\sinh \xi z + \xi z \cosh \xi z\} + C\xi^2 \sinh \xi z + D\xi\{2\cosh \xi z + \xi z \sinh \xi z\}] + (1-\nu)\xi^2[A\cosh \xi z + Bz\cosh \xi z + C\sinh \xi z + Dz\sinh \xi z]$$

$$= A\xi^2 \cosh \xi z + Bz\xi^2 \cosh \xi z + C\xi^2 \sinh \xi z + Dz\xi^2 \sinh \xi z + 2\nu\xi B \sinh \xi z + 2\nu\xi D \cosh \xi z$$

$$= A\xi^2 \cosh \xi z + B\{z\xi^2 \cosh \xi z + 2\nu C \sinh \xi z\} + C\xi^2 \sinh \xi z + D\{z\xi^2 \sinh \xi z + 2\nu\xi \cosh \xi z\}$$

Using the boundary conditions, we find

$$0 = A\xi \cosh \xi h + B\{\xi h \cosh \xi h + 2\nu \sinh \xi h\} + C\xi \sinh \xi h + D\{\xi h \sinh \xi h + 2\nu \cosh \xi h\} \quad ...(6)$$

$$0 = A\xi \cosh \xi h - B\{\xi h \cosh \xi h + 2\nu \sinh \xi h\} - C\xi \sinh \xi h + D\{\xi h \sinh \xi h + 2\nu \cosh \xi h\} \quad ...(7)$$

Solving Eqs. (4) to (7) for A, B, C and D and inserting in Eq. (3), we get

$$G(\xi, z) = \left(\frac{\overline{p}_1 - \overline{p}_2}{\xi^3}\right)\left[\frac{\xi h \sinh \xi h \cosh \xi z + 2\nu \cosh \xi h \cosh \xi z - z\xi \sinh \xi z \cosh \xi h}{\sinh 2\xi h - 2\xi h}\right] + \left(\frac{\overline{p}_1 + \overline{p}_2}{\xi^3}\right)\left[\frac{\{\xi h \cosh \xi h + 2\nu \sinh \xi h\}\sinh \xi z - \xi z \cosh \xi z \sinh \xi h}{\sinh 2\xi h + 2\xi h}\right] \quad ...(8)$$

Eq. (8) may be substituted in Eqs. (8.34), (8.35), (8.37) and (8.38) to obtain the stress components and the displacement components may be obtained from Eqs. (8.32) and (8.33).

Let $p_1(r) = p_2(r) = p(r)$

then Eq. (8) becomes

$$G(\xi, z) = \frac{2\bar{p}}{\xi^3}$$

$$\times \left[\frac{\{\xi h \cosh \xi h + 2\nu \sinh \xi h\} \sinh \xi z - \xi z \cosh \xi z \sinh \xi h}{\sinh 2\xi h + 2\xi h} \right] \qquad \text{...(9)}$$

From Eq. (8.34), we have

$$\sigma_z = \int_0^\infty \xi \left[(1-\nu) \frac{d^3G}{dz^3} - (2-\nu)\, \xi^2 \frac{dG}{dz} \right] J_0(\xi r)\, d\xi$$

$$= -\int_0^\infty \xi \left[\frac{2\bar{p}(\xi)}{\sinh 2\xi h + 2\xi h} \{(\xi h \cosh \xi h + 2\nu \sinh \xi h) \cosh \xi z - \xi z \sinh \xi h \sinh \xi z + (1-2\nu) \sinh \xi h \cosh \xi z\} \right] \times J_0(\xi r)\, d\xi$$

$$= -2 \int_0^\infty \left[\frac{\xi h \cosh \xi h \cosh \xi z - \xi z \sinh \xi h \sinh \xi z + \sinh \xi h \cosh \xi z}{\sinh 2\xi h + 2\xi h} \right] \times \xi\, \bar{p}(\xi)\, J_0(\xi r) d\xi.$$

At $\quad z = 0$

$$\sigma_z = -2 \left[\frac{\xi h \cosh \xi h + \sinh \xi h}{\sinh 2\xi h + 2\xi h} \right] \xi \bar{p}(\xi)\, J_0(\xi r)\, d\xi$$

Let $\quad \xi h = \eta$

so that $\quad d\xi = \dfrac{d\eta}{h}$

and $\quad \lambda = \dfrac{r}{h}$

$$\therefore \quad \sigma_z = -2 \int_0^\infty \frac{\eta \cosh \eta + \sinh \eta}{\sinh 2\eta + 2\eta} \cdot \frac{\eta}{h} \cdot \bar{p}\left(\frac{\eta}{h}\right) J_0(\lambda\eta) \frac{d\eta}{h}$$

$$= -\frac{2}{h^2} \int_0^\infty \eta\, \bar{p}\left(\frac{\eta}{h}\right) \left\{ \frac{\eta \cosh \eta + \sinh \eta}{\sinh 2\eta + 2\eta} \right\} . J_0(\lambda\eta)\, d\eta. \qquad \text{...(10)}$$

Similar expressions may be obtained for other stress components.

For a point force

$$\bar{p}\,(\xi) = \frac{P}{2\pi}$$

and for a force P distributed over circular area of radius a

$$\bar{p}\,(\xi) = \frac{P\,J_1(\xi a)}{\pi \xi a}$$

After obtaining the expressions for stress components, these integrals may be evaluated numerically.

Example 8.8 *Stress distribution in a long circular cylinder subjected to a discontinuous pressure band.*

Solution. Consider a long circular cylinder of radius $r = a$ deformed by the application of unit pressure to the part $z > 0$ while the remainder, $z < 0$ of the boundary is unloaded. The shear stresses over the curved surface is taken to be zero for the whole length of cylinder. Thus the boundary conditions become

$$\sigma_r = \begin{cases} -f(z)\,; & 0 < z < \infty \\ 0\;; & -\infty < z < 0 \end{cases} \qquad \text{...(1)}$$

$$\tau_{rz} = 0,\ r = a,\ -\infty < z < \infty \qquad \text{...(2)}$$

The stress components given by

$$\left.\begin{aligned} \sigma_r &= \frac{\partial}{\partial z}\left(\nu\,\nabla^2 - \frac{\partial^2}{\partial r^2}\right)\Phi \\ \sigma_\theta &= \frac{\partial}{\partial z}\left(\upsilon\,\nabla^2 - \frac{1}{r}\frac{\partial}{\partial r}\right)\Phi \\ \sigma_z &= \frac{\partial}{\partial z}\left[(2-\nu)\,\nabla^2 - \frac{\partial^2}{\partial z^2}\right]\Phi \\ \tau_{rz} &= \frac{\partial}{\partial r}\left[(1-\nu)\,\nabla^2 - \frac{\partial^2}{\partial z^2}\right]\Phi \end{aligned}\right\} \qquad \text{...(3)}$$

satisfy the compatibility equation.

$$\nabla^4\Phi = 0$$

where
$$\nabla^2 = \frac{\partial^2}{\partial r^2} + \frac{1}{r}\frac{\partial}{\partial r} + \frac{\partial^2}{\partial z^2} \qquad \text{...(4)}$$

Then
$$\left.\begin{aligned} u_r &= -\left(\frac{1+\nu}{E}\right)\frac{\partial^2\Phi}{\partial r\,\partial z} \\ u_z &= \frac{1+\nu}{E}\left[(1-2\nu)\,\nabla^2 + \frac{\partial^2}{\partial r^2} + \frac{1}{r}\frac{\partial}{\partial r}\right]\Phi \end{aligned}\right\} \qquad \text{...(5)}$$

Thus boundary conditions become

$$\frac{\partial}{\partial z}\left(\nu \nabla^2 - \frac{\partial^2}{\partial r^2}\right)\Phi = \begin{cases} -f(z)\,; & 0 < z < \infty\,, r = a \\ 0 & ; -\infty < z < 0, r = a \end{cases} \quad ...(6)$$

$$\frac{\partial}{\partial r}\left[(1-\nu)\nabla^2 - \frac{\partial}{\partial z}\right]\Phi = 0\,; -\infty < z < \infty, r = a \quad ...(7)$$

Multiply Eq. (4) by $e^{i\xi z}$ and integrate w.r.t. z from $-\infty$ to ∞

$$\therefore \quad \int_{-\infty}^{\infty} \nabla^4\Phi.\, e^{i\xi z}\, dz = 0$$

Let $$\chi = \int_{-\infty}^{\infty} \Phi\, e^{i\xi z}\, dz \quad ...(8)$$

Thus we get $\overline{\nabla}^4 \chi = 0$ ...(9)

where $\overline{\nabla}^2 = \dfrac{d^2}{dr^2} + \dfrac{1}{r}\dfrac{d}{dr} - \xi^2$

Also let $F(\xi) = \int_0^{\infty} f(z)\, e^{i\xi z}\, dz$

then Eq. (6) can be written as

$$\left(\nu\,\overline{\nabla}^2 - \frac{d^2}{dr^2}\right)\chi = -\frac{iF(\xi)}{\xi}\,; r = a \quad ...(10)$$

and Eq. (7) as

$$\frac{d}{dr}\left[(1-\nu)\,\overline{\nabla}^2 + \xi^2\right]\chi = 0\,; r = a \quad ...(11)$$

The solution of Eq. (9) is

$$\chi = A(\xi)\, I_0(\xi r) + B(\xi)\, \xi r\, I_1(\xi r) \quad ...(12)$$

Now $\overline{\nabla}^2 \chi = 2\xi^2 B(\xi)\, I_0(\xi r)$ ...(13)

Substituting Eq. (13) in Eqs. (10) and (11), we get

$$A\left[I_0(\xi a) - \frac{1}{\xi a} I_1(\xi a)\right] + B[(1-2\nu)\, I_0(\xi a) + \xi a\, I_1(\xi a)] = -\frac{i\,F(\xi)}{\xi^3}$$

$$AI_1(\xi a) + B[2(1-\nu)\, I_1(\xi a) + \xi a\, I_0(\xi a)] = 0$$

Solving for A and B, we get

$$A(\xi) = \frac{ai}{\xi^3 D(\xi a)}[2(1-\nu)\, I_1(\xi a) + \xi a\, I_0(\xi a)]\, F(\xi)$$

$$B(\xi) = -\frac{ai}{\xi^3 D(\xi a)} I_1(\xi a)\, F(\xi)$$

where $D(z) = [z^2 + 2(1-\nu)]\, I_1^2(z) - z^2 I_0^2\,(z)$

Thus the auxiliarly function χ can be evaluated. Now multiply first of Eqs. (3) by $e^{i\xi z}$ and integrate w.r.t. z from $-\infty$ to ∞, we obtain

$$\bar{\sigma}_r = \int_0^{\infty} \sigma_r\, e^{i\xi z}\, dz$$

$$= -i\xi\left(\nu\,\bar{\nabla}^2 - \frac{d^2}{dr^2}\right)\chi$$

$$\therefore \qquad \bar{\sigma}_r = \frac{aF(\xi)}{rD(\xi a)}\,[ar\xi^2 I_0\,(\xi a)\, I_0(\xi r) + \xi r\, I_0(\xi r)\, I_1(\xi a)$$
$$- a\xi\, I_0(\xi a)\, I_1(\xi r) - \{r^2\xi^2 + 2(1-\nu)\}\, I_1(\xi a)\, I_1(\xi r)]$$

where $\bar{\sigma}_r$ is the Fourier transform of σ_r.

$$\bar{\sigma}_r = \frac{1}{2\pi}\int_{-\infty}^{\infty} \bar{\sigma}_r\, e^{-i\xi z}\, d\xi$$

Let $\quad f(z) = 1.$

Then $\quad F(\xi) = 2\pi\delta_+\,(\xi)$

where $\delta_+\,(\xi) = \dfrac{1}{2}\,\delta(\xi) - \dfrac{1}{2\pi i\xi}$

is the Heisenberg delta function.

and
$$\int_{-\infty}^{\infty} \delta_+(\xi)\, e^{-i\xi z}\, d\xi = \frac{1}{2} + \frac{1}{\pi}\int_0^{\infty} \frac{\sin\,(\xi z)}{\xi}\, d\xi$$
$$= \begin{cases} 1 \text{ if } z > 0 \\ 0 \text{ if } z < 0 \end{cases} \qquad \text{...(14)}$$

Substituting this value of $F(\xi)$ and putting $\eta = \xi a$, we find that

$$\sigma_r = -\frac{1}{2} + \frac{a}{\pi r}\int_0^{\infty}\left\{\frac{r}{a}\,\eta^2\, I_0(\eta)\, I_0\left(\frac{r\eta}{a}\right) + \frac{r}{a}\cdot\eta\, I_0\left(\frac{r\eta}{a}\right) I_1\,(\eta)\right.$$
$$\left. - I_0(\eta)\, I_1\left(\frac{r\eta}{a}\right) - \left[\frac{r^2}{a^2}\cdot\eta^2 + 2(1-\nu)\right] I_1(\eta)\, I_1\left(\frac{r\eta}{a}\right)\right\}$$
$$\times\, \frac{\sin\left(\dfrac{z\eta}{a}\right) d\eta}{\eta D(\eta)} \qquad \text{...(15)}$$

Similarly

$$\bar{\sigma}_\theta = 2\pi\,\delta_+\,(\xi)\,\frac{a}{r}\,\frac{1}{D(\xi a)}\,[a\xi\, I_0(\xi a)\, I_1(r\xi)$$
$$+ 2(1-\nu)\, I_1(\xi a)\, I_1(r\xi) + (2\nu - 1)\,\xi r\, I_1(\xi a)\, I_1(\xi r)]$$

$$\bar{\sigma}_z = \frac{1}{i\xi}\frac{a\xi}{D(\xi a)}[a\xi\, I_0(\xi a)\, I_0(r\xi) - \xi r\, I_1(\xi a)\, I_1(r\xi) - 2\, I_0(r\xi)\, I_1(\xi a)]$$

$$\tau_{rz} = \frac{1}{\xi}\frac{a}{D(\xi a)}[a\xi\, I_1(\xi r)\, I_0(\xi a) - \xi r\, I_1(\xi a)\, I_0(\xi r)]$$

Inverting we get

$$\sigma_\theta = -\frac{1}{2} + \frac{a}{\pi r}\int_0^\infty \left[\eta I_0(\eta) + I_1\left(\eta\frac{r}{a}\right) + 2(1-\nu)\, I_1(\eta)\left(\eta\frac{r}{a}\right) + (2\nu - 1)\frac{r}{a}\eta\, I_1(\eta)\, I_0\left(\eta\frac{r}{a}\right)\right]\frac{\sin\left(\eta\frac{z}{a}\right)d\eta}{\eta D(\eta)} \quad ...(16)$$

$$\sigma_z = \frac{1}{\pi}\int_0^\infty \left[2I_0\left(\eta\frac{r}{a}\right)I_1(\eta) + \frac{r}{a}\eta\, I_1(\eta)\, I_1\left(\eta\frac{r}{a}\right) - \eta\, I_0(\eta)\, I_0\left(\eta\frac{r}{a}\right)\right]\frac{\sin\left(\eta\frac{\eta}{a}\right)d\eta}{D(\eta)} \quad ...(17)$$

$$\tau_{rz} = \frac{1}{\pi}\int_0^\infty \left[\eta\, I_1\left(\eta\frac{r}{a}\right) I_0(\eta) - \frac{r}{a}\eta\, I_1(\eta)\, I_0\left(\eta\frac{r}{a}\right)\right] \times \frac{\cos\left(\eta\frac{\eta}{a}\right)d\eta}{D(\eta)} \quad ...(18)$$

These integrals can be evaluated numerically.

EXERCISES

8.1 A circular disc of radius 200 mm is 25 mm thick. If this is rotating at a speed of 3000 r.p.m., determine the radial and hoop stresses in the disc at the mid-thickness and on the surface. Assume $\rho = 7470$ kg/m^3, $\nu = 0.28$.

8.2 A force of 800 N is applied at a point in an infinite solid. Determine the stress distribution in the solid at a depth of 20 mm at an angle of 30° with the line of action of the force. Take $\nu = 0.30$.

8.3 A spherical shell has an internal diameter of 400 mm and external diameter of 500 mm. It is subjected to an internal pressure of 12 MPa. Determine the radial and hoop stresses in the shell.

8.4 A point force of 500 N is applied normal to a semi-infinite solid. Determine the stress distribution at a depth of 50 mm and angular location of 15° with the line of action of the force.

8.5 By adopting stress function approach, determine the expressions for stresses in a solid circular cylinder of radius 'a' carrying a band of pressure of intensity p.

8.6 A thick hollow cylinder of circular cross-section of inner radius a and outer radius b is subjected to a uniform band of pressure of intensity p. Formulate this problem and present a solution by using stress function approach.

8.7 Formulate the problem as given in exercise 8.5 by using integral transform method and derive expressions for the stress components.

8.8 A thick plate of thickness $2h$ is subjected to uniform pressure intensities of p_1 and p_2 applied over circular areas of radius a and b on the top and bottom surfaces respectively. Derive expression for the stress components using integral transform method.

8.9 Determine the solution of Exercise 8.8 by using stress-function approach.

8.10 A thick hollow cylinder of circular cross section of inner radius a and outer radius b is subjected to a semi-infinite band pressure of intensity p. Determine expressions for the stress components developed in the cylinder.

9

Bending of Plates

9.1 Introduction

In general a plate may be defined as a flat structure of which thickness is small compared with other two dimensions. Plates are capable of taking bending moments and shear forces perpendicular to their own planes, as well as forces lying in their own plane [Fig. 9.1 (*a*)]. Otherwise we can say that it is a two-dimensional beam.

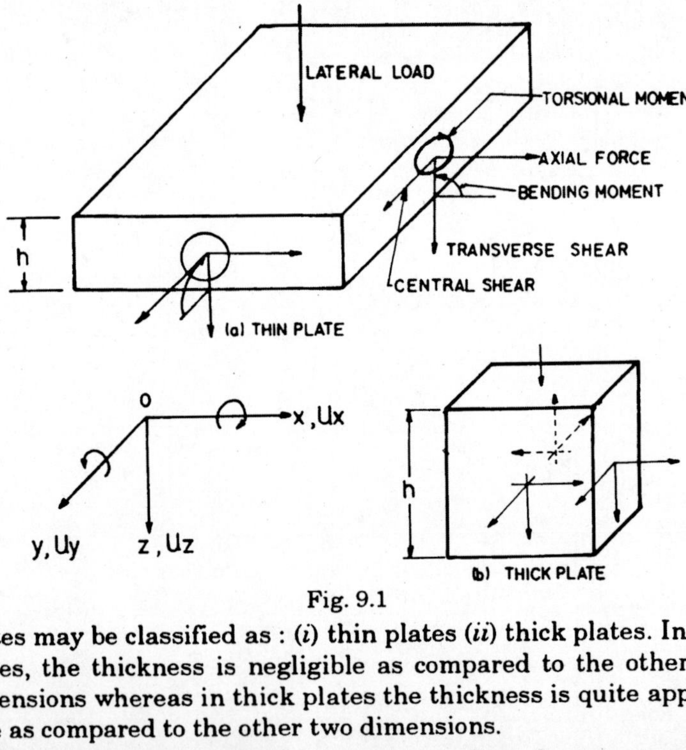

Fig. 9.1

Plates may be classified as : (*i*) thin plates (*ii*) thick plates. In thin plates, the thickness is negligible as compared to the other two dimensions whereas in thick plates the thickness is quite appreciable as compared to the other two dimensions.

As shown in Fig. 9.1 (*a*), transverse shear forces act perpendicularly to the plane of the plate whereas central shear forces act in the plane of the plate. A thick plate has been shown in Fig. 9.1 (*b*). In this chapter, we shall study the bending of thin plates of various shapes having various end conditions assuming small deflection.

9.2 Cylindrical Bending of Rectangular Plates

If a long rectangular plate is subjected to a transverse load that does not vary along the length of the plate then the deflected surface of such a plate at a considerable distance from the ends can be assumed cylindrical, with the axis of the cylinder parallel to the length of the plate.

To determine the differential equation for the deflection of the plate, consider a plate of uniform thickness, equal to h, and take the xy plane as the middle plane of the plate before loading and positive direction of z-axis be downward. Then if we take width of the plate as l, then the stresses acting on an element of the plate at a distance z from the middle plane and of thickness dz are as shown in Fig. 9.2.

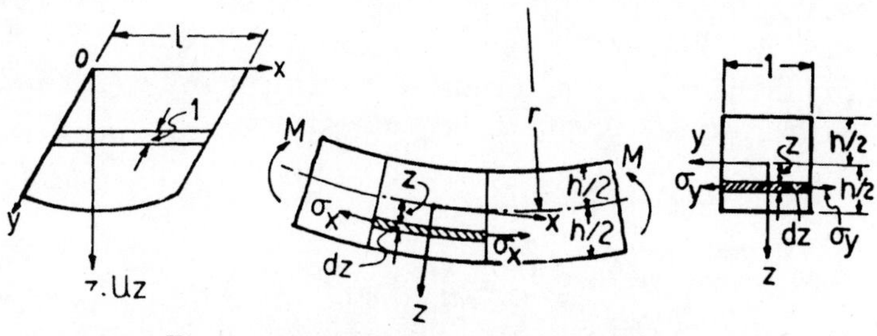

Fig. 9.2. Bending of a rectangular plate.

The following assumptions are made in calculating the bending stresses :

(*i*) There is no distortion in the cross-section of the strip during bending.

(*ii*) Plane cross-sections of the strip remain plane during bending.

Using Hooke's law, the strains are

$$\left.\begin{aligned} \varepsilon_x &= \frac{1}{E}(\sigma_x - \nu\sigma_y) \\ \varepsilon_y &= \frac{1}{E}(\sigma_y - \nu\sigma_x) = 0 \end{aligned}\right\} \qquad ...(1)$$

The lateral strain in the y-direction must be zero in order to maintain continuity. in the plate during bending. Hence, we get

$\sigma_y = \nu\sigma_x$ from second of Eq. (1). Substituting in first of Eq. (1), we find

$$\varepsilon_x = \frac{\sigma_x}{E}(1 - \nu^2)$$

$$\therefore \quad \sigma_x = \frac{E\varepsilon_x}{1-\nu^2} = \left(\frac{E}{1-\nu^2}\right)\cdot\frac{z}{r} = -\frac{Ez}{(1-\nu^2)}\cdot\frac{d^2u_z}{dx^2} \quad \text{...(2)}$$

where r is the radius of curvature of the plate after bending and

$$\sigma_y = -\frac{\nu\, Ez}{(1-\nu^2)}\frac{d^2u_z}{dx^2} \quad \text{...(3)}$$

The bending moment M on the elemental strip is given by :

$$M = \int_{-h/2}^{+h/2} \sigma_x \,.\, zdz = -\int_{-h/2}^{h/2} \frac{Ez^2}{(1-\nu^2)}\frac{d^2u_z}{dx^2}dz$$

$$= -\frac{Eh^3}{12(1-\nu^2)}\frac{d^2u_z}{dx^2}$$

Now $$\frac{Eh^3}{12(1-\nu^2)} = EI = D \quad \text{...(4)}$$

where I is the moment of inertia of the strip of unit width and D is called the flexural rigidity of the plate. Hence

$$D\frac{d^2u_z}{dx^2} = -M$$

or $$\frac{d^2u_z}{dx^2} = \frac{1}{r} = -\frac{M}{D} \quad \text{...(9.1)}$$

Thus to determine the deflection of the plate, we are required to integrate Eq. (9.1).

9.2.1 Uniformly loaded Rectangular Plate with Simply Supported Edges

(i) **Exact Solution.** Consider a long rectangular plate simply supported at the edges and carrying a uniformly distributed load of intensity q whose elemental strip is as shown in Fig. 9.3.

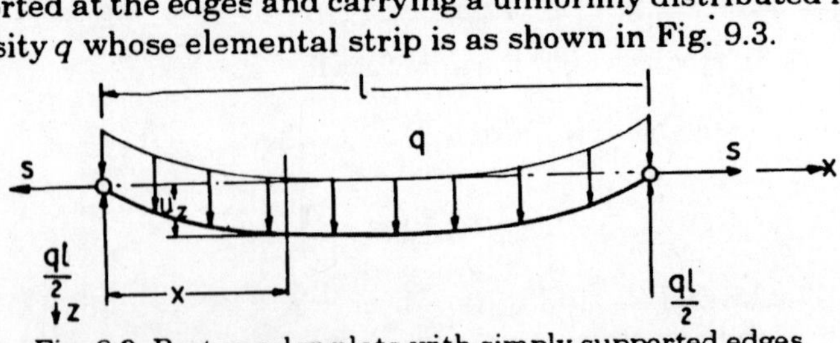

Fig. 9.3. Rectangular plate with simply supported edges.

The elemental strip is also subjected to an axial force S which is such as to prevent the ends of the bar from moving along the x-axis. Now

$$M = \frac{ql}{2}x - \frac{qx^2}{2} - Su_z \qquad ...(1)$$

Substituting in Eq. (9.1), we get

$$\frac{d^2u_z}{dx^2} - \frac{Su_z}{D} = -\frac{qlx}{2D} + \frac{qx^2}{2D} \qquad ...(2)$$

Let $$\frac{Sl^2}{4D} = \alpha^2 \qquad ...(3)$$

then the general solution of Eq. (2) can be written as :

$$u_z = C_1 \sinh\frac{2\alpha x}{l} + C_2 \cosh\frac{2\alpha x}{l} + \frac{ql^3x}{8\alpha^2 D} - \frac{ql^2x^2}{8\alpha^2 D} - \frac{ql^4}{16\alpha^4 D} \qquad ...(4)$$

where C_1, C_2 are constants of integration.

Now $u_z = 0$ for $x = 0$ and $x = l$

$$\therefore \quad C_1 = \frac{ql^4}{16\alpha^4 D}\left(\frac{1-\cosh 2\alpha}{\sinh 2\alpha}\right), \quad C_2 = \frac{ql^4}{16\alpha^4 D}$$

Thus Eq. (4) becomes :

$$u_z = \frac{ql^4}{16\alpha^4 D}\left(\frac{1-\cosh 2\alpha}{\sinh 2\alpha} . \sinh\frac{2\alpha x}{l} + \cosh\frac{2\alpha x}{l} - 1\right) + \frac{ql^3x}{8\alpha^2 D} - \frac{ql^2x^2}{8\alpha^2 D}$$

$$= \frac{ql^4}{16\alpha^4 D}\left[\frac{-\sinh\alpha\sinh\frac{2\alpha x}{l} + \cosh\alpha\cosh\frac{2\alpha x}{l}}{\cosh\alpha} - 1\right] + \frac{ql^2x}{8\alpha^2 D}(l-x)$$

$$= \frac{ql^4}{16\alpha^4 D}\left[\frac{\cosh\alpha\left(1-\frac{2x}{l}\right)}{\cosh\alpha} - 1\right] + \frac{ql^2x}{8\alpha^2 D}(l-x) \qquad ...(5)$$

Now extension of the centre line of the strip is

$$\lambda = \frac{1}{2}\int_0^l \left(\frac{du_z}{dx}\right)^2 dx$$

$$= \frac{S(1-\nu^2)}{hE}\,l \qquad ...(6)$$

Assuming that the lateral strain of the strip in the y-direction is prevented. Substituting Eq. (5) and (6), we get

$$\frac{S(1-\nu^2)l}{hE} = \frac{q^2l^7}{D^2}\left[\frac{5}{256}\frac{\tanh\alpha}{\alpha^7} + \frac{1}{256}\frac{\tanh^2\alpha}{\alpha^6} - \frac{5}{256\,\alpha^6} + \frac{1}{384\,\alpha^4}\right]$$

Substituting for S from Eq. (3), we get

$$\frac{E^2h^8}{(1-\nu^2)^2q^2l^8} = \frac{135}{16}\frac{\tanh\alpha}{\alpha^9} + \frac{27}{16}\frac{\tanh^2\alpha}{\alpha^8} - \frac{135}{16\alpha^8} + \frac{9}{8\alpha^6} \qquad ...(7)$$

For a given value of $\frac{h}{l}$ and q, the value of α may be determined by hit and trial from Eq. (7) and then the value of u_z can be obtained from Eq. (5). Now

$$M_{max} = -D\left(\frac{d^2u_z}{dx^2}\right)_{x=\frac{l}{2}}$$

Substituting for u_z from Eq. (5), we get

$$M_{max} = \frac{ql^2}{8}\,\psi_0(\alpha) \qquad ...(9.2)$$

where
$$\psi_0(\alpha) = \frac{1-\text{sech}\,\alpha}{\frac{\alpha^2}{2}} \qquad ...(9.3)$$

Now direct tensile stress σ_d is

$$\sigma_d = \frac{S}{h} = \frac{4\alpha^2D}{hl^2} = \frac{E\alpha^2}{3(1-\nu^2)}\left(\frac{h}{l}\right)^2 \qquad ...(9.4)$$

and the bending stress σ_b is

$$\sigma_b = \pm\frac{6}{h^2}M_{max} = \frac{3}{4}q\left(\frac{l}{h}\right)^2\psi_0 \qquad ...(9.5)$$

$$\sigma_{max} = \sigma_d + \sigma_b \qquad ...(9.6)$$

$$(u_z)_{max} = \frac{5ql^4}{384D}f_0(\alpha) \qquad ...(9.7)$$

where
$$f_0(\alpha) = \frac{\text{sech}\,\alpha - 1 + \frac{\alpha^2}{2}}{\frac{5}{24}\alpha^4} \qquad ...(9.8)$$

(ii) **Approximate Solution.** The deflection curve for the elemental strip may be represented by

$$u_z = \delta \sin \frac{\pi x}{l} \qquad \text{...(8)}$$

where δ = deflection at the middle of the strip.

Extension of the middle line of the strip is

$$\lambda = \frac{1}{2}\int_0^l \left(\frac{du_z}{dx}\right)^2 dx = \frac{\pi^2\delta^2}{4l} \qquad \text{...(9)}$$

$$\delta = \frac{\delta_0}{1+\beta}$$

where $\delta_0 = \dfrac{5ql^4}{384\,D}$, $\beta = \dfrac{S}{S_{cr}} = \dfrac{Sl^2}{\pi^2 D}$

where S_{cr} is the Euler critical load.

$$\therefore \qquad \lambda = \frac{\pi^2}{4l}\cdot\frac{\delta_0^{\,2}}{(1+\beta)^2} = \frac{Sl(1-\nu^2)}{Eh} = \frac{\pi^2\beta h^2}{12l}$$

$$\therefore \qquad \beta(1+\beta)^2 = 3\,\frac{\delta_0^{\,2}}{h^2} \qquad \text{...(10)}$$

From Eq. (10) we can determine β and the parameter α is now determined from the equation :

$$\alpha^2 = \frac{Sl^2}{4D} = \frac{\pi^2\beta}{4} \qquad \text{...(9.9)}$$

Then by knowing α, the value of u_z can be determined from Eq. (5) and stresses from Eqs. (9.4) and (9.5). If we let

$1 + \beta = x$ then Eq. (10) becomes

$$x^3 - x^2 = \frac{3\delta_0^{\,2}}{h^2} \qquad \text{...(9.10)}$$

Eq. (9.10) is more easier to evaluate.

9.2.2 Uniformly Loaded Plate with clamped Edges

(*i*) **Exact Solution.** Consider now that the longitudinal edges of the plate are clamped in such a manner that they cannot rotate.

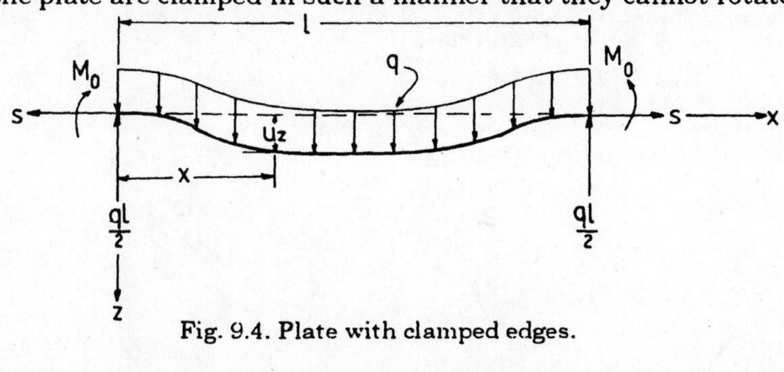

Fig. 9.4. Plate with clamped edges.

We take an elemental strip of unit width of the plate so that M_0 is the bending moment per unit length acting on the longitudinal edges as shown in Fig. 9.4.

The bending moment at any cross-section of the strip is

$$M = \frac{ql}{2}x - \frac{qx^2}{2} - Su_z + M_0 \qquad \text{...(1)}$$

Substituting in Eq. (9.1), we get

$$\frac{d^2u_z}{dx^2} - \frac{S}{D}u_z = -\frac{qlx}{2D} + \frac{qx^2}{2D} - \frac{M_0}{D} \qquad \text{...(2)}$$

The general solution of Eq. (2) is

$$u_z = C_1 \sinh\frac{2\alpha x}{l} + C_2 \cosh\frac{2\alpha x}{l} + \frac{ql^3x}{8x^2D} - \frac{ql^2x^2}{8\alpha^2D}$$

$$- \frac{ql^4}{16\alpha^4D} + \frac{M_0l^2}{4\alpha^2D} \qquad \text{...(3)}$$

where C_1, C_2 are constants and $\alpha^2 = \dfrac{Sl^2}{4D}$.

Now $\quad \dfrac{du_z}{dx} = 0$ for $x = 0$ and $x = \dfrac{l}{2}$

$u_z = 0$ for $x = 0$

Substituting in Eq. (3), we get

$$C_1 = -\frac{ql^4}{1\alpha^3D},$$

$$C_2 = \frac{ql^4}{16\alpha^3D}\coth\alpha$$

$$M_0 = \frac{ql^2}{4\alpha^2} - \frac{ql^2}{4\alpha}\coth\alpha$$

$$= -\frac{ql^2}{12}\psi_1(\alpha)$$

where $\psi_1(\alpha) = \dfrac{3(\alpha - \tanh\alpha)}{\alpha^2\tanh\alpha}$

Hence, Eq. (3) becomes

$$u_z = -\frac{ql^4}{16\alpha^3D}\sinh\frac{2\alpha x}{l} + \frac{ql^4}{16\alpha^3D}\coth\alpha\cosh\frac{2\alpha x}{l}$$

$$+ \frac{ql^3x}{8\alpha^2D} - \frac{ql^2x^2}{8\alpha^2D} - \frac{ql^4}{16\alpha^3D}\coth\alpha$$

$$= \frac{ql^4}{16\alpha^3 D \tanh\alpha}\left[\frac{\cosh\left\{\alpha\left(1-\frac{2x}{l}\right)\right\}}{\cosh\alpha} - 1\right] + \frac{ql^2(l-x)x}{8\alpha^2 D} \qquad ...(4)$$

Now extension of the centre line of the strip is

$$\lambda = \frac{1}{2}\int_0^l \left(\frac{du_z}{dx}\right)^2 du$$

$$= \frac{S(1-\nu^2)\,l}{hE}$$

$$= \frac{q^2 l^7}{D^2}\left[-\frac{3}{256\,\alpha^5 \tanh\alpha} - \frac{1}{256\,\alpha^4 \sinh^2\alpha} + \frac{1}{64\,\alpha^6} + \frac{1}{384\,\alpha^4}\right]$$

Substituting for S and D, we get

$$\frac{E^2 h^8}{(1-\nu^2)^2 q^2 l^8} = -\frac{81}{16\alpha^7 \tanh\alpha} - \frac{27}{16\alpha^6 \sinh^2\alpha} + \frac{27}{4\alpha^8} + \frac{9}{8\alpha^6} \qquad ...(5)$$

Eq. (5) can be solved by hit and trial for α for any given values of $\frac{h}{l}$ and q. Then deflection can be known from Eq. (4).

The stresses become

$$\sigma_d = \frac{E\alpha^2}{3(1-\nu^2)}\cdot\left(\frac{h}{l}\right)^2 \qquad ...(9.11)$$

$$\sigma_b = -\frac{6M_0}{h^2}$$

$$= \frac{q}{2}\left(\frac{l}{h}\right)^2 \psi_1(\alpha) \qquad ...(9.12)$$

$$\sigma_{max} = \sigma_d + \sigma_b \qquad ...(9.13)$$

$$(u_z)_{max} = (u_z)_{x=\frac{l}{2}}$$

$$= \frac{ql^4}{384D} f_1(\alpha) \qquad ...(9.14)$$

where $f_1(\alpha) = \frac{24}{\alpha^4}\left[\frac{\alpha^2}{2} + \frac{\alpha}{\sinh\alpha} - \frac{\alpha}{\tanh\alpha}\right]$

(*ii*) **Approximate solution.** We assume

$$u_z = \frac{\delta}{2}\left(1 - \cos\frac{2\pi x}{l}\right) \qquad \text{...(6)}$$

$$\lambda = \frac{1}{2}\int_0^l \left(\frac{du_z}{dx}\right)^2 dx$$

$$= \frac{\pi^2 \delta^2}{4l} \qquad \text{...(7)}$$

$$\delta = \frac{\delta_0}{1 + \frac{\beta}{4}}$$

where $\delta_0 = \dfrac{5ql^4}{384D}$

$$\beta = \frac{S}{S_{cr}} = \frac{Sl^2}{\pi^2 D}$$

$$\therefore \qquad \beta\left(1 + \frac{\beta}{4}\right)^2 = \frac{3\delta_0^2}{h}$$

Put $\qquad 1 + \dfrac{\beta}{4} = x$

$$\therefore \qquad x^3 - x^2 = \frac{3}{4}\frac{\delta_0^2}{h^2} \qquad \text{...(9.15)}$$

Example 9.1 *A circular plate 50 mm diameter is clamped at the edge. Its deflection at the centre is limited to 1 mm when a pressure of 1 MPa is applied. Calculate the thickness of the plate and the maximum stress developed in it. E = 200 GPa, ν = 0.3.*

Solution.

$$D = \frac{Et^3}{12(1 - \nu^2)} = \frac{200 \times 10^9 \times t^3}{12(1 - 0.09)} = 18.315 \times 10^9 \times t^3$$

$$(u_z)_{max} = \frac{pR^4}{64D}$$

$$1 \times 10^{-3} = \frac{1 \times 10^6 \times 25^4 \times 10^{-12}}{64 \times 18.315 \times 10^9 \times t^3}$$

$$t = 6.933 \times 10^{-4} \text{ m or } 0.6933 \text{ mm}$$

$$M_{max} = \frac{pR^2}{8} = \frac{1 \times 10^6 \times 25^2 \times 10^{-6}}{8} = 78.125 \text{ N.m.}$$

$$\sigma_{max} = \frac{6M_{max}}{t^2} = \frac{6 \times 78.125}{(0.6933)^2 \times 10^{-6}} = 975.2 \text{ MPa}$$

Example 9.2 *A flat plate is used to cover the end of a cylinder. Calculate the plate thickness if the cylinder is subjected to an internal pressure of 1.35 MPa. The diameter of the cylinder is 300 mm and maximum permissible stress is limited to 70 MPa. The edge of the cover fitting with the cylinder end may be treated as (a) rigidly clamped, and (b) simply supported.* $\nu = 0.33$.

Solution.

(a)
$$M_{max} = \frac{pa^2}{8}$$
$$= \frac{1.35 \times 10^6 \times (150)^2 \times 10^{-6}}{8}$$
$$= 3796.875 \text{ N.m}$$
$$\sigma_{max} = \frac{3pa^2}{4t^2}$$
$$70 \times 10^6 = \frac{3 \times 1.35 \times 10^6 \times (150)^2 \times 10^{-6}}{4t^2}$$
$$t = 18.04 \times 10^{-3} \text{ m or } 18.04 \text{ mm}$$

(b)
$$\sigma_{max} = \frac{3}{8}(3 + \nu)\frac{pa^2}{t^2}$$
$$70 \times 10^6 = \frac{3}{8} \times 3.33 \times \frac{1.35 \times 10^6 \times (150)^2 \times 10^{-6}}{t^2}$$
$$t = 23.28 \text{ mm}$$

Example 9.3 *A rectangular plate 375 mm × 125 mm is used as a cover for a rectangular tank. If the tank is to carry a fluid under a pressure of 2.7 MPa, find the minimum thickness of the plate required if the permissible stress for the material is 70 MPa. Consider the plate to be simply supported at the edges.*

Solution.

$$\beta_1 = 0.1189 \text{ for } \frac{a}{b} = \frac{375}{125} = 3$$
$$M_{max} = \beta_1 q b^2$$
$$= 0.1189 \times 2.7 \times 10^6 \times 125^2 \times 10^{-6}$$
$$= 5016.1 \text{ N.m}$$
$$\sigma_{max} = \frac{6M_{max}}{t^2}$$
$$t = \sqrt{\frac{6 \times 5016.1}{70 \times 10^6}} = 20.73 \text{ mm}$$

9.3 Slope and Curvatures

We refer the plate to a system of orthogonal coordinate system *oxyz* with the horizontal *xy* plane coincident with the middle surface of the undeformed plate and the vertical *z*-axis directed positively downward as shown in Fig. 9.5. Components of the elastic displacement in the *x*, *y* and *z* directions will be indicated, respectively, by the letters u_x, u_y and u_z.

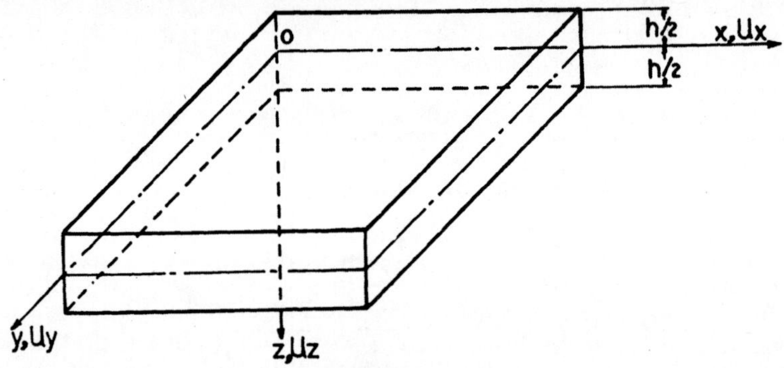

Fig. 9.5

Taking a normal section of the plate parallel to the *xz* plane as shown in Fig. 9.6 (*a*), the slope of the middle surface in the *x*-direction is $\phi_x = \frac{\partial u_z}{\partial x}$. Similarly slope in the *y*-direction is $\phi_y = \frac{\partial u_z}{\partial y}$. Now taking any direction *n* along *AB* in the *xy* plane making an

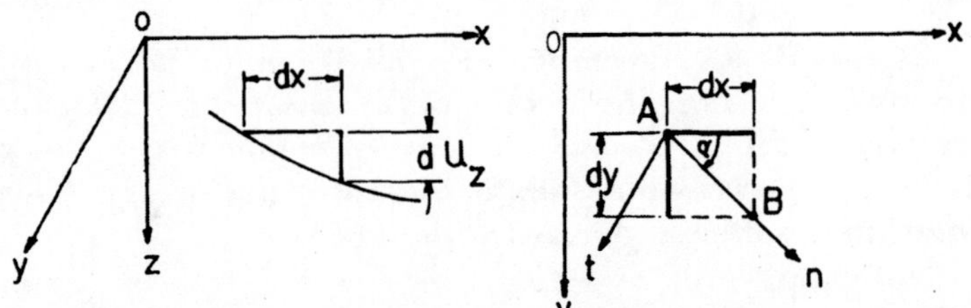

(*a*) (*b*)

Fig. 9.6

angle α with the *x*-axis [Fig. 9.6 (*b*)], the difference in the deflectors of two adjacent points *A* and *B* is

$$du_z = \frac{\partial u_z}{\partial x} dx + \frac{\partial u_z}{\partial y} dy$$

since $u_z = f(x, y)$

and the corresponding slope is

$$\frac{\partial u_z}{\partial n} = \frac{\partial u_z}{\partial x} \cdot \frac{dx}{dn} + \frac{\partial u_z}{\partial y} \cdot \frac{dy}{dn}$$

Now $\frac{dx}{dn} = \cos\alpha,$

$$\frac{dy}{dn} = \sin\alpha$$

$$\therefore \quad \frac{\partial u_z}{\partial n} = \frac{\partial u_z}{\partial x}\,.\cos\alpha + \frac{\partial u_z}{\partial y}\,.\sin\alpha$$

For $(\alpha + 90°)$ *i.e.* t direction

$$\frac{\partial u_z}{\partial t} = \frac{\partial u_z}{\partial x}\cos(\alpha + 90°) + \frac{\partial u_z}{\partial y}\sin(\alpha + 90°)$$

$$= -\frac{\partial u_z}{\partial x}\sin\alpha + \frac{\partial u_z}{\partial y}\cos\alpha$$

$$\therefore \quad \frac{\partial}{\partial n} = \cos\alpha\,.\frac{\partial}{\partial x} + \sin\alpha\,.\frac{\partial}{\partial y}$$

$$\frac{\partial}{\partial t} = -\sin\alpha\frac{\partial}{\partial x} + \cos\alpha\frac{\partial}{\partial y}$$

For small deflections, the slope of the surface in any direction can be taken equal to the angle that the tangent to the surface in that direction makes with the xy plane. The curvature is considered positive if it is convex downward. Thus the curvature of the surface in a plane parallel to the xz plane is given by

$$\frac{1}{r_x} = -\frac{\partial}{\partial x}\left(\frac{\partial u_z}{\partial x}\right) = -\frac{\partial^2 u_z}{\partial x^2} \qquad ...(9.16)$$

Similarly the curvature in a plane parallel to the yz plane is

$$\frac{1}{r_y} = -\frac{\partial}{\partial y}\left(\frac{\partial u_z}{\partial y}\right) = -\frac{\partial^2 u_z}{\partial y^2} \qquad ...(9.17)$$

The curvature in the n-direction is

$$\frac{1}{r_n} = -\frac{\partial}{\partial n}\left(\frac{\partial u_z}{\partial n}\right)$$

$$= -\left(\cos\alpha\frac{\partial}{\partial x} + \sin\alpha\frac{\partial}{\partial y}\right)\left(\frac{\partial u_z}{\partial x}\cos\alpha + \frac{\partial u_z}{\partial y}\sin\alpha\right)$$

$$= \frac{1}{r_x}\cos^2\alpha - \frac{1}{r_{xy}}\sin 2\alpha + \frac{1}{r_y}\sin^2\alpha \qquad ...(9.18)$$

where $\frac{1}{r_{xy}} = \frac{\partial^2 u_z}{\partial x\,\partial y}$ is called the twist of the surface w.r.t. the x and y axes.

Similarly,

$$\frac{1}{r_t} = -\frac{\partial}{\partial t}\left(\frac{\partial u_z}{\partial t}\right)$$

$$= -\left(-\sin\alpha\frac{\partial}{\partial x} + \cos\alpha\frac{\partial}{\partial y}\right)\left(-\sin\alpha\frac{\partial u_z}{\partial x} + \cos\alpha\frac{\partial u_z}{\partial y}\right)$$

$$= \frac{1}{r_x}\sin^2\alpha + \frac{1}{r_{xy}}\sin 2\alpha + \frac{1}{r_y}\cos^2\alpha \qquad ...(9.19)$$

Adding we find that

$$\frac{1}{r_n} + \frac{1}{r_t} = \frac{1}{r_x} + \frac{1}{r_y} \qquad ...(9.20)$$

Thus this sum is independent of the angle α and is called the average curvature.

Now $$\frac{1}{r_{nt}} = \frac{\partial}{\partial t}\left(\frac{\partial u_z}{\partial n}\right)$$

$$= \left(-\sin\alpha\frac{\partial}{\partial x} + \cos\alpha\frac{\partial}{\partial y}\right)\left(\cos\alpha\frac{\partial u_z}{\partial x} + \sin\alpha\frac{\partial u_z}{\partial y}\right)$$

$$= -\sin\alpha\cos\alpha\frac{\partial^2 u_z}{\partial x^2} + \frac{\partial^2 u_z}{\partial x\partial y}(\cos^2\alpha - \sin^2\alpha) + \sin\alpha\cos\alpha\frac{\partial^2 u_z}{\partial y^2}$$

$$= \frac{1}{2}\sin 2\alpha\left(\frac{1}{r_x} - \frac{1}{r_y}\right) + \cos 2\alpha \cdot \frac{1}{r_{xy}} \qquad ...(9.21)$$

For $\frac{1}{r_{nt}}$ to be zero, the coordinate system is said to be laid along the principal directions of curvature at the point A. Therefore, setting $\frac{1}{r_{nt}} = 0$, we get

$$\tan 2\alpha = -\frac{\dfrac{2}{r_{xy}}}{\dfrac{1}{r_x} - \dfrac{1}{r_y}} \qquad ...(9.22)$$

Substituting Eq. (9.22) in Eq. (9.18), we find two values of $1/r_n$, one representing the maximum and the other the minimum curvature at a point A of the surface. These two curvatures are called the principal curvatures of the surface and the corresponding planes the principal planes of curvature.

We can draw the Mohr's circle for the curvature and twist at a point of a curved surface as shown in Fig. 9.7.

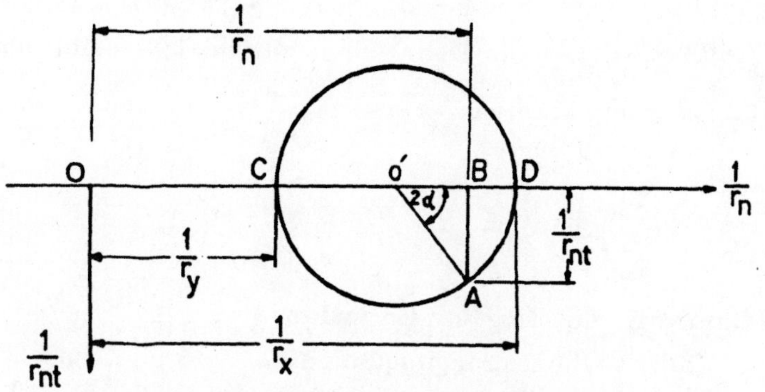

Fig. 9.7. Mohr's circle for curvature and twist.

If the coordinate planes xz and yz are taken parallel to the principal planes of curvature at point A, then $1/r_{xy} = 0$ and Eqs. (9.18) and (9.21) give

$$\left.\begin{aligned} \frac{1}{r_n} &= \frac{1}{r_x}\cos^2\alpha + \frac{1}{r_y}\sin^2\alpha \\ &= \frac{1}{2}\left(\frac{1}{r_x}+\frac{1}{r_y}\right)+\frac{1}{2}\left(\frac{1}{r_x}-\frac{1}{r_y}\right)\cos 2\alpha \\ \frac{1}{r_{nt}} &= \frac{1}{2}\left(\frac{1}{r_x}-\frac{1}{r_y}\right)\sin 2\alpha \end{aligned}\right\} \quad \text{...(9.23)}$$

From Fig. 9.7 we find

$$OB = OO' + O'B$$

$$= \frac{1}{2}\left(\frac{1}{r_x}+\frac{1}{r_y}\right)+\frac{1}{2}\left(\frac{1}{r_x}-\frac{1}{r_y}\right)\cos 2\alpha$$

$$= \frac{1}{r_n}$$

$$AB = \frac{1}{2}\left(\frac{1}{r_x}-\frac{1}{r_y}\right)\sin 2\alpha = \frac{1}{r_{nt}}$$

If the curvatures $\frac{1}{r_x}$ and $\frac{1}{r_y}$ are both positive then the surface is bent convex downward and if they are both negative then the surface is bent convex upward. Surfaces in which the curvatures in all planes have like signs are called *synclastic*.

And when the curvatures have opposite signs then such surfaces are called *anticlastic*.

When $\frac{1}{r_x} = \frac{1}{r_y}$ the Mohr's circle shrinks to a point and the surface is spherical when

$$\frac{1}{r_x} = -\frac{1}{r_y}$$

i.e.
$$\frac{1}{r_x} + \frac{1}{r_y} = 0.$$

9.4 Lagrange Equilibrium Equation

To derive the Lagrange equation of equilibrium, the following assumptions are made :

1. It is assumed that a straight line initially normal to a middle surface of the plate remains straight and normal to the deformed middle surface of the plate and unchanged in length. Further, transverse shear and transverse normal strains are neglected.

2. The normal stresses σ_x and σ_y are assumed to be zero at the middle surface of the plate. This assumption is valid, if the deflections u_z are small as compared to the plate thickness h.

3. The displacement u_z is assumed to be small, consequently, the curvatures $\frac{1}{r_x}$ and $\frac{1}{r_y}$ of the middle surface can be expressed as

$$\frac{1}{r_x} = -\frac{\partial^2 u_z}{\partial x^2};$$

$$\frac{1}{r_y} = -\frac{\partial^2 u_z}{\partial y^2}$$

and the slopes ϕ_x and ϕ_y that the tangent plane to the middle surface makes with the x and y axes can be expressed as :

$$\phi_x \simeq \tan\phi_x = \frac{\partial u_z}{\partial x}$$

$$\phi_y \simeq \tan\phi_y = \frac{\partial u_z}{\partial y}.$$

As a consequence of the first two assumptions, stresses σ_x and σ_y at a point within the plate are proportional to the distance z of the point from the middle surface of the plate as shown in Fig. 9.8.

Under the third assumption, points on the middle surface of the plate can be assumed to be displaced only in the z-direction. For other points of the plate, the corresponding u_x and u_y displacements in the x and y directions are proportional to their distances from the middle surface. Thus a point is displaced by the amount

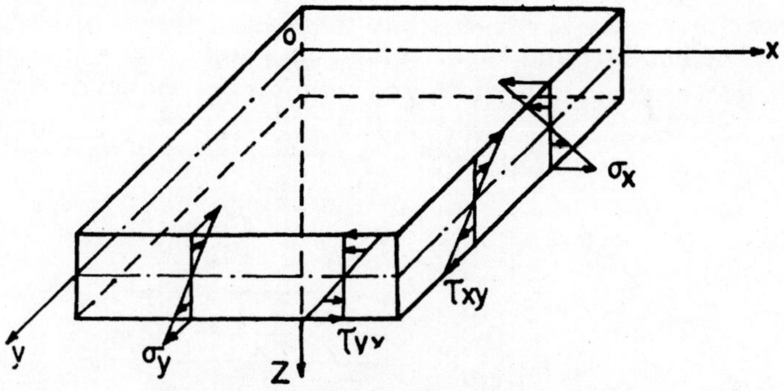

Fig. 9.8. Stresses on a rectangular plate.

$$u_x = -z\phi_x = -z\frac{\partial u_z}{\partial x}$$

in the x-direction, and by the amount

$$u_y = -z\phi_y = -z\frac{\partial u_z}{\partial y}$$

in the y-direction.

The components of strain are

$$\left.\begin{aligned} \varepsilon_x &= \frac{\partial u_x}{\partial x} = -z\frac{\partial^2 u_z}{\partial x^2} = \frac{z}{r_x} \\ \varepsilon_y &= \frac{\partial u_y}{\partial y} = -z\frac{\partial^2 u_z}{\partial y^2} = \frac{z}{r_y} \\ \gamma_{xy} &= \frac{\partial u_x}{\partial y} + \frac{\partial u_y}{\partial x} = -2z\frac{\partial^2 u_z}{\partial x \partial y} \end{aligned}\right\} \quad \text{...(9.24)}$$

and the corresponding stresses are :

$$\left.\begin{aligned} \sigma_x &= \frac{E}{1-\nu^2}(\varepsilon_x + \nu\varepsilon_y) = \frac{-Ez}{1-\nu^2}\left(\frac{\partial^2 u_z}{\partial x^2} + \nu\frac{\partial^2 u_z}{\partial y^2}\right) \\ &= \frac{Ez}{1-\nu^2}\left(\frac{1}{r_x} + \frac{\nu}{r_y}\right) \\ \sigma_y &= \frac{E}{1-\nu^2}(\varepsilon_y + \nu\varepsilon_x) = \frac{-Ez}{1-\nu^2}\left(\frac{\partial^2 u_z}{\partial y^2} + \nu\frac{\partial^2 u_z}{\partial x^2}\right) \\ &= \frac{Ez}{1-\nu^2}\left(\frac{1}{r_y} + \frac{\nu}{r_x}\right) \\ \tau_{xy} &= G\gamma_{xy} = -2Gz\frac{\partial^2 u_z}{\partial x \partial y} = -\frac{Ez}{(1+\nu)}\cdot\frac{\partial^2 u_z}{\partial x \partial y} \end{aligned}\right\}$$

...(9.25)

The bending and twisting moments and shear forces acting on the plate area of unit width and height h are shown in Fig. 9.9.

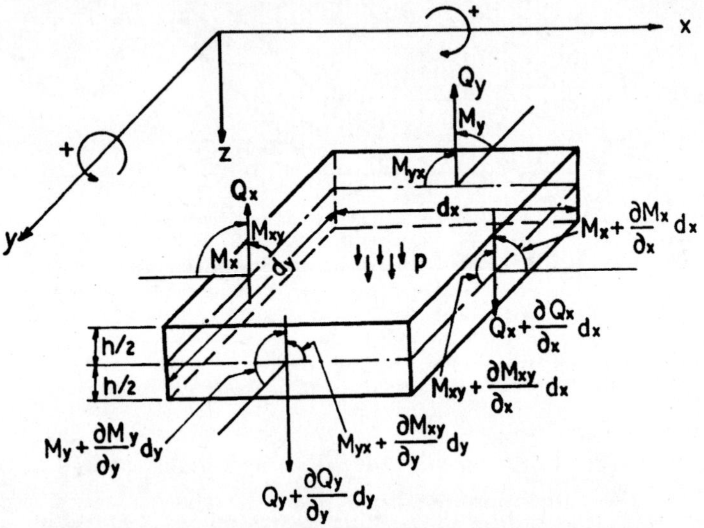

Fig. 9.9. Bending and twisting moments on the plate.

where M_x, M_y = bending moments on a plane normal to the x and y axes respectively.

M_{xy} = twisting moment on a plane normal to the x-axis in the direction y.

Q_x, Q_y = transverse shear forces on planes normal to the x and y axes respectively.

It is assumed that positive moments produce tension in the fibres located at the bottom part of the plate.

$$\left.\begin{aligned} \text{Now} \quad M_x &= \int_{-h/2}^{h/2} \sigma_x \,.\, (1) \,.\, z dz \\ M_y &= \int_{-h/2}^{h/2} \sigma_y \,.\, (1) \,.\, z \,.\, dz \\ M_{xy} = M_{yx} &= \int_{-h/2}^{h/2} \tau_{xy} \,.\, (1) . z dz = \int_{-h/2}^{h/2} \tau_{yx} \,.\, (1) \,.\, z dz \end{aligned}\right\} \quad ...(9.26)$$

$$\text{Also} \quad I = \int_{-h/2}^{h/2} z^2 . 1 . dz = \frac{h^2}{12}$$

and $$D = EI = \frac{Eh^3}{12(1-\nu^2)}$$

Substituting the values of σ_x, σ_y and τ_{xy} from Eqs. (9.25) in (9.26), we get

$$\left.\begin{aligned} M_x &= -D\left(\frac{\partial^2 u_z}{\partial x^2} + \nu \frac{\partial^2 u_z}{\partial y^2}\right) \\ M_y &= -D\left(\frac{\partial^2 u_z}{\partial y^2} + \nu \frac{\partial^2 u_z}{\partial x^2}\right) \\ M_{xy} &= -(1-\nu) D \frac{\partial^2 u_z}{\partial x \partial y} = M_{yz} \end{aligned}\right\} \quad ...(9.27)$$

The condition of moment equilibrium about the y-axes is expressed by the equation

$$Q_x\, dxdy - \frac{\partial M_x}{\partial x} dxdy - \frac{\partial M_{yx}}{\partial y} dxdy = 0$$

$$\therefore \quad Q_x = \frac{\partial M_x}{\partial x} + \frac{\partial M_{yx}}{\partial y} \quad ...(9.28)$$

Similarly considering moment equilibrium about the x-axis, we find

$$Q_y = \frac{\partial M_y}{\partial y} + \frac{\partial M_{xy}}{\partial x} \quad ...(9.29)$$

Substituting Eqs. (9.27) in (9.28) and (9.29), we get

$$\left.\begin{aligned} Q_x &= -D \frac{\partial}{\partial x}\left[\frac{\partial^2 u_z}{\partial x^2} + \frac{\partial^2 u_z}{\partial y^2}\right] \\ Q_y &= -D \frac{\partial}{\partial y}\left[\frac{\partial^2 u_z}{\partial x^2} + \frac{\partial^2 u_z}{\partial y^2}\right] \end{aligned}\right\} \quad ...(9.30)$$

If p = intensity of distributed external load on the plate surface, then considering equilibrium of forces in the z-direction, we get

$$p dxdy + \frac{\partial Q_x}{\partial x} dxdy + \frac{\partial Q_y}{\partial y} dxdy = 0$$

$$\therefore \quad \frac{\partial Q_x}{\partial x} + \frac{\partial Q_y}{\partial y} = -p \quad ...(9.31)$$

Substituting Eqs. (9.27) in (9.28), we get

$$\frac{\partial^4 u_z}{\partial x^4} + \frac{2\partial^4 u_z}{\partial x^2 \partial y^2} + \frac{\partial^4 u_z}{\partial y^4} = \frac{(1-\nu^2)p}{EI} = \frac{p}{D}$$

or $$\nabla^4 u_z = \frac{p}{D}, \quad \nabla^2 = \frac{\partial^2}{\partial x^2} + \frac{\partial^2}{\partial y^2} \qquad \text{...(9.32)}$$

which is the Lagrange equilibrium equation for the bending of a thin plate with constant thickness. Therefore the solution of the problem of bending of plates by a lateral load reduces to the integration of Eq. (9.32). If, for a particular case, a solution of this equation is found that satisfies the conditions at the boundaries of the plate, then bending and twisting moments can be calculated from Eqs. (9.27). The corresponding normal and shearing stresses are obtained from Eqs. (9.25). Now

$$(\tau_{xy})_{max} = \frac{6M_{xy}}{h^2} \qquad \text{...(9.33)}$$

The shearing stresses τ_{xz} and τ_{yz} can now be determined by assuming that they are distributed across the thickness of the plate according to the parabolic law.

Then $$(\tau_{xz})_{max} = \frac{3}{2}\frac{Q_x}{h}, \quad (\tau_{yz})_{max} = \frac{3}{2}\frac{Q_y}{h} \qquad \text{...(9.34)}$$

9.5 Boundary Conditions

(1) *Built-in edge.* Assuming the built-in edge to be given by $x = a$, the boundary conditions are

$$(u_z)_{x=a} = 0, \quad \left(\frac{\partial u_z}{\partial x}\right)_{x=a} = 0 \qquad \text{...(9.35)}$$

(2) *Simply-supported edge.* At the simply supported edge $x = a$, the deflection is zero and the bending moment M_x along this edge is zero, *i.e.*,

$$(u_z)_{x=a} = 0, \quad \left(\frac{\partial^2 u_z}{\partial x^2} + \nu \frac{\partial^2 u_z}{\partial y^2}\right)_{x=a} = 0$$

But the condition $u_z = 0$ along the edge $x = a$ means also that

$$\frac{\partial u_z}{\partial y} = \frac{\partial^2 u_z}{\partial y^2}$$

along the edge. Hence, we find that

$$(u_z)_{x=a} = 0, \quad \left(\frac{\partial^2 u_z}{\partial x^2}\right)_{x=a} = 0 \qquad \text{...(9.36)}$$

(3) *Free edge.* At the free edge $x = a$

$$(M_x)_{x=a} = 0, \quad (M_{xy})_{x=a} = 0, \quad (Q_x)_{x=a} = 0 \qquad \text{...(9.37)}$$

Eqs. (9.37) were expressed by Poisson. Later on Kirchhoff proved that three boundary conditions are too many and that two conditions are sufficient for the complete determination of the deflection u_z satisfying Eq. (9.32). Kirchhoff pointed out that the two conditions prescribing M_{xy} and Q_x can be replaced by a single one. The reason is that the twisting moment acting on an element of the edge of the plate may be replaced by two statically equivalent vertical forces, which can then be combined with the vertical shearing forces. Owing to such a replacement, the stress distribution in the immediate neighbourhood of the edge will naturally be affected, but the stress distribution in the rest of the plate will remain essentially the same.

Consider the edge $x = a$. On an element of length dy we have the twisting moment $M_{xy}\,dy$ which may be replaced by two vertical forces of the magnitude M_{xy} at a distance dy part, as shown in Fig. 9.10. At the next element of length dy, the twisting moment will be of magnitude $\left[M_{xy} + \frac{\partial M_{xy}}{\partial y}dy\right]dy$, which may be replaced by the

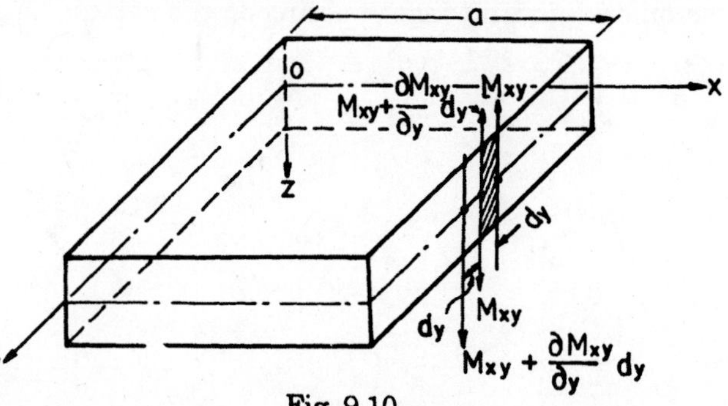

Fig. 9.10

vertical forces of the magnitude $M_{xy} + \frac{\partial M_{xy}}{\partial y}dy$ as shown in Fig. 9.10. Because of the foregoing replacements of statically equivalent systems we find that there is a net vertical force of the magnitude $\frac{\partial M_{xy}}{\partial y}$ per unit length. For positive M_{xy}, the statically equivalent forces are directed upward and are, therefore, negative according to our sign convention. Thus, for a free edge, instead of requiring both M_{xy} and Q_x to be zero along the edge, we may require that the statically equivalent vertical force V_x be zero. Thus

$$(V_x)_{x=a} = \left(Q_x - \frac{\partial M_{xy}}{\partial y}\right)_{x=a} = 0.$$

Substituting for Q_x and M_{xy} from Eqs. (9.30) and (9.26) respectively, we find

$$\left[\frac{\partial^3 u_z}{\partial x^3} + (2 - \nu)\frac{\partial^3 u_z}{\partial x \partial y^2}\right]_{x=a} = 0 \quad ...(9.38)$$

The condition that $M_x = 0$, requires that

$$\left(\frac{\partial^2 u_z}{\partial x^2} + \nu\frac{\partial^2 u_z}{\partial y^2}\right)_{x=a} = 0 \quad ...(9.39)$$

Eqs. (9.38) and (9.39) may be taken as the boundary conditions for a free edge.

9.6 Determination of Bending and Twisting Moment on any Plane

The stresses acting on the side AC as shown in Fig. 9.11 (*a*) are

$$\sigma_n = \sigma_x \cos^2\alpha + \sigma_y \sin^2\alpha + \tau_{xy}\sin 2\alpha$$

$$\tau_{nt} = \frac{1}{2}(\sigma_y - \sigma_x)\sin 2\alpha + \tau_{xy}\cos 2\alpha$$

where α is the angle between the normal n and the x-axis. The bending moment per unit length acting on the side AC is given by

$$M_n = \int_{-h/2}^{h/2} \sigma_n.1.zdz.$$

$$= M_x\cos^2\alpha + M_y\sin^2\alpha + M_{xy}\sin 2\alpha$$

$$= \frac{M_x + M_y}{2} + \frac{M_x - M_y}{2}\cos 2\alpha + M_{xy}\sin 2\alpha \quad ...(9.40)$$

$$M_{nt} = \int_{-h/2}^{h/2} \tau_{nt}.1.z.dz$$

$$= M_{xy}\cos 2\alpha - \frac{M_x - M_y}{2}\sin 2\alpha \quad ...(9.41)$$

and $\quad Q_n = Q_x\cos\alpha + Q_y\sin\alpha \quad ...(9.42)$

These have been shown in Fig. 9.11 (*b*).

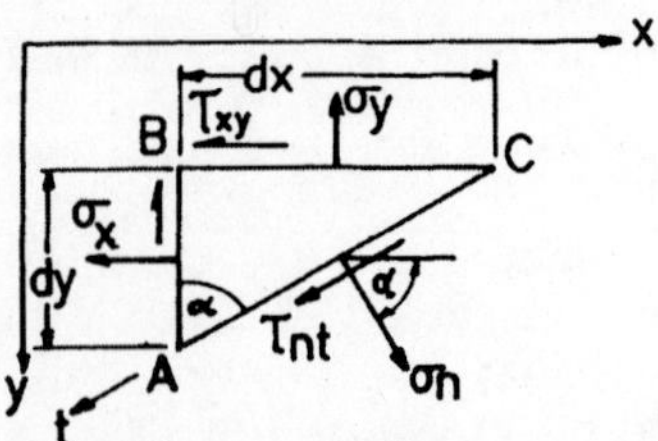

(*a*)

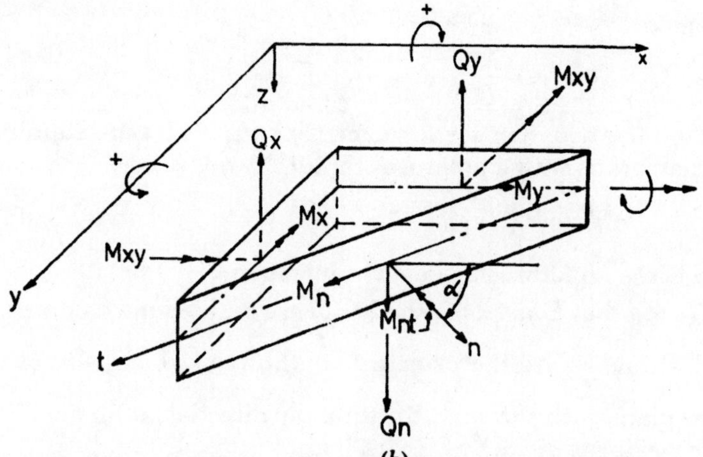

(b)

Fig. 9.11

From Eqs. (9.40) and (9.41), Mohr's circle can be constructed as shown in Fig. 9.12.

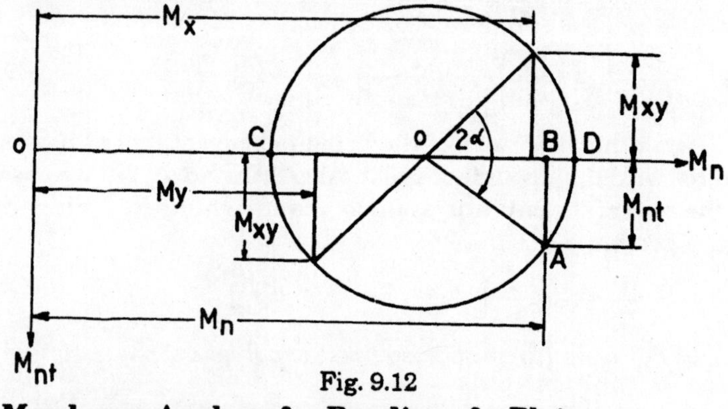

Fig. 9.12

9.7 Membrane Analogy for Bending of a Plate

The Lagrange Eq. (9.32) is

$$\nabla^2.\nabla^2 u_z = \frac{p}{D}$$

where $\nabla^2 = \frac{\partial^2}{\partial x^2} + \frac{\partial^2}{\partial y^2}$

Adding M_x and M_y from Eq. (9.27), we find

$$M_x + M_y = -D(1 + \nu)\, \nabla^2 u_z$$

Let $$M = \frac{M_x + M_y}{1 + \nu} = -D\nabla^2 u_z \qquad ...(9.43)$$

Then
$$\left.\begin{aligned}\nabla^2 M &= -p \\ \nabla^2 u_z &= -\frac{M}{D}\end{aligned}\right\} \qquad ...(9.44)$$

Now the equation for a uniformly stretched and laterally loaded membrane with a pressure of intensity q is

$$\nabla^2 z = -\frac{q}{S} \qquad ...(9.45)$$

where S is the uniform tension in the membrane.

We see that Eqs. (9.44) and (9.45) are of the same kind.

If $\frac{1}{r_x}$ and $\frac{1}{r_y}$ are the principal curvatures at any point of a deflected plate, then the curvatures in any direction at an angle α with the x-axis are given by Eq. (9.23) as

$$\frac{1}{r_n} = \frac{1}{r_x}\cos^2\alpha + \frac{1}{r_y}\sin^2\alpha$$

$$\frac{1}{r_{nt}} = \frac{1}{2}\left(\frac{1}{r_x} - \frac{1}{r_y}\right)\sin 2\alpha$$

$$\therefore \qquad 2\left(\frac{1}{r_{nt}}\right)_{max} = \frac{1}{r_x} - \frac{1}{r_y} \qquad ...(9.46)$$

Thus the difference of the principal curvatures at a point is equal to twice the maximum twist. Also from Eq. (9.20) we observe that the sum of the curvatures in planes at right angles is invariant in any direction, *i.e.*

$$\frac{1}{r_n} + \frac{1}{r_t} = \frac{1}{r_x} + \frac{1}{r_y}$$

(*i*) For a simply supported plate of polygonal shape, for which along each rectilinear portion of the boundary, we have $\frac{\partial^2 u_z}{\partial s^2} = 0$ since $u_z = 0$ at the boundary. Since $M_n = 0$ at the simply supported boundary, hence $\frac{\partial^2 u_z}{\partial n^2} = 0$ at the boundary. Hence, we have

$$\frac{\partial^2 u_z}{\partial s^2} + \frac{\partial^2 u_z}{\partial n^2} = \frac{\partial^2 u_z}{\partial x^2} + \frac{\partial^2 u_z}{\partial y^2} = -\frac{M}{D} = 0 \qquad ...(9.47)$$

Thus the solution of the plate problem reduces to the integration of the two Eqs. (9.44). We begin with the first of these equations and find a solution satisfying the condition $M = 0$ at the boundary. Substituting this solution in the second equation and integrating, we find the deflection u_z. Both these problems are of the same kind as the problem of the deflection of a uniformly stretched and laterally loaded membrane having zero deflection at the boundary, which is easier to solve by either Ritz's or finite-difference method.

(*ii*) In the case of a simply supported plate of polygonal shape bent by moments M_n uniformly distributed along the boundary, Eqs. (9.33) become

$$\left.\begin{aligned} \nabla^2 M &= 0 \\ \nabla^2 u_z &= -\frac{M}{D} \end{aligned}\right\} \qquad \text{...(9.48)}$$

Along a rectilinear edge, we have $\dfrac{\partial^2 u_z}{\partial s^2} = 0$. Hence

$$M_n = -D\frac{\partial^2 u_z}{\partial n^2}$$

and at the boundary

$$\nabla^2 u_z = \frac{\partial^2 u_z}{\partial n^2} = -\frac{M_n}{D} = -\frac{M}{D} \qquad \text{...(9.49)}$$

Boundary condition (9.49) and the first of Eqs. (9.48) will be satisfied if we take $M = M_n$ at all points of the plate, which means that the sum of the bending moments M_x and M_y remains constant over the entire surface of the plate. Then the second of Eqs. (9.48) becomes

$$\nabla^2 u_z = -\frac{M_n}{D} \qquad \text{...(9.50)}$$

from which the deflections u_z of the plate can be determined. From Eq. (9.50) it may be observed that the deflection surface of the plate is the same as that of the uniformly stretched membrane with a uniformly distributed load.

Example 9.4 *Bending of a triangular plate by uniform bending moment.*

Solution. Consider the bending of a triangular plate as shown in Fig. 9.13 by moments M_n uniformly distributed along the boundary. The deflection surface of the plate is the same as that of a uniformly stretched and uniformly loaded membrane. If we take,

$$u_z = m\left[\frac{x^3 - 3y^2x}{3} - \frac{a(x^2 + y^2)}{3} + \frac{4a^3}{81}\right] \qquad \text{...(1)}$$

then u_z becomes zero at the boundary. Hence the boundary condition $u_z = 0$ for the membrane is satisfied. The constant m is determined from Eq. (9.38), giving finally

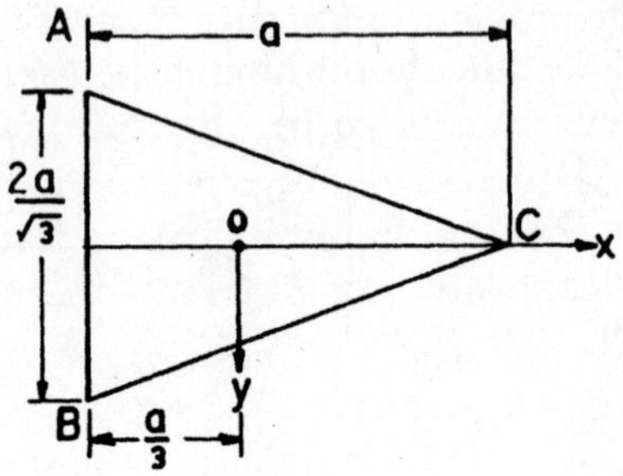

Fig. 9.13

$$u_z = \frac{M_n}{4aD}\left[x^3 - 3y^2x - a(x^2 + y^2) + \frac{4a^3}{27}\right] \quad ...(2)$$

At the centroid $x = y = 0$ and we get

$$(u_z)_0 = \frac{M_n a^2}{27D} \quad ...(3)$$

Now

$$\left.\begin{aligned} M_x &= -D\left(\frac{\partial^2 u_z}{\partial x^2} + \nu\frac{\partial^2 u_z}{\partial y^2}\right) \\ &= \frac{M_n}{2}\left[1 + \nu - (1-\nu)\frac{3x}{a}\right] \\ M_y &= -D\left(\frac{\partial^2 u_z}{\partial y^2} + \nu\frac{\partial^2 u_z}{\partial x^2}\right) \\ &= \frac{M_n}{2}\left[1 + \nu + (1-\nu)\frac{3x}{a}\right] \\ M_{xy} &= -(1-\nu)D\frac{\partial^2 u_z}{\partial x \partial y} \\ &= -3(1-\nu)M_n\frac{y}{2a} \end{aligned}\right\} \quad ...(4)$$

and

$$Q_x = Q_y = 0 \quad (5)$$

Now from Eq. (9.30), the twisting moment along the side BC is

$$M_{nt} = M_{xy}\cos 2\alpha - \frac{M_x - M_y}{2}\sin 2\alpha$$

$$= \frac{3(1-\nu)M_n}{4a}(y - \sqrt{3}\,x) \quad ...(6)$$

and the vertical reactions acting on the plate along the side BC are

$$V_n = Q_n - \frac{\partial M_{nt}}{\partial s} = -\frac{3(1-\nu)}{2a}M_n \quad ...(7)$$

By symmetry the same uniformly distributed reactions also act along the two other sides of the plate. These forces are balanced by the concentrated reactions at the corners of the plate, given by

$$R = 2\left(M_{nt}\right)_{\substack{x = \frac{2}{3}a \\ y = 0}} = (1 - \nu)\sqrt{3}\, M_n \qquad \ldots(8)$$

Also $$(M_y)_{max} = \left(M_y\right)_{x = \frac{2}{3}a} = \frac{(3 - \nu)}{2} M_n \qquad \ldots(9)$$

9.8 Symmetrical Bending of Circular Plates

Consider the symmetrical bending of a circular plate. Let us take the origin of co-ordinates O at the centre of the undeflected plate as shown in Fig. 9.14. Let r denoted the radial distances of points in the middle plane of the plate and u_z be the deflection of the plate in the z-direction at any point A. Then for small u_z, the maximum slope of the deflection surface at A is given by

$$\phi = -\frac{du_z}{dr}$$

and the curvature of the plate in diametral section rz is

$$\frac{1}{r_n} = -\frac{d^2u_z}{dr^2} = \frac{d\phi}{dr} \qquad \ldots(1)$$

where ϕ is the small angle between the normal to the deflection surface at A and the axis of symmetry OB.

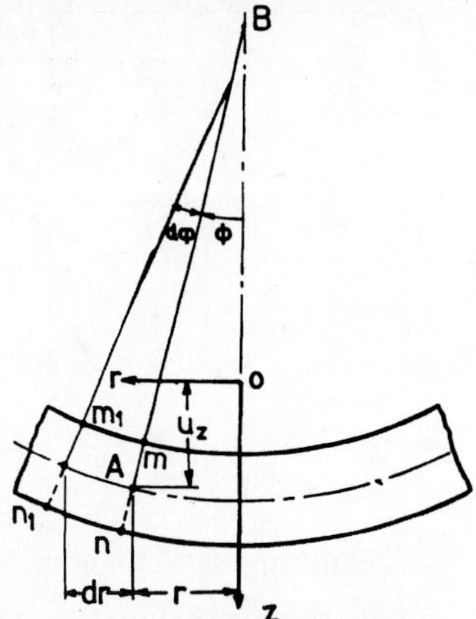

Fig. 9.14. Bending of a circular plate.

From symmetry we conclude that $\frac{1}{r_n}$ is one of the principal curvatures of the deflection surface at A. The second principal curvature r_θ is equal to the length AB given by

$$\frac{1}{r_\theta} = -\frac{1}{r}\frac{du_z}{dr} = \frac{\phi}{r} \qquad \ldots(2)$$

Now bending moment per unit length along mn is

$$M_r = -D\left(\frac{d^2u_z}{dr^2} + \frac{\nu}{r}\frac{du_z}{dr}\right)$$

$$= D\left(\frac{d\phi}{dr} + \frac{\nu}{r}\phi\right) \qquad ...(3)$$

and along rz is

$$M_\theta = -D\left(\frac{1}{r}\frac{du_z}{dr} + \nu\frac{\partial^2 u_z}{dr^2}\right)$$

$$= D\left(\frac{\phi}{r} + \nu\frac{d\phi}{dr}\right) \qquad ...(4)$$

To determine ϕ or u_z, consider the equilibrium of an element of the plate such as $abcd$ as shown in Fig. 9.15.

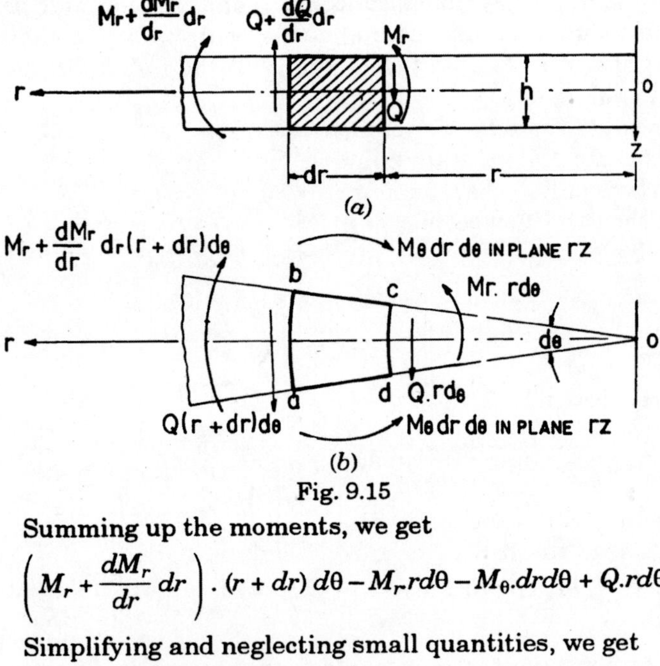

Fig. 9.15

Summing up the moments, we get

$$\left(M_r + \frac{dM_r}{dr}dr\right).(r + dr)\,d\theta - M_r r d\theta - M_\theta . dr d\theta + Q.r d\theta dr = 0$$

Simplifying and neglecting small quantities, we get

$$M_r + \frac{dM_r}{dr}.r - M_\theta + Qr = 0 \qquad ...(5)$$

Substituting from Eqs. (9.41) and (9.42), we get

$$\therefore \quad \frac{d^2\phi}{dr^2} + \frac{1}{r}\frac{d\phi}{dr} - \frac{\phi}{r^2} = -\frac{Q}{D} \qquad ...(6)$$

or

$$\frac{d^3u_z}{dr^3} + \frac{1}{r}\frac{\partial^2 u_z}{dr^2} - \frac{1}{r^2}\frac{du_z}{dr} = \frac{Q}{D} \qquad ...(7)$$

These equations can also be written as

$$\frac{d}{dr}\left[\frac{1}{r}\frac{d}{dr}(r\phi)\right] = -\frac{Q}{D} \quad ...(8)$$

$$\frac{d}{dr}\left[\frac{1}{r}\frac{d}{dr}\left(r\frac{du_z}{dr}\right)\right] = \frac{Q}{D} \quad ...(9.51)$$

Multiplying both sides by $2\pi r$ and observing that

$$2\pi rQ = \int_0^r 2\pi rq.dr$$

where q is the intensity of load distributed over the plate surface. Hence

$$r\frac{d}{dr}\left[\frac{1}{r}\frac{d}{dr}\left(r\frac{du_z}{dr}\right)\right] = \frac{1}{D}\int_0^r qrdr$$

Differentiating both sides w.r.t. r and dividing by r, we get

$$\frac{1}{r}\frac{d}{dr}\left[r\frac{d}{dr}\left\{\frac{1}{r}\frac{d}{dr}\left(r\frac{du_z}{dr}\right)\right\}\right] = \frac{q}{D} \quad ...(9.52)$$

9.8.1 Uniformly Loaded Circular Plate

Consider a circular plate of radius a carrying a uniformly distributed load of intensity q over its entire surface. Now

$$2\pi rQ = \pi r^2 q$$

$$\therefore \quad Q = \frac{qr}{2} \quad ...(1)$$

Substituting in Eq. (9.47), we get

$$\frac{d}{dr}\left[\frac{1}{r}\frac{d}{dr}\left(r\frac{du_z}{dr}\right)\right] = \frac{qr}{2D} \quad ...(2)$$

Integrating, we get

$$\frac{1}{r}\frac{d}{dr}\left(r\frac{du_z}{dr}\right) = \frac{qr^2}{4D} + C_1 \quad ...(3)$$

where C_1 is an arbitrary constant of integration. Eq. (3) can be written as

$$\frac{d}{dr}\left(r\frac{du_z}{dr}\right) = \frac{qr^3}{4D} + C_1 r$$

Integrating again, we get

$$\frac{du_z}{dr} = \frac{qr^3}{16D} + \frac{C_1 r}{2} + \frac{C_2}{r} \quad ...(4)$$

Integrating once more, we get

$$u_z = \frac{qr^4}{64D} + \frac{C_1 r^2}{4} + C_2 \log r + C_3 \qquad \text{...(5)}$$

(*i*) *Clamped edges.* In this case the slope

$$\phi = 0 \text{ at } r = 0 \text{ and } r = a.$$

Hence from Eq. (4), we get

$$\left[\frac{qr^3}{16D} + \frac{C_1 r}{2} + \frac{C_2}{r}\right]_{r=0} = 0$$

$$\left[\frac{qr^3}{16D} + \frac{C_1 r}{2} + \frac{C_2}{r}\right]_{r=a} = 0$$

$$\therefore \qquad C_2 = 0 \text{ and } C_1 = -\frac{qa^2}{8D}$$

$$\therefore \qquad \phi = -\frac{du_z}{dr}$$

$$= \frac{qr}{16D}(a^2 - r^2) \qquad \text{...(6)}$$

Eq. (5) gives $u_z = \frac{qr^4}{64D} - \frac{qa^2r^2}{32D} + C_3$

At $\qquad r = a,\ u_z = 0$

$$\therefore \qquad C_3 = \frac{qa^4}{64D}$$

$$\therefore \qquad u_z = \frac{q}{64D}(a^2 - r^2)^2 \qquad \text{...(7)}$$

$$(u_z)_{max} = (u_z)_{r=0}$$

$$= \frac{qa^4}{64D} \qquad \text{...(9.53)}$$

Substituting for ϕ in Eqs. (9.41) and (9.42), we get

$$M_r = \frac{q}{16}[a^2(1+\nu) - r^2(3+\nu)] \qquad \text{...(8)}$$

$$M_\theta = \frac{q}{16}[a^2(1+\nu) - r^2(1+3\nu)] \qquad \text{...(9)}$$

At the boundary

$$\left.\begin{aligned} (M_r)_{r=a} &= -\frac{qa^2}{8} \\ (M_\theta)_{r=a} &= -\frac{\nu qa^2}{8} \end{aligned}\right\} \qquad \text{...(10)}$$

At the centre of the plate

$$(M_r)_{r=0} = (M_\theta)_{r=0} = \frac{qa^2}{16}(1+\nu) \qquad ...(11)$$

$$(\sigma_r)_{max} = (\sigma_r)_{r=a} = -\frac{6M_r}{h^2} = \frac{3}{4}\frac{qa^2}{h^2} \qquad ...(9.54)$$

(*ii*) *Simply supported edges.* In this case

$$(u_z)_{r=a} = 0, \left(\frac{du_z}{dr}\right)_{r=0} = 0$$

and
$$(M_r)_{r=a} = -D\left(\frac{d^2u_z}{dr^2} + \frac{\nu}{r}\frac{du_z}{dr}\right)_{r=a} = 0$$

$\therefore$
$$C_2 = 0$$

$$C_1 = -\frac{qa^2}{8D}\left(\frac{3+\nu}{1+\nu}\right)$$

$$C_3 = \frac{qa^4}{64D}\left(\frac{5+\nu}{1+\nu}\right)$$

$\therefore$
$$u_z = \frac{q(a^2-r^2)}{64\,D}\left[\left(\frac{5+\nu}{1+\nu}\right)a^2 - r^2\right] \qquad ...(12)$$

At $r = 0$

$$(u_z)_{max} = \frac{(5+\nu)qa^4}{64(1+\nu)D} \qquad ...(9.55)$$

$$M_r = \frac{q}{16}(3+\nu)(a^2-r^2) \qquad ...(13)$$

$$M_\theta = \frac{q}{16}[a^2(3+\nu) - r^2(1+3\nu)] \qquad ...(14)$$

At $r = 0$

$$M_r = M_\theta = \frac{3+\nu}{16}qa^2$$

and
$$(\sigma_r)_{max} = (\sigma_\theta)_{max} = \frac{6M_r}{h^2} = \frac{3(3+\nu)qa^2}{8h^2} \qquad ...(9.56)$$

9.8.2 Circular Plate with a Circular Hole at the Centre

Consider the bending of a plate by moments M_1 and M_2 uniformly distributed along the inner and outer boundaries as shown in Fig. 9.16. In this case $Q = 0$ and Eq. (9.51) becomes

$$\frac{d}{dr}\left[\frac{1}{r}\frac{d}{dr}\left(r\frac{du_z}{dr}\right)\right] = 0$$

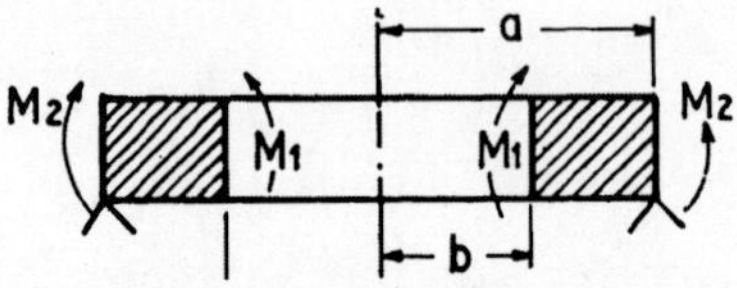

Fig. 9.16

Integrating twice, we get

$$-\frac{du_z}{dr} = \phi = \frac{C_1 r}{2} + \frac{C_2}{r} \quad \text{...(1)}$$

Integrating again, we find

$$u_z = -C_1 \frac{r^2}{4} - C_2 \log \frac{r}{a} + C_3 \quad \text{...(2)}$$

At $r = a$, $M_r = M_2$ and at $r = b$, $M_r = M_1$.

$$\therefore \qquad M_r = -D\left[\frac{d^2u_z}{dr^2} + \frac{\nu}{r}\frac{du_z}{dr}\right]$$

$$M_r = D\left[\frac{C_1}{2} - \frac{C_2}{r^2} + \nu\left(\frac{C_1}{2} + \frac{C_2}{r^2}\right)\right] \quad \text{...(3)}$$

Hence $\quad M_1 = D\left[\frac{C_1}{2}(1+\nu) - \frac{C_2}{b^2}(1-\nu)\right]$

$$M_2 = D\left[\frac{C_1}{2}(1+\nu) - \frac{C_2}{a^2}(1-\nu)\right]$$

From which we get

$$C_1 = \frac{2(a^2M_2 - b^2M_1)}{(1+\nu)D(a^2 - b^2)}$$

$$C_2 = \frac{a^2b^2(M_2 - M_1)}{(1+\nu)D(a^2 - b^2)}$$

If the plate is supported along the outer edge, then $u_z = 0$ at $r = a$.

$$\therefore \qquad C_3 = \frac{a^2(a^2 M_2 - b^2M_1)}{2(1+\nu)D(a^2 - b^2)}$$

when $\quad M_2 = 0$, then

$$C_1 = -\frac{2b^2M_1}{(1+\nu)D(a^2 - b^2)}$$

$$C_2 = -\frac{a^2b^2M_1}{(1-\nu)D(a^2-b^2)}$$

$$C_3 = -\frac{a^2b^2M_1}{2(1+\nu)D(a^2-b^2)}$$

$$\therefore \quad \frac{du_z}{dr} = \frac{a^2b^2M_1}{D(1-\nu)(a^2-b^2)}\left[\frac{1}{r}+\frac{1-\nu}{1+\nu}\frac{r}{a^2}\right] \quad ...(4)$$

$$u_z = \frac{b^2M_1}{2(1+\nu)D(a^2-b^2)}(a^2-r^2) + \frac{a^2b^2M_1}{(1-\nu)D(a^2-b^2)}\log\frac{r}{a} \quad ...(5)$$

Now consider the case of bending of a plate by shearing forces Q_0 uniformly distributed along the inner edge as shown in Fig. 9.17. The shearing force per unit length of a circumference of radius r is

$$Q = \frac{Q_0 b}{r} = \frac{P}{2\pi r} \quad \text{where} \quad P = 2\pi b Q_0$$

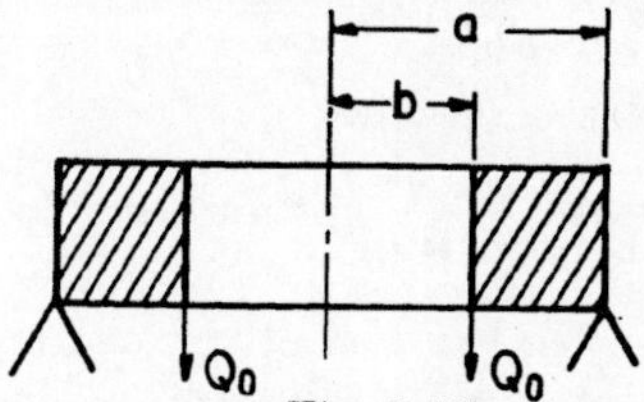

Fig. 9.17

Substituting in Eq. (9.51) and integrating, we get

$$\frac{du_z}{dr} = \frac{Pr}{8\pi D}\left(2\log\frac{r}{a}-1\right) - \frac{C_1 r}{2} - \frac{C_2}{r} \quad ...(6)$$

$$u_z = \frac{Pr^2}{8\pi D}\left(\log\frac{r}{a}-1\right) - \frac{C_1 r^2}{4} - C_2\log\frac{r}{a} + C_3 \quad ...(7)$$

If the plate is simply supported at the outer edge, then

$$(u_z)_{r=a} = 0, \ (M_r)_{r=a} = 0 \quad ...(8)$$

and $$(M_r)_{r=b} = 0 \quad ...(9)$$

Now $$M_r = -D\left[\frac{d^2u_z}{dr^2}+\frac{\nu}{r}\frac{du_z}{dr}\right]$$

Using these conditions, we get

$$C_1 = \frac{P}{4\pi D}\left[\frac{1-\nu}{1+\nu} - \frac{2b^2}{a^2-b^2}\log\frac{b}{a}\right]$$

$$C_2 = -\frac{(1+\nu)P}{(1-\nu)\,4\pi D}\cdot\frac{a^2b^2}{a^2-b^2}\log\frac{b}{a}$$

$$C_3 = \frac{Pa^2}{8\pi D}\left(1+\frac{1}{2}\frac{1-\nu}{1+\nu}-\frac{b^2}{a^2-b^2}\log\frac{b}{a}\right)$$

Substituting these values in Eqs. (6) and (7) we can determine the slope and deflection

$$\left(\frac{du_z}{dr}\right)_{r=b} = \frac{Pb}{8\pi D}\left[2\log\frac{b}{a}-1-\frac{1-\nu}{1+\nu}\right.$$

$$\left.+\frac{2b^2}{a^2-b^2}\log\frac{b}{a}\left(1+\frac{a^2}{b^2}\frac{1+\nu}{1-\nu}\right)\right] \quad ...(10)$$

When b is infinitely small, then $\log\frac{b}{a}\to 0$.

Then we find $C_1 = \frac{1-\nu}{1+\nu}\cdot\frac{P}{4\pi D}$, $C_2 = 0$,

$$C_3 = \frac{Pa^2}{8\pi D}\left(1+\frac{1}{2}\frac{1-\nu}{1+\nu}\right)$$

Substituting in Eq. (7), we get

$$u_z = \frac{P}{8\pi D}\left[\frac{3+\nu}{2(1+\nu)}(a^2-r^2)+r^2\log\frac{r}{a}\right] \quad ...(9.57)$$

9.8.3 Circular plate loaded at the centre

Consider a circular plate of radius a carrying a concentrated load P at the centre. Then $Q = \frac{P}{2\pi r}$ and Eq. (9.51) becomes

$$\frac{d}{dr}\left[\frac{1}{r}\frac{d}{dr}\left(r\frac{du_z}{dr}\right)\right] = \frac{Q}{D} = \frac{P}{2\pi rD}$$

Integrating, we get

$$\frac{1}{r}\frac{d}{dr}\left(r\frac{du_z}{dr}\right) = \frac{P}{2\pi D}\log r + C_1$$

$$\frac{d}{dr}\left(r\frac{du_z}{dr}\right) = \frac{Pr}{2\pi D}\log r + C_1 r$$

Integrating again, we get

$$r\frac{du_z}{dr} = \frac{Pr^2}{8\pi D}(2\log r-1)+C_1\frac{r^2}{2}+C_2$$

$$\frac{du_z}{dr} = \frac{Pr}{8\pi D}(2\log r-1)+C_1\frac{r}{2}+\frac{C_2}{r}$$

Integrating once again, we get

$$\therefore \quad u_z = \frac{Pr^2}{8\pi D}(\log r - 1) + C_1 \frac{r^2}{4} + C_2 \log r + C_3.$$

(*i*) *Built in edges.* In this case

$$(u_z)_{r=a} = 0,$$

$$\left(\frac{du_z}{dr}\right)_{r=0} = 0, \left(\frac{du_z}{dr}\right)_{r=a} = 0$$

$$\therefore \quad C_2 = 0$$

$$C_1 = -\frac{P}{4\pi D}(2 \log a - 1)$$

$$C_3 = \frac{Pa^2}{16\pi D}$$

$$\therefore \quad u_z = \frac{Pr^2}{8\pi D} \log \frac{r}{a} + \frac{P}{16\pi D}(a^2 - r^2) \qquad \text{...(9.58)}$$

$$(u_z)_{max} = (u_z)_{r=0} = \frac{Pa^2}{16\pi D} \qquad \text{...(9.59)}$$

$$M_r = \frac{P}{4\pi}\left[(1 + \nu) \log \frac{a}{r} - 1\right]$$

$$M_\theta = \frac{P}{4\pi}\left[(1 + \nu) \log \frac{a}{r} - \nu\right]$$

At $r = a$

$$M_r = -\frac{P}{4\pi}, M_\theta = -\frac{\nu P}{4\pi}$$

$$(\sigma_r)_{max} = \frac{3}{2}\frac{P}{\pi h^2} \qquad \text{...(9.60)}$$

$$(\sigma_\theta)_{max} = \frac{3}{2}\frac{\nu P}{\pi h^2} \qquad \text{...(9.61)}$$

We observe that at $r = 0$, σ_r and σ_θ both tend to infinity. Therefore, this analysis is not valid near the centre of the plate.

(*ii*) *Simply supported edges*

Here $(u_z)_{r=a} = 0, (M_r)_{r=a} = 0$

and $$\left(\frac{du_z}{dr}\right)_{r=0} = 0$$

$$\therefore \quad C_2 = 0$$

$$M_r = -D\left(\frac{d^2u_z}{dr^2} + \frac{\nu}{r}\frac{du_z}{dr}\right)$$

$$\therefore \qquad C_1 = -\frac{P}{4\pi D}\left[2 \log a + \frac{1-\nu}{1+\nu}\right]$$

$$C_3 = \frac{Pa^2}{8\pi D}\left[1 + \frac{1}{2}\left(\frac{1-\nu}{1+\nu}\right)\right]$$

$$\therefore \qquad u_z = \frac{P}{16\pi D}\left[2r^2 \log\frac{r}{a} + \left(\frac{3+\nu}{1+\nu}\right)(a^2 - r^2)\right] \quad ...(9.62)$$

$$(u_z)_{max} = (u_z)_{r=0} = \left(\frac{3+\nu}{1+\nu}\right)\frac{Pa^2}{16\pi D} \quad ...(9.63)$$

$$M_r = \frac{P}{4\pi}(1+\nu)\log\frac{a}{r}$$

$$M_\theta = \frac{P}{4\pi}\left[(1+\nu)\log\frac{a}{r} + 1 - \nu\right]$$

$$\sigma_r = \frac{Pz}{4\pi I}(1+\nu)\log\frac{a}{r} \quad ...(9.64)$$

$$\sigma_\theta = \frac{Pz}{4\pi I}\left[(1+\nu)\log\frac{a}{r} + 1 - \nu\right] \quad ...(9.65)$$

Again at $r = 0$, σ_r and σ_θ becomes infinite.

9.9 Navier's Solution for Simply Supported Rectangular Plates

Consider a rectangular plate of sides a and b which is simply supported on the four sides and subjected to a distributed load p. Suppose that p, which is a function of the two variables x and y can be expressed in the domain $0 \le x \le a$, $0 \le y \le b$ by a double trigonometric series given by :

$$p = \sum_{m=1}^{\infty}\sum_{n=1}^{\infty} p_{mn} \sin\frac{m\pi x}{a} . \sin\frac{n\pi y}{b} \quad ...(1)$$

where $p_{mn} = \dfrac{4}{ab}\displaystyle\int_0^a\int_0^b p \sin\frac{m\pi x}{a} \sin\frac{n\pi y}{b}\, dxdy$

The Lagrange's equation becomes in this case

$$\nabla^4 u_z = \frac{1}{D}\sum_{m=1}^{\infty}\sum_{n=1}^{\infty} p_{mn} \sin\frac{m\pi x}{a} . \sin\frac{n\pi y}{b} \quad ...(2)$$

Let $$u_z = \sum_{m=1}^{\infty}\sum_{n=1}^{\infty} A_{mn} \sin\frac{m\pi x}{a} \sin\frac{n\pi y}{b}$$

Then substituting in Eq. (2), we get

$$A_{mn} = \frac{p_{mn}}{D\pi^4\left(\frac{m^2}{a^2}+\frac{n^2}{b^2}\right)^2}$$

$$\therefore \quad u_z = \frac{1}{\pi^4 D}\sum_{m=1}^{\infty}\sum_{n=1}^{\infty}\frac{p_{mn}}{\left(\frac{m^2}{a^2}+\frac{n^2}{b^2}\right)^2}\sin\frac{m\pi x}{a}\sin\frac{n\pi y}{b} \qquad \text{...(9.66)}$$

(*i*) Plate carrying a uniformly distributed load of intensity p_0.

$$\therefore \quad p_{mn} = \frac{4p_0}{ab}\int_0^a\int_0^b \sin\frac{m\pi x}{a}\sin\frac{n\pi y}{b}\,dx\,dy$$

$$= \frac{4p_0}{\pi^2 mn}(1-\cos m\pi)(1-\cos n\pi)$$

$$= \begin{cases} \dfrac{16p_0}{\pi^2 mn} & \text{for } m \text{ and } n \text{ to be odd} \\ 0 & \text{for } m \text{ and } n \text{ to be even} \end{cases} \qquad \text{...(9.67)}$$

$$u_z = \frac{16p_0}{\pi^6 D}\sum_{m=1,3,5,\dots}^{\infty}\sum_{n=1,3,5,\dots}^{\infty}\frac{1}{\left(\frac{m^2}{a^2}+\frac{n^2}{b^2}\right)^2} \times \sin\frac{m\pi x}{a}\sin\frac{n\pi y}{b}$$

$$= \frac{16p_0a^4}{\pi^6 D}\sum_{m=1,3,5}^{\infty}\sum_{n=1,3,5}^{\infty}\frac{\sin\frac{m\pi x}{a}\,.\,\sin\frac{n\pi y}{b}}{mn(m^2+n^2\rho^2)^2} \qquad \text{...(9.68)}$$

where $\rho = \dfrac{a}{b}$

$$(u_z)_{max}\Big|_{\substack{x=a/2\\ y=b/2}} = \frac{16p_0a^4}{\pi^6 D}\sum_{m=1,3,5}^{\infty}\sum_{n=1,3,5}^{\infty}\frac{(-1)^{\frac{m+n-2}{2}}}{mn(m^2+n^2\rho^2)^2} \qquad \text{...(9.69)}$$

$$\left.\begin{aligned} M_x &= \frac{16p_0}{\pi^4}\sum_{m=1,3,5}^{\infty}\sum_{n=1,3,5}^{\infty}\frac{\left(\dfrac{m^2}{a^2}+\nu\dfrac{n^2}{b^2}\right)}{mn\left(\dfrac{m^2}{a^2}+\dfrac{n^2}{b^2}\right)^2}\sin\frac{m\pi x}{a}\sin\frac{n\pi y}{b} \\ M_y &= \frac{16p_0}{\pi^4}\sum_{m=1,3,5}^{\infty}\sum_{n=1,3,5}^{\infty}\frac{\left(\nu\dfrac{m^2}{a^2}+\dfrac{n^2}{b^2}\right)}{\left(\dfrac{m^2}{a^2}+\dfrac{n^2}{b^2}\right)^2}\sin\frac{m\pi x}{a}\sin\frac{n\pi y}{b} \end{aligned}\right\} \qquad \ldots(9.70)$$

The bending stresses are

$$\sigma_x = 12M_x\frac{z}{h^3},\ \sigma_y = 12M_y\frac{z}{h^3} \qquad \ldots(9.71)$$

$$(\sigma_x)_{z=h/2} = \frac{8p_0h}{\pi^4 I}\left[a^2\sum\frac{m}{n(m^2+n^2\rho^2)^2}\sin\frac{m\pi x}{a}\sin\frac{n\pi y}{b} + \nu b^2\sum\frac{n}{m\left(\dfrac{m^2}{\rho^2}+n^2\right)^2}\sin\frac{m\pi x}{a}\sin\frac{n\pi y}{b}\right] \qquad \ldots(9.72)$$

At the centre of the plate $x = \frac{a}{2}$, $y = \frac{b}{2}$, this stress will be maximum.

$$(\sigma_x)_{max} = \frac{8p_0h}{\pi^4 I}\left[a^2\sum\frac{m(-1)^{\frac{m+n-2}{2}}}{n(m^2+n^2\rho^2)^2} + \nu b^2\sum\frac{n(-1)^{\frac{m+n-2}{2}}}{m\left(\dfrac{m^2}{\rho^2}+n^2\right)^2}\right]$$

Let $\quad N(\rho) = \sum\dfrac{m(-1)^{\frac{m+n-2}{2}}}{n(m^2+n^2\rho^2)^2}$

$$\left.\begin{aligned} &\text{Then} \quad (\sigma_x)_{max} = \frac{8}{\pi^4}\frac{p_0h}{I}\left[a^2N(\rho)+\nu b^2N\left(\frac{1}{\rho}\right)\right] \\ &\text{Similarly,} \quad (\sigma_y)_{max} = \frac{8}{\pi^4}\frac{p_0h}{I}\left[\nu a^2N(\rho)+b^2N\left(\frac{1}{\rho}\right)\right] \end{aligned}\right\} \qquad \ldots(9.73)$$

$$(\sigma_x)_{max}-(\sigma_y)_{max} = \frac{8}{\pi^4}\frac{p_0\,b^2h(1-\nu)}{I}\left[\rho^2N(\rho)-N\left(\frac{1}{\rho}\right)\right] \qquad \ldots(9.74)$$

For a square plate $\rho = 1$

and $\qquad (\sigma_x)_{max} = (\sigma_y)_{max}$

Now $\tau_{xy} = -\dfrac{Ez}{(1+\nu)}\dfrac{d^2u_z}{\partial x \partial y}$

$$(\tau_{xy})_{z=h/2} = -\frac{8}{\pi^4}\cdot\frac{a^3}{b}\frac{p_0h(1-\nu)}{I}\sum\frac{\cos\dfrac{m\pi x}{a}\cdot\cos\dfrac{n\pi y}{b}}{(m^2+n^2\rho^2)^2} \quad ..(9.75)$$

τ_{xy} is zero at the centre of the plate and along the axis of symmetry of the plate for which $x = \dfrac{a}{2}, y = \dfrac{b}{2}$. At the four corners, the absolute value of τ_{xy} is the same, where it is maximum.

$$(\tau_{xy})_{max}\Bigg|_{x=y=0} = \frac{8p_0h}{\pi^4 I}(1-\nu)a^2\rho\sum\frac{1}{(m^2+n^2\rho^2)^2} \quad ...(9.76)$$

9.9.1 Concentrated Load at (ξ, η)

Consider a rectangular plate of sides a and b simply supported at the edges and carrying a concentrated load P at (ξ, η). In this case

$$p_{mn} = \frac{4P}{ab}\sin\frac{m\pi\xi}{a}\cdot\sin\frac{n\pi\eta}{b}$$

and Navier's solution becomes

$$u_z = \frac{4P}{\pi^4 abD}\sum_{m=1}^{\infty}\sum_{n=1}^{\infty}\frac{\sin\dfrac{m\pi\xi}{a}\sin\dfrac{n\pi\eta}{b}}{\left(\dfrac{m^2}{a^2}+\dfrac{n^2}{b^2}\right)^2} \times\sin\frac{m\pi x}{a}\cdot\sin\frac{n\pi y}{b} \quad ...(9.77)$$

when the load P is in the centre then $\xi = \dfrac{a}{2}, \eta = \dfrac{b}{2}$.

$$u_z = \frac{4P}{\pi^4 abD}\sum_{m=1,3,5}^{\infty}\sum_{n=1,3,5}^{\infty}\frac{\sin\dfrac{m\pi x}{a}\sin\dfrac{n\pi y}{b}}{\left(\dfrac{m^2}{a^2}+\dfrac{n^2}{b^2}\right)^2} \quad ...(9.78)$$

For a square plate $a = b$

$$(u_z)_{max} = \frac{4Pa^2}{\pi^4 D}\sum_{m=1,3,5}^{\infty}\sum_{n=1,3,5}^{\infty}\frac{1}{(m^2+n^2)^2}$$

$$\simeq 0.01142 \frac{Pa^2}{D} \text{ by taking the first nine terms} \qquad ..(9.79)$$

The exact solution is

$$(u_z)_{max} = 0.01159 \frac{Pa^2}{D} \qquad ...(9.80)$$

and the error is 1.47 per cent.

9.9.2 Bending of an Elliptic Plate

Consider the bending of an elliptic plate built up at the edges and subjected to a uniformly distributed load of intensity p as shown

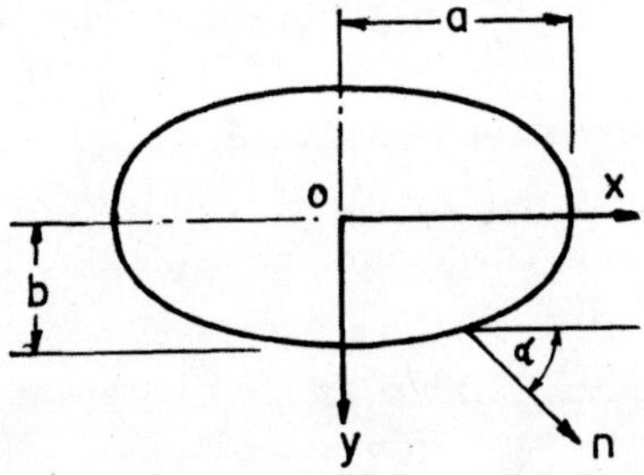

Fig. 9.18

in Fig. 9.18. Taking the co-ordinate axes as shown, the equation of the boundary of the plate is,

$$\frac{x^2}{a^2} + \frac{y^2}{b^2} - 1 = 0$$

Now the Navier's differential equation for the bending of plate is

$$\nabla^4 u_z = \frac{p}{D}$$

where $\nabla^4 = \frac{\partial^4}{\partial x^4} + 2\frac{\partial^4}{\partial x^2 \partial y^2} + \frac{\partial^4}{\partial y^4}$

Let $$u_z = C\left(1 - \frac{x^2}{a^2} - \frac{u^2}{b^2}\right)^2$$

Then $$\nabla^4 u_z = \frac{24\,C}{a^4} + \frac{24C}{b^4} + \frac{8C}{a^2b^2}$$

$$\therefore \quad C = \frac{p}{8D} \cdot \frac{a^4b^4}{3a^4 + 2a^2b^2 + 3b^4}$$

and $$u_z = \frac{p}{8D}\left(\frac{a^4b^4}{3a^4 + 2a^2b^2 + 3b^4}\right)\left(1 - \frac{x^2}{a^2} - \frac{y^2}{b^2}\right)^2 \qquad ...(9.81)$$

The boundary conditions are

$$\left.\begin{aligned} u_z &= 0 \\ \frac{\partial u_z}{\partial n} &= 0 \end{aligned}\right\} \text{ for } \frac{x^2}{a^2} + \frac{y^2}{b^2} = 1$$

where n is the normal to the plate boundary. We observe that both the boundary conditions are satisfied.

$$(u_z)_{max}\Big|_{x=y=0} = \frac{pa^4b^4}{8D(3a^4 + 2a^2b^2 + 3b^4)} \qquad ...(9.82)$$

The bending moments are

$$M_x = -D\left(\frac{\partial^2 u_z}{\partial x^2} + \nu\frac{\partial^2 u_z}{\partial y^2}\right)$$

$$= 4DC\left[\left\{\frac{1}{a^2} - \frac{3x^2}{a^4} - \frac{y^2}{a^2b^2}\right\} + \nu\left\{\frac{1}{b^2} - \frac{3y^2}{b^4} - \frac{x^2}{a^2b^2}\right\}\right] \qquad ...(9.83)$$

$$M_y = -D\left(\frac{\partial^2 u_z}{\partial y^2} + \nu\frac{\partial^2 u_z}{\partial x^2}\right)$$

$$= 4DC\left[\left\{\frac{1}{b^2} - \frac{3y^2}{b^4} - \frac{x^2}{a^2b^2}\right\} + \nu\left\{\frac{1}{a^2} - \frac{3x^2}{a^4} - \frac{y^2}{a^2b^2}\right\}\right] \qquad ...(9.84)$$

$$M_{xy} = -(1-\nu)D\frac{\partial^2 u_z}{\partial x\,\partial y} = -(1-\nu)\frac{8DC}{a^2b^2}xy \qquad ...(9.85)$$

$$\left.\begin{aligned} (M_x)_{x=y=0} &= 4DC\left(\frac{1}{a^2} + \frac{\nu}{b^2}\right) \\ (M_y)_{x=y=0} &= 4DC\left(\frac{1}{b^2} + \frac{\nu}{a^2}\right) \\ (M_x)_{x=\pm a,\, y=0} &= -\frac{8DC}{a^2} \\ (M_y)_{x=\pm a,\, y=0} &= -\nu\frac{8DC}{a^2} \\ (M_x)_{x=0,\, y=\pm b} &= -\frac{8\nu DC}{b^2} \\ (M_y)_{x=0,\, y=\pm b} &= -\frac{8DC}{b^2} \end{aligned}\right\} \qquad ...(9.86)$$

9.9.3 Bending of Circular Plates

If (r, θ) coordinates are taken for the bending of circular plates as shown in Fig. 9.19, then the relation between the polar and Cartesian co-ordinates is given by the equations

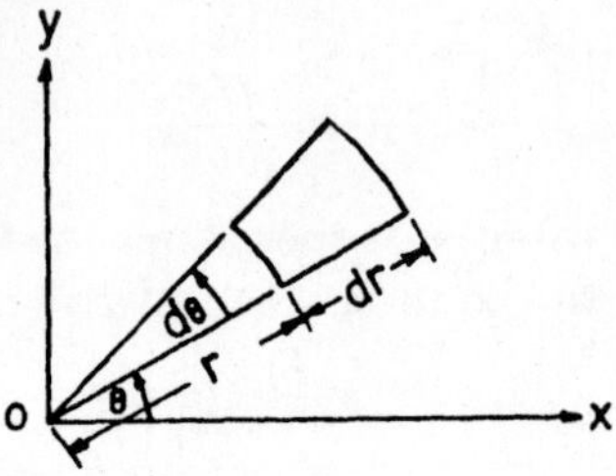

Fig. 9.19

$$r^2 = x^2 + y^2$$

$$\tan\theta = \frac{y}{x},$$

from which we find

$$\frac{\partial r}{\partial x} = \frac{x}{r} = \cos\theta$$

$$\frac{\partial r}{\partial y} = \frac{y}{r} = \sin\theta$$

$$\frac{\partial \theta}{\partial x} = -\frac{y}{r^2} = -\frac{\sin\theta}{r}, \quad \frac{\partial \theta}{\partial y} = \frac{x}{r^2} = \frac{\cos\theta}{r}$$

Now $\quad u_z = u_z(r, \theta)$

$$\therefore \quad \frac{\partial u_z}{\partial x} = \frac{\partial u_z}{\partial r}\frac{\partial r}{\partial x} + \frac{\partial u_z}{\partial \theta}\cdot\frac{\partial \theta}{\partial x}$$

$$= \frac{\partial u_z}{\partial r}\cdot\cos\theta - \frac{1}{r}\frac{\partial u_z}{\partial \theta}\cdot\sin\theta$$

$$= \left(\cos\theta\cdot\frac{\partial}{\partial r} - \frac{1}{r}\sin\theta\frac{\partial}{\partial \theta}\right)u_z$$

$$\frac{\partial u_z}{\partial y} = \frac{\partial u_z}{\partial r}\cdot\frac{\partial r}{\partial y} + \frac{\partial u_z}{d\theta}\cdot\frac{\partial \theta}{\partial y}$$

$$= \frac{\partial u_z}{\partial r}\cdot\sin\theta + \frac{1}{r}\frac{\partial u_z}{\partial \theta}\cos\theta$$

$$= \left(\sin\theta\cdot\frac{\partial}{\partial r} + \frac{1}{r}\cos\theta\cdot\frac{\partial}{\partial \theta}\right)u_z$$

$$\frac{\partial^2 u_z}{\partial x^2} = \left(\cos\theta\cdot\frac{\partial}{\partial r} - \frac{1}{r}\sin\theta\cdot\frac{\partial}{\partial \theta}\right)^2 u_z$$

$$= \frac{\partial^2 u_z}{\partial r^2}\cos^2\theta - 2\frac{\partial^2 u_z}{\partial \theta\partial r}\cdot\frac{\sin\theta\cos\theta}{r}$$

$$+ \frac{\partial u_z}{\partial r}\cdot\frac{\sin^2\theta}{r} + 2\frac{\partial u_z}{\partial \theta}\frac{\sin\theta\cos\theta}{r^2} + \frac{\partial^2 u_z}{\partial \theta^2}\frac{\sin^2\theta}{r^2}$$

$$\text{Similarly } \frac{\partial^2 u_z}{\partial y^2} = \frac{\partial^2 u_z}{\partial r^2} \sin^2\theta + 2\frac{\partial^2 u_z}{\partial\theta\partial r}\frac{\sin\theta\cos\theta}{r}$$

$$+ \frac{\partial u_z}{\partial r}\frac{\cos^2\theta}{r} - 2\frac{\partial u_z}{\partial\theta}\cdot\frac{\sin\theta\cos\theta}{r^2} + \frac{\partial^2 u_z}{\partial\theta^2}\frac{\cos^2\theta}{r^2}$$

$$\frac{\partial^2 u_z}{\partial x^2 \partial y^2} = \frac{\partial^2 u_z}{\partial r^2}\sin\theta\cos\theta + \frac{\partial^2 u_z}{\partial r\partial\theta}\left(\frac{\cos^2\theta - \sin^2\theta}{r}\right)$$

$$- \frac{\partial u_z}{\partial\theta}\left(\frac{\cos^2\theta - \sin^2\theta}{r^2}\right)$$

$$- \frac{\partial u_z}{\partial r}\frac{\sin\theta\cos\theta}{r} - \frac{\partial^2 u_z}{\partial\theta^2}\cdot\frac{\sin\theta\cos\theta}{r^2}$$

$$\therefore \quad \nabla^2 u_z = \left(\frac{\partial^2}{\partial x^2} + \frac{\partial^2}{\partial y^2}\right)u_z$$

$$= \left(\frac{\partial^2}{\partial r^2} + \frac{1}{r}\frac{\partial}{\partial r} + \frac{1}{r^2}\frac{\partial^2}{\partial\theta^2}\right)u_z$$

$$\therefore \quad \nabla^4 u_z = \nabla^2 \; \nabla^2 u_z$$

$$= \left(\frac{\partial^2}{\partial r^2} + \frac{1}{r}\frac{\partial}{\partial r} + \frac{1}{r^2}\frac{\partial^2}{\partial\theta^2}\right)\left(\frac{\partial^2 u_z}{\partial r^2} + \frac{1}{r}\frac{\partial u_z}{\partial r} + \frac{1}{r^2}\frac{\partial^2 u_z}{\partial\theta^2}\right)$$

$$= \frac{p}{D} \qquad \text{...(9.87)}$$

To derive the expressions for the moments and shearing forces in polar co-ordinates, let us consider an element of the plate bounded by two adjacent axial planes forming an angle $d\theta$ and by two cylindrical surfaces of radii r and $r + dr$ respectively as shown in Fig. 9.20.

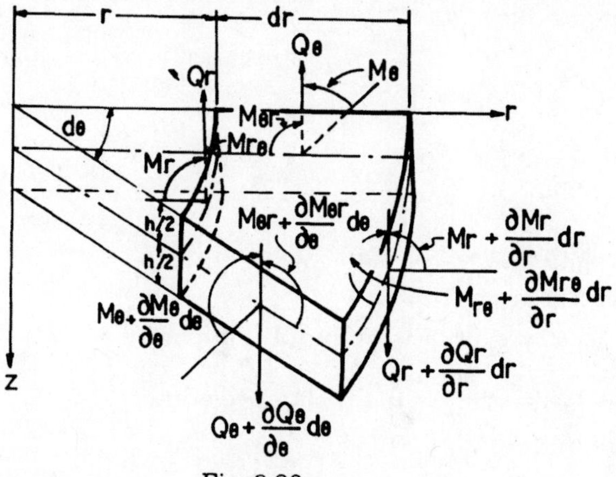

Fig. 9.20

Assume that the x-axis coincides with the radius r. The moments M_r, M_θ, $M_{r\theta}$ and the shears Q_r, Q_θ then have the same values as the moment M_x, M_y, M_{xy} and the shears Q_x, Q_y at the same point. By substituting $\theta = 0$ in the expressions for

$$\frac{\partial^2 u_z}{\partial x^2}, \frac{\partial^2 u_z}{\partial y^2}, \frac{\partial^2 u_z}{\partial x \partial y}$$

we obtain

$$M_r = -D\left(\frac{\partial^2 u_z}{\partial x^2} + \nu \frac{\partial u_z}{\partial y^2}\right)_{\theta=0}$$

$$= -D\left[\frac{\partial^2 u_z}{\partial r^2} + \nu\left(\frac{1}{r}\frac{\partial u_z}{\partial r} + \frac{1}{r^2}\frac{\partial^2 u_z}{\partial \theta^2}\right)\right] \quad \text{...(9.88 a)}$$

$$M_\theta = -D\left(\frac{\partial^2 u_z}{\partial y^2} + \nu \frac{\partial^2 u_z}{\partial x^2}\right)_{\theta=0}$$

$$= -D\left(\frac{1}{r}\frac{\partial u_z}{\partial r} + \frac{1}{r^2}\frac{\partial^2 u_z}{\partial \theta^2} + \nu \frac{\partial^2 u_z}{\partial r^2}\right) \quad \text{...(9.88 b)}$$

$$M_{r\theta} = M_{\theta r} = (1-\nu)D\left(\frac{\partial^2 u_z}{\partial x \partial y}\right)_{\theta=0}$$

$$= (1-\nu)D\left(\frac{1}{r}\frac{\partial^2 u_z}{\partial r \partial \theta} - \frac{1}{r^2}\frac{\partial u_z}{\partial \theta}\right) \quad \text{...(9.88 c)}$$

$$Q_r = -D\frac{\partial}{\partial x}\left(\frac{\partial^2 u_z}{\partial x^2} + \frac{\partial^2 u_z}{\partial y^2}\right)_{\theta=0}$$

$$= -D\frac{\partial}{\partial r}\left(\frac{\partial^2 u_z}{\partial r^2} + \frac{1}{r}\frac{\partial u_z}{\partial r} + \frac{1}{r^2}\frac{\partial^2 u_z}{\partial \theta^2}\right) \quad \text{...(9.89 a)}$$

$$Q_\theta = -D\frac{\partial}{\partial y}\left(\frac{\partial^2 u_z}{\partial x^2} + \frac{\partial^2 u_z}{\partial y^2}\right)_{\theta=0}$$

$$= -D\frac{1}{r}\frac{\partial}{\partial \theta}\left(\frac{\partial^2 u_z}{\partial r^2} + \frac{1}{r}\frac{\partial u_z}{\partial r} + \frac{1}{r^2}\frac{\partial^2 u_z}{\partial \theta^2}\right) \quad \text{...(9.89 b)}$$

The boundary conditions at the edge of a circular plate of radius $r = a$ are as follows :

(*i*) $u_z = 0$, $M_r = 0$ for simply supported edge.

(*ii*) $u_z = 0$, $\frac{\partial u_z}{\partial r} = 0$, for clamped edge.

(*iii*) $M_r = 0, V = Q_r - \frac{\partial M_{r\theta}}{r\partial\theta} = 0$ for a free edge.

The general solution of Eq. (9.87) can be taken in the form

$$u_z = (u_z)_c + (u_z)_p$$

where $(u_z)_c$ is the complimentary solution of the homogeneous equation

$$\left(\frac{\partial^2}{\partial r^2} + \frac{1}{r}\frac{\partial}{\partial r} + \frac{1}{r^2}\frac{\partial^2}{\partial \theta^2}\right)\left(\frac{\partial^2 u_z}{\partial r^2} + \frac{1}{r}\frac{\partial u_z}{\partial r} + \frac{1}{r^2}\frac{\partial^2 u_z}{\partial \theta^2}\right) = 0 \quad ...(9.90)$$

and $(u_z)_p$ is the particular solution of Eq. (9.87).

The complimentary solution can be assumed in the form of the following series :

$$(u_z)_c = R_\theta + \sum_{m=1}^{\infty} R_m \cos m\theta + \sum_{m=1}^{\infty} R'_m \sin m\theta \quad ...(9.91)$$

where R_0 ; R_1 .....; R_1', ... are functions of the radial distance r only. Substituting Eq. (9.91) in Eq. (9.90), we get

$$\left(\frac{d^2}{dr^2} + \frac{1}{r}\frac{d}{dr} - \frac{m^2}{r^2}\right)\left(\frac{d^2 R_m}{dr^2} + \frac{1}{r}\frac{dR_m}{dr} - \frac{m^2 R_m}{r}\right) = 0$$

The general solution of this equation is for $m > 1$

$$R_m = A_m r^m + B_m r^{-m} + C_m r^{m+2} + D_m r^{-m+2}$$

For $m = 0$ and $m = 1$ the solutions are

$$R_0 = A_0 + B_0 r^2 + C_0 \log r + D_0 r^2 \log r$$

$$R_1 = A_1 r + B_1 r^3 + C_1 r^{-1} + D_1 r \log r \quad ...(9.92)$$

Similar expressions can be written for the function R'_m. Substituting these expressions for the functions R_m and R'_m in series (9.91), we obtain the general solution of Eq. (9.90). The constants A_m, B_m, ..., D_m in each particular case must be determined so as to satisfy the boundary conditions.

9.9.4 Circular Plate Under a Linearly Varying Load

Consider a circular plates as shown in Fig. 9.21 of radius a and carrying a linearly varying load of intensity q_1 to q_2. This load can be divided into two parts : (*i*) a uniformly distributed load of intensity $\frac{1}{2}(q_1 + q_2)$ and (*ii*) a linearly varying load having zero intensity along the diameter CD of the plate and the intensities $-q$ and $+q$ at the ends A and B of the diameter AB.

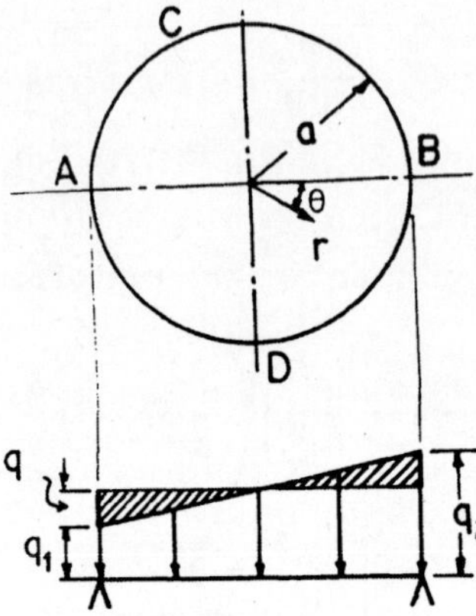

Fig. 9.21

The intensity of the non-uniform load q at any point with co-ordinates r and θ is

$$p = \frac{qr \cos \theta}{a}, q = \tfrac{1}{2}(q_2 - q_1)$$

The particular solution of Eq. (9.87) can thus be taken in the following form :

$$(u_z)_p = A \frac{qr^5 \cos \theta}{a}$$

which gives $\quad A = \dfrac{1}{192\, D}$

$$\therefore \quad (u_z)_p = \frac{qr^5 \cos \theta}{192\, aD}$$

The complimentary solution can be taken as

$$(u_z)_c = (A_1 r + B_1 r^3 + C_1 r^{-1} + D_1 r \log r) \cos \theta$$

Let $\quad \rho = \dfrac{r}{a}$, then

$$u_z = (u_z)_c + (u_z)_p$$

$$= \frac{qa^4}{192\, D}(\rho^5 + A_1\rho + B_1\rho^3 + C_1\rho^{-1} + D_1\rho \log \rho) \cos \theta$$

For a simply supported plate

$$(u_z)_{\rho=1} = 0, (M_r)_{\rho=1} = 0$$

At $\quad \rho = 0$, u_z and M_r must be finite.

Thus $\quad C_1 = D_1 = 0$

$$(u_z)_{\rho=1} = \frac{q\,a^4}{192\,D}(1 + A_1 + B_1)\cos\theta = 0$$

$$(M_r)_{\rho=1} = -\frac{qa^2}{192}[4(5+\nu) + 2(3+\nu)B_1]\cos\theta = 0$$

$$\therefore \qquad 1 + A_1 + B_1 = 0$$

$$4(5+\nu) + 2(3+\nu)B_1 = 0$$

$$B_1 = -\frac{2(5+\nu)}{3+\nu}$$

$$A_1 = \frac{7+\nu}{3+\nu}.$$

Substituting in Eq. (4), we get

$$u_r = \frac{qa^4\,\rho(1-\rho^2)}{192\,(3+\nu)\,D}[7 + \nu - (3+\nu)\rho^2]\cos\theta \qquad ...(9.93)$$

Substituting Eq. (9.93) in Eqs. (9.88), we get

$$\left.\begin{aligned} M_r &= \frac{qa^2}{48}(5+\nu)\rho(1-\rho^2)\cos\theta \\ M_\theta &= \frac{qa^2}{48(3+\nu)}\rho[(5+\nu)(1+3\nu) \\ &\quad - (1+5\nu)(3+\nu)\rho^2]\cos\theta \end{aligned}\right\} \qquad ...(9.94)$$

$$Q_r = \frac{qa}{24(3+\nu)}[2\,(5+\nu) - 9(3+\nu)\rho^2] \times \cos\theta$$

$$Q_\theta = -\frac{qa}{24(3+\nu)}\rho[2(5+\nu) - 3\,(3+\nu)\rho^2]\sin\theta$$

At $\rho = \frac{1}{\sqrt{3}}$, M_r is maximum and is equal to

$$(M_r)_{max} = \frac{qa^2\,(5+\nu)}{72\sqrt{3}}$$

The maximum value of M_θ occurs at

$$\rho = \sqrt{(5+\nu)(1+3\nu)}/\sqrt{3(1+5\nu)(3+\nu)}$$

and is equal to

$$(M_\theta)_{max} = \frac{qa^2}{72}\frac{(5+\nu)(1+3\nu)}{(3+\nu)}$$

Intensity of the vertical reaction at the boundary is

$$-V = -Q_r + \frac{\partial M_{r\theta}}{r\partial\theta} = \frac{qa\cos\theta}{4}$$

For the uniform load of intensity $\frac{1}{2}(q_1 + q_2)$, we have

$$u_z = \frac{(q_1 + q_2)}{128D}(a^2 - r^2)\left[\frac{5+\nu}{1+\nu}a^2 - r^2\right] \quad ...(9.95)$$

$$\left.\begin{aligned} M_r &= \frac{(q_1+q_2)}{32}[(a^2 - r^2)(3+\nu)] \\ M_\theta &= \frac{(q_1+q_2)}{32}[a^2(3+\nu) - r^2(1+3\nu)] \end{aligned}\right\} \quad ...(9.96)$$

Eqs. (9.93) and (9.95) may be added to obtain the resultant deflection and Eq. (9.94) may be added to Eq. (9.96) to obtain the resultant moments.

9.10 Combined Bending and Stretching of Rectangular Plates

If there are forces acting in the middle plane of the plate in addition to the lateral loads, then the governing differential equation for the bending of the plate must be modified. Let us denote the in-plane forces per unit length by N_x, N_y, $N_{xy} = N_{yx}$ as shown in Fig. 9.22. Consider the equilibrium of an element of the plate with

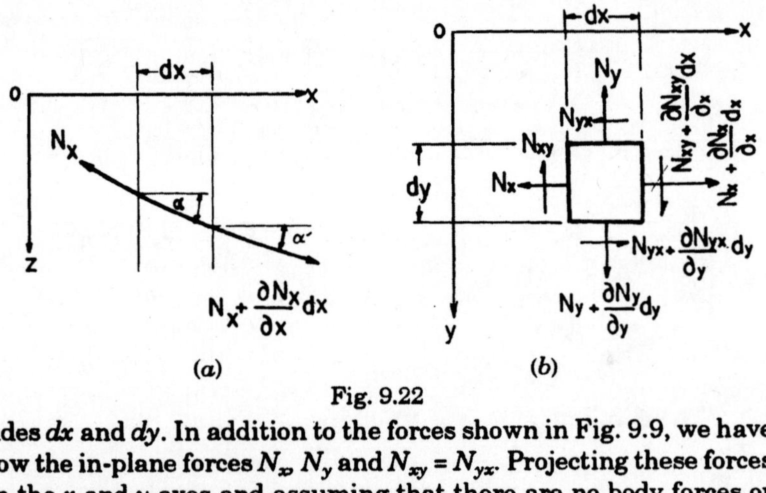

Fig. 9.22

sides dx and dy. In addition to the forces shown in Fig. 9.9, we have now the in-plane forces N_x, N_y and $N_{xy} = N_{yx}$. Projecting these forces on the x and y axes and assuming that there are no body forces or tangential forces acting in those directions at the faces of the plate, we have for $\Sigma F_x = 0$.

$$\left(N_x + \frac{\partial N_x}{\partial x}dx\right)dy\cos\alpha' - N_x\,dy\cos\alpha$$

$$+\left(N_{yx} + \frac{\partial N_{yx}}{\partial y}dy\right)dx\cos\alpha' - N_{yx}\,dx\cos\alpha = 0.$$

Now $\alpha' = \alpha + \frac{\partial \alpha}{\partial x} dx$

$$\cos \alpha = 1 - \frac{\alpha^2}{2} + \ldots$$

For small α, $\cos \alpha \simeq 1$ and $\cos \alpha' \simeq 1$.

Hence, we get

$$\frac{\partial N_x}{\partial x} + \frac{\partial N_{yx}}{\partial y} = 0 \qquad \text{...(9.97)}$$

Similarly, the condition $\Sigma F_y = 0$ leads to

$$\frac{\partial N_{xy}}{\partial x} + \frac{\partial N_{yx}}{\partial y} = 0 \qquad \text{...(9.98)}$$

Now let us find the equation of equilibrium in the z-direction. In addition to the z components of the forces shown in Fig. 9.22, we have to consider also the z-components of the forces as shown in Fig. 9.9. The z components of the forces $N_x\, dy$ and

$\left[N_x + \frac{\partial N_x}{\partial x} dx\right] dy$ give a resultant

$$\left[N_x + \frac{\partial N_x}{\partial x} dx\right] dy \sin \alpha' - N_x\, dy \sin \alpha.$$

For small α and α', we find $\sin \alpha \simeq \alpha \simeq \frac{\partial u_z}{\partial x}$

and $\sin \alpha^n \simeq \alpha' \simeq \alpha + \frac{\partial \alpha}{\partial x} dx = \frac{\partial u_z}{\partial x} + \frac{\partial^2 u_z}{\partial x^2} dx.$

Thus the above expression becomes

$$\left[N_x + \frac{\partial N_x}{\partial x} dx\right] dy \left(\frac{\partial u_z}{\partial x} + \frac{\partial^2 u_z}{\partial x^2} dx\right) - N_x\, dy . \frac{\partial u_z}{\partial x}$$

$$= N_x \frac{\partial^2 u_z}{\partial x^2} dx\, dy + \frac{\partial N_x}{\partial x} \frac{\partial u_z}{\partial x} dx\, dy \qquad \text{...}(a)$$

where higher order terms are neglected. In a similar manner, the resultant of the z-components of the forces $N_y dx$ and $\left[N_y + \frac{\partial N_y}{\partial y} dy\right] dx$ can be shown to be

$$N_y \frac{\partial^2 u_z}{\partial y^2} dxdy + \frac{\partial N_y}{\partial y} . \frac{\partial u_z}{\partial y} dxdy \qquad \text{...}(b)$$

To find the z components of the forces $N_{xy}\,dy$ and

$$\left[N_{xy} + \frac{\partial N_{xy}}{\partial x}\,dx\right]dy,$$

we refer to Fig. 9.23. After bending, the deflection of O' is u_z and deflection of B' is $u_z + \frac{\partial u_z}{\partial y}\,dy$. The line $O'B'$ is, therefore, inclined downward at an angle $\frac{\partial u_z}{\partial y}$ with the y-axis. Similarly we can show

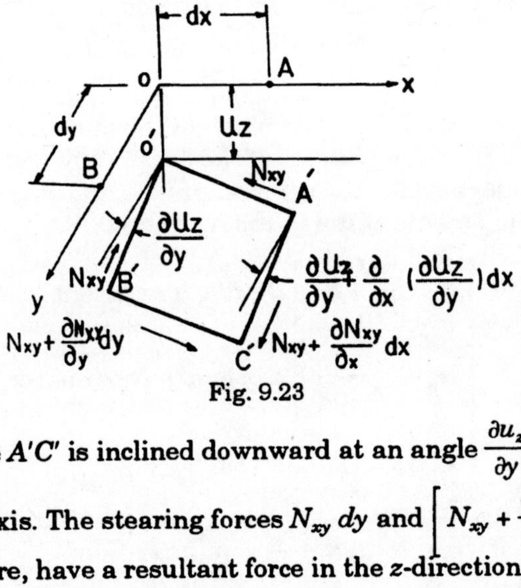

Fig. 9.23

that the line $A'C'$ is inclined downward at an angle $\frac{\partial u_z}{\partial y} + \frac{\partial^2 u_z}{\partial x\,\partial y}\,.\,dx$ with the y-axis. The stearing forces $N_{xy}\,dy$ and $\left[N_{xy} + \frac{\partial N_{xy}}{\partial x}\,dx\right]dy$ will, therefore, have a resultant force in the z-direction equal to

$$\left(N_{xy} + \frac{\partial N_{xy}}{\partial x}\,dx\right)dy\left(\frac{\partial u_z}{\partial y} + \frac{\partial^2 u_z}{\partial x\,\partial y}\,dx\right) - N_{xy}\,.\,dy\,.\,\frac{\partial u_z}{\partial y}$$

$$= N_{xy}\frac{\partial^2 u_z}{\partial x\,\partial y}\,dxdy + \frac{\partial N_{xy}}{\partial x}\frac{\partial u_z}{\partial y}\,dx\,dy \qquad ...(c)$$

An analogous expression can be obtained for the z projection of the forces $N_{yx}\,dx$ and $\left[N_{yx} + \frac{\partial N_{yx}}{\partial y}\,dy\right]dy$ and is equal to

$$N_{yx}\frac{\partial^2 y_z}{\partial x\,\partial y}\,dxdy + \frac{\partial N_{yx}}{\partial y}\frac{\partial u_z}{\partial x}\,dxdy \qquad ...(d)$$

Adding all the forces in the z-direction given by Eqs. (a), (b), (c) and (d) and dividing by $dxdy$, we get

$$N_x \frac{\partial^2 u_z}{\partial x^2} + N_y \frac{\partial^2 u_z}{\partial y^2} + 2N_{xy} \frac{\partial^2 u_z}{\partial x \partial y}$$
$$+ \left(\frac{\partial N_x}{\partial x} + \frac{\partial N_{yx}}{\partial y} \right) \frac{\partial u_z}{\partial x} + \left(\frac{\partial N_y}{\partial y} + \frac{\partial N_{xy}}{\partial y} \right) \frac{\partial u_z}{\partial y}$$

Using Eqs. (9.97) and (9.98), we find

$$N_x \frac{\partial^2 u_z}{\partial x^2} + N_y \frac{\partial^2 u_z}{\partial y^2} + 2N_{xy} \frac{\partial^2 u_z}{\partial x \partial y} \qquad ...(e)$$

Adding Eq. (*e*) to the load *p dxdy* acting on the element, we get

$$\frac{\partial Q_x}{\partial x} + \frac{\partial Q_y}{\partial x} + p + N_x \frac{\partial^2 u_z}{\partial x^2} + N_y \frac{\partial^2 u_z}{\partial y^2} + 2N_{xy} \frac{\partial^2 u_z}{\partial x \partial y} = 0$$

or
$$\frac{\partial^2 M_x}{\partial x^2} + 2 \frac{\partial^2 M_{xy}}{\partial x \partial y} + \frac{\partial^2 M_y}{\partial y^2} = -\left(p + N_x \frac{\partial^2 u_z}{\partial x^2} + N_y \frac{\partial^2 u_z}{\partial y^2} + 2N_{xy} \frac{\partial^2 u_z}{\partial x \partial y} \right]$$

or
$$\frac{\partial^4 u_z}{\partial x^4} + 2 \frac{\partial^4 u_z}{\partial x^2 \partial y^2} + \frac{\partial^4 u_z}{\partial y^4} = \frac{1}{D} \left(p + N_x \frac{\partial^2 u_z}{\partial x^2} + N_y \frac{\partial^2 u_z}{\partial y^2} + 2N_{xy} \frac{\partial^2 u_z}{\partial x \partial y} \right]$$

or
$$\nabla^2 . \nabla^2 u_z = \frac{1}{D} \left(p + N_x \frac{\partial^2 u_z}{\partial x^2} + N_y \frac{\partial^2 u_z}{\partial y^2} + 2N_{xy} \frac{\partial^2 u_z}{\partial x \partial y} \right) \qquad ...(9.99)$$

If body forces are also present in the middle plane of the plate or tangential forces distributed over the surface of the plate, then Eqs. (9.97) and (9.98) become

$$\left. \begin{aligned} \frac{\partial N_x}{\partial x} + \frac{\partial N_{xy}}{\partial y} + X = 0 \\ \frac{\partial N_{xy}}{\partial x} + \frac{\partial N_y}{\partial y} + Y = 0 \end{aligned} \right\} \qquad ...(9.100)$$

where X and Y are the components of the body forces or of the tangential forces per unit area of the middle plane of the plate.

Eq. (9.99) then becomes

$$\nabla^4 u_z = \frac{1}{D} \left(p + N_x \frac{\partial^2 u_z}{\partial x^2} + N_y \frac{\partial^2 u_z}{\partial y^2} + 2N_{xy} \frac{\partial^2 u_z}{\partial x \partial y} - X \frac{\partial u_z}{\partial x} - Y \frac{\partial u_z}{\partial y} \right) \qquad ...(9.101)$$

9.10.1 Rectangular Plate with Simply Supported Edges under the combined action of Uniform Lateral Load and uniform Tension

Let $\quad N_x = N_0, N_y = N_{xy} = 0$ ...(1)

Then Eqs. (9.97) and (9.98) are satisfied identically. Now uniform lateral load p can be represented by the double Fourier series as :

$$p = \frac{16p_0}{\pi^2} \sum_{m=1,3,5}^{\infty} \sum_{n=1,3,5}^{\infty} \frac{1}{mn} \sin\frac{m\pi x}{a} \sin\frac{n\pi y}{b}$$

...(2)

Eq. (9.99) then becomes

$$\nabla^4 u_z - \frac{N_0}{D}\frac{\partial^2 u_z}{\partial x^2} = \frac{16p_0}{\pi^2} \sum_{m=1,3,5}^{\infty} \sum_{n=1,3,5}^{\infty} \frac{1}{mn} \sin\frac{m\pi x}{a} \sin\frac{n\pi y}{b}$$

...(3)

The boundary conditions are

$$\left.\begin{aligned} (u_z)_{\substack{x=0\\x=a}} &= \left(\frac{\partial^2 u_z}{\partial x^2}\right)_{\substack{x=0\\x=a}} = 0 \\ (u_z)_{\substack{x=0\\x=b}} &= \left(\frac{\partial^2 u_z}{\partial y^2}\right)_{\substack{x=0\\x=b}} = 0 \end{aligned}\right\} \qquad ...(4)$$

The boundaries conditions will be satisfied if we assume u_z in the following double series :

$$u_z = \sum_{m=1}^{\infty} \sum_{n=1}^{\infty} A_{mn} \sin\frac{m\pi x}{a} \sin\frac{n\pi y}{b} \qquad ...(5)$$

Substituting Eq. (5) in (3), we get

$$A_{mn} = -\frac{16p_0}{D\pi^6\, mn\left[\left(\frac{m^2}{a^2}+\frac{n^2}{b^2}\right)^2 + \frac{N_0 m^2}{\pi^2 D a^2}\right]}$$

for odd m and n

$= 0$ for even m or n.

$$u_z = \frac{16p_0}{\pi^6 D} \sum_{m=1,3,5}^{\infty} \sum_{n=1,3,5}^{\infty} \frac{\sin\frac{m\pi x}{a} \cdot \sin\frac{n\pi y}{b}}{mn\left[\left(\frac{m^2}{a^2}+\frac{n^2}{b^2}\right)^2 + \frac{N_0 m^2}{\pi^2 D a^2}\right]}$$

...(9.102)

9.11 The Effect of Transverse Shear Deformation on the Bending of Thin Plates

We have seen that the Navier theory for bending of thin plates leads to a fourth order differential equation. To determine the deflection we need to satisfy two boundary conditions at each edge. For a plate of finite thickness, it appears more logical to satisfy three boundary conditions. The formal reason for the impossibility of satisfying more than two boundary conditions lies in the fact that the distortion of the plate element due to transverse shear forces Q_x and Q_y has been neglected in Navier's theory. The inaccuracy of Navier's thin plate theory becomes of practical interest in the edge zones of the plates and around holes that have a diameter which is not large as compared to the plate thickness.

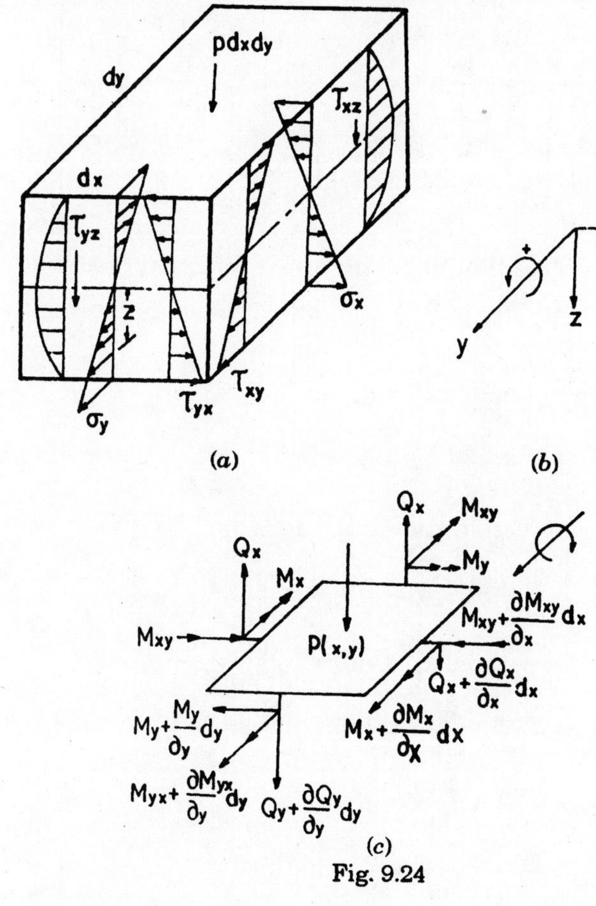

Fig. 9.24

Let us consider an element of the plate subjected to the external transverse load $pdxdy$ and to a system of stress components as shown in Fig. 9.24. The effect of transverse shear deformation was studied first of all by Reissner who assumed a linear law for the distribution of stress components σ_x, σ_y and τ_{xy} through the thickness of the plate. By using equations of equilibrium in Cartesian coordinates with no body forces, the distribution of the components τ_{xz} and τ_{yz} then follows a parabolic law. The stress component σ_z can then be obtained from the third equilibrium equation by taking into account the condition.

$$(\sigma_z)_{z=-h/2} = -p, \quad (\sigma_z)_{z=h/2} = 0$$

Thus the stress components become

$$\left.\begin{aligned} \sigma_x &= \frac{12M_x \cdot z}{h^3}, \quad \sigma_y = \frac{12M_y \cdot z}{h^3}, \quad \tau_{xy} = \frac{12M_{xy} \cdot z}{h^3} \\ \tau_{xz} &= \frac{3Q_x}{2h}\left[1-\left(\frac{2z}{h}\right)^2\right], \quad \tau_{yz} = \frac{3Q_y}{2h}\left[1-\left(\frac{2z}{h}\right)^2\right] \\ \sigma_z &= -\frac{3p}{4}\left[\frac{2}{3}-\frac{2z}{h}+\frac{1}{3}\left(\frac{2z}{h}\right)^3\right] \end{aligned}\right\} \quad ...(9.103)$$

Conditions of equilibrium of the stress resultants become

$$\left.\begin{aligned} \frac{\partial M_x}{\partial x} + \frac{\partial M_{xy}}{\partial y} - Q_x &= 0 \\ \frac{\partial M_{xy}}{\partial x} + \frac{\partial M_y}{\partial y} - Q_y &= 0 \\ \frac{\partial Q_x}{\partial x} + \frac{\partial Q_y}{\partial y} + p &= 0 \end{aligned}\right\} \quad ...(9.104)$$

The general stress-strain relations are :

$$\left.\begin{aligned} \varepsilon_x &= \frac{\partial u_{x0}}{\partial x} = \frac{1}{E}[\sigma_x - \nu(\sigma_y + \sigma_z)] \\ \varepsilon_y &= \frac{\partial u_{y0}}{\partial y} = \frac{1}{E}[\sigma_y - \nu(\sigma_x + \sigma_z)] \\ \gamma_{xy} &= \frac{\partial u_{x0}}{\partial y} + \frac{\partial u_{y0}}{\partial x} = \frac{\tau_{xy}}{G} \\ \gamma_{yz} &= \frac{\partial u_{y0}}{\partial z} + \frac{\partial u_{z0}}{\partial y} = \frac{\tau_{yz}}{G} \\ \gamma_{xz} &= \frac{\partial u_{x0}}{\partial z} + \frac{\partial u_{z0}}{\partial x} = \frac{\tau_{xz}}{G} \end{aligned}\right\} \quad ...(9.105)$$

We do not use the relation

$$\varepsilon_z = \frac{\partial u_{z0}}{\partial z} = \frac{1}{E}[\sigma_z - \nu(\sigma_x + \sigma_y)]$$

where u_{x0}, u_{y0}, u_{z0} are the displacement components. Since this is in contradiction with the assumed linear law for the distribution of the stress components $\sigma_x, \sigma_y, \tau_{xy}$.

Now the average displacement resultant

$$u_z = \frac{3}{2h}\int_{-h/2}^{h/2} u_{z0}\left[1 - \left(\frac{2z}{h}\right)^2\right] dz$$

and the average rotation resultants are

$$\omega_x = \frac{6}{h^2}\int_{-h/2}^{h/2} u_{x0} \cdot \frac{2z}{h}\, dz$$

$$\omega_y = \frac{6}{h^2}\int_{-h/2}^{h/2} u_{y0} \cdot \frac{2z}{h}\, dz \qquad \text{...(9.106)}$$

Now $$\int_{-h/2}^{h/2} \sigma_x \cdot u_{x0}\, dz = M_x \omega_x$$

$$\int_{-h/2}^{h/2} \sigma_y \cdot u_{y0}\, dz = M_y \cdot \omega_y$$

$$\int_{-h/2}^{h/2} \tau_{xz} \cdot u_{z0}\, dz = Q_x \cdot u_z$$

$$\int_{-h/2}^{h/2} \tau_{xy} \cdot u_{y0}\, dz = M_{xy}\, \omega_y$$

$$\int_{-h/2}^{h/2} \tau_{xy} \cdot u_{x0}\, dz = M_{xy} \cdot \omega_x$$

$$\int_{-h/2}^{h/2} \tau_{yz} \cdot u_{z0}\, dz = Q_y \cdot u_z \qquad \text{...(9.107)}$$

Hence

$$\left.\begin{aligned}\frac{\partial \omega_x}{\partial x} &= \frac{12}{Eh^3}\left[M_x - \nu M_y - \frac{h^2}{10}\nu p\right]\\ \frac{\partial \omega_y}{\partial y} &= \frac{12}{Eh^3}\left[-\nu M_x + M_y - \frac{h^2}{10}\nu p\right]\\ \frac{\partial \omega_x}{\partial y} + \frac{\partial \omega_y}{\partial x} &= \frac{12}{h^3 G} M_{xy}\end{aligned}\right\} \quad \text{...(9.108)}$$

Substituting Eqs. (9.105) in (9.103), we get

$$\left.\begin{aligned}\sigma_x &= \frac{E}{(1-\nu^2)}\left(\frac{\partial u_{x0}}{\partial x} + \nu\frac{\partial u_{y0}}{\partial y}\right) - \frac{3p\nu}{4(1-\nu)}\left[\frac{2}{3} - \frac{2z}{h} + \frac{1}{3}\left(\frac{2z}{h}\right)^3\right]\\ \sigma_y &= \frac{E}{(1-\nu^2)}\left(\frac{\partial u_{y0}}{\partial y} + \nu\frac{\partial u_{x0}}{\partial x}\right) - \frac{3p\nu}{4(1-\nu)}\left[\frac{2}{3} - \frac{2z}{h} + \frac{1}{3}\left(\frac{2z}{h}\right)^3\right]\\ \tau_{xy} &= \frac{E}{2(1+\nu)}\left(\frac{\partial u_{x0}}{\partial y} + \frac{\partial u_{y0}}{\partial x}\right)\end{aligned}\right\} \quad \text{...(9.109)}$$

Substituting the values of τ_{xz} and τ_{yz} from Eqs. (9.103) in the last two equations of Eqs. (9.105) and multiplying the result by $\frac{3}{2h}\left[1-\left(\frac{2z}{h}\right)^2\right]dz$ and integrating between the limits $z = \pm\frac{h}{2}$, we obtain

$$\left.\begin{aligned}\omega_x &= -\frac{\partial u_z}{\partial x} + \frac{6}{5h}\frac{Q_x}{G}\\ \omega_y &= -\frac{\partial u_z}{\partial y} + \frac{6}{5h}\frac{Q_y}{G}\end{aligned}\right\} \quad \text{...(9.110)}$$

Eliminating ω_x and ω_y from Eqs. (9.108) and (9.110), we find

$$\left.\begin{aligned}M_x &= -D\left(\frac{\partial^2 u_z}{\partial x^2} + \nu\frac{\partial^2 u_z}{\partial y^2}\right) + \frac{h^2}{5}\frac{\partial Q_x}{\partial x} - \frac{h^2 p}{10}\left(\frac{\nu}{1-\nu}\right)\\ M_y &= -D\left(\frac{\partial^2 u_z}{\partial y^2} + \nu\frac{\partial^2 u_z}{\partial x^2}\right) + \frac{h^2}{5}\frac{\partial Q_y}{\partial y} - \frac{h^2 p}{10}\left(\frac{\nu}{1-\nu}\right)\\ M_{xy} &= -D(1-\nu)\frac{\partial^2 u_z}{\partial x \partial y} + \frac{h^2}{10}\left(\frac{\partial Q_y}{\partial x} + \frac{\partial Q_x}{\partial y}\right)\end{aligned}\right\} \quad \text{...(9.111)}$$

or

$$\left.\begin{aligned} M_x &= D\left[\frac{\partial \omega_x}{\partial x} + \nu \frac{\partial \omega_y}{\partial y} + \frac{6\nu(1+\nu)}{5Eh} p\right] \\ M_y &= D\left[\frac{\partial \omega_y}{\partial y} + \nu \frac{\partial \omega_x}{\partial x} + \frac{6\nu(1+\nu)}{5Eh} p\right] \\ M_{xy} &= D\frac{(1-\nu)}{2}\left(\frac{\partial \omega_x}{\partial y} + \frac{\partial \omega_y}{\partial x}\right) \end{aligned}\right\} \quad \text{...(9.112)}$$

Substituting Eqs. (9.111) in the first two of Eqs. (9.104), we get

$$\left.\begin{aligned} Q_x - \frac{h^2}{10}\nabla^2 Q_x &= -D\frac{\partial(\nabla^2 u_z)}{\partial x} - \frac{h^2}{10(1-\nu)}\frac{\partial p}{\partial x} \\ Q_y - \frac{h^2}{10}\nabla^2 Q_y &= -D\frac{\partial(\nabla^2 u_z)}{\partial y} - \frac{h^2}{10(1-\nu)}\frac{\partial p}{\partial y} \end{aligned}\right\} \quad \text{...(9.113)}$$

where $\nabla^2 = \frac{\partial^2}{\partial x^2} + \frac{\partial^2}{\partial y^2}$, $D = \frac{Eh^3}{12(1-\nu^2)}$

Substituting Eqs. (9.113) in the third of Eqs. (9.104), we find

$$D\nabla^4 u_z = p - \frac{h^2}{10}\left(\frac{2-\nu}{1-\nu}\right)\nabla^2 p \quad \text{...(9.114)}$$

Solution of Eq. (9.114) can be taken in the form

$$u_z = (u_z)_c + (u_z)_p \quad \text{...(9.115)}$$

where $(u_z)_p$ is the particular solution and $(u_z)_c$ is the auxiliary solution of the equation $\nabla^4 u_z = 0$. Eq. (9.114) requires to satisfy four boundary conditions.

Now the third of equilibrium equations (9.104) is satisfied if we take

$$\left.\begin{aligned} Q_x &= -D\frac{\partial}{\partial x}(\nabla^2 u_z) + \frac{\partial \psi}{\partial y} \\ Q_y &= -D\frac{\partial(\nabla^2 u_z)}{\partial y} - \frac{\partial \psi}{\partial x} \end{aligned}\right\} \quad \text{...(9.116)}$$

or

$$\left.\begin{aligned} Q_x &= Q'_x - D\frac{\partial[\nabla^2 (u_z)_c]}{\partial x} + \frac{\partial \psi}{\partial y} \\ Q_y &= Q'_y - D\frac{\partial[\nabla^2 (u_z)_c]}{\partial y} + \frac{\partial \psi}{\partial x} \end{aligned}\right\} \quad \text{...(9.117)}$$

where ψ = stress function and Q'_x and Q'_y must satisfy the relations

$$\left.\begin{aligned} Q'_x - \frac{h^2}{10}\nabla^2 Q'_x &= -D\frac{\partial}{\partial x}[\nabla^2(u_z)_p] - \frac{h^2}{10(1-\nu)}\frac{\partial p}{\partial x} \\ Q'_y - \frac{h^2}{10}\nabla^2 Q'_y &= -D\frac{\partial}{\partial y}[\nabla^2(u_z)_p] - \frac{h^2}{10(1-\nu)}\frac{\partial p}{\partial y} \end{aligned}\right\} \quad \text{...(9.118)}$$

Differentiating Eqs. (9.118) w.r.t. x and y and adding the results, we get

$$\frac{\partial Q'_x}{\partial x} + \frac{\partial Q'_y}{\partial y} + p = 0 \quad \text{...(9.119)}$$

Substituting Eqs. (9.117) in (9.113), we get

$$\frac{\partial}{\partial y}\left(\psi - \frac{h^2}{10}\nabla^2\psi\right) = -\frac{\partial}{\partial x}\left(\psi - \frac{h^2}{10}\nabla^2\psi\right) = 0 \quad \text{...(9.120)}$$

$$\therefore \quad \psi - \frac{h^2}{10}\nabla^2\psi = \text{const.} = 0 \quad \text{...(9.121)}$$

Eq. (9.121) is the second fundamental equation of the generalized theory of bending in addition to Eq. (9.114). Eq. (9.114) is of fourth order whereas Eq. (9.121) is of second order and we can satisfy three boundary conditions instead of only two.

EXERCISES

9.1 A rectangular plate of length 500 mm is simply supported at the edges and carries a uniformly distributed load of intensity 20 N/mm. Determine the maximum stress in the plate. Also calculate the maximum deflection. Assume $E = 2 \times 10^5$ N/mm^2, $\nu = 0.28$.

9.2 If the plate in exercise 9.1 is clamped at the edges, then, determine the maximum stress and the maximum deflection of the plate.

9.3 (*a*) State clearly the assumptions made in deriving Lagrange's equilibrium equations for the bending of a plate subject to lateral load.

(*b*) Derive Lagrange's equilibrium equation for the bending of a thin plate of uniform thickness.

9.4 Describe membrane analogy for the bending of thin plates. How this analogy can be used to determine the deflection of a plate ?

9.5 A circular steel plate 10 mm thick is of 300 mm outside diameter and the inside diameter is 100 mm. It is clamped at its outer edge and loaded by a uniform edge moment of 50 N/mm of the circumference at its inner edge. Calculate the radial stress distribution assuming $\nu = 0.3$ and $E = 2 \times 10^5$ N/mm^2.

9.6 A circular plate of outside radius 250 mm and inside radius 50 mm carries a shear force of 800 N at the inner edge. If the thickness of the plate is 10 mm and $\nu = 0.28$, determine the maximum stresses and deflection of the plate when (*i*) the plate is simply supported at the outer edge (*ii*) the outer edges are clamped. Take $E = 2.1 \times 10^5$ N/mm^2.

9.7 A circular plate of radius 300 mm is 8 mm thick. It carries a concentrated load of 600 N at the centre. Determine the maximum deflection and stresses in the plate, when (*i*) the plate is simply supported at the edges (*ii*) the edges of the plate are clamped. Take $\nu = 0.30$, $E = 2.1 \times 10^5$ N/mm^2.

9.8 A solid circular steel plate 10 mm thick is clamped at its outer edge and carries a ring of load at a radius of 50 mm. If the radius of the plate is 150 mm and the total load on the plate is 12500 N, calculate the maximum deflection of the plate. $E = 2 \times 10^5$ N/mm^2.

9.9 A simply supported rectangular plate of sides a and b is under the action of hydrostatic pressure so that the load is distributed according to the formula $p(x, y) = p_0 \dfrac{x}{a}$.

Find the deflection, moments and stresses in the plate.

9.10 A uniformly loaded circular plate with radius a has its edge simply supported. Assume that there is not hole in the plate. Find the maximum deflection and bending stress.

9.11 A circular plate with radius a has a hole with radius b at the middle of the plate. The plate is subjected to uniformly distributed load of intensity p_0 and has its inner edge clamped and external edge free. Find the maximum deflection, moment and bending stress if $b = \dfrac{a}{4}$.

9.12 A simply supported rectangular plate of sides 400 mm and 600 mm carries a uniformly distributed load of intensity 10 N/mm^2. Determine the maximum deflection and the maximum stresses in the plate. Assume $\nu = 0.30$, $E = 2 \times 10^5$ N/mm^2 and thickness of plate 15 mm.

9.13 A rectangular plate of sides 300 mm and 540 mm is simply supported at the edges. The thickness of the plate is 10 mm.

It carries a concentrated load of 5000 N at a point whose coordinates are (150, 200) mm with reference to origin at one corner. Determine the deflection of the plate at the centre. Assume $E = 2 \times 10^5$ N/mm^2.

9.14 An elliptic plate semi-major axis 200 mm and semi-minor axis 150 mm is subjected to a uniformly distributed load of intensity 12 N/mm^2. If the edges of the plate are clamped, determine the maximum deflection and the maximum bending stress. Assume $E = 2 \times 10^5$ N/mm^2 and $\nu = 0.3$. The plate is 15 mm thick.

9.15 Derive Navier's equation for the bending of a circular plate. State clearly the assumptions made.

9.16 A simply supported rectangular plate of sides a and b is under the action of hydrostatic pressure given by the formula $p(x, y) = p_0 \dfrac{x}{a}$.

If the plate is also under a uniform tensile force N_x along the sides $x = 0$ and $x = a$, find the deflection, moments and stresses in the plate.

9.17 A simply supported rectangular plate is under the combined action of uniform lateral pressure p_0 and uniform in-plane forces N along the four edges. Find the deflection, moments and stresses in the plate.

9.18 What do you understand by transverse shear in plates ? How would you account for the effect of transverse shear deformation on the bending of thin plates ?

9.19 A rectangular steel plate (E = 200 GPa, ν = 0.30, σ_0 = 280 MPa) has the length of 2 m, a width of 1 m and fixed edges. The plate is subjected to a uniform pressure p = 270 kPa. Determine the plate thickness and maximum deflection taking factor of safety of 2.

9.20 A mild steel plate (E = 200 GPa, ν = 0.30, σ_0 = 320 MPa) has a thickness of 10 mm and covers a circular opening having a diameter of 200 mm. The plate is fixed at the edges and is subjected to a uniform pressure p. Determine the working pressure based on a factor of safety of 2.

10

Thin Shells

10.1 Introduction

A shell may be defined as a curved structure of which one dimension, the thickness is 'small' in comparison with the other two dimensions. Shells bear the same relation to plates as curved beams to straight beams. In general shells may be termed as curved plates. There are two different classes of shells—thick shells and thin shells. A shell will be called thin, if the maximum value of the ratio h/R (where h is the shell thickness and R is the radius of curvature of the middle surface) can be neglected in comparison with unity. Correspondingly, shells will be called thick whenever such terms cannot be neglected. For ordinary technical calculations, it is admissible to have a relative error of 5 per cent and for thin shells this corresponds to

$$\max\left(\frac{h}{R}\right) \le \frac{1}{20}$$

Hence for shells for which this inequality is violated will be regarded as thick shells. In this chapter, we shall study the distribution of stresses in thin shells by considering bending and without bending *i.e.* membrane stresses.

10.2 The Equilibrium Equations

The location of any point on a shell can be determined by three parameters, two of which vary on the middle surface of the shell while third varies along the normal to the middle surface. If we choose the lines of curvature at a point on the middle surface as the parametric curves given by

$$x = f(\xi_1, \xi_2)$$
$$y = f(\xi_1, \xi_2)$$
$$z = f(\xi_1, \xi_2)$$

then we have a three-dimensional orthogonal coordinate system. Let us assume that on the middle surface ξ_1 = constant and ξ_2 = constant. Let z be the distance of the point from the middle surface. A point on the shell will then have the orthogonal curvilinear co-ordinates (ξ_1, ξ_2, z).

Let us consider a space element separated from a shell by means of four cuts, normal to its middle surface and tangent to the lines

$$\xi_1, \xi_1 + d\xi_1, \xi_2, \xi_2 + d\xi_2, \text{ and } z = \pm\frac{h}{2}.$$

The sides of the element will have the lengths $\alpha_1 d\xi_1$ and $\alpha_2 d\xi_2$ where α_1 and α_2 are factors so that the square of a line element ds on the middle surface of the shell has the form

$$ds^2 = \alpha_1^2 d\xi_1^2 + \alpha_2^2 \xi_2^2$$

For example in polar coordinates r and θ, let $\xi_1 = r$, $\xi_2 = \theta$, then

$$ds^2 = (dr)^2 + r^2(d\theta)^2$$

so that $\alpha_1 = 1$ and $\alpha_2 = r$.

Let us denote by x and y the directions of the tangents to the curvilinear coordinates ξ_1 and at the point O as shown in Fig. 10.1. Let R_1 and R_2 be the principal radii of curvature at O which lie in the xz and yz planes respectively. The angle subtended by the arc length $\alpha_1 d\xi_2$ is then $(\alpha_2 d\xi_2)/R_2$ and the length of the element which is at a distance z from the middle surface in the yz-plane is

$$\frac{(R_2 + z)\,\alpha_2 d\xi_2}{R_2} = \left(1 + \frac{z}{R_2}\right)\alpha_2 d\xi_2$$

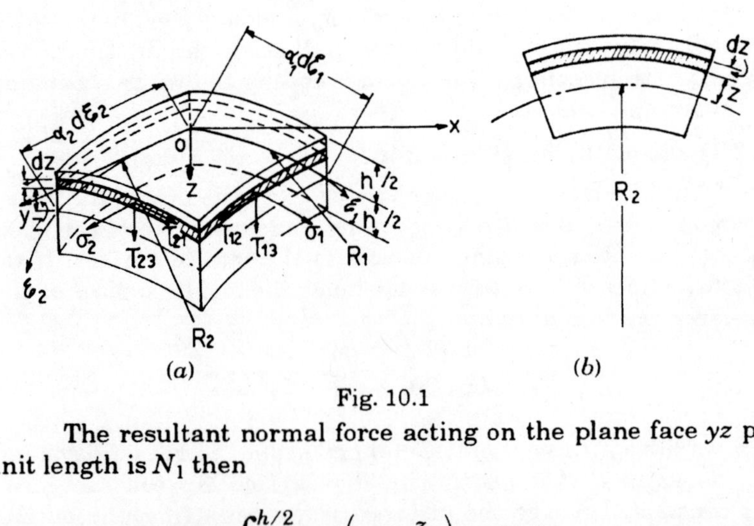

(a) (b)

Fig. 10.1

The resultant normal force acting on the plane face yz per unit length is N_1 then

$$N_1 \alpha_2 d\xi_2 = \int_{-h/2}^{h/2} \sigma_1 \left(1 + \frac{z}{R_2}\right)\alpha_2 d\xi_2 dz$$

or $$N_2 = \int_{-h/2}^{h/2} \sigma_1 \left(1 + \frac{z}{R_2}\right) dz$$

Similarly other resultant forces and moments per unit length of the normal section as shown in Fig. 10.2 are

$$\left.\begin{aligned}
N_2 &= \int_{-h/2}^{h/2} \sigma_2 \left(1 + \frac{z}{R_1}\right) dz \\
N_{12} &= \int_{-h/2}^{h/2} \tau_{12} \left(1 + \frac{z}{R_2}\right) dz \\
N_{21} &= \int_{-h/2}^{h/2} \tau_{21} \left(1 + \frac{z}{R_1}\right) dz \\
Q_1 &= \int_{-h/2}^{h/2} \tau_{13} \left(1 + \frac{z}{R_2}\right) dz \\
Q_2 &= \int_{-h/2}^{h/2} \tau_{23} \left(1 + \frac{z}{R_1}\right) dz \\
M_1 &= \int_{-h/2}^{h/2} \sigma_1 \left(1 + \frac{z}{R_2}\right) zdz \\
M_2 &= \int_{-h/2}^{h/2} \sigma_2 \left(1 + \frac{z}{R_1}\right) zdz \\
M_{12} &= -\int_{-h/2}^{h/2} \tau_{12} \left(1 + \frac{z}{R_2}\right) zdz \\
M_{21} &= -\int_{-h/2}^{h/2} \tau_{21} \left(1 + \frac{z}{R_2}\right) zdz
\end{aligned}\right\} \quad ...(10.1)$$

We see that although $\tau_{12} = \tau_{21}$, $N_{12} \neq N_{21}$, $M_{12} \neq M_{21}$, because R_1 is not generally equal to R_2. However, for thin shells, h is small in comparison with R_1 and R_2 and the terms $\frac{z}{R_1}$ and $\frac{z}{R_2}$ can be neglected compared with 1. Thus $N_{12} = N_{21}$ and $M_{12} = M_{21}$.

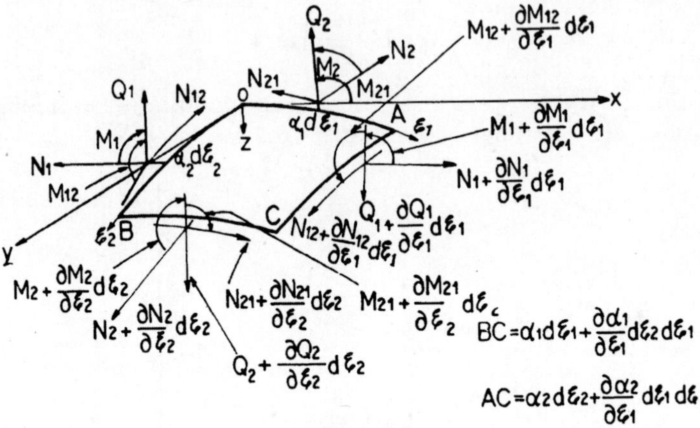

Fig. 10.2

Now consider the equilibrium of the x-component of the forces as shown in Fig. 10.3 (a).

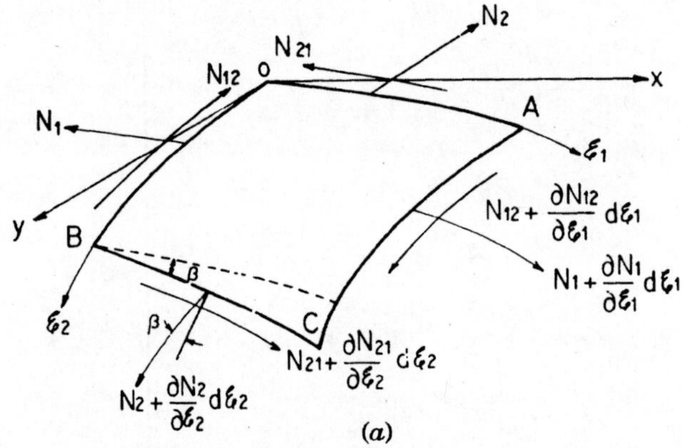

(a)

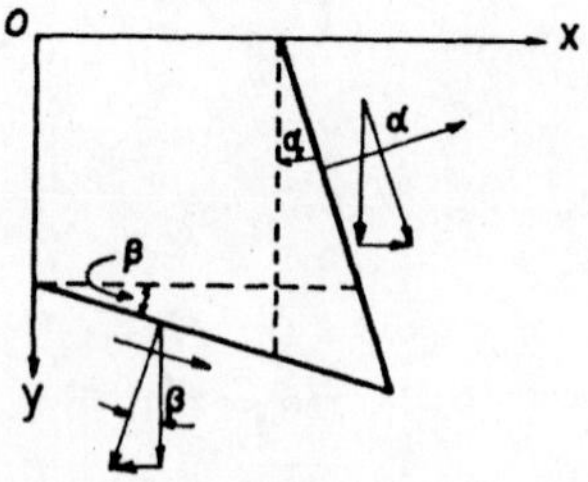

(b)

Fig. 10.3

The x-component of forces N_1 and $N_1 + \frac{\partial N_1}{\partial \xi_1} d\xi_1$ is

$$\left(N_1 + \frac{\partial N_1}{\partial \xi_1} d\xi_1\right)\left(\alpha_2\, d\xi_2 + \frac{\partial \alpha_2}{\partial \xi_1} d\xi_1 d\xi_2\right) - N_1\alpha_2 d\xi_2$$

$$= \left(\frac{\partial N_1}{\partial \xi_1}\alpha_2 + N_1 \frac{\partial \alpha_2}{\partial \xi_1}\right) d\xi_1 d\xi_2$$

$$= \frac{\partial(\alpha_2 N_1)}{\partial \xi_1} d\xi_1\, d\xi_2 \qquad \text{...(a)}$$

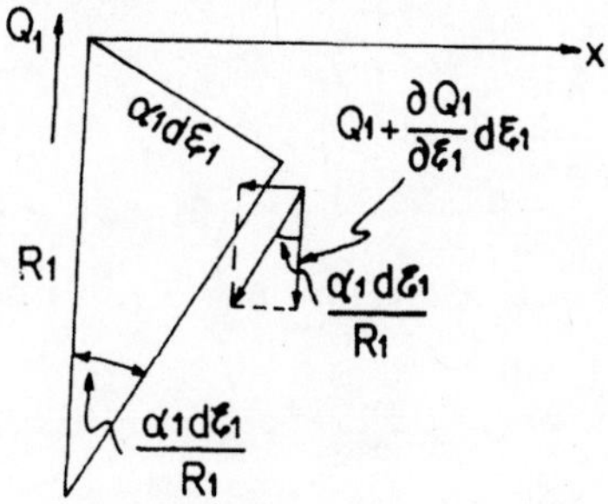

Fig. 10.4

Similarly, the x-component of the forces N_{21} and

$N_{21} + \frac{\partial N_{21}}{\partial \xi_2} d\xi_2$ is

$$\left(\frac{\partial N_{21}}{\partial \xi_2}\alpha_1 + N_{21}\frac{\partial \alpha_1}{\partial \xi_2}\right) d\xi_1 d\xi_2 = \frac{\partial(\alpha_1 N_{21})}{\partial \xi_2} d\xi_1 d\xi_2 \qquad \text{...(b)}$$

Since α_2 varies from point to point, the force $N_2 + \frac{\partial N_2}{\partial \xi_2} d\xi_2$, which is normal to the face BC, is inclined to the y-axis by a small angle β as shown in Fig. 10.3 (b), where the curvature of the sides is neglected.

$$\beta \simeq \frac{\left(\frac{\partial \alpha_2}{\partial \xi_1}\right) d\xi_1 d\xi_2}{\alpha_1 d\xi_1 + \frac{\partial \alpha_1}{\partial \xi_2} \cdot d\xi_1 d\xi_2}$$

Thus the x-component of this force is

$$-\left(N_2 + \frac{\partial N_2}{\partial \xi_2} d\xi_2\right)\left(\alpha_1 d\xi_1 + \frac{d\alpha_1}{\partial \xi_2} d\xi_1 d\xi_2\right)\sin\beta$$

Since β is small, therefore, $\sin\beta \simeq \beta$ and the above component after simplifying and neglecting small quantities becomes

$$-N_2 \frac{\partial \alpha_2}{\partial \xi_1} d\xi_1\, d\xi_2 \qquad ...(c)$$

Similarly the x-component of the force $N_{12} + \frac{\partial N_{12}}{\partial \xi_1} d\xi_1$ is

$$\left(N_{12} + \frac{\partial N_{12}}{\partial \xi_1} d\xi_1\right)\left(\alpha_2 d\xi_2 + \frac{\partial \alpha_2}{\partial \xi_1} d\xi_1\, \partial\xi_2\right) \sin\alpha$$

where $\alpha \simeq \dfrac{\left(\dfrac{\partial \alpha_1}{\partial \xi_2}\right) d\xi_1 d\xi_2}{\alpha_2 d\xi_2 + \dfrac{\partial \alpha_2}{d\xi_1} d\xi_1 d\xi_2}$

Since α is small, therefore $\sin\alpha \simeq \alpha$ and above component becomes

$$N_{12} \frac{\partial \alpha_1}{\partial \xi_1} \cdot d\xi_1 d\xi_2 \qquad ...(d)$$

For Fig. 10.4, we find that the shearing force $Q_1 + \frac{\partial Q_1}{\partial \xi_1} d\xi_1$ has an x-component equal to

$$-\left(Q_1 + \frac{\partial Q_1}{\partial \xi_1} \partial\xi_1\right)\left(\alpha_2 d\xi_2 + \frac{\partial \alpha_2}{\partial \xi_1} d\xi_1 d\xi_2\right)\frac{\alpha_1 d\xi_1}{R_1}$$

$$= -Q_1 \frac{\alpha_1 \alpha_2}{R} d\xi_1 d\xi_2 \qquad ...(e)$$

If the lateral pressure has a component p_1 in the x-direction, the total force is

$$p_1 \alpha_1 \alpha_2 d\xi_1 d\xi_2 \qquad ...(f)$$

Summing up forces given by Eqs. (*a*) to (*f*) and dividing by $d\xi_1 d\xi_2$, we find that the condition $\Sigma F_x = 0$ gives

$$\frac{\partial(\alpha_2 N_1)}{\partial \xi_1} + \frac{\partial(\alpha_1 N_{21})}{\partial \xi_2} + N_{12}\frac{\partial \alpha_1}{\partial \xi_2} - N_2 \frac{\partial \alpha_2}{\partial \xi_1} - Q_1 \frac{\alpha_1 \alpha_2}{R_1} + \alpha_1 \alpha_2 p_1 = 0 \qquad ...(10.2)$$

Similarly the condition $\Sigma F_y = 0$ gives

$$\frac{\partial(\alpha_1 N_2)}{\partial \xi_2} + \frac{\partial(\alpha_2 N_{12})}{\partial \xi_1} + N_{21}\frac{\partial \alpha_2}{\partial \xi_1} - N_1 \frac{\partial \alpha_1}{\partial \xi_2} - Q_2 \frac{\alpha_1 \alpha_2}{R_2} + \alpha_1 \alpha_2 p_2 = 0 \qquad ...(10.3)$$

and the condition $\Sigma F_z = 0$ leads to

$$\frac{\partial(\alpha_2 Q_1)}{\partial \xi_1} + \frac{\partial(\alpha_1 Q_2)}{\partial \xi_2} + N_1 \frac{\alpha_1 \alpha_2}{R_1} + N_2 \frac{\alpha_1 \alpha_2}{R_2} + \alpha_1 \alpha_2 p_3 = 0 \quad ...(10.4)$$

where p_1, p_2, p_3 are the x, y and z components of the lateral load per unit area.

To find the equilibrium equations for the moments, let us represent the moments by vectors according to the right hand rule as shown in Fig. 10.5. The components of the moments about the

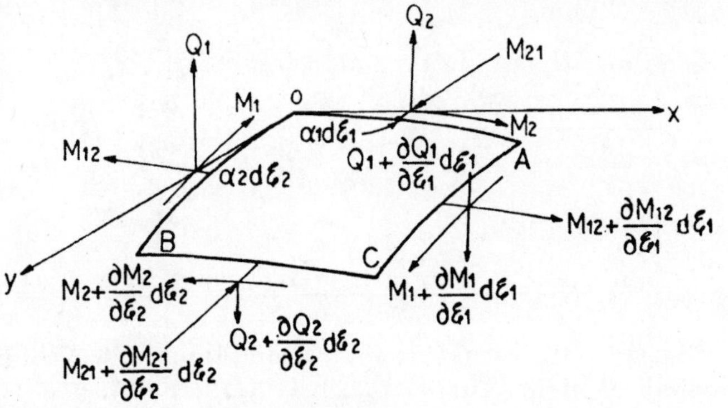

Fig. 10.5

x-axis can now be found by projecting these moment vectors on the x-axis. Thus, we have

$$-M_{12}\alpha_2 d\xi_2 + \left(M_{12} + \frac{\partial M_{12}}{\partial \xi_1} d\xi_1\right)\left(\alpha_2 + \frac{\partial \alpha_2}{\partial \xi_1} d\xi_1\right) d\xi_2$$

$$+ M_2 \alpha_1 d\xi_1 - \left(M_2 + \frac{\partial M_2}{\partial \xi_2} d\xi_2\right)\left(\alpha_1 + \frac{\partial \alpha_1}{\partial \xi_2} d\xi_2\right) d\xi_1$$

$$+ \left(M_1 + \frac{\partial M_1}{\partial \xi_1} d\xi_1\right)\left(\alpha_2 + \frac{\partial \alpha_2}{\partial \xi_1} d\xi_1\right) d\xi_2 \left\{\frac{\frac{\partial \alpha_1}{\partial \xi_2} d\xi_2 d\xi_1}{\left(\alpha_2 + \frac{\partial \alpha_2}{\partial \xi_1} d\xi_1\right)}\right.$$

$$+ \left(M_{21} + \frac{\partial M_{21}}{\partial \xi_2} d\xi_2\right)\left(\alpha_1 + \frac{\partial \alpha_1}{\partial \xi_2} d\xi_2\right) d\xi_1 \left\{\frac{\frac{\partial \alpha_2}{\partial \xi_1} d\xi_1 d\xi_2}{\left(\alpha_1 + \frac{\partial \alpha_1}{\partial \xi_2} d\xi_2\right) d\xi}\right.$$

$$= \left(\frac{\partial(\alpha_2 \, . \, M_{12})}{\partial \xi_1} - \frac{\partial(\alpha_1 M_2)}{\partial \xi} + M_1 \frac{\partial \alpha_1}{\partial \xi_2} + M_{21} \frac{\partial \alpha_2}{\partial \xi_1}\right) d\xi$$

The shearing force Q_2 has a moment about the x-axis magnitude

$$\left(Q_2 + \frac{\partial Q_2}{\partial \xi_2} d\xi_2\right) \alpha_1 d\xi_1 . \alpha_2 d\xi_2$$

$$= Q_2 \alpha_1 \alpha_2 d\xi_1 d\xi_2$$

Hence the condition $\Sigma M_x = 0$ gives

$$\frac{\partial(\alpha_2 M_{12})}{\partial \xi_1} - \frac{\partial(\alpha_1 M_2)}{\partial \xi_2} + M_1 \frac{\partial \alpha_1}{\partial \xi_2} + M_{21} \frac{\partial \alpha_2}{\partial \xi_1} + Q_2 \alpha_1 \alpha_2 = 0 \quad ...(10.5)$$

Similarly the condition $\Sigma M_y = 0$ gives

$$\frac{\partial(\alpha_1 M_{21})}{\partial \xi_2} - \frac{\partial(\alpha_2 M_1)}{\partial \xi_1} + M_2 \frac{\partial \alpha_2}{\partial \xi_1} + M_{12} \frac{\partial \alpha_1}{\partial \xi_2} + Q_1 \alpha_1 \alpha_2 = 0 \quad ...(10.6)$$

and $\Sigma M_z = 0$ leads to

$$N_{12} - N_{21} + \frac{M_{12}}{R_1} - \frac{M_{21}}{R_2} = 0 \quad ...(10.7)$$

The Eqs. (10.2) to (10.7) are the conditions for the equilibrium of the shell element. The condition (10.7) is identically satisfied, since substituting their values from Eqs. (10.1) we find that Eq. (10.7) is equivalent to

$$\int_{-h/2}^{h/2} \left(1 + \frac{Z}{R_1}\right)\left(1 + \frac{Z}{R_2}\right)(\tau_{12} - \tau_{21})\, dZ = 0$$

Since $\tau_{12} = \tau_{21}$

The above six equations of equilibrium involve ten unknowns, *i.e.*, the problem of the theory of shells is 'internally' statically indeterminate.

10.3 Membrane Theory of Shells

In many problems of thin shells, the deformations are such that the stresses due to bending can be neglected and the deformations are such that the stresses are mainly due to the middle surface forces N_1, N_2, N_{12}. In such cases, good approximations can be obtained by neglecting the bending stresses completely. By letting the moments and shearing forces go to zero, we find that there are three equilibrium equations with three unknown quantities N_1, N_2 and N_{12}. If the external forces acting on the shell are given, the problem becomes statically determinate, and these unknown forces can be determined without using the strain relations. The problem of stress analysis is therefore greatly simplified. The forces N_1, N_2,

N_{12} obtained in this manner are sometimes called membrane forces. The theory of thin shells based on the assumption of no bending stresses is called membrane theory. By letting shear forces to be zero in Eqs. (10.2) to Eq. (10.4) we get the equilibrium equations for membrane theory as :

$$\left.\begin{aligned}
&\frac{\partial(\alpha_2 N_1)}{\partial \xi_1} + \frac{\partial(\alpha_1 N_{21})}{\partial \xi_2} + N_{12}\frac{\partial \alpha_1}{\partial \xi_2} - N_2 \frac{\partial \alpha_2}{\partial \xi_1} + \alpha_1\alpha_2 p_1 = 0 \\
&\frac{\partial(\alpha_1 N_2)}{\partial \xi_2} + \frac{\partial(\alpha_2 N_{12})}{\partial \xi_1} + N_{21}\frac{\partial \alpha_2}{\partial \xi_1} - N_1 \frac{\partial \alpha_1}{\partial \xi_2} + \alpha_1\alpha_2 p_2 = 0 \\
&N_1 \frac{\alpha_1\alpha_2}{R_1} + N_2 \frac{\alpha_1\alpha_2}{R_2} + \alpha_1\alpha_2 p_3 = 0
\end{aligned}\right\} \quad ...(10.8)$$

10.4 Geometry of Shells of Revolution

Most pressure vessels are shells of revolution, *i.e.* spheres, cylinders, cones, etc. Such a surface is obtained by rotating a plane curve about an axis lying in the plane of the curve. This curve is called the meridian, and its plane is a meridian plane as shown in Fig. 10.6. At a point P the radius of curvature of the meridian is R_1

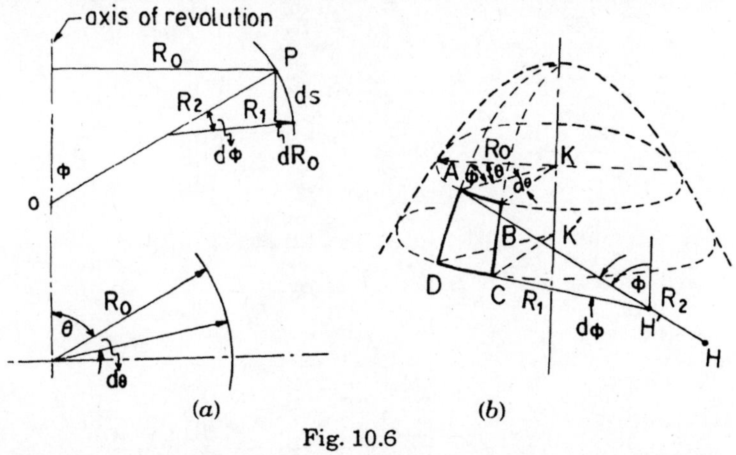

(a) (b)

Fig. 10.6

and the circle of radius R_0 in the plane view of Fig. 10.6 is called a parallel circle. Thus an element of shell ($R_0 d\theta \times R_1 d\phi$) is cut by two meridians and two parallel circles. OP is the normal to this element of shell and obviously R_1 is one radius of principal curvature. The second radius of principal curvature R_2 equals OP since by considering two adjacent points on the parallel circle at P, the normals from these points will intersect the axis of rotation at O. From Fig. 10.6, we obtain

$$R_0 = R_2 \sin \phi$$
$$dR_0 = ds \cos \phi$$
$$ds = R_1 d\phi$$

from which $$\frac{dR_0}{d\phi} = R_1 \cos \phi$$

10.5 Equations of Equilibrium for Shells of Revolution

Consider an element $R_0 d\theta \times R_1 d\phi$ of the shell of revolution. Fig. 10.7 shows the direct and shear stress resultants in the plane of the surface of the shell and also the transverse shear stress resultants. Fig. 10.8 shows the bending and twisting moments in

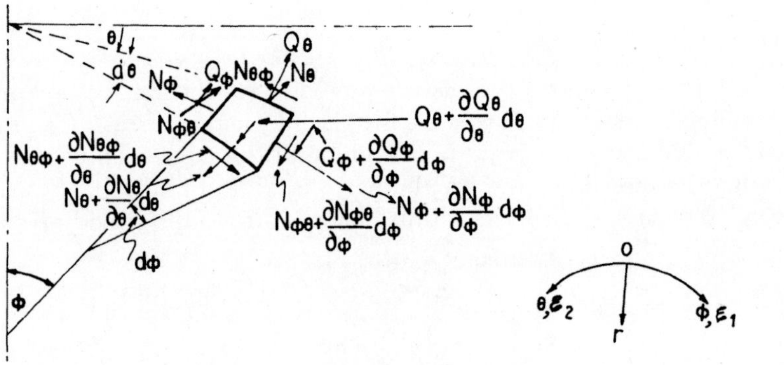

Fig. 10.7

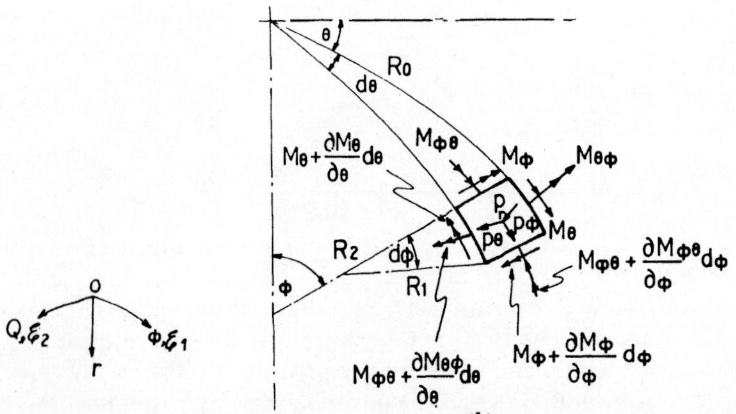

Fig. 10.8

the element. All stress resultants are shown in the positive directions, the right hand screw rule being used for the moments in Fig. 10.8. The external forces on the plate element are p_r, p_θ, p_ϕ per unit area of shell and are shown in Fig. 10.8.

A line element ds on the middle surface of the shell will be given by the formula

$$ds^2 = R_0{}^2 d\theta^2 + R_1{}^2 d\phi^2$$

Here $\alpha_1 = R_1,\ \alpha_2 = R_0 = R_2 \sin\phi$

$\xi_1 = \phi,\ \xi_2 = \theta$

Hence we get from Eqs. (10.2) to (10.7), the six equilibrium equations.

$$\frac{\partial(R_0 N_\phi)}{\partial\phi} + R_1 \frac{\partial N_{\theta\phi}}{\partial\theta} - N_\theta R_1 \cos\phi - Q_\phi R_0 + R_1 R_0 p_\phi = 0 \quad \text{...(10.9)}$$

$$\frac{\partial}{\partial\phi}(R_0 N_{\phi\theta}) + R_1 \frac{\partial N_\theta}{\partial\theta} + R_1 N_{\theta\phi} \cos\phi - R_1 Q_\theta \sin\phi + R_1 R_0 p_\theta = 0 \quad \text{...(10.10)}$$

$$R_1 N_\theta \sin\phi + R_0 N_\phi + R_1 \frac{\partial Q_\theta}{\partial\theta} + \frac{\partial}{\partial\phi}(R_0 Q_\phi) + R_1 R_0 p_r = 0 \quad \text{...(10.11)}$$

$$\frac{\partial}{\partial\phi}(R_0 M_\phi) + R_1 \frac{\partial M_{\theta\phi}}{\partial\theta} - R_1 M_\theta \cos\phi - R_1 R_\theta Q_\phi = 0 \quad \text{...(10.12)}$$

$$\frac{\partial}{\partial\phi}(R_0 M_{\phi\theta}) + R_1 \frac{\partial M_\theta}{\partial\theta} + R_1 M_{\theta\phi} \cos\phi - R_1 R_0 Q_\theta = 0 \quad \text{...(10.13)}$$

$$R_0 R_1 N_{\theta\phi} - R_0 R_1 N_{\phi\theta} - R_1 M_{\theta\phi} \sin\phi + R_0 M_{\theta\phi} = 0 \quad \text{...(10.14)}$$

If the loading is rotationally symmetric, *i.e.* if $p_\theta = 0$, then $Q_\theta = N_{\theta\phi} = N_{\phi\theta} = M_{\theta\phi} = M_{\phi\theta} = 0$ and N_θ and M_θ are independent of θ. The equations of equilibrium then take the form

$$\frac{d}{d\phi}(R_0 N_\phi) - R_1 N_\theta \cos\phi - R_0 Q_\phi + R_1 R_0 p_\phi = 0 \quad \text{...(10.15)}$$

$$\frac{d}{d\phi}(R_0 Q_\phi) + R_1 N_\theta \sin\phi + R_0 N_\phi + R_1 R_0 p_r = 0 \quad \text{...(10.16)}$$

$$\frac{d}{d\phi}(R_0 M_\phi) - R_1 M_\theta \cos\phi - R_1 R_0 Q_\phi = 0 \quad \text{...(10.17)}$$

Eliminating N_θ from Eqs. (10.15) and (10.16) we obtain

$$\frac{d}{d\phi}(R_0 N_\phi) \sin\phi - R_0 Q_\phi \sin\phi + R_1 R_0 p_\phi \sin\phi + R_0 N_\phi \cos\phi$$

$$+\frac{d}{d\phi}(R_0 Q_\phi)\cos\phi + R_0 R_1 p_r \cos\phi = 0$$

$$\frac{d}{d\phi}(R_0 N_\phi \sin\phi) + \frac{d}{d\phi}(R_0 Q_\phi \cos\phi) + R_0 R_1 (p_\phi \sin\phi + p_r \cos\phi) = 0.$$

Hence on integration from 0 to 2π, we get

$$2\pi[R_0 N_\phi \sin\phi + R_0 Q_\phi \cos\phi] = -2\pi \int R_0 R_1 (p_r \cos\phi + p_\phi \sin\varphi)\, d\phi$$

$= -P$, the total load on the shell above the parallel circle ...(10.18)

This is a very useful but not independent equation of equilibrium and follows from Fig. 10.9.

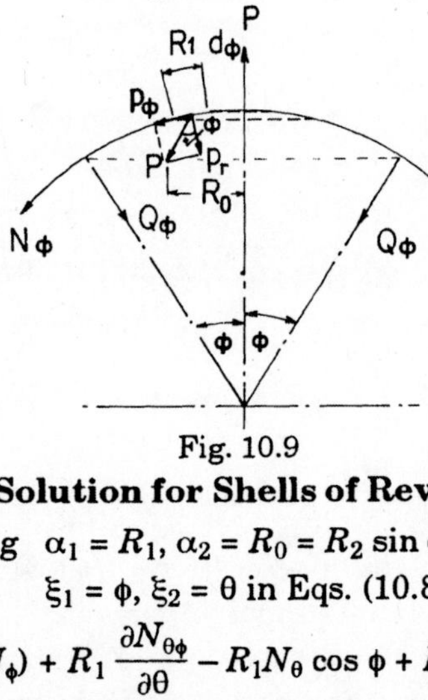

Fig. 10.9

10.6 Membrane Solution for Shells of Revolution

Substituting $\alpha_1 = R_1$, $\alpha_2 = R_0 = R_2 \sin\phi$

and $\xi_1 = \phi$, $\xi_2 = \theta$ in Eqs. (10.8), we get

$$\frac{\partial}{\partial\phi}(R_\theta N_\phi) + R_1 \frac{\partial N_{\theta\phi}}{\partial\theta} - R_1 N_\theta \cos\phi + R_0 R_1 p_\phi = 0 \quad ...(10.19)$$

$$\frac{\partial}{\partial\phi}(R_0 N_{\phi\theta}) + R_1 \frac{\partial N_\theta}{\partial\theta} + R_1 N_{\theta\phi} \cos\phi + R_0 R_1 p_\theta = 0 \quad ...(10.20)$$

$$R_1 N_\theta \sin\phi + R_0 N_\phi + R_0 R_1 p_r = 0 \quad ...(10.21)$$

Now $R_0 = R_2 \sin\phi$. Therefore, Eq. (10.21) becomes

$$R_1 N_\theta \sin\phi + R_2 N_\phi \sin\phi + R_1 R_2 \sin\phi\, p_r = 0$$

or

$$\frac{N_\theta}{R_2} + \frac{N_\phi}{R_1} = -p_r \quad ...(10.22)$$

For rotationally symmetric loading, we get from Eq. (10.18) by setting $Q_\phi = 0$.

$$2\pi\, R_0 N_\phi \sin\phi = -P$$

or $$2\pi\, R_0 N_\phi \sin\phi + P = 0 \qquad ...(10.23)$$

where P is the sum of all the external forces applied to the shell above the parallel circle.

Example 10.1 *Spherical dome of constant thickness under its own weight.*

Solution. Consider a spherical dome of constant thickness under its own weight (Fig. 10.10). Let q be the constant weight of the dome per unit area and a the radius of the sphere so that $R_0 = a \sin\phi$. The total load P on that part of the spherical dome subtended by angle ϕ is given by

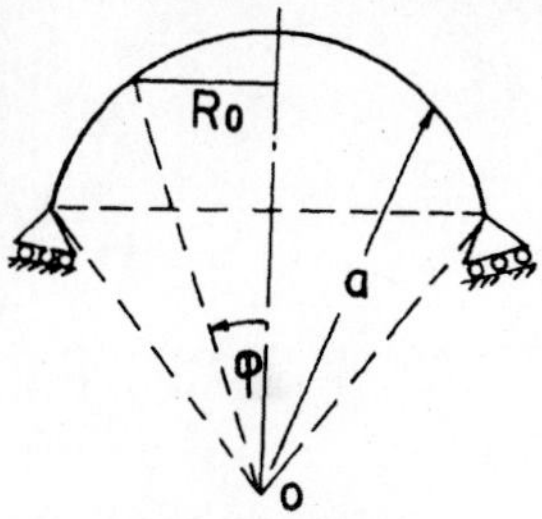

Fig. 10.10

$$P = 2\pi \int_0^\phi aq \sin\phi \,.\, a d\phi$$

$$= 2\pi a^2 q\,(1 - \cos\phi)$$

For spherical dome $R_1 = R_2 = a$.

$\therefore$ From Eq. (10.23), we get

$$N_\phi = \frac{-P}{2\pi R_0 \sin\phi} = \frac{-P}{2\pi a \sin^2\phi}$$

$$= -\frac{aq(1-\cos\phi)}{\sin^2\phi} = \frac{-aq}{1+\cos\phi} \qquad ...(1)$$

Now substituting N_ϕ in Eq. (10.22), we get

$$\frac{N_\theta}{R_2} - \frac{qa}{(1+\cos\phi)R_1} = -p_r$$

$$\therefore \qquad N_\theta = \left(\frac{qa}{1+\cos\phi}\right)\frac{R_2}{R_1} - R_2 p_r$$

$$= \frac{qa}{1+\cos\phi} - a p_r \qquad [\because\ p_r = q\cos\phi]$$

$$= -qa\left[\frac{-1}{1+\cos\phi} + \cos\phi\right] \qquad ...(2)$$

The negative sign in Eqs. (1) and (2) indicate that N_θ and N_ϕ are compressive forces. N_ϕ remains compressive throughout while N_θ becomes tensile when ϕ is greater than 51° 50′.

Example 10.2 *Shell in the form of an Ellipsoid of Revolution.*

Solution. Let a and b be the lengths of the semi-major and semi-minor axes of the ellipse respectively as shown in Fig. 10.11. Generally one-half of ellipsoid is used. The equation of the ellipse is

$$\frac{x^2}{a^2} + \frac{y^2}{b^2} = 1.$$

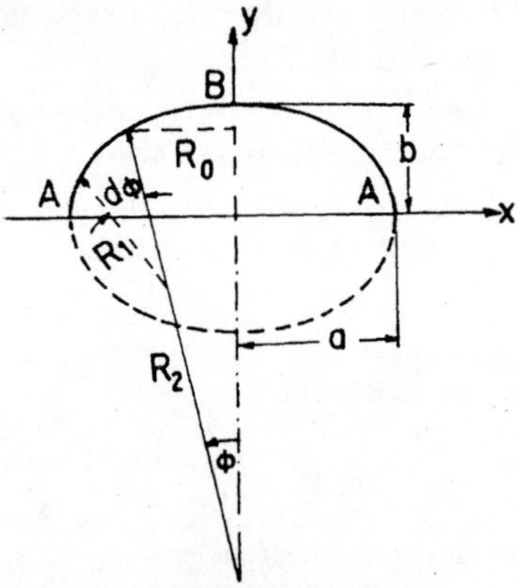

Fig. 10.11

The principal radii of curvature are

$$R_1 = \frac{a^2b^2}{(a^2\sin^2\phi + b^2\cos^2\phi)^{1/2}}$$

$$= \frac{(a^4y^2 + b^4x^2)^{3/2}}{a^4b^4}$$

$$R_2 = \frac{a^2}{(a^2\sin^2\phi + b^2\cos^2\phi)^{1/2}}$$

$$= \frac{(a^4y^2 + b^4x^2)^{1/2}}{b^2}$$

If p is the uniform pressure acting on the shell, then the resultant force P corresponding to an angle ϕ is

$$P = -p\pi R_0^2 = -\pi p R_2^2 \sin^2\phi$$

Since $p_r = -p$

$$\therefore \quad N_\phi = \frac{pR_0}{2\sin\phi} = \frac{pR_2}{2}$$

$$N_\theta = R_2 p - \frac{R_2}{R_1} N_\phi$$

$$= p\left(R_2 - \frac{R_2^2}{2R_1}\right)$$

At the top of the shell, we have

$$R_1 = R_2 = \frac{a^2}{b}$$

$$\therefore \qquad N_\phi = N_\theta = \frac{pa^2}{2b}$$

At the equator A-A, we have

$$R_1 = \frac{b^2}{a} \text{ and } R_2 = a.$$

$$N_\phi = \frac{pa}{2}, N_\theta = pa\left(1 - \frac{a^2}{2b^2}\right).$$

Example 10.3 *Spherical tank filled with liquid.*

Solution. Consider a spherical tank supported along a parallel circle A-A as shown in Fig. 10.12 and filled with liquid of specific weight γ.

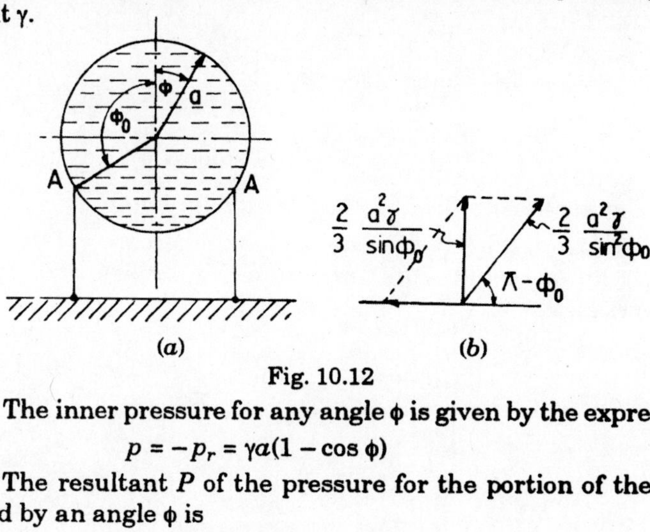

(a) (b)

Fig. 10.12

The inner pressure for any angle ϕ is given by the expression.

$$p = -p_r = \gamma a(1 - \cos\phi)$$

The resultant P of the pressure for the portion of the shell defined by an angle ϕ is

$$P = -2\pi a^2 \int_0^\phi \gamma a\,(1 - \cos\phi)\,.\,\sin\phi\cos\phi\,d\phi$$

$$= -2\pi a^2\gamma\left[\frac{1}{6} - \frac{1}{2}\cos^2\phi\left(1 - \frac{2}{3}\cos\phi\right)\right]$$

Substituting in Eq. (10.23), we find

$$N_\phi = \frac{\gamma a^2}{6\sin^2\phi}[1 - \cos^2\phi(3 - 2\cos\phi)]$$

$$= \frac{\gamma a^2}{6}\left(1 - \frac{2\cos^2\phi}{1+\cos\phi}\right) \text{ for } \phi < \phi_0$$

and Eq. (10.22) gives

$$N_\theta = \frac{\gamma a^2}{6}\left(5 - 6\cos\phi + \frac{2\cos^2\phi}{1+\cos\phi}\right) \text{ for } \phi < \phi_0$$

For calculating the resultant P for $\phi > \phi_0$, we must take into account the sum of the vertical reactions along the ring A-A also in addition to the internal pressure.

Now total weight of the liquid

$$= \frac{4}{3}\pi a^3\gamma$$

$$\therefore \quad P = -\frac{4}{3}\pi a^3\gamma - 2\pi a^3\gamma\left[\frac{1}{6} - \frac{1}{2}\cos^2\phi\left(1 - \frac{2}{3}\cos\phi\right)\right]$$

Hence $$N_\varphi = \frac{\gamma a^2}{6}\left(5 + \frac{2\cos^2\varphi}{1-\cos\varphi}\right)$$

$$N_\theta = \frac{\gamma a^2}{6}\left(1 - 6\cos\phi - \frac{2\cos^2\phi}{1-\cos\phi}\right).$$

Example 10.4 *Conical shell filled with liquid.*

Solution. If a force P is applied in the direction of the axis of cone as shown in Fig. 10.13 (a), then the stress distribution is symmetrical. We obtain

$$N_\phi = \frac{-P}{2\pi R_0 \cos\alpha} \quad \text{...(1)}$$

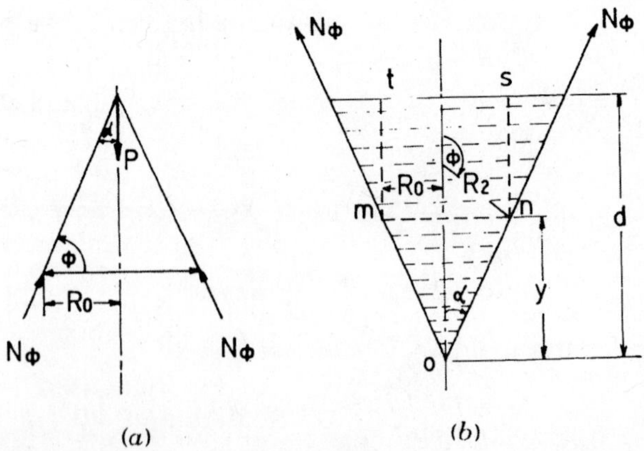

Fig. 10.13

Then Eq. (10.22) gives

$$N_\theta = 0.$$

Since the curvature of the meridian in the case of a cone is zero, $R_1 = \infty$, therefore,

$$\left.\begin{aligned} N_\phi &= \frac{-P}{2\pi R_\theta \sin\phi} \\ N_\theta &= -p_r R_2 = -p_r \cdot \frac{R_\theta}{\sin\phi} \end{aligned}\right\} \qquad \text{...(2)}$$

Let the conical tank be filled with a liquid of specific weight γ as shown in Fig. 10.13 (*b*). Measuring the distances y from the bottom of the tank and denoting by d the total depth of the liquid in the tank, the pressure at any parallel circle mn is

$$p = -p_r = \gamma(d - y)$$

Also for such a tank $\phi = \frac{\pi}{2} + \alpha$ and $R_0 = y \tan\alpha$, where α is the semi-cone angle of the tank. Substituting in the second of Eqs. (2), we get

$$N_\theta = \frac{\gamma(d-y)\, y \tan\alpha}{\cos\alpha} \qquad \text{...(3)}$$

$$(N_\theta)_{max}\Bigg|_{y = d/2} = \frac{\gamma d^2 \tan\alpha}{4\cos\alpha} \qquad \text{...(4)}$$

Total load P = Weight of liquid in the conical part mno + Weight of the liquid in the cylindrical part $mnst$.

$$= -\pi\gamma y^2\left(d - y + \frac{1}{3}y\right)\tan^2\alpha$$

$$\therefore \qquad N_\phi = \frac{\gamma y\left(d - \frac{2}{3}y\right)\tan\alpha}{2\cos\alpha} \qquad \text{...(5)}$$

$$(N_\phi)_{max}\Bigg|_{y = 3/4d} = \frac{3}{16}\,\frac{d^2\gamma\tan\alpha}{\cos\alpha} \qquad \text{...(6)}$$

Example 10.5 *Shell in the form of a Torus.*

Solution. A toroidal shell or doughtnut is obtained by rotation of a circle of radius a about a vertical axis as shown in Fig. 10.14. The forces N_ϕ are obtained by considering the equilibrium of the ring shaped portion of the shell represented by line AB. Since the forces N_ϕ along the parallel circle BB are horizontal, we need consider only

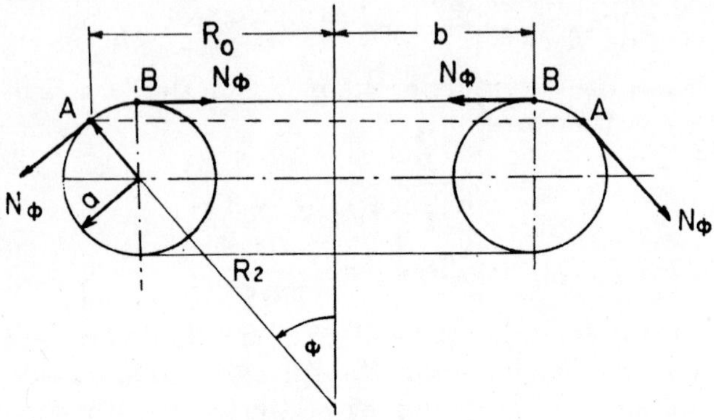

Fig. 10.14

the forces N_ϕ along the circle AA and the external forces acting on the ring when considering equilibrium in the vertical direction. Assuming that the shell is subjected to the action of uniform internal pressure p, the equation of equilibrium becomes

$$2\pi R_0 N_\phi \sin\phi = \pi p(R_0^2 - b^2)$$

$$\therefore \quad N_\phi = \frac{p(R_0^2 - b^2)}{2R_0 \sin\phi} = \frac{pa(R_0 + b)}{2R_0} \qquad \text{...(1)}$$

Substituting in Eq. (10.22), we get

$$N_\theta = \frac{pR_2\,(R_0 - b)}{2R_0} = \frac{pa}{2}.$$

10.7 Equations of Equilibrium for Cylindrical Shells

Let us consider the case of a circular cylindrical shell. We choose the co-ordinate axes so that x is in direction of the generatrix, θ is in the direction of the tangent to the circumference, and z is in the direction of the normal to the middle surface of the shell. Thus $\xi_1 = x$, $\xi_2 = \theta$. Let a be the radius of the circular cylinder. A line element ds on the middle surface is given in this case by the formula

$$ds^2 = dx^2 + a^2 d\theta^2$$

Hence $\quad \alpha_1 = 1, \alpha_2 = a.$

If we denote the tangent to the co-ordinate θ by y, we find that the radii of curvature in the xz and yz planes arc the principal radii of curvature.

Thus $R_1 = \infty$, $R_2 = a$.

The force resultants and moments have been shown in Figs. 10.15 and 10.16 respectively.

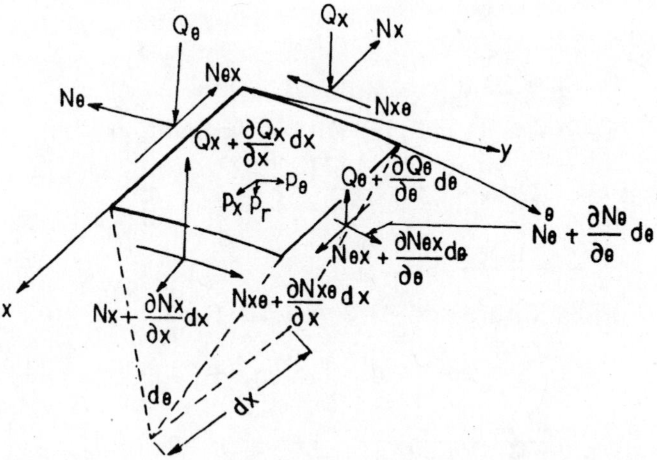

Fig. 10.15

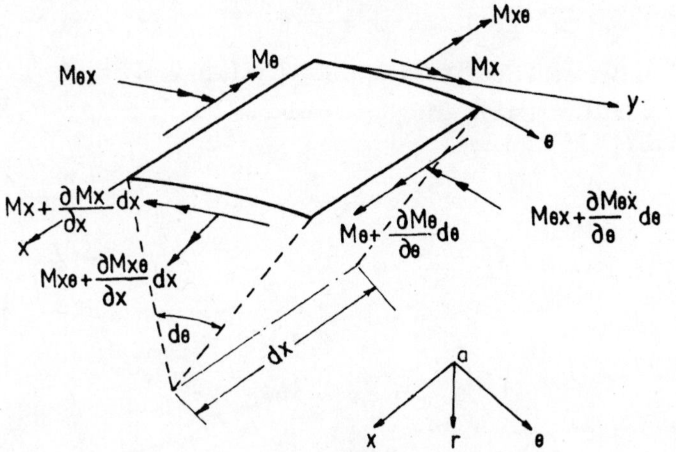

(*b*)

Fig. 10.16

Substituting in Eq. (10.2) to (10.7), we get

$$a\frac{\partial N_x}{\partial x} + \frac{\partial N_{\theta x}}{\partial \theta} + ap_x = 0 \qquad \text{...(10.24)}$$

$$a\frac{\partial N_{x\theta}}{\partial x} + \frac{\partial N_\theta}{\partial \theta} - Q_\theta + ap_\theta = 0 \qquad \text{...(10.25)}$$

$$a\frac{\partial Q_x}{\partial x} + \frac{\partial Q_\theta}{\partial \theta} + N_\theta + ap_r = 0 \qquad \text{...(10.26)}$$

$$a\frac{\partial M_{x\theta}}{\partial x} - \frac{\partial M_\theta}{\partial \theta} + aQ_\theta = 0 \qquad \text{...(10.27)}$$

$$\frac{\partial M_{\theta x}}{\partial \theta} - a\frac{\partial M_x}{\partial x} + Q_x a = 0 \qquad ...(10.28)$$

$$a(N_{x\theta} - N_{\theta x}) + aM_{x\theta} - M_{\theta x} = 0 \qquad ...(10.29)$$

From Eqs. (10.27) and (10.28), we find that

$$\left.\begin{aligned} Q_\theta &= \frac{1}{a}\frac{\partial M_\theta}{\partial \theta} - \frac{\partial M_{x\theta}}{\partial x} \\ Q_x &= \frac{\partial M_x}{\partial x} - \frac{1}{a}\frac{\partial M_{x\theta}}{\partial \theta} \end{aligned}\right\} \qquad ...(10.30)$$

Substituting Eqs. (10.30) in Eqs. (10.24) to (10.26), we get

$$\left.\begin{aligned} a\frac{\partial N_x}{\partial x} + \frac{\partial N_{\theta x}}{\partial \theta} + ap_x &= 0 \\ a\frac{\partial N_{x\theta}}{\partial x} + \frac{\partial N_\theta}{\partial \theta} - \frac{1}{a}\frac{\partial M_\theta}{\partial \theta} + \frac{\partial M_{x\theta}}{\partial x} + ap_\theta &= 0 \\ a\frac{\partial^2 M_x}{\partial x^2} - 2\frac{\partial^2 M_{x\theta}}{\partial x \partial \theta} + \frac{1}{a}\frac{\partial^2 M_\theta}{\partial \theta^2} + N_\theta + ap_r &= 0 \end{aligned}\right\} \qquad ...(10.31)$$

10.8 Membrane Theory of Circular Cylindrical Shells

By letting the shearing forces and moments go to zero, Eqs. (10.24) to (10.29) become :

$$\left.\begin{aligned} a\frac{\partial N_x}{\partial x} + \frac{\partial N_{\theta x}}{\partial \theta} + ap_x &= 0 \\ a\frac{\partial N_{x\theta}}{\partial x} + \frac{\partial N_\theta}{\partial \theta} + ap_\theta &= 0 \\ N_\theta + ap_r &= 0 \end{aligned}\right\} \qquad ...(10.32)$$

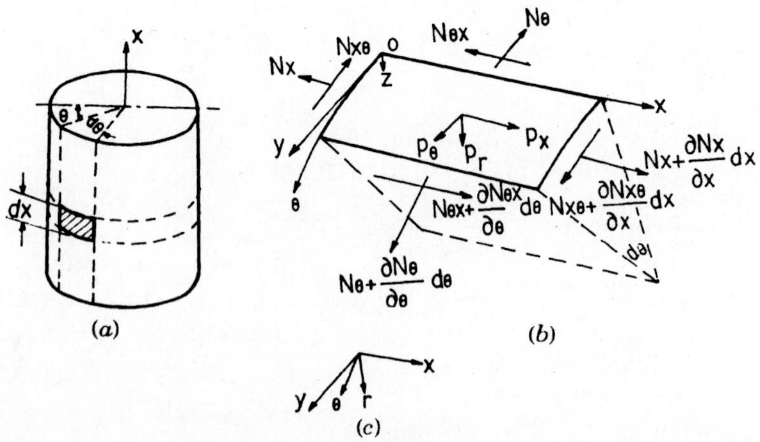

Fig. 10.17

The forces on the circular cylindrical shell have been shown in Fig. 10.17.

Example 10.6 *Horizontal circular cylindrical shell filled with liquid and supported at the ends.*

Solution. Consider the circular cylindrical shell filled with liquid as shown in Fig. 10.18. Measuring the angle θ as shown and

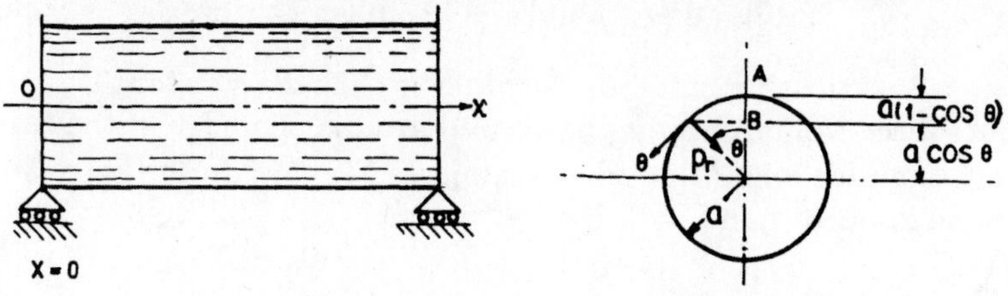

Fig. 10.18

denoting by p_0 the pressure at the axis of the tube, the pressure at any point is $p_0 - \gamma a \cos\theta$. We thus get

$$p_x = 0,\ p_\theta = 0,\ p_r = -p_0 + \gamma a \cos\theta \qquad \text{...(1)}$$

Substituting in Eqs. (10.32), we get

$$N_\theta + a(-p_0 + \gamma a \cos\theta) = 0$$

$$\therefore \quad N_\theta = ap_0 - \gamma a^2 \cos\theta \qquad \text{...(2)}$$

$$\frac{\partial N_{x\theta}}{\partial x} = -\frac{1}{a}\frac{\partial N_\theta}{\partial \theta}$$

Integrating, we get

$$N_{x\theta} = -\frac{1}{a}\int \frac{\partial}{\partial\theta}(p_0 a - \gamma a^2 \cos\theta)\, dx + C_1(\theta)$$

$$= -\int \gamma a \sin\theta\, dx + C_1(\theta)$$

$$= -\gamma a x \sin\theta + C_1(\theta) \qquad \text{...(3)}$$

Now
$$\frac{\partial N_x}{\partial x} = -\frac{1}{a}\frac{\partial N_{\theta x}}{\partial \theta}$$

$$= -\frac{1}{a}\frac{\partial}{\partial\theta}[-\gamma a x \sin\theta + C_1(\theta)]$$

$$= -\frac{1}{a}\left[-\gamma a x \cos\theta + \frac{\partial C_1(\theta)}{\partial\theta}\right]$$

Integrating, we get

$$N_x = \int \gamma x \cos\theta\, dx - \frac{1}{a}\int \frac{dC_1(\theta)}{d\theta}\, dx + C_2(\theta)$$

$$= \frac{\gamma x^2}{2} \cos\theta - \frac{x}{a}\frac{dC_1(\theta)}{d\theta} + C_2(\theta) \qquad ...(4)$$

The functions $C_1(\theta)$ and $C_2(\theta)$ are to be determined from the conditions at the edges.

If $(N_x)_{\substack{x=0\\x=L}} = 0$, then

$$C_2(\theta) = 0 \text{ and } C_1(\theta) = \frac{a\gamma L}{2}\sin\theta + C$$

Substituting in Eq. (3) we observe that the constant C represents force $N_{x\theta}$ uniformly distributed around the edge of the tube, as in the case when the tube is subjected to torsion. If there is no torque applied then $C = 0$. Hence, we get

$$N_\theta = ap_0 - \gamma a^2 \cos\theta$$

$$N_{x\theta} = -\gamma a x \sin\theta + \frac{\gamma a L}{2}\sin\theta$$

$$= \gamma a\left(\frac{L}{2} - x\right)\sin\theta$$

$$N_x = -\frac{\gamma x}{2}(L - x)\cos\theta.$$

10.9 Determination of Strain Components

Let ξ_1, ξ_2 and ξ_3 be the curvilinear co-ordinates. Consider two points $P(\xi_1, \xi_2, \xi_3)$ and $Q(\xi_1 + d\xi_1, \xi_2 + d\xi_2, \xi_3 + d\xi_3)$ at a small distance ds apart. Let the direction cosines of PQ, referred to the curvilinear co-ordinates ξ_1, ξ_2, ξ_3 be l, m, n respectively, shown as x_1, x_2, x_3 in Fig. 10.19. If A_1, A_2, A_3 are the factors at P which transform the curvilinear coordinates into linear distances, then we find that the projections of ds on x_1, x_2, x_3 are :

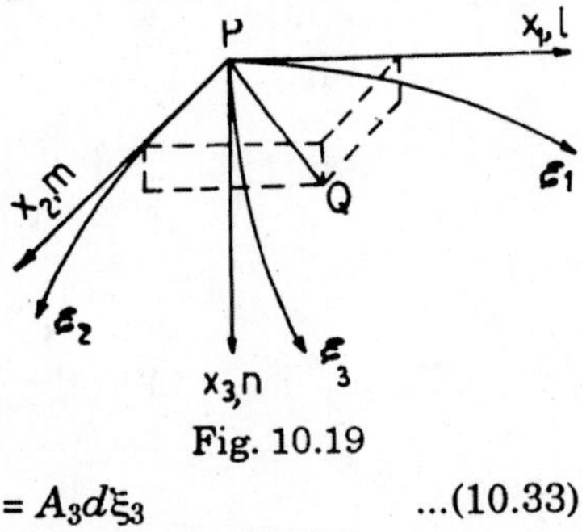

Fig. 10.19

$$l ds = A_1 d\xi_1,\ m ds = A_2 d\xi_2,\ n ds = A_3 d\xi_3 \qquad ...(10.33)$$

Squaring and adding, we get

$$ds^2 = A_1^2\, d\xi_1^2 + A_2^2\, d\xi_2^2 + A_3^2\, d\xi_3^2$$

After deformation, let the particles which were at P, Q in the unstrained state be displaced to P', Q'. Let u_1, u_2, u_3 be the projections of the displacement PP' on x_1, x_2, x_3 and let $\xi_1' = \xi_1 + \mu_1$, $\xi_2' = \xi_2 + \mu_1$, $\xi_3' = \xi_3 + \mu_3$, be the curvilinear coordinates of P'. Assuming small displacements, then

$$u_1 = A_1\mu_1,\ u_2 = A_2\mu_2,\ u_3 = A_3\mu_3 \qquad \text{...(10.34)}$$

The curvilinear co-ordinates of $Q'(\xi_1'', \xi_2'', \xi_3'')$ can be expressed to the first orders as,

$$\xi_1'' = \xi_1' + d\xi_1 + d\mu_1$$
$$= \xi_1 + \mu_1 + d\xi_1 + \frac{\partial \mu_1}{\partial \xi_1} d\xi_1 + \frac{\partial \mu_1}{\partial \xi_2} d\xi_2 + \frac{\partial \mu_1}{\partial \xi_3} d\xi_3$$
$$\xi_2'' = \xi_2' + d\xi_2 + d\mu_2$$
$$= \xi_2 + \mu_2 + d\xi_2 + \frac{\partial \mu_2}{\partial \xi_1} d\xi_1 + \frac{\partial \mu_2}{\partial \xi_2} d\xi_2 + \frac{\partial \mu_2}{\partial \xi_3} d\xi_3$$
$$\xi_3'' = \xi_3' + d\xi_3 + d\mu_3$$
$$= \xi_3 + \mu_3 + d\xi_3 + \frac{\partial \mu_3}{\partial \xi_1} d\xi_1 + \frac{\partial \mu_3}{\partial \xi_2} d\xi_2 + \frac{\partial \mu_3}{\partial \xi_3} d\xi_3$$

Similarly the quantities A_1', A_2', A_3' at P' become :

$$A_1' = A_1 + \frac{\partial A_1}{\partial \xi_1}\mu_1 + \frac{\partial A_1}{\partial \xi_2}\mu_2 + \frac{\partial A_1}{\partial \xi_3}\mu_3$$

$$A_2' = A_2 + \frac{\partial A_2}{\partial \xi_1}\mu_1 + \frac{\partial A_2}{\partial \xi_2}\mu_2 + \frac{\partial A_2}{\partial \xi_3}\mu_3.$$

Let ds' be the length of $P'Q'$ and l', m', n' be the direction cosines of $P'Q'$ referred to the curvilinear coordinates ξ_1, ξ_2, ξ_3 which pass through P'. The projections of $P'Q'$ on these tangents can be expressed with sufficient accuracy by

$$l'd's' = A_1'(\xi_1'' - \xi_1')$$
$$= \left(A_1 + \frac{\partial A_1}{\partial \xi_1}\mu_1 + \frac{\partial A_1}{\partial \xi_2}\mu_2 + \frac{\partial A_1}{\partial \xi_3}\mu_3\right)$$
$$\times \left(d\xi_1 + \frac{\partial \mu_1}{\partial \xi_1} d\xi_1 + \frac{\partial \mu_1}{\partial \xi_2} d\xi_2 + \frac{\partial \mu_1}{\partial \xi_3} d\xi_3\right)$$

Neglecting higher-order terms, we get

$$l'ds' = A_1 d\xi_1 + \frac{\partial A_1}{\partial \xi_1}\mu_1 d\xi_1 + \frac{\partial A_1}{\partial \xi_2}\mu_2 d\xi_1$$
$$+ \frac{\partial A_1}{\partial \xi_3}\mu_3 d\xi_1 + \frac{\partial \mu_1}{\partial \xi_1} A_1 d\xi_1 + \frac{\partial \mu_1}{\partial \xi_2} A_1 d\xi_2 + \frac{\partial \mu_1}{\partial \xi_3} A_1 d\xi_3$$

Substituting Eqs. (10.33) and (10.34) into the above equation and noting that

$$\frac{\partial A_1}{\partial \xi_1}\mu_1 d\xi_1 + \frac{\partial \mu_1}{\partial \xi_1} A_1 d\xi_1 = \frac{\partial (A_1 \mu_1)}{\partial \xi_1} d\xi_1$$

we find that $l'ds' = lds\left(1 + \frac{1}{A_1}\frac{\partial u_1}{\partial \xi_1} + \frac{u_2}{A_1A_2}\frac{\partial A_1}{\partial \xi_2} + \frac{u_3}{A_1A_3}\frac{\partial A_1}{\partial \xi_3}\right)$

$$+ mds\frac{A_1}{A_2}\frac{\partial}{\partial \xi_2}\left(\frac{u_1}{A_1}\right) + nds\frac{A_1}{A_3}\frac{\partial}{\partial \xi_3}\left(\frac{u_1}{A_1}\right)$$

Similarly, we find that

$$m'ds' = lds\frac{A_2}{A_1}\frac{\partial}{\partial \xi_1}\left(\frac{u_2}{A_2}\right) + mds\left(1 + \frac{1}{A_2}\frac{\partial u_2}{\partial \xi_2}\right.$$

$$\left. + \frac{u_1}{A_1A_2}\frac{\partial A_2}{\partial \xi_1} + \frac{u_3}{A_2A_3}\frac{\partial A_2}{\partial \xi_3}\right) + nds\frac{A_2}{A_3}\frac{\partial}{\partial \xi_2}\left(\frac{u_2}{A_2}\right)$$

$$n'ds' = lds\frac{A_3}{A_1}\frac{\partial}{\partial \xi_1}\left(\frac{u_3}{A_3}\right) + mds\frac{A_3}{A_2}\frac{\partial}{\partial \xi_2}\left(\frac{u_3}{A_3}\right)$$

$$+ nds\left(1 + \frac{1}{A_3}\frac{\partial u_3}{\partial \xi_3} + \frac{u_2}{A_2A_3}\frac{\partial A_3}{\partial \xi_2} + \frac{u_1}{A_1A_3}\frac{\partial A_3}{\partial \xi_1}\right)$$

If ε is the longitudinal strain of a linear element along PQ, then we have

$$\varepsilon = \frac{P'Q' - PQ}{PQ}$$

or
$$P'Q' = PQ(1 + \varepsilon) = ds(1 + \varepsilon)$$

But $(P'Q')^2 = (l'ds')^2 + (m'ds')^2 + (n'ds')^2$

$$\therefore \quad (1 + \varepsilon)^2 = \left[l\left(1 + \frac{1}{A_1}\frac{\partial u_1}{\partial \xi_1} + \frac{u_2}{A_1A_2}\frac{\partial A_1}{\partial \xi_2} + \frac{u_3}{A_1A_3}\frac{\partial A_1}{\partial \xi_3}\right)\right.$$

$$\left. + m\frac{A_1}{A_2}\frac{\partial}{\partial \xi_2}\left(\frac{u_1}{A_1}\right) + n\frac{A_1}{A_3}\frac{\partial}{\partial \xi_3}\left(\frac{u_1}{A_1}\right)\right]^2$$

+ Higher-order terms, which are neglected.

This can be written as :

$$\varepsilon = \varepsilon_1 l^2 + \varepsilon_2 m^2 + \varepsilon_2 n^2 + \varepsilon_3 n^2 + \gamma_{12} lm + \gamma_{13} ln + \gamma_{23} mn \qquad ...(10.35)$$

where
$$\varepsilon_1 = \frac{1}{A_1}\frac{\partial u_1}{\partial \xi_1} + \frac{u_2}{A_1A_2}\frac{\partial A_1}{\partial \xi_2} + \frac{u_3}{A_1A_3}\frac{\partial A_1}{\partial \xi_3} \qquad ...(10.36a)$$

$$\varepsilon_2 = \frac{1}{A_2}\frac{\partial u_2}{\partial \xi_2} + \frac{u_1}{A_1A_2}\frac{\partial A_3}{\partial \xi_1} + \frac{u_3}{A_2A_3}\frac{\partial A_2}{\partial \xi_3} \qquad ...(10.36b)$$

$$\varepsilon_3 = \frac{1}{A_3}\frac{\partial u_3}{\partial \xi_3} + \frac{u_2}{A_2A_3}\frac{\partial A_3}{\partial \xi_2} + \frac{u_1}{A_1A_3}\frac{\partial A_3}{\partial \xi_1} \qquad ...(10.36c)$$

$$\gamma_{12} = \frac{A_2}{A_1}\frac{\partial}{\partial \xi_1}\left(\frac{u_2}{A_2}\right) + \frac{A_1}{A_2}\frac{\partial}{\partial \xi_2}\left(\frac{u_1}{A_1}\right) \qquad ...(10.36d)$$

$$\gamma_{13} = \frac{A_1}{A_3}\frac{\partial}{\partial \xi_3}\left(\frac{u_1}{A_1}\right) + \frac{A_3}{A_1}\frac{\partial}{\partial \xi_1}\left(\frac{u_3}{A_3}\right) \qquad ...(10.36e)$$

$$\gamma_{23} = \frac{A_3}{A_2}\frac{\partial}{\partial \xi_2}\left(\frac{u_3}{A_3}\right) + \frac{A_2}{A_3}\frac{\partial}{\partial \xi_3}\left(\frac{u_2}{A_2}\right) \qquad ...(10.36f)$$

These are the expressions for strain components in curvilinear coordinates. To determine the strain components for thin shells, the following assumptions are made.

(*i*) The normals of the undeformed middle surface are deformed into normals of the middle surface.

(*ii*) The stress component normal to the middle surface are small compared with the other stress components and may be neglected in the stress-strain relations.

For thin shell we take $\xi_1 = \xi_1$, $\xi_2 = \xi_2$, $\xi_3 = z$. Since the middle surface of the shell is also strained unlike plates, the displacements at any point in the directions of ξ_1 and ξ_2 are, to the first order of z,

$$u_1 = u_{10} + z\left(\frac{\partial u_1}{\partial z}\right)_{z=0}$$

$$u_2 = u_{20} + z\left(\frac{\partial u_2}{\partial z}\right)_{z=0} \qquad ...(10.37)$$

where u_{10} and u_{20} are the displacements u_1 and u_2 on the middle surface at $z = 0$

The first assumption is equivalent to $\gamma_{1z} = \gamma_{2z} = 0$.

From Eq. (10.36), we find that

$$\left.\begin{aligned} \gamma_{1z} &= \frac{1}{A_3}\frac{\partial u_1}{\partial z} - \frac{u_1}{A_1 A_3}\frac{\partial A_1}{\partial z} + \frac{A_3}{A_1}\frac{\partial}{\partial \xi_1}\left(\frac{u_3}{A_3}\right) \\ \gamma_{2z} &= \frac{A_3}{A_2}\frac{\partial}{\partial \xi_2}\left(\frac{u_3}{A_3}\right) + \frac{1}{A_3}\frac{\partial u_2}{\partial z} - \frac{u_2}{A_2 A_3}\frac{\partial A_2}{\partial z} \end{aligned}\right\} \qquad ...(10.38)$$

If α_1 and α_2 are the factors which transform the curvilinear coordinates to linear distances on the middle surface of the shell, we find then that

$$A_1 = \alpha_1\left(1 + \frac{z}{R_1}\right)$$

$$A_2 = \alpha_2\left(1 + \frac{z}{R_2}\right), A_3 = 1$$

Substituting into Eqs. (10.38), at $z = 0$, we find that

$$\left(\frac{\partial u_1}{\partial z}\right)_{z=0} = -\frac{u_{10}}{R_1} - \frac{1}{\alpha_1}\frac{\partial u_z}{\partial \xi_1}$$

$$\left(\frac{\partial u_2}{\partial z}\right)_{z=0} = -\frac{u_{20}}{R_2} - \frac{1}{\alpha_2}\frac{\partial u_z}{\partial \xi_2}$$

where u_z is the displacement u_3 on the middle surface of the shell. Hence

$$\left.\begin{aligned} u_1 &= u_{10} - z\left(\frac{u_{10}}{R_1} + \frac{1}{\alpha_1}\frac{\partial u_z}{\partial \xi_1}\right) \\ u_2 &= u_{20} - z\left(\frac{u_{20}}{R_2} + \frac{1}{\alpha_2}\frac{\partial u_z}{\partial \xi_2}\right) \end{aligned}\right\} \quad \text{...(10.39)}$$

Neglecting $\frac{z}{R_1}$ and $\frac{z}{R_2}$ as compared to 1, we get

$$A_1 = \alpha_1, A_2 = \alpha_2$$

Also $$\frac{\partial A_1}{\partial z} = \frac{\alpha_1}{R_1}$$

$$\frac{\partial A_2}{\partial z} = \frac{\alpha_2}{R_2}, \; u_3 = u_z$$

Hence the strain components become

$$\varepsilon_1 = \left(\frac{1}{\alpha_1}\frac{\partial u_{10}}{\partial \xi_1} + \frac{u_{20}}{\alpha_1\alpha_2}\frac{\partial \alpha_1}{\partial \xi_2} + \frac{u_z}{R_1}\right) - z\left[\frac{1}{\alpha_1}\frac{\partial}{\partial \xi_1}\left(\frac{u_{10}}{R_1} + \frac{1}{\alpha_1}\frac{\partial u_z}{\partial \xi_1}\right) + \frac{1}{\alpha_1\alpha_2}\left(\frac{u_{20}}{R_2} + \frac{1}{\alpha_2}\frac{\partial u_z}{\partial \xi_2}\right)\frac{\partial \alpha_1}{\partial \xi_2}\right]$$

$$\varepsilon_2 = \left(\frac{1}{\alpha_2}\frac{\partial u_{20}}{\partial \xi_2} + \frac{u_{10}}{\alpha_1\alpha_2}\frac{\partial \alpha_2}{\partial \xi_1} + \frac{u_z}{R_2}\right) - z\left[\frac{1}{\alpha_2}\frac{\partial}{\partial \xi_2}\left(\frac{u_{20}}{R_2} + \frac{1}{\alpha_2}\frac{\partial u_z}{\partial \xi_2}\right) + \frac{1}{\alpha_1\alpha_2}\left(\frac{u_{10}}{R_1} + \frac{1}{\alpha_1}\frac{\partial u_z}{\partial \xi_1}\right)\frac{\partial \alpha_2}{\partial \xi_1}\right]$$

$$\gamma_{12} = \left[\frac{\alpha_2}{\alpha_1}\frac{\partial}{\partial\xi_1}\left(\frac{u_{20}}{\alpha_2}\right) + \frac{\alpha_1}{\alpha_2}\frac{\partial}{\partial\xi_2}\left(\frac{u_{10}}{\alpha_1}\right)\right]$$

$$-z\left[\frac{\alpha_2}{\alpha_1}\frac{\partial}{\partial\xi_1}\left(\frac{u_{20}}{\alpha_2 R_2} + \frac{1}{{\alpha_2}^2} + \frac{\partial u_z}{\partial\xi_2}\right)\right.$$

$$\left.+\frac{\alpha_1}{\alpha_2}\frac{\partial}{\partial\xi_2}\left(\frac{u_{10}}{\alpha_1 R_1} + \frac{1}{{\alpha_1}^2}\frac{\partial u_z}{\partial\xi_1}\right)\right] \qquad ...(10.40)$$

Eqs. (10.40) may be written as

$$\left.\begin{aligned} \varepsilon_1 &= \varepsilon_{10} - z\chi_1 \\ \varepsilon_2 &= \varepsilon_{20} - z\chi_2 \\ \gamma_{12} &= \gamma_0 - 2z\chi_{12} \end{aligned}\right\} \qquad ...(10.41)$$

where $\varepsilon_{10} = \dfrac{1}{\alpha_1}\dfrac{\partial u_{10}}{\partial\xi_1} + \dfrac{u_{20}}{\alpha_1\alpha_2}\dfrac{\partial\alpha_1}{\partial\xi_2} + \dfrac{u_z}{R_1}$

$$\varepsilon_{20} = \frac{1}{\alpha_2}\frac{\partial u_{20}}{\partial\xi_2} + \frac{u_{10}}{\alpha_1\alpha_2}\frac{\partial\alpha_2}{\partial\xi_1} + \frac{u_z}{R_2}$$

$$\gamma_0 = \frac{\alpha_2}{\alpha_1}\frac{\partial}{\partial\xi_1}\left(\frac{u_{20}}{\alpha_2}\right) + \frac{\alpha_1}{\alpha_2}\frac{\partial}{\partial\xi_2}\left(\frac{u_{10}}{\alpha_1}\right)$$

$$\chi_1 = \frac{1}{\alpha_1}\frac{\partial}{\partial\xi_1}\left(\frac{u_{10}}{R_1} + \frac{1}{\alpha_1}\frac{\partial u_z}{\partial\xi_1}\right) + \frac{1}{\alpha_1\alpha_2}\left(\frac{u_{20}}{R_2} + \frac{1}{\alpha_2}\frac{\partial u_z}{\partial\xi_2}\right)\frac{\partial\alpha_1}{\partial\xi_2}$$

$$\chi_2 = \frac{1}{\alpha_2}\frac{\partial}{\partial\xi_2}\left(\frac{u_{20}}{R_2} + \frac{1}{\alpha_2}\frac{\partial u_z}{\partial\xi_2}\right) + \frac{1}{\alpha_1\alpha_2}\left(\frac{u_{10}}{R_1} + \frac{1}{\alpha_1}\frac{\partial u_z}{\partial\xi_1}\right)\frac{\partial\alpha_2}{\partial\xi_1}$$

$$\chi_{12} = \frac{1}{2}\left[\frac{\alpha_2}{\alpha_1}\frac{\partial}{\partial\xi_1}\left(\frac{u_{20}}{\alpha_2 R_2} + \frac{1}{{\alpha_2}^2}\frac{\partial u_z}{\partial\xi_2}\right) + \frac{\alpha_1}{\alpha_2}\frac{\partial}{\partial\xi_2}\left(\frac{u_{10}}{\alpha_1 R_1} + \frac{1}{{\alpha_1}^2}\frac{\partial u_z}{\partial\xi_1}\right)\right]$$

ε_{10}, ε_{20}, γ_0 are the strains in the middle surface of the shell, while χ_1, χ_2, χ_{12} are the changes in curvatures.

From the second assumption, the stresses become

$$\left.\begin{aligned} \sigma_1 &= \frac{E}{1-\nu^2}[\varepsilon_{10} + \nu\varepsilon_{20} - z(\chi_1 + \nu\chi_2) \\ \sigma_2 &= \frac{E}{1-\nu^2}[\varepsilon_{20} + \nu\varepsilon_{10} - z(\chi_2 + \nu\xi_1)] \\ \tau_{12} &= \frac{E}{2(1+\nu)}(\gamma_0 - 2z\chi_{12}) \end{aligned}\right\} \qquad ...(10.42)$$

Substituting these expressions into Eqs. (10.1), we get after neglecting $\frac{z}{R}$ and $\frac{z}{R_2}$,

$$\left.\begin{aligned}
N_1 &= \frac{Eh}{1-\nu^2}(\varepsilon_{10} + \nu\varepsilon_{20}) \\
N_2 &= \frac{Eh}{1-\nu^2}(\varepsilon_{20} + \nu\varepsilon_{10}) \\
N_{12} = N_{21} &= \frac{Eh\gamma_0}{2(1+\nu)} \\
M_1 &= -D(\chi_1 + \nu\chi_2) \\
M_2 &= -D(\chi_2 + \nu\chi_1) \\
M_{12} = M_{21} &= D(1-\nu)\chi_{12}
\end{aligned}\right\} \quad ...(10.43)$$

10.10 Strains and Displacements in Shells of Revolution

For shells of revolution, we have

$$\alpha_1 = R_1,\ \alpha_2 = R_0 = R_2 \sin\phi$$

$$\xi_1 = \phi,\ \xi_2 = \theta,\ \xi_3 = z.$$

Let the displacements of a point P in the middle surface of the shell with coordinates θ, ϕ has displacements u_θ, u_ϕ, u_z as shown in Fig. 10.20. Then from Eqs. (10.40), we get

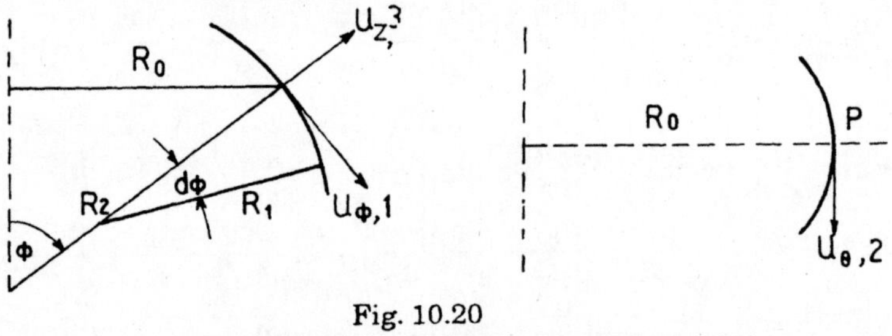

Fig. 10.20

$$\left.\begin{aligned}
\varepsilon_\theta &= \frac{1}{R_0}\frac{\partial u_\theta}{\partial\theta} + \frac{u_\phi}{R_1 R_0}\frac{\partial R_0}{\partial\phi} + \frac{u_z}{R_2} \\
&= \frac{1}{R_0}\left[\frac{\partial u_\theta}{\partial\theta} + u_\phi\cos\phi + u_z\sin\phi\right] \\
\varepsilon_\phi &= \frac{1}{R_1}\frac{\partial u_\phi}{\partial\phi} + \frac{u_z}{R_1} \\
\gamma_{\theta\phi} &= \frac{1}{R_1}\frac{\partial u_\theta}{\partial\phi} - \frac{u_\theta\cos\phi}{R_0} + \frac{1}{R_0}\frac{\partial u_\phi}{\partial\theta}
\end{aligned}\right\} \quad ...(10.44)$$

$$\left.\begin{aligned}
\chi_\theta &= \frac{1}{R_0}\left[\frac{\partial}{\partial\theta}\left\{\frac{1}{R_0}\left(\frac{\partial u_z}{\partial\theta} - u_\theta \sin\phi\right)\right\} + \frac{\cos\phi}{R_1}\left(\frac{\partial u_z}{\partial\phi} - u_\phi\right)\right] \\
&= \frac{1}{R_2^2}\frac{\partial^2 u_z}{\partial\theta^2} - \frac{1}{R_0 R_2}\frac{\partial u_\theta}{\partial\theta} + \frac{\cot\phi}{R_2}\left[\frac{1}{R_1}\frac{\partial u_z}{\partial\phi} - \frac{u_\phi}{R_1}\right] \\
\chi_\phi &= \frac{1}{R_1}\frac{\partial}{\partial\phi}\left[\frac{1}{R_1}\frac{\partial u_z}{\partial\phi} - \frac{u_\phi}{R_1}\right] \\
2\chi_{\theta\phi} &= \frac{R_0}{R_1}\frac{\partial}{\partial\phi}\left[\frac{1}{R_0^2}\left(\frac{\partial u_z}{\partial\theta} - u_\theta \sin\phi\right)\right] \\
&\quad + \frac{1}{R_0}\frac{\partial}{\partial\theta}\left[\frac{1}{R_1}\frac{\partial u_z}{\partial\phi} - \frac{u_\phi}{R_1}\right] \\
&= \frac{2}{R_0 R_1}\frac{\partial^2 u_z}{\partial\phi\partial\theta} - \frac{2\cos\phi}{R_0^2}\frac{\partial u_z}{\partial\theta} - \frac{1}{R_0 R_1}\frac{\partial u_\phi}{\partial\theta} \\
&\quad - \frac{\sin\phi}{R_0 R_1}\frac{\partial u_\theta}{\partial\phi} - \frac{u_\theta \cot\phi}{R_1 R_2^2}(R_2 - 2R_1)
\end{aligned}\right\} \quad \text{...(10.45)}$$

The direct and shear strains at a distance z from the middle surface of shell are given by Eq. (10.41) as :

$$\left.\begin{aligned}
\bar{\varepsilon}_\theta &= \varepsilon_\theta - z\chi_\theta \\
\bar{\varepsilon}_\phi &= \varepsilon_\phi - z\chi_\phi \\
\bar{\gamma}_{\theta\phi} &= \gamma_{\theta\phi} - 2z\chi_{\theta\phi}
\end{aligned}\right\} \quad \text{...(10.46)}$$

10.11 Strains and Displacement in Cylindrical Shells

For a cylindrical shell $\alpha_1 = 1$, $\alpha_2 = a$, $R_1 = \infty$, $R_2 = a$. Hence Eqs. (10.44) and (10.45) become (Fig. 10.21).

$$\left.\begin{aligned}
\varepsilon_x &= \frac{\partial u_x}{\partial x} \\
\varepsilon_\theta &= \frac{1}{a}\frac{\partial u_\theta}{\partial\theta} + \frac{u_z}{a} \\
\gamma_{\theta x} &= \frac{1}{a}\frac{\partial u_x}{\partial\theta} + \frac{\partial u_\theta}{\partial x}
\end{aligned}\right\} \quad \text{...(10.47)}$$

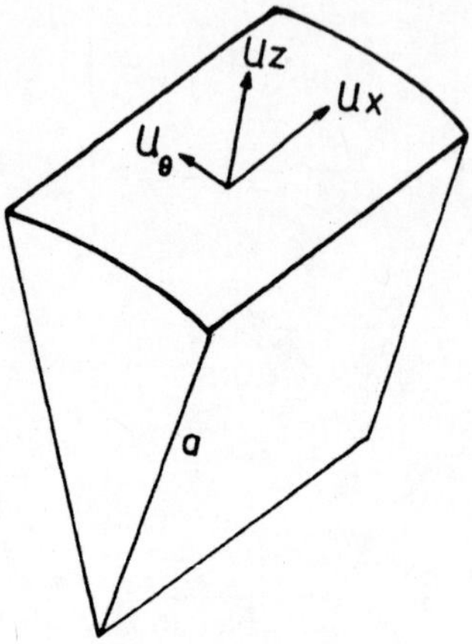

Fig. 10.21

$$\left.\begin{aligned} \chi_x &= \frac{\partial^2 u_z}{\partial x^2} \\ \chi_\theta &= \frac{1}{a^2}\left[\frac{\partial^2 u_z}{\partial \theta^2} - \frac{\partial u_\theta}{\partial \theta}\right] \\ 2\chi_{\theta x} &= \frac{1}{a}\left[\frac{2\partial^2 u_z}{\partial x \partial \theta} - \frac{\partial u_\theta}{\partial x}\right] \end{aligned}\right\} \quad \text{...(10.48)}$$

The strains at a distance z (+ve in direction of outward normal) from the middle surface are

$$\left.\begin{aligned} \bar{\varepsilon}_x &= \varepsilon_x - z\chi_x \\ \bar{\varepsilon}_\theta &= \varepsilon_\theta - z\chi_\theta \\ \bar{\gamma}_{\theta x} &= \gamma_{\theta x} - 2z\chi_{\theta x} \end{aligned}\right\} \quad \text{...(10.49)}$$

10.12 Donnell Equations for Cylindrical Shells

For a circular cylindrical shell Eqs. (10.43) become :

$$\left.\begin{aligned} N_\theta &= K(\varepsilon_\theta + \nu\varepsilon_x) \\ N_x &= K(\varepsilon_x + \nu\varepsilon_\theta) \\ N_{\theta x} &= K\left(\frac{1-\nu}{2}\right)\gamma_{\theta x} \\ M_\theta &= -D(\chi_\theta + \nu\chi_x) \\ M_x &= -D(\chi_x + \nu\chi_\theta) \\ M_{\theta x} &= D(1-\nu)\chi_{\theta x} \end{aligned}\right\} \quad \text{...(10.50)}$$

where $K = \dfrac{Eh}{1 - \nu^2}$, $D = \dfrac{Eh^3}{12(1 - \nu^2)}$

Now equilibrium equations for a circular cylindrical shell are given by Eqs. (10.24) to (10.29). Dropping the terms in u_θ in the expressions for χ_θ and $\chi_{\theta x}$ in Eqs. (10.48) and ignoring the effect of Q_θ in Eq. (10.25) and ignoring Eq. (10.29), we get after substituting the expressions for the stress resultants in terms of u_x, u_θ and u_z in the equations of equilibrium,

$$\left.\begin{aligned}
&\frac{\partial^2 u_x}{\partial x^2} + \frac{1-\nu}{2a^2}\frac{\partial^2 u_x}{\partial \theta^2} + \frac{1+\nu}{2a}\frac{\partial^2 u_\theta}{\partial x \partial \theta} - \frac{\nu}{a}\frac{\partial u_z}{\partial x} + \frac{p_x}{K} = 0 \\
&\frac{1+\nu}{2a}\frac{\partial^2 u_x}{\partial x \partial \theta} + \frac{1-\nu}{2}\frac{\partial^2 u_\theta}{\partial x^2} - \frac{1}{a^2}\frac{\partial^2 u_\theta}{\partial \theta^2} + \frac{1}{a^2}\frac{\partial u_z}{\partial \theta} + \frac{p_\theta}{K} = 0 \\
&\frac{\nu \partial u_x}{\partial x} + \frac{1}{a}\frac{\partial u_\theta}{\partial \theta} - \frac{u_z}{a} - \frac{h^2}{12}\left[\frac{a\partial^4 u_z}{\partial x^4} + \frac{2}{a}\frac{\partial^4 u_z}{\partial x^2 \partial \theta^2} + \frac{1}{a^3}\frac{\partial^4 u_z}{\partial \theta^4}\right] + \frac{ap_r}{K} = 0
\end{aligned}\right\} \quad ...(10.51)$$

For $p_r = p_\theta = p_x = 0$, Eqs. (10.51) can be witten as :

$$\left.\begin{aligned}
&\nabla^8 u_z + \frac{12(1-\nu^2)}{r^2h^2}\frac{\partial^4 u_z}{\partial x^4} = 0 \\
&\nabla^4 u_\theta = -\left(\frac{2+\nu}{a^2}\right)\frac{\partial^3 u_z}{\partial x^2 \partial \theta} - \frac{1}{a^4}\frac{\partial^3 u_z}{\partial \theta^3} \\
&\nabla^4 u_x = -\frac{\nu}{a}\frac{\partial^3 u_z}{\partial x^3} + \frac{1}{a^3}\frac{\partial^3 u_z}{\partial x \partial \theta^2}
\end{aligned}\right\} \quad ...(10.52)$$

where $\nabla^2 = \dfrac{\partial^2}{\partial x^2} + \dfrac{1}{r^2}\dfrac{\partial^2}{\partial \theta^2}$

These are known as the Donnell equations.

10.13 Symmetrically Loaded Circular Cylindrical Shell

For symmetrically loaded circular cylindrical shell of radius a, Eqs. (10.51) become

$$\left.\begin{aligned}
&\frac{\partial^2 u_x}{\partial x^2} - \frac{\nu}{a}\frac{du_z}{dx} + \frac{p_x}{K} = 0 \\
&\nu\frac{du_x}{dx} - \frac{u_z}{a} - \frac{h^2 a}{12}\frac{d^4 u_z}{dx^4} + \frac{ap_r(1-\nu^2)}{Eh} = 0
\end{aligned}\right\} \quad ...(10.53)$$

If the cylinder is loaded by lateral load only, then $p_x = 0$ and the first of Eqs. (10.53) can be integrated to give

$$\frac{du_x}{dx} = \frac{\nu u_z}{a} + C \qquad ...(10.54)$$

where C is a constant of integration. From Eq. (10.48) to (10.50), we find that

$$\frac{du_x}{dx} = \frac{\nu u_z}{a} + \frac{N_x(1-\nu^2)}{Eh} \qquad ...(10.55)$$

Therefore from Eqs. (10.54) and (10.55) it appears that N_x are constant.

Eliminating $\frac{du_x}{dx}$ from the second equation in (10.53), we obtain,

$$\frac{d^4u_z}{dx^4} + 4\beta^4 u_z = \frac{p_r}{D} + \frac{\nu N_x}{aD} \qquad ...(10.56)$$

where $\beta^4 = \frac{3(1-\nu^2)}{a^2h^2}$

This is an ordinary differential equation with constant coefficients, whose complementary solution is

$$u_z = e^{-\beta x}(C_1 e^{i\beta x} + C_2 e^{-i\beta x}) + e^{\beta x}(C_3 e^{i\beta x} + C_4 e^{-i\beta x})$$

If the particular integral is $f(x)$, the general solution of Eq. (10.56) can be written as

$$u_z = e^{-\beta x}(C_1 e^{i\beta x} + C_2 e^{-i\beta x}) + e^{\beta x}(C_3 e^{i\beta x} + C_4 e^{-i\beta x}) + f(x) \qquad ...(10.57)$$

where C_1, C_2, C_3, C_4 are arbitrary constants.

Example 10.7 *Circular pipe subjected to the action of bending and shear.*

Solution. Consider a circular pipe subjected to the action of bending moments M_0 and shearing forces Q_0, both uniformly distributed along the edge $x = 0$ as shown in Fig. 10.22. In this case $p_r = 0$ and if the end condition is such that $N_x = 0$, then $f(x) = 0$. Since u_z is to remain finite as x tends to infinity, therefore $C_3 = C_4 = 0$.

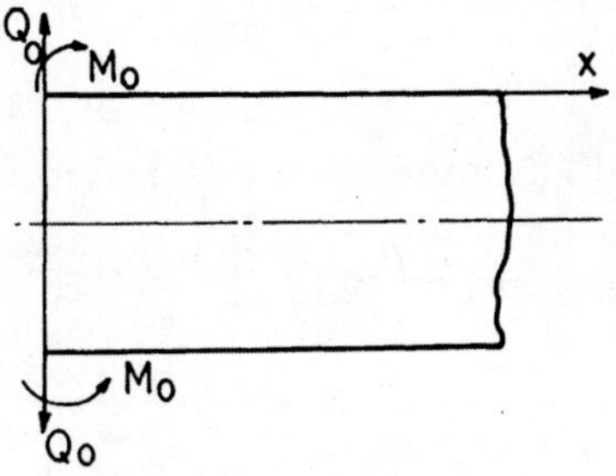

Fig. 10.22

$$\therefore \qquad u_z = e^{-\beta x}(C_1 e^{i\beta x} + C_2 e^{-i\beta x}) \qquad ...(1)$$

$$u_z = e^{-\beta x}(C_1 \cos\beta x + C_2 \sin\beta x)$$

$$\left.\begin{aligned} \text{Now} \quad (M_x)_{x=0} &= -D\left(\frac{d^2u_z}{dx^2}\right)_{x=0} = M_0 \\ (Q_x)_{x=0} &= \left(\frac{dM_x}{dx}\right)_{x=0} = -D\left(\frac{d^3u_z}{dx^3}\right)_{x=0} = Q_0 \end{aligned}\right\} \quad ...(2)$$

Substituting Eq. (1) in Eqs. (2), we obtain

$$C_1 = -\frac{1}{2\beta^3 D}(Q_0 + \beta M_0),\ C_2 = \frac{M_0}{2\beta^2 D}$$

$$\therefore \qquad u_z = \frac{e^{-\beta x}}{2\beta^3 D}[\beta M_0(\sin\beta x - \cos\beta x) - Q_0\cos\beta x] \quad ...(3)$$

$$(u_z)_{x=0} = (u_z)_{max} = -\frac{1}{2\beta^3 D}(\beta M_0 + Q_0) \quad ...(4)$$

Example 10.8 *Bending of a long circular pipe by a load uniformly distributed along a circular section.*

Solution. Choosing the coordinates as shown in Fig. 10.23, consider the cylinder on the right side of z-axis. Since $p_3 = 0$ and if

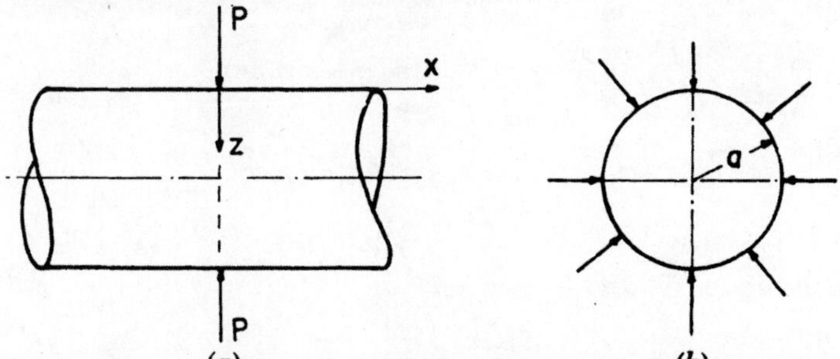

(a) (b)

Fig. 10.23

the end condition is such that $N_x = 0$, then $f(x) = 0$. Since the cylinder is long, at the support, which is at a large value of x, u_z must be zero. This is possible only if $C_3 = C_4 = 0$ in Eq. (10.57). We have, therefore

$$u_z = e^{-\beta x}(C_1\cos\beta x + C_2\sin\beta x) \quad ...(1)$$

Now $\qquad N_{x\theta} = M_{x\theta} = 0$

and

$$\left.\begin{aligned} N_\theta &= \frac{Eh}{1-\nu^2}\left(\nu\frac{du_x}{dx} - \frac{u_z}{a}\right) \\ M_x &= -D\frac{d^2u_z}{dx^2} \\ M_\theta &= -D\nu\frac{d^2u_z}{dx^2} \end{aligned}\right\} \quad ...(2)$$

Since $N_x = 0, \dfrac{du_x}{dx} = -\dfrac{\nu u_z}{a}$

$$\therefore \quad N_\theta = -\frac{Ehu_z}{a}$$

Now $$Q_x = \frac{dM_x}{dx} = -D\frac{d^3u_z}{dx^3} \qquad \text{...(3)}$$

Since each half of the cylinder will take one-half of the external load, at $x = 0$, we have $Q_x = -\dfrac{P}{2}$. For a long cylinder u_z must be symmetrical about $x = 0$, which requires that $\dfrac{du_z}{dx} = 0$ there. The integration constants C_1 and C_2 can be determined from the conditions :

$$\left(\frac{du_z}{dx}\right)_{x=0} = [-\beta C_1 e^{-\beta x}(\cos\beta x + \sin\beta x) + \beta C_2 e^{-\beta x}(\cos\beta x - \sin\beta x)]_{x=0}$$

$$= -C_1 + C_2 = 0$$

$$\left(\frac{d^3u_z}{dx^3}\right)_{x=0} = [2\beta^3 C_1 e^{-\beta x}(\cos\beta x - \sin\beta x) + 2\beta^3 C_2 e^{-\beta x}(\cos\beta x + \sin\beta x)]_{x=0}$$

$$= 2\beta^3(C_1 + C_2) = \frac{P}{2D}$$

Solving for C_1 and C_2, we get

$$C_1 = C_2 = \frac{P}{8\beta^3 D}$$

$$\therefore \quad u_z = \frac{Pe^{-\beta x}}{8\beta^3 D}(\sin\beta x + \cos\beta x) \qquad \text{...(4)}$$

Substituting in Eqs. (2) and (3), we get

$$N_\theta = -\frac{Ehu_z}{a} = -\frac{EhPe^{-\beta x}}{8\beta^3 Da}(\sin\beta x + \cos\beta x)$$

$$M_x = -D\frac{d^2u_z}{dx^2} = \frac{Pe^{-\beta x}}{4\beta}(\cos\beta x - \sin\beta x)$$

$$M_\theta = -D\nu\frac{d^2u_z}{dx^2} = \frac{nPe^{-\beta x}}{4\beta}(\cos\beta x - \sin\beta x)$$

$$Q_x = -D\frac{d^3u_z}{dx^3} = -\frac{Pe^{-\beta x}}{2}\cos\beta x$$

$$(u_z)_{max} = (u_z)_{x=0} = \frac{P}{8\beta^3 D} = \frac{Pa^2\beta}{2Eh}$$

$$(M_x)_{max} = (M_x)_{x=0} = \frac{P}{4\beta}$$

We observe that for $x > \dfrac{\pi}{\beta}$, all the above quantities are negligible indicating that the effect of P is of local character and a cylinder with length $L = \dfrac{2\pi}{\beta}$ loaded at the middle will have practically the same maximum deflection and bending moments as those for a long cylinder.

Now the bending stresses are

$$\sigma_x = \frac{12M_x z}{h^3}, \; \sigma_\theta = \frac{N_\theta}{h} + \frac{12M_\theta \cdot z}{h^3}$$

$$(\sigma_x)_{max}\Big|_{\substack{x=0\\z=h/2}} = \frac{3P}{2\beta h^2}$$

$$(\sigma_\theta)_{max}\Big|_{\substack{x=0\\z=h/2}} = -\frac{Pa\beta}{2} - \frac{3\nu P}{2\beta h^2}$$

Example 10.9 *Bending of a long cylinder by a load distributed along a certain length of the cylinder.*

Solution. Consider the case of a load of intensity q uniformly distributed along a length l of the cylinder as shown in Fig. 10.24. Assuming that the load is at a considerable distance from the ends of the cylinder, the deflection at a point A produced by an elementary ring load of an intensity $qd\xi$ at a distance ξ from A is obtained by substituting $qd\xi$ for P and ξ for x in the expression

$$u_z = \frac{Pe^{-\beta x}}{8\beta^3 D}(\cos\beta x + \sin\beta x)$$

obtained in example 10.8 for the concentrated load.

$$\therefore \quad du_z = \frac{qd\xi}{8\beta^3 D} e^{-\beta\xi}(\cos\beta\xi + \sin\beta\xi) \quad \text{...(1)}$$

The deflection produced at A by the total load distributed over the length l is then

$$u_z = \int_0^b \frac{qd\xi}{8\beta^3 D} e^{-\beta\xi}(\cos\beta\xi + \sin\beta\xi)$$

$$+ \int_0^c \frac{qd\xi}{8\beta^3 D} e^{-\beta\xi}(\cos\beta\xi + \sin\beta\xi)$$

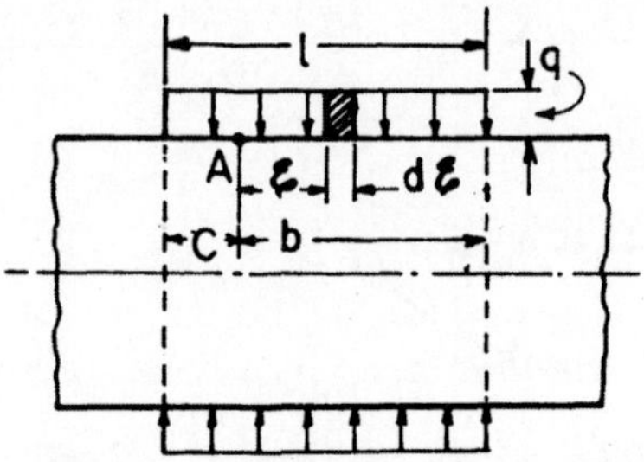

Fig. 10.24

$$= \frac{qa^2}{2Eh}\,(2 - e^{-\beta b}\cos\beta b - e^{-\beta c}\cos\beta c)$$

If $\qquad b = c = l/2$, then

$$u_z = \frac{qa^2}{Eh}\left(1 - e^{-\beta\frac{l}{2}}\cos\beta\frac{l}{2}\right)$$

Similarly bending moments, shear force and stresses can be calculated.

Example 10.10 *Cylindrical shell subjected to uniform internal pressure.*

Solution. If the edges of the shell are free then

$$\sigma_\theta = \frac{pa}{h},\ \sigma_x = \frac{pa}{2h}$$

and the radius of the cylinder increases by the amount

$$\delta = \sigma_\theta \cdot \frac{a}{E} = \frac{pa^2}{Eh} \qquad ...(1)$$

If the ends of the shell are clamped as shown in Fig. 10.25 (*a*) they can not move out, and local bending occurs at the edges. If the length L of the shell is sufficiently large then at $x = 0$, $u_z = \frac{du_z}{dx} = 0$.

Using the results from example 10.7, we find

$$-\frac{1}{2\beta^3 D}(\beta M_0 + Q_0) = \delta$$

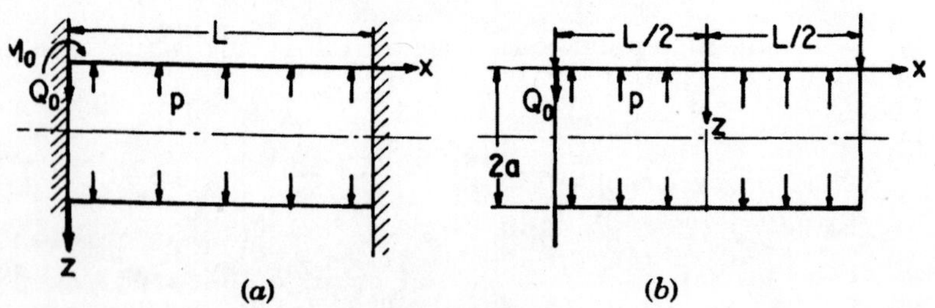

Fig. 10.25

$$\frac{1}{2\beta^2 D}(2\beta M_0 + Q_0) = 0$$

$$\therefore \quad \left.\begin{aligned} M_0 &= 2\beta^2 D\delta = \frac{p}{2\beta^2} \\ Q_0 &= -3\beta^3 D\delta = -\frac{p}{\beta} \end{aligned}\right\} \quad \text{...(2)}$$

If we have simply supported edges as shown in Fig. 10.25 (*b*) then at $x = 0$, $u_x = \frac{d^2u_z}{dx^2} = 0$, and we get

$$Q_0 = 2\beta^2 D\delta \quad \text{...(3)}$$

Then the bending moments and stresses can be calculated.

If the cylinder is of finite length then the bending at one end can not be considered as independent of the conditions at the other end. Then $p_r = -p$ and the general solution becomes

$$u_z = -\frac{pa^2}{Eh} + C_1 \sin\beta x \sinh\beta x + C_2 \sin\beta x \times \cosh\beta x + C_3 \cos\beta x \sinh\beta x + C_4 \cos\beta x \times \cosh\beta x \quad \text{...(4)}$$

Choosing the coordinates at the centre of the cylinder as shown in Fig. 10.25 (*b*), the expression (4) must be an even function of x. Hence

$$C_2 = C_3 = 0 \quad \text{...(5)}$$

The constants C_1 and C_4 must now be selected so as to satisfy the conditions at the ends.

For simply supported ends

$$(u_z)_{x=\frac{L}{2}} = \left(\frac{d^2u_z}{dx^2}\right)_{x=\frac{L}{2}} = 0 \quad \text{...(6)}$$

Substituting in Eq. (4) and using Eq. (5), we find

$$\frac{-pa^2}{Eh} + C_1 \sin\alpha \sinh\alpha + C_4 \cos\alpha \cosh\alpha = 0$$

$$C_1 \cos\alpha \cosh\alpha - C_4 \sin\alpha \sinh\alpha = 0$$

where $\alpha = \beta\frac{L}{2}$

Solving for C_1 and C_4, we get

$$C_1 = \frac{pa^2}{Eh}\left[\frac{2\sin\alpha \sinh\alpha}{\cos 2\alpha + \cosh\alpha}\right]$$

$$C_4 = \frac{pa^2}{Eh}\left[\frac{2\cos\alpha\cosh\alpha}{\cos 2\alpha + \cosh 2\alpha}\right]$$

Substituting in Eq. (4), we find that

$$\therefore \qquad u_z = -\frac{pL^4}{64D\alpha^4}\left[1 - \frac{2\sin\alpha\sinh\alpha}{\cos 2\alpha + \cosh 2\alpha}\sin\beta x\sinh\beta x - \frac{2\cos\alpha\cosh\alpha}{\cos 2\alpha + \cosh 2\alpha}.\cos\beta x\cosh\beta x\right] \qquad ...(7)$$

$$(u_z)_{x=0} = -\frac{pL^4}{64D\alpha^4}\left(1 - \frac{2\cos\alpha\cosh\alpha}{\cos 2\alpha + \cosh 2\alpha}\right) \qquad ...(8)$$

$$M_x = -D\frac{d^2u_z}{dx^2}$$

$$= -\frac{pL^4}{4\alpha^4}\left[\frac{\sin\alpha\sinh\alpha}{\cos 2\alpha + \cosh 2\alpha}\cosh\beta x\cos\beta x - \frac{\cos\alpha\cosh\alpha}{\cos 2\alpha + \cosh 2\alpha}.\sin\beta x\sinh\beta x\right] \qquad ...(9)$$

$$(M_x)_{x=0} = -\frac{pL^4}{4\alpha^4}\left(\frac{\sin\alpha\sinh\alpha}{\cos 2\alpha + \cosh 2\alpha}\right) \qquad ...(10)$$

For a cylinder with clamped edges, proceeding as above we can find that the bending moment M_0 acting along the clamped edge is

$$M_0 = \frac{p}{2\beta^2}\left(\frac{\sinh 2\alpha - \sin 2\alpha}{\sinh 2\alpha + \sin 2\alpha}\right) \qquad ...(11)$$

For cylinders of finite length the following results can be obtained for the case of bending by uniformly distributed shearing force Q_0 along the edges.

$$\left.\begin{aligned}(u_z)_{\substack{x=0\\x=L}} &= -\frac{2Q_0\beta a^2}{Eh}\left[\frac{\cosh 2\alpha + \cos 2\alpha}{\sinh 2\alpha + \sin 2\alpha}\right]\\ \left(\frac{du_z}{dx}\right)_{\substack{x=0\\x=L}} &= \pm\frac{2Q_0\beta^2a^2}{Eh}\left[\frac{\sinh 2\alpha - \sin 2\alpha}{\sinh 2\alpha + \sin 2\alpha}\right]\end{aligned}\right\} \qquad ...(12)$$

and in the case of bending by moments M_0, we have

$$\left.\begin{aligned}(u_z)_{\substack{x=0\\x=L}} &= -\frac{2M_0\beta^2a^2}{Eh}\left[\frac{\sinh 2\alpha - \sin 2\alpha}{\sinh 2\alpha + \sin 2\alpha}\right]\\ \left(\frac{du_z}{dx}\right)_{\substack{x=0\\x=L}} &= \pm\frac{4M_0\beta^2a^2}{Eh}\left[\frac{\cosh 2\alpha - \cos 2\alpha}{\sinh 2\alpha + \sin 2\alpha}\right]\end{aligned}\right\} \qquad ...(13)$$

Example 10.11 *Stresses in a long pipe reinforced by equidistant rings and carrying internal pressure.*

Solution. In the absence of rings

$$\sigma_\theta = \frac{pa}{h} \text{ and } \delta = \frac{pa^2}{Eh},$$

where δ is the increase in radius a.

Now assume that the pipe is reinforced by rigid rings. As a result of it, reactive force will be produced between each ring and

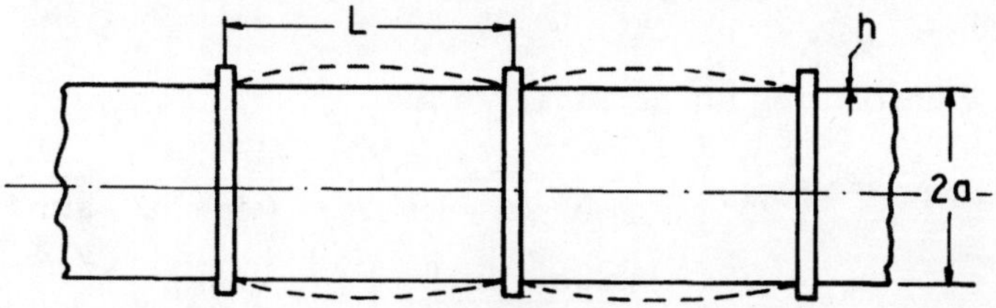

Fig. 10.26

the pipe. Let P be the force per unit length of the circumference of the tube. Since there is no extension of the cylinder under the rings therefore the deflection of the pipe produced by P under the ring is equal to the expansion δ created by the internal pressure p. As shown in Fig. 10.26, each portion of the tube between two adjacent rings may be considered as shell subjected to the action of shear forces Q_0 and bending moments M_0. In this case $Q_0 = -\frac{P}{2}$ and the bending moment M_0 is determined from the condition that $\frac{du_z}{dx} = 0$ under the ring. Hence from Eqs. (12) and (13) of Example 10.10, we find

$$-\frac{P\beta^2 a^2}{Eh}\left[\frac{\sinh 2\alpha - \sin 2\alpha}{\sinh 2\alpha + \sin 2\alpha}\right] + \frac{4M_0\beta^3 a^2}{Eh}\left[\frac{\cosh 2\alpha - \cos 2\alpha}{\sinh 2\alpha + \sin 2\alpha}\right] = 0$$

$$\therefore \quad M_0 = \frac{P}{4\beta}\left[\frac{\sinh 2\alpha - \sin 2\alpha}{\cosh 2\alpha - \cos 2\alpha}\right] \quad \text{...(1)}$$

For large values of α

$$M_0 = P/4\beta$$

For calculating the force P, we have

$$\frac{P\beta a^2}{Eh}\left[\frac{\cosh 2\alpha + \cos 2\alpha}{\sinh 2\alpha + \sin 2\alpha}\right] - \frac{4M_0\beta^2 a^2}{Eh}\left[\frac{\sinh 2\alpha - \sin 2\alpha}{\sinh 2\alpha + \sin 2\alpha}\right] = \delta = \frac{pa^2}{Eh}$$

Substituting the value of M_0 from Eq. (1), we get

$$\frac{P\beta a^2}{Eh}\left[\frac{\cosh 2\alpha + \cos 2\alpha}{\sinh 2\alpha + \sin 2\alpha} - \frac{1}{2}\frac{(\sinh 2\alpha - \sin 2\alpha)^2}{(\sinh 2\alpha + \sin 2\alpha)(\cosh 2\alpha - \cos 2\alpha)}\right] = \frac{pa^2}{Eh}$$

$$P = \frac{2p}{\beta}\left[\frac{(\sinh 2\alpha + \sin 2\alpha)(\cosh 2\alpha - \cos 2\alpha)}{2(\cosh^2 2\alpha - \cos^2 2\alpha) - (\sinh 2\alpha - \sin 2\alpha)^2}\right] \quad \text{...(2)}$$

For large values of α

$$P = 2p/\beta$$

Substituting Eq. (2) in (1), we find

$$M_0 = \frac{p}{2\beta^2}\left[\frac{\sinh^2 2\alpha - \sin^2 2\alpha}{2(\cosh^2 2\alpha - \cos^2 2\alpha) - (\sinh 2\alpha - \sin 2\alpha)^2}\right]$$

$$= \frac{p}{2\beta^2}\left[\frac{\sinh 2\alpha - \sin 2\alpha}{\sinh 2\alpha + \sin 2\alpha}\right] \quad \text{...(3)}$$

Knowing P and M_0 stresses in the pipe can be determined.

Example 10.12 *Cylindrical tank filled with liquid.*

Solution. Consider a cylindrical tank filled with liquid of specific weight γ as shown in Fig. 10.27. Taking the coordinates as shown the pressure acting at any point x on the wall of the cylindrical tank is

$$p_r = -\gamma(L - x) \quad \text{...(1)}$$

where the –ve sign indicates that p_r is directed outward and is opposite to the + ve direction. Thus the equilibrium equation becomes

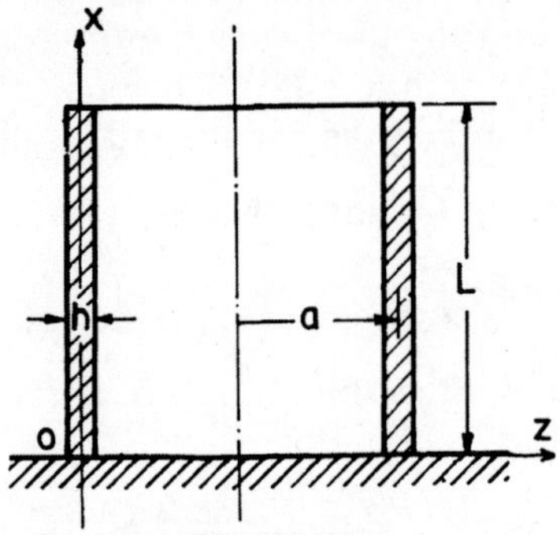

Fig. 10.27

$$\frac{d^4u_z}{dx^4} + 4\beta^4 u_z = -\frac{\gamma(L-x)}{D} \quad \text{...(2)}$$

The particular integral $f(x)$ becomes

$$f(x) = -\frac{\gamma(L-x)}{4\beta^4 D} = -\frac{\gamma(L-x)a^2}{Eh}$$

and the general solution is

$$u_z = e^{-\beta x}(C_1 \cos \beta x + C_2 \sin \beta x) + e^{\beta x}(C_3 \cos \beta x + C_4 \sin \beta x) - \frac{\gamma(L-x)a^2}{Eh} \quad \text{...(3)}$$

If L is large compared with $\sqrt{ah}$, we may regard the cylinder as infinitely long. The constants C_3 and C_4 are then zero. If the tank is rigidly fixed at the bottom, then

$$u_z = \frac{du_z}{dx} = 0 \text{ at } x = 0$$

we find $\quad (u_z)_{x=0} = C_1 = \frac{\gamma a^2 L}{Eh} = 0$

$$\left(\frac{du_z}{dx}\right)_{x=0} = \beta(C_1 - C_2) + \frac{\gamma a^2}{Eh} = 0$$

$$\therefore \quad C_1 = \frac{\gamma a^2 L}{Eh}$$

$$C_2 = \frac{\gamma a^2}{Eh}\left(L - \frac{1}{\beta}\right)$$

$$\therefore \quad u_z = -\frac{\gamma a^2}{Eh}\left[L - x - e^{-\beta x}\left\{L\cos\beta x + \left(L - \frac{1}{\beta}\right)\sin\beta x\right\}\right] \quad \text{...(4)}$$

$$N_\theta = -\frac{Ehu_z}{a}$$

$$= \gamma a\left[L - x - e^{-\beta x}\left\{L\cos\beta x + \left(L - \frac{1}{\beta}\right)\sin\beta x\right\}\right] \quad \text{...(5)}$$

$$M_x = -D\frac{d^2u_z}{dx^2}$$

$$= \frac{\gamma aLh}{\sqrt{12(1-\nu^2)}}e^{-\beta x}\left[-\sin\beta x + \left(1 - \frac{1}{\beta L}\right)\cos\beta x\right] \quad \text{...(6)}$$

$$M_\theta = -D\nu\frac{d^2u_z}{dx^2} = \frac{\nu\gamma aLh}{\sqrt{12(1-\nu^2)}}e^{-\beta x} \times\left[-\sin\beta x + \left(1 - \frac{1}{\beta L}\right)\cos\beta x\right] \quad \text{...(7)}$$

$$(M_x)_{max} = (M_x)_{x=0} = \left(1 - \frac{1}{\beta L}\right)\frac{\gamma aLh}{\sqrt{12(1-\nu^2)}} \quad \text{...(8)}$$

Example 10.13 *Cylindrical shells with supported edges.*

Solution. Consider a cylindrical shell supported at the ends and subjected to pressure due to the enclosed liquid as shown in Fig.

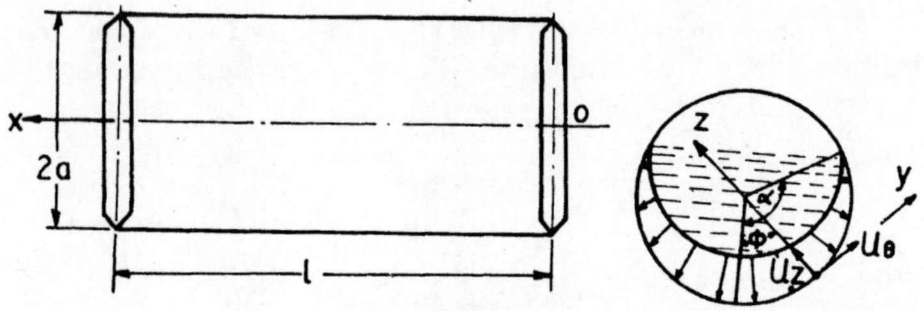

Fig. 10.28

10.28. The conditions at the supports and the conditions of symmetry of deformation will be satisfied if we assume

$$\left.\begin{aligned} u_x &= \sum\sum A_{mn} \cos n\varphi \cos \frac{m\pi x}{l} \\ u_\theta &= \sum\sum A_{mn} \sin n\varphi \sin \frac{m\pi x}{l} \\ u_z &= \sum\sum C_{mn} \cos n\varphi \sin \frac{m\pi x}{l} \end{aligned}\right\} \quad \text{...(1)}$$

where l is the length of the cylinder and φ is the angle measured as shown.

The intensity of the load p_r is represented by

$$\begin{aligned} p_r &= -\gamma a (\cos\varphi - \cos\alpha) \text{ for } \varphi < \alpha \quad \text{...(2)} \\ &= 0 \text{ for } \varphi > \alpha \end{aligned}$$

where γ is the specific weight of the liquid and the angle α defines the level of the liquid. The load p_r can be expressed as

$$p_r = \sum\sum D_{mn} \cos n\varphi \sin \frac{m\pi x}{l} \quad \text{...(3)}$$

$$\left.\begin{aligned} &\text{where } D_{mn} = \frac{8\gamma a}{mn\pi^2 (n^2 - 1)} (\cos\alpha \sin n\alpha \\ &\qquad\qquad - n \cos n\alpha \sin\alpha) \\ &\quad m \text{ is odd and } n \text{ is even.} \\ &\text{and} \quad D_{m0} = -\frac{4\gamma a}{m\pi^2} (\sin\alpha - \alpha\cos\alpha) \\ &\qquad D_{m1} = -\frac{2\gamma a}{m\pi^2} (2\alpha - \sin 2\alpha) \end{aligned}\right\} \quad \text{...(4)}$$

For a completely filled cylinder $\alpha = 180°$ and

$$D_{mn} = 0,\ D_{m0} = -\frac{4\gamma a}{m\pi},\ D_{m1} = -\frac{4\gamma a}{m\pi}.$$

To obtain the deformation of the shell, we substitute Eqs. (1) and (3) in Eqs. (10.51), we obtain

$$A_{mn}\,[2m^2\pi^2 + (1-\nu)\lambda^2 n^2] - B_{mn}\,(1-\nu)\lambda mn\pi + C_{mn}\,2\nu\lambda m\pi = 0$$

$$A_{mn}\,3(1+\nu)\,\lambda mn\pi - B_{mn}[3(1-\nu)m^2\pi^2 + 6\lambda^2 n^2] - C_{mn}\,2\lambda^2 n[3 + \eta^2(m^2\pi^2 + \lambda^2 n^2)] = 0$$

$$A_{mn}\,3\nu\,\lambda m\pi - B_{mn}\lambda^2 n^2[3 + \eta^2(m^2\pi^2 + \lambda^2 n^2)] - C_{mn}\,[3\lambda^2 + \eta^2(m^2\pi^2 + \eta^2 n^2)^2] = -\frac{D_{mn} l^2 h}{2D}$$

where $\lambda = \dfrac{l}{a}$, $\eta = \dfrac{h}{2l}$.

Since D_{mn} is known from Eqs. (4), the coefficients A_{mn}, B_{mn}, C_{mn} can be calculated for any value of m and n and, therefore, the resultant forces, moments and components of displacement can be found for any point of the cylindrical shell by using Eqs. (10.50) and Eqs. (1).

10.14 Analysis of a Spherical Shell

For a spherical shell $R_1 = R_2 = R$ and Eqs. (10.9) to (10.14) become

$$\left.\begin{aligned}
&\frac{\partial(N_\varphi \sin\varphi)}{\partial\varphi} + \frac{\partial N_{\theta\varphi}}{\partial\theta} - N_\theta\cos\varphi - Q_\varphi\sin\varphi + R\sin\varphi . p_\varphi = 0\\
&\frac{\partial}{\partial\varphi}(N_{\varphi\theta}\sin\varphi) + \frac{\partial N_\theta}{\partial\theta} + N_{\theta\varphi}\cos\varphi - Q_\theta\sin\varphi + R\sin\varphi . p\theta = 0\\
&(N_\theta + N_\varphi)\sin\varphi + \frac{\partial Q_\theta}{\partial\theta} + \frac{\partial}{\partial\varphi}(Q_\varphi\sin\varphi) + R\sin\varphi . p_r = 0\\
&\frac{\partial}{\partial\varphi}(M_\varphi\sin\varphi) + \frac{\partial M_{\theta\varphi}}{\partial\theta} - M_\theta\cos\varphi - R\sin\varphi . Q_\varphi = 0\\
&\frac{\partial}{\partial\varphi}(M_{\varphi\theta}\sin\varphi) + \frac{\partial M_\theta}{\partial\theta} + M_{\theta\varphi}\cos\varphi - R\sin\varphi . Q_\theta = 0\\
&R(N_{\theta\varphi} - N_{\varphi\theta})\sin\varphi - M_{\theta\varphi}\sin\varphi + \sin\varphi\, M_{\varphi\theta} = 0
\end{aligned}\right\} \quad ...(10.58)$$

The expressions for direct strains and curvatures become from Eqs. (10.44) and (10.45) by changing z into r as

$$\left.\begin{aligned}
\varepsilon_\varphi &= \frac{1}{R}\left(\frac{\partial u_\varphi}{\partial\varphi} + u_r\right)\\
\varepsilon_\theta &= \frac{1}{R}\left[\frac{1}{\sin\varphi}\frac{\partial u_\theta}{\partial\theta} + u_\varphi\cot\varphi + u_r\right]\\
\gamma_{\theta\varphi} &= \frac{1}{R}\frac{\partial u_\theta}{\partial\varphi} - \frac{u_\theta}{R}\cot\varphi + \frac{1}{R\sin\varphi}\frac{\partial u_\varphi}{\partial\theta}
\end{aligned}\right\} \quad ...(10.59)$$

$$
\left.\begin{aligned}
\chi_\varphi &= \frac{1}{R^2}\frac{\partial}{\partial\varphi}\left[\frac{\partial u_r}{\partial\varphi} - u_\varphi\right] \\
\chi_\theta &= \frac{1}{R^2}\left[\frac{1}{\sin^2\varphi}\frac{\partial^2 u_r}{\partial\theta^2} + \cot\varphi\frac{\partial u_r}{\partial\varphi} - u_\varphi\cot\varphi - \frac{1}{\sin\varphi}\frac{\partial u_\theta}{\partial\theta}\right] \\
\chi_{\varphi\theta} &= \frac{1}{R^2}\left[\frac{1}{\sin\varphi}\frac{\partial^2 u_r}{\partial\varphi\partial\theta} - \frac{\cos\varphi}{\sin^2\varphi}\frac{\partial u_r}{\partial\theta}\right] - \frac{1}{2R^2}\left[\frac{1}{\sin\varphi}\frac{\partial u_\varphi}{\partial\theta} - u_\theta\cot\varphi + \frac{\partial u_\theta}{\partial\varphi}\right]
\end{aligned}\right\} \quad \text{...(10.60)}
$$

The expressions for the curvatures can also be written as :

$$
\left.\begin{aligned}
\chi_\varphi &= \frac{1}{R^2}\left(\frac{\partial^2 u_r}{\partial\varphi^2} + u_r\right) - \frac{\varepsilon_\varphi}{R} \\
\chi_\theta &= \frac{1}{R^2}\left(\frac{1}{\sin^2\varphi}\cdot\frac{\partial^2 u_r}{\partial\theta^2} + \cot\varphi\frac{\partial u_r}{\partial\varphi} + u_r\right) - \frac{\varepsilon_\theta}{R} \\
\chi_{\varphi\theta} &= \frac{1}{R^2}\left(\frac{1}{\sin\varphi}\frac{\partial^2 u_r}{\partial\varphi\partial\theta} - \frac{\cos\varphi}{\sin^2\varphi}\frac{\partial u_r}{\partial\theta}\right) - \frac{\gamma_{\theta\varphi}}{2R}
\end{aligned}\right\} \quad \text{...(10.61)}
$$

Using the first three of Eqs. (10.58) the following compatibility equations can be derived

$$
\frac{\partial}{\partial\varphi}\left(\sin\varphi\frac{\partial\gamma_{\theta\varphi}}{\partial\theta}\right) = \frac{\partial}{\partial\varphi}\left(\sin^2\varphi\frac{\partial\varepsilon_\varphi}{\partial\theta}\right) + \frac{\partial^2\varepsilon_\varphi}{\partial\theta^2} - \sin\varphi\cos\varphi\frac{\partial\varepsilon_\varphi}{\partial\varphi} + 2\sin^2\varphi\,\varepsilon_\varphi - \frac{\sin^2\varphi}{R}(\nabla^2 + 2)\,u_r \quad \text{...(10.62)}
$$

where $\nabla^2 = \dfrac{\partial^2}{\partial\varphi^2} + \cot\varphi\dfrac{\partial}{\partial\varphi} + \dfrac{1}{\sin^2\varphi}\dfrac{\partial^2}{\partial\theta^2}$.

Now stress resultants are

$$
N_\theta = \int_{-h/2}^{h/2}\sigma_\theta\,dz = K[\varepsilon_\theta + \nu\varepsilon_\varphi] \quad \text{...(10.63a)}
$$

$$
N_\varphi = \int_{-h/2}^{h/2}\sigma_\varphi\,dz = K[\varepsilon_\varphi + \nu\varepsilon_\theta] \quad \text{...(10.63b)}
$$

$$N_{\theta\varphi} = N_{\varphi\theta} = \int_{-h/2}^{h/2} \tau_{\theta\varphi}\, dz = \frac{K(1-\nu)}{2}\gamma_{\theta\varphi} \qquad ...(10.63c)$$

$$M_{\theta} = \int_{-h/2}^{h/2} \sigma_{\theta}\, z\, dz = -D[\chi_{\theta} + \nu\chi_{\varphi}] \qquad ...(10.63d)$$

$$M_{\varphi} = \int_{-h/2}^{h/2} \sigma_{\varphi}\, z\, dz = -D[\chi_{\varphi} + \nu\chi_{\theta}] \qquad ...(10.63e)$$

$$M_{\theta\varphi} = M_{\varphi\theta} = -\int_{-h/2}^{h/2} \tau_{\theta\varphi}\, z\, dz = D(1-\nu)\chi_{\theta\varphi} \qquad ...(10.63f)$$

where $K = \dfrac{Eh}{1-\nu^2}$, $D = \dfrac{Eh^3}{12(1-\nu^2)}$

and h is the shell thickness.

Substituting for the strain terms in terms of the direct stress resultants and using Eqs. (10.58), we get

$$(\nabla^2 + 2)(N_{\varphi} + N_{\theta}) - \frac{Eh}{R}(\nabla^2 + 2)u_r$$

$$= -(1+\nu)\left\{\frac{1}{\sin\varphi}\frac{\partial}{\partial[\text{illegible}]}([\text{illegible}]R\sin\varphi) + \frac{1}{\sin^2\varphi}\frac{\partial}{\partial\theta}[\text{illegible}]\varphi) - p_r\right\} \qquad ...(10.64)$$

Also

$$\left.\begin{aligned} \bar{\varepsilon}_{\theta} &= \varepsilon_{\theta} - z\chi_{\theta} \\ \bar{\varepsilon}_{\varphi} &= \varepsilon_{\varphi} - z\chi_{\varphi} \\ \bar{\gamma}_{\theta\varphi} &= \gamma_{\theta\varphi} - 2z\chi_{\theta\varphi} \end{aligned}\right\} \qquad ...(10.65)$$

and

$$\left.\begin{aligned} \bar{\varepsilon}_{\theta} &= \frac{\sigma_{\theta}}{E} - \frac{\nu\sigma_{\varphi}}{E} \\ \bar{\varepsilon}_{\varphi} &= \frac{\sigma_{\varphi}}{E} - \frac{\nu\sigma_{\theta}}{E} \\ \bar{\gamma}_{\theta\varphi} &= \frac{\tau_{\theta\varphi}}{G} \end{aligned}\right\} \qquad ...(10.66)$$

or

$$\sigma_{\theta} = \frac{E}{1-\nu^2}\left[\bar{\varepsilon}_{\theta} + \nu\,\bar{\varepsilon}_{\varphi}\right] \qquad ...(10.67a)$$

$$\sigma_{\varphi} = \frac{E}{1-\nu^2}\left[\bar{\varepsilon}_{\varphi} + \nu\,\bar{\varepsilon}_{\theta}\right] \qquad ...(10.67b)$$

$$\tau_{\theta\varphi} = \frac{E}{2(1+\nu)} \bar{\gamma}_{\theta\varphi} \qquad ...(10.67c)$$

By means of fourth of Eqs. (10.58) and Eqs. (10.67) the lateral shear force resultant Q_φ can be expressed in terms of u_r as

$$Q_\varphi = \frac{D}{R^3} \frac{\partial}{\partial \varphi} (\nabla^2 + 2) u_r \Big/ \left(1 + \frac{h^2}{12 R^2}\right)$$

Similarly from the fifth of Eqs. (10.58), we find

$$Q_\theta = \frac{D}{R^3} \cdot \frac{1}{\sin\varphi} \cdot \frac{\partial}{\partial \theta} (\nabla^2 + 2) u_r \Big/ \left(1 + \frac{h^2}{12 R^2}\right)$$

If $\frac{h^2}{12R^2} << 1$, then we find

$$\left.\begin{aligned} \sigma_\varphi &= \frac{D}{R^3} \frac{\partial}{\partial \varphi} (\nabla^2 + 2) u_r \\ Q_\theta &= \frac{D}{R^3} \frac{1}{\sin\varphi} \frac{\partial}{\partial \theta} (\nabla^2 + 2) u_r \end{aligned}\right\} \qquad ...(10.68)$$

The first and second equilibrium equations (10.58) are satisfied if we express the stress resultants in terms of a stress function Φ, the radial displacement u_r, and a potential function V defining the loading as follows :

$$\left.\begin{aligned} N_\varphi &= \frac{1}{\sin^2\varphi} \frac{\partial^2 \Phi}{\partial \theta^2} + \cot\varphi \frac{\partial \Phi}{\partial \varphi} + \Phi + \frac{D}{R^3} (\nabla^2 + 2) u_r + V \\ N_\theta &= \frac{\partial^2 \Phi}{\partial \varphi^2} + \Phi + \frac{D}{R^3} (\nabla^2 + 2) u_r + V \\ N_{\theta\varphi} &= \frac{\cot\varphi}{\sin\varphi} \frac{\partial \Phi}{\partial \theta} - \frac{1}{\sin\varphi} \frac{\partial^2 \Phi}{\partial \varphi \partial \theta} \end{aligned}\right\} \qquad ...(10.69)$$

where V defines the tangential loading components as :

$$p_\theta = \frac{-1}{R\sin\varphi} \frac{\partial V}{\partial \theta}, \; p_\varphi = -\frac{1}{R} \frac{\partial V}{\partial \varphi} \qquad ...(10.70)$$

Using these definitions in the third equilibrium Eq. (10.58), yields

$$(\nabla^2 + 2)\Phi + \frac{D}{R^3} (\nabla^2 + 2) u_r = -2V + p_r R \qquad ...(10.71)$$

Also by substituting for N_φ and N_θ from Eqs. (10.69) in the compatibility equation (10.64) :

$$\nabla^2(\nabla^2 + 2)\,\Phi - \frac{Eh}{R}(\nabla^2 + 2)u_r = -(1-\nu)\,\nabla^2 V - (1-\nu)p_r R \qquad \text{...(10.72)}$$

Eqs. (10.71) and (10.72) define the problem of a spherical shell.

10.15 Membrane Solution for a Spherical Shell

Neglecting moments and shears, Eqs. (10.58) reduce to

$$\left.\begin{aligned} \frac{\partial}{\partial \varphi}(N_\varphi \sin\varphi) + \frac{\partial N_{\theta\varphi}}{\partial \theta} - N_\theta \cos\varphi + R\sin\varphi\, p_\varphi = 0 \\ \frac{\partial}{\partial \varphi}(N_{\varphi\theta} \sin\varphi) + \frac{\partial N_\theta}{\partial \theta} + N_{\theta\varphi}\cos\varphi + R\sin\varphi\, p_\theta = 0 \\ (N_\theta + N_\varphi)\sin\varphi + R\sin\varphi\, p_r = 0 \end{aligned}\right\} \qquad \text{...(10.73)}$$

EXERCISES

(*Note* : Assume $E = 2 \times 10^5$ N/mm^2 ; $\nu = 0.30$)

10.1 (*a*) Differentiate between thin and thick shells.

(*b*) Explain membrane stresses.

10.2 Derive equations of equilibrium for a thin shell of general shape. State clearly the assumptions made.

10.3 Explain the membrane theory of thin shells and derive the equilibrium equations based upon this theory.

10.4 (*a*) What do you understand by a shell of revolution ?

(*b*) Derive equilibrium equations for a shell of revolution.

10.5 By neglecting bending stresses, derive the equations of equilibrium for shells of revolution.

10.6 A spherical dome of constant thickness 8 mm is having radius of curvature 1 m. If the density of the dome material is (7470 kg/m^3), determine the stresses in the dome at an angular locations of 30° and 60° with the line of symmetry.

10.7 A shell in the form of ellipsoid of revolution is used as an end plate for a cylindrical boiler. The semi-major and semi-minor axes for the shell are 1.5 m and 1.2 m respectively. If the steam pressure is 1.5 MPa determine the stresses in the shell at the top and at the equator.

10.8 A spherical tank of diameter 2 m is completely filled with a liquid of density 1150 kg/m^3. The tank is supported along a parallel circle whose radius is 0.5 m. Determine the stresses in the tank at parallel circles whose radii are 0.8 m and 0.4 m.

10.9 A conical shell of vertical height 1.5 m and semi-vertex angle 60° is completely filled with a liquid of density 1050 kg/m^3. Find the stresses in the shell.

10.10 A thin shell in the form of a torus is obtained by rotation of a circle of radius 0.5 m about a vertical axis. The radius of rotation is 2 m. This shell is filled with a liquid which exerts a uniform pressure of 2 MPa. Determine the maximum stresses in the shell.

10.11 (*a*) Derive equilibrium equations for a cylindrical shell.

(*b*) Reduce the above equations to obtain the membrane solution for a circular cylindrical shell.

10.12 A horizontal thin circular cylinder is filled with steam at a constant pressure of 2 MPa. The radius of the cylinder is 0.4 m and is 1.2 m long. Assuming thickness of the cylinder wall as 10 mm, determine the membrane forces in the cylinder, when (*i*) the ends are simply supported (*ii*) the ends are assumed to be rigidly built in.

10.13 A horizontal thin circular cylinder is under its own weight. If the density of the material from which the cylinder is made is 7470 kg/m^3, radius of cylinder 0.5 m, length 2 m and wall thickness 8 mm ; find the membrane forces in the cylinder. Assume the ends to be rigidly built in.

10.14 (*a*) Derive expressions for strains and displacements in shells of revolution.

(*b*) Reduce the above expressions for cylindrical shells.

10.15 (*a*) Derive Donnell's equations for cylindrical shells.

(*b*) Reduce these equations for symmetrically loaded circular cylindrical shells.

10.16 A long circular pipe of radius 300 mm and wall thickness 10 mm is subjected to a uniform bending moment of 10 N-m and a shear force of 1600 N at one end. Determine the maximum deflection of the pipe.

10.17 A long circular pipe of radius 250 mm and wall thickness 12 mm is subjected to a uniformly distributed load of 7500 N along a circular section. Determine the maximum stresses in the pipe and the maximum deflection.

10.18 A long circular cylinder of radius 400 mm and wall thickness 10 mm is subjected to a uniform band of pressure of intensity 2 N/mm^2 over a length of 200 mm. Determine the maximum deflection of the cylinder and the stresses at the centre and end of the pressure band.

10.19 A vertical cylindrical taken is 600 mm in radius and the wall thickness is 15 mm. The length of the cylinder is 5 m. It is completely filled with water. Determine the maximum stresses in the tank wall.

10.20 A horizontal circular cylindrical tank of radius 500 mm and wall thickness 12 mm is $\frac{3}{4}$th full with water. The length of the tank is 3 m and it is simply supported at the ends. Determine the stresses in the tank wall.

11

Beams on Elastic Foundation

11.1 Introduction

In this chapter, we shall analyse the elastically supported beams. The elastic support is provided by a load-bearing medium, referred to as the 'foundation', distributed continuously along the length of the beam. Such conditions of support can be found in a large variety of technical problems. In some of these problems the identity of the beam and the foundation can be easily established, as in the case of actual foundation structures or in the case of railroad track. In other problems, however, the concept of beam and foundation is more of an abstract nature as in the case of thin-walled tubes, shells and domes, where the elastic foundation for the beam part is supplied by the resilience of the adjoining portions of a continuous elastic structure.

There are two basic types of elastic foundation. The first type is characterized by the fact that the pressure in the foundation is proportional at every point to the deflection occurring at that point and is independent of pressures or deflections produced elsewhere in the foundation. Such a foundation is called Winkler type foundation. The second type of foundation is furnished by an elastic solid which, in contrast to the first one, represents the case of complete continuity in the supporting medium. Foundations of the first type have by far the wider field of applications in physical sciences. We shall study, in ths chapter, both of these type of foundations.

11.2 Differential Equation for a Beam Resting on Elastic Foundation

Let us consider a straight beam supported along its entire length by an elastic medium and subjected to vertical forces acting in the principal plane of the symmetrical cross-section as shown in Fig. 11.1. Because of this action the beam will deflect, producing continuously distributed reaction forces in the supporting medium whose intensity p at any point is proportional to the deflection of the beam y at that point, *i.e.* $p = ky$, where k is a constant of the supporting medium, called the modulus of the foundation. If we consider unit width of the beam then the units of k will be N/m^2.

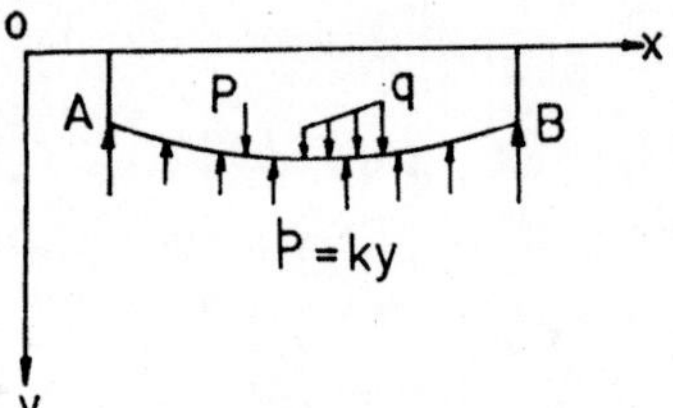

Fig. 11.1

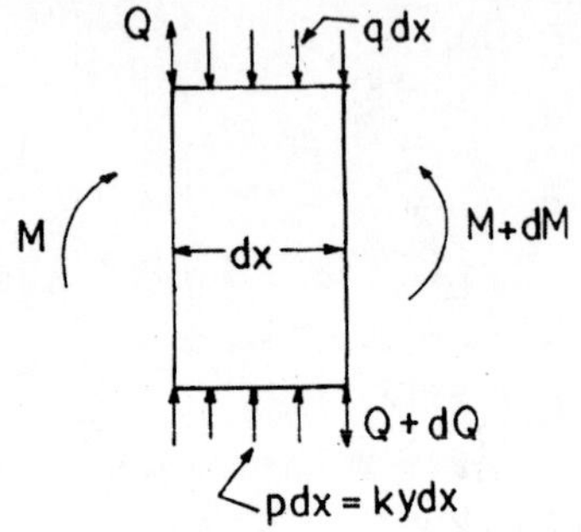

Fig. 11.2

Let us consider an infinitely small element enclosed between two vertical cross-sections a distance dx apart on the beam under consideration where the beam is acted upon by a distributed loading q N/m. The forces acting on this element are shown in Fig. 11.2. The upward acting shearing force, Q, to the left of the cross-section is considered positive and clockwise bending moment M is taken positive. For the equilibrium of the element, we have

$$Q - (Q + dQ) + kydx - qdx = 0$$

or
$$\frac{dQ}{dx} = ky - q$$

Now
$$Q = \frac{dM}{dx}$$

$$\therefore \quad \frac{dQ}{dx} = \frac{d^2M}{dx^2} = ky - q$$

Also for the bending of a beam

$$EI\frac{d^2y}{dx^2} = -M$$

$$\therefore \quad EI\frac{d^4y}{dx^4} = -ky + q \qquad ...(11.1)$$

This is the differential equation for the deflection curve of a beam supported on an elastic foundation. On the unloaded part of the beam where $q = 0$, we have

$$EI\frac{d^4y}{dx^4} = -ky \qquad ...(11.2)$$

Let $EI = D$, the flexural rigidity of the beam, then Eq. (11.1) becomes

$$\frac{d^4y}{dx^4} + \frac{k}{D}y = \frac{q}{D} \qquad ...(11.3)$$

The general solution of Eq. (11.3) may be written as

$$y = y_c + y_p \qquad ...(11.4)$$

where y_c is the complimentary solution of the auxiliary equation

$$\frac{d^4y}{dx^4} + 4\lambda^4 y = 0 \qquad ...(11.5)$$

where $\lambda = \sqrt[4]{\frac{k}{4D}}$,

is called the characteristic of the system.

The complimentary solution of Eq. (11.5) is

$$y_c = e^{\lambda x}(C_1 \cos \lambda x + C_2 \sin \lambda x) + e^{-\lambda x}(C_3 \cos \lambda x + C_4 \sin \lambda x) \qquad ...(11.6)$$

The particular solution y_p is

$$y_p = \frac{q}{k} \qquad ...(11.7)$$

Hence the complete solution becomes

$$y = e^{\lambda x}(C_1 \cos \lambda x + C_2 \sin \lambda x) + e^{-\lambda x}(C_3 \cos \lambda x + C_4 \sin \lambda x) + \frac{q}{k} \qquad ...(11.8)$$

Eq. (11.8) represents the general solution for the deflection line of a straight prismatic bar supported on an elastic foundation.

By differentiation of Eq. (11.8), we find

$$\frac{1}{\lambda}\frac{dy}{dx} = e^{\lambda x}(C_1(\cos \lambda x - \sin \lambda x) + C_2(\cos \lambda x + \sin \lambda x)] - e^{-\lambda x}[C_3(\cos \lambda x + \sin \lambda x) - C_4(\cos \lambda x - \sin \lambda x)] \qquad ...(11.9)$$

$$\frac{1}{2\lambda^2}\frac{d^2y}{dx^2} = -e^{-\lambda x}(C_1 \sin \lambda x - C_2 \cos \lambda x) + e^{-\lambda x}(C_3 \sin \lambda x - C_4 \cos \lambda x) \qquad ...(11.10)$$

$$\frac{1}{2\lambda^3}\frac{d^3y}{dx^3} = -e^{\lambda x}[C_1(\cos \lambda x + \sin \lambda x) - C_2(\cos \lambda x - \sin \lambda x)] + e^{-\lambda x}[C_3(\cos \lambda x - \sin \lambda x) + C_4(\cos \lambda x + \sin \lambda x)] \qquad ...(11.11)$$

Now $\frac{dy}{dx} = \tan \theta, -EI\frac{d^2y}{dx^2} = M, -EI\frac{d^3y}{dx^3} = Q \qquad ...(11.12)$

Hence we can obtain the general expressions for the slope θ of the deflection line, bending moment M and shearing force Q. The constants C_1, C_2, C_3 and C_4 can be obtained from the conditions existing at the two ends of the beam.

From Eqs. (11.8) to (11.11), we get at $x = 0$

$$\left.\begin{aligned} (y)_{x=0} &= y_0 = C_1 + C_3 \\ \left(\frac{dy}{dx}\right)_{x=0} &= \theta_0 = \lambda(C_1 + C_2 - C_3 + C_4) \\ \left(-EI\frac{d^2y}{dx^2}\right)_{x=0} &= M_0 = 2\lambda^2 EI(-C_2 + C_4) \\ \left(-EI\frac{d^3y}{dx^3}\right)_{x=0} &= Q_0 = 2\lambda^3 EI(C_1 - C_2 - C_3 - C_4) \end{aligned}\right\} \quad ...(11.13)$$

Solving for C_1 to C_4, we find

$$\left.\begin{aligned} C_1 &= \frac{1}{2}y_0 + \frac{1}{4\lambda}\theta_0 + \frac{1}{8\lambda^3 EI}Q_0 \\ C_2 &= \frac{1}{4\lambda}\theta_0 - \frac{1}{4\lambda^2}EIM_0 - \frac{1}{8\lambda^3 EI}Q_0 \\ C_3 &= \frac{1}{2}y_0 - \frac{1}{4\lambda}\theta_0 - \frac{1}{8\lambda^3 EI}Q_0 \\ C_4 &= \frac{1}{4\lambda}\theta_0 + \frac{1}{4\lambda^2 EI}M_0 - \frac{1}{8\lambda^3 EI}Q_0 \end{aligned}\right\} \quad ...(11.14)$$

Substituting in Eq. (11.8), we get

$$y_x = y_0F_1(\lambda x) + \frac{1}{\lambda}\theta_0F_2(\lambda x) - \frac{1}{\lambda^2 EI}M_0F_3(\lambda x) - \frac{1}{\lambda^3 EI}Q_0F_4(\lambda x) + \frac{q}{k} \quad ...(11.15)$$

where $F_1(\lambda x) = \cosh \lambda x \cos \lambda x$

$$F_2(\lambda x) = \frac{1}{2}(\cosh \lambda x \sin \lambda x + \sinh \lambda x \cos \lambda x)$$

$$F_3(\lambda x) = \frac{1}{2}\sinh \lambda x \sin \lambda x$$

$$F_4(\lambda x) = \frac{1}{4}(\cosh \lambda x \sin \lambda x - \sinh \lambda x \cos \lambda x)$$

When the beam carries concentrated and distributed forces and moments as shown in Fig. 11.3 along its length, then we can write Eq. (11.15) as :

$$y_x = y_0F_1(\lambda x) + \frac{1}{\lambda}\theta_0F_2(\lambda x) - \frac{1}{\lambda^3 EI}M_0F_3(\lambda x) - \frac{1}{\lambda^3 EI}Q_0F_4(\lambda x) - \frac{1}{\lambda^3 EI}MF_3[\lambda(x - x_a)]$$

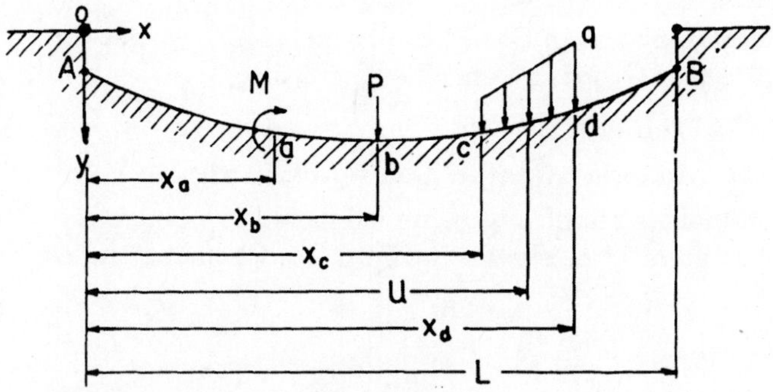

Fig. 11.3

$$+\frac{1}{\lambda^3 EI}PF_4[\lambda(x-x_b)]+\frac{1}{\lambda^3 EI}\int_{x_c}^{x_d} qE_4\,[(x-u)]du \quad ...(11.16)$$

$$\left.\begin{aligned}
\frac{dy_x}{dx} &= \theta_x = \theta_0 F_1(\lambda x)-\frac{1}{\lambda EI}M_0F_2(\lambda x)-\frac{1}{\lambda^2 EI}Q_0F_3(\lambda x)\\
&-4\lambda y_0F_4(\lambda x)-\frac{1}{\lambda EI}MF_2\,[\lambda(x-x_a)]\\
&+\frac{1}{\lambda^2 EI}PF_3[\lambda(x-x_b)]+\frac{1}{\lambda^2 EI}\int_{x_c}^{x_d} qF_3\,[\lambda(x-u)]\,du\\
-EI\frac{d^2y_x}{dx^2} &= M_x = M_0F_1(\lambda x)+\frac{1}{\lambda}Q_0F_2(\lambda x)+\frac{k}{\lambda^2}y_0F_3(\lambda x)\\
&+\frac{k}{\lambda^3}\theta_0F_4(\lambda x)+MF_1[\lambda(x-x_a)]\\
&-\frac{1}{\lambda}PF_2[\lambda(x-x_b)]-\frac{1}{\lambda}\int_{x_c}^{x_d} qF_2[\lambda(x-u)]\,du\\
-EI\frac{d^3y_x}{dx^3} &= Q_x = Q_0F_1(\lambda x)+\frac{k}{\lambda}y_0F_2(\lambda x)+\frac{k}{\lambda^2}\theta_0F_3(\lambda x)\\
&-4\lambda M_0F_4(\lambda x)-4\lambda MF_4[\lambda(x-x_a)]\\
&-PF_1[\lambda(x-x_b)]-\int_{x_c}^{x_d} qF_1[\lambda(x-u)]\,du
\end{aligned}\right\}$$

...(11.17)

where $\frac{dF_1}{dx}=-4\lambda F$; $\quad \frac{dF_2}{dx}=\lambda F_1$

$\frac{dF_3}{dx}=\lambda F_2$; $\quad \frac{dF_4}{dx}=\lambda F_3$

Putting $x = L$ in Eqs. (11.16) and (11.17), we get the values of y_L, θ_L, M_L and Q_L for the end B of the beam.

11.3 Infinite Beam

11.3.1 Concentrated Load on an Infinite Beam

Consider an infinite beam subjected to a single concentrated load P at point O as shown in Fig. 11.4. Choosing the co-ordinate

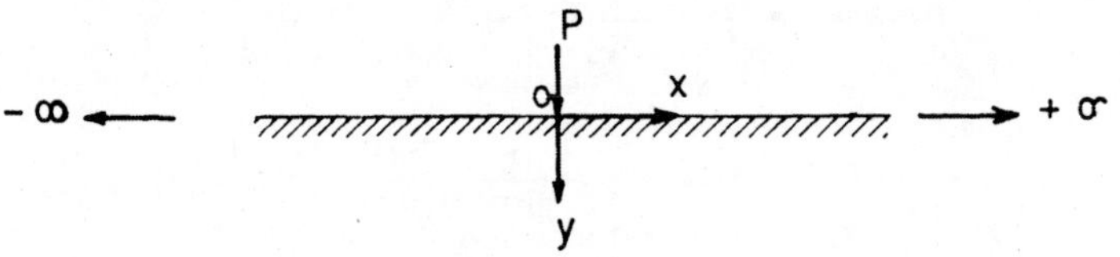

Fig. 11.4

system as shown, due to the symmetry of the deflection curve, we consider only the right half of the beam.

As x and y tends to infinity, the deflection of the beam must be zero. Hence $C_1 = C_2 = 0$ and Eq. (11.8) becomes

$$y = e^{-\lambda x} (C_3 \cos \lambda x + C_4 \sin \lambda x)$$

From the condition of symmetry, as have

$$\left(\frac{dy}{dx}\right)_{x=0} = 0$$

$$\therefore \quad C_3 - C_4 = 0$$

or $$C_3 = C_4 = C \text{ (say)}$$

Hence $$y = Ce^{-\lambda x} (\cos \lambda x + \sin \lambda x).$$

The constant C can be obtained from the condition that

$$2\int_0^{\infty} ky \, dx = P$$

$$\therefore \quad 2\int_0^{\infty} kCe^{-\lambda x} (\cos \lambda x + \sin \lambda x) \, dx = P$$

$$2kC \cdot \frac{1}{\lambda} = P$$

$$\therefore \quad C = \frac{P\lambda}{2k}$$

$$\therefore \quad y = \frac{P\lambda}{2k} e^{-\lambda x} (\cos \lambda x + \sin \lambda x) \qquad ...(11.18)$$

$$(y)_{x=0} = \frac{P\lambda}{2k}$$

$$y = 0,$$

where $\cos \lambda x + \sin \lambda x = 0$.

or $$\lambda x = \frac{3\pi}{4}, \frac{7\pi}{4}, \text{ etc.}$$

Now $$\frac{dy}{dx} = \theta = -\frac{P\lambda^2}{k} e^{-\lambda x} \sin \lambda x \qquad \text{...(11.19)}$$

$$-EI\frac{d^2y}{dx^2} = M = \frac{P}{4\lambda} e^{-\lambda x} (\cos \lambda x - \sin \lambda x) \qquad \text{...(11.20)}$$

$$-EI\frac{d^3y}{dx^3} = Q = -\frac{P}{2} e^{-\lambda x} \cos \lambda x \qquad \text{...(11.21)}$$

The curves represented by Eqs. (11.18) to (11.21) are given in Fig. 11.5.

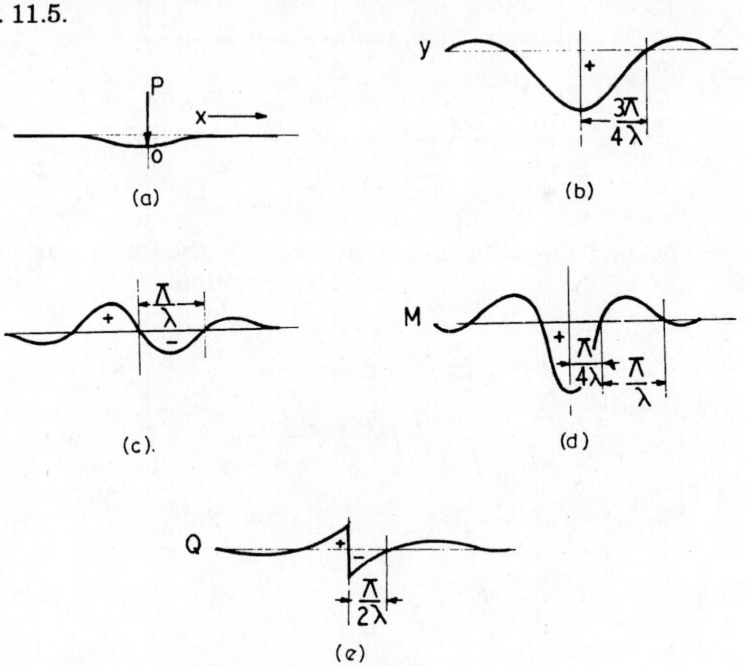

Fig. 11.5

Let $$e^{-\lambda x} (\cos \lambda x + \sin \lambda x) = A_{\lambda x}$$

$$e^{-\lambda x} \sin \lambda x = B_{\lambda x}$$

$$e^{-\lambda x} (\cos \lambda x - \sin \lambda x) = C_{\lambda x}$$

$$e^{-\lambda x} \cos \lambda x = D_{\lambda x}.$$

Then these equations become

$$\left.\begin{aligned} y &= \frac{P\lambda}{2k} A_{\lambda x} \\ \theta &= -\frac{P\lambda^2}{k} B_{\lambda x} \\ M &= \frac{P}{4\lambda} C_{\lambda x} \\ Q &= -\frac{P}{2} D_{\lambda x} \end{aligned}\right\} \quad \text{...(11.22)}$$

11.3.2 Concentrated Moment on an Infinite Beam

Let a concentrated moment M_0 be applied at point O on the infinitely long beam, as shown in Fig. 11.6 (*a*). The concentrated

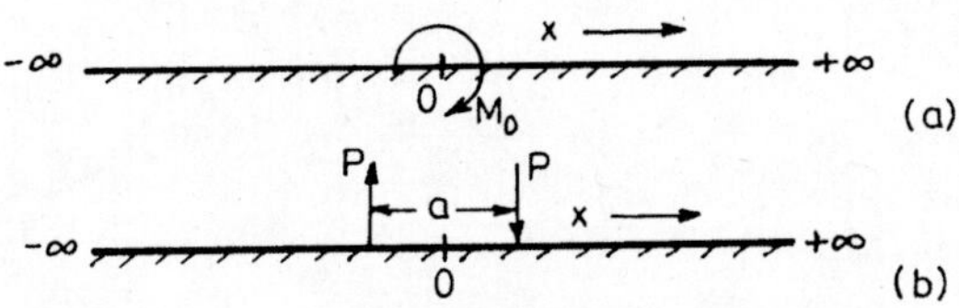

Fig. 11.6

moment can be regarded as a limiting case of the loading shown in Fig. 11.6 (*b*). As $a \to 0$, $Pa \to M_0$. Hence from Eq. (11.22), we find that

$$y = \frac{P\lambda}{2k} [-A_{\lambda(x+a)} + A_{\lambda x}]$$

$$= -\frac{Pa\lambda}{2k} \left[\frac{A_{\lambda(x+a)} + A_{\lambda x}}{a} \right] \quad \text{for } x > 0.$$

$$\text{Now} \left[\frac{A_{\lambda(x+a)} + A_{\lambda x}}{a} \right]_{a \to 0} = \frac{d}{dx} A_{\lambda x} = -2\lambda B_{\lambda x}$$

and $\quad (Pa)_{a \to 0} = M_0$

$$y = \frac{M_0 \lambda^2}{k} B_{\lambda x} \quad \text{...(11.23a)}$$

and

$$\frac{dy}{dx} = \theta = \frac{M_0 \lambda^3}{k} C_{\lambda x} \quad \text{...(11.23b)}$$

$$-EI \frac{d^2y}{dx^2} = M = \frac{M_0}{2} D_{\lambda x} \quad \text{...(11.23c)}$$

$$-EI\frac{d^3y}{dx^3} = Q = -\frac{M_0\lambda}{2}A_{\lambda x} \qquad ...(11.23d)$$

The curves represented by Eq. (11.23) are shown in Fig. 11.7.

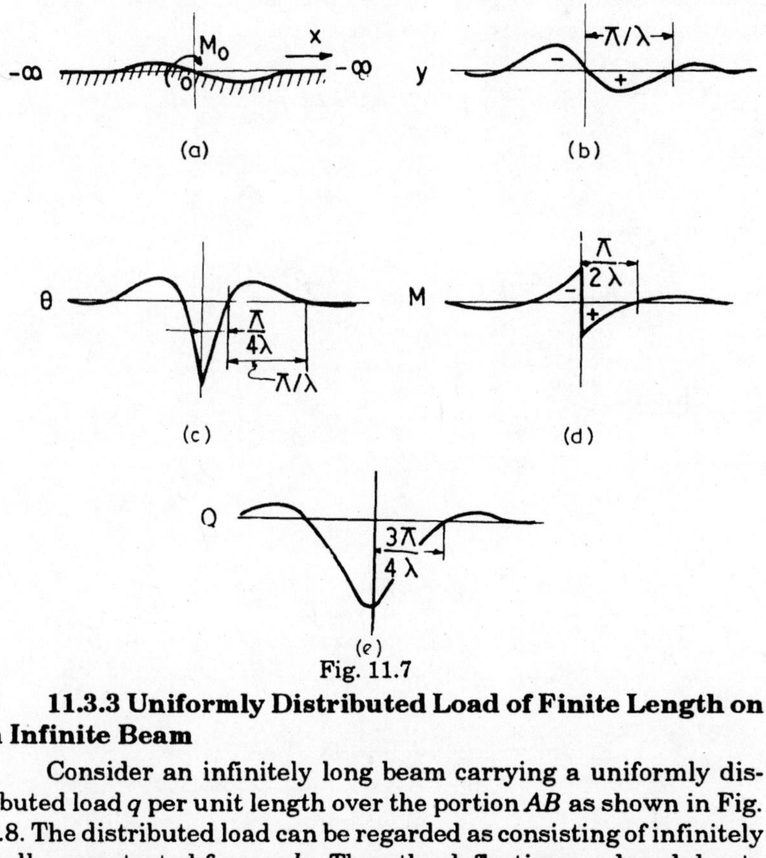

Fig. 11.7

11.3.3 Uniformly Distributed Load of Finite Length on an Infinite Beam

Consider an infinitely long beam carrying a uniformly distributed load q per unit length over the portion AB as shown in Fig. 11.8. The distributed load can be regarded as consisting of infinitely small concentrated force qdx. Then the deflection produced due to this load at any point C at a distance x is given by Eq. (11.18) as :

$$\delta y = qdx \cdot \frac{\lambda}{2k} \cdot e^{-\lambda x}(\cos \lambda x + \sin \lambda x)$$

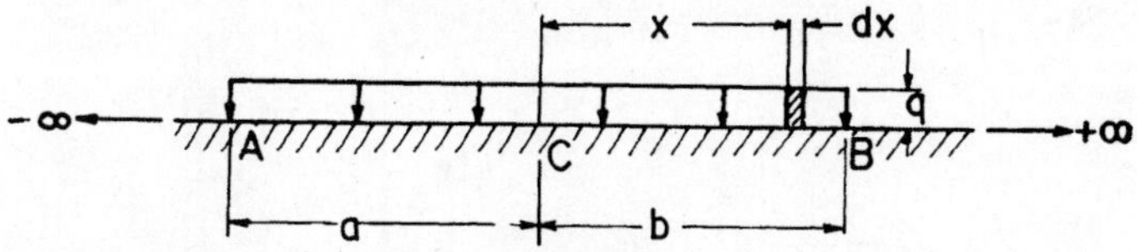

Fig. 11.8

The total effect of the distributed loading can be obtained by integrating the above expression within the $a-b$ limits of the loaded art of the beam.

(*i*) *Point C under the loading.* If the point C is under the loading as shown in Fig. 11.8, then the deflection at C becomes by noting that a is negative and b is positive.

$$y_c = \frac{q\lambda}{2k}\left[\int_0^a e^{-\lambda x}(\cos \lambda x + \sin \lambda x)\,dx + \int_0^b e^{-\lambda x}(\cos \lambda x + \sin \lambda x)\,dx\right]$$

$$= \frac{q}{2k}\left[(1 - e^{-\lambda a}\cos \lambda a) + (1 - e^{-\lambda b}\cos \lambda b)\right]$$

$$= \frac{q}{2k}(2 - D_{\lambda a} - D_{\lambda b})$$

Similarly,

$$\left.\begin{aligned}\theta_c &= \frac{q\lambda}{2k}(A_{\lambda a} - A_{\lambda b})\\ M_e &= \frac{q}{4\lambda^2}(B_{\lambda a} + B_{\lambda b})\\ Q_c &= \frac{q}{4\lambda}(C_{\lambda a} - C_{\lambda b})\end{aligned}\right\} \qquad \text{...(11.24)}$$

(*ii*) *Point C to the left of the loading.* When the point C lies to the left of the loading as shown in Fig. 11.9, then a and b are both positive. Hence on integration, we get

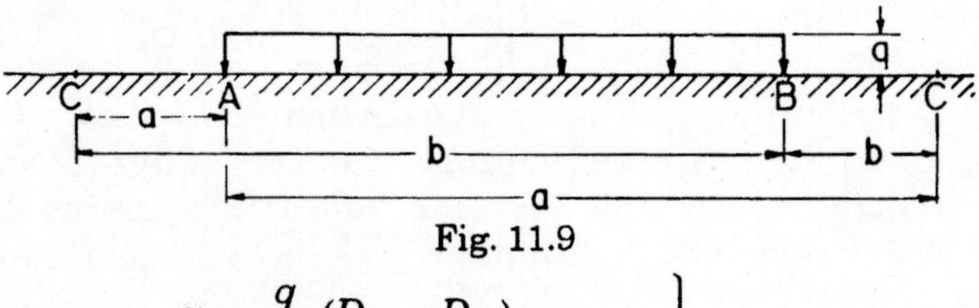

Fig. 11.9

$$\left.\begin{aligned}y_c &= \frac{q}{2k}(D_{\lambda a} - D_{\lambda b})\\ \theta_c &= \frac{q\lambda}{2k}(A_{\lambda a} - A_{\lambda b})\\ M_c &= -\frac{q}{4\lambda^2}(B_{\lambda a} - B_{\lambda b})\\ Q_c &= \frac{q}{4\lambda}(C_{\lambda a} - C_{\lambda b})\end{aligned}\right\} \qquad \text{...(11.25)}$$

(*iii*) *Point C to the right of the loading.* In this case a and b are both negative and we get on integration (Fig. 11.9).

$$\left.\begin{aligned} y_c &= -\frac{q}{2k}(D_{\lambda a} - D_{\lambda b}) \\ \theta_c &= \frac{q\lambda}{2k}(A_{\lambda a} - A_{\lambda b}) \\ M_c &= \frac{q}{4\lambda^2}(B_{\lambda a} - B_{\lambda b}) \\ Q_c &= \frac{q}{4\lambda}(C_{\lambda a} - C_{\lambda b}) \end{aligned}\right\} \qquad ...(11.26)$$

11.3.4 Infinite Beam Loaded Sinusoidally

Let the loading be represented by

$$q = q_0 \sin \frac{2\pi x}{l}$$

where q_0 is the peak intensity of the load.

Let us also assume that

$$y = A \sin \frac{2\pi x}{l}$$

Then the differential equation becomes

$$\left(\frac{2\pi}{l}\right)^4 A + \frac{k}{EI} A = \frac{q_0}{EI}$$

$$\therefore \qquad A = \frac{q_0}{\left(k + \dfrac{16\pi^4 EI}{l^4}\right)}$$

$$\therefore \qquad y = e^{-\beta x}(C_3 \cos \beta x + C_4 \sin \beta x) + \frac{q_0}{\left(k + \dfrac{16\pi^4 EI}{l^4}\right)} \sin \frac{2\pi x}{l}$$

For a repeated load

$$C_3 = C_4 = 0$$

$$\therefore \qquad y = \frac{q_0}{k} \cdot \frac{1}{\left(1 + \dfrac{\sqrt{2}\pi}{\lambda l}\right)^4} \sin \frac{2\pi x}{l} \qquad ...(11.27)$$

If the beam is carrying periodic load as shown in Fig. 11.10 (*a*) then this type of loading can be considered as the sum of loading shown in Fig. 11.10 (*b*) and (*c*).

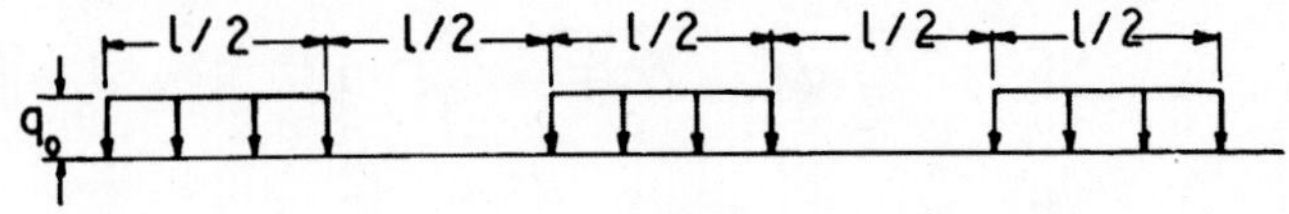

(a)

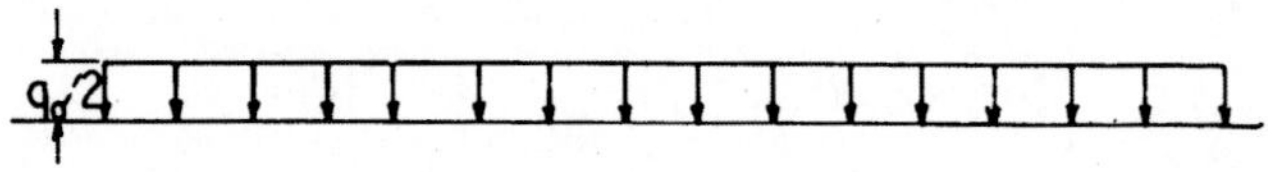

(b)

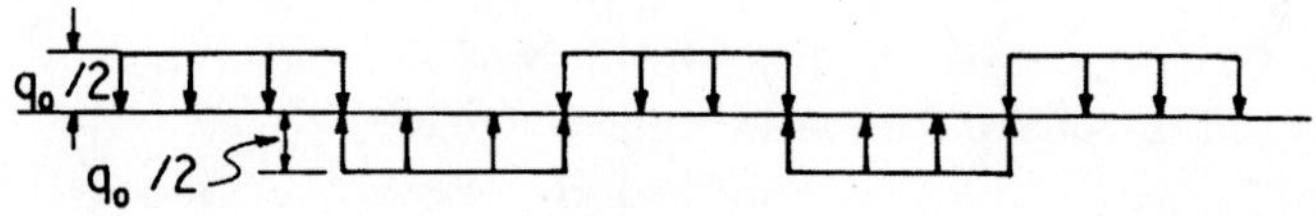

(c)

Fig. 11.10

This loading is equal to

$$= \frac{q_0}{2}\left[1 + \frac{4}{\pi}\left(\sin\frac{2\pi x}{l} + \frac{1}{3}\sin\frac{6\pi x}{l} + \frac{1}{5}\sin\frac{10\pi x}{l} + \ldots\right)\right]$$

Hence $$y_{max} = \frac{q_0}{2k}\left[1 + \frac{4}{\pi}\left\{\frac{1}{1+\left(\frac{\sqrt{2}\pi}{\lambda l}\right)^4} + \frac{1}{1+\left(\frac{3\pi\sqrt{2}}{\lambda l}\right)^4} + \ldots\right\}\right] \qquad \ldots(11.28)$$

11.3.5 General Loading Conditions on the Infinite Beam

When the infinitely long beam is subjected to a group of concentrated forces, then

$$y = \frac{\lambda}{2k}\sum_{n=1}^{n} P_n A_\lambda x_n$$

where x_n = absolute distance of the force P_n from the cross-section where the deflection is sought.

Similarly, under the action of group of concentrated moments, we have

$$y = \frac{\lambda^2}{k} \sum_{n=1}^{n} M_n B_\lambda x_n$$

Therefore, the resultant deflection under the action of a group of concentrated forces and moments become

$$\left.\begin{aligned} y &= \frac{\lambda}{2k} \sum_{n=1}^{n} P_n A_\lambda x_n + \frac{\lambda^2}{k} \sum_{n=1}^{n} M_n B_\lambda x_n \\ \text{Similarly, } M &= \frac{1}{4\lambda} \sum_{n=1}^{n} P_n C_{\lambda x} + \frac{1}{2} \sum_{n=1}^{n} M_n D_{\lambda x} \end{aligned}\right\} \quad \text{...(11.29)}$$

If there are concentrated reactions also then

$$y = \frac{\lambda}{2k}\left[\sum_{n=1}^{n} P_n A_\lambda x_n - \sum_{m=1}^{m} R_m A_\lambda x_m\right] + \frac{\lambda^2}{k} \sum_{n=1}^{n} M_n B_\lambda x_n \quad \text{...(11.30)}$$

11.4 Semi-Infinite Beam

A semi-infinite beam may be defined as a beam which has unlimited extension in only one direction, having at point A a finite end.

11.4.1 Semi-infinite Beam with Free End

Consider an infinitely long beam subjected to some loading due to which we have at point A (finite end) a bending moment M_A and shear force Q_A. If this moment and shear is made to vanish then the infinite beam would become a semi-infinite beam, since the end A of the semi-infinite beam has been assumed free. This can be achieved by applying at A a moment M_0 and force P_0 which will cause a bending moment $-M_A$ and a shearing force $-Q_A$ at that section as shown in Fig. 11.11.

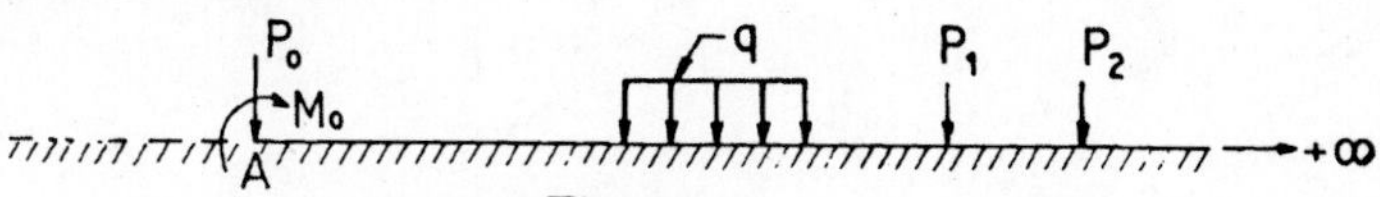

Fig. 11.11

As the end A is free, hence

$$M_A + \frac{P_0}{4\lambda} + \frac{M_0}{2} = 0$$

$$Q_A - \frac{P_0}{2} - \frac{M_0 \lambda}{2} = 0$$

$$\therefore \quad \left.\begin{aligned} P_0 &= 4(\lambda M_A + Q_A) \\ M_0 &= -\frac{2}{\lambda}(2\lambda M_A + Q_A) \end{aligned}\right\} \qquad \text{...(11.31)}$$

Hence, under the action of applied loads, together with P_0 and M_0, the part of the infinitely long beam which is to the right of point A will behave in every respect as if there was a free end at A. The forces P_0 and M_0 are called the end-conditioning forces.

Let us consider that a force P_1 is applied at end A (Fig. 11.12) so that $M_A = 0$ and $Q_A = P_1$. Substituting in Eqs. (11.31), we get

$$P_0 = 4Q_A = 4P_1$$

$$M_0 = -\frac{2Q_A}{\lambda} = -\frac{2P_1}{\lambda}$$

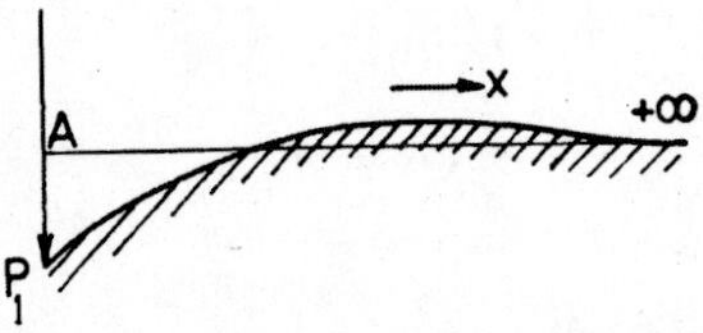

Fig. 11.12

Now if we apply these on the infinitely long beam then using Eqs. (11.22) and (11.23), we find

$$y = \frac{2P_1\lambda}{k} A_{\lambda x} - \frac{2P_1\lambda}{k} B_{\lambda x}$$

$$= \frac{2P_1\lambda}{k}(A_{\lambda x} - B_{\lambda x})$$

$$= \frac{2P_1\lambda}{k} D_{\lambda x}$$

$$\text{Similarly,} \quad \left.\begin{aligned} \theta &= -\frac{2P_1\lambda^2}{k} A_{\lambda x} \\ M &= -\frac{P_1}{\lambda} B_{\lambda x} \\ Q &= -P_1 C_{\lambda x} \end{aligned}\right\} \qquad \text{...(11.32)}$$

Now if we apply amoment M_1 at A (Fig. 11.13), then

$$M_A = -M_1, \; Q_A = 0.$$

Substituting in Eq. (11.31), we get

$$P_0 = -4\lambda M_1$$

$$M_0 = 4M_1$$

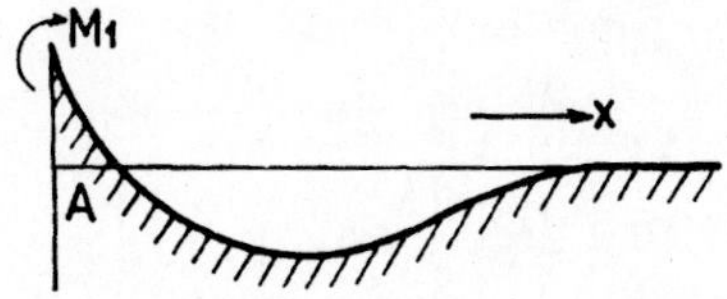

Fig. 11.13

Using again Eqs. (11.22) and (11.23), we get

$$\left.\begin{aligned} y &= -\frac{2M_1\lambda^2}{k} C_{\lambda x} \\ \theta &= \frac{4M_1\lambda^3}{k} D_{\lambda x} \\ M &= M_1 A_{\lambda x} \\ Q &= -2M_1\lambda B_{\lambda x} \end{aligned}\right\} \quad \text{...(11.33)}$$

11.4.2 Semi-infinite Beam with Hinged End

At the hinged end A, we have

$$y = 0, M = 0$$

Hence in this case, we have at end A

$$y_A + \frac{P_0\lambda}{2k} = 0$$

$$M_A + \frac{P_0}{4\lambda} + \frac{M_0}{2} = 0$$

$$\therefore \quad \left.\begin{aligned} P_0 &= -\frac{2k}{\lambda} y_A \\ M_0 &= \frac{k}{\lambda^2} y_A - 2M_A \end{aligned}\right\} \quad \text{...(11.34)}$$

Let us apply a moment M_1 at end A which is hinged as shown in Fig. 11.14. Then putting $y_A = 0$ and $M_A = -M_1$ in Eq. (11.34), we get

$$P_0 = 0, M_0 = 2M_1$$

Fig. 11.14

Hence, we find

$$\left.\begin{aligned} y &= 2M_1 \frac{\lambda^2}{k} B_{\lambda x} \\ \theta &= 2M_1 \frac{\lambda^3}{k} C_{\lambda x} \\ M &= M_1 D_{\lambda x} \\ Q &= -M_1 \lambda A_{\lambda x} \end{aligned}\right\} \quad \text{...(11.35)}$$

11.4.3 Semi-infinite, beam with free end subjected to uniformly distributed load over a finite length

Consider a semi-infinite beam with free end carrying a uniformly distributed load as shown in Fig. 11.15. For an infinite beam from Eqs. (11.24), we have

$$M = \frac{q}{4\lambda^2} B_{\lambda l}$$

$$Q = \frac{q}{4\lambda} (1 - C_{\lambda l})$$

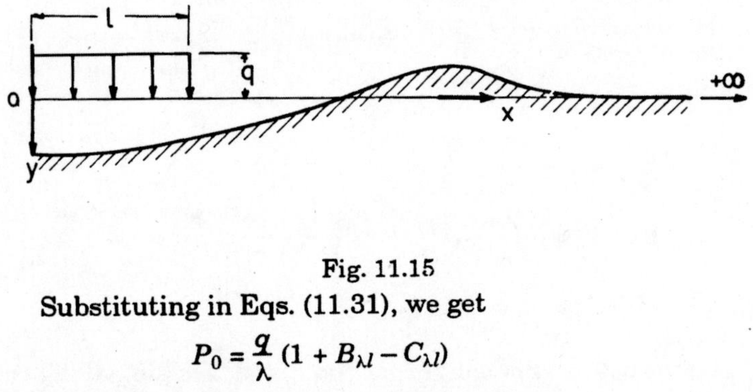

Fig. 11.15

Substituting in Eqs. (11.31), we get

$$P_0 = \frac{q}{\lambda} (1 + B_{\lambda l} - C_{\lambda l})$$

$$M_0 = -\frac{q}{2\lambda^2} (1 + 2B_{\lambda l} - C_{\lambda l})$$

Applying P_0 and M_0 on the infinite beam together with q, we get

$$y = \frac{q}{2k} [(1 + B_{\lambda l} - C_{\lambda l}) A_{\lambda x} - (1 + 2B_{\lambda l} - C_{\lambda l} + B_{\lambda x}) + (2 - D_{\lambda x} - D_{\lambda(l-x)})] \text{ for } 0 < x < l \quad \text{...(11.36)}$$

and
$$y = \frac{q}{2k} [(1 + B_{\lambda l} - C_{\lambda l})A_{\lambda x} - (1 + 2B_{\lambda l} - C_{\lambda l})B_{\lambda x} - (D_{\lambda x} - D_{\lambda(x-l)})] \text{ for } x > l \quad \text{...(11.37)}$$

The expressions for θ, M and Q can be determined by differentiation.

11.5 Beams of Finite Length

For a beam of finite length the correct solution is the one which, besides fulfilling the differential equation of the elastic line, also satisfies the required conditions at both ends of the beam.

Assume that an infinite beam is subjected to a given loading P and q as shown in Fig. 11.16. Because of this loading certain values

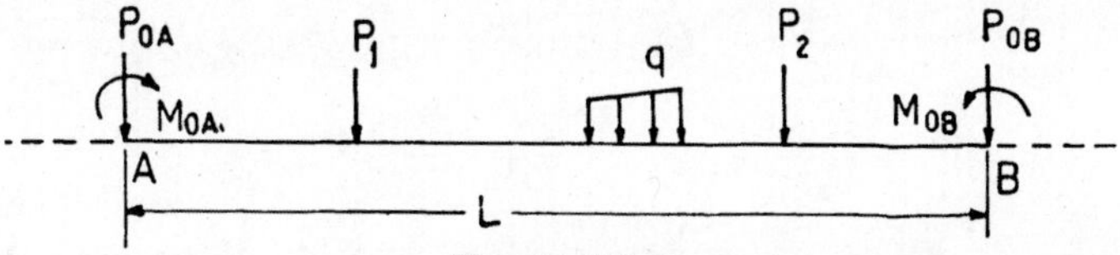

Fig. 11.16

of y, θ, M and Q will be produced at points A and B. By superposing on this loaded beam two pairs of concentrated forces and moments (M_{0A}, P_{0A} and M_{0B}, P_{0B}), we can modify the elastic curve in such a way that at points A and B the required end conditions will be fulfilled. These forces are known as the end conditioning forces.

11.5.1 Beams with Free Ends

Let us consider a beam AB carrying a concentrated load P and uniformly distributed load q as shown in Fig. 11.17 (*a*), whose ends are free. To determine the solution for the beam of finite length, we consider a beam of infinite length subjected to the same loads P and q as shown in Fig. 11.17 (*b*). In the beam shown in Fig. 11.17 (*b*)

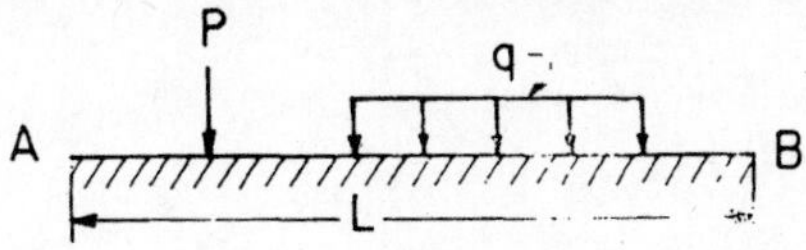

(*a*)

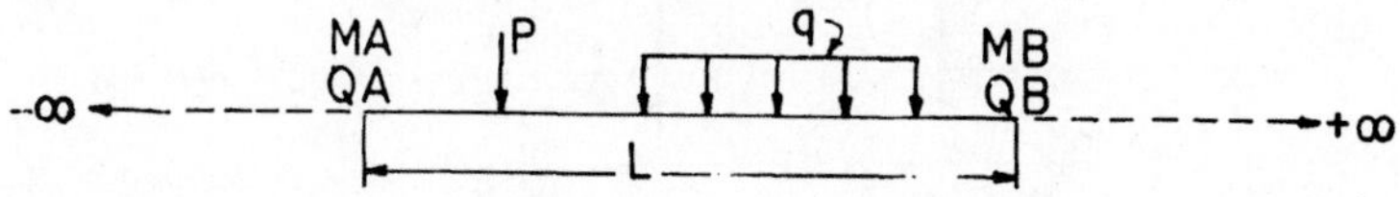

(*b*)

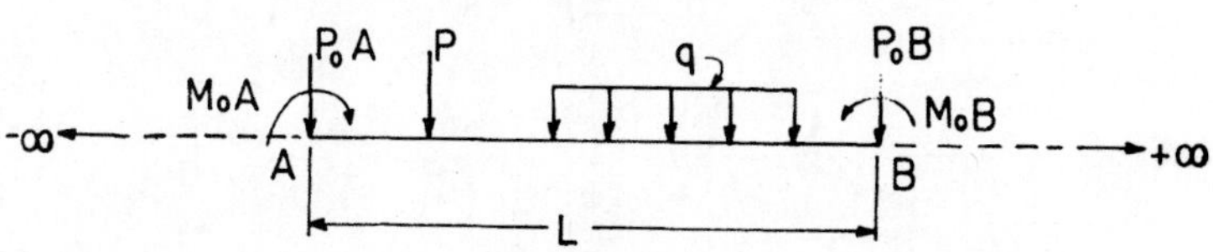

(*c*)

Fig. 11.17

there are bending moments and shear forces at A and B, M_A, Q_A and M_B, Q_B respectively. To make these moments and shear forces vanish, we apply end-conditioning forces P_{0A}, M_{0A} and P_{0B}, M_{0B} at A and B respectively as shown in Fig. 11.17 (*c*).

The end-conditioning forces must produce $-M_A$, $-Q_A$, $-M_B$ and $-Q_B$ at the points A and B.

Since the ends A and B are free, hence we get from Eqs. (11.22) and (11.23),

$$\left.\begin{aligned} M_A + \frac{P_{0A}}{4\lambda} + \frac{P_{0B}}{4\lambda} C_{\lambda L} + \frac{M_{0A}}{2} + \frac{M_{0B}}{2} D_{\lambda L} &= 0 \\ Q_A - \frac{P_{0A}}{2} + \frac{P_{0B}}{2} D_{\lambda L} - \lambda \frac{M_{0A}}{2} + \lambda \frac{M_{0B}}{2} A_{\lambda L} &= 0 \\ M_B + \frac{P_{0A}}{4\lambda} C_{\lambda L} + \frac{P_{0B}}{4\lambda} + \frac{M_{0A}}{2} D_{\lambda L} + \frac{M_{0B}}{2} &= 0 \\ Q_B - \frac{P_{0A}}{2} D_{\lambda L} + \frac{P_{0B}}{2} - \lambda \frac{M_{0A}}{2} A_{\lambda L} + \lambda \frac{M_{0B}}{2} &= 0 \end{aligned}\right\} \quad \text{...(11.38)}$$

Solving these four equations, we can determine the four end-conditioning forces. However, this leads to complicated expressions. To simplify the solution,

(*i*) we resolve the original loading into two parts, a symmetrical part and an antisymmetrical part as shown in Fig. 11.18 ;

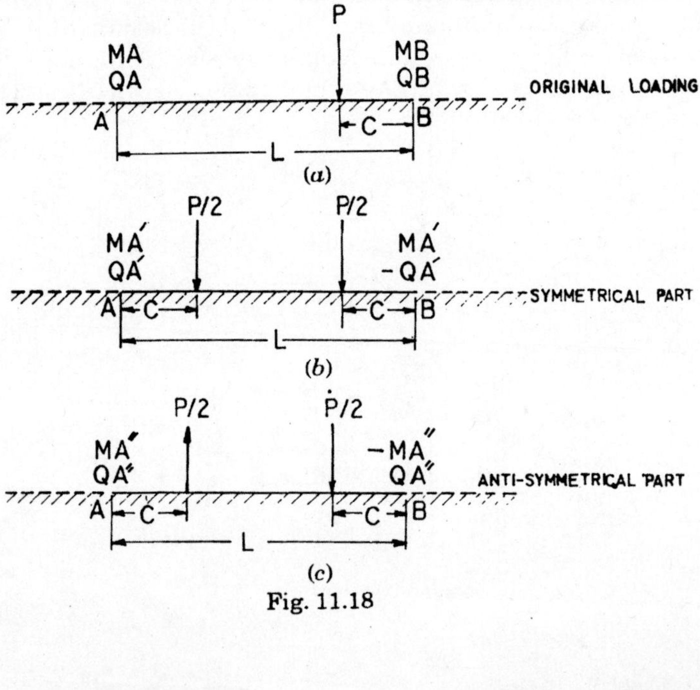

Fig. 11.18

(*ii*) determine the end-conditioning forces in each of these parts ; and

(*iii*) adding them together afterward at each end.

From Eq. 11.18, we find that

$$M_A = M_A + M_A''$$
$$M_B = M_A' - M_A''$$
$$Q_A = Q_A' + Q_A''$$
$$Q_B = -Q_A' + Q_A''$$

Hence, we get

$$\left.\begin{aligned} M_A' &= \tfrac{1}{2}(M_A + M_B) \\ Q_A' &= \tfrac{1}{2}(Q_A - Q_B) \\ M_A'' &= \tfrac{1}{2}(M_A - M_B) \\ Q_A'' &= \tfrac{1}{2}(Q_A + Q_B) \end{aligned}\right\} \qquad \text{...(11.39)}$$

Knowing M_A, M_B and Q_A, Q_B on the infinite beam due to original loading we can determine the symmetrical and antisymmetrical loadings.

Now we make these moments and shearing forces vanish at the point A and B by applying P_0', M_0' in the symmetrical case and P_0'', M_0'' in the antisymmetrical case as shown in Fig. 11.19. By adding the symmetrical and anti-symmetrical components we obtain the end-conditioning forces for the original loading.

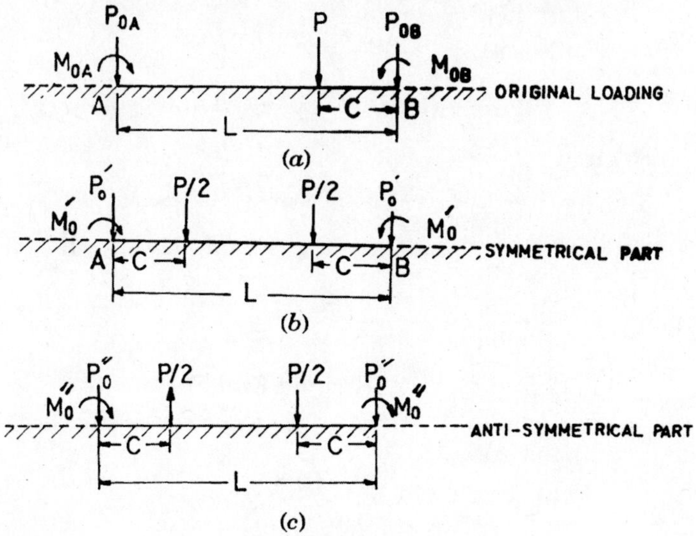

Fig. 11.19

Now,
$$P_{0A} = P_0' + P_0''$$
$$P_{0B} = P_0' - P_0''$$
$$M_{0A} = M_0' + M_0''$$
$$M_{0B} = M_0' - M_0''$$

In order to remove the moments and shears at A and B on the infinite beam we shall require that P_{0A}, M_{0A}, P_{0B} and M_{0B} should produce altogether $-M_A$, $-Q_A$ at A and $-M_B$, $-Q_B$ at B, which they will do if ; P_0' and M_0' produce $-M_A'$, $-Q_A'$ at A $-M_A'$, $+Q_A'$ at B and if P_0'' and M_0'' produce $-M_A''$, $-Q_A''$ at A and $+M_A''$, $-Q_A''$ at B.

(*a*) *Symmetrical case.* From Eqs. (11.12) and (11.23), we get

$$\frac{P_0'}{4\lambda}(1 + C_{\lambda L}) + \frac{M_0'}{2}(1 + D_{\lambda L}) = -M_A'$$

$$-\frac{P_0'}{2}(1 - D_{\lambda L}) - \lambda\frac{M_0'}{2}(1 - A_{\lambda L}) = -Q_A'$$

which gives

$$\left.\begin{aligned} P_0' &= 4\,E_I\,[Q_A'(1 + D_{\lambda L}) + \lambda M_A'(1 - A_{\lambda L})] \\ M_0' &= -\frac{2}{\lambda}E_I\,[Q_A'(1 + C_{\lambda L}) + 2\lambda M_A'(1 - D_{\lambda L})] \end{aligned}\right\} \quad ...(11.40)$$

where $E_I = \dfrac{1}{2}\dfrac{e^{\lambda L}}{\sinh \lambda L + \sin \lambda L}$.

(*b*) *Anti-symmetrical case.* In this case, we get

$$\frac{P_0''}{4\lambda}(1 - C_{\lambda L}) + \frac{M_0''}{2}(1 - D_{\lambda L}) = -M_A''$$

$$-\frac{P_0''}{2}(1 + D_{\lambda L}) - \lambda\frac{M_0''}{2}(1 + A_{\lambda L}) = -Q_A''$$

which gives

$$\left.\begin{aligned} P_0'' &= 4E_{II}[Q_A''(1 - D_{\lambda L}) + \lambda M_A''(1 + A_{\lambda L})] \\ M_0'' &= -\frac{2}{\lambda}E_{II}[Q_A''(1 - C_{\lambda L}) + 2\lambda M_A''(1 + D_{\lambda L})] \end{aligned}\right\} \quad ...(11.41)$$

where $E_{II} = \dfrac{1}{2}\dfrac{e^{\lambda L}}{\sinh \lambda L - \sin \lambda L}$.

11.5.2 Beams with Hinged Ends

For a beam with hinged ends as shown in Fig. 11.20 (*a*), the end conditions are

$$y_A = y_B = 0$$

and
$$M_A = M_B = 0$$

Again using single dashes for symmetrical case and double dashes for anti-symmetrical case, we get

$$y_A = y_A' + y_A''$$
$$y_B = y_A' - y_A''$$
$$M_A = M_A' + M_A''$$
$$M_B = M_A' - M_A''$$

which gives

$$\left.\begin{aligned} y_A' &= \tfrac{1}{2}(y_A + y_B) \\ y_A'' &= \tfrac{1}{2}(y_A - y_B) \\ M_A' &= \tfrac{1}{2}(M_A + M_B) \\ M_A'' &= \tfrac{1}{2}(M_A - M_B) \end{aligned}\right\} \quad \text{...(11.42)}$$

Now the end conditioning forces can be determined for the symmetrical [Fig. 11.20 (*b*)] and anti-symmetrical [Fig. 11.20 (*c*)] loading separately.

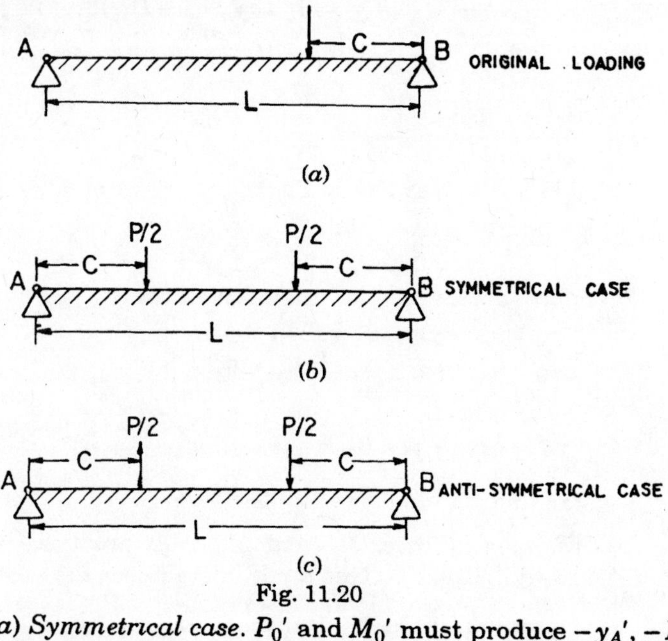

Fig. 11.20

(*a*) *Symmetrical case.* P_0' and M_0' must produce $-y_A'$, $-M_A'$ at points A and B on the infinite beam. Hence using Eqs. (11.22) and (11.23), we get

$$\frac{P_0'\lambda}{2k}(1 + A_{\lambda L}) + \frac{M_0'\lambda^2}{k} B_{\lambda L} = -y_A'$$

$$\frac{P_0'}{4\lambda}(1 + C_{\lambda L}) + \frac{M_0'}{2}(1 + D_{\lambda L}) = -M_A'$$

which gives

$$\left.\begin{aligned} P_0' &= 4\lambda F_I[M_A' B_{\lambda L} - 2\lambda^2 EI\, y_A'(1 + D_{\lambda L})] \\ M_0' &= 2F_I[-M_A'(1 + A_{\lambda L}) + 2\lambda^2 EI y_A'(1 + C_{\lambda L})] \end{aligned}\right\} \quad ...(11.43)$$

where $F_I = \dfrac{1}{2}\dfrac{e^{\lambda L}}{\cosh \lambda L + \cos \lambda L}$.

(b) Anti-symmetrical case. P_0'' and M_0'' must reproduce $-y_A''$ and $-M_A''$ at A and $+y_A''$ and M_A'' at B. Using Eqs. (11.22) and (11.23), we get

$$\frac{P_0''\lambda}{2k}(1 - A_{\lambda L}) - \frac{M_0''\lambda^2}{k} B_{\lambda L} = -y_A''$$

$$\frac{P_0''}{4\lambda}(1 - C_{\lambda L}) + \frac{M_0''}{2}(1 - D_{\lambda L}) = -M_A''$$

from which

$$\left.\begin{aligned} P_0'' &= -4\lambda F_{II}[M_A'' B_{\lambda L} + 2\lambda^2 EI y_A''(1 - D_{\lambda L})] \\ &- 2F_{II}[M_A''(1 - A_{\lambda L}) - 2\lambda^2 EI\, y_A''(1 - C_{\lambda L})] \end{aligned}\right\} \quad ...(11.44)$$

where $F_{II} = \dfrac{1}{2}\dfrac{e^{\lambda L}}{\cosh \lambda L - \cos \lambda L}$.

11.5.3 Beams with Fixed Ends

For a beam with fixed ends

$$y = 0 \text{ and } \theta = 0.$$

Therefore, in this case for the infinite beam

$$y_A = y_A' + y_A'', \quad \theta_A = \theta_A' + \theta_A''$$
$$y_B = y_A' - y_A'', \quad \theta_B = -\theta_A' + \theta_A''$$

Hence

$$\left.\begin{aligned} y_A' &= \tfrac{1}{2}(y_A + y_B), \quad \theta_A' = \tfrac{1}{2}(\theta_A - \theta_B) \\ y_A'' &= \tfrac{1}{2}(y_A - y_B), \quad \theta_A'' = \tfrac{1}{2}(\theta_A + \theta_B) \end{aligned}\right\} \quad ...(11.45)$$

(a) Symmetrical case. P_0' and M_0' must produce $-y_A'$ and $-\theta_A'$ at A and $-y_A'$, θ_A' at B. This condition requires that

$$P_0'\frac{\lambda}{2k}(1 + A_{\lambda L}) + M_0'\frac{\lambda^2}{k} B_{\lambda L} = -y_A'$$

$$P_0'\frac{\lambda^2}{k} B_{\lambda L} + M_0'\frac{\lambda^3}{k}(1 - C_{\lambda L}) = -\theta_A'$$

From this, we have

$$\left.\begin{aligned} P_0' &= 8\lambda^2 EI\, E_I\, [\theta_A'\, B_{\lambda L} - y_A'\, \lambda\, (1 - C_{\lambda L})] \\ M_0' &= -4\lambda EI\, E_I\, [\theta_A'(1 + A_{\lambda L}) - y_A' 2\lambda B_{\lambda L}] \end{aligned}\right\} \quad ...(11.46)$$

(*b*) *Antisymmetrical case.* P_0'' and M_0'' win produce $-y_A''$ and $-\theta_A''$ at A and $+y_A''$ and $-\theta_A''$ at B. This condition required that

$$P_0'' \frac{\lambda}{2k}(1 - A_{\lambda L}) - M_0'' \frac{y^2}{k} B_{\lambda L} = -y_A''$$

$$-P_0'' \frac{\lambda^2}{k} B_{\lambda L} + M_0'' \frac{\lambda^3}{k}(1 + C_{\lambda L}) = -\theta_A''$$

From which, we get

$$\left.\begin{aligned} P_0'' &= -8\lambda^2 EI\, E_{II}[\theta_A''\, B_{\lambda L} + \lambda y_A''\, (1 + C_{\lambda L})] \\ M_0'' &= -4\lambda EI\, E_{II}\, [\theta_A''\, (1 - A_{\lambda L}) + y_A''\, 2\lambda\, B_{\lambda L}] \end{aligned}\right\} \quad ...(11.47)$$

where E_I and E_{II} denote the same expressions as given in art. 11.5.1.

Example 11.1 *Beam with free ends carrying equal concentrated forces at both ends.*

Solution. Consider a beam of length L carrying concentrated forces equal to P at both ends as shown in Fig. 11.21.

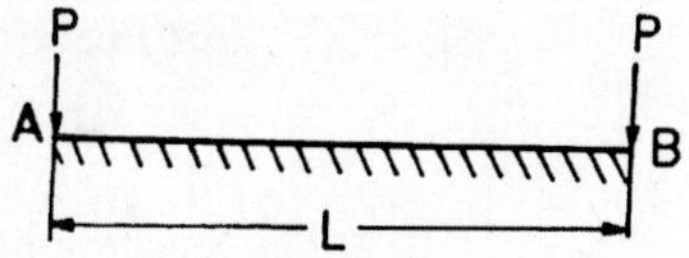

Fig. 11.21

Since the ends of the beam are free, hence according to Eq. (11.29), we have for the infinite beam

$$M_A = \frac{P}{4\lambda} + \frac{P}{4\lambda} e^{-\lambda L} (\cos \lambda L - \sin \lambda L) = M_B$$

$$Q_A = -\frac{P}{2} - \frac{P}{2} e^{-\lambda L} \cos \lambda L = -Q_B$$

From Eqs. (11.39), we get

$$M_A' = \frac{1}{2}(M_A + M_B) = M_A \quad Q_A' = \frac{1}{2},\ (Q_A - Q_B) = Q_A$$

$$M_A'' = \frac{1}{2}(M_A - M_B) = 0$$

$$Q_A'' = \frac{1}{2}(Q_A + Q_B) = 0$$

Now from Eqs. (11.40), we have

$$\begin{aligned} P_0' &= 4E_I\, [Q_A'\, (1 + D_{\lambda L}) + \lambda M_A'\, (1 - A_{\lambda L})] \\ &= 4E_I\, [Q_A(1 + D_{\lambda L}) + \lambda M_A\, (1 - A_{\lambda L})] \end{aligned}$$

$$\begin{aligned} M_0' &= -\frac{2}{\lambda} E_I\, [Q_A'\, (1 + C_{\lambda L}) + 2\lambda M_A'\, (1 - D_{\lambda L})] \\ &= -\frac{2}{\lambda} E_I\, [Q_A(1 + C_{\lambda L}) + 2\lambda M_A\, (1 - D_{\lambda L})] \end{aligned}$$

where $E_I = \dfrac{1}{2} \dfrac{e^{\lambda L}}{\sinh \lambda L + \sin \lambda L}$

Obviously $= P_0'' = M_0'' = 0$.

Therefore, the end-conditioning forces become

$$P_{0A} = P_0' + P_0'' = P_0'$$
$$P_{0B} = P_0' - P_0'' = P_0'$$
$$M_{0A} = M_0' + M_0'' = M_0'$$
$$M_{0B} = M_0' - M_0'' = M_0'$$

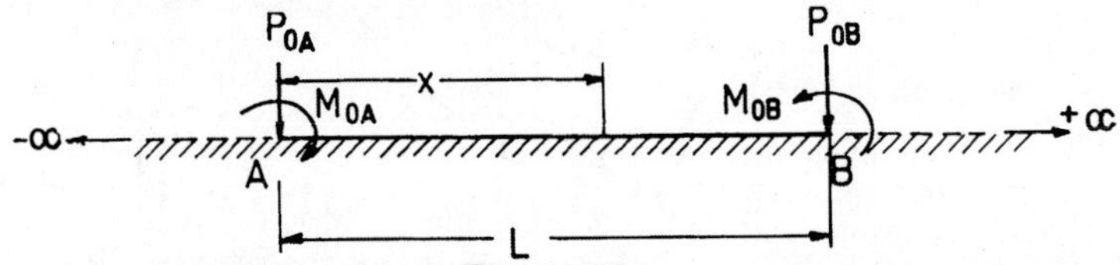

Fig. 11.22

These end-conditioning forces have been shown in Fig. 11.22 on the infinite beam.

Using Eqs. (11.29), we get at any distance x from end A,

$$y = \frac{\lambda}{2k} [\, P_{0A}\, e^{-\lambda x} (\cos \lambda x - \sin \lambda x) + P_{0B}\, e^{-\lambda(L-x)} \{\cos \lambda (L-x) - \sin \lambda (L-x)\} + P\, e^{-\lambda x} (\cos \lambda x - \sin \lambda x) + P\, e^{-\lambda(L-x)} \{\cos \lambda (L-x) - \sin \lambda (L-x)\}]$$

$$= \frac{\lambda}{2k} [(P_{0A} + P)\, e^{-\lambda x} (\cos \lambda x - \sin \lambda x) + (P_{0B} + P)\, e^{-\lambda(L-x)} \{\cos \lambda(L-x) - \sin \lambda (L-x)\}]$$

Now $P_{0A} + P = P_0' + P$

$$= 4E_I\, [Q_A (1 + D_{\lambda L}) + \lambda M_A (1 + A_{\lambda L})]$$

$$= 4E_I \left[-\frac{P}{2} (1 + D_{\lambda L})^2 + \frac{P}{4} \{(1 + A_{\lambda L}) \times (1 + C_{\lambda L})\} \right]$$

$$= 2PE_I\, [-(1 + D_{\lambda L}) + \tfrac{1}{2} (1 + A_{\lambda L})(1 + C_{\lambda L})]$$

$$P_{0B} = P_{0A} + P$$

$$\therefore \quad y = \frac{\lambda P}{k} E_I\, [\{1 - (1 + D_{\lambda L})^2 + \tfrac{1}{2} (1 + A_{\lambda L}) \times (1 + C_{\lambda L})\} (C_{\lambda x} + C_{\lambda(L-x)})]$$

$$= \frac{P\lambda}{2k} \left(\frac{e^{\lambda L}}{\sinh \lambda L + \sin \lambda L} \right) [\{-(1 + D_{\lambda L})^2 + \tfrac{1}{2} (1 + A_{\lambda L}) \times (1 + C_{\lambda L})\} (C_{\lambda x} + C_{\lambda(L-x)})]$$

The deflection due to moments is

$$y = \frac{\lambda^2}{k}\,[M_{0A}\,B_{\lambda x} + M_{0B}\,B_{\lambda(L-x)}]$$

$$= \frac{\lambda^2}{k}\,[M_0'\,B_{\lambda x} + M_0'\,B_{\lambda(L-x)}]$$

$$= \frac{\lambda^2}{k}\left[-\frac{2}{\lambda}E_I\,\{Q_A\,(1 + C_{\lambda L}) + 2\lambda M_A\,(1 - D_{\lambda L})\}\,(B_{\lambda x} + B_{\lambda(L-x)})\right]$$

$$= -\frac{2\lambda}{k}E_I\left[\left\{-\frac{P}{2}\,(1 + D_{\lambda L})\,(1 + C_{\lambda L}) + \frac{P}{2}\,(1 + C_{\lambda L}) \times (1 - D_{\lambda L})\right\} \times (B_{\lambda x} + B_{\lambda(L-x)})\right]$$

$$= -\frac{P\lambda}{k}E_I\,[\{-(1 + D_{\lambda L})(1 + C_{\lambda L}) + (1 + C_{\lambda L})(1 - D_{\lambda L})\} \times (B_{\lambda x} + B_{\lambda(L-x)})]$$

$$= -\frac{P\lambda}{2k}\left(\frac{e^{\lambda L}}{\sin L_{\lambda L} + \sin \lambda_L}\right)[\{-(1 + D_{\lambda L})(1 + C_{\lambda L}) + (1 + C_{\lambda L})(1 - D_{\lambda L})\} \times (B_{\lambda x} + B_{\lambda(L-x)})]$$

Adding we get

$$y = \frac{P\lambda}{2k}\left(\frac{4}{\sinh \lambda L + \sin \lambda L}\right)[\cosh \lambda x \cos \lambda\,(L - x) + \cosh \lambda\,(L - x) \cos \lambda x]$$

$$= \frac{2P\lambda}{k}\left[\frac{\cosh \lambda x \cos \lambda\,(L - x) + \cos \lambda(L - x) \cos \lambda x}{\sinh \lambda L + \sin \lambda L}\right]$$

$$y_A = y_B = \frac{2P\lambda}{k}\left(\frac{\cosh \lambda L + \cos \lambda L}{\sinh \lambda L + \sin \lambda L}\right)$$

$$(y)_{x=L/2} = \frac{4P\lambda}{k}\left(\frac{\cos \dfrac{\lambda L}{2} \cos \dfrac{\lambda L}{2}}{\sinh \lambda L + \sin \lambda L}\right)$$

Differentiating we can get

$$M = -\frac{P}{\lambda}\left(\frac{\sinh \lambda x \sin \lambda(L - x) + \sinh \lambda(L - x) \sin \lambda x}{\sinh \lambda L + \sin \lambda L}\right).$$

11.6 Thin Cylindrical Shells

Consider a thin cylindrical shell subjected to radial forces uniformly distributed along an arbitrary circle on the shell as shown in Fig. 11.23.

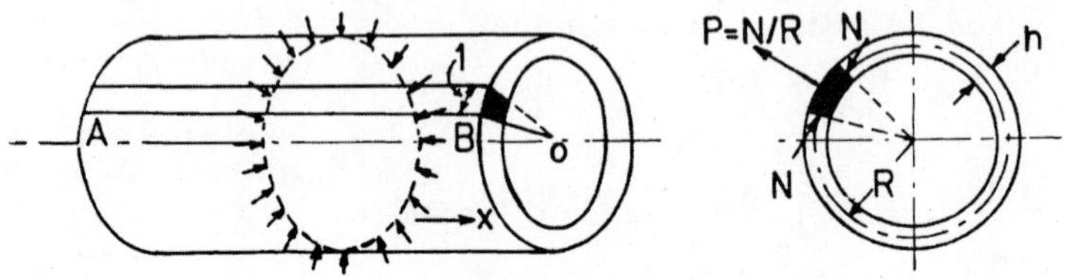

Fig. 11.23

Because of the symmetry of the loading every section normal to the axis will remain circular, while the radius R will undergo a change, $\Delta R = y$, different for each cross-section. The radial displacement y can be regarded as deflection for a longitudinal element of the shell. On account of the symmetry, let us consider the deformation of only one element AB as shown in the Fig. 11.23 and having unit width. The radial displacement y must be accompanied by circumferential compression $\frac{y}{R}$ of the shell, which, in turn will give rise to compressive hoop forces N having a magnitude of

$$N = \frac{Eh}{R} y$$

per unit length of the $A - B$ element. R is the mean radius and h is the thickness of the shell wall. The resultant of these forces will have a radial direction and its value is

$$P = N \cdot \frac{1}{R} = \frac{Eh}{R^2} y.$$

Therefore the force P, opposing the deflection, will be proportional to the deflection, $\frac{Eh}{R^2}$ being the proportionality factor. Hence it may be concluded that a longitudinal element of a cylindrical shell loaded symmetrically with respect to its axis can be regarded as a beam resting on an elastic foundation, whose modulus is given by

$$k = \frac{Eh}{R^2} \qquad \text{...(11.48)}$$

the modulus of foundation depends on the cross-sectional dimensions of the shell and its material.

On account of the axial symmetry of the deformation of the shell due to bending, the sides of each longitudinal element will not be able to rotate in order to permit the lateral extension or compression of the longitudinal fibres, but will have to remain parallel with their original directions. This restraining influence is equivalent to a bending moment

$$M_\theta = \nu M_x$$

in the circumferential ring where M_x denotes the bending moment in the longitudinal beam. The stiffening effect of M_θ on the bending deformation of the longitudinal beam can be taken into account by increasing the moment of inertia of each beam of unit width in the ratio $\left(\frac{1}{1-\nu^2}\right)$, and thus we have

$$I = \frac{h^3}{12(1-\nu^2)} \qquad ...(11.49)$$

$$\therefore \quad \lambda = \sqrt[4]{\frac{k}{4EI}} = \sqrt[4]{3(1-\nu^2)}\,\frac{1}{\sqrt{Rh}} \qquad ...(11.50)$$

Then the same formulae which have been derived for the elastically supported beams may be used to calculate y_x, θ_x, M_x, Q_x by taking k and λ as given by Eqs. (11.48) and (11.50) respectively.

The stresses in the shell wall are

$$\left.\begin{aligned} \sigma_x &= \pm\frac{6}{h^2}M_x \\ \sigma_\theta &= -\frac{E}{R}y_x \pm \frac{6}{h^2}\nu M_x \end{aligned}\right\} \qquad ...(11.51)$$

where the plus and minus signs on the right side give the fibre stresses at the inside and at the outside surface of the shell respectively, tensile stress always being regarded as positive.

Example 11.2 *Thin cylindrical shell subjected to uniform band of pressure.*

Solution. Consider a thin cylinder of infinite length subjected to a uniform band of pressure of intensity q and of length l as shown in Fig. 11.24.

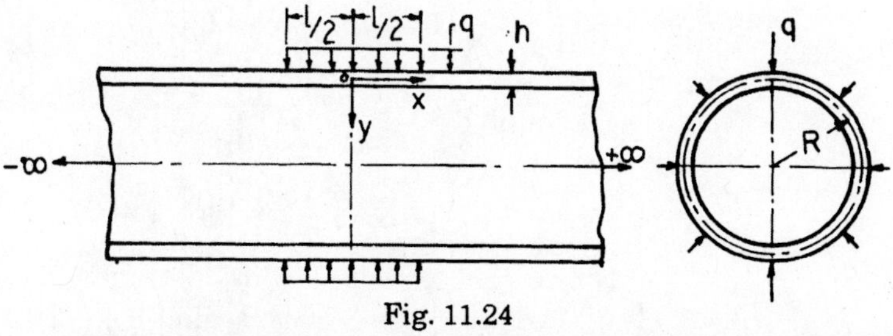

Fig. 11.24

Since the loading is symmetrical, we need to consider only the right half of the shell. Choosing coordinate axes as shown in the figure and denoting the loaded part of the cylinder by subscript 1 and unloaded part by 2, we can write the equations for the elastic lines as :

$$\left.\begin{aligned}\frac{d^4y_1}{dx^4} + 4\lambda^4 y_1 &= \frac{q}{D}, -\frac{l}{2} \le x \le \frac{l}{2} \\ \frac{d^4y_2}{dx^4} + 4\lambda^4 y_2 &= 0, \frac{l}{2} \le x < \infty\end{aligned}\right\} \quad ...(1)$$

Since the loading is symmetric, in general

$$y(x) = y(-x), -\infty < x < \infty$$

$$\therefore \quad \frac{d^2y_1}{dx^2} = \frac{d^3y_1}{dx^3} = 0 \quad ...(2)$$

For the continuity of the cylinder material at

$$x = \frac{l}{2}$$

the following condition must be satisfied :

$$\left.\begin{aligned}y_1 &= y_2 \\ \frac{dy_1}{dx} &= \frac{dy_2}{dx} \\ \frac{d^2y_1}{dx^2} &= \frac{d^2y_2}{dx^2} \\ \frac{d^3y_1}{dx^3} &= \frac{d^3y_2}{dx^3}\end{aligned}\right\} \quad ...(3)$$

Also as x tends to infinity, y_2 tends to zero.

Hence the solutions of Eqs. (1) may be written as :

$$\begin{aligned}\frac{y_1(\xi)}{q/k} = A_1 &\left[\sin\left(\frac{1}{2} - \xi\right)\theta \cosh\left(\frac{1}{2} + \xi\right)\theta + \sin\left(\frac{1}{2} + \xi\right)\theta \cosh\left(\frac{1}{2} - \xi\right)\theta \right] \\ &+ B_1 \left[\sinh\left(\frac{1}{2} - \xi\right)\theta \cos\left(\frac{1}{2} + \xi\right)\theta + \sinh\left(\frac{1}{2} + \xi\right)\theta \cos\left(\frac{1}{2} - \xi\right)\theta \right] + 1\end{aligned} \quad ...(4)$$

$$\frac{y_2(\xi)}{q/k} = e^{-\left(\xi - \frac{1}{2}\right)\theta} \left[A_2 \cos\left(\xi - \frac{1}{2}\right)\theta + B_2 \sin\left(\xi - \frac{1}{2}\right)\theta \right] \quad ...(5)$$

where $\xi = x/l$. $\theta = \lambda l$, and the constants A_1, B_1, A_2, B_2 are to be determined from Eqs. (3). Substituting Eqs. (4) and (5) in Eqs. (3) we get

$$\begin{bmatrix} 1 & 0 & -b & -d \\ 1 & -1 & a-c & -(a-c) \\ 0 & 1 & -d & b \\ 1 & 1 & -(a-c) & a-c \end{bmatrix} \begin{Bmatrix} A_2 \\ B_2 \\ A_1 \\ B_1 \end{Bmatrix} = \begin{Bmatrix} 1 \\ 0 \\ 0 \\ 0 \end{Bmatrix} \qquad \text{...(6)}$$

where $a = \cos\theta$, $b = \sin\theta$, $c = \cosh\theta$, $d = \sinh\theta$. Solving Eqs. (6), we get

$$A_1 = \frac{2}{\delta} b$$

$$B_1 = \frac{2}{\delta} (d - a + c)$$

$$A_2 = \frac{2}{\delta} [d(a-c) - (a-c)^2]$$

$$B_2 = \frac{2}{\delta} b(a-c)$$

$$\delta = -2 + 4a(c+d) - 2(c+d)^2$$

Thus by knowing the constants, the values of deflection for the cylinder can be determined from Eqs. (4) and (5)

$$M_1(x) = -\frac{Dd^2y_1}{dx^2}$$

$$\therefore \quad \frac{M_1(\xi)}{(q/2\lambda^2)} = A_1\left[\cos\left(\frac{1}{2}-\xi\right)\theta \sinh\left(\frac{1}{2}+\xi\right)\theta + \cos\left(\frac{1}{2}+\xi\right)\theta \sinh\left(\frac{1}{2}-\xi\right)\theta\right] - B_1\left[\sin\left(\frac{1}{2}-\xi\right)\theta \cosh\left(\frac{1}{2}+\xi\right)\theta + \sin\left(\frac{1}{2}+\xi\right)\theta \cosh\left(\frac{1}{2}-\xi\right)\theta\right] \qquad \text{...(7)}$$

Similarly

$$\frac{M_2(\xi)}{(q/2\lambda^2)} = -e^{-\left(\xi-\frac{1}{2}\right)\theta}\left[A_2 \sin\left(\xi-\frac{1}{2}\right)\theta - B_2\cos\left(\xi-\frac{1}{2}\right)\theta\right] \qquad \text{...(8)}$$

Hence the stresses in the cylinder become

$$\left.\begin{aligned} (\sigma_x)_1 &= \pm\frac{6}{h^2} M_1(\xi), \; -\frac{1}{2} \le \xi < \frac{1}{2} \\ (\sigma_x)_2 &= \pm\frac{6}{h^2} M_2(\xi), \; \frac{1}{2} \le \xi < \infty \end{aligned}\right\} \qquad \text{...(9)}$$

$$(\sigma_\theta)_1 = -\frac{E}{R} y_1(\xi) \pm \frac{6}{h^2} \nu M_1(\xi)$$

$$(\sigma_\theta)_2 = -\frac{E}{R} y_2(\xi) \pm \frac{6}{h^2} \nu M_2(\xi)$$

In the similar manner the other thin cylindrical shell problems can be solved under various types of symmetrical loading conditions. Problems involving thin cylinders of finite length may also be solved by adopting the same procedure as followed for beams of finite length.

Example 11.3 *An aluminium alloy I-beam (depth 100 mm, $I_x = 2.45 \times 10^6$ mm^4, $E = 72$ GPa) has a length $L = 7$ m and is supported by 8 springs ($k = 100$ N/mm) spaced at distance $l = 1.0$ m centre to centre along the beam. A load $P = 15$ kN is applied at the centre of the bean over one of the springs. Determine the deflection of the beam under the load, the maximum bending moment and the maximum bending stress in the beam.*

Solution. $k = \dfrac{k}{l} = \dfrac{100}{1.0 \times 10^3} = 0.1 \text{ N/mm}^2$

$$\lambda = \sqrt[4]{\frac{k}{4EI}} = \sqrt[4]{\frac{0.1}{4(72 \times 10^3)(2.45 \times 10^6)}} = 0.000614 \text{ mm}^{-1}$$

Now $\dfrac{\pi}{4\lambda} = \dfrac{\pi}{4 \times 0.000614} = 1279$ mm

Hence $l = 1000$ mm is less than $\pi/4\lambda$.

Also $\dfrac{3\pi}{2\lambda} = \dfrac{3\pi}{2 \times 0.000614} = 7675$ mm

Number of springs $m = 8$

$ml = 8 \times 1000 = 8000$ mm

Hence $ml > \dfrac{3\pi}{2\lambda}$.

Therefore, approximate solution for a spring supported infinite beam may be used for a spring supported finite beam of length L.

$$y_{max} = \frac{P\lambda}{2k} = \frac{15 \times 10^3(0.000614)}{2 \times 0.1} = 46.05 \text{ mm}$$

$$M_{max} = \frac{P}{4\lambda} = \frac{15 \times 10^3}{4 \times 0.000614} = 6.1075 \times 10^6 \text{ N-mm}$$

$$\sigma_{max} = \frac{M_{max} \times \sigma}{I_x} = \frac{6.1075 \times 10^6 \times 50}{2.45 \times 10^6} = 124.64 \text{ N/mm}^2.$$

11.7 Straight Bars under Simultaneous Axial and Transverse Loading

Consider a straight bar subjected to transverse and axial tensile loading as shown in Fig. 11.25. The horizontal tensile forces

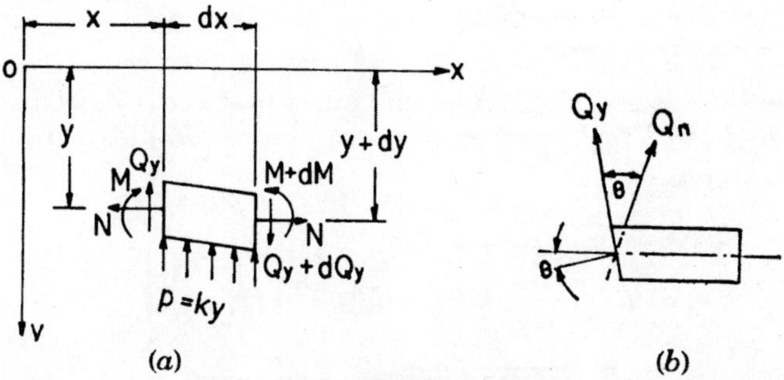

Fig. 11.25

N are assumed to act through the centre of gravity of the end cross-section of the bar. Let us now consider the moments equilibrium of an infinitely small unloaded element bounded by two verticals a distance dx apart, we get

$$M + dM - M + Ndy - Q_y dx = 0$$

$$\frac{dM}{dx} + N\frac{dy}{dx} - Q_y = 0$$

The normal shear force Q_n acting in the plane of the section normal to the deflection line is

$$Q_n = Q_y \cos\theta - N\sin\theta$$

For θ to be small cos θ = 1,

$$\sin\theta = \theta = \tan\theta = \frac{dy}{dx}$$

$$\therefore \quad Q_n = Q_y - N\frac{dy}{dx} = \frac{dM}{dx}$$

Putting $M = -EI\dfrac{d^2y}{dx^2}$

and substituting $\dfrac{dQ_y}{dx} = ky$, we obtain

$$EI\frac{d^4y}{dx^4} - N\frac{d^2y}{dx^2} + ky = 0 \qquad \text{...(11.52)}$$

Let $y = e^{mx}$, then the characteristic equation is

$$EIm^4 - Nm^2 + k = 0$$

whose roots are

$$m_{1,2,3,4} = \pm\sqrt{\frac{N}{2EI} \pm i\sqrt{\frac{k}{EI} - \left(\frac{N}{2EI}\right)^2}}$$

Thus the general solution of Eq. (11.52) is

$$y = A_1 e^{m_1 x} + A_2 e^{m_2 x} + A_3 e^{m_3 x} + A_4 e^{m_4 x}$$

Now we discuss different cases

(*i*) $N < 2\sqrt{kEI}$

Then the expression under the inner bracket is positive and the four roots will form two pairs of conjugate complex numbers, which can be put as

$$m_{1,2,3,4} = \pm(\alpha \pm i\beta),$$

where
$$\left.\begin{aligned}\alpha &= \sqrt{\sqrt{\frac{k}{4EI}} + \frac{N}{4EI}} = \sqrt{\lambda^2 + \frac{N}{4EI}}\\ \beta &= \sqrt{\sqrt{\frac{k}{4EI}} - \frac{N}{4EI}} = \sqrt{\lambda^2 - \frac{N}{4EI}}\end{aligned}\right\} \quad ...(11.53)$$

Then the general solution becomes

$$y = (C_1 e^{\alpha x} + C_2 e^{-\alpha x}) \cos \beta x + (C_3 e^{\alpha x} + C_4 e^{-\alpha x}) \sin \beta x \quad ...(11.54)$$

When the infinitely long bar carries a concentrated load P at the centre and axial tensile force N then as $x \to \infty, y \to 0$ and $M \to 0$. Hence $C_1 = C_3 = 0$ and we get for $x > 0$.

$$\left.\begin{aligned}y &= \frac{P}{2k}\frac{\lambda^2}{\alpha\beta} e^{-\alpha x}(\beta \cos \beta x + \alpha \sin \beta x)\\ \theta &= -\frac{P}{4EI}\cdot\frac{1}{\alpha\beta} e^{-\alpha x} \sin \beta x\\ M &= \frac{P}{4}\cdot\frac{1}{\alpha\beta} e^{-\alpha x}(\beta \cos \beta x - \alpha \sin \beta x)\\ Qy &= -\frac{P}{4}\cdot\frac{1}{\alpha\beta}\cdot e^{-\alpha x}[2\alpha\beta \cos \beta x + (\alpha^2 - \beta^2)\sin \beta x]\end{aligned}\right\} \quad ...(11.55)$$

Similarly a clockwise moment M_0 at the origin leads to for $x > 0$.

$$\left.\begin{aligned}y &= \frac{M_0}{4EI}\cdot\frac{1}{\alpha\beta} e^{\alpha x} \sin \beta x\\ \theta &= \frac{M_o}{4EI}\cdot\frac{1}{\alpha\beta} e^{-\alpha x}(\beta \cos \beta x - \alpha \sin \beta x)\\ M &= \frac{M_o}{4}\cdot\frac{1}{\alpha\beta} e^{-\alpha x}[2\alpha\beta \cos \beta x - (\alpha^2 - \beta^2)\sin \beta x]\\ Qy &= -\frac{M_0}{2}\frac{\lambda^2}{\alpha\beta}(\beta \cos \beta x + \alpha \sin \beta x)\end{aligned}\right\} \quad ...(11.56)$$

From Eqs. (11.55) and (11.56) it may be observed that although an axial force N is present, the deflection line and its derivatives remain proportional to the transverse loading P and M_0. The solution for the bars of finite length can thus be obtained by using the principle of superposition as was applied in the case of infinite length beams where no axial force was applied. However in the present case there will be two types of shearing forces Q_n and Q_y and accordingly the end conditions have to be stated involving shear.

(*ii*) $N = 2\sqrt{kEI}$

In this case $\alpha = \sqrt{\dfrac{N}{2EI}} = a,\ \beta = 0$

Hence the general solution becomes

$$y = C_1 e^{ax} + C_2 e^{-ax} \qquad ...(11.57)$$

and Eqs. (11.55) and (11.56) become for deflection :

$$y = \frac{P}{4k} a\, e^{-ax}(1 + ax) \qquad ...(11.58)$$

$$y = \frac{M_0 a^3}{4k} e^{-ax} . x \qquad ...(11.59)$$

(*iii*) $N > 2\sqrt{kEI}$

In this case we will have four real roots of the characteristic equation. Thus the solution obtained in case (*i*) can be transformed into case (*iii*) by putting $i\beta$ instead of the term β, where

$$\bar{\beta} = \sqrt{\frac{N}{4EI} - \sqrt{\frac{k}{4EI}}} = \sqrt{\frac{N}{4EI} - \lambda^2} \qquad ...(11.60)$$

The general solution then becomes

$$y = (C_1 e^{ax} + C_2 e^{-ax}) \cosh \bar{\beta}x + (C_3 e^{ax} + C_4 e^{-ax})\, i \sinh \bar{\beta}x. \qquad ...(11.61)$$

The equations for the elastic line for an infinitely long bar carrying a concentrated load P or moment M_0 at the origin becomes from Eqs. (11.55) and (11.56) respectively for $x > 0$.

$$y = \frac{P}{2k}\frac{\lambda^2}{\alpha\bar{\beta}} e^{-ax}(\bar{\beta}\cosh \beta x + \alpha \sinh \bar{\beta}x) \qquad ...(11.62)$$

$$y = \frac{M_0}{4EI} \cdot \frac{1}{\alpha\bar{\beta}} e^{-ax} \sinh \bar{\beta}x \qquad ...(11.63)$$

If the bar carries compressive axial force then α and β may be interchanged in the above solutions.

11.8 Curved Beam (arch) on an Elastic Foundation

Consider a prismatic beam whose neutral axis in the undeflected state forms an arc of a circle of radius r is subjected to bending forces acting in the plane of curvature of the beam, as shown in Fig. 11.26 (*a*).

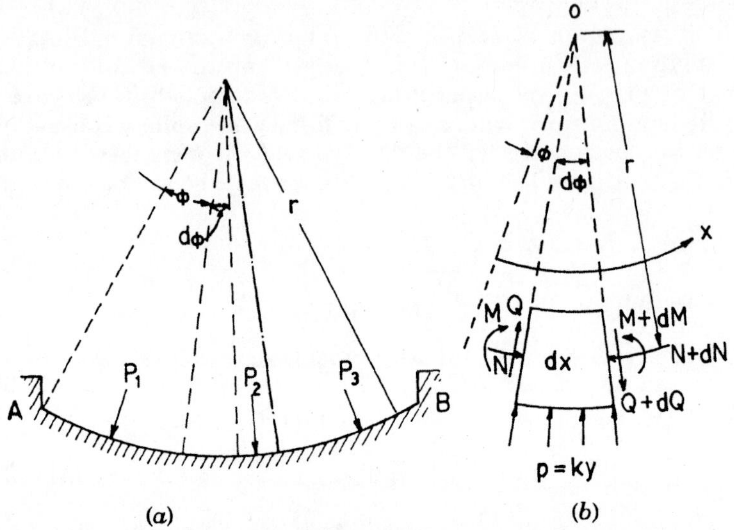

Fig. 11.26

An infinitesimal element of this beam will be acted upon by shearing force Q, normal force N, bending moment M and reaction of the foundation pdx, as shown in Fig. 11.26 (*b*). For the equilibrium of the forces on this element in the radial direction, we have

$$pdx - Nd\varphi = dQ$$

and in the tangential direction

$$Qd\varphi = dN$$

The moment equilibrium condition gives

$$dM = Q.\, rd\varphi$$

Putting $p = ky$ and $dx = rd\varphi$, we find

$$ky - \frac{N}{r} = \frac{dQ}{dx} = \frac{d^2M}{dx^2}$$

and

$$\frac{Q}{r} = \frac{dN}{dx}$$

Eliminating N and Q, we get

$$k\frac{dy}{dx} - \frac{1}{r^2}\frac{dM}{dx} = \frac{d^3M}{dx^3} \quad \text{...}(a)$$

Now the differential equation of bending of a circular arch of radius of curvature r and flexural rigidity EI is

$$EI\left(\frac{d^2y}{dx^2} + \frac{y}{r^2}\right) = -M$$

$$\therefore \quad \frac{dM}{dx} = -EI\left(\frac{d^3y}{dx^3} + \frac{1}{r^2}\frac{dy}{dx}\right)$$

$$\frac{d^3M}{dx^3} = -EI\left(\frac{d^5y}{dx^5} + \frac{1}{r^2}\frac{d^3y}{dx^3}\right) \qquad ...(b)$$

Comparing Eqs. (*a*) and (*b*), we get

$$\frac{d^5y}{dx^5} + \frac{2}{r^2}\frac{d^3y}{dx^3} + \left(\frac{k}{EI} + \frac{1}{r^4}\right)\frac{dy}{dx} = 0 \qquad ...(11.64)$$

which is the fundamental differential equation of bending of circular arches or curved beams supported on an elastic foundation.

In terms of $\varphi = \frac{x}{r}$, Eq. (11.64) becomes

$$\frac{d^5y}{d\varphi^5} + 2\frac{d^3y}{d\varphi^3} + \eta^2\frac{dy}{d\varphi} = 0 \qquad ...(11.65)$$

where $\eta = \sqrt{\frac{r^4k}{EI} + 1}$

Let $y = e^{mx}$ then Eq. (11.65) becomes

$$m^5 + 2m^3 + \eta^2 = 0$$

whose roots are $m_1 = 0, m_{2,3,4,5} = \pm(\alpha \pm i\beta)$

where $$\alpha = \sqrt{\frac{\eta - 1}{2}}, \beta = \sqrt{\frac{\eta + 1}{2}} \qquad ...(11.66)$$

The general solution of Eq. (11.65) can be written as :

$$y = C_0 + (C_1 \cosh \alpha\varphi + C_2 \sinh \alpha\varphi)\cos \beta\varphi + (C_3 \cosh \alpha\varphi + C_4 \sinh \alpha\varphi)\sin \beta\varphi \qquad ...(11.67)$$

Taking the derivatives we obtain the following expressions :

$$M = -\frac{EI}{r^2}[C_0 - 2\alpha\beta\{(C_1 \sinh \alpha\varphi + C_2 \cosh \alpha\varphi)\sin \beta\varphi - (C_3 \sinh \alpha\varphi + C_4 \cosh \alpha\varphi)\cos \beta\varphi\}] \qquad ...(11.68)$$

$$Q = 2\alpha\beta\frac{EI}{r^3}[(\alpha C_1 + \beta C_4)\cos \alpha\varphi \sin \beta\varphi + (\beta C_1 - \alpha C_4)\sinh \alpha\varphi \cos \beta\varphi + (\alpha C_2 + \beta C_3)\sinh \alpha\varphi \sin \beta\varphi + (\beta C_2 - \alpha C_3)\cosh \alpha\varphi \cos \beta\varphi] \qquad ...(11.69)$$

$$N = rkC_0 + 2\alpha\beta\frac{EI}{r^3}[(C_1 \sinh \alpha\varphi + C_2 \cosh \alpha\varphi)\sin \beta\varphi - (C_3 \sinh \alpha\varphi + C_4 \cosh \alpha\varphi)\cos \beta\varphi] \qquad ...(11.70)$$

11.9 Plates on Elastic Foundation

Consider a rectangular plate as shown in Fig. 11.27 resting on an elastic foundation so that $p = ku_z$ where k is the modulus of the foundation and u_z is the vertical deflection of the plate.

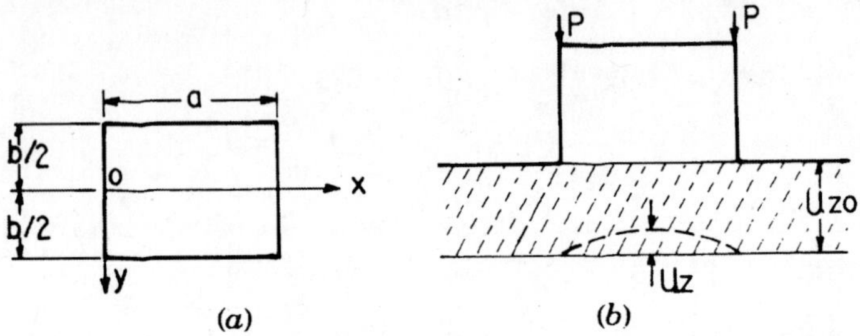

Fig. 11.27

The differential equation for the deflection of the plate can be written as :

$$\nabla^4 u_z = \frac{q - ku_z}{D} \quad ...(11.71)$$

where q is the intensity of the lateral loading

and $$\nabla^4 = \frac{\partial^4}{\partial x^4} + \frac{2\partial^4}{\partial x^2 \partial y^2} + \frac{\partial^4}{\partial y^4}.$$

Let u_{z0} denote the deflection of the edges of the bottom plate and u_z the deflection of the plate with respect to the plane of its boundary, then $p = k\,(u_{z0} - u_z)$. Hence Eq. (11.71) becomes

$$\nabla^4 u_z = \frac{k}{D}\,(u_{z0} - z_z) \quad ...(11.72)$$

Let the edges of the plate parallel to the y-axis be simply supported and the other two edges be clamped, the boundary conditions are

$$(u_z)_{x=0,\, x=a} = 0, \left(\frac{\partial^2 u_z}{\partial x^2}\right)_{x=0,\, x=a} = 0$$

$$(u_z)_{y=\pm\frac{b}{2}} = 0, \left(\frac{\partial u_z}{\partial y}\right)_{y=\pm\frac{b}{2}} = 0$$

The deflection u_z can be taken in the form of a series

$$u_z = \frac{4ku_{z0}}{D\pi} \sum_{m=1,3,5,\ldots}^{\infty} \frac{\sin\dfrac{m\pi x}{a}}{m\left(\dfrac{m^4\pi^4}{a^4} + \dfrac{k}{D}\right)} + \sum_{m=1,3,5\ldots}^{\infty} Y_m \sin\frac{m\pi x}{a} \quad ...(11.73)$$

The first series is a particular solution of Eq. (11.72) and the second series is the solution of the equation

$$\nabla^4 u_z + \frac{k}{D} u_z = 0 \qquad \text{...(11.74)}$$

Hence the functions Y_m have to satisfy the ordinary differential equation

$$\frac{d^4 Y_m}{dx^4} - 2\frac{m^2\pi^2}{a^2}\frac{d^2 Y_m}{dx^2} + \left(\frac{m^4\pi^4}{a^4} + \frac{k}{D}\right) Y_m = 0 \qquad \text{...(11.75)}$$

Let $\frac{m\pi}{a} = \mu_m, \frac{k}{D} = \lambda^4$

$$2\beta_m{}^2 = \sqrt{\mu_m^4 + \lambda^4} + \mu_m^2$$

$$2\gamma_m{}^2 = \sqrt{\mu_m^4 + \lambda^4} - \mu_m^2$$

Solution of Eq. (11.75) can be written as

$$Y_m = A_m \cosh \beta_m y \cosh \gamma_m y + B_m \sinh \beta_m y \sin \gamma_m y.$$

The deflection of the plate is

$$u_z = \sum_{m=1,3,5,\ldots}^{\infty} \sin\frac{m\pi x}{a}\left[\frac{4ku_{z0}}{D\pi}\cdot\frac{1}{m\left(\frac{m^4\pi^4}{a^4}+\frac{k}{D}\right)} + A_m \cosh \beta_m y \cos \gamma_m y + B_m \sinh \beta_m y \sin \gamma_m y\right] \qquad \text{...(11.76)}$$

From the slope boundary condition at the fixed edge, we get

$$\frac{4ku_{z0}}{D\pi}\cdot\frac{1}{m\left(\frac{m^4\pi^4}{a^4}+\frac{k}{D}\right)} + A_m \cosh\frac{\beta_m b}{2}\cos\frac{\gamma_m b}{2} + B_m \sinh\frac{\beta_m b}{2}\cdot\sin\frac{\gamma_m b}{2} = 0$$

$$(A_m \beta_m + B_m\gamma_m)\sinh\frac{\beta_m b}{2}\cdot\cos\frac{\gamma_m b}{2} - (A_m\gamma_m - B_m\beta_m)\cosh\frac{\beta_m b}{2}\sin\frac{\gamma_m b}{2} = 0$$

From which we can get A_m and B_m. Substituting these values in Eq. (11.76), we get the deflection of the plate.

Similarly for a plate having a load P at (ξ, η) and all the four edges simply supported we can obtain that

$$u_z = \frac{4P}{ab} \sum_{m=1}^{\infty} \sum_{n=1}^{\infty} \frac{\dfrac{\sin m\pi\xi}{a} \sin \dfrac{n\pi\eta}{b}}{\pi^4 D \left(\dfrac{m^2}{a^2} + \dfrac{n^2}{b^2} \right)^2 + k} \sin \frac{m\pi x}{a} \sin \frac{n\pi y}{b} .$$

EXERCISES

11.1 (*a*) What are the different types of elastic foundations ? Give examples.

(*b*) Derive the differential equation for the elastic line of a beam resting on an elastic foundation.

11.2 An infinitely long steel beam of unit width and 200 mm thick is resting on an elastic foundation whose modulus of foundation is 10 N/mm^2. A concentrated load of 8 kN is applied at a point on the beam. Determine the maximum deflection of the beam and the maximum bending stresses. Take $E = 2 \times 10^5$ N/mm^2, $\nu = 0.28$.

11.3 An infinitely long steel beam 150 mm wide and 250 mm thick is resting on an elastic foundation whose modulus of foundation is 15.75 N/mm^3. This beam is subjected to a concentrated moment of 25 N-m. Determine the maximum deflection and the maximum bending stresses in the beam. Assume $E = 2.1 \times 10^5$ N/mm^2, $\nu = 0.30$.

11.4 An infinitely long steel beam 100 mm wide and 150 mm thick is resting on an elastic foundation whose modulus of foundation is 10.50 N/mm^3. This beam is subjected to a uniformly distributed load of intensity 10 N/m over a length of 400 mm.

Determine the deflection and stresses at a point at mid length of this load and at two locations 100 mm to the left and 200 mm to the right of this load.

11.5 A semi-infinite beam resting on an elastic foundation is hinged at one end and a moment of 12 N-m is applied at the hinged end. If the beam is 80 mm wide and 60 mm thick, determine the maximum deflection and stresses in the beam. Assume $E = 2 \times 10^5$ N/mm^2, $\nu = 0.30$ and modulus of foundation as 8.4 N/mm^3.

11.6 A semi-infinite beam with free ends is resting on an elastic foundation. The beam is 100 mm wide and 150 mm thick. Determine the maximum deflection and moment in the beam if (*i*) a force of 800 N is applied at one end, (*ii*) a moment of 8 N-m is applied at one end. Assume $E = 2 \times 10^5$ N/mm^2, $\nu = 0.28$, and modulus of foundation as 10.50 N/mm^3.

11.7 A semi-infinite beam with free ends is resting on an elastic foundation. The beam is 60 mm wide and 80 mm thick. It carries a uniformly distributed load of 50 N/mm over a length of 0.5 m at one end. Determine the maximum deflection and the stresses in the beam. Assume $E = 2 \times 10^5$ N/mm^2, $\nu = 0.30$ and modulus of elastic foundation as 63 N/mm^3.

11.8 (*a*) Explain the two approaches generally adopted to solve problems of beams of finite length resting on an elastic foundation. Out of these two approaches which one is generally preferred and why ?

(*b*) Explain the term 'end-conditioning forces'.

11.9 A steel beam of length 2 m is resting on an elastic foundation and has free ends. The beam is 10 mm wide and 120 mm thick and carries two concentrated forces of 1 kN, one at each end. Determine the maximum bending stresses developed in the beam. Assume $E = 2 \times 10^5$ N/mm^2, $\nu = 0.28$ and modulus of foundation as 105 N/mm^3.

11.10 If the beam in Exercise 11.9 is simply supported at the ends, then determine the maximum bending stresses.

11.11 If the beam in Exercise 11.9 is having fixed ends, then determine the maximum bending stresses.

11.12 A steel beam of length 5 m is resting on an elastic foundation and has free ends. The beam is 100 mm wide and 200 mm thick and carries a uniformly distributed load of 50 N/mm over its entire length. Determine the maximum bending stresses developed in the beam. Assume $E = 2 \times 10^5$ N/mm^2 $\nu = 0.3$ and modulus of elastic foundation as 10.5 N/mm^2.

11.13 If the beam in Exercise 11.12 is clamped at the ends, then determine the maximum bending stresses.

11.14 A steel beam of 3 m length is resting on an elastic foundation. It carries a uniformly distributed load of 20 N/mm of length 0.5 m placed symmetrically with respect to the mid length of the beam. The beam is 80 mm wide and 120 mm thick and has free ends. Determine the maximum bending stresses developed in the beam. Assume $E = 2 \times 10^5$ N/mm^2, $\nu = 0.30$ and modulus of elastic foundation is 8 N/mm^3.

11.15 If the beam in Exercise 11.14 is simply supported at the ends, then determine the maximum bending stresses.

11.16 (*a*) Show that a longitudinal element of a thin cylindrical shell subjected to radial forces uniformly distributed along the circumference can be considered as a beam resting on an elastic foundation.

(*b*) Determine the expressions for bending stresses in a thin cylindrical shell subjected to symmetrical loading.

11.17 A thin infinitely long cylinder of mean radius, R, wall thickness, h, is subjected to a uniform band of external pressure of intensity q over a length l. Determine the expressions for bending stresses developed in the cylinder.

11.18 A steel bar 100 mm wide and 100 mm thick is resting on an elastic foundation. It carries a concentrated load of 1 kN at the centre together with an axial force of 2 kN. Determine the maximum deflection and the bending moment developed in the bar. Assume $E = 2 \times 10^5$ N/mm^2, $\nu = 0.3$ and modulus of elastic foundation as 10.5 N/mm^3.

11.19 If the bar in Exercise 11.18 carries a clockwise moment at the origin instead of the concentrated force of magnitude 12 N-m, then determine the maximum deflection and bending moment in the bar.

11.20 Derive an expression for the elastic line of a curved beam (arch) resting on an elastic foundation and determine expressions for various quantities of interest.

11.21 A rail road uses steel rails (E = 200 GPa) with a depth of 184 mm. The distance from the top of the rail to its centroid is 99.1 mm, and the moment of inertia of the rail is 36.9×10^6 mm^4. The spring constant of the road bed is 14 N/mm^2. Determine the maximum deflection, maximum bending moment, and maximum flexure stress in the rail for a single wheel load of 170 kN.

[**Ans.** 5.039 mm, 51.21 kN m, 137.5 MPa]

12

Beam Columns

12.1 Introduction

In this chapter, we shall study the buckling behaviour of beam columns. A beam column may be defined as a column which is subjected to the action of axial compression and simultaneously supporting lateral loads.

12.2 Beam-Column with a Concentrated Lateral Load

Let us consider a beam column of length l, simply supported at the ends and carrying a single concentrated lateral load Q at a distance C from the right end and axial compressive load P, as shown in Fig. 12.1.

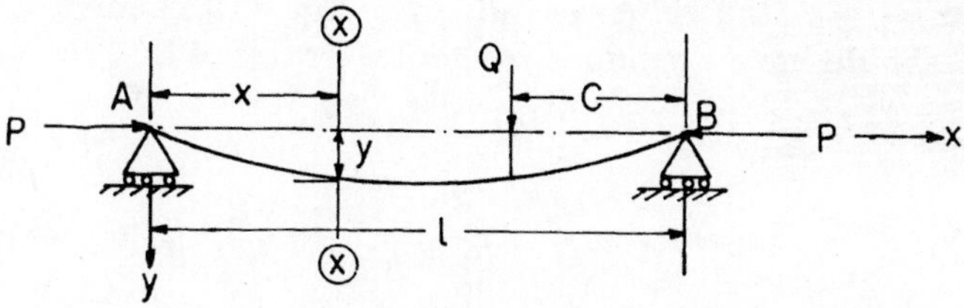

Fig. 12.1

Considering any cross-section x-x, the bending moments in the left and right-hand portions of the beam are respectively.

$$M = \frac{Qc}{l} x + Py \quad \text{for} \quad 0 \le x \le l - c$$

$$M = \frac{Q(l-c)}{l}(l-x) + Py \quad \text{for} \quad l - c \le x \le l$$

Hence, we obtain

$$\left.\begin{aligned} EI\frac{d^2y}{dx^2} &= -\frac{Qcx}{l} - Py \\ EI\frac{d^2y}{dx^2} &= -\frac{Q(l-c)(l-x)}{l} - Py \end{aligned}\right\} \quad \text{...(12.1)}$$

Let $$k^2 = \frac{P}{EI}$$

Hence, we find that

$$\frac{d^2y}{dx^2} + k^2y = -\frac{Qc\,x}{EI\,l}$$

$$\frac{d^2y}{dx^2} + k^2y = -\frac{Q(l-c)(l-x)}{EI\,l}$$

The general solutions of these equations are

$$\left.\begin{aligned} y &= A\cos kx + B\sin kx - \frac{Qc}{Pl}x \\ y &= C\cos kx + D\sin kx - \frac{Q(l-c)}{Pl}(l-x) \end{aligned}\right\} \qquad ...(12.2)$$

The constants of integration A, B, C and D are determined from the conditions at the ends of the beam and at the point of application of the load Q.

Now $$\Big[\, y \,\Big]_{\substack{x=0\\x=l}} = 0$$

therefore, $$A = 0,\ C = -D\tan kl.$$

At $x = (l-c)$, the values of y given by Eqs. (12.2) must be the same and should have a common tangent. Hence, we find

$$B\sin k(l-c) - \frac{Qc}{Pl}(l-c)$$

$$= D[\sin k(l-c) - \tan kl\cos k(l-c)] - \frac{Qc}{Pl}(l-c)$$

$$Bk\cos k(l-c) - \frac{Qc}{Pl}$$

$$= Dk[\cos k(l-c) + \tan kl\sin k(l-c)] + \frac{Q(l-c)}{Pl}$$

Solving for B and D, we get

$$B = \frac{Q\sin kc}{Pk\sin kl},\ D = -\frac{Q\sin k(l-c)}{Pk\tan kl}$$

Substituting into Eqs. (12.2), we get

$$y = \frac{Q\sin kc}{Pk\sin kl}\sin kx - \frac{Qc}{Pl}x,\ 0 \le x \le l-c \qquad ...(12.3)$$

$$y = \frac{Q\sin k(l-c)}{Pk\sin kl}\sin k(l-x) - \frac{Q(l-c)(l-x)}{Pl},\ l-c \le x \le l.$$

By differentiation, Eqs. (12.3) give

$$\frac{dy}{dx} = \frac{Q \sin kc}{P \sin kl} \cos kx - \frac{Qc}{Pl}, 0 \le x \le l - c \quad ...(12.4)$$

$$\frac{dy}{dx} = -\frac{Q \sin k(l-c)}{P \sin kl} \cos k(l-x) + \frac{Q(l-c)}{Pl}, l - c \le x \le l$$

$$\left.\begin{aligned} \frac{d^2y}{dx^2} &= -\frac{Qk \sin kc}{P \sin kl} \sin kx, 0 \le x \le l - c \\ \frac{d^2y}{dx^2} &= -\frac{Qk \sin k(l-c)}{P \sin kl} \sin k(l-x), l - c \le x \le l \end{aligned}\right\} \quad ...(12.5)$$

when the load is applied at the centre of the beam,

$x = c = \frac{l}{2}$, the maximum deflection becomes

$$\delta = \Big[y \Big]_{x = l/2} = \frac{Q}{2Pk} \left[\tan \frac{kl}{2} - \frac{kl}{2} \right] \quad ...(12.6)$$

Let $u = \frac{kl}{2} = \frac{l}{2} \cdot \sqrt{\frac{P}{EI}}$

then Eq. (12.6) becomes $\delta = \frac{Ql^3}{48\,EI} \frac{3(\tan u - u)}{u^3}$ ...(12.7)

When P is small then u will also be small, hence

$$\tan u \simeq u + \frac{u^3}{3} + ...$$

$$\therefore \quad \frac{\tan u - u}{u^3} = \frac{1}{3}$$

Hence $\delta = \frac{Ql^3}{48EI}$ ...(12.8)

When $u = \frac{\pi}{2}$

then δ becomes infinite. The value of P corresponding to the value of δ equal to infinity is called the critical load and is given by

$$P_{cr} = \frac{\pi^2 EI}{l^2} \quad ...(12.9)$$

and $u = \frac{\pi}{2}\sqrt{\frac{P}{P_{cr}}} = \frac{l}{2}\sqrt{\frac{P}{EI}}$

Now $\left(\frac{dy}{dx}\right)_{x=0} = \frac{Q}{2P}\left[\frac{1}{\cos \frac{kl}{2}} - 1\right]$

$$= \frac{Ql^2}{16EI} \cdot \frac{2(1 - \cos u)}{u^2 \cos u} \qquad \text{...(12.10)}$$

and
$$(M)_{max} = -EI\left(\frac{d^2y}{dx^2}\right)_{x=\frac{l}{2}}$$

$$= \frac{Qk\,EI}{2P} \tan \frac{kl}{2}$$

$$= \frac{Ql}{4} \frac{\tan u}{u} \qquad \text{...(12.11)}$$

Maximum bending stress is given by

$$(\sigma_b)_{max} = \frac{M_{max}}{I} \cdot y_c \qquad \text{...(12.12)}$$

where y_c = distance of the extreme fibre from the beam neutral axis.

$$\sigma_d = \frac{P}{A} \qquad \text{...(12.13)}$$

where A is the cross-sectional area of the beam. The resultant stress becomes

$$\sigma_r = (\sigma_b)_{max} \pm \sigma_d \qquad \text{...(12.14)}$$

We have considered here simply supported ends. For other end conditions, equivalent length of the column may be used in calculating the above quantities.

12.3 Several Concentrated Loads

Since the deflection is proportional to the lateral load Q, hence the principle of superposition may be used to determine the deflection under several lateral loads. Assume that there are n lateral forces $Q_1, Q_2, ..., Q_n$ and that their distances from the right support of the beam are $c_1, c_2, ..., c_n$, where $c_1 < c_2 < ... < c_n$, then from Eqs. (12.3) the deflection curve between the loads Q_m and Q_{m+1} is given by the equation

$$y = \frac{\sin kx}{Pk \sin kl} \sum_{i=1}^{i=m} Q_i \sin kc_i - \frac{x}{Pl} \sum_{i=1}^{i=m} Q_i \,.\, c_i$$

$$+ \frac{\sin k(l-x)}{Pk \sin kl} \sum_{i=m+1}^{i=n} Q_i \,.\, \sin k(l - c_i) - \frac{l-x}{Pl} \sum_{i=m+1}^{i=n} Q_i . (l - c_i) \qquad \text{...(12.15)}$$

Similarly the bending moment becomes from Eqs. (12.5)

$$M = EI\left[\frac{k \sin kx}{P \sin kl}\sum_{i=1}^{i=m} Q_i \sin kc_i + \frac{k \sin k(l-x)}{P \sin kl}\sum_{i=m+1}^{i=n} Q_i \sin k\,(l-c_i)\right] \quad ...(12.16)$$

Thus the stresses in the beam column can be determined at any cross-section.

12.4 Uniformly Distributed Lateral Load

Let us consider a beam column carrying a uniformly distributed load of intensity q as shown in Fig. 12.2.

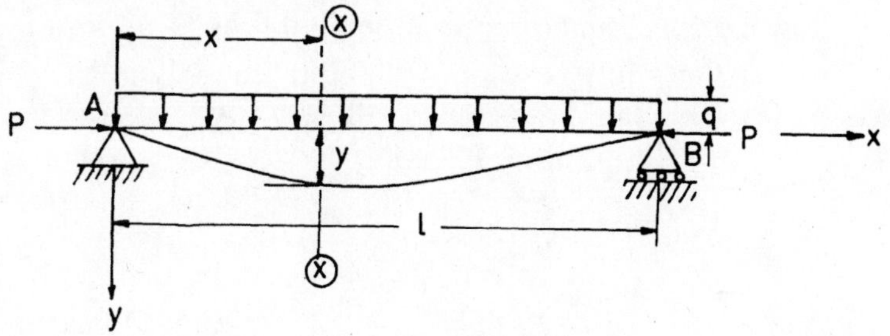

Fig. 12.2

Taking origin at A, the bending moment at any cross-section x-x is

$$M = Py + \frac{ql}{2}x - \frac{qx^2}{2}$$

$$\therefore \quad EI\frac{d^2y}{dx^2} = -Py - \frac{ql}{2}x + \frac{qx^2}{2} \quad ...(12.17)$$

The solution of this equation is of the form

$$y = A\cos kx + B\sin kx + a + bx + cx^2 \quad ...(12.18)$$

where A, B, a, b, c are constants.

Differentiating, we get

$$\frac{dy}{dx} = -Ak\sin kx + Bk\cos kx + b + 2cx$$

$$\frac{d^2y}{dx^2} = -Ak^2\cos kx - Bk^2\sin kx + 2c \quad ...(12.19)$$

Substituting the value of y from Eq. (12.18) in (12.17) and comparing with Eq. (12.19), we find

$$-\frac{P}{EI}(a + bx + cx^2) - \frac{qlx}{2EI} + \frac{qx^2}{2EI} = 2c$$

Comparing the coefficients, we get

$$c = \frac{q}{2P}$$

$$b = -\frac{ql}{2P}$$

$$a = -\frac{qEI}{P^2}$$

Hence, Eq. (12.18) becomes

$$y = A\cos kx + B\sin kx + \frac{q}{2P}x^2 - \frac{ql}{2P}x - \frac{qEI}{p^2}$$

Now using the end conditions, we find that

at $x = 0, y = 0$

and at $x = l, y = 0$.

Hence, we get $A = \frac{qEI}{P^2}$

$$B = \frac{qEI}{P^2}\tan\frac{kl}{2}$$

$$\therefore \quad y = \frac{qEI}{P^2}\left[\cos k\left(\frac{l}{2} - x\right)\sec\frac{kl}{2} - 1\right] - \frac{qx}{2P}(l - x) \quad \text{...(12.20)}$$

$$y_{max}\bigg|_{x=\frac{l}{2}} = \frac{qEI}{P^2}\left(\sec\frac{kl}{2} - 1\right) - \frac{ql^2}{8P} \quad \text{...(12.21)}$$

Eqs. (12.20) and (12.21) can be written as

$$y = \frac{ql^4}{16EI\,u^4}\left[\frac{\cos\left(u - \frac{2ux}{l}\right)}{\cos u} - 1\right] - \frac{ql^2x}{8EI\,u^4}(l - x) \quad \text{...(12.22)}$$

$$y_{max} = \frac{5ql^4}{384\,EI}\left[\frac{12(2\sec u - 2 - u^2)}{5u^4}\right] \quad \text{...(12.23)}$$

where $k = \sqrt{\frac{P}{EI}}$

$$u = \frac{kl}{2}$$

Differentiating Eq. (12.22), we get the slope of the deflection curve. At the origin

$$\left(\frac{dy}{dx}\right)_{x=0} = \frac{ql^3}{24\,IE}\left[\frac{3(\tan u - u)}{u^3}\right] \qquad ...(12.24)$$

and $$M_{max} = \frac{qEI}{P}\left[\cos\frac{kl}{2} - 1\right] = \frac{ql^2}{8}\left[\frac{2(1-\cos u)}{u^2\cos u}\right] \qquad ...(12.25)$$

The maximum compressive stress becomes

$$\sigma_r = \frac{P}{A} + \frac{qEy_c}{P}\left[\sec\frac{kl}{2} - 1\right] \qquad ...(12.26)$$

12.5 Bending of a Beam Column by Couples

Suppose now that a moment M_B is acting at the end B of the beam column as shown in Fig. 12.3.

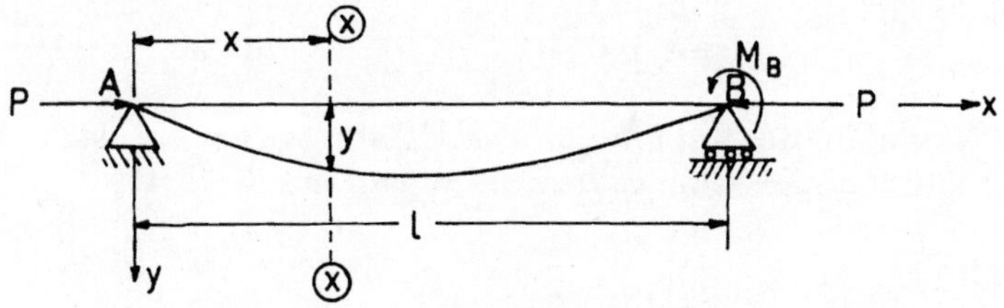

Fig. 12.3.

At any cross-section x-x at a distance x from origin A, the bending moment is

$$M = Py + \frac{M_B}{l}\,.\,x$$

$$\therefore \quad EI\frac{d^2y}{dx^2} = -Py - \frac{M_B}{l}y$$

$$\frac{d^2y}{dx^2} + k^2y = -\frac{M_B}{EIl}x \qquad ...(12.27)$$

where $$k = \sqrt{\frac{P}{EI}}$$

The solution of Eq. (12.27) is

$$y = A\cos kx + B\sin kx - \frac{M_Bx}{Pl}$$

At $x = 0, y = 0$, hence $A = 0$.

At $x = l, y = 0$.

$$\therefore \qquad B = \frac{M_B}{P \sin kl}$$

$$\therefore \qquad y = \frac{M_B}{P}\left[\frac{\sin kx}{\sin kl} - \frac{x}{l}\right] \qquad ...(12.28)$$

$$\frac{dy}{dx} = \frac{M_B}{P}\left[\frac{k \cos kx}{\sin kl} - \frac{1}{l}\right]$$

$$\therefore \qquad \left(\frac{dy}{dx}\right)_{x=0} = \frac{M_B}{P}\left[\frac{k}{\sin kl} - \frac{1}{l}\right]$$

$$= \frac{M_B \,.\, l}{6\,EI} \cdot \frac{3}{u}\left[\frac{1}{\sin 2u} - \frac{1}{2u}\right] \qquad ...(12.29)$$

$$\left(\frac{dy}{dx}\right)_{x=l} = \frac{M_B}{P}\left[k \cot kl - \frac{1}{l}\right]$$

$$= \frac{M_B \,.\, l}{3\,EI} \cdot \frac{3}{2u}\left[\frac{1}{\tan 2u} - \frac{1}{2u}\right] \qquad ...(12.30)$$

Now suppose that two couples M_A and M_B are applied at the ends A and B of the beam column as shown in Fig. 12.4.

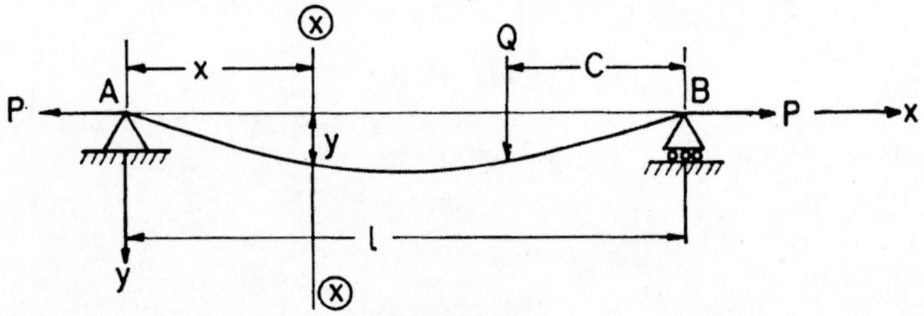

Fig. 12.4

Let R_A be the reaction at the end A. Then bending moment at any cross-section x-x is

$$M = R_A . x + M_A + Py \qquad ...(12.31)$$

$$M_B = R_A . l + M_A$$

$$\frac{dM}{dx} = R_A + P\frac{dy}{dx}$$

$$\frac{d^2M}{dx^2} = P\frac{d^2y}{dx^2}$$

$$= -\frac{PM}{EI} = -k^2 M$$

$$\therefore \qquad \frac{d^2M}{dx^2} + k^2M = 0.$$

The solution of this equation is

$$M = A \sin kx + B \cos kx$$

At $\quad x = 0$

$$M = M_A$$

$$\therefore \qquad B = M_A$$

At $\quad x = l$

$$M = -M_B$$

$$\therefore \qquad -M_B = M_A \cos kl + A \sin kl$$

$$A = \frac{-M_B - M_A \cos kl}{\sin kl}$$

$$\therefore \qquad M = M_A \cos kx - (M_B + M_A \cos kl)\frac{\sin kx}{\sin kl}$$

$$\frac{dM}{dx} = -kM_A \sin kx - k(M_B + M_A \cos kl)\frac{\cos kx}{\sin kl}$$

At $\quad x = 0$

$$\frac{dM}{dx} = R_A$$

$$\therefore \qquad R_A = -k(M_B + M_A \cos kl) \cdot \frac{1}{\sin kl} \qquad \text{...(12.32)}$$

$$\therefore \qquad M_B = -kl\left[\frac{M_B + M_A \cos kl}{\sin kl}\right] + M_A$$

$$M_B\left[1 + \frac{kl}{\sin kl}\right] = -M_A\left[\frac{kl \cos kl}{\sin kl} - 1\right]$$

$$M_B = M_A\left[\frac{\sin kl - kl \cos kl}{kl + \sin kl}\right] \qquad \text{...(12.33)}$$

Now Eq. (12.31) becomes

$$M = -kx\left[\frac{M_B + M_A \cos kl}{\sin kl}\right] + M_A + Py$$

$$\therefore \qquad EI\frac{d^2y}{dx^2} = kx\left[\frac{M_B + M_A \cos kl}{\sin kl}\right] - M_A - Py$$

$$\frac{d^2y}{dx^2} + \frac{P}{EI}y = \frac{kx}{EI}\left[\frac{M_B + M_A \cos kl}{\sin kl}\right] - \frac{M_A}{EI}$$

The solution of this equation can be written as

$$y = C \cos kx + D \sin kx + \left[\frac{M_B + M_A \cos kl}{\sin kl}\right]\frac{kx}{P} - \frac{M_A}{P}$$

At $\quad x = 0, y = 0$

$$0 = C - \frac{M_A}{P}$$

$$\therefore \quad C = \frac{M_A}{P}$$

At $\quad x = l, y = 0$

$$\therefore \quad 0 = \frac{M_A}{P}\cos kl + D \sin kl + \left[\frac{M_B + M_A \cos kl}{\sin kl}\right]\frac{kl}{P} - \frac{M_A}{P}$$

$$\therefore \quad D \sin kl = -\frac{M_A}{P}\cos kl + \frac{M_A}{P} - \left[\frac{M_B + M_A \cos kl}{\sin kl}\right]\frac{kl}{P}$$

$$D = -\frac{M_A}{P}\frac{\cos kl}{\sin kl} + \frac{M_A}{P}\cdot\frac{1}{\sin kl} - \left[\frac{M_B + M_A \cos kl}{\sin^2 kl}\right]\frac{kl}{P}$$

$$\therefore \quad y = \frac{M_A}{P}\cos kx + \frac{M_A}{P}\left[\frac{1}{\sin kl} - \frac{\cos kl}{\sin kl} - \frac{kl \cos kl}{\sin^2 kl}\right] \times \sin kx - \frac{kl}{\sin^2 kl}\frac{M_B}{P}\,.\sin kx + \frac{M_A}{P}\cdot\frac{kx \cos kl}{\sin kl} + \frac{M_B}{p}\cdot\frac{kx}{\sin kl} - \frac{M_A}{P}$$

$$= \frac{M_A}{P}\left[\cos kx + \frac{\sin kx}{\sin kl}\left\{1 - \cos kl - \frac{kl \cos kl}{\sin kl}\right\} + \frac{kx \cos kl}{\sin kl} - 1\right] + \frac{M_B}{P}\left[\frac{kx}{\sin kl} - \frac{kl \sin kx}{\sin^2 kl}\right]$$

$$= \frac{M_A}{P}\left[\frac{1}{\sin kl}\left\{\sin kl \cos kx + \sin kx - \sin kx \cos kl - kl\frac{\sin kx \cos kl}{\sin kl} + kx \cos kl - \sin kl\right\}\right] + \frac{M_B}{P}\cdot\frac{kl}{\sin kl}\left[\frac{x}{l} - \frac{\sin kx}{\sin xl}\right]$$

$$= \frac{M_A}{P}\left[\frac{1}{\sin kl}\left\{\sin k(l-x) + \sin kx - \sin kl - kl\cos kl\left(\frac{\sin kx}{\sin kl} - \frac{x}{l}\right)\right\}\right] + \frac{M_B}{P}\cdot\frac{kl}{\sin kl}\left[\frac{x}{l} - \frac{\sin kx}{\sin xl}\right] \quad ...(12.34)$$

12.6 Uniformly Distributed Lateral Load and Eccentric Axial Load

Consider a beam column carrying load as shown in Fig. 12.5.

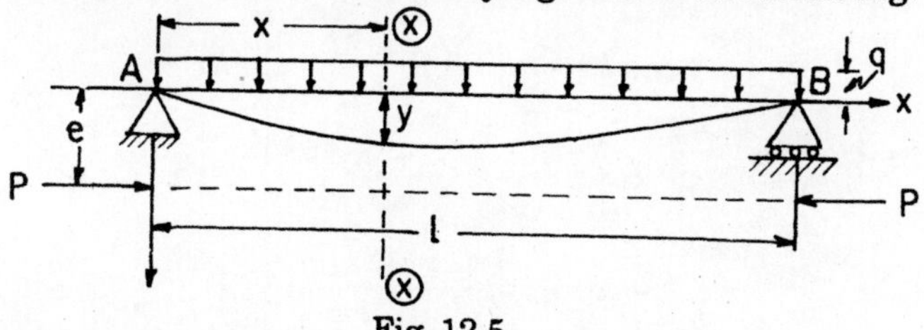

Fig. 12.5

The bending moment at any cross-section x-x is

$$M = \frac{ql}{2}\cdot x - \frac{qx^2}{2} - P(e-y) \quad ...(12.35)$$

$$\therefore \quad EI\frac{d^2y}{dx^2} = -\frac{qlx}{2} + \frac{qx^2}{2} + P(e-y)$$

$$\frac{d^2y}{dx^2} + \frac{P}{EI}y = \frac{qx^2}{2EI} - \frac{qlx}{2EI} + \frac{Pe}{EI}.$$

The solution of this equation may be written as

$$y = A\cos kx + B\sin kx + \frac{qx^2}{2P} - \frac{qEI}{P^2} - \frac{qlx}{2P} + e$$

At $\quad x = 0, y = 0.$

$$\therefore \quad 0 = A - \frac{qEI}{P^2} + e$$

$$A = \frac{qEI}{P^2} - e$$

At $\quad x = l, y = 0$

$$0 = \left(\frac{qEI}{P^2} - e\right)\cos kl + B\sin kl + \frac{ql^2}{2P} - \frac{qEI}{P^2} - \frac{2l^2}{2P} + e$$

$$= \frac{qEI}{P^2}(\cos kl - 1) + e(1-\cos kl) + B\sin kl$$

$$\therefore \qquad B = \frac{qEI}{P^2}\left(\frac{1-\cos kl}{\sin kl}\right) - e\left(\frac{1-\cos kl}{\sin kl}\right)$$

$$= \left(\frac{qEI}{P^2} - e\right)\left(\frac{1-\cos kl}{\sin kl}\right)$$

$$y = \left(\frac{qEI}{P^2} - e\right)\cos kx + \left(\frac{qEI}{P^2} - e\right)\left(\frac{1-\cos kl}{\sin kl}\right)\sin kx$$

$$+ \frac{qx^2}{2P} - \frac{qEI}{P^2} - \frac{qlx}{2P} + e \qquad \text{...(12.36)}$$

If $q = 0$, then

$$y = -e\cos kx - e\left(\frac{1-\cos kl}{\sin kl}\right)\sin kx + e$$

$$= e(1-\cos kx) - e\left(\frac{1-\cos kl}{\sin kl}\right)\sin kx$$

$$= e\left[1 - \cos kx - \frac{1-\cos kl}{\sin kl}\sin kx\right]$$

$$= \frac{e}{\sin kl}[\sin kl - \cos kx \sin kl - \sin kx + \cos kl \sin kx]$$

$$= e\left[1 - \frac{\sin kx}{\sin kl} - \frac{\sin k(l-x)}{\sin kl}\right] \qquad \text{...(12.37)}$$

Substituting Eq. (12.36) in (12.35), we find

$$M = \frac{qlx}{2} - \frac{qx^2}{2} - P\left[e - \left(\frac{qEI}{P^2} - e\right)\cos kx\right.$$

$$\left. - \left(\frac{qEI}{P^2} - e\right)\left(\frac{1-\cos kl}{\sin kl}\right) \times \sin kx - \frac{qx^2}{2P} + \frac{qEI}{P^2} + \frac{qlx}{2P} - e\right]$$

$$= P\left(\frac{qEI}{P^2} - e\right)\cos kx$$

$$+ P\left(\frac{qEI}{P^2} - e\right)\left(\frac{1-\cos kl}{\sin kl}\right)\sin kx - \frac{qEI}{P}$$

$$= \left(\frac{qEI}{P} - Pe\right)\left[\cos kx + \frac{1-\cos kl}{\sin kl} . \sin kx\right] - \frac{qEI}{P}$$

At $x = \dfrac{l}{2}$, M is maximum.

$$M_{max} = \left(\frac{qEI}{P} - Pe\right)\left[\cos\frac{kl}{2} + \frac{1-\cos kl}{\sin kl} . \sin\frac{kl}{2}\right] - \frac{qEI}{P}$$

$$= \left(\frac{qEI}{P} - Pe \right) \left[\sec \frac{kl}{2} \right] - \frac{qEI}{P} \qquad \text{...(12.38)}$$

The resultant stress is

$$\sigma_r = \frac{P}{A} + \frac{M_{max}}{I} \cdot y_c$$

$$\sigma_r = \frac{P}{A} + \frac{\left(\frac{qEI}{P} - Pe \right) \sec \frac{kl}{2} - \frac{qEI}{P}}{1} \cdot y_c \qquad \text{...(12.39)}$$

12.7 Tie Rod With Lateral Loading

Let us consider a tie rod carrying a lateral load Q at a distance C from the right hand support as shown in Fig. 12.6.

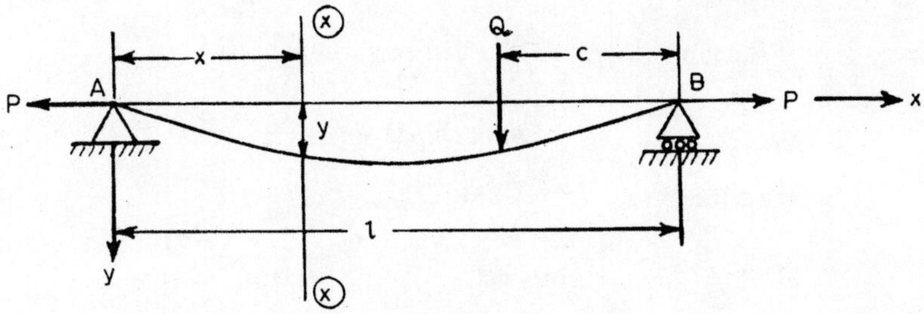

Fig. 12.6

The bending moment at any cross-section x-x is

$$M = \frac{Qc}{l} \cdot x - Py$$

$$\therefore \quad EI \frac{d^2y}{dx^2} = Py - \frac{Qcx}{l} \text{ for } 0 \le x \le l - c$$

and $$EI \frac{d^2y}{dx^2} = Py - \frac{Q(l-c)}{l} \cdot (l - x) \text{ for } l - c \le x \le l$$

or $$\frac{d^2y}{dx^2} - k^2y = -\frac{Qcx}{EIl}$$

and $$\frac{d^2y}{dx^2} - k^2y = -\frac{Q(l-c)}{EIl} (l - x).$$

The solutions of these equations are

$$y = A \cosh kx + B \sinh kx + \frac{Qcx}{Pl}$$

and $$y = C \cosh kx + D \sinh kx + \frac{Q(l-c)}{Pl} (l - x)$$

At $x = 0$ and $x = l, y = 0$

$\therefore\ A = 0, C = -D \tanh kl$

At $x = l - c$,

the values of y obtained from both the equations must be the same and the deflection curve should have a common tangent.

$$\therefore\ B \sinh k(l - C) + \frac{Qc(l-c)}{Pl} = -D \tanh kl \cosh k(l-c) + D \sinh k(l-c) + \frac{Qc(l-c)}{Pl}.$$

$$Bk \cosh k(l-c) + \frac{Qc}{Pl} = -\left[-Dk \tanh kl \sinh k(l-c) + Dk \cosh k(1-c) - \frac{Q(l-c)}{Pl} \right]$$

or

$$Bk \cosh k(l-c) + \frac{Qc}{Pl} = Dk \tanh kl \sin k(l-c) - Dk \cosh k(l-c) + \frac{Q(l-c)}{Pl}$$

Solving for B and D, we get

$$B = -\frac{Q \sinh kc}{Pk \sinh kl}, \quad D = \frac{Q \sinh k(l-c)}{Pk \tanh kl}$$

$$\therefore\quad y = -\frac{Q \sinh kc}{Pk \sinh kl} \sinh kx + \frac{Qcx}{Pl} \quad \text{for } 0 \le x \le l - c \qquad ...(12.40)$$

$$y = -\frac{Q \sinh k(l-c)}{Pk \sinh kl} \sinh k(l-x) + \frac{Q(l-c)(l-x)}{Pl}, \text{ for } l - c \le x \le l \qquad ...(12.41)$$

Differentiating Eq. (12.40), we get

$$\frac{dy}{dx} = -\frac{Q \sinh kc}{P \sinh kl} \cosh kx + \frac{Qc}{Pl}$$

$$\frac{d^2y}{dx^2} = -\frac{Qk \sinh kc}{P \sinh kl} . \sinh kx$$

$$y_{max}\Big|_{x = l/2} = -\frac{Q \sinh \frac{kl}{2}}{Pk \sinh kl} \sinh \frac{kl}{2} + \frac{Ql}{4P} = \frac{Q}{2Pk}\left[\frac{kl}{2} - \tanh \frac{kl}{2}\right]$$

$$= \frac{Q}{2Pk}[u - \tanh u] = \frac{Ql^3}{48EI} \cdot \left[\frac{u - 3 \tanh u}{u^3}\right] \qquad ...(12.42)$$

$$M_{max} = \frac{Ql}{4} \cdot \left(\frac{\tanh u}{u} \right) \qquad ...(12.43)$$

Similar results may be derived for other type of lateral loadings and axial tension.

EXERCISES

12.1 A beam column of rectangular cross-section 80 mm wide and 100 mm deep is 12 m long. It carries an axial compressive load 45 kN together with a concentrated lateral load of 1.5 kN at a distance of 300 mm from the right hand support. If the column has hinged ends, determine the maximum deflection of the column and the maximum stress at mid-span. Assume $E = 2 \times 10^5$ N/mm^2.

12.2 If the concentrated lateral load in Exercise 12.1 is acting at the mid-span, then determine the maximum deflection and the maximum stress developed in the column.

12.3 A beam column of circular cross-section is 100 mm in diameter. It carries an axial compressive load of 5 kN. The column is 2 m long and has hinged ends. It also carries three concentrated lateral loads each equal to 2 kN at 0.5 m spacing. Determine the deflection of the column at mid length and the maximum stress at the mid-span. Assume $E = 2.1 \times 10^5$ N/mm^2.

12.4 A beam column of square cross-section is of 50 mm side length. It carries an axial compressive load of 4 kN and a uniformly distributed load of intensity 150 N/mm is applied on its lateral surface. The column is 3 m in length and the ends are hinged. Determine the maximum deflection and the stress developed in the column. Take $E = 2 \times 10^5$ N/mm^2.

12.5 A beam column of circular cross-section is 50 mm in diameter and 2 m long. It is simply supported at the ends and carries an axial compressive load of 3.5 kN. If a couple of 15 N-m is applied at the right end in the anti-clockwise direction, determine the deflection and maximum stress developed in the column at mid-span. Take $E = 2 \times 10^5$ N/mm^2.

12.6 If in Exercise 12.5 a couple of 10 N-m is also applied on the left end support in the clockwise direction, then determine the deflection and maximum stress in the column at mid-span.

12.7 A beam column of circular cross-section of diameter 150 mm is subjected to an eccentric axial compressive load of 6.5 kN. It also carried a lateral load of 60 N/mm distributed uniformly over its length of 2 m. It has fixed ends. Determine the deflection and maximum stress at mid-span. Take $E = 2 \times 10^5$ N/mm^2.

12.8 If the axial load in Exercise 12.1 is tensile instead of being compressive, then determine the maximum deflection and stress in the column.

13

Numerical and Energy Methods

13.1 Introduction

Sometimes it is not possible or rather difficult to obtain the solution of the governing differential equations of an elasticity problem by adopting the methods as described in the previous chapters. In those cases the solution can be obtained by minimization of a certain energy expression by the variational methods. These methods are, therefore, referred to as the energy or variational methods. The technique of using these methods shall be discussed in this chapter.

Out of the numerical methods, the finite difference and finite element methods have been extensively used to solve elasticity problems. We shall also study the basic principles of these methods.

13.2 Strain Energy

When an elastic body is under the action of external forces, the body deforms and work is done by these forces. If a strained, perfectly elastic body is allowed to recover slowly to its unstrained state, it is capable of giving back all the work done by these external forces. This work done in straining such a body may be regarded as energy stored in the body and is called the strain energy of deformation. We shall use U to represent the strain energy for a given body and u to represent strain energy per unit volume.

In order to calculate the amount of strain energy stored in a strained body let us first consider an infinitesimal rectangular parallelopiped of material under the action of simple tension, as shown in Fig. 13.1. The increment of work done by the

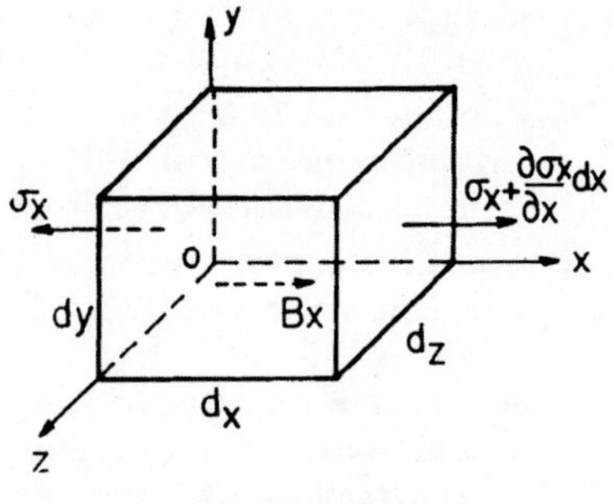

Fig. 13.1

stresses in the x-direction during infinitesimal displacement du_x is :

$$\left(\sigma_x + \frac{\partial \sigma_x}{\partial x} dx\right) d\left(u_x + \frac{\partial u_x}{\partial x} dx\right) dydz - \sigma_x \, . \, du_x \, dydz$$

$$+ B_x \, dxdydz.d\left(u_x + \frac{1}{2}\frac{\partial u_x}{\partial x} dx\right)$$

Solving and neglecting higher order terms, we get

$$\left[\sigma_x \, d\left(\frac{\partial u_x}{\partial x}\right) + \left(\frac{\partial \sigma_x}{\partial x} + B_x\right) du_x\right] dxdydz$$

Now $\frac{\partial u_x}{\partial x} = \varepsilon_x$ and $\frac{\partial \sigma_x}{\partial x} + B_x = 0$

Hence the above expression becomes

$$du_x = \sigma_x d\varepsilon_x \, dxdydz + \sigma_x \, d\varepsilon_x \, dv$$

where $dv = dxdydz$ is the elementary volume.

By considering normal stresses and strains in the y and z directions, we get

$$dU_y = \sigma_y . d\varepsilon_y \, dv$$

$$dU_z = \sigma_z . d\varepsilon_z \, dv$$

Therefore, the strain energy resulting from normal strains is

$$dU_\varepsilon = (\sigma_x \, d\varepsilon_x + \sigma_y d\varepsilon_y + \sigma_z d\varepsilon_z) dv \quad ...(13.1)$$

To determine the strain energy associated with shear strain, let us consider for this purpose an infinitesimal parallelopiped of this material subjected to shear stress τ_{xy} as shown in Fig. 13.2.

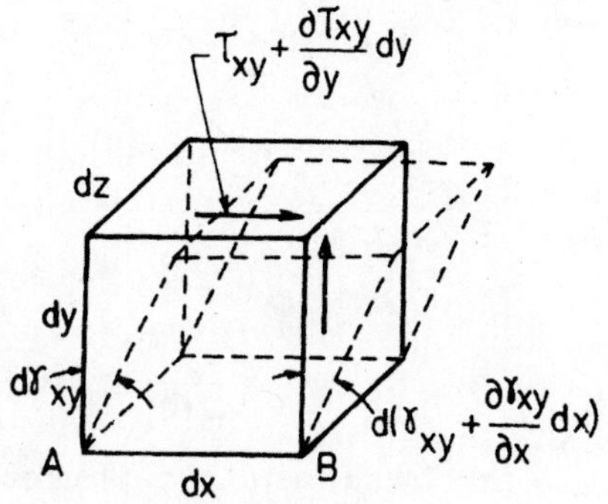

Fig. 13.2

The work done for the strain differentials $d\gamma_{xy}$ at A and $d\left[\gamma_{xy} + \frac{\partial \gamma_{xy}}{\partial x} dx\right]$ and B is :

$$\left[\left\{\tau_{xy} + \frac{\partial \tau_{xy}}{\partial y} dy\right\} dxdz\right] \times \left[d\left\{\gamma_{xy} + \frac{\partial \gamma_{xy}}{\partial x} \; \beta dx\right\} dy\right]$$

where β is some fraction.

Expanding the expressions and neglecting quantities of higher order, we get

$$\tau_{xy}\, d\gamma_{xy}.dxdydz$$

By similarly considering the differentials of the work done by shear stresses τ_{xz} and τ_{yz} and adding, we get

$$dU_\gamma = (\tau_{xy}\, d\gamma_{xy} + \tau_{yz}\, d\gamma_{yz} + \tau_{zx} d\gamma_{zx})\, dv \quad ...(13.2)$$

Therefore, for an elastic material, the total strain energy increment dU for an element under a general state of stress becomes

$$dU = (\sigma_x d\varepsilon_x + \sigma_y d\varepsilon_y + \sigma_z d\varepsilon_z + \tau_{xy} d\gamma_{xy} + \tau_{yz} d\gamma_{yz} + \tau_{zx} d\gamma_{zx})\, dv \quad ...(13.3)$$

or $$du = \sigma_x d\varepsilon_x + \sigma_y d\varepsilon_y + \sigma_z d\varepsilon_z + \tau_{xy} d\gamma_{xy} + \tau_{yz} d\gamma_{yz} + \tau_{zx} d\gamma_{zx} \quad ...(13.4)$$

$$= \tau_{ij} \,.\, d\varepsilon_{ij}$$

The total strain energy stored in a deformed elastic body, U, can be found by integration dU over the whole volume V, thus

$$U = \int [\sigma_x d\varepsilon_x + \sigma_y d\varepsilon_y + \sigma_z d\varepsilon_z + \tau_{xy} d\gamma_{xy} + \tau_{yz} d\gamma_{yz} + \tau_{zx} d\gamma_{zx}]\, dv$$

$$= \frac{1}{2}(\sigma_x \varepsilon_x + \sigma_y \varepsilon_y + \sigma_z \varepsilon_z + \tau_{xy}\gamma_{xy} + \tau_{yz}\gamma_{yz} + \tau_{zx}\gamma_{zx})\, dv \quad ...(13.5)$$

$$\therefore \quad U = \frac{1}{2} \iiint_V (\sigma_x \varepsilon_x + \sigma_y \varepsilon_y + \sigma_z \varepsilon_z + \tau_{xy}\gamma_{xy} + \tau_{yz}\gamma_{yz} + \tau_{zx}\gamma_{zx}) dv \quad ...(13.6)$$

Substituting strains in terms of stresses by using Hooke's law, we get

$$U = \iiint_V \left[\frac{1}{2E} (\sigma_x^2 + \sigma_y^2 + \sigma_z^2) - \frac{\nu}{E} (\sigma_x\sigma_y + \sigma_y\sigma_z + \sigma_z\sigma_x) + \frac{1}{2G} (\tau_{xy}^2 + \tau_{yz}^2 + \tau_{zx}^2) \right] dxdydz \quad ...(13.7)$$

And by substituting in terms of strains, we get

$$U = \iiint_V \left[\frac{E\nu}{2(1+\nu)(1-2\nu)} (\varepsilon_x + \varepsilon_y + \varepsilon_z)^2 + G(\varepsilon_x^2 + \varepsilon_y^2 + \varepsilon_z^2) + \frac{1}{2}G(\gamma_{xy}^2 + \gamma_{yz}^2 + \gamma_{zx}^2) \right] dxdydz \quad ...(13.8)$$

(i) In case of plane stress

$$\sigma_z = \tau_{yz} = \tau_{zx} = 0$$

and we have $$U = \frac{1}{2} \iiint_V \left[\frac{1}{E} (\sigma_x^2 + \sigma_y^2 - 2\nu\, \sigma_x\sigma_y) + \frac{1}{G} \tau_{xy}^2 \right] dv \quad ...(13.9)$$

or $$U = \frac{E}{2(1+\nu)} \iiint_V \left[\frac{1}{1-\nu} (\varepsilon_x{}^2 + \varepsilon_y{}^2 + 2\nu\, \varepsilon_{xy}) + \frac{1}{2}\gamma_{xy}^2 \right] dv \quad ...(13.10)$$

(ii) Strain-energy under simple tension

$$U = \iiint_V \frac{1}{2E} \sigma_x{}^2\, dv \quad ...(13.11)$$

For a straight bar of length l, area of cross-section A and pulled with a force P

$$U = \frac{1}{2E} \int_0^l \left(\frac{P}{A} \right)^2 A dx = \frac{P^2 l}{2AE} \quad ...(13.12)$$

or in terms of deflection δ

$$U = \frac{EA\delta^2}{2l} \quad ...(13.13)$$

(iii) Strain-energy under symmetrical bending

$$U_\varepsilon = \iiint_V \frac{M^2 y^2}{2EI^2} dv$$

$$= \int_0^l \int_0^A \frac{M^2 y^2}{2EI^2} dA dx$$

$$= \int_0^l \frac{M^2}{2EI} dx = \frac{M^2 l}{2EI}$$

$$U_\gamma = \iiint_V \frac{1}{2G} \left(\frac{FA\bar{y}}{Ib} \right)^2 dx dy dz$$

where F = Shear force in the section

A = Area of the bar above the cross-section considered

$\bar{y}$ = Distance of the centre of gravity of the area from the neutral axis

b = Width of the bar

I = Moment of inertia

M = Bending moment

$$U_\gamma = \frac{1}{2I^2} \iint_V \left[\left(\frac{A\bar{y}}{b} \right)^2 \int_0^l \frac{F^2}{G} dx \right] dA$$

$$= \frac{1}{2\,GI^2} \left[\iint_A \left(\frac{A\bar{y}}{b} \right)^2 dA \right] \int_0^l F^2 dx$$

For a rectangular beam of width b and depth h

$$A\bar{y} = b\left(\frac{h}{2} - y\right)\cdot\left(\frac{y}{2} + \frac{h}{4}\right)$$

$$= \frac{b}{2}\left(\frac{h^2}{4} - y^2\right)$$

$$\therefore \qquad U = U_\varepsilon + U_\gamma \qquad \text{...(13.14)}$$

(*iv*) *Strain-energy in torsion.*

$$U = \frac{1}{2G}\iiint_V \tau^2_{xy}\, dv$$

$$= \frac{1}{2G}\iiint_V \left(\frac{Tr}{J}\right)^2 dv$$

$$= \frac{T^2}{2GJ^2}\int_A r^2 dA \int_0^l dx$$

$$= \frac{T^2 l}{2GJ} = \frac{\theta^2 J}{2Gl} \qquad \text{...(13.15)}$$

where T = Torque applied

J = Polar moment of inertia

θ = Angle of twist.

13.3 The Maxwell-Betti Reciprocal Theorem

Consider an elastic body constrained in space and acted on by point forces as shown in Fig. 13.3.

Let δ_i be the displacement of the point of application of load P_i due to the total action of all the n forces. Let us imagine that the loads are applied in a quasi-static way and at any stage of loading process each load can be given as $\lambda(P_i)_f$, where λ is a fraction between zero and one and $(P_i)_f$ is the final value of the load. Because of the linear elastic behaviour of the material, the deflection $\delta_i = \lambda(\delta_i)_f$. The work done W_k by the forces become :

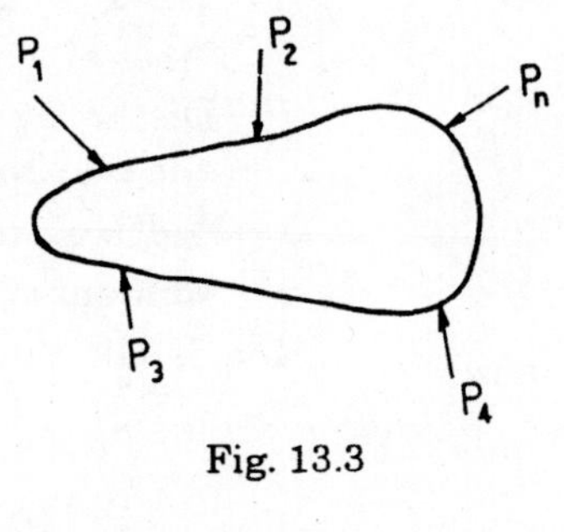

Fig. 13.3

$$W_k = \sum_{i=1}^{n}\int_0^{(\delta_i)_f} P_i \,.\, d\delta_i$$

$$= \sum_{i=1}^{n} \int_0^{(\delta_i)_f} \lambda (P_i)_f \,.\, d[\lambda (\delta_i)_f]$$

$$= \int_0^1 \sum_{i=1}^{n} (P_i)_f \,.\, (\delta_i)_f \,\lambda d\lambda$$

$$= \sum_{i=1}^{n} (P_i)_f \,.\, (\delta_i)_f \int_0^1 \lambda\, d\lambda = \sum_{i=1}^{n} \frac{1}{2} (P_i)_f \,.\, (\delta_i)_f$$

$$\therefore \qquad U_I = \sum_{i=1}^{n} \frac{1}{2} P_i \,.\, \delta_i \qquad \text{...(13.16)}$$

Now consider another loading process for the body wherein a new set of point forces denoted with the subscript II, is applied. Let there be now w forces for this loading process. Therefore, the work done in this case is

$$U_{II} = \sum_{j=1}^{m} \frac{1}{2} P_j \,.\, \delta_j \qquad \text{...(13.17)}$$

To determine the strain energy of the body under the combined action of both force systems I and II, we apply loading system I first producing strain energy U_I given by

$$U_I = \frac{1}{2} \sum_{i=1}^{n} (P_i)_I \,.\, (\delta_i)_I \qquad \text{...(13.18)}$$

Now apply load II to the system. The work done by force system I as a result of deformations induced by force system II, $U_{I,\,II}$, during which force system I remains constant is

$$U_{I,\,II} = \sum_{i=1}^{n} (P_i)_I \,.\, (\delta_i)_{II} \qquad \text{...(13.19)}$$

where $(\delta_i)_{II}$ = Displacement due to force system II at the position of the point of application of force $(P_i)_I$ of system I.

and

$$U_{II} = \frac{1}{2} \sum_{j=1}^{m} (P_j)_{II} \,.\, (\delta_j)_{II} \qquad \text{...(13.20)}$$

Total strain energy becomes

$$U = U_I + U_{I,\,II} + U_{II} \qquad \text{...(13.21)}$$

If we apply loading system II first, then

$$U' = U_{II} + U_{II,\,I} + U_I \qquad \text{...(13.22)}$$

But $\qquad U = U'$, hence

$$U_{I,\,II} = U_{II,\,I} \qquad \text{...(13.23)}$$

or

$$\sum_{i=1}^{n} (P_i)_I \cdot (\delta_i)_{II} = \sum_{j=1}^{m} (P_j)_{II} \cdot (\delta_j)_I \qquad \text{...(13.24)}$$

This is the Maxwell-Betti reciprocal theorem. It says that the work done by one system of loads due to displacements caused by a second system of loads equals the work done by the second system of loads due to displacements caused by the first system of loads.

13.4 Castigliano's Theorem

Let us consider an elastic body loaded by n concentrated external forces P_i as shown in Fig. 13.4. The concentrated external forces P_i and supporting forces N_i from the loading system I. If we increase the load P_k by a small amount ΔP_k, to maintain equilibrium, some or all of the supporting forces will change slightly as a result of ΔP_k. The load increment ΔP_k and the associated force increments at the supports form loading system II.

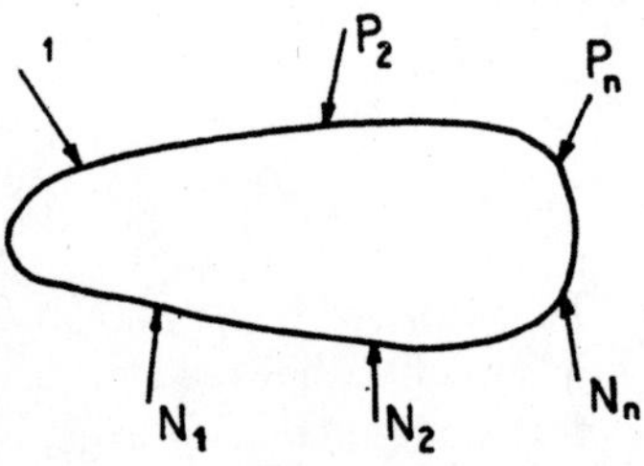

Fig. 13.4

Applying Maxwell-Betti-reciprocal theorem, we get

$$\sum_{i=1}^{n} (\bar{P}_i)_I \cdot (\Delta\,\bar{\delta}_i)_{II} = (\Delta \bar{P}_k)_{II} \cdot (\bar{\delta}_k)_I \qquad \text{...(13.25)}$$

where $(\Delta\bar{\delta}_i)_{II}$ = Deflection of the point of application of force P_i due to the force increment ΔP_k and the corresponding reactive force increments.

$(\bar{\delta}_k)_I$ = Deflection of the point of application of force P_k due to the original set of forces.

Increase in strain-energy resulting from the application of the load increment ΔP_k is

$$\Delta U = \frac{1}{2} (\Delta \bar{P}_k)_{II} \cdot (\Delta\bar{\delta}_k)_{II} + \sum_{i=1}^{n} (\bar{P}_i)_I \cdot (\Delta\bar{\delta}_i)_{II} \qquad \text{...(13.26)}$$

where the first and second terms on the right side represents work done by loading system II and I respectively.

Using Eq. (13.25), we get

$$\Delta U = \frac{1}{2}(\Delta\bar{P}_k)_{II} \cdot (\Delta\bar{\delta}_k)_{II} + (\Delta\bar{P}_k)_{II} \cdot (\bar{\delta}_k)_I \qquad \text{...(13.27)}$$

Dividing throughout by ΔP_k, we get after omitting subscripts II and I whose meaning is obvious,

$$\lim_{\Delta P_k \to 0} \frac{\Delta U}{\Delta P_k} = \lim_{\Delta P_k \to 0} \frac{1}{2}\frac{\Delta\bar{P}_k}{\Delta P_k} \cdot \Delta\bar{\delta}_k + \lim_{\Delta P_k \to 0} \frac{\Delta\bar{P}_k}{\Delta P_k} \cdot \bar{\delta}_k$$

Since $\Delta\bar{\delta}_k \to 0$ as $\Delta P_k \to 0$, hence

$$\frac{\partial U}{\partial P_k} = 0 + \lim_{\Delta P_k \to 0}\left(\frac{\Delta\bar{P}_k}{\Delta P_k} \cdot \bar{\delta}_k\right)$$

$$\frac{\partial U}{\partial P_k} = \delta_k, \text{ because } \frac{\Delta\bar{P}_k}{\Delta P_k} = 1 \qquad \text{...(13.28)}$$

This is Castigliano's theorem. It says that the rate of change of the strain energy for a body with respect to statically independent force P_k gives the component deflection of the point of application of this force in the direction of the force.

In a similar manner we can show what

$$\theta = \frac{\partial U}{\partial M} \qquad \text{...(13.29)}$$

where θ is the angle of rotation about the axis about which the couple M is applied.

13.5 Principle of Virtual Work

The principle of virtual work was enunciated by John Bernoulli in 1717. The virtual displacement is a hypothetical, infinitesimal displacement of a particle consistent with the constraints. The virtual work is the work done by the forces acting on the particle during a virtual displacement.

Consider a particle at a point O and imagine that it is acted upon by certain forces. Let one of these forces be P. Let the particle undergoes an arbitrary small virtual displacement δr at an angle θ with P. Then the virtual work done is

$$\delta W = P\delta r.\cos\theta$$
$$= P_r \cdot \delta r \qquad \text{...(13.30)}$$

where P_r is the component of P in the direction of δr.

Now let us assume that the particle O is in equilibrium under the n forces $P_1, P_2, \ldots, P_n$. The principle of virtual work states that if a particle is in equilibrium the total virtual work done by these n forces during any arbitrary virtual displacement of the particle is zero.

To verify this, the work done by any forces P_i $(i = 1, 2, \ldots, n)$ is given by Eq. (13.30) as

$$\delta W_i = P_{ir} \cdot \delta r$$

where P_{ir} = Components of forces P_i in the direction of δr.

The total virtual work is therefore

$$\delta W = \sum_{i=1}^{n} \delta W_i$$

$$= P_{1r} \, \delta r + P_{2r} \, \delta r + \ldots\ldots + P_{nr} \, . \, \delta r$$

$$= \left(\sum_{i=1}^{n} P_{ir} \right) \delta r$$

But the equilibrium of these static forces require that

$$\sum_{i=1}^{n} P_{ir} = 0$$

$$\therefore \quad \delta W = 0 \qquad \ldots(13.31)$$

i.e. the total virtual work done during any virtual displacement is zero.

13.6 Principle of Stationary Potential Energy

An elastic body can be considered as a system consisting of an infinite number of particles connected through internal forces and acted on by external body forces and surface tractions. We must give to each particle a virtual displacement field $\delta \bar{u}\,(x, y, z)$ rather than discrete displacements. The displacement field must not violate the constraints on the body and must be a continuous function.

Let u_x, u_y, u_z be the x, y, z components respectively of the actual displacement due to the external loads in an elastic body and δu_x, δu_y, δu_z the components of the virtual displacement. Let us consider an element of the elastic body $dxdydz$ acted on by the following forces :

(*i*) internal surface forces from surrounding particles.

(*ii*) body forces, and

(*iii*) surface tractions if the particle is at the boundary of the elastic body.

Imposing a virtual displacement field on a body means that there is also imposed a strain field increment on the body while maintaining the stress as constant.

This amounts to imposing a first-order variation of strain energy on the body given as :

$$\delta U = \iiint_V \tau_{ij}\, \delta\varepsilon_{ij} \,.\, dv$$

Now the virtual work of the internal forces is the negative of the variation of strain energy. Hence

$$\delta W_I = -\,\delta U$$

$$= -\iiint_V \tau_{ij} \,.\, \delta\varepsilon_{ij}\, dv \qquad \text{...(13.32)}$$

The virtual work done by body force distribution $\overline{B}$ can be given as

$$\delta W_B = \iiint_V \overline{B} \,.\, \delta\overline{u}\, dv \qquad \text{...(13.33)}$$

and the virtual work done by surface tractions $\overline{T}$ is

$$\delta W_T = \iint_A \overline{T} \,.\, \delta\overline{u}\, dA \qquad \text{...(13.34)}$$

Hence by the principle of virtual work

$$\delta W_B + \delta W_T + \delta W_I = 0$$

or

$$\delta W_B + \delta W_T - \delta U = 0$$

or

$$\delta\left[U - \iint_A \overline{T} \,.\, \overline{u}\, dA - \iiint_V \overline{B} \,.\, \overline{u}\, dv\right] = 0 \qquad \text{...(13.35)}$$

or

$$\delta[U - W] = 0 \qquad \text{...(13.36)}$$

where

$$W = \iint_A \overline{T} \,.\, \overline{u}\, dA + \iiint_V \overline{B} .\, \overline{u}\, dv$$

$$= \iint_A (T_x u_x + T_y u_y + T_z u_z)\, dA$$

$$+ \iiint_V (B_x \,.\, u_x + B_y \,.\, u_y + B_z \,.\, u_z)\, dv$$

and
$$T_x = \sigma_x \cdot a_{nx} + \tau_{xy} \cdot a_{ny} + \tau_{xz} \cdot a_{nz}$$
$$T_y = \tau_{xy} \cdot a_{nx} + \sigma_y \cdot a_{ny} + \tau_{yz} \cdot a_{nz}$$
$$T_z = \tau_{xz} \cdot a_{nx} + \tau_{yz} \cdot a_{ny} + \sigma_z \cdot a_{nz}$$

The quantity $\pi = U - W$ is called the potential energy of the system since U is the potential energy of deformation and $-W$ represents the potential energy of the external forces acting on the body if the potential energy of these forces for the unstressed condition is taken as zero.

Eq. (13.36) states that, in comparing various values of the displacements, u_x, u_y and u_z the displacements which actually occur in an elastic system under the action of given external forces are those which lead to zero variation of the total energy of the system for any virtual displacement from the position of equilibrium.

The expression $\delta\pi = \delta(U - W) = 0$

indicates that the potential energy of the system at the position of equilibrium is an extremum. For an elastic body with small displacements and strains, it can be shown that the potential energy at equilibrium is a minimum. The principle of potential energy can be stated as follows :

Of all displacements satisfying given boundary conditions, those which satisfy the equilibrium conditions make the potential energy π assume a stationary value. For stable equilibrium, π is a minimum.

13.7 Principle of Complementary Energy or Least Work

The complementary work performed by a gradually applied force P undergoing a deflection δ is given by

$$W_c = \int_0^{\delta_1} \delta \, dP$$

where δ is the component of the total deflection in the direction of P. The complementary work is illustrated by the shaded area shown in Fig. 13.5. Defining the complementary energy of the system by π^* as being equal to the complementary work, we have

$$\pi^* = \int_0^{P_1} \delta \, dP$$

If a number of forces P_i are applied to a structure and each P_i deflects δ_i, where δ_i is the component of the total deflection of P, in the direction of the line of action of the force, the total complementary energy of the system is

$$\pi^* = \sum_{i=1}^{n} \int \delta_i \, dP_i \qquad \text{...(13.37)}$$

where n is the total number of forces.

Now if instead of varying the strain components we vary the stress distribution under certain prescribed conditions then the principle which permits one to deduce the state of stress corresponding to equilibrium by this technique is called the principle of least work or complementary energy. When we are varying the stress components then in addition to the equilibrium equations, the compatibility equations must also be taken into account.

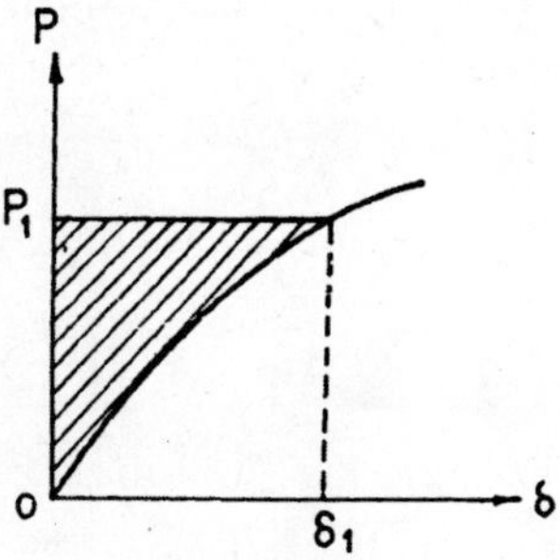

Fig. 13.5

Let τ_{ij} represent the true state of stress, *i.e.* one which satisfies compatibility, equilibrium and boundary conditions. Thus we have equilibrium equations

$$\frac{\partial \tau_{ij}}{\partial x_j} + B_i = 0 \qquad \text{...(13.38)}$$

and boundary conditions

$$\tau_{ij}\, a_{nj} = T_i \qquad \text{...(13.39)}$$

Next consider a stress variation $\delta\tau_{ij}$ such that the total stress components $(\tau_{ij} + \delta\tau_{ij})$ also satisfy the (Eqs. 13.38) and (13.39). Thus we can say that

$$\frac{\partial(\tau_{ij} + \delta\tau_{ij})}{\partial x_j} + B_i = 0 \qquad \text{...(13.40)}$$

$$(\tau_{ij} + \delta\tau_{ij})a_{nj} = T_i \qquad \text{...(13.41)}$$

Subtracting Eqs. (13.38) from (13.40) and (13.39) from (13.41), we find that

$$\frac{\partial(\delta\tau_{ij})}{\partial x_j} = 0 \qquad \text{...(13.42)}$$

$$(\delta\tau_{ij})a_{nj} = 0 \qquad \text{...(13.43)}$$

We have assumed that the surface tractions and body forces are kept constant during the variation of stress.

Now the strain energy of a body per unit volume at a point for a linear elastic material is given from Eq. (13.5) as :

$$u = \left(\frac{1+\nu}{2E}\right)\tau_{ij}\tau_{ij} - \frac{\nu}{2E}(\tau_{kk})^2$$

The variation of u achieved by varying stress can be expressed as

$$\delta u = \frac{\partial u}{\partial \tau_{ij}}\,\delta\tau_{ij} \qquad \text{...(13.44)}$$

Now $$\frac{\partial u}{\partial \tau_{ij}} = \left(\frac{1+\nu}{E}\right)\tau_{ij} - \frac{\nu}{E}\tau_{kk}\,.\,\delta_{ij}$$

$$= \varepsilon_{ij} \qquad \text{...(13.45)}$$

$$= \frac{1}{2}\left(\frac{\partial u_i}{\partial x_j} + \frac{\partial u_j}{\partial x_i}\right)$$

Hence Eq. (13.44) becomes

$$\delta u = \frac{1}{2}\left(\frac{\partial u_i}{\partial x_j} + \frac{\partial u_j}{\partial x_i}\right)\delta\tau_{ij} \qquad \text{...(13.46)}$$

The variation of strain energy for the entire body is then

$$\delta U = \iiint_V \delta u\, dv$$

$$= \frac{1}{2}\iiint_V \left(\frac{\partial u_i}{\partial x_j} + \frac{\partial u_j}{\partial x_i}\right)\delta\tau_{ij}\, dv$$

$$= \frac{1}{2}\iiint_V \frac{\partial u_i}{\partial x_j}\,\delta\tau_{ij}\, dv$$

$$+ \frac{1}{2}\iiint_V \frac{\partial u_j}{\partial x_i}\,\delta\tau_{ij}\, dv \qquad \text{...(13.47)}$$

Now $$\iiint_V \frac{\partial u_i}{\partial x_j}\,\delta\tau_{ij}\, dv = \iiint_V \frac{\partial(\overline{u_i}\,\delta\tau_{ij})}{\partial x_j}\, dv$$

$$-\iiint_V u_i \frac{\partial(\delta\tau_{ij})}{\partial x_j}\, dv \qquad \text{...(13.48)}$$

By using Eq. (13.42) we see that the second integral in Eq. (13.48) is zero and the first integral by Stoke's theorem may be written as

$$\iiint_V \frac{\partial(u_i\,\partial\tau_{ij})}{\partial x_j}\, dv = \iint_A u_i\delta\tau_{ij}\, a_{nj}\, dA \qquad \text{...(13.49)}$$

By using Eq. (13.43), we see that this integral is also equal to zero. Similarly it can be shown that the second integral in Eq. (13.47) is also zero. Hence

$$\delta U = 0 \qquad \text{...(13.50)}$$

If the body forces are kept constant but the surface tractions change, then we will get

$$\delta[U - \iint_A (u_x T_x + u_y T_y + u_z T_z) dA] = 0$$

or
$$\delta(U - W^*) = 0 \qquad \text{...(13.51)}$$

or
$$\delta\pi^* = 0$$

where π^* is called the complementary energy and

$$W^* = \iint_A (u_x T_x + u_y T_y + u_z T_z)\, dA$$

Eq. (13.51) gives the principle of complementary energy which states that for all stresses satisfying the equilibrium conditions in the interior and on that part of the boundary surface where the surface forces are prescribed, the actual state of stress *i.e.*, the stresses which satisfy the compatibility equations, is such that the complementary energy π^* assumes a stationary value. For problems with small strains and displacements, we can further show that π^* is a minimum.

13.8 The Dummy Load Method

The dummy load method is based on the complementary energy theorem and is applicable to nonlinear load-displacement problems.

When an infinitesimal force ΔP_i is placed on a loaded structure, the complementary work, due to ΔP_i is

$$\Delta W_c = \delta_i \, \Delta P_i$$

or
$$\delta_i = \frac{\Delta\pi^*}{\Delta P_i} = \frac{\partial\pi^*}{\partial P_i} \qquad \text{...(13.52)}$$

where δ_i is the deflection of point i on the structure in the direction of ΔP_i. The increase $\Delta\pi^*$ in complementary energy of the system is equal to ΔW_c. The increase in π^* of the system is equal to the sum of the increase in the complementary energy of all members. For each member, the increase in complementary energy is equal to that part of ΔP_i which performs complementary work on the member. If the load applied is equal to unity then this is called unit load method.

(*i*) *Axial loading.* Let the deflection of an infinitesimal dx element due to the applied loads be $d(\delta)$. Consider ΔF to be the internal force within the element due to a dummy load placed on the structure at a given point ; then the work performed by ΔF only is $\Delta F\, d(\delta)$. This is the increase in the complementary energy of the dx element.

The total increase in complementary energy of the member is

$$\Delta\pi^* = \int_0^l \Delta F\, d(\delta)$$

For a linear elastic material $d(\delta) = \dfrac{F dx}{AE}$, where F is the internal force in the element due to the applied loads. Hence

$$\Delta\pi^* = \int_0^l \Delta F \frac{F}{AE}\, dx = \Delta F \cdot \frac{Fl}{AE} \qquad \text{...(13.53)}$$

(*ii*) *Bending.* In this case

$$\Delta\pi^* = \int_0^l \Delta M\, d\theta$$

where ΔM is the bending moment due to dummy load and $d\theta$ the rotation of the element due to applied loads.

Now $\qquad d\theta = \dfrac{M}{EI}\, dx$

where M is the internal bending moment due to the applied loads. Thus for a linear material

$$\Delta\pi^* = \int_0^l \Delta M . \frac{M}{EI}\, dx = \Delta M . \frac{Ml}{EI} \qquad \text{...(13.54)}$$

(*iii*) *Torsional loading*. Here, we have

$$\Delta\pi^* = \int_0^l \Delta T d\theta$$

where $\qquad \Delta T$ = torque being transmitted at position x due to the dummy load.

$d\theta$ = angle of twist of dx element due to applied loads.

For a linear elastic material,

$$d\theta = (1 + \nu) \frac{T}{EJ}\, dx,$$

where T is the torque being transmitted at x due to the applied loads. Thus

$$\Delta\pi^* = \int_0^l \Delta T(1 + \nu) \frac{T}{EJ} dx$$

$$= \Delta(1 + \nu) \frac{Tl}{EJ} \qquad \text{...(13.55)}$$

13.9 Rayleigh's Method

Rayleigh's method is often employed to determine the deflection of beams. The technique assumes a deflection shape, and the strain energy is formulated based on the assumed shape and then equated to the work done by the applied forces.

Consider a load distribution $q(x)$ applied to a beam. If a dx element is isolated at position x, the net force on the element is qdx. If the beam deflects in the positive y direction in a linear fashion a distance y_c, the net work performed is $-\frac{1}{2}qu_c dx$. The work is negative since the force and deflection are in opposite directions. The total work performed by the application of q is

$$W = -\frac{1}{2}\int_o^l qy_c \, dx$$

The increase in strain energy of the beam is

$$U = \frac{1}{2}\int EI \, (y_c'')^2 \, dx$$

$$\therefore \qquad W = U \qquad \text{...(13.56)}$$

The shape for the deflection curve is in general a function of x and must be consistent with the geometric boundary conditions of the beam. From Eq. (13.56) the value of the unknown constant can be determined.

Example 13.1 *Strain energy for a simply supported rectangular beam.*

Solution. Consider a rectangular beam loaded as shown in Fig. 13.6.

$$U_\varepsilon = \int_0^l \frac{M_z^2}{2EI_{zz}} dx$$

$$M_z = 0.5x \text{ kN-m for } 0 \le x \le 5 \text{ m}$$

$$= [0.5x - (x - 5)] \text{ kN-m for } 5 \le x \le 10 \text{ m}$$

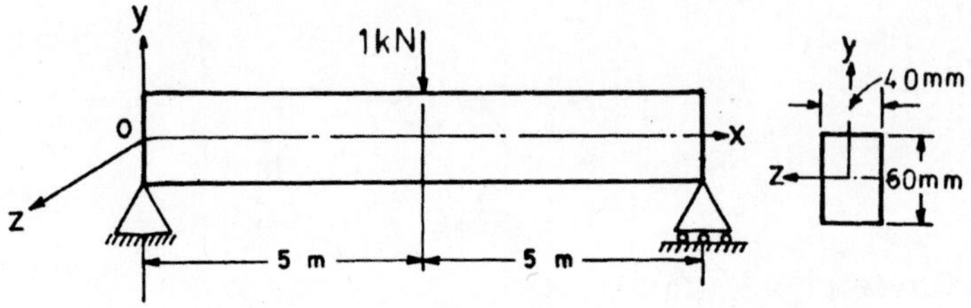

(a) (b)

Fig. 13.6

$$U_\varepsilon = \int_0^5 \frac{(0.5)x^2 \times 10^6}{2EI_{zz}}\,dx + \int_0^{10} \frac{[0.5x - (x-5)]^2 \times 10^6}{2EI_{zz}}\,dx$$

$$= \frac{1.041 \times 10^6}{EI_{zz}} \text{ N-m.}$$

Taking $E = 2 \times 10^5$ N/mm^2

and $$I = \frac{4 \times 6^2 \times 10^4}{12} = 72 \times 10^4 \text{ mm}^4$$

$\therefore$ $U_\varepsilon = 7.23$ N-m

$$U_\gamma = \iiint_V \frac{1}{2G}\left[\frac{FA\bar{y}}{Ib}\right]^2 dx\,dy\,dz$$

$$A\bar{y} = b\left(\frac{a}{2} - y\right)\left(\frac{y}{2} + \frac{a}{4}\right) = \frac{b}{2}\left(\frac{a^2}{4} - y^2\right)$$

Here $b = 40$ mm, $a = 60$ mm

$\therefore$ $A\bar{y} = 20(900 - y^2)$

$$U_\gamma = \frac{1}{2GI_{zz}^2}\left[\iint_A \left(\frac{A\bar{y}}{b}\right)^2 dA\right]\left(\int_0^l F^2 dx\right)$$

$F = -\,0.5$ kN for $0 \le x \le 5$ m

$= +\,0.5$ kN for $5 \le x \le 10$ m.

$$\therefore \quad U_\gamma = \frac{1}{2GI_{zz}^2}\left[\int_{-a/2}^{+a/2} \frac{\frac{b^2}{4}\left(\frac{a^2}{4} - y^2\right)^2}{b^2}\,b\,dy\right] \times \left[\int_0^5 (-\,500)^2\,dx + \int_0^{10} (500)^2\,dx\right]$$

$= 536$ N.mm.

$$\therefore \quad U = U_\varepsilon + U_\gamma$$
$$= 7230 + 536 = 7766 \text{ N.mm.}$$

Example 13.2 *Deflection of a beam by dummy load method.*

Solution. Let us consider a rectangular beam carrying load of 1 kN at the centre. Let us determine the deflection of the beam at points A and B as shown in Fig. 13.7.

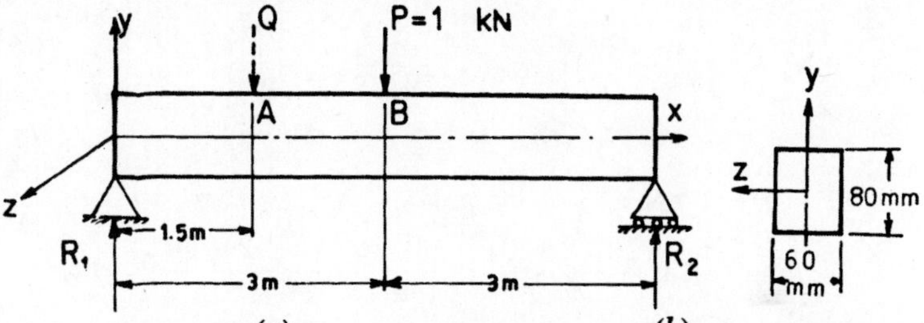

(*a*) (*b*)

Fig. 13.7

To determine the deflection at A we apply a dummy load Q at A. Then

$$M_x = \left(\frac{P}{2} + \frac{3Q}{4}\right)x - Q(x - 1.5) - P(x - 3)$$

$$U = \int_0^l \frac{M_z^2}{2EI_{zz}}\,dx$$

$$= \frac{1}{2EI_{zz}}\left[\int_0^{1.5}\left(\frac{P}{2} + \frac{3Q}{4}\right)^2 x^2\,dx + \int_{1.5}^{3}\left[\left(\frac{P}{2} + \frac{3Q}{4}\right)x - Q(x - 1.5)\right]^2 dx + \int_3^6\left[\left(\frac{P}{2} + \frac{3Q}{4}\right)x - Q(x - 1.5) - P(x - 3)\right]^2 dx\right]$$

$$\frac{\partial U}{\partial P} = \frac{1}{2EI_{zz}}\left[\int_0^{1.5}\left(\frac{P}{2} + \frac{3Q}{4}\right)x^2\,dx + \int_{1.5}^{3}\left[\left(\frac{P}{2} + \frac{3Q}{4}\right)x - Q(x - 1.5)\right]^2 \times x\,dx + \int_3^6\left[\left(\frac{P}{2} + \frac{3Q}{4}\right)x - Q(x - 1.5) - P(x - 3)\right]2\left(\frac{x}{2} - x + 3\right) \times dx\right]$$

Put $Q = 0$ and $P = 1$ kN, we get $\dfrac{\partial U}{\partial P} = \dfrac{450 \times 10^6}{EI_{zz}}$

$$\therefore \quad \delta_P = \delta_B = \frac{450 \times 10^6}{EI_{zz}} = \frac{450 \times 12 \times 10^6}{2 \times 10^5 \times 6 \times 512 \times 10^4}$$
$$= 0.878 \times 10^{-3} \text{ mm}$$

Now

$$\frac{\partial U}{\partial Q}=\frac{1}{EI_{zz}}\left[\begin{aligned}&\int_0^{1.5}\left(\frac{P}{2}+\frac{3Q}{4}\right)x^2\cdot\frac{3}{4}\,dx\\&+\int_{1.5}^{3}\left[\left(\frac{P}{2}+\frac{3}{4}Q\right)x-Q(x-1.5)\right]\left(\frac{3}{4}x-x+1.5\right)dx\\&+\int_3^6\left[\left(\frac{P}{2}+\frac{3Q}{4}\right)x-Q(x-1.5)-P(x-3)\right]\left(\frac{3}{4}x-x+1.5\right)\\&\qquad\times dx\end{aligned}\right]$$

Put $P = 1$ kN, $Q = 0$, we get

$$\delta_Q=\delta_A=\frac{267.2\times10^6}{EI_{zz}}=\frac{267.2\times12\times10^6}{2\times10^5\times6\times512\times10^4}$$

$$=0.521\times10^{-2}\text{ mm}$$

Example 13.3 *Rotation of a beam.*

Solution. Consider a beam carrying uniformly distributed load of intensity q per unit length, as shown in Fig. 13.8. In order to determine the change of slope of the beam at B (say), we apply a dummy couple C at B.

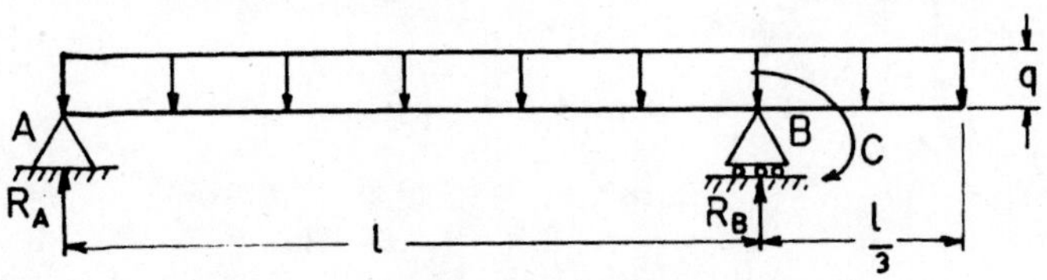

Fig. 13.8

$$\theta=\int_0^l\frac{M_1}{EI}\frac{\partial M_1}{\partial C}\,dx+\int_l^{4l/3}\frac{M_2}{EI}\frac{\partial M_2}{\partial C}\,dx$$

where

$$M_1=\left(R_A+\frac{C}{l}\right)x-\frac{qx^2}{2}\text{ for }0\le x\le l$$

$$M_2=\left(R_A+\frac{C}{l}\right)x-\frac{qx^2}{2}+\left(R_B-\frac{C}{l}\right)(x-l)-C\quad\text{for }l\le x\le\frac{4l}{3}$$

$$\frac{\partial M_1}{\partial C}=\frac{x}{l},\quad\frac{\partial M_2}{\partial C}=0$$

$$\therefore \qquad \theta = \frac{1}{EI}\left[\frac{R_A l^2}{3} - \frac{ql^3}{8}\right]$$

$$R_A = \frac{4}{9} ql$$

$$\therefore \qquad \theta = \frac{5}{216}\frac{ql^3}{EI}$$

13.10 Rayleigh-Ritz Method

This method may be carried out as explained below. First assume the solution in the form of a series which satisfies the boundary conditions but with undetermined parameters C_i. Second, insert these functions into the expression of the potential energy or the complementary energy, and carry out any required integration. The resulting expressions are functions of the undetermined parameters C_i, i = 1, 2, ... Since the potential energy or the complementary energy must be a minimum at equilibrium, these parameters can be determined from the minimizing conditions.

$$\frac{\partial \pi}{\partial C_1} = 0, \frac{\partial \pi}{\partial C_2} = 0$$

or

$$\frac{\partial \pi^*}{\partial C_1} = 0, \frac{\partial \pi^*}{\partial C_2} = 0 \qquad \text{...(13.57)}$$

If n parameters are taken, Eq. (13.57) gives n simultaneous equations from which these parameters may be solved. Substituting these values of the parameters into the assumed form of the function, we get an approximate solution of the given problem.

Example 13.4 *Torsion of a rectangular bar.*

Solution. Let us consider the torsion of a rectangular bar with sides $2a$ and $2b$. To satisfy the boundary condition $\Phi = 0$ on sides $x = \pm a$ and $y = \pm b$, we may take the polynomial solution as

$$\Phi = (x^2 - a^2)(y^2 - b^2)(C_1 + C_2x^2 + C_3y^2 + C_4x^2y^2 + \ldots\ldots) \qquad \text{...(1)}$$

where C_1, C_2, C_3, C_4, ... are undetermined parameters. Since we know that Φ is symmetrical with respect to the x and y axes, we have taken only terms with even powers in x and y. As a first approximation, let us take

$$\Phi = C_1(x^2 - a^2)(y^2 - b^2)$$

Now $$\tau_{yz} = -G\theta \frac{\partial \Phi}{\partial x}, \tau_{zx} = G\theta \frac{\partial \Phi}{\partial y}$$

and $$\sigma_x = \sigma_y = \sigma_z = \tau_{xy} = 0$$

$$U = \frac{1}{2G}\iiint_V (\tau_{yz}^2 + \tau_{zx}^2)dxdydz$$

$$= \frac{G\theta^2 l}{2}\iint_A \left[\left(\frac{\partial \Phi}{\partial x}\right)^2 + \left(\frac{\partial \Phi}{\partial y}\right)^2\right] dxdy$$

where A is the cross-sectional area of the bar. W^* is zero over the lateral surfaces of the bar because the boundary force is prescribed there. On the two ends we have

$$W^* = \left[\iint_A (\tau_{zy} u_x + \tau_{yz} u_y)\, dxdy\right]_{\substack{z=0\\ z=l}}$$

Now $u_z = -\theta zy,\ u_y = \theta zx$

At $z = 0,\ u_x = u_y = 0$ and at $z = l$

$u_x = -\theta ly,\ u_y = \theta lx.$

Hence $$W^* = \theta l \iint_A \left(-y\,\tau_{zx} + x\,\tau_{yz}\right) dxdy$$

$$= G\theta^2 l \iint_A \left(-y\frac{\partial \Phi}{\partial y} - x\frac{\partial \Phi}{\partial x}\right) dxdy$$

$$= 2G\theta^2 l \iint_A \Phi dxdy - G\theta^2 l \oint_S (\Phi x dy - \Phi y dx)$$

$\therefore$ $$\pi^* = U - W^*$$

$$= \frac{G\theta^2 l}{2}\iint_A \left[\left(\frac{\partial \Phi}{\partial x}\right)^2 + \left(\frac{\partial \Phi}{\partial y}\right)^2 - 4\Phi\right] dxdy - G\theta^2 l \oint_S (\Phi x dy - \Phi y dx)$$

with $\Phi = 0$ on the boundary, we get

$$\pi^* = \frac{G\theta^2 l}{2}\iint_A \left[\left(\frac{\partial \Phi}{\partial x}\right)^2 + \left(\frac{\partial \Phi}{\partial y}\right)^2 - 4\Phi\right] dxdy$$

$$= \frac{G\theta^2 l}{2}\int_{x=-a}^{+a}\int_{y=-b}^{+b} [4C_1^2\{x^2(y^2-b^2)^2 + (x^2-a^2)^2y^2\} - 4C_1(x^2-a^2)(y^2-b^2)]dxdy$$

$$= \frac{G\theta^2 l}{2}\cdot\frac{64}{45}[2C_1^2a^3b^3(a^2+b^2) - 5C_1a^3b^3] \quad \text{...(2)}$$

Now $\frac{\partial \pi^*}{\partial C_1} = 0$, which gives

$$C_1 = \frac{5}{4(a^2 + b^2)}$$

$$\therefore \qquad \Phi = \frac{5}{4(a^2 + b^2)} (x^2 - a^2)(y^2 - b^2) \qquad ...(3)$$

Now torque $T = 2\int_A \Phi \, dxdy$

$$T = 2C_1 \int_{x=-a}^{+a} \int_{y=-b}^{+b} (x^2 - a^2)(y^2 - b^2) \, dxdy$$

$$= \frac{32}{9} a^3 b^3 C_1 = \frac{40}{9} \frac{(b/a)^3}{1 + \left(\frac{b}{a}\right)^2} . a^4 \qquad ...(4)$$

For $\frac{b}{a} = 1$, $T = \frac{20}{9} a^4 = 2.222\, a^4$ compared with the exact value of $T = 2.250\, a^4$.

$$\tau_{max} = G\theta \left(\frac{\partial \Phi}{\partial x}\right)_{\substack{x=a \\ y=0}} = -G\theta . \frac{5}{4(a^2 + b^2)} . 2a(-b^2)$$

$$= 2.5\, G\theta . \frac{ab^2}{(a^2 + b^2)} = 2.5\, G\theta . \frac{\left(\frac{b^2}{a^2}\right) . a}{\left(1 + \frac{b^2}{a^2}\right)} \qquad ...(5)$$

For $\quad \frac{b}{a} = 1$

$$\tau_{max} = 1.25\, G\theta\, a$$

whereas exact value is $\tau_{max} = 1.350\, G\theta\, a$

If we assume

$$\Phi = (x^2 - a^2)(y^2 - b^2)(C_1 + C_2 x^2 + C_3 y^2) \qquad ...(6)$$

$$\text{Then } \pi^* = \frac{G\theta^2 l}{2} \iint_A \left[\left(\frac{\partial \Phi}{\partial x}\right)^2 + \left(\frac{\partial \Phi}{\partial y}\right)^2 - 4\Phi \right] dxdy$$

$$= \frac{G\theta^2 l}{2} . \frac{64}{4725} a^3 b^3 [210(a^2 + b^2) C_1^2$$
$$+ a^4(66b^2 + 10a^2) C_2^2 + b^4(66a^2 + 10b^2) C_3^2$$
$$+ a^2(84b^2 + 60a^2) C_1 C_2 + b^2(84a^2 + 60b^2) C_1 C_3$$
$$+ 12a^2 b^2 (a^2 + b^2) C_2 C_3 - 525 C_1 - 105 a^2 C_2$$
$$- 105 b^2 C_3]$$

The conditions $\frac{\partial \pi^*}{\partial C_1} = 0, \frac{\partial \pi^*}{\partial C_2} = 0, \frac{\partial \pi^*}{\partial C_3} = 0$

give, respectively

$$140(a^2 + b^2)C_1 + a^2(28b^2 + 20a^2)C_2 + b^2(28a^2 + 20b^2)C_3 = 175$$
$$(84b^2 + 60a^2)C_1 + a^2(132b^2 + 20a^2)C_2 + 12b^2(a^2 + b^2)C_3 = 105$$
$$(84a^2 + 60b^2)C_1 + 12(a^2 + b^2)C_2 + b^2(132a^2 + 20b^2)C_3 = 105$$

From these equations we can find the values of C_1, C_2, C_3. For a square bar, it can be determined that

$$C_1 = \frac{1295}{2216a^2}, C_2 = C_3 = \frac{525}{4432a^4}$$

Now $T = 2\iint \Phi \, dxdy$

$$= \frac{32}{9} a^3b^3 \left(C_1 + \frac{a^2}{5} C_2 + \frac{b^2}{5} C_3 \right)$$

$$\tau_{max} = -G0 \left(\frac{\partial \Phi}{\partial x} \right)_{\substack{x = a \\ y = 0}}$$

$$= G\theta.2ab^2(C_1 + C_2a^2)$$

Hence for $\frac{b}{a} = 1$, we find

$$T = \frac{32}{9} a^6 \left[\frac{1295}{2216a^2} + \frac{105}{4432a^2} + \frac{105}{4432a^2} \right]$$

$$= \frac{32}{9} a^4 \left[\frac{2590 + 210}{4432} \right] = \frac{32}{9} . a^4 . \frac{2800}{4432}$$

$$= 2.246a^4 \qquad \text{...(7)}$$

and $$\tau_{max} = G\theta \,.\, 2a^3 \left[\frac{1295}{2216a^2} + \frac{525}{4432a^3} \right]$$

$$= 2G\theta \,.\, a \left[\frac{2590 + 525}{4432} \right] = 2G\theta a \,.\, \frac{3115}{4432} = \frac{6230}{4432} G\theta a$$

$$= 1.405 \, G\theta a \qquad \text{...(8)}$$

Example 13.5 *Deflection of a simply supported beam.*

Solution. Let us consider the deflection of a beam carrying uniformly distributed load of intensity q per unit length and simply supported at the ends. The boundary conditions require that

$$y = \frac{d^2y}{dx^2} = 0$$

at $x = 0$ and l. To satisfy the boundary conditions we take the solution in the form of a series

$$y = \sum_{n=1}^{\infty} C_n \sin \frac{n\pi x}{l} \qquad \text{...(1)}$$

where C_n are undetermined parameters.

Now $$U = \iiint_V \frac{\sigma_x^2}{2E}\, dxdydz$$

$$= \frac{1}{2}\int_0^l EI\left(\frac{d^2y}{dx^2}\right)^2 dx = \frac{EI}{2}\int_0^l \left(\frac{d^2y}{dx^2}\right)^2 dx$$

$$W = \int_0^l q\, ydx$$

$\therefore$ $$\pi = U - W$$

$$= \frac{EI}{2}\int_0^l \left(\frac{d^2y}{dx^2}\right)^2 dx - \int_0^l qydx$$

$$= \frac{EI\pi^4}{4l^3} \cdot \sum_{n=1}^{\infty} n^4 C_n^2 - \frac{2ql}{\pi} \sum_{n=1,3,5}^{\infty} \frac{C_n}{n}$$

$$\frac{\partial \pi}{\partial C_n} = 0, \text{ gives}$$

$$\frac{EI\pi^4}{4l^3} \cdot 2n^4 C_n - \frac{2ql}{\pi} \cdot \frac{1}{n} = 0 \text{ for odd } n$$

$$\frac{EI\pi^4}{4l^3} \cdot 2n^4 C_n = 0 \text{ if } n \text{ is even}$$

Hence $$C_n = \frac{4ql^4}{EI\pi^5} \cdot \frac{1}{n^5} \text{ for odd } n$$

$$C_n = 0 \text{ for even } n$$

$\therefore$ $$y = \frac{4ql^4}{EI\pi^5} \sum_{n=1,3,5}^{\infty} \frac{1}{n^5} \sin \frac{n\pi x}{l}$$

$$y_{max}\Big|_{x=l/2} = \frac{4ql^4}{EI\pi^5}\left[1 - \frac{1}{3^5} + \frac{1}{5^5} - \ldots\right]$$

$$\simeq \frac{ql^4}{76.6\, EI} \text{ by taking only first term}$$

where as exact value is $\frac{ql^4}{76.8\, EI}$.

Example 13.6 *Deflection of a rectangular plate by the principal of virtual work.*

Solution. Now $U = \frac{1}{2}\iiint_V (\sigma_x\varepsilon_x + \sigma_y\varepsilon_y + \sigma_z\varepsilon_z + \tau_{xy}\gamma_{xy} + \tau_{yz}\gamma_{yz} + \tau_{zx}\gamma_{zx})\, dv$

For thin plates $\sigma_z = \tau_{xz} = \tau_{yz} = 0$.

$$\text{Hence } U = \iiint_V \left[\frac{1}{2E}(\sigma_x^2 + \sigma_y^2 - 2\nu\sigma_x\sigma_y) + \frac{1+\nu}{E}\tau_{xy}^2 \right] dxdydz$$

Now
$$\sigma_x = -\frac{Ez}{1-\nu^2}\left[\frac{\partial^2 u_z}{\partial x^2} + \nu\frac{\partial^2 u_z}{\partial y^2}\right]$$

$$\sigma_y = -\frac{Ez}{1-\nu^2}\left[\frac{\partial^2 u_z}{\partial y^2} + \nu\frac{\partial^2 u_z}{\partial x^2}\right]$$

$$\tau_{xy} = -\frac{Ez}{1+\nu}\frac{\partial^2 u_z}{\partial x \partial y}$$

$$\therefore \quad U = \frac{D}{2}\iint_A \left[\left(\frac{\partial^2 u_z}{\partial x^2}\right)^2 + \left(\frac{\partial^2 u_z}{\partial y^2}\right)^2 + 2\nu\frac{\partial^2 u_z}{\partial y^2}\cdot\frac{\partial^2 u_z}{\partial x^2} + 2(1-\nu)\left(\frac{\partial^2 u_z}{\partial x\partial y}\right)^2\right] dxdy$$

$$= \frac{D}{2}\iint_A \left[\left(\frac{\partial^2 u_z}{\partial x^2} + \frac{\partial^2 u_z}{\partial y^2}\right) - 2(1-\nu)\left\{\frac{\partial^2 u_z}{\partial x^2}\cdot\frac{\partial^2 u_z}{\partial y^2} - \left(\frac{\partial^2 u_z}{\partial x\partial y}\right)^2\right\}\right] \times dxdy \quad \text{...(1)}$$

Let the deflection of the plate be represented by

$$u_z = \sum_{m=1}^{\infty}\sum_{n=1}^{\infty} a_{mn} \sin\frac{m\pi x}{a} \,.\, \sin\frac{n\pi y}{b} \quad \text{...(2)}$$

Substituting Eq. (2) in (1), the first term becomes

$$\frac{D}{2}\int_0^a\int_0^b \left[\sum_{m=1}^{\infty}\sum_{n=1}^{\infty} a_{mn}\left(\frac{m^2\pi^2}{a^2} + \frac{n^2\pi^2}{b^2}\right)\sin\frac{m\pi x}{a}\sin\frac{n\pi y}{b}\right]^2 dxdy$$

$$= \frac{\pi^4 ab}{8} D \sum_{m=1}^{\infty}\sum_{n=1}^{\infty} a^2{}_{mn}\left(\frac{m^2}{a^2} + \frac{n^2}{b^2}\right)^2$$

Similarly the second term in Eq. (1) after integration becomes equal to zero.

$$\therefore \quad U = \frac{\pi^4 ab}{8} D \sum_{m=1}^{\infty} \sum_{n=1}^{\infty} a^2{}_{mn} \left(\frac{m^2}{a^2} + \frac{n^2}{b^2} \right)^2 \qquad ...(3)$$

To obtain the deflection of a plate by a concentrated force P at (ξ, η), we give a virtual displacement satisfying boundary conditions. To obtain this virtual displacement we give to any coefficient $a_{m'n'}$ of series give by Eq. (2) an infinitely small variation $\delta a_{m'n'}$. As a result of this the variation in deflection u_z is

$$du_z = \delta a_{m'n'} \sin \frac{m'\pi x}{a} \sin \frac{n'\pi y}{b}.$$

and the virtual work done by load P is

$$\delta W = P.\delta a_{m'n'} \sin \frac{m'\pi\xi}{a} \sin \frac{n'\pi\eta}{b}$$

By the principle of virtual work.

$$\delta W = \delta U$$

$$\therefore \quad P.\delta a_{m'n'} \sin \frac{m'\pi\xi}{a} . \sin \frac{n'\pi\eta}{b} = \frac{\partial U}{\partial a_{m'n'}} \delta a_{m'n'}$$

or

$$P.\delta a_{m'n'} \sin \frac{m'\pi\xi}{a} . \sin \frac{n'\pi\eta}{b}$$

$$= \frac{\pi^4 ab}{4} D\, a_{m'n'} \left(\frac{m'^2}{a^2} + \frac{n'^2}{b^2} \right)^2 \delta a_{m'n'}$$

$$a_{m'n'} = \frac{4P \sin \dfrac{m'\pi\xi}{a} \sin \dfrac{n'\pi\eta}{b}}{\pi^4 abD \left(\dfrac{m'^2}{a^2} + \dfrac{n'^2}{b^2} \right)^2}$$

$$\therefore \quad u_z = \frac{4P}{\pi^4 abD} \sum_{m=1}^{\infty} \sum_{n=1}^{\infty} \frac{\sin \dfrac{m\pi\xi}{a} \sin \dfrac{n\pi\eta}{b}}{\left(\dfrac{m^2}{a^2} + \dfrac{n^2}{b^2} \right)^2} \sin \frac{m\pi x}{a} \sin . \frac{n\pi y}{b}.$$

Example 13.7 *Bending of a rectangular plate by Rayleigh-Ritz method.*

Solution. Consider a rectangular plate with clamped edges. For this case $u_z = 0$ along the edges. Now strain energy for the bending of plates is

$$U = \frac{D}{2}\iint_A \left[\left(\frac{\partial^2 u_z}{\partial x^2} + \frac{\partial^2 u_z}{\partial y^2}\right)^2 - 2(1-\nu)\left[\frac{\partial^2 u_z}{\partial x^2}\cdot\frac{\partial^2 u_z}{\partial y^2} - \left(\frac{\partial^2 u_z}{\partial x \partial y}\right)^2\right\}\right] dxdy \qquad ...(1)$$

where A is the area of the plate.

Now integrating by parts, we find

$$\iint_A \frac{\partial^2 u_z}{\partial x \partial y}\frac{\partial^2 u_z}{\partial x \partial y}\, dxdy = -\oint_S \frac{\partial^2 u_z}{\partial x \partial y}\frac{\partial u_z}{\partial x}\, dx - \iint_A \frac{\partial u_z}{\partial x}\cdot\frac{\partial^3 u_z}{\partial x \partial y^2}\, dxdy$$

$$= \oint_S \frac{\partial^2 u_z}{\partial x \partial y}\frac{\partial u_z}{\partial x}\, dx - \oint_S \frac{\partial u_z}{\partial x}\frac{\partial^2 u_z}{\partial y^2}\, dy + \iint_A \frac{\partial^2 u_z}{\partial x^2}\frac{\partial^2 u_z}{\partial y^2}\, dxdy.$$

For plates with clamped edges, $\frac{\partial u_z}{\partial x} = \frac{\partial u_z}{\partial y} = 0$ along the edges.

For a rectangular plate $u_z = 0$ along the edges, $\frac{\partial u_z}{\partial x} = 0$ along the edges y = const. and $\frac{\partial u_z}{\partial y} = \frac{\partial^2 u_z}{\partial y^2} = 0$ along the edges x = constant. Hence, the first two integrals in the above expression become identically zero, and we have

$$\iint_A \left[\frac{\partial^2 u_z}{\partial x^2}\frac{\partial^2 u_z}{\partial y^2} - \left(\frac{\partial^2 u_z}{\partial x \partial y}\right)^2\right] dxdy = 0 \qquad ...(2)$$

$$\therefore \qquad U = \frac{D}{2}\iint_A \left(\frac{\partial^2 u_z}{\partial x^2} + \frac{\partial^2 u_z}{\partial y^2}\right)^2 dxdy \qquad ...(3)$$

For a rectangular plate of sides $2a$ and $2b$ the boundary conditions are

$$u_z = 0, \frac{\partial u_z}{\partial x} = 0 \text{ at } x = 0 \text{ and } a.$$

$$u_z = 0, \frac{\partial u_z}{\partial y} = 0 \text{ at } y = 0 \text{ and } b.$$

These boundary conditions are satisfied if we assume u_z in the following form

$$u_z = \sum_{m=1}^{\infty} \sum_{n=1}^{\infty} A_{mn} \left(1 - \cos \frac{2m\pi x}{a}\right)\left(1 - \cos \frac{2n\pi y}{b}\right) \quad \text{...(4)}$$

where the parameters A_{mn} are to be determined from the condition that the potential energy of the system.

$$\pi = U - W$$

is a minimum with respect to these parameters.

Substituting Eq. (4) in (3), we get

$$U = \frac{D}{2}\int_0^a\int_0^b \left[\sum_{m=1}^{\infty} \sum_{n=1}^{\infty} 4\pi^2 A_{mn} \left\{ \frac{m^2}{a^2} \cos \frac{2m\pi x}{a} \times \left(1 - \cos \frac{2n\pi y}{b}\right) + \frac{n^2}{b^2} \cos \frac{2n\pi y}{b} \left(1 - \cos \frac{2m\pi x}{b}\right)\right\}\right]^2 \times dxdy$$

$$= 2D\pi^4 ab \left[\sum_{m=1}^{\infty} \sum_{n=1}^{\infty} \left\{ 3\left(\frac{m^4}{a^4}\right) + 3\left(\frac{n^4}{b^4}\right) + 2\left(\frac{m^2}{a^2}\right)\left(\frac{n^2}{b^2}\right)\right\} A_{mn}^2 + \sum_{m=1}^{\infty} \sum_{\substack{r=1 \\ r \neq s}}^{\infty} \sum_{s=1}^{\infty} 2\left(\frac{m^4}{a^4}\right) A_{mr} A_{ms} + \sum_{r=1}^{\infty} \sum_{\substack{s=1 \\ r \neq s}}^{\infty} \sum_{n=1}^{\infty} 2\left(\frac{n^4}{b^4}\right) A_{rn} A_{sn} \right]$$

If the plate is carrying uniformly distributed load of intensity p_0 then the potential energy of the external force is

$$-W = -\int_0^a\int_0^b p_0 u_z \, dxdy$$

$$= -p_0 \int_0^a\int_0^b \left[\sum_{m=1}^{\infty} \sum_{n=1}^{\infty} A_{mn} \left(1 - \cos \frac{2m\pi x}{a}\right) \times \left(1 - \cos \frac{2n\pi y}{b}\right)\right] dxdy$$

$$= -p_0 \, ab \sum_{m=1}^{\infty} \sum_{n=1}^{\infty} A_{mn}$$

The condition $\dfrac{\partial \pi}{\partial A_{mn}} = 0$ gives

$$4D\pi^4 ab\left[\left\{3\left(\frac{m^4}{a^4}\right)+3\left(\frac{n^4}{b^4}\right)+2\left(\frac{m^2}{a^2}\right)\left(\frac{n^2}{b^2}\right)\right\}A_{mn}\right.$$

$$\left.+\sum_{\substack{r=1\\ (\text{except } r=n)}}^{\infty} 2\left(\frac{m^4}{a^4}\right)A_{mr}+\sum_{\substack{r=1\\ (\text{except } r=m)}}^{\infty} 2\left(\frac{n^4}{b^4}\right)A_{rn}\right]-p_0ab=0 \quad ...(5)$$

By taking $m = n = 1$, we get

$$A_{11}=\frac{p_0a^4}{4D\pi^4}\,\frac{1}{3+3\left(\frac{a}{b}\right)^4+2\left(\frac{a}{b}\right)^2}$$

$$\therefore \quad u_z\Big|_{\substack{x=a/2\\ y=a/2}}=\frac{p_0a^4}{\pi^4 D}\left[\frac{1}{3+3\left(\frac{a}{b}\right)^4+2\left(\frac{a}{b}\right)^2}\right]$$

For a square plate, $u_s = \dfrac{p_0a^4}{8\pi^4 D}$

Better accuracy can be obtained by taking more parameters.

Example 13.8 *Deflection of a beam by Rayleigh's method.*

Solution. Consider a cantilever beam of length l carrying a concentrated load P at the free end. In order to determine the deflection of the beam at the free end, let deflection curve for the bean be given by

$$y=\delta\left(1-\cos\frac{\pi x}{2l}\right)$$

so that $$\frac{d^2y}{dx^2}=-\delta\left(\frac{\pi}{2l}\right)^2\cos\frac{\pi x}{2l} \quad \therefore \quad U=\int_0^l\frac{EI}{2}\left(\frac{d^2y}{dx^2}\right)^2 dx$$

$$=\frac{EI}{2}\,.\,\delta^2\left(\frac{\pi}{2l}\right)^4\int_0^l\cos^2\frac{\pi x}{2l}\,dx=\frac{\pi^4 EI\,\delta^2}{64\,l^3}$$

Also $$W=\frac{1}{2}P\delta$$

$$\therefore \quad \frac{\pi^2 EI\delta^2}{64\,l^3}=\frac{1}{2}P\delta$$

$$\delta=\frac{32}{\pi^4}\cdot\frac{Pl^3}{EI}=0.327\,\frac{Pl^3}{EI}$$

Example 13.9 *Two members AB and BC of lengths l_1 and l_2 respectively of a pin-joined truss are attached to a rigid foundation at points A and C as shown in Fig. 13.9. Under the action of forces P and Q, pin B undergoes finite horizontal and vertical displacement components u and v respectively. Derive formulae for P and Q in terms of u and v by using the principle of stationary potential energy.*

Solution. Let the elongations of members AB and BC be δ_1 and δ_2 respectively. Then

$$\left.\begin{aligned}(l_1+\delta_1)^2 &= (b_1+u)^2+(h+v)^2\\ l_1^2 &= b_1^2+h^2\\ (l_2+\delta_2)^2 &= (b_2-u)^2+(h+v)^2\\ l_2^2 &= b_2^2+h^2\end{aligned}\right\} \quad ...(1)$$

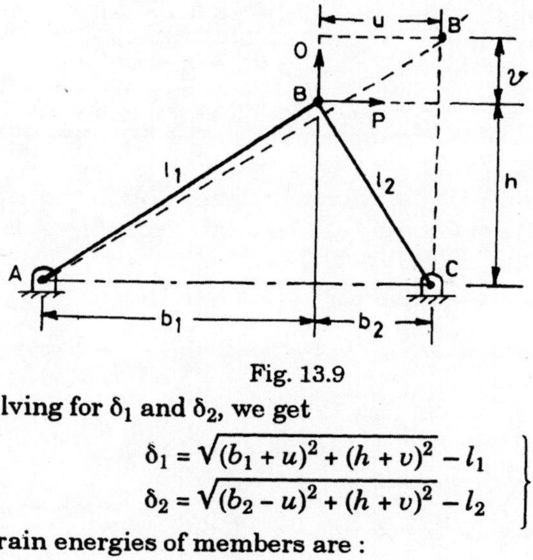

Fig. 13.9

Solving for δ_1 and δ_2, we get

$$\left.\begin{aligned}\delta_1 &= \sqrt{(b_1+u)^2+(h+v)^2}-l_1\\ \delta_2 &= \sqrt{(b_2-u)^2+(h+v)^2}-l_2\end{aligned}\right\} \quad ...(2)$$

Strain energies of members are :

$$\left.\begin{aligned}U_1 &= \frac{1}{2}F_1\delta_1\\ U_2 &= \frac{1}{2}F_2\delta_2\end{aligned}\right\} \quad ...(3)$$

where F_1 and F_2 are the tensions in two members.

Now $\delta_1 = F_1 l_1/E_1A_1$

$\delta_2 = F_2 l_2/E_2A_2.$

Total strain energy

$$U = U_1 + U_2$$

$$= \frac{E_1A_1}{2l_1}\delta_1^2 + \frac{E_2A_2}{2l_2}\delta_2^2$$

$$\left.\begin{aligned} \text{Now} \quad P &= \frac{\partial U}{\partial u} = \frac{E_1A_1\delta_1}{l_1}\cdot\frac{\partial\delta_1}{\partial u} + \frac{E_2A_2\delta_2}{l_2}\cdot\frac{\partial\delta_2}{\partial u} \\ Q &= \frac{\partial U}{\partial v} = \frac{E_1A_1\delta_1}{l_1}\cdot\frac{\partial\delta_1}{\partial v} + \frac{E_2A_2\delta_2}{l_2}\cdot\frac{\partial\delta_2}{\partial v} \end{aligned}\right\} \quad ...(4)$$

Substituting Eqs. (2) in (4), we get

$$P = \frac{E_1A_1\,(b_1+u)}{l_1}\cdot\frac{\sqrt{(b_1+u)^2+(h+v)^2}-l_1}{\sqrt{(b_1+u)^2+(h+v)^2}}$$

$$-\frac{E_2A_2\,(b_2-u)}{l_2}\cdot\frac{\sqrt{(b_2-u)^2+(h+v)^2}-l_2}{\sqrt{(b_2-u)^2+(h+v)^2}}$$

$$Q = \frac{E_1A_1\,(h+v)}{l_1}\cdot\frac{\sqrt{(b_1+u)^2+(h+v)^2}-l_1}{\sqrt{(b_1+u)^2+(h+v)^2}}$$

$$+\frac{E_2A_2\,(h_2+v)}{l_2}\cdot\frac{\sqrt{(b_2-u)^2+(h+v)^2}-l_2}{\sqrt{(b_2-u)^2+(h+v)^2}}$$

Example 13.10 *Calculate the vertical deflection of joint A (Fig. 13.10) by the principle of virtual work. Area of cross-section of AB and AC is 5 cm² and E = 200 GPa.*

Solution. $F_{AC} \cos 60° = P$

$$F_{AC} = \frac{10}{\cos 60°} = 10 \times 2 = 20 \text{ kN} \qquad \text{(compressions)}$$

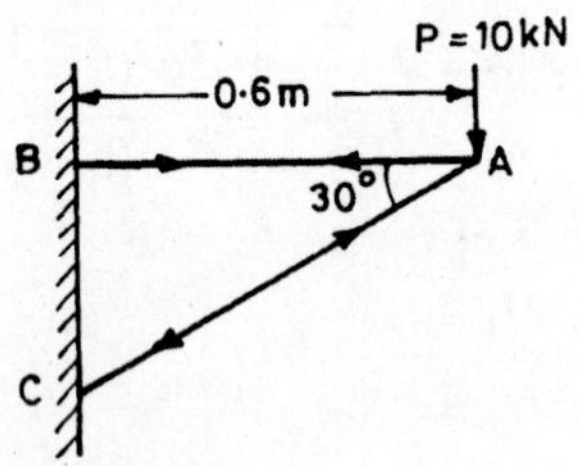

Fig. 13.10

$F_{AC} \cos 30° = F_{AB}$

$$F_{AB} = 20 \times \frac{\sqrt{3}}{2} = 17.32 \text{ kN (tensile)}$$

$$AC = \frac{AB}{\cos 30°} = \frac{0.6}{\sqrt{3}} \times 2 = 0.6928 \text{ m}$$

Total strain energy,

$$U = \frac{F_{AB}^2 \times AB}{2EA} + \frac{F_{AC}^2 \times AC}{2EA}$$

$$= \frac{10^6}{2 \times 200 \times 10^9 \times 5 \times 10^{-4}} \times [(17.32)^2 \times 0.6 + (-20)^2 \times 0.6928]$$

$$= 2.2855 \text{ N.m}$$

Let δ be the vertical deflection of A.

$$\text{Virtual work done} = \frac{1}{2} P \delta$$

$$= \frac{1}{2} \times 10 \times \delta = 5 \times \delta \text{ kN.m.}$$

$$5 \times \delta \times 10^3 = 2.2855$$

$$\delta = 0.457 \text{ mm}$$

Example 13.11 *Calculate the virtual displacement of joint C of the structure shown in Fig. 13.11. Area of cross-section of all members is 5 cm² and length of member 1 to 4 is 0.5 m. E = 200 GPa.*

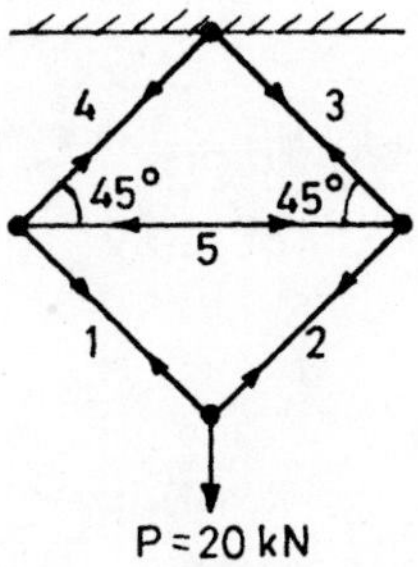

Fig. 13.11

Solution. Force in the members are :

$$F_1 = F_2 = F_3 = F_4$$

$$= \frac{P}{\sqrt{2}} = \frac{20}{\sqrt{2}} = 14.14 \text{ kN}$$

$$F_5 = -P = -20 \text{ kN}$$

$$l_1 = l_2 = l_3 = l_4 = 0.5 \text{ m}$$

$$l_5 = 0.5\sqrt{2} = 0.707 \text{ m}$$

Strain energy stored,

$$U = \sum_{i=1}^{5} \frac{F_i^2 l_i}{2EA}$$

$$= \frac{10^6}{2 \times 200 \times 10^9 \times 5 \times 10^{-4}} \times \left[4 \times (14.14)^2 \times 0.5 + (-20)^2 \times 0.707 \right]$$

$$= 5.4128 \text{ N.m}$$

Let δ be the vertical deflection under load P

$$\text{Virtual work done} = \frac{1}{2} P\delta$$

$$= \frac{1}{2} \times 20 \times 10^3 \times \delta$$

$$= 10^4 \times \delta \text{ N.m}$$

$$10^4 \times \delta = 5.4128$$

$$\delta = 0.5418 \text{ mm}$$

Example 13.12 *Calculate the vertical and horizontal deflections of the end A of the bar shown in Fig. 13.12. Cross-sectional area of AB and BC is 30 mm × 20 mm, AB = 0.6 m and BC = 0.4 m. Take E = 200 GPa.*

Solution. Let us apply a horizontal dummy load Q at A.

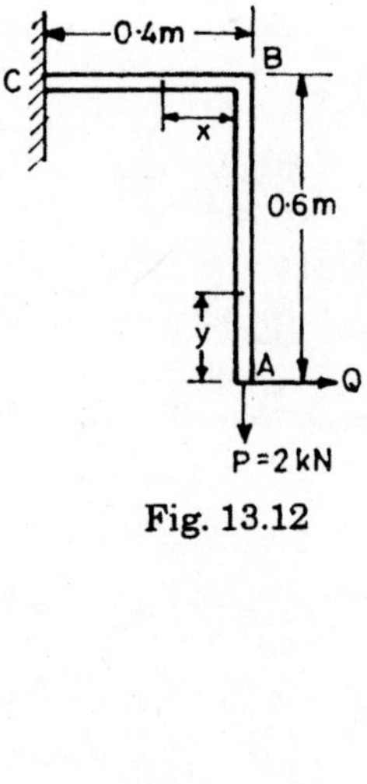

Fig. 13.12

Strain energy,

$$U = \int_0^{0.6} \frac{P^2 dy}{2AE} + \int_0^{0.6} \frac{(Qy)^2 \, dy}{2EI} + \int_0^{0.4} \frac{Q^2 \, dx}{2AE} + \int_0^{0.4} \frac{(Px + 0.6\,Q)^2}{2EI} dx$$

$$\delta_V = \left. \frac{\partial U}{\partial P} \right|_{Q = 0}$$

$$= \int_0^{0.6} \frac{Pdy}{AE} + \int_0^{0.4} \frac{Px^2 \, dx}{EI}$$

$$= \frac{P}{AE} \left| y \right|_0^{0.6} + \frac{P}{EI} \left| \frac{x^3}{3} \right|_0^{0.4}$$

$$= \frac{2 \times 10^3 \times 0.6}{400 \times 10^{-6} \times 200 \times 10^9} + \frac{2 \times 10^3 \times (0.4)^3 \times 12}{3 \times 200 \times 10^9 \times 20 \times 20^3 \times 10^{-12}}$$

$$= 0.000015 + 0.016$$

$$= 0.016015 \text{ m or } 16 \text{ mm}$$

$$\delta_H = \left. \frac{\partial U}{\partial Q} \right|_{Q = 0} = \int_0^{0.6} \frac{Qy^2}{EI} dy + \int_0^{0.4} \frac{Qdx}{AE}$$

$$+ \int_0^{0.4} \frac{(Px + 0.6\,Q) \times 0.6}{EI} dx$$

$$= \int_0^{0.4} \frac{0.6\,Px\,dx}{EI} = \frac{0.6P}{EI} \left| \frac{x^2}{2} \right|_0^{0.4}$$

$$= \frac{0.6 \times 2 \times 10^3 \times (0.4)^2 \times 12}{2 \times 200 \times 10^9 \times 20^4 \times 10^{-12}} = 0.036 \text{ m or } 36 \text{ mm}$$

Example 13.13 *A load of 100 N is suspended from end A of a semi-circular ring fixed at end B in horizontal plane, as shown in Fig. 13.13. Calculate the vertical displacement of end A. The cross-section of the ring is of 15 mm diameter and its mean radius is 200 mm. E = 200 GPa and G = 80 GPa.*

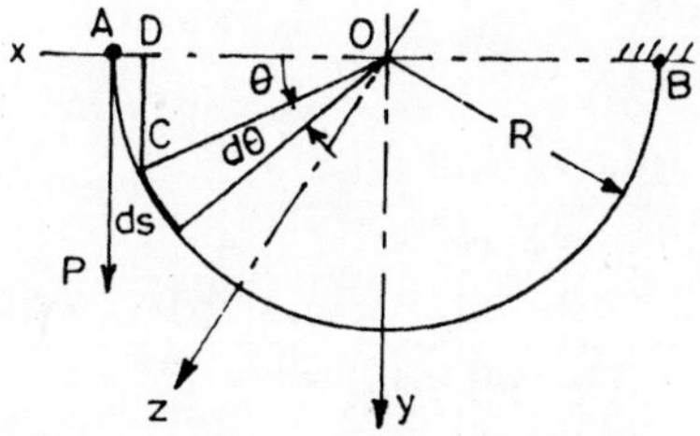

Fig. 13.13

Solution. Consider an elementary length $ds = Rd\theta$ of the ring en θ and $\theta + d\theta$.

$$AD = R - R\cos\theta$$

$$= R(1 - \cos\theta)$$

$$CD = R\sin\theta$$

Bending moment on ds, $M = P \times CD = PR\sin\theta$

Torque on ds, $T = P \times AD$

$$= PR(1 - \cos\theta)$$

$$U = \int_0^{\pi} \left[\frac{M^2\,ds}{2EI} + \frac{T^2\,ds}{2GJ} \right]$$

$$= \int_0^{\pi} \left[\frac{P^2R^2 \sin^2\theta \,.\, Rd\theta}{2EI} + \frac{P^2R^2(1-\cos\theta)^2\,Rd\theta}{2GJ} \right]$$

$$\delta_V = \frac{\partial U}{\partial P} = \int_0^{\pi} \left[\frac{PR^3 \sin^2\theta}{EI} + \frac{PR^3(1-\cos\theta)^2}{GJ} \right] d\theta$$

$$= \frac{PR^3}{EI}\int_0^{\pi}\left(\frac{1-\cos 2\theta}{2}\right)d\theta + \frac{PR^3}{GJ}\int_0^{\pi}(1+\cos^2\theta - 2\cos\theta)\,d\theta$$

$$= \frac{PR^3}{2EI}\int_0^{\pi}(1-\cos 2\theta)\,d\theta$$

$$+ \frac{PR^3}{GJ}\int_0^{\pi}\left(1+\frac{1+\cos 2\theta}{2} - 2\cos\theta\right)d\theta$$

$$= \frac{PR^3}{2EI}\left|\theta - \frac{\sin 2\theta}{2}\right|_0^{\pi} + \frac{PR^3}{GJ}\left|\theta + \frac{1}{2}\left(\theta + \frac{\sin 2\theta}{2}\right) - 2\sin\theta\right|_0^{\pi}$$

$$= \frac{PR^3}{2EI}(\pi) + \frac{PR^3}{GJ}\left[\pi + \frac{1}{2}(\pi)\right]$$

$$= \frac{\pi \times 100 \times 200^3 \times 10^{-9} \times 64}{2 \times 200 \times 10^9 \times \pi \times 15^4 \times 10^{-12}}$$

$$+ \frac{3\pi \times 100 \times 200^3 \times 10^{-9} \times 32}{2 \times 80 \times 10^9 \times \pi \times 15^4 \times 10^{-12}}$$

$$= 0.0025284 + 0.009481$$

$$= 0.012 \text{ m or } 12 \text{ mm}$$

Example 13.14 *A beam of 50 mm × 100 mm cross-section and 2 m long is fixed at one end and supported by a 1 m long wire of 5 mm diameter at the free end (Fig. 13.14). A load of 5 kN acts at the mid-span of the beam. Calculate support reactions at free end and fixed end, support moment at fixed end and the deflection at the free end and mid-span. Take E = 200 GPa.*

Solution.

$$I = \frac{50 \times 100^3}{12} \times 10^{-12}$$

$$= 4.167 \times 10^{-6} \text{ m}^4$$

$$A_w = \frac{\pi \times 25}{4} \times 10^{-6}$$

$$= 19.635 \times 10^{-6} \text{ m}^2$$

$$l_w = 1 \text{ m}$$

$$M_x = R_B x \text{ for } 0 < x < \frac{l}{2}$$

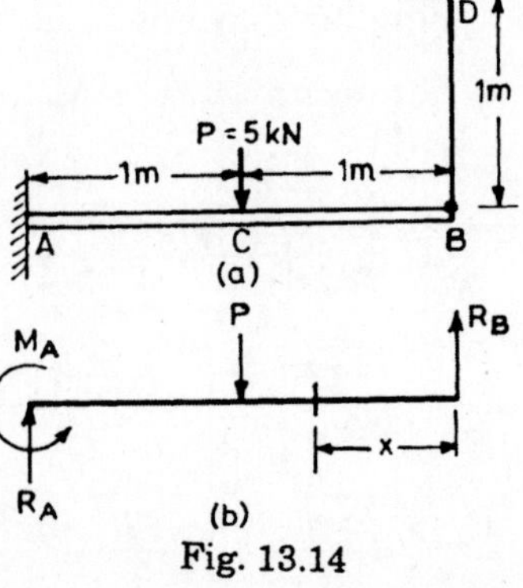

Fig. 13.14

$$= R_B x - P\left(x - \frac{l}{2}\right) \text{ for } \frac{l}{2} < x < l$$

Deflection of free end = Elongation of wire.

$$\delta_B = \frac{R_B l_w}{A_w E}$$

$$U_{beam} = \int_0^{l/2} \frac{(R_B x)^2 dx}{2EI} + \int_{l/2}^{l} \frac{[R_B x - P(x - l/2)]^2 dx}{2EI}$$

$$\delta_B = \frac{\partial U_{beam}}{\partial R_B} = \int_0^{l/2} \frac{R_B x^2 dx}{EI} + \int_{l/2}^{l} \frac{\left[R_B x - P\left(x - \frac{l}{2}\right)\right] x\, dx}{EI}$$

$$= \frac{R_B}{EI}\left|\frac{x^3}{3}\right|_0^{l/2} + \frac{1}{EI}\left|\frac{R_B x^3}{3} - \frac{Px^3}{3} + \frac{Plx^2}{4}\right|_{l/2}^{l}$$

$$= \frac{R_B l^3}{24EI} + \frac{1}{EI}\left[\frac{R_B}{3}\left(l^3 - \frac{l^3}{8}\right) - \frac{P}{3}\left(l^3 - \frac{l^3}{8}\right) + \frac{Pl}{4}\left(l^2 - \frac{l^2}{4}\right)\right]$$

$$= \frac{R_B l^3}{24EI} + \frac{1}{EI}\left[\frac{7R_B l^3}{24} - \frac{7Pl^3}{24} + \frac{3Pl^3}{16}\right]$$

$$= \frac{R_B l^3}{24\,EI} + \frac{7}{24}\frac{R_B l^3}{EI} - \frac{5}{48}\frac{Pl^3}{EI}$$

$$= \frac{R_B l^3}{3EI} - \frac{5}{48}\frac{Pl^3}{EI} = -\frac{R_B l_w}{E A_w}$$

or

$$\frac{R_B}{E}\left(\frac{l^3}{3I} + \frac{l_w}{A_w}\right) = \frac{5}{48}\frac{Pl^3}{EI}$$

$$R_B\left(\frac{8}{3 \times 4.167 \times 10^{-6}} + \frac{1}{19.635 \times 10^{-6}}\right) = \frac{5}{48} \times \frac{5 \times 10^3 \times 8}{4.167 \times 10^{-6}}$$

$$R_B = 1.447 \text{ kN}$$

$$R_A = 5 - 1.447 = 3.553 \text{ kN}$$

$$M_A = R_B l - \frac{Pl}{2}$$

$$= 1.447 \times 2 - \frac{5 \times 2}{2} = 2.894 - 5 = 2.106 \text{ kN.m}$$

$$\delta_B = \frac{1.447 \times 10^3 \times 1}{200 \times 10^9 \times 19.635 \times 10^{-6}} = 0.3685 \text{ mm}$$

$$\delta_C = \frac{\partial U}{\partial P} = \frac{1}{EI}\int_{l/2}^{l} [R_B x - P(x - l/2)] (-x + l/2)\, dx$$

$$= \frac{l}{EI}\int_{l/2}^{l}\left[-R_B x^2 + P\left(x^2 - \frac{lx}{2}\right) + R_B x\frac{l}{2} - \frac{Pl}{2} - \left(x - \frac{l}{2}\right)\right] dx$$

$$= \frac{1}{EI}\left| -\frac{R_B x^3}{3} + P\left(\frac{x^3}{3} - \frac{lx^2}{4}\right) + \frac{R_B lx^2}{4} - \frac{Pl}{2}\left(\frac{x^2}{2} - \frac{l}{2}x\right)\right|_{l/2}^{l}$$

$$= \frac{1}{EI}\left[-\frac{R_B}{3}\left(l^3 - \frac{l^3}{8}\right) + P\left(\frac{l^3}{3} - \frac{l^3}{24} - \frac{l^3}{4} + \frac{l^3}{16}\right)\right.$$

$$\left. + \frac{R_B l}{4}\left(l^2 - \frac{l^2}{4}\right) - \frac{Pl}{2}\left(\frac{l^2}{2} - \frac{l^2}{8} - \frac{l^2}{2} + \frac{l^2}{4}\right)\right]$$

$$= \frac{1}{EI}\left[-\frac{7}{24}R_B l^3 + \frac{5}{48}Pl^3 + \frac{3R_B l^3}{16} - \frac{Pl^3}{16}\right]$$

$$= \frac{1}{EI}\left[-\frac{5}{48}R_B l^3 + \frac{Pl^3}{24}\right]$$

$$= \frac{1}{200 \times 10^9 \times 4.167 \times 10^{-6}}$$

$$\times\left[-\frac{5}{48} \times 1.447 \times 10^3 \times 8 + \frac{5 \times 10^3 \times 8}{24}\right]$$

$= 0.553$ mm

Example 13.15 *Calculate the vertical deflection of point A (Fig. 13.15) $R = 8$ m, $d_0 = 150$ mm, $d_i = 110$ mm, and $E = 200$ GPa.*

Solution.

$$I = \frac{\pi}{64}(150^4 - 110^4) \times 10^{-12} = 17663605 \times 10^{-12}\ \text{m}^4$$

$$M = PR \sin\theta$$

$$U = \int_0^{\pi/2} \frac{P^2R^2 \sin^2\theta\, Rd\theta}{EI}$$

$$\delta_A = \frac{\partial U}{\partial P} = \frac{PR^3}{EI}\int_0^{\pi/2}\left(\frac{1 - \cos 2\theta}{2}\right) d\theta$$

$$= \frac{PR^3}{EI}\left|\frac{\theta}{2} - \frac{\sin 2\theta}{4}\right|_0^{\pi/2} = \frac{PR^3}{EI}\left[\frac{\pi}{4}\right] = \frac{\pi PR^3}{4\,EI}$$

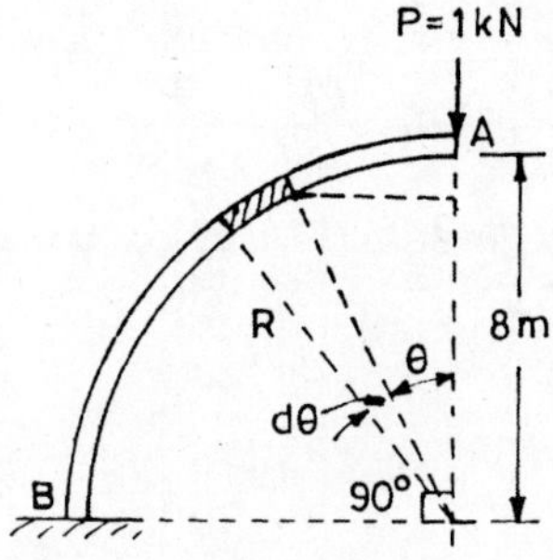

Fig. 13.15

$$= \frac{\pi \times 10^3 \times 8^3}{4 \times 200 \times 10^9 \times 17663605 \times 10^{-12}} = 11.38 \text{ mm}$$

Example 13.16 *Calculate the vertical deflection of point A of the frame shown in Fig. 13.16.*

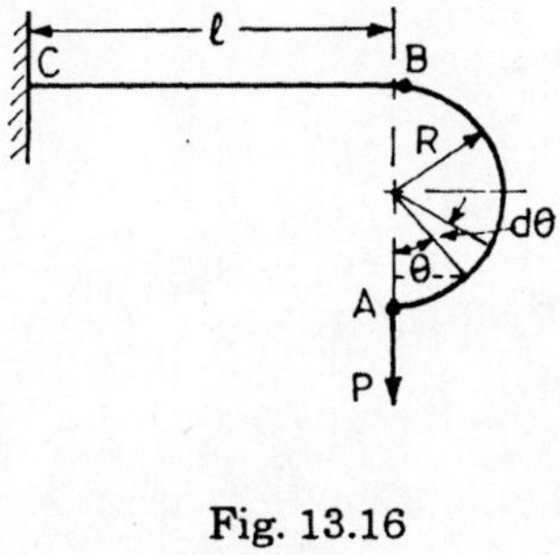

Fig. 13.16

Solution. $U = \frac{1}{2EI}\int_0^{\pi} P^2R^2 \sin^2\theta \; Rd\theta$

$$+ \frac{1}{2EI}\int_0^{l} P^2x^2\,dx$$

$$\delta_A = \frac{\partial U}{\partial P} = \frac{PR^3}{EI}\int_0^{\pi} \sin^2\theta\, d\theta + \frac{P}{EI}\int_0^{l} x^2dx$$

$$= \frac{PR^3}{EI}\int_0^{\pi} \left(\frac{1 - \cos 2\theta}{2}\right) d\theta + \frac{P}{EI}\int_0^{l} x^2dx$$

$$= \frac{PR^3}{2EI}\left|\theta - \frac{\sin 2\theta}{2}\right|_0^{\pi} + \frac{P}{3EI}\left|x^3\right|_0^{l}$$

$$= \frac{PR^3}{2EI}(\pi) + \frac{Pl^3}{3EI} = \frac{P}{6EI}(3\pi R^3 + 2l^3)$$

Example 13.17 *Calculate the vertical deflection of point A of the bar shown in Fig. 13.17.*

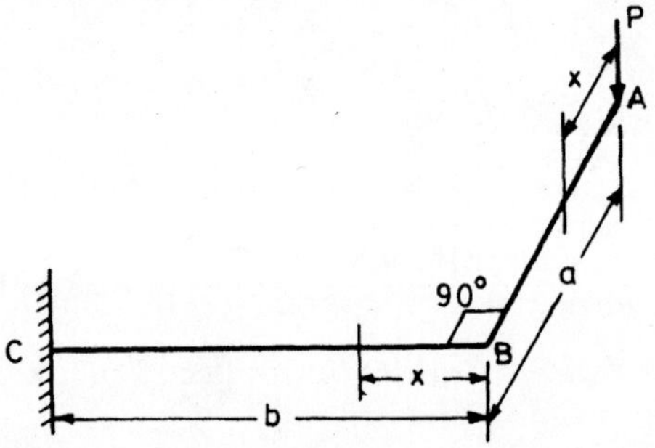

Fig. 13.17

Solution.

AB : $M_x = Px$

BC : $M_x = Px$

$T = Pa$

$$U = \frac{1}{2EI}\int_0^a P^2x^2dx + \frac{1}{2EI}\int_0^b P^2x^2dx + \frac{1}{2GJ}\int_0^b P^2a^2dx$$

$$\delta_A = \frac{\partial U}{\partial P} = \frac{P}{EI}\int_0^a x^2dx + \frac{P}{EI}\int_0^b x^2dx + \frac{Pa^2}{GJ}\int_0^b dx$$

$$= \frac{P}{3EI}(a^3 + b^3) + \frac{Pa^2b}{GJ}$$

Example 13.18 *Determine the support reaction for the propped cantilever shown in Fig. 13.18.*

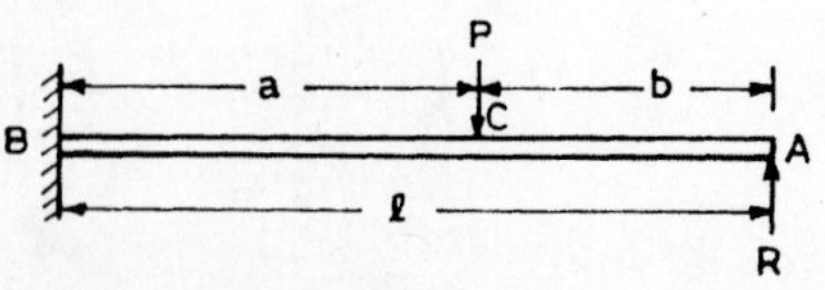

Fig. 13.18

Solution. Let R be the reaction at the prop. The deflection at the prop is zero

$$U = \int_0^b \frac{(Rx)^2\,dx}{2EI} \int_0^a \frac{[R(b + x) - Px]^2\,dx}{2EI}$$

$$\delta_A = \frac{\partial U}{\partial R} = \frac{R}{EI}\int_0^b x^2 dx + \frac{1}{EI}\int_0^a [R(b + x) - Px]\,(b + x)\,dx$$

$$= \frac{R}{EI}\left|\frac{x^3}{3}\right|_0^b + \frac{1}{EI}\int_0^a [R(b + x)^2 - Px(b + x)]\,dx$$

$$= \frac{Rb^3}{3EI} + \frac{1}{EI}\left| R\left(b^2x + \frac{x^3}{3} + bx^2\right) - P\left(\frac{bx^2}{2} + \frac{x^3}{3}\right)\right|_0^a$$

$$= \frac{Rb^3}{3EI} + \frac{1}{EI}\left[R\left(ab^2 + \frac{a^3}{3} + a^2b\right) - P\left(\frac{a^2b}{2} + \frac{a^3}{3}\right)\right] = 0$$

$$\frac{R}{EI}\left(\frac{b^3}{3} + ab^2 + \frac{a^3}{3} + a^2b\right) = \frac{P}{EI}\left(\frac{a^2b}{2} + \frac{a^3}{3}\right)$$

$$R = \frac{P}{2}\left(\frac{3a^2b + 2a^3}{a^3 + b^3 + 3ab^2 + 3a^2b}\right)$$

$$= \frac{P}{2}\left[\frac{a^2(2a + 3b)}{(a + b)^3}\right] = \frac{Pa^2}{2l^3}\,(2a + 3b)$$

Example 13.19 *Calculate the reaction at A for the spring supported cantilever shown in Fig. 13.19.*

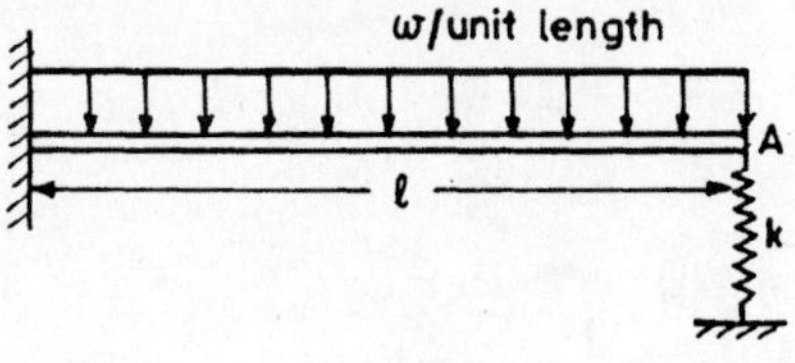

Fig. 13.19

Solution.

Let R be the reaction at A.

Strain energy in the spring,

$$U_1 = \frac{1}{2}R\delta = \frac{1}{2}\frac{R^2}{k}$$

Strain energy in the beam,

$$U_2 = \int_0^l \frac{M^2\,dx}{2EI} = \int_0^l \frac{\left(Rx - \frac{\omega x^2}{2}\right)^2 dx}{2EI}$$

$$U = U_1 + U_2$$

$$\delta_A = \frac{\partial U}{\partial R} = \frac{R}{k} + \frac{1}{EI}\int_0^l \left(Rx - \frac{\omega x^2}{2}\right) x\,dx$$

$$= \frac{R}{k} + \frac{1}{EI}\left|\frac{Rx^3}{3} - \frac{\omega x^4}{8}\right|_0^l = \frac{R}{k} + \frac{1}{EI}\left[\frac{Rl^3}{3} - \frac{\omega l^4}{8}\right] = 0$$

$$R\left(\frac{1}{k} + \frac{l^3}{3EI}\right) = \frac{\omega l^4}{8EI}$$

$$R = \frac{\omega l^4/8EI}{\frac{1}{k} + \frac{l^3}{3EI}} = \frac{3k\,\omega l^4}{8\,(3EI + kl^3)}$$

Example 13.20. *Determine the vertical deflection of the point of application of force P in Fig. 13.20.*

Solution.

AB : $M_x = Px$

$SF = P$

BC : $F = P$

$m = Pa$

CD : $M_x = Px,\ T = Pa,\ SF = P$

$$U = \int_0^a \frac{(Px)^2\,dx}{2EI} + \int_0^b \frac{P^2\,dx}{2EA}$$

$$+ \int_0^b \frac{(Pa)^2\,dx}{2EI} + \int_0^c \frac{(Px)^2\,dx}{2EI}$$

$$+ \int_0^c \frac{(Pa)^2\,dx}{2GJ} + \int_0^a \frac{kP^2 ds}{2AG}$$

$$+ \int_0^c \frac{kP^2 ds}{2AG}$$

where k is a factor for cross-section.

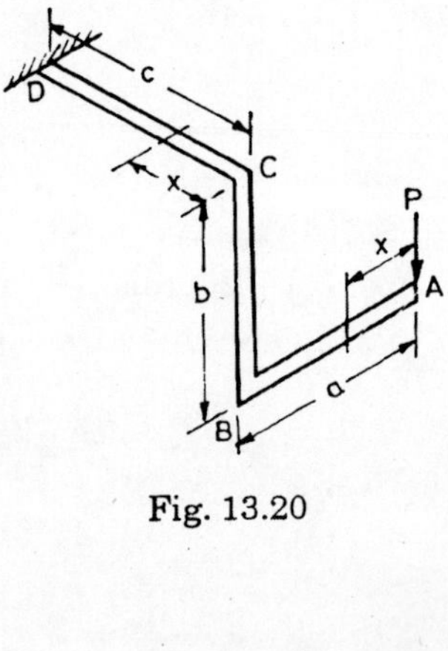

Fig. 13.20

$$\delta_A = \frac{\partial U}{\partial P} = \frac{P}{EI}\int_0^a x^2 dx + \frac{P}{EA}\int_0^b dx + \frac{Pa^2}{EI}\int_0^b dx$$

$$+ \frac{P}{EI}\int_0^c x^2 dx + \frac{Pa^2}{GJ}\int_0^c dx + \frac{kP}{AG}\int_0^a dx + \frac{kP}{2AG}\int_0^c dx$$

$$= \frac{P}{EI}\left(\frac{a^3}{3}\right) + \frac{Pb}{EA} + \frac{Pa^2b}{EI} + \frac{Pc^3}{3EI} + \frac{Pa^2c}{GJ} + \frac{kPa}{AG} + \frac{kPc}{AG}$$

$$= P\left[\frac{1}{E}\left\{\frac{b}{A} + \frac{1}{I}\left(\frac{a^3}{3} + \frac{c^3}{3} + a^2b\right)\right\} + \frac{1}{G}\left\{\frac{k}{A}(a+c) + \frac{a^2c}{J}\right\}\right]$$

Example 13.21 *A rectangular frame of constant EI is loaded as shown in Fig. 13.21. Assuming the strain energy to be attributable to bending alone, determine the increase in distance between the points of application of the load.*

Solution.

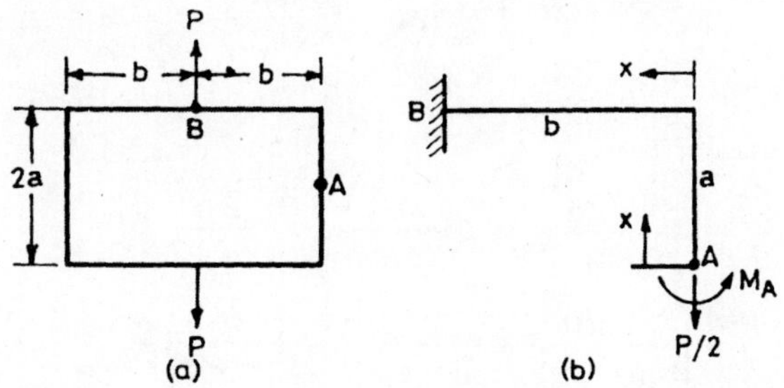

Fig. 13.21

Because of symmetry, we need to analyze only one quadrant. Since the slope at B is zero before and after application of the load, the segment may be treated as fixed at B. The moment distributions are :

$$M_1 = -M_A \qquad 0 \le x \le a$$

$$M_2 = -M_A + \frac{1}{2}Px \qquad 0 \le x \le b$$

Since the slope is zero at A, we have

$$\theta_A = \frac{1}{EI}\int M\frac{\partial M}{\partial M_A}dx$$

$$= \frac{1}{EI}\left[\int_0^a M_A\,dx + \int_0^b \left(M_A - \frac{1}{2}Px\right)dx\right] = 0$$

$$M_A = \frac{Pb^2}{4(a+b)}$$

$$U = 4\left[\int_0^a \frac{M_A^2\,dx}{2EI} + \int_0^b \frac{\left(M_A - \frac{1}{2}Px\right)^2 dx}{2EI}\right]$$

$$= \frac{2}{EI}\left[\int_0^a \frac{P^2b^4\,dx}{16(a+b)^2} + \int_0^b \left(\frac{Pb^2}{4(a+b)} - \frac{1}{2}Px\right)^2 dx\right]$$

$$\delta = \frac{\partial U}{\partial P} = \frac{4}{EI}\left[\int_0^a \frac{Pb^4\,dx}{16(a+b)^2} + \int_0^b \left(\frac{Pb^2}{4(a+b)} - \frac{1}{2}Px\right) \times \left(\frac{b^2}{4(a+b)} - \frac{1}{2}x\right) dx\right]$$

$$= \frac{4}{EI}\left[\frac{Pb^4}{16(a+b)^2}\Big|\, x \Big|_0^a + \frac{Pb^4}{16(a+b)^2}\Big|\, x \Big|_0^b - \frac{Pb^2}{8(a+b)}\Big|\, \frac{x^2}{2}\Big|_0^b \times 2 + \frac{P}{4}\Big|\, \frac{x^3}{3}\Big|_0^b\right]$$

$$= \frac{4}{EI}\left[\frac{Pb^4a}{16(a+b)^2} + \frac{Pb^5}{16(a+b)^2} - \frac{Pb^4}{8(a+b)} + \frac{Pb^3}{12}\right]$$

$$= \frac{Pb^3}{EI}\left[\frac{ab}{4(a+b)^2} + \frac{b^2}{4(a+b)^2} - \frac{b}{2(a+b)} + \frac{1}{3}\right]$$

$$= \frac{Pb^3}{EI}\left[\frac{3ab + 3b^2 - 6b(a+b) + 4(a+b)^2}{12(a+b)^2}\right]$$

$$= \frac{Pb^3}{EI}\left[\frac{3ab + 3b^2 - 6ab - 6b^2 + 4a^2 + 4b^2 + 8ab}{12(a+b)^2}\right]$$

$$= \frac{Pb^3}{EI}\left[\frac{4a^2 + b^2 + 5ab}{12(a+b)^2}\right] = \frac{Pb^3}{EI}\left[\frac{4a(a+b) + b(a+b)}{12(a+b)^2}\right]$$

$$= \frac{Pb^3}{EI}\left(\frac{4a+b}{12(a+b)}\right) = \frac{Pb^3}{12EI}\left(\frac{4a+b}{a+b}\right)$$

Example 13.22 *Determine the vertical and horizontal deflections and angular rotation of the free end by Castigliano's theorem, (Fig. 13.22).*

Solution. **BC :** $M_x = Px$ **AB :** $M_x = Pb + Qx$

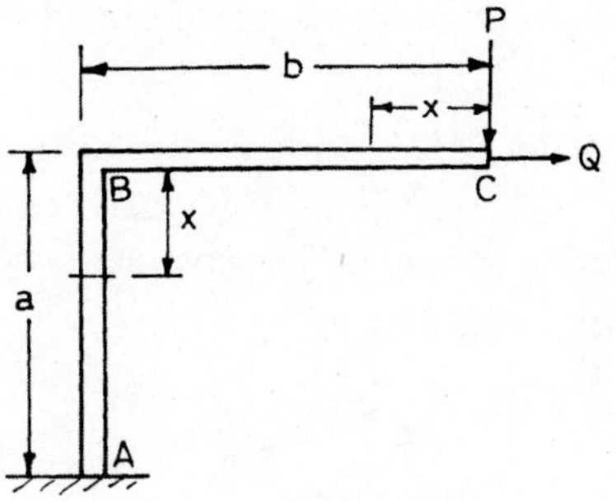

Fig. 13.22

$$U = \frac{1}{2EI}\left[\int_0^b (Px)^2\,dx + \int_0^a (Pb + Qx)^2\,dx\right]$$

$$\delta_V = \frac{\partial U}{\partial P} = \frac{1}{EI}\left[P\int_0^b x^2\,dx + \int_0^a (Pb + Qx)\,b\,dx\right]$$

$$= \frac{1}{EI}\left[P\left|\frac{x^3}{3}\right|_0^b + Pb^2\left|x\right|_0^a + Qb\left|\frac{x^2}{2}\right|_0^a\right]$$

$$= \frac{1}{EI}\left[\frac{Pb^3}{3} + Pb^2a + \frac{Qb\,a^2}{2}\right]$$

$$\delta_H = \frac{\partial U}{\partial Q} = \frac{1}{EI}\left[0 + \int_0^a (Pb + Qx)\,x\,dx\right]$$

$$= \frac{1}{EI}\left[Pb\left|\frac{x^2}{2}\right|_0^a + Q\left|\frac{x^3}{3}\right|_0^a\right]$$

$$= \frac{1}{EI}\left[Pb\frac{a^2}{2} + \frac{Qa^3}{3}\right] = \frac{a^2}{6EI}[3Pb + 2Qa]$$

13.11 Finite Difference Method

This is the numerical method to obtain the solution of elasticity equations consistent with the boundary conditions. This method consists in replacing the governing partial differential equations of elasticity and the equations defining the boundary conditions by the corresponding finite difference equations. This reduces the problem to a set of simultaneous algebraic equations which can be solved without mathematical difficulty. Once the governing equations have been derived, no further reference is made to the actual physical problem.

13.11.1 Finite Difference Equations

The finite differences may be defined in three ways, *i.e.* forward, backward and central, as given below.

First forward difference,

$$\overline{\Delta} f = \frac{1}{h}[f(x+h) - f(x)]$$

Second forward difference,

$$\overline{\Delta}^2 f = f(x+2h) - 2f(x+h) + f(x)$$

Similarly, First backward difference,

$$\underline{\Delta} f = \frac{1}{h}[f(x) - f(x-h)]$$

Second backward difference,

$$\underline{\Delta}^2 f = f(x) - 2f(x-h) + f(x-2h)$$

and First central difference,

$$\Delta f = \frac{1}{2h}[f(x+h) - f(x-h)]$$

Second central difference,

$$\Delta^2 f = f(x+h) - 2f(x) + f(x-h)$$

where $\Delta x = h$ is the interval.

In general, the derivatives at any point can be better approximated by the corresponding central differences because these differences involve points on both sides of the point in question along the curve and, therefore, give better average values.

Hence we would use central differences only.

Likewise, third central difference

$$\Delta^3 f = \tfrac{1}{2}[f(x+2h) - 2f(x+h) + 2f(x-h) - f(x-2h)]$$

and fourth central difference

$$\Delta^4 f = f(x+2h) - 4f(x+h) + 6f(x) - 4f(x-h) + f(x-2h)$$

If f is a function of both x and y then taking

$$\Delta x = \Delta y = h$$

and dividing the domain into a square mesh as shown in Fig. 13.23 at the point O we obtain.

$$\frac{\partial f}{\partial x} \simeq \frac{1}{2h}[f(x+h, y) - f(x-h, y)] = \frac{1}{2h}(f_1 - f_3)$$

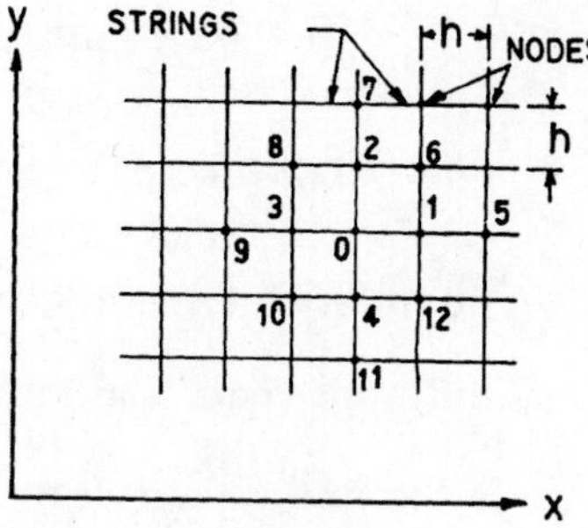

Fig. 13.23

$$\frac{\partial^2 f}{\partial x^2} \simeq \frac{1}{h^2} [f(x+h, y) - 2f(x, y) + f(x-h, y)]$$

$$= \frac{1}{h^2} (f_1 - 2f_0 + f_3)$$

$$\frac{\partial^2 f}{\partial y^2} \simeq \frac{1}{h} [f(x, y+h) - 2f(x, y) + f(x, y-h)]$$

$$= \frac{1}{h^2} (f_2 - 2f_0 + f_4).$$

Now $\nabla^2 f = \dfrac{\partial^2 f}{\partial x^2} + \dfrac{\partial^2 f}{\partial y^2} = \dfrac{1}{h^2} (f_1 + f_2 + f_3 + f_4 - 4f_0)$

Hence the Laplace equation $\nabla^2 f = 0$ becomes

$$f_1 + f_2 + f_3 + f_4 - 4f_0 = 0 \qquad \text{...(13.58)}$$

This is the corresponding finite difference equation at the point 0. At every point of the domain a finite difference equation of the form (13.58) can be written. If the domain is divided into a net with n inner points, n difference equations are obtained. Together with the boundary conditions, we can solve for the values of the function at every point in the domain.

If the governing equation is of the Poisson type

$$\nabla^2 f = -2$$

then the corresponding difference equation is

$$f_1 + f_2 + f_3 + f_4 - 4f_0 = -2h^2 \qquad \text{...(13.59)}$$

Now consider the biharmonic equation

$$\nabla^4 f = \frac{\partial^4 f}{\partial x^4} + 2\frac{\partial^4 f}{\partial x^2 \partial y^2} + \frac{\partial^4 f}{\partial y^4} = 0$$

Now $$\frac{\partial^4 f}{\partial x^4} = \frac{1}{h^4}(f_5 - 4f_1 + 6f_0 - 4f_3 + f_9)$$

$$\frac{\partial^4 f}{\partial y^4} = \frac{1}{h^4}(f_7 - 4f_2 + 6f_0 - 4f_4 + f_{11})$$

$$\frac{\partial^4 f}{\partial x^2 \partial y^2} = \frac{1}{h^4}(f_6 + f_8 + f_{10} + f_{12} - 2f_1 - 2f_2 - 2f_3 - 2f_4 + 4f_0)$$

Hence the finite difference equation at 0 becomes :

$$\frac{1}{h^4}[20f_0 - 8(f_1 + f_2 + f_3 + f_4) + 2(f_6 + f_8 + f_{10} + f_{12}) + f_5 + f_7 + f_9 + f_{11}] = 0 \quad ...(13.60)$$

13.11.2 Solution of the Finite-Difference Equations

To obtain the solution of the finite difference equations the relaxation method due to R.V. Southwell may be used. In this method the finite difference equation is written as

$$f_1 + f_2 + f_3 + f_4 - 4f_0 = Q_0 \quad ...(13.61)$$

in the case of Laplace equation and

$$f_1 + f_2 + f_3 + f_4 - 4f_0 + 2h^2 = Q_0 \quad ...(13.62)$$

in the case of Poisson's equation.

If the values f_1, f_2, f_3, f_4, f_0 are the correct ones then the residue Q_0 is equal to zero. Conversely, if the values are not the true values then the residue Q_0 does not vanish. In terms of the residues the numerical solution of Laplace or Poisson's equation may be said to consist of such adjustments in the initial values of f as would make the residues zero. When the residual Q_0 vanishes the point is said to be relaxed.

In order to obtain some interesting results for relaxation patterns, let $(Q_0)_0$ and $(Q_0)_1$ denote the residuals at 0 corresponding to values of f_0 and $(f_0 + \Delta f_0)$ respectively.

Then $$(Q_0)_0 = f_1 + f_2 + f_3 + f_4 - 4f_0$$
$$(Q_0)_1 = f_1 + f_2 + f_3 + f_4 - 4(f_0 + \Delta f_0)$$

The change in the residual at 0 *i.e.* ΔQ_0 is then given by the expression

$$\Delta Q_0 = (Q_0)_1 - (Q_0)_0 = -4\Delta f_0$$

As shown in Fig. 13.24

$$(Q_1)_0 = f_0 + f_5 + f_6 + f_7 - 4f_1$$
$$(Q_1)_1 = (f_0 + 4f_0) + f_5 + f_6 + f_7 + 4f_1$$

and $$\Delta Q_1 = (Q_1)_1 - (Q_1)_0 = \Delta f_0$$

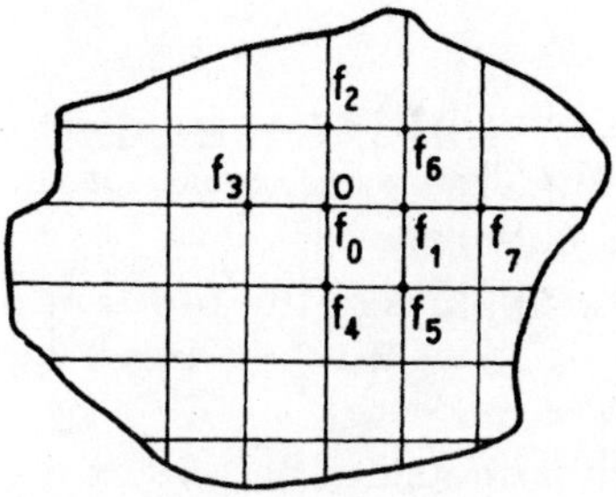

Fig. 13.24

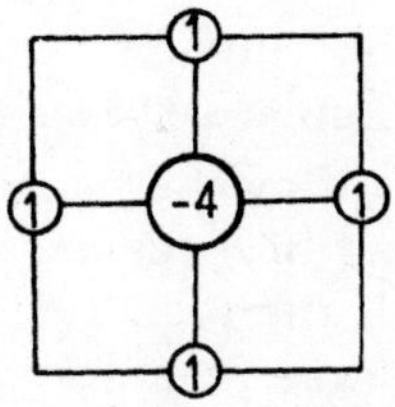

Fig. 13.25. Relaxation pattern.

By analogy $\Delta Q_1 = \Delta Q_2 = \Delta Q_3 = \Delta Q_4 = \Delta f_0 = -\frac{\Delta Q_0}{4}$

There are thus three significant facts to remember :

(*i*) The change in f_0 affects all five residuals, Q_0 Q_1, Q_2, Q_3, Q_4 but does not affect any other residuals.

(*ii*) A change in Q_0 equals four times the change Δf_0 and is of oposite sign.

(*iii*) The effect on Q_1, Q_2, Q_3, Q_4 equals the change Δf_0.

The relaxation pattern has been shown in Fig. 13.25.

A set of five residuals at a point O which satisfies the expression

$$Q_1 = Q_2 = Q_3 = Q_4 = \frac{Q_0}{4}$$

is called a vanishing set.

Generally several changes in the value of f_0 must be made before a vanishing set is obtained. In some cases it is desirable to introduce a value of Δf_0 which is greater than that necessary to make Q_0 vanish. In other cases the opposite choice is made. This is often referred to as overshooting and undershooting, respectively.

The following operation instructions may be used :

(*i*) Calculate and record all residuals in the network.

(*ii*) Select the point at which the residual is numerically the greatest, and relax it approximately over shooting or undershooting as may be indicated.

(*iii*) Calculate all the resulting changes in the residuals and as well as their final values, resulting from the relaxation of the largest residue.

(*iv*) Proceed to the point of the next greatest residual and repeat the process.

13.11.3 Block Relaxation

Sometimes it is desirable to relax a block of points simultaneously. The rule for writing the line and block-relaxation operators for Laplace or Poisson's equations is as follows :

By simultaneously altering the values of the function at a group of points in a region along a line or within a block by an amount of + 1, the residuals at all points which are directly connected with three points (*e.g.* point *a*) outside the line or block-relaxation region are altered by an amount of – 3 as shown in Figs. 13.26 and 13.27. The residuals at all points which are connected with two outside points (*e.g.* point *b*) are altered by an amount of – 2. The residual at a point such as *C* is altered by an amount of – 1 when the point is connected with one outside point. There are no changes in the residuals at points such as *d* which are not directly connected with any outside points. The residuals at points like *e*, which are outside the line or block-relaxation region but directly connected to one point within the region are altered by + 1. The residual at any

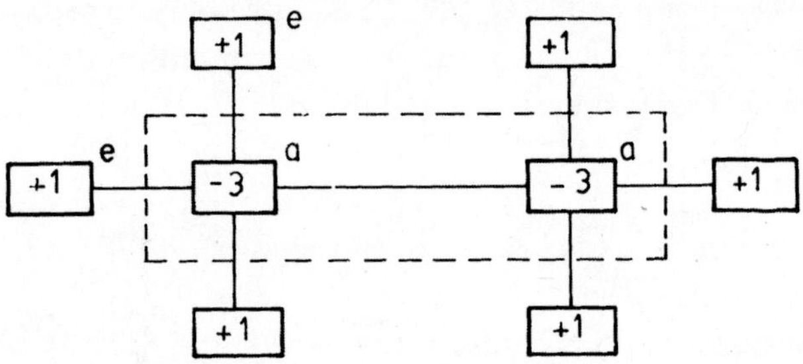

(*a*)

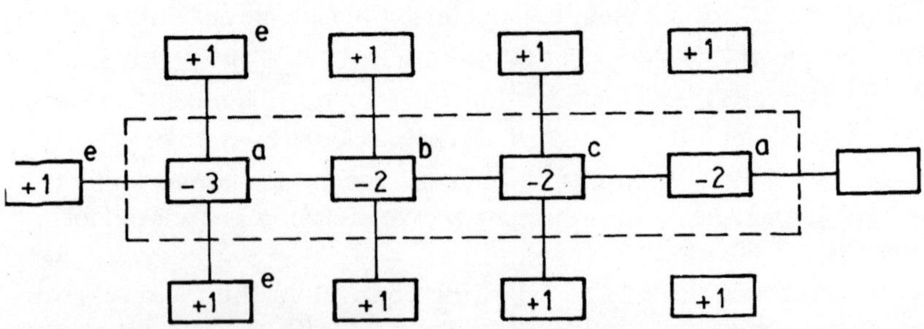

(*b*)

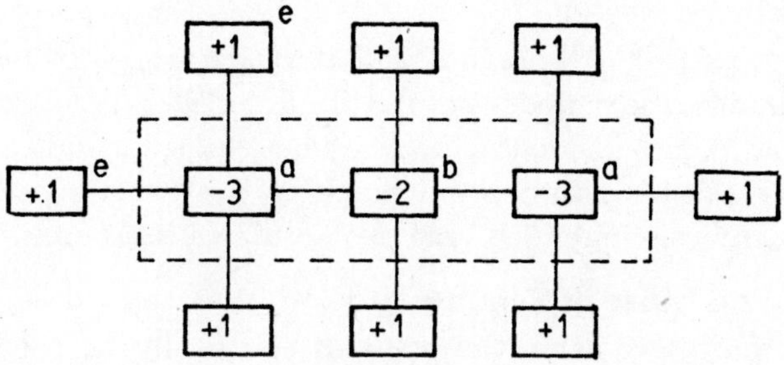

(c)

Fig. 13.26

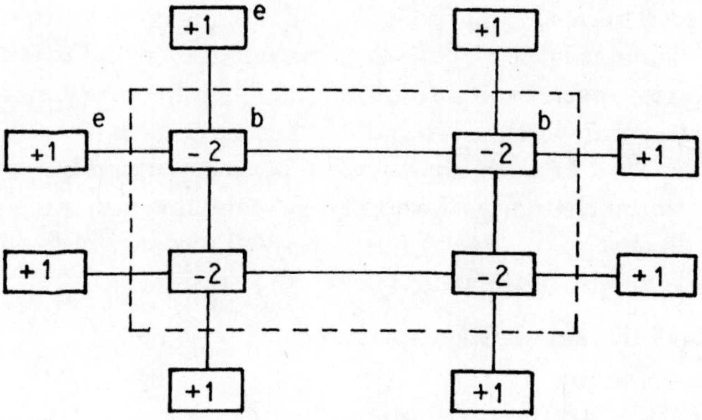

(a)

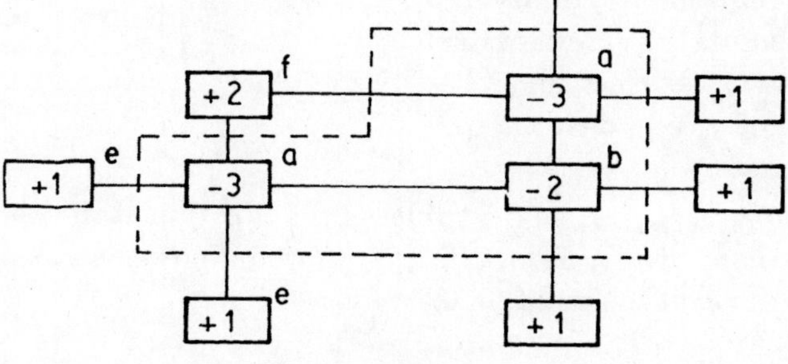

(b)

Fig. 13.27

outside point such as f which is directly connected to two inside points is altered by + 2.

The greatest advantage of block iteration is that the residuals at the points on the boundary of the block are altered, whereas the residuals at points inside the block are not changed. Judicious use of block relaxation can prevent much or the "washing back" of residuals, thus saving much time in obtaining a solution.

13.11.4 Lines of Symmetry

In many problems the solution can easily be seen to be symmetrical with respect to one or more lines because of the symmetry of the domain and boundary conditions. In solving such problems, it is unnecessary to find the unknown function over the entire area. There is no markedly new technique involved in solving such problems. It is merely necessary to remember that to preserve the symmetry, each time a point adjacent to a line of symmetry is altered, the point which is symmetrical to this point is altered at the same time. That is, such an operation is accompanied by an automatically equal change on the other side of the line of symmetry. As a result, a point on the line of symmetry will receive a change in its residual from both of the points being altered.

13.11.5 Irregular Boundaries

In many practical problems, the boundaries are curved. On dividing the region into a square mesh, points like that marked 0 in Fig. 13.28 occur, in which one or more of the associated arms, 01, 02, 03, 04 are less than the standard interval h. For such points the standard unit operators are incorrect, and it becomes necessary to develop special parameters.

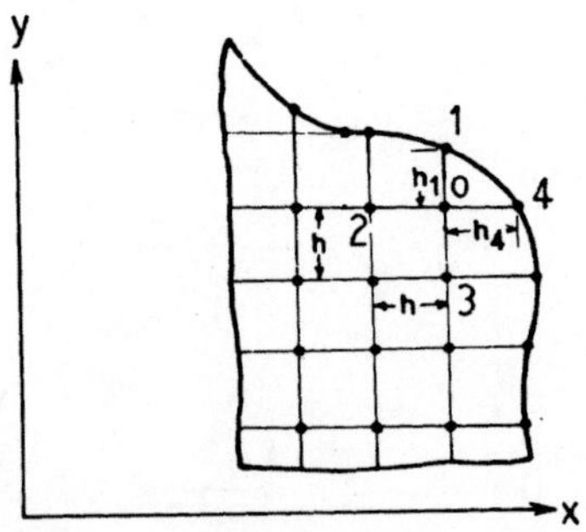

Fig. 13.28

Let f_0, f_1, f_2, f_3, f_4 be the values of the function at the points 0, 1, 2, 3, 4 respectively. Expanding $f(x, y)$ into a power series in the neighbourhood of the point (0, 0), we get

$$f(x, y) = f_0 + a_1x + a_2y + a_3x^2 + a_4y^2 + a_5xy + ...$$

At $x = 0, y = h_1$, we have $f = f_1$ and at $x = 0, y = -h, f = f_3$. Hence it follows from above equation that

$$f_1 = f_0 + a_2 h_1 + a_4 h_1^2$$
$$f_3 = f_0 - a_2 h + a_4 h^2$$

Solving, we find $$a_2 = \frac{h^2(f_1 - f_0) + h_1^2 (f_0 - f_3)}{hh_1(h + h_1)}$$

$$a_4 = \frac{h(f_1 - f_0) - h_1 (f_0 - f_3)}{hh_1 (h + h_1)}$$

Similarly considering points 2 and 4 get

$$a_1 = \frac{h^2(f_4 - f_0) + h_1^2(f_0 - f_2)}{hh_4(h + h_4)}$$

$$a_3 = \frac{h(f_4 - f_0) - h_4(f_0 - f_2)}{hh_4(h + h_4)}$$

At the point 0 ($x = 0, y = 0$) we have

$$\left(\frac{\partial^2 f}{\partial x^2}\right)_0 = 2a_3 = 2\frac{h(f_4 - f_0) - h_4(f_0 - f_2)}{hh_4(h + h_4)}$$

$$\left(\frac{\partial^2 f}{\partial y^2}\right)_0 = 2a_4 = 2\frac{h(f_1 - f_0) - h(f_0 - f_3)}{hh_1(h + h_1)}$$

So that $(\nabla^2 f)_0 = \left(\frac{\partial^2 f}{\partial x^2} + \frac{\partial^2 f}{\partial y^2}\right)$ can be written as

$$h^2\left(\frac{\partial^2 f}{\partial x^2} + \frac{\partial^2 f}{\partial y^2}\right)_0 = \frac{2}{\alpha_1(1 + \alpha_1)} f_1 + \frac{2}{1 + \alpha_4} f_2 + \frac{2}{1 + \alpha_1} f_3 + \frac{2}{\alpha_4(1 + \alpha_4)} f_4 - \left(\frac{2}{\alpha_1} + \frac{2}{\alpha_4}\right) f_0$$

where $$\alpha_1 = \frac{h_1}{h}, \alpha_4 = \frac{h_4}{h} \qquad \text{...(13.63)}$$

When $0 < \alpha_1 < 1, 0 < \alpha_4 < 1$ then Eq. (13.63) may we written as

$$h^2(\nabla^2 f)_0 \simeq Af_1 + Bf_2 + Cf_2 + Df_4 - (E + F)f_0 \qquad \text{...(13.64)}$$

where $A = \frac{2}{\alpha_1(1 + \alpha_1)}, B = \frac{2}{1 + \alpha_4}, C = \frac{2}{1 + \alpha_1}$,

and $D = \frac{2}{\alpha_4(1 + \alpha_4)}, E = \frac{2}{\alpha_1}, F = \frac{2}{\alpha_4}$

The finite-difference approximation formulas for other differential equations near a curved boundary can be derived in a like manner.

The finite difference equivalent for the Laplace operator may then be used in the same manner as the standard (– 4, 1, 1, 1, 1) operator. For the point 0, the residual operator, *i.e.* the operator from which the residual can be computed, is as shown in Fig. 13.29, where the numbers A_0, ..., F_0 will all be known.

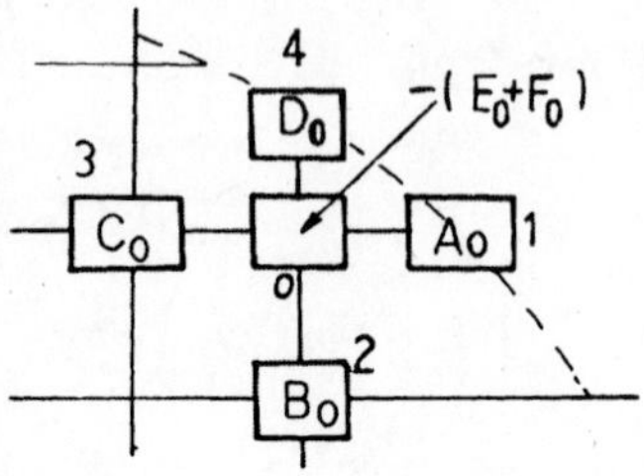

Fig. 13.29

If relaxation operators can be obtained as follows. If the value of the function is altered by + 1 at point 0, the residual there is altered by $-(E_0 + F_0)$. At the same time at position 2 and 3 as shown in Fig. 13.30, the residual alters by + 1. No residuals are recorded at the boundary points 1 and 4.

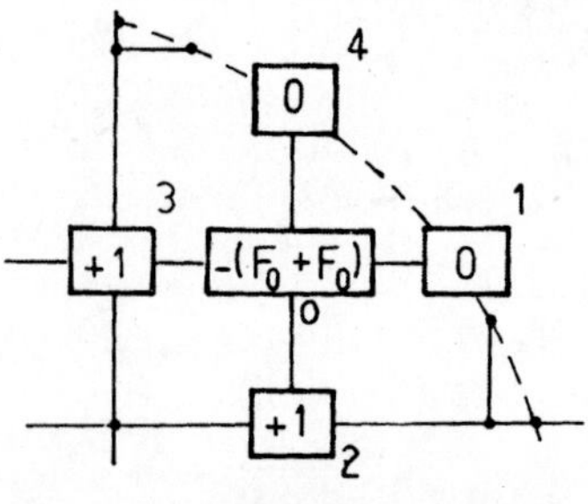

Fig. 13.30

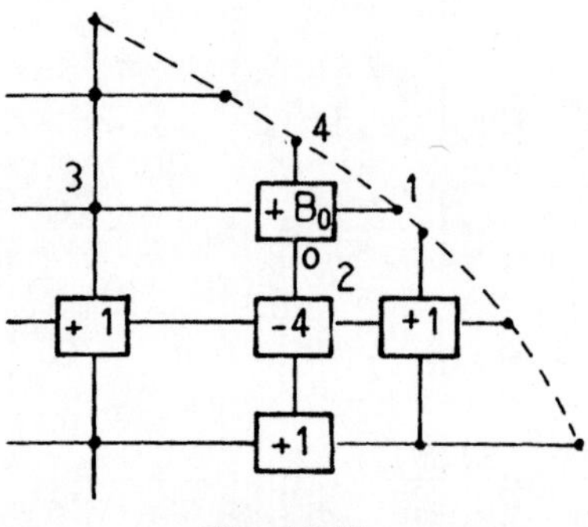

Fig. 13.31

If we alter the value of the function at 2, the residual at 0 will change B_0 times the value altered at 2 as shown in Fig. 13.31.

The relaxation process may then be carried out in the manner described earlier.

Example 13.23 *Analyze the torsion of a bar of square section (Fig. 13.32) using finite difference technique.*

Solution. The governing partial differential equation is

$$\frac{\partial^2 \phi}{\partial x^2} + \frac{\partial^2 \phi}{\partial y^2} = -2G\theta$$

ϕ is assigned the value of zero at boundaries.

Finite difference equation about the point 0 is :

$$\phi_1 + \phi_2 + \phi_3 + \phi_4 - 4\,\phi_0 = -2G\theta h^2$$

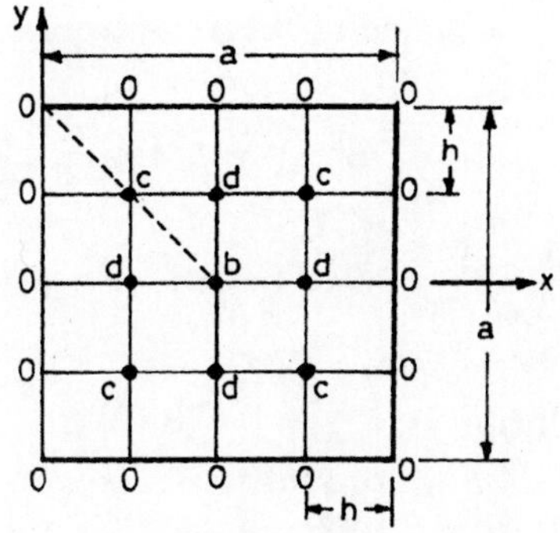

Fig. 13.32

Let the square shaft be divided into 16 small squares of size h. We consider only one fourth of the shaft to take into account the symmetry. For the nodes b, c and d, we get

$$4\phi_d - 4\phi_b = -2\,G\theta h^2$$
$$2\phi_d - 4\phi_c = -2\,G\theta h^2$$
$$2\phi_c + \phi_b - 4\phi_d = -2G\theta h^2$$

Solving simultaneous equations, we get

$$\phi_b = 2.25\,G\theta h^2,\ \phi_c = 1.375\,G\theta h^2,\ \phi_d = 1.175\,G\theta h^2$$

x	$\phi / G\theta h^2$	$\Delta / G\theta h^2$	$\Delta^2 / G\theta h^2$	$\Delta^3 / G\theta h^2$	$\Delta^4 / G\theta h^2$
0	0				
		1.75			
h	1.75		– 1.25		
		0.5		0.25	
$2h$	2.25		– 1.0		– 0.5
		– 0.5		– 0.25	
$3h$	1.75		– 1.25		
		– 1.75			
$4h$	0				

Using Newton's interpolation formula,

$$\phi(x) = \phi_0 + \frac{x}{h}\Delta\phi_0 + \frac{x(x-h)}{2!\,h^2}\Delta^2\phi_0 + \frac{x(x-h)(x-2h)}{3!\,h^3}\Delta^3\phi_0 + \ldots\ldots$$

where $\Delta\phi_0$'s are the forward differences, calculated at $x = 0$

$$\Delta\phi_0 = \phi_d - \phi_0 = (1.75 - 0)\,G\theta h^2 = 1.75\,G\theta h^2$$
$$\Delta^2\phi_0 = \phi_b - 2\,\phi_d + \phi_0 = -1.25\,G\theta h^2$$
$$\Delta^3\phi_0 = \phi_d - 3\phi_b + 3\phi_d - \phi_0 = 0.25\,G\theta h^2$$
$$\Delta^4\phi_0 = \phi_0 - 4\,\phi_d + 6\phi_b - 4\phi_d + \phi_0 = -0.5\,G\theta h^2$$

τ_{max} occurs at $x = 0$ and is obtained from $\left(\frac{\partial\phi}{\partial x}\right)_0$

$$\left(\frac{\partial\phi}{\partial x}\right)_0 = \frac{1}{h}\left(\Delta\phi_0 - \frac{1}{2}\Delta^2\phi_0 + \frac{1}{3}\Delta^3\phi_0 - \frac{1}{4}\Delta^4\phi_0 + \ldots\right)$$

$$= \frac{31}{12} G\theta h = 0.646\, G\theta a$$

$$\tau_{max} = 0.646\, G\theta a$$

The exact solution is, $\tau_{max} = 0.678\, G\theta\, a$

Example 13.24 *Find the shear stresses at the points A and B of an elliptical shaft (Fig. 13.33) for which a = 15 mm, b = 10 mm and h = 5 mm.*

Solution.

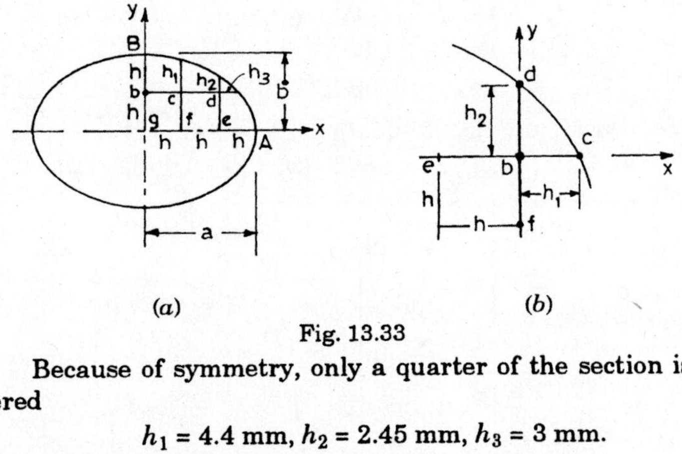

(a) (b)

Fig. 13.33

Because of symmetry, only a quarter of the section is considered

$$h_1 = 4.4 \text{ mm}, h_2 = 2.45 \text{ mm}, h_3 = 3 \text{ mm}.$$

For a curved boundary,

$$h^2(\nabla^2\phi)_b = \frac{2\,\phi_c}{\alpha_1(1+\alpha_1)} + \frac{\phi_d}{\alpha_2(1+\alpha_2)} + \frac{2\phi_e}{1+\alpha_1} + \frac{2\phi_f}{1+\alpha_2} - \left(\frac{2}{\alpha_1} + \frac{2}{\alpha_2}\right)\phi_b$$

where $\alpha_1 = \frac{h_1}{h}, \alpha_2 = \frac{h_2}{h}$

$$h^2\left(\frac{\partial^2\phi}{\partial x^2}\right)_b = \frac{2\,\phi_c}{\alpha_1(1+\alpha_1)} + \frac{2\alpha_e}{1+\alpha_1} - \frac{2\phi_b}{\alpha_1}$$

$$h^2\left(\frac{\partial\phi}{\partial y^2}\right)_b = \frac{\phi_d}{\alpha_2(1+\alpha_2)} + \frac{2\phi_f}{1+\alpha_2} - \frac{2\phi_b}{\alpha_1}$$

At points *b, e, f, g* and *c, d*, the six equations are :

$$\begin{bmatrix} 2 & 0 & 0 & 0 & 2 & -4 \\ 0 & 2 & 0 & 1 & -4 & 1 \\ 0 & 0 & 2 & -4 & 1 & 0 \\ -4 & 2 & 0 & 0 & 0 & 1 \\ 1 & -4.27 & 1 & 0 & 1.06 & 0 \\ 0 & 1.25 & -7.41 & 1.34 & 0 & 0 \end{bmatrix} \begin{Bmatrix} \phi_b \\ \phi_c \\ \phi_d \\ f_e \\ f_f \\ f_g \end{Bmatrix} = (-2h^2 G\theta) \begin{Bmatrix} 1 \\ 1 \\ 1 \\ 1 \\ 1 \\ 1 \end{Bmatrix}$$

Solving these equations, we get

$$\phi_b = 2.075\ G\theta h^2\ ;\ \phi_c = 1.767\ G\theta h^2,\ \phi_d = 0.843\ G\theta h^2$$

$$\phi_e = 1.536\ G\theta h^2,\ \phi_f = 2.459\ G\theta h^2,\ \phi_g = 2.767\ G\theta h^2$$

$$\Delta\phi_B = 2.075\ G\theta h^2,\ \Delta^2\phi_B = -1.383 G\theta h^2$$

$$\Delta^3\phi_B = -0.001\ G\theta h^2,\ \Delta^4\phi_B = 0.002\ G\theta h^2$$

For point *A*,

$$\Delta\phi_A = 1.536\ G\theta h^2,\ \Delta^2\phi_A = -0.613\ G\theta h^2$$

$$\Delta^3\phi_A = -0.002\ G\theta h^2,\ \Delta^4\phi_A = 0.001\ G\theta h^2$$

$$\Delta^5\phi_A = 0.001\ G\theta h^2,\ \Delta^6\phi_A = -0.002\ G\theta h^2$$

Thus

$$\tau_A = \left(\frac{\partial\phi}{\partial x}\right)_A = \left(1.536 + \frac{0.613}{2} - \frac{0.002}{3} - \frac{0.001}{4} + \frac{0.001}{5} + \frac{0.002}{6}\right) G\theta h$$

$$= 1.842\ G\theta h = 0.614\ G\theta a = 0.00921\ G\theta$$

$$\tau_B = \left(\frac{\partial\phi}{\partial y}\right)_B = \left(2.075 + \frac{1.383}{2} - \frac{0.001}{3} + \frac{0.002}{4}\right) G\theta h$$

$$= 2.766\ G\theta h = 1.383\ G\theta b = 0.01383\ G\theta$$

13.12 Finite Element Method

This is a relatively new technique in which the complete solid under consideration is assumed to consist of a number of small elements, whose behaviour can be readily formulated. The whole assembly is then analysed by considering the equilibrium and continuity of displacement at the interfaces between adjoining elements. This has the advantage of retaining the physical characteristics of the problem throughout the process of its solution. This leads to a large number of simultaneous linear equations that have to be solved in order to find the values of the stress resultants at the edges of each element. The method has become of immense use due to the availability of high-speed electronic digital computers.

13.12.1 Finite Element Concept

The basis of the finite element method is the representation of a body or a structure by an assemblage of sub-divisions called Finite Elements. These finite elements are obtained by means of fictitious cuts through the original structure. Adjoining elements may be thought of as being connected at common points, termed as nodes or nodal points. Then simple functions are chosen to approximate the variation of the actual displacements over each finite element. Such assumed functions are called Displacement Functions of Displacement Models. The unknown magnitudes of the displacement functions are the displacements at the nodal points. Hence, the final solution will yield the approximate displacements at the nodal points. The displacement model can be expressed in various simple forms, such as polynomials and trigonometric functions.

Instead of the displacement model, other type of models like the Stress model or Hybride model may also be used. In the stress model, either stresses or stress functions are approximated over individual elements. The stresses need not be continuous over interelement boundaries. In the Hybrid model, the stresses are assumed as in the stress model ; however, instead of consideration of equilibrium of boundary tractions, the boundary displacements are now interpolated in terms of a finite number of generalized boundary displacements.

The idealized model is now analysed by using the principle of mechanics, such as the principle of minimum potential energy (for Displacement model) or the principle of minimum complementary energy (for Stress model), to obtain the set of equilibrium equations for each element. The displacements or stresses of the nodes are introduced as the primary indeterminate quantities. Thus nodal point forces are expressed in terms of nodal point displacements or stresses. The equilibrium equations for the entire body are then obtained by combining the equations for the individual elements in such a way that continuity of displacements is preserved at the interconnecting nodes. These equations are modified for the given displacement boundary conditions and then solved to obtain the unknown displacements.

13.12.2 Basic Steps

The basic six steps to be followed for the analysis of the problem are as follows :

1. *Discretization of the continuum.* The discretization may be simply described as the process in which the given body is subdivided into an equivalent system of finite elements. The finite elements may be triangles or quadrilaterals for a two-dimensional continuum or tetrahedra, rectangular prisms or hexahedra for there dimensional bodies.

2. *Selection of the displacement models.* The assumed displacement functions or models represent only approximately the actual or exact distribution of the displacements. Commonly used displacement function is a polynomial. The following three factors are considered while selecting displacement models.

(*i*) The degree of polynomial used in the displacement function must be chosen.

(*ii*) The particular displacement magnitudes that describe the model must be selected.

(*iii*) The model should satisfy certain requirements such that the numerical results approach the correct solution.

3. *Derivation of the element stiffness matrix.* The stiffness matrix consists of the coefficients of the equilibrium equations derived from the material and geometric properties of an element and obtained by using the principle of minimum potential energy or virtual work principle. The stiffness relates the displacement at the nodal points to the applied forces at the nodes. The distributed forces applied to the structure are converted into equivalent nodal forces. The equilibrium relation between the element stiffness matrix $[k]$, nodal force vector $\{F\}^e$ and nodal. displacement vector $\{\delta\}^e$ is expressed as a set of simultaneous linear algebraic equations :

$$[k]\,\{\delta\}^e = \{F\}^e$$

The stiffness matrix for an element depends upon the displacement model, the geometry of the element and the local material properties or constitutive relations.

4. *Assembly for the overall discretized continuum.* In this the global stiffness of the entire body $[K]$ and the global load vector for the continuum $\{R\}$ is obtained from the element stiffness matrix $[k]$ and load vector $\{F\}^e$ using the direct stiffness method. The basis for

the assembly method is that the nodal interconnections require the displacements at a node to be the same for all elements adjacent to that node. Thus the overall equilibrium equation is

$$[K]\{\delta\} = \{R\}$$

where $\{\delta\}$ is the nodal displacement vector for the entire body.

5. *Solution for the unknown displacements.* To solve for the unknown displacements, generally the Gauss elimination method is used.

6. *Computation of the element strains and stresses.* The strains and stresses can be obtained from the derivatives of the displacements.

In order to understand the theory and applications of the finite element method, the working knowledge in the following areas is essential.

(*i*) Matrix algebra
(*ii*) Solid mechanics
(*iii*) Variational methods and
(*iv*) Computer skills.

13.12.3 Displacement Approach

Now the procedure of displacement approach will be explained in brief for plane stress analysis.

(*i*) *Displacement models.* The most common form of the displacement model is that of a polynomial. Three are two principal reasons for this, namely (*a*) It is easy to handle mathematics of polynomial in formulating the desired properties of an element and (*b*) by controlling the order of the polynomial, desired approximation of the solution can be achieved. The general form of the polynomial is

$$u(x) = \alpha_1 + \alpha_2 x + \alpha_3 x^2 + \ldots + \alpha_{n+1} x^n$$

where α's are known as the generalised coordinates or "generalised displacement amplitudes". The shape of the displacement model is determined by the number of terms in the polynomial and the amplitude is governed by the magnitudes of the generalised displacements. For example for a three-dimensional displacement model of nth order the displacement components are expressed as :

$$\{f\} = \begin{Bmatrix} u_x(x, y, z) \\ u_y(x, y, z) \\ u_z(x, y, z) \end{Bmatrix} = [\phi]\{\alpha\}$$

$$= \begin{bmatrix} \{\phi_1\}^T & \{0\}^T & \{0\}^T \\ \{0\}^T & \{\phi_1\}^T & \{0\}^T \\ \{0\}^T & \{0\}^T & \{\phi_1\}^T \end{bmatrix} \{\alpha\}$$

where $\{\phi_1\}^T = [1\ x\ y\ z\ zx\ x^2\ xy\ y^2\ yz\ z^2\ \dots\ z^n]$

$$\{\alpha\}^T = [\alpha_1\ \alpha_2\ \alpha_3\ \dots\ \alpha_{3m}]$$

By truncation the polynomial of desired degree can be obtained.

For two-dimensional analysis, a linear model is

$$\{\phi_1\}^T = [1\ x\ y]$$

$$\{\alpha_1\}^T = [\alpha_1\ \alpha_2\ \dots\ \alpha_6]$$

and a quadratic model is

$$\{\phi_1\}^T = [1\ x\ y\ x^2\ xy\ y^2]$$

$$\{\alpha\}^T = [\alpha_1\ \alpha_2\ \dots\ \alpha_{12}]$$

(*ii*) *Displacement function.* Substituting the coordinates of the nodes of an element, we can determine the α's. Thus we get

$$\{f\} = [N]\{\delta\}^e$$

where $[N]$ is generally called the shape function matrix and $\{\delta\}^e$ is the displacement at the nodes.

In the case of a two-dimensional triangular element

$$\{f\} = \begin{Bmatrix} u_x \\ u_y \end{Bmatrix} \begin{bmatrix} [I]N_i, & [I]N_j, & [I]N_m \end{bmatrix} \begin{Bmatrix} \delta_i \\ \delta_j \\ \delta_m \end{Bmatrix}$$

where $[I] = \begin{bmatrix} 1 & 0 \\ 0 & 1 \end{bmatrix}$ is the identity matrix.

In general, $\{f\} = [N_i,\ N_j,\ N_m,\ \dots.] \begin{Bmatrix} \delta_i \\ \delta_j \\ \delta_m \\ \vdots \end{Bmatrix}$

$$N_i(x_i, y_i) = I$$

and $N_i(x_j, y_j) = N_i(x_m, y_m) = 0$

(*iii*) *Strains.* Strains are necessary to be calculated to determine the potential energy to be used in the variational approach to calculate the element properties. In general the strain-displacement relationship can be written as

$$\{\varepsilon\} = [B]\{\delta\}^e$$

(*iv*) *Stresses.* In general elastic relationships between stresses and strains is

$$\{\sigma\} = [D][\{\varepsilon\} - \{\varepsilon_0\}] + \{\sigma_0\}$$

where $[D]$ is the elasticity matrix, $\{\varepsilon_0\}$ is the initial strain due to temperature changes, shrinkage etc. $\{\sigma_0\}$ is the initial stress like residual stresses.

(*v*) *Equivalent nodal forces.* The load force vector is

$$\{F\}^e = \begin{Bmatrix} F_i \\ F_j \\ F_m \\ \vdots \end{Bmatrix}$$

Each of the forces $\{F_j\}$ contain the same number of components as the corresponding nodal displacements $\{\delta_i\}$. Thus for a two dimensional element

$$\{F_i\} = \begin{Bmatrix} F_x \\ F_y \end{Bmatrix}$$

The distributed loads $\{p\}$ are defined as those acting on a unit volume of material within the element with directions corresponding to those of the displacements $\{f\}$ at the point. In the direct method of formulation, the distributed load on the surface of an element are divided among these nodes corresponding to the loads acting on a region tributary to the node. They are called the lumped loads. In the second method, the variational principle itself is used to calculate the nodal loads.

(*vi*) *Element stiffness matrix.* The potential energy can be written as

$$\pi = \frac{1}{2}\iiint_V (\varepsilon^T \sigma - 2u^T \overline{B})\, dV - \iint_{S_1} u^T T dS_1$$

where $\overline{B}$ is the body force distribution and T is the tractive forces. S_1 is the surface area. Substituting the relations between the strains and displacements and expressing the displacements in terms of shape functions, we obtain

$$\pi = \frac{1}{2}\iiint_V \Big[\{\delta\}^{eT}[B]^T[D][B]\{\delta\}^e - \{\delta\}^{eT}[B]^T\,[D]\{\varepsilon_0\} + \{\delta\}^{eT}[B]^T\{\sigma_0\} - 2\{\delta\}^{eT}\,[N]^T[\overline{B}]\,\Big]\, dV - \iint_{S_1} \{\delta\}^T[N]^T[T]dS_1$$

Applying the variational principle

$$d\{\delta\}^{eT}\left\{\left[\iiint_V [B]^T[D][B]\{\delta\}^e - [B]^T[D]\{\varepsilon_0\} + [B]^T\{\sigma_0\}\right] dV - \iiint_V [N]^T[\overline{B}]\, dV - \iint_{S_1} [N]^T[T]ds_1\right\} = 0$$

Since the variations of the nodal displacements $d\{\delta\}^e$ is arbitrary, the expression in the parenthesis must vanish. Hence we get the equilibrium equation

$$[k]^e\{\delta\}^e = \{F\}^e$$

where

$$[k]^e = \iiint_V [B]^T[D][B]dV \quad \text{...(13.64)}$$

is the element stiffness matrix.

$$\{F\}^e = \int_V B^T D\varepsilon_0 dV - \int_V B^T\sigma_0 dV + \int_V N^T\,\overline{B}\,dV + \int_{S_1} N^T T dS_1 \quad \text{...(13.65)}$$

$\int_V B^T D\varepsilon_0\, dV$ = Equivalent load due to initial strains

$\int_V B^T\sigma_0 dV$ = Initial stresses

$\int_V N^T\,\overline{B}\, dV$ = Load due to body forces

$\int_{S_1} NT^T dS_1$ = Load due to tractive forces.

For external concentrated forces acting at the nodes, the matrix

$$\{R\} = \begin{Bmatrix} R_1 \\ R_2 \\ \vdots \\ R_n \end{Bmatrix}$$

is added to the consideration of equilibrium at the nodes. For a two-dimensional plane stress case

$$\{\varepsilon_0\} = \begin{Bmatrix} \varepsilon_{x0} \\ \varepsilon_{y0} \\ \gamma_{xy0} \end{Bmatrix}$$

and for strains associated with linear thermal expansion, we get

$$\{\varepsilon_0\} = \begin{Bmatrix} \alpha\theta^e \\ \alpha\theta^e \\ 0 \end{Bmatrix}$$

where θ^e is the temperature rise and α is the coefficient of linear expansion.

If the initial stress system is self equilibriating then

$$\int B^T \sigma_0 \, dV = 0.$$

(vii) Convergence criteria. For the approximate displacement solution to converge, the following three conditions must be satisfied.

(a) The displacement function chosen should be such that it does not permit straining of an element to occur when the nodal displacements are caused by a rigid body displacement. This amounts to adding a constant term in the displacement model.

(b) The displacement models must include constant strain states of the element

(c) The displacement functions should be so chosen that the strains at interface between elements are finite. Elements satisfying this condition are called compatible or conforming type of elements.

(viii) Selection of the order of the polynomial. The polynomials selected should be independent of the orientation of the local coordinate system. This property of the model is known as Geometric Invariance. One way of achieving this is to consider Pascal triangle for the variable terms of the two-dimensional polynomial.

1 ⟶ Constant terms

$x \; \vdots \; y$ ⟶ Linear terms

$x^2 \; xy \; y^2$ ⟶ Quadratic terms

$x^3 x^2y \; \vdots \; xy^2 y^3$ ⟶ Cubic terms

$x^4 x^3y x^2y^2 \; xy^3 y^4$ ⟶ Quartic terms

$x^5 \; x^4y \; x^3y^2 \; \vdots \; x^2y^3 \; xy^4 \; y^5$ ⟶ Quintic terms

Axis of symmetry

Terms from both sides of the axis of symmetry should be included simultaneously. Further the total number of generalised coordinates for an element must be equal to or greater than the number of joints or external degrees of freedom of the element.

13.12.4 Transformation of coordinate axes

In order to transform the stiffness matrices from the local to the datum or global coordinate system the relationship between the element displacements $\{\delta\}$ in the local system and the element displacements $\{\bar{\delta}\}$ in the datum system can be expressed as :

$$\{\delta\} = [\lambda]\,\{\bar{\delta}\} \qquad ...(13.66)$$

where $[\lambda]$ is a matrix of coefficients obtained by resolving datum displacements in the directions of local coordinates. As the corresponding force components must perform the same amount of work in either system.

$$\{F\}^T\,\{\delta\} = \{\bar{F}\}^T\,\{\bar{\delta}\}$$

or
$$\{F\}^T[\lambda]\{\bar{\delta}\} = \{\bar{F}\}^T\{\bar{\delta}\}$$

or
$$\{\bar{F}\} = [\lambda]^T\{F\} \qquad ...(13.67)$$

Transformation of stiffness takes place as

$$\{F\} = [K]\{\delta\}$$

$$\therefore \quad \{\bar{F}\} = [\lambda]^T\,[K]\{\delta\} = [\lambda]^T[K][\lambda]\{\bar{\delta}\} = [\bar{K}]\,\{\bar{\delta}\}$$

where
$$[\bar{K}] = [\lambda]^T[K][\lambda] \qquad ...(13.68)$$

13.12.5 Plane Stress Analysis

For plane stress problems $\sigma_z = \tau_{yz} = \tau_{zx} = 0$ and the displacement components are u, v. Two types of plane stress and plane strain elements are generally used, *viz.*, the triangular and quadrilateral elements, both with straight and curved sides.

(*a*) *Stiffness of triangular elements.* This element is also called constant strain triangular element. The triangular element of arbitrary shape and with nodal point at the corners is shown in Fig. 13.34. The element is located arbitrarily in relation to a rectangular coordinate system.

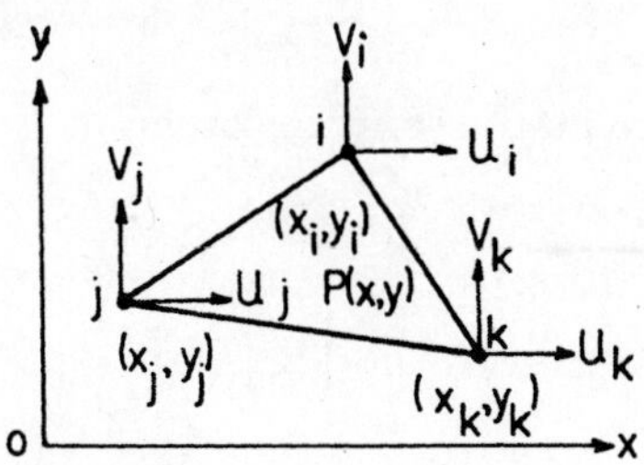

Fig. 13.34

The six components of element displacements are

$$\{\delta\}^e = \{u_i\; v_i\; u_j\; v_j\; u_k\; v_k\}^T \qquad ...(13.69)$$

For linear transformation, we assume

$$u = \alpha_1 + \alpha_2 x + \alpha_3 y$$

$$v = \alpha_4 + \alpha_5 x + \alpha_6 y$$

where $\alpha_1, \alpha_2, ..., \alpha_6$ may be expressed as functions of the displacements at the u_i, v_i ; u_j, v_j and u_k, v_k and

$$\begin{Bmatrix} u_i \\ u_j \\ u_k \\ v_i \\ v_j \\ v_k \end{Bmatrix} = \begin{bmatrix} 1 & x_i & y_i & 0 & 0 & 0 \\ 1 & x_j & y_j & 0 & 0 & 0 \\ 1 & x_k & y_k & 0 & 0 & 0 \\ 0 & 0 & 0 & 1 & x_i & y_i \\ 0 & 0 & 0 & 1 & x_j & y_j \\ 0 & 0 & 0 & 1 & x_k & y_k \end{bmatrix} \begin{Bmatrix} \alpha_1 \\ \alpha_2 \\ \alpha_3 \\ \alpha_4 \\ \alpha_5 \\ \alpha_6 \end{Bmatrix}$$

or $$\{\delta\}^e = [A]\{\alpha\}$$

Solving this system of equations, we get

$$\alpha_1 = [u_i(x_jy_k - x_ky_i) + u_j(x_ky_i - x_iy_k) + u_k(x_iy_j - x_jy_i)]/\Delta$$
$$\alpha_2 = u_i(y_j - y_k) + u_j(y_k - y_i) + u_k(y_i - y_k)]/\Delta$$
$$\alpha_3 = [u_i(u_k - x_j) + u_j(x_i - x_k) + u_k(y_j - y_i)]/\Delta$$
$$\alpha_4 = [v_i(x_jy_k - x_ky_i) + v_j(x_ky_i - x_iy_k) + v_k(x_iy_i - x_jy_j)]/\Delta$$
$$\alpha_5 = [v_i(y_j - y_k) + v_j(y_k - y_i) + v_k(y_i - y_j)]/\Delta$$
$$\alpha_6 = [v_i(x_k - x_j) + v_j(x_i - x_k) + v_k(x_j - x_i)]/\Delta$$
$$\Delta = (x_j y_k + x_iy_j + x_k y_i) - (x_jy_i + x_iy_k + x_ky_j)$$
$$= \text{area of the triangle.}$$

The general displacement of a point P becomes

$$u = \frac{u_j}{\Delta}[(x_jy_k - x_ky_i) + (y_i - y_k)x + (x_k - x_i)y]$$
$$+ \frac{u_i}{\Delta}(x_ky_i - x_iy_k) + (y_k - y_i)x + (x_i - x_k)y$$
$$+ \frac{u_k}{\Delta}[(x_iy_j - x_jy_i) + (y_i - y_j)x + (x_j - x_i)y]$$

and a similar expression for v.

Or we can write

$$\{f\} = \begin{Bmatrix} u \\ v \end{Bmatrix}$$
$$= \begin{bmatrix} N_1 & 0 & N_2 & 0 & N_3 & 0 \\ 0 & N_1 & 0 & N_2 & 0 & N_3 \end{bmatrix} \{\delta\}^e \quad ...(13.70)$$

where $$N_1 = \frac{1}{\Delta}[x_jy_k - x_ky_i) + (y_i - y_k)x + (x_k - x_j)y]$$

$$N_2 = \frac{1}{\Delta}[x_ky_j - x_iy_k) + (y_k - y_i)x + (x_i - x_k)y]$$

$$N_3 = \frac{1}{\Delta}\left[(x_i y_j - x_j y_i) + (y_i - y_j)x + (x_j - x_i)y\right]$$

are called the shape functions.

(*b*) *Element stresses and strains.*

Now $$\{\varepsilon\} = \begin{Bmatrix} \varepsilon_x \\ \varepsilon_y \\ \gamma_{xy} \end{Bmatrix}$$

$$= \begin{Bmatrix} \dfrac{\partial u}{\partial x} \\ \dfrac{\partial v}{\partial y} \\ \dfrac{\partial u}{\partial y} + \dfrac{\partial v}{\partial x} \end{Bmatrix} \qquad \text{...(13.71)}$$

$$\varepsilon_x = \frac{\partial u}{\partial x} = [u_i(y_j - y_k) + u_j(y_k - y_i) + u_k(y_i - y_j)]/\Delta$$

$$\varepsilon_y = \frac{\partial v}{\partial y} = [v_i(x_k - x_j) + v_j(x_i - x_k) + v_k(x_j - x_i)]/\Delta$$

$$\gamma_{xy} = \frac{\partial u}{\partial y} + \frac{\partial v}{\partial x}$$

$$= [u_i(x_k - x_j) + u_j(x_i - x_k) + u_k(x_j - x_i) + v_i(y_j - y_k) + v_j(y_k - y_i) + v_k(x_j - x_i)]/\Delta$$

Thus $$\{\varepsilon\} = [B]\{\delta\}^e$$

where $$B = \frac{1}{\Delta}\begin{bmatrix} (y_j - y_k) & 0 & (y_k - y_i) & 0 & (y_i - y_j) & 0 \\ 0 & (x_k - x_j) & 0 & (x_i - x_k) & 0 & (x_j - x_i) \\ (x_k - x_j) & (y_j - y_k) & (x_i - x_k) & (y_k - y_i) & (x_j - x_i) & (y_i - y_j) \end{bmatrix}$$

and $$\{\delta\}^e = \begin{Bmatrix} u_i \\ v_i \\ u_j \\ v_j \\ u_k \\ v_k \end{Bmatrix}$$

If there are some initial strains due to temperature rise θ^e then

$$\{\varepsilon_0\} = \begin{Bmatrix} \varepsilon_{x_0} \\ \varepsilon_{y_0} \\ \gamma_{xy_0} \end{Bmatrix} = \begin{Bmatrix} \alpha\theta^e \\ \alpha\theta^e \\ 0 \end{Bmatrix} \text{ for the plane stress case.}$$

For the plane strain case we should multiply by $(1 + \nu)$.

By defining an angle β between the x' and x axes, the transformation matrix is

$$[\lambda] = \begin{bmatrix} \cos^2\beta & \sin^2\beta & -\sin 2\beta \\ \sin^2\beta & \cos^2\beta & \sin 2\beta \\ \sin\beta\cos\beta & -\sin\beta\cos\beta & \cos^2\beta - \sin^2\beta \end{bmatrix}$$

The general relationship between stresses and strains can be written as

$$\{\sigma\} = \begin{Bmatrix} \sigma_x \\ \sigma_y \\ \tau_{xy} \end{Bmatrix} = [D][\{\varepsilon\} - \{\varepsilon_0\}] + \{\sigma_0\} \qquad \text{...(13.72)}$$

where $[D]$ is the elasticity matrix.

$$[D] = \frac{E}{1-\nu^2}\begin{bmatrix} 1 & \nu & 0 \\ \nu & 1 & 0 \\ 0 & 0 & \dfrac{1-\nu}{2} \end{bmatrix} \text{ for plane stress case} \qquad \text{...(13.73)}$$

$$[D] = \frac{E(1-\nu)}{(1+\nu)(1-2\nu)}\begin{bmatrix} 1 & \dfrac{\nu}{1-\nu} & 0 \\ \dfrac{\nu}{1-\nu} & 1 & 0 \\ 0 & 0 & \dfrac{1-2\nu}{2(1-\nu)} \end{bmatrix} \qquad \text{...(13.74)}$$

for plane strain case.

(c) *The stiffness matrix.* If body forces are absent then work done by external forces is

$$W = \frac{1}{2}[u_i\ v_i\ u_j\ v_j\ v_k\ v_k]\begin{Bmatrix} X_i \\ Y_i \\ X_j \\ Y_j \\ X_k \\ Y_k \end{Bmatrix}$$

$$= \frac{1}{2}\{\delta\}^{e^T}\{F\}^e$$

where X and Y are the components of external forces in x and y directions respectively. Strain energy stored per unit volume

$$u = \frac{1}{2}(\sigma_x\,\varepsilon_x + \sigma_y\,\varepsilon_y + \tau_{xy}\,\gamma_{xy}) = \{\varepsilon\}^T\{\sigma\}$$

$$= \frac{1}{2}[B]^T\{\delta\}^{e^T}[D][B]\{\delta\}^e$$

$\therefore$ $U = W$ gives

$$\{F\}^e = \int_V [B]^T \{\delta\}^{e^T} [D][B]dv = [K]\{\delta\}^e$$

where
$$[K] = \int_V [B]^T [D][B]dv$$

$$[K] = [B]^T [D][B] \times \text{(Area of triangle)} \times \text{Thickness}$$

or
$$K = B^T DB.\Delta.t.$$

(*d*) *Applying force to the element.* For the plane stress condition, only the force that lie in the *x-y* plane are considered. For example in Fig. 13.35, a solid is divided into elements with numbered nodes and there is a force acting at node 9 of element 4. Since

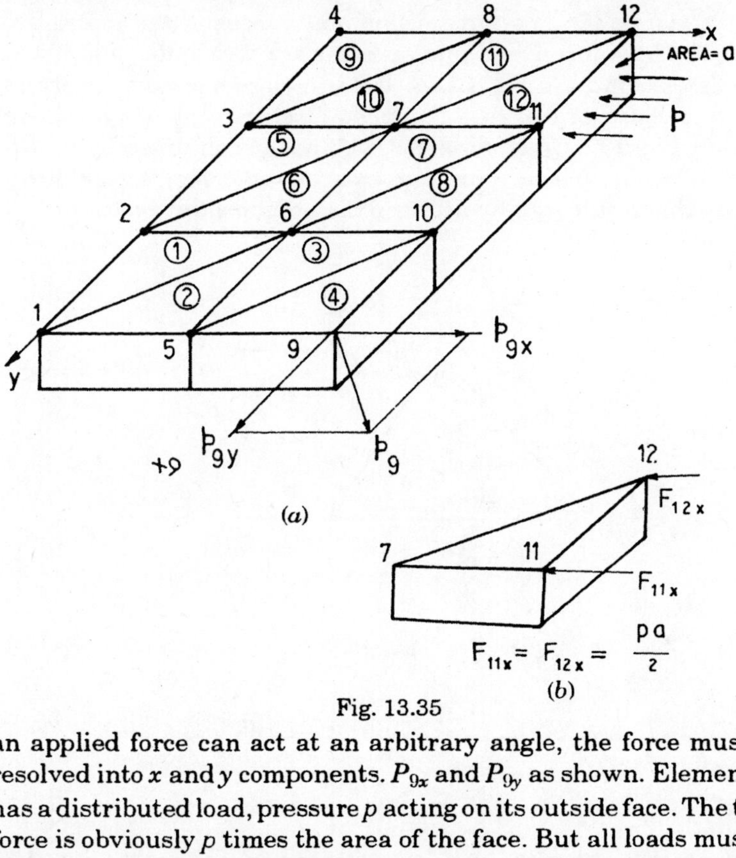

(*a*) (*b*)

Fig. 13.35

an applied force can act at an arbitrary angle, the force must be resolved into x and y components. P_{9x} and P_{9y} as shown. Element 12 has a distributed load, pressure p acting on its outside face. The total force is obviously p times the area of the face. But all loads must be applied at the nodes. We divide the load due to the pressure equally between the nodes 11 and 12.

Thus the load vector for the whole system becomes

$$\begin{Bmatrix} F_{1x} \\ F_{1y} \\ \vdots \\ \vdots \\ \vdots \\ F_{12x} \\ F_{12y} \end{Bmatrix} \begin{Bmatrix} 0 \\ 0 \\ \vdots \\ P_{9x} \\ P_{9y} \\ \vdots \\ pa/2 \\ 0 \\ pa/2 \end{Bmatrix}$$

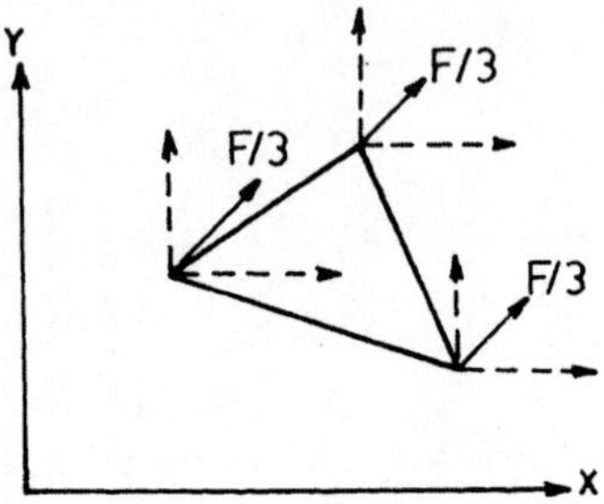

Fig. 13.36

If more than one force acts simultaneously, the forces are added. The body forces and inertia forces are distributed equally at the nodes as shown in Fig. 13.36. The translation of uniform stress fields into concentrated forces applied at element nodes has been shown in Fig. 13.37. The force for each node is obtained by multiplying stresses by the projected area of the triangle, and then dividing this result equally between the appropriate nodes.

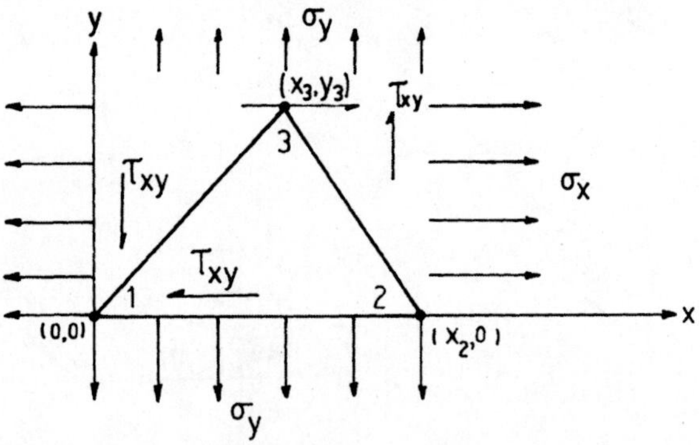

(a)

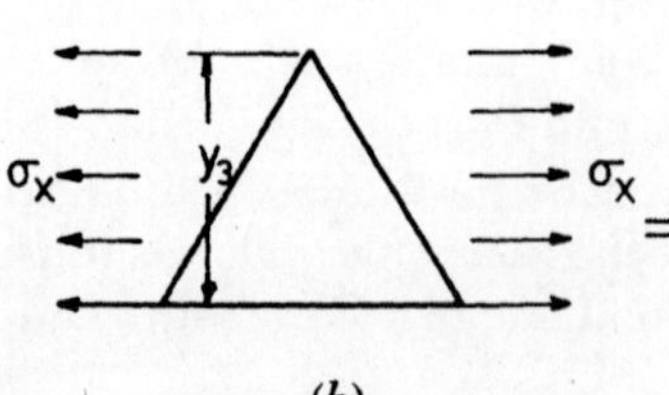

(b)

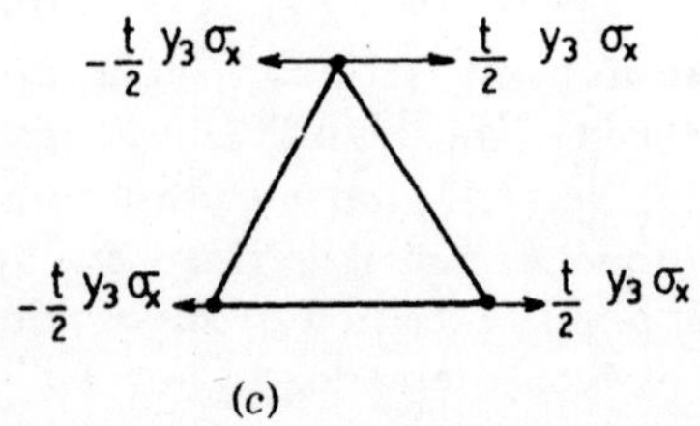

(c)

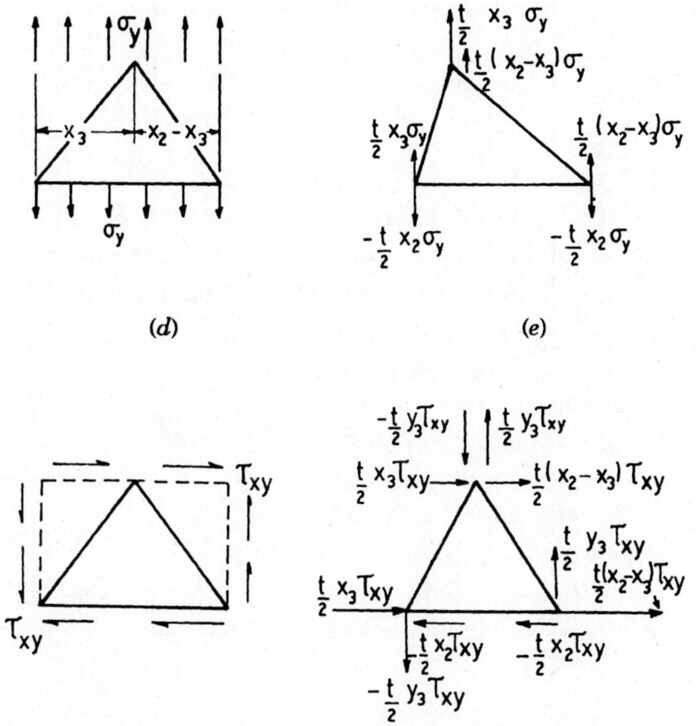

(*f*) Constant stress. (*g*) Nodal force equivalents.

Fig. 13.37

(*e*) *Assembly of the complete stiffness matrix*. The final equations in the solution are based on a simple static equilibrium of forces. At each node, all forces in the x-direction must balance, and all forces in the y-direction must balance, *i.e.*

$$F_{ix} = \sum\nolimits^{*} f_{ix}, \quad F_{iy} = \sum\nolimits^{*} f_{iy}$$

The asterisk means that forces are to be summed over all the elements that produce an equivalent nodal force at i in the x or y direction. For an externally loaded structure (without body forces from acceleration), this procedure merely reduces to equating externally applied loads to the net internal nodal forces acting in opposition. At all other nodes where no external forces act, the net force is zero, so entries in the force vector $\{F\}$ for these nodes are zero.

$$\therefore \quad \{F\}_{system} = [K]_{system} \{\delta\}_{system} \qquad ...(13.76)$$

Thus the finite element method may be summarised as :

Step		Procedure
1.	For each element	Input the co-ordinates of the element nodes, the E and ν values of the element material and the thickness t.
2.		Form the $[B]$ matrix.
3.		Form the $[D]$ matrix.
4.		Form the element stiffness matrix $[k]^e$
5.		Assemble the master stiffness matrix $[K]$.

6. Form the $[F]$ vector by direct substitution for external forces or by input boundary displacement through adjustments to F and K.
7. Solve for u and v of all nodes.
8. Determine stresses for an element.
9. Determine strains for an element.

13.12.6 Axi-symmetric stress analysis

(*a*) *Displacement function.* Using the triangular element with nodes i, j, m numbered in anti-clockwise sense, the nodal displacements are defined as

$$\{\delta_i\} = \begin{Bmatrix} u_i \\ v_i \end{Bmatrix} \quad \text{...(13.77)}$$

and element displacements by the vector

$$\{\delta\}^e = \begin{Bmatrix} \delta_i \\ \delta_j \\ \delta_m \end{Bmatrix} \quad \text{...(13.78)}$$

A linear polynomial can be used to define uniquely the displacements within the element. The displacement field is given by

$$\{f\} = \begin{Bmatrix} u \\ v \end{Bmatrix} = [IN_i', IN_j', IN_m'] \{\delta\}^e \quad \text{...(13.79)}$$

where $N_i' = (a_i + b_i r + c_i z)/2\Delta$ etc.

and I is a two-by-two identity matrix.

$$\begin{aligned} a_i &= r_j z_m - r_m z_j \\ b &= z_j - z_m = z_{jm} \\ c_i &= r_m - r_j = r_{mi} \end{aligned} \quad \text{...(13.80)}$$

etc. in cyclic order. Δ is the area of the triangle.

(*b*) *Strain* (*total*)

$$\{\varepsilon\} = \begin{Bmatrix} \varepsilon_z \\ \varepsilon_r \\ \varepsilon_\theta \\ \gamma_{rz} \end{Bmatrix} = \begin{Bmatrix} \dfrac{\partial v}{\partial z} \\ \dfrac{\partial u}{\partial r} \\ \dfrac{u}{r} \\ \dfrac{\partial u}{\partial z} + \dfrac{\partial v}{\partial r} \end{Bmatrix} \quad ...(13.81)$$

Using the displacement functions defined by Eq. (13.79) and (13.80), we get

$$\{\varepsilon\} = [B]\{\delta\}^e = [B_i, B_j, B_m]\,\{\delta\}^e$$

in which

$$B_i = \begin{bmatrix} 0 & \dfrac{\partial N_i'}{\partial z} \\ \dfrac{\partial N_i'}{\partial r} & 0 \\ \dfrac{N_i'}{r} & 0 \\ \dfrac{\partial N_i'}{\partial z} & \dfrac{\partial N_i'}{\partial r} \end{bmatrix} = \frac{1}{2\Delta}\begin{bmatrix} 0 & 0 \\ b_i & 0 \\ \dfrac{a_i}{r} + b_i + \dfrac{c_i z}{r} & 0 \\ c_i & b_i \end{bmatrix} \quad ...(13.82)$$

(*c*) *Initial strains.* $\{\varepsilon_0\} = \begin{Bmatrix} \varepsilon_{z0} \\ \varepsilon_{r0} \\ \varepsilon_{\theta 0} \\ \gamma_{rz0} \end{Bmatrix}$

For thermal expansion $\{\varepsilon_0\} = \begin{Bmatrix} \alpha_\theta^{\,e} \\ \alpha_\theta^{\,e} \\ \alpha_\theta^{\,e} \\ 0 \end{Bmatrix}$

where θ^e is the average temperature rise and α is the coefficient of linear expansion.

(*d*) *Elasticity matrix*

$$\{\sigma\} = \begin{Bmatrix} \sigma_z \\ \sigma_r \\ \sigma_\theta \\ \tau_{rz} \end{Bmatrix} = [D](\{\varepsilon\} - \{\varepsilon_0\}) + \{\sigma_0\}$$

For an isotropic material

$$[D] = \frac{E(1-\nu)}{(1+\nu)(1-2\nu)} \begin{bmatrix} 1 & \frac{\nu}{1-\nu} & \frac{\nu}{1-\nu} & 0 \\ \frac{\nu}{1-\nu} & 1 & \frac{\nu}{1-\nu} & 0 \\ \frac{\nu}{1-\nu} & \frac{\nu}{1-\nu} & 1 & 0 \\ 0 & 0 & 1 & \frac{1-2\nu}{2(1-\nu)} \end{bmatrix} \quad ...(13.83)$$

(*e*) *The stiffness matrix*

$$[k]^e = 2\pi \int [B]^T [D][B] r dr dz \quad ...(13.84)$$

Now the $[B]$ matrix depends upon the co-ordinates, so the integration has to be carried out either by numerical integration or by explicit multiplication and term by term integration.

The simplest procedure is to evaluate $[\overline{B}]$ for a centroidal point.

$$\bar{r} = \frac{r_i + r_j + r_m}{3}, \bar{z} = \frac{z_i + z_j + z_m}{3}$$

As a first approximation, we have

$$[k]^e = 2\pi [\overline{B}]^T [D][\overline{B}]\, \bar{r}\, \Delta$$

where Δ is the area of the triangle.

(*f*) *External nodal forces.* In this case the nodal forces represent a combined effect of the force acting along the whole circumference of the circle forming the element node. Thus if $\overline{R}$ represents the radial component of force per unit length of the circumference of a node of a radius r, the external force to be introduced for computation is $2\pi r\overline{R}$. In the axial direction we will have $2\pi r\overline{Z}$, to represent the combined effect of axial forces.

(*g*) *Nodal forces due to initial strains*

$$\left\{ F \right\}_{\varepsilon_0}^{e} = -2\pi \int [B]^T [D]\{\varepsilon_0\} r dr dz \quad ...(13.85)$$

or partitioning and noting that ε_0 is constant

$$\left\{ F_i \right\}_{\varepsilon_0}^{e} = -2\pi \left(\int [B_i]^T r dr dz \right) [D]\, \{\varepsilon_0\} \quad ...(13.86)$$

Matrix $[B_i]$ may be splitted into constant and variable parts, such as

$$[B_i] = [\overline{B}_i] + [B_i']$$

in which $[\overline{B}_i]$ is the value of $[B_i]$ at the centroid and the $[B_i']$ accounts for the variation from this value, *i.e.*, given by

$$[B_i'] \begin{bmatrix} 0 & 0 \\ 0 & 0 \\ 1 & 0 \\ 0 & 0. \end{bmatrix} \left\{ \frac{a_i + c_i z}{r} - \frac{a_i + c_i z}{r} \right\} \Big/ 2\Delta \qquad \text{...(13.87)}$$

It can be seen that

$$\int [B_i]\, r dr dz = 0 \qquad \text{...(13.88)}$$

Thus Eq. (13.86) becomes

$$\left\{ F_i \right\}_{\varepsilon_0}^{e} = -2\pi\, [\overline{B}_i]^T [D]\, \{\varepsilon_0\} \bar{r}\, \Delta \qquad \text{...(13.89)}$$

Initial stress forces are treated in an identical manner.

(*h*) *Distributed body forces.* Distributed body forces such as those due to gravity, centrifugal force in rotating machine parts often occur in axi-symmetric problems. Let such forces be denoted by

$$\{p\} = \begin{Bmatrix} R \\ Z \end{Bmatrix} \qquad \text{...(13.90)}$$

per unit volume of material in r and z directions respectively. Thus

$$\{F\}^e = -2\pi \int [N]^T \begin{Bmatrix} R \\ Z \end{Bmatrix} r dr dz \qquad \text{...(13.91)}$$

or partitioning

$$\{F_i\}_p^{\,e} = -2\pi \int \begin{Bmatrix} R \\ Z \end{Bmatrix} N_i\, r dr dz$$

Shifting the co-ordinates to the centroid, the first approximation give

$$\left\{ \overline{F}_i \right\}_p^{e} = \left\{ \overline{F}_i \right\}_p^{e} = \left\{ \overline{F}_m \right\}_p^{e} = -2\pi \begin{Bmatrix} R \\ Z \end{Bmatrix} \bar{r} \frac{\Delta}{3} \qquad \text{...(13.92)}$$

(*i*) *Evaluation of stresses*. The stresses vary throughout the element. It is generally convenient to evaluate the average stresses at centroid of the element. The stress matrix resulting from Eqs. (13.82) and

$$\{\varepsilon\} = [B]\, \{\delta\}^e$$

is given by

$$\{\, \overline{\sigma}\, \}^e = [D\}\, [\overline{B}]\, \{\delta\}^e - [D]\, \{\varepsilon_0\} + \{\sigma_0\} \qquad \text{...(13.93)}$$

13.12.7 Three-Dimensional Stress Analysis

The most widely used elements for three dimensional stress analysis are the tetrahedron and hexahedron. We will study the characteristics of tetrahedral element.

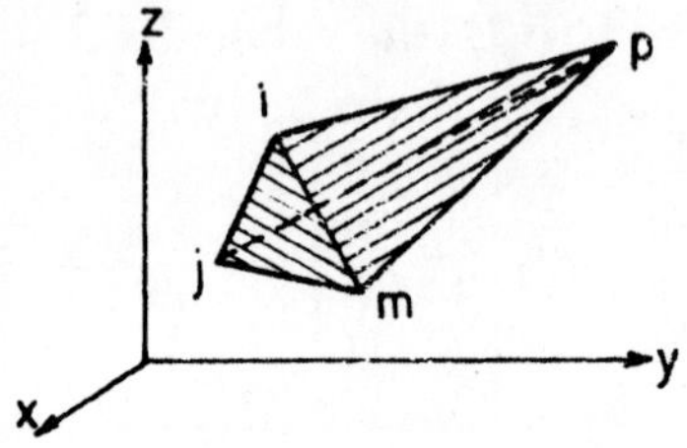

Fig. 13.38

(*a*) *Displacement functions.* Fig. 13.38 shows a tetrahedral element i, j, m, p in space defined by the x, y and z coordinates. The displacement at a point is defined as u, v, w in the directions x, y, z respectively.

Thus $$\{f\} = \begin{Bmatrix} u \\ v \\ w \end{Bmatrix} \qquad ...(13.94)$$

A linear variation for the displacements at the nodes is given as

$$\left.\begin{aligned} u &= \alpha_1 + \alpha_2 x + \alpha_3 y + \alpha_4 z \\ v &= \alpha_5 + \alpha_6 x + \alpha_7 y + \alpha_8 z \\ w &= \alpha_9 + \alpha_{10} x + \alpha_{11} y + \alpha_{12} z \end{aligned}\right\} \qquad ...(13.95)$$

Equating the values of the displacements at the nodes we have four equations of the type

$$u_i = \alpha_1 + \alpha_2 x_i + \alpha_3 y_i + \alpha_4 z_i \qquad ...(13.96)$$

from which α_1 to α_4 can be determined. Similarly other α's can be determined.

Again it is possible to write the displacements in the form of shape functions thus

$$\{f\} = [IN_i', IN_j', IN_m', IN_p']\{\delta\}^e \qquad ...(13.97)$$

where I is the three-by-three identity matrix.

$$\{\delta\}^e = \begin{Bmatrix} \delta_i \\ \delta_j \\ \delta_m \\ \delta_p \end{Bmatrix} \text{ with } \{\delta_i\} = \begin{Bmatrix} u_i \\ v_i \\ w_i \end{Bmatrix} \text{ etc.}$$

and $$N_i' = (a_i + b_i x + c_i y + d_i z)/6V$$

where $$6V = \det \begin{vmatrix} 1 & x_i & y_i & z_i \\ 1 & x_j & y_j & z_j \\ 1 & x_m & y_m & z_m \\ 1 & x_p & y_p & z_p \end{vmatrix}$$

$$a_i = \begin{vmatrix} x_j & y_j & z_j \\ x_m & y_m & z_m \\ x_p & y_p & z_p \end{vmatrix}$$

$$b_i = -\det \begin{vmatrix} 1 & y_j & z_j \\ 1 & y_m & z_m \\ 1 & y_p & z_p \end{vmatrix}$$

$$c_i = \det \begin{vmatrix} x_j & 1 & z_j \\ x_m & 1 & z_m \\ x_p & 1 & z_p \end{vmatrix}$$

$$d_i = -\det \begin{vmatrix} x_j & y_j & 1 \\ x_m & y_m & 1 \\ x_p & y_p & 1 \end{vmatrix}$$

$$\begin{aligned} u = \frac{1}{6V} \{(a_i + b_i x + c_i y + d_i z)\, u_i & \\ & + (a_j + b_j x + c_j y + d_j z) u_j \\ & + (a_m + b_m x + c_m y + d_m z) u_m \\ & + (a_p + b_p x + c_p y + d_p z) u_p\} \end{aligned}$$

(b) Strain matrix

$$\{\varepsilon\} = \begin{Bmatrix} \varepsilon_x \\ \varepsilon_y \\ \varepsilon_z \\ \gamma_{xy} \\ \gamma_{yz} \\ \gamma_{zx} \end{Bmatrix} = \begin{Bmatrix} \dfrac{\partial u}{\partial x} \\ \dfrac{\partial v}{\partial y} \\ \dfrac{\partial w}{\partial z} \\ \dfrac{\partial u}{\partial y} + \dfrac{\partial v}{\partial x} \\ \dfrac{\partial v}{\partial z} + \dfrac{\partial w}{\partial y} \\ \dfrac{\partial w}{\partial x} + \dfrac{\partial u}{\partial z} \end{Bmatrix}$$

$$= [B]\,\{\delta\}^e$$

$$= [B_i, B_j, B_m, B_p]\,\{\delta\}^e \qquad \text{...(13.98)}$$

where
$$[B_i] = \begin{bmatrix} \frac{\partial N_i'}{\partial x} & 0 & 0 \\ 0 & \frac{\partial N_i'}{\partial y} & 0 \\ 0 & 0 & \frac{\partial N_i'}{\partial z} \\ \frac{\partial N_i'}{\partial y} & \frac{\partial N_i'}{\partial x} & 0 \\ 0 & \frac{\partial N_i'}{\partial z} & \frac{\partial N_i'}{\partial y} \\ \frac{\partial N_i'}{\partial z} & 0 & \frac{\partial N_i'}{\partial x} \end{bmatrix} = \frac{1}{6V} \begin{bmatrix} b_i & 0 & 0 \\ 0 & c_i & 0 \\ 0 & 0 & d_i \\ c_i & b_i & 0 \\ 0 & d_i & c_i \\ d_i & 0 & b_i \end{bmatrix}$$

Initial strains such as that due to thermal expansion can be written as

$$\{\varepsilon_0\} = \begin{Bmatrix} \alpha\theta^e \\ \alpha\theta^e \\ \alpha\theta^e \\ 0 \\ \nu \\ 0 \end{Bmatrix}$$

(c) *Elasticity matrix.*
$$\{\sigma\} = \begin{Bmatrix} \sigma_x \\ \sigma_y \\ \sigma_z \\ \tau_{xy} \\ \tau_{yz} \\ \tau_{zx} \end{Bmatrix}$$

$$= [D]\,(\{\varepsilon\} - \{\varepsilon_0\}) + \{\sigma_0\} \qquad \text{...(13.99)}$$

For elastic isotropic material

$$[D] = \frac{E(1-\nu)}{(1+\nu)(1-2\nu)} \begin{bmatrix} 1 & \frac{\nu}{1-\nu} & \frac{\nu}{1-\nu} & 0 & 0 & 0 \\ & 1 & \frac{\nu}{1-\nu} & 0 & 0 & 0 \\ & & 1 & 0 & 0 & 0 \\ & & & \frac{1-2\nu}{2(1-\nu)} & 0 & 0 \\ \text{Symmetric} & & & & \frac{1-2\nu}{2(1-\nu)} & 0 \\ & & & & & \frac{1-2\nu}{2(1-\nu)} \end{bmatrix}$$

(*d*) *Stiffness, stress and load matrices*. The stiffness matrix is given by

$$[K] = [B]^T [D][B]\, dv \qquad \text{...(13.100)}$$

and may be integrated explicitly.

The stress matrix is given as

$$[D]\,[B]\,\{\delta\}^e$$

and the constant load matrix due to initial strains is given as

$$\left\{ F \right\}_{\varepsilon_0}^{e} = -[B]^T [D]\{\varepsilon_0\}\, V$$

with similar expression for forces due to initial stresses.

13.12.8 Isoparametric Elements

An interpolation function or a shape function is a function which has unit value at one nodal point and zero value at all other nodal points.

When the geometry and displacements of the element are described in terms of the same parameters and are of the same order then these elements are called isoparametric elements.

(*a*) *Rectangular elements of the 'serendipity' family*. It is most convenient to make the functions depend on nodal values placed on the element boundary. The various elements in this family are shown in Fig. 13.39.

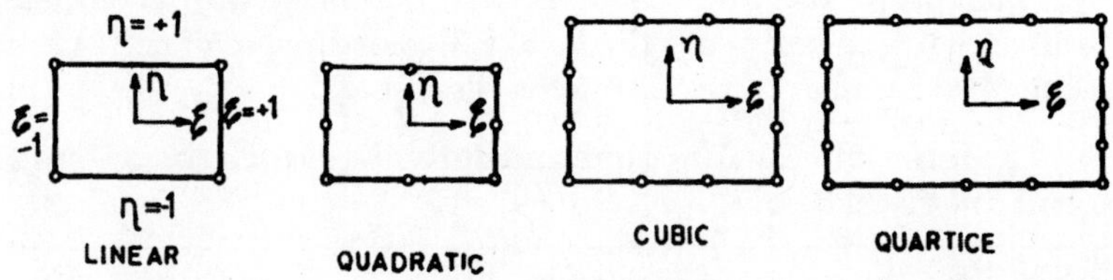

(*a*) (*b*) (*c*) (*d*)

Fig. 13.39

To achieve the shape function for the first element it is obvious that a product of the form

$$\frac{1}{2}(\xi + 1)\,(\eta + 1)$$

gives unity at top right corners where $\xi = 1$, $\eta = 1$ and zero at all other corners. Further, a linear variation of the shape function of all sides exists and hence continuity is satisfied. Introducing new variables

$$\xi_0 = \xi\xi_1,\ \eta_0 = \eta\eta_1$$

the form $$N_i = \frac{1}{4}\,(1 + \xi_0)\,(1 + \eta_0)$$

allows all shape functions to be written down in one expression.

For quadratic element : (8 nodes)

Corner nodes $\quad N_i = \frac{1}{4}(1+\xi_0)(1+\eta_0)(\xi_0+\eta_0-1)$

Mid-side nodes $\quad \xi_i = 0, N_i = \frac{1}{2}(1-\xi^2)(1+\eta_0)$

$\eta_i = 0, N_i = \frac{1}{2}(1+\xi_0)(1-\eta^2)$

Cubic element : (12 nodes)

Corner nodes $\quad N_i = \frac{1}{32}(1+\xi_0)(1+\eta_0)[-10+9(\xi^2+\eta^2)]$

Mid-side nodes $\quad \xi_i = \pm 1$ and $\eta_i = \pm \frac{1}{3}$

$$N_i = \frac{9}{32}(1+\xi_0)(1-\eta^2)(1+9\eta_0).$$

(*b*) *Quadrilateral.* For the quadrilateral elements as shown in Fig. 13.40, the shape of the membrane element can be expressed in terms of interpolation functions and its nodal co-ordinates as

$$x = \sum_{i-1}^{n} N_i(\xi, \eta)\, x_i \; ; \; y = \sum_{i-1}^{n} N_i(\xi, \eta)\, y_i$$

where $N_i(\xi, \eta)$, $i = 1$ to n are interpolation functions in the curvilinear co-ordinates ξ, η, x_i and y_i are the local x, y co-ordinates of node i and N is the total number of nodes for the element.

The interpolation functions and derivatives for a linear quadrilateral are given below.

Node i	ξ_i	η_i	*Interpolation function* $N_i(\xi, \eta)$	*Derivatives*	
				$\frac{\partial N_i}{\partial \xi}$	$\frac{\partial N_i}{\partial \eta}$
1	−1	−1	$\frac{1}{4}(1-\xi)(1-\eta)$	$-\frac{1}{4}(1-\eta)$	$-\frac{1}{4}(1-\xi)$
2	1	−1	$\frac{1}{4}(1+\xi)(1-\eta)$	$\frac{1}{4}(1-\eta)$	$-\frac{1}{4}(1+\xi)$
3	1	1	$\frac{1}{4}(1+\xi)(1+\eta)$	$\frac{1}{4}(1+\eta)$	$\frac{1}{4}(1+\xi)$
4	−1	1	$\frac{1}{4}(1-\xi)(1+\eta)$	$-\frac{1}{4}(1+\eta)$	$\frac{1}{4}(1-\xi)$

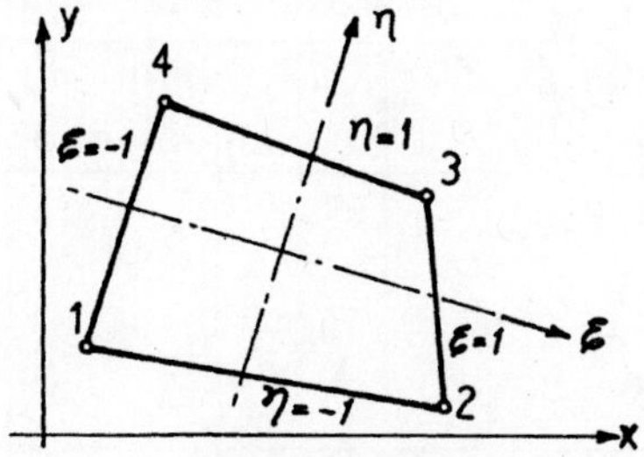

Fig. 13.40

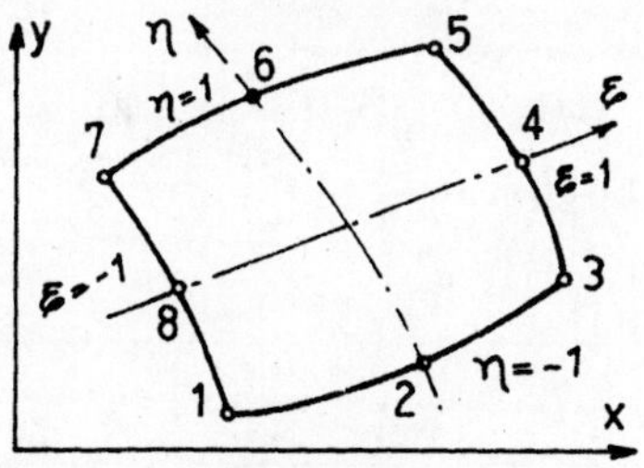

Fig. 13.41

In general for the linear quadrilateral

$$N_i = \frac{1}{4}(1 + \xi\xi_i)(1 + \eta\eta_i)$$

where ξ_i, η_i take their nodal values.

For the quadratic quadrilateral corner nodes.

$$N_i = \frac{1}{4}(1 + \xi\xi_i)(1 + \eta\eta_i) - \frac{1}{4}(1 - \xi^2)(1 + \eta\eta_i) - \frac{1}{4}(1 + \xi\xi_i)(1 - \eta^2)$$

Mid side nodes

$$\xi_i = 0, \eta_i = \pm 1$$

$$N_i = \frac{1}{2}(1 - \xi^2)(1 + \eta\eta_i)$$

Mid side nodes

$$\xi_i = \pm 1, \eta_i = 0.$$

$$N_i = \frac{1}{2}(1 + \xi\xi_i)(1 - \eta^2)$$

For the quadratic quadrilateral as shown in Fig. 13.41, the interpolation functions and derivatives are given below.

Node	ξ_i	η_i	$N_i(\xi, \eta)$	$\dfrac{\partial N_i}{\partial \xi}$	$\dfrac{\partial N_i}{\partial \eta}$
1	−1	−1	$\frac{1}{4}(1-\xi)(1-\eta)$ $-\frac{1}{4}(1-\xi^2)(1-\eta)$ $-\frac{1}{4}(1-\xi)(1-\eta^2)$	$-\frac{1}{4}(1-\eta)+\frac{1}{2}\xi(1-\eta)$ $+\frac{1}{4}(1-\eta^2)$	$-\frac{1}{4}(1-\xi)$ $+\frac{1}{4}(1-\xi^2)$ $+\frac{1}{2}\eta(1-\xi)$
2	0	−1	$\frac{1}{2}(1-\xi^2)(1-\eta)$	$-\xi(1-\eta)$	$-\frac{1}{2}(1-\xi^2)$
3	1	−1	$\frac{1}{4}(1+\xi)(1-\eta)$ $-\frac{1}{4}(1-\xi^2)(1-\eta)$ $-\frac{1}{4}(1+\xi)(1-\eta^2)$	$\frac{1}{4}(1-\eta)+\frac{1}{2}\xi(1-\eta)$ $-\frac{1}{4}(1-\eta^2)$	$-\frac{1}{4}(1+\xi)$ $+\frac{1}{4}(1-\xi^2)$ $+\frac{1}{2}\eta(1+\xi)$
4	1	0	$\frac{1}{2}(1+\xi)(1-\eta^2)$	$\frac{1}{2}(1-\eta^2)$	$-\eta(1+\xi)$
5	1	1	$\frac{1}{4}(1+\xi)(1+\eta)$ $-\frac{1}{4}(1-\xi^2)(1+\eta)$ $-\frac{1}{4}(1+\xi)(1-\eta^2)$	$\frac{1}{4}(1+\eta)+\frac{1}{2}\xi(1+\eta)$ $-\frac{1}{4}(1-\eta^2)$	$\frac{1}{4}(1+\xi)$ $-\frac{1}{4}(1-\xi^2)$ $+\frac{1}{2}\eta(1+\xi)$
6	0	1	$\frac{1}{2}(1-\xi^2)(1+\eta)$	$-\xi(1+\eta)$	$\frac{1}{2}(1-\xi^2)$
7	−1	1	$\frac{1}{4}(1-\xi)(1+\eta)$ $-\frac{1}{4}(1-\xi^2)(1+\eta)$ $-\frac{1}{2}(1-\xi)(1-\eta^2)$	$-\frac{1}{4}(1+\eta)+\frac{1}{2}\xi(1+\eta)$ $+\frac{1}{4}(1-\eta^2)$	$\frac{1}{4}(1-\xi)$ $-\frac{1}{4}(1-\xi^2)$ $+\frac{1}{2}\eta(1-\xi)$
8	−1	0	$\frac{1}{2}(1-\xi)(1-\eta^2)$	$-\frac{1}{2}(1-\eta^2)$	$-\eta(1-\xi)$

The interpolation function have special property that at node j

$$N_i = \begin{cases} 1, & i = j \\ 0, & i \neq j \end{cases}$$

Displacement functions :

$$u = \sum_{i=1}^{n} N_i(\xi, \eta)\, u_i$$

$$v = \sum_{i=1}^{n} N_i(\xi, \eta)\, v_i$$

Stress and strain fields.

$$\{\varepsilon\} = \begin{Bmatrix} \varepsilon_x \\ \varepsilon_y \\ \gamma_{xy} \end{Bmatrix}$$

$$\{\varepsilon\} = [B] \begin{Bmatrix} u_1 \\ v_1 \\ u_2 \\ v_2 \\ \vdots \end{Bmatrix} \quad \text{where} \quad [B] = [B_1, B_2, \ldots]$$

$$B_i = \begin{bmatrix} \frac{\partial N_i}{\partial x} & 0 \\ 0 & \frac{\partial N_i}{\partial y} \\ \frac{\partial N}{\partial y} & \frac{\partial N_i}{\partial x} \end{bmatrix}$$

The interpolation functions are defined in terms of curvilinear co-ordinates ξ and η. The Cartesian and curvilinear derivatives are related by :

$$\begin{bmatrix} \frac{\partial}{\partial x} \\ \frac{\partial}{\partial y} \end{bmatrix} = [J]^{-1} \begin{bmatrix} \frac{\partial}{\partial \xi} \\ \frac{\partial}{\partial \eta} \end{bmatrix}$$

where $[J]$ is the Jacobian matrix.

$$[J] = \begin{bmatrix} \frac{\partial x}{\partial \xi} & \frac{\partial y}{\partial \xi} \\ \frac{\partial \eta}{\partial \eta} & \frac{\partial y}{\partial \eta} \end{bmatrix}$$

$$[J] = \begin{bmatrix} \frac{\partial N_1}{\partial \xi} & \frac{\partial N_2}{\partial \xi} & \cdots & \frac{\partial N_i}{\partial \xi} & \cdots & \frac{\partial N_n}{\partial \xi} \\ \frac{\partial N_1}{\partial \eta} & \frac{\partial N_2}{\partial \eta} & \cdots & \frac{\partial N_i}{\partial \eta} & \cdots & \frac{\partial N_n}{\partial \eta} \end{bmatrix} \begin{bmatrix} x_1 & y_1 \\ x_2 & y_2 \\ \vdots & \vdots \\ x_i & y_i \\ \vdots & \vdots \\ x_n & y_n \end{bmatrix}$$

$$\therefore \quad \begin{bmatrix} \frac{\partial N_i}{\partial x} \\ \frac{\partial N_i}{\partial y} \end{bmatrix} = [J]^{-1} \begin{bmatrix} \frac{\partial N_i}{\partial \xi} \\ \frac{\partial N_i}{\partial \eta} \end{bmatrix}.$$

Now $\{\sigma\} = [D]\{\varepsilon\}$

While integrating the following four-point Gauss-Quadrature rule may be used.

$$\iint (......)\, dx\, dy = \int_{-1}^{+1}\int_{-1}^{+1} (......) \det [J]\, d\xi\, d\eta$$

$$I = \int_{-1}^{+1}\int_{-1}^{+1} f(\xi, \eta)\, d\xi d\eta$$

$$= \sum_i \sum_j w_i\, w_i\, f(\xi_i, \eta_j)$$

$$= (1.)\,(1.)\, f(\xi_1, \eta_1) + ... + (1.)(1.1)\, f(\xi_4, \eta_4)$$

where the four sampling points are :

$$\xi_i, \eta_i = \pm\, 0.57735 = \pm \frac{1}{\sqrt{3}}.$$

The integration points have been shown in Fig. 13.42.

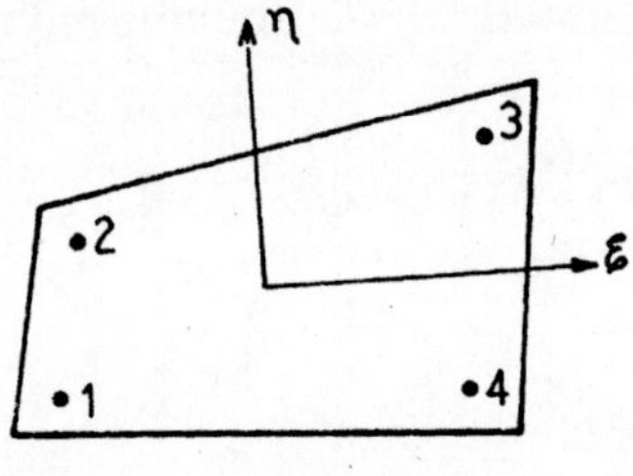

Fig. 13.42

(c) *Eight node hexahedron.* For the eight node hexahedron shown in Fig. 13.43, we have

Shape functions :

$$x = \sum_{i=1}^{n} N_i(\xi, \eta, \zeta)\, x_i$$

$$y = \sum_{i=1}^{n} N_i(\xi, \eta, \zeta)\, y_i$$

$$z = \sum_{i=1}^{n} N_i(\xi, \eta, \zeta)\, z_i.$$

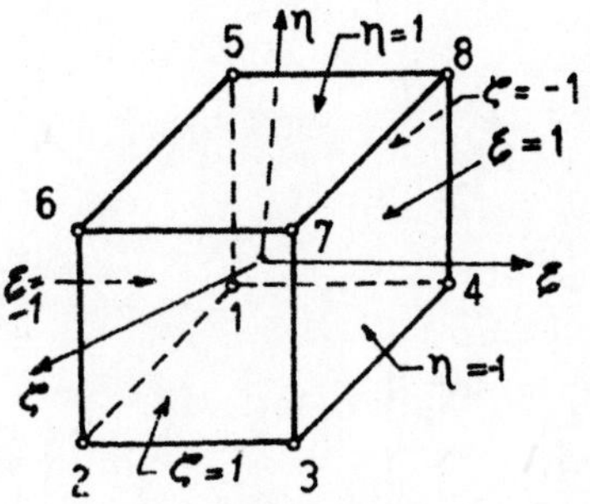

Fig. 13.43

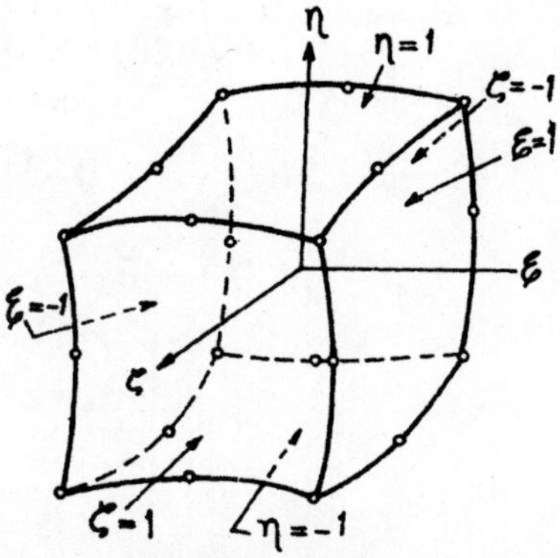

Fig. 13.44

(*d*) ***Twenty node (quadratic) hexahedron***. For the twenty node hexahedron as shown in Fig. 13.44, we have for the, corner nodes

$$N_i = \frac{1}{8}(1 + \xi\xi_i)(1 + \eta\eta_i)(1 + \zeta\zeta_i)(\xi\xi_i + \eta\eta_i + \zeta\zeta_i - 2)$$

Mid side nodes

$$\xi_i = 0,\ \eta_i = \pm 1,\ \zeta_i = \pm 1$$

$$N_i = \frac{1}{4}(1 - \xi^2)(1 + \eta\eta_i)(1 + \zeta\zeta_i)$$

Mid side nodes

$$\xi_i = \pm 1,\ \eta_i = \pm 1,\ \zeta_i = 0$$

$$N_i = \frac{1}{4}(1 + \xi\xi_i)(1 + \eta\eta_i)(1 - \zeta^2)$$

Displacement functions

$$u = \sum_{i=1}^{n} N_i(\xi, \eta, \zeta)\, u_i$$

$$v = \sum_{i=1}^{n} N_i(\xi, \eta, \zeta)\, v_i$$

$$w = \sum_{i=1}^{n} N_i(\xi, \eta, \zeta)\, w_i$$

$$\{\varepsilon\} = \{\varepsilon_x, \varepsilon_y, \varepsilon_z, \gamma_{xy}, \gamma_{yz}, \gamma_{zx}\}$$

$$= \left\{ \frac{\partial u}{\partial x}, \frac{\partial v}{\partial y}, \frac{\partial w}{\partial z}, \frac{\partial u}{\partial y} + \frac{\partial v}{\partial x}, \frac{\partial v}{\partial z} + \frac{\partial w}{\partial y}, \frac{\partial w}{\partial x} + \frac{\partial u}{\partial z} \right\}$$

$$\{\varepsilon\} = [B]\begin{Bmatrix} u_1 \\ v_1 \\ w_1 \\ u_2 \\ v_2 \\ w_2 \\ \vdots \end{Bmatrix}$$

$$[B] = [B_1, B_2 \ldots\ldots, B_n]$$

$$[B_i] = \begin{bmatrix} \frac{\partial N_i}{\partial x} & 0 & 0 \\ 0 & \frac{\partial N_i}{\partial y} & 0 \\ 0 & 0 & \frac{\partial N_i}{\partial z} \\ \frac{\partial N_i}{\partial y} & \frac{\partial N_i}{\partial x} & 0 \\ 0 & \frac{\partial N_i}{\partial z} & \frac{\partial N_i}{\partial y} \\ \frac{\partial N_i}{\partial z} & 0 & \frac{\partial N_i}{\partial x} \end{bmatrix}$$

$$\begin{Bmatrix} \frac{\partial}{\partial x} \\ \frac{\partial}{\partial y} \\ \frac{\partial}{\partial z} \end{Bmatrix} = [J]^{-1}\begin{Bmatrix} \frac{\partial}{\partial \xi} \\ \frac{\partial}{\partial \eta} \\ \frac{\partial}{\partial \zeta} \end{Bmatrix}$$

$$[J] = \begin{bmatrix} \frac{\partial x}{\partial \xi} & \frac{\partial y}{\partial \xi} & \frac{\partial z}{\partial \xi} \\ \frac{\partial x}{\partial \eta} & \frac{\partial y}{\partial \eta} & \frac{\partial z}{\partial \eta} \\ \frac{\partial x}{\partial \zeta} & \frac{\partial y}{\partial \zeta} & \frac{\partial z}{\partial \zeta} \end{bmatrix}$$

$$[J] = \begin{bmatrix} \frac{\partial N_1}{\partial \xi} & \frac{\partial N_2}{\partial \xi} & \cdots & \frac{\partial N_i}{\partial \xi} & \cdots & \frac{\partial N_n}{\partial \xi} \\ \frac{\partial N_1}{\partial \eta} & \frac{\partial N_2}{\partial \eta} & \cdots & \frac{\partial N_i}{\partial \eta} & \cdots & \frac{\partial N_n}{\partial \eta} \\ \frac{\partial N_1}{\partial \zeta} & \frac{\partial N_2}{\partial \zeta} & \cdots & \frac{\partial N_i}{\partial \zeta} & \cdots & \frac{\partial N_n}{\partial \zeta} \end{bmatrix}\begin{bmatrix} x_1 & y_1 & z_1 \\ x_2 & y_2 & z_2 \\ \ldots & \ldots & \ldots \\ x_i & y_i & z_i \\ \ldots & \ldots & \ldots \\ x_n & y_n & z_n \end{bmatrix}$$

or
$$[J] = \sum_{i=1}^{n} \begin{bmatrix} \dfrac{\partial N_i}{\partial \xi} \\ \dfrac{\partial N_i}{\partial \eta} \\ \dfrac{\partial N_i}{\partial \zeta} \end{bmatrix} [x_i \; y_i \; z_i]$$

$$\begin{bmatrix} \partial N_i/\partial x \\ \partial N_i/\partial y \\ \partial N_i/\partial z \end{bmatrix} = [J]^{-1} \begin{bmatrix} \partial N_i/\partial \xi \\ \partial N_i/\partial \eta \\ \partial N_i/\partial \zeta \end{bmatrix}$$

$$\{\sigma\} = [D]\{\varepsilon\}$$

where
$$\{\sigma\} = \begin{Bmatrix} \sigma_x \\ \sigma_y \\ \sigma_z \\ \tau_{xy} \\ \tau_{yz} \\ \tau_{zx} \end{Bmatrix}$$

$$[D] = \frac{E}{(1+\nu)(1-2\nu)} \begin{bmatrix} 1-\nu & \nu & \nu & 0 & 0 & 0 \\ & 1-\nu & \nu & 0 & 0 & 0 \\ & & 1-\nu & 0 & 0 & 0 \\ \text{Symmetric} & & & \dfrac{1-2\nu}{2} & 0 & 0 \\ & & & & \dfrac{1-2\nu}{2} & 0 \\ & & & & & \dfrac{1-2\nu}{2} \end{bmatrix}$$

13.13 Variational Methods

Let us consider the problem of determining function of $y(x)$ which makes the integral

$$I(y) = \int_a^b F(x, y, y')\, dx \qquad \text{...(13.101)}$$

an extremum *i.e.* stationary and which satisfies the prescribed end conditions,

$$y(x)_{x=a} = y(a) = x_a \qquad \text{...(13.102)}$$

$$y(x)_{x=b} = y(b) = y_b \qquad \text{...(13.103)}$$

The value of the integral in Eq. (13.101) depends on the choice of $y = y(x)$, hence the notation $I(y)$. The integral $F(x, y, y')$ is a function of x, y and the first derivative $dy/dx = y'$. Let us assume that $y = y(x)$ is the actual minimizing function (Fig. 13.45), and find out what happens to the integral I when neighbouring functions are used in F. These functions are constructed by adding to $y(x)$ a function $\varepsilon\eta(x)$ where ε is a constant which can take different quantitative values and $\eta(x)$ is an arbitrary function of x which vanishes at $x = a$ and $x = b$, that is

$$\eta(a) = 0 \text{ and } \eta(b) = 0 \qquad ...(13.104)$$

Therefore, $y(x) + \varepsilon\eta(x)$ will satisfy the end conditions. If we now replace $y(x)$ by $y(x) + \varepsilon\eta(x)$, the integral I becomes a function of ε once $\eta(x)$ is chosen, since y is the actual minimizing function. therefore, we can write :

$$I(\varepsilon) = \int_a^b F(x, y + \varepsilon\eta, y' + \varepsilon\eta')\, dx \qquad ...(13.105)$$

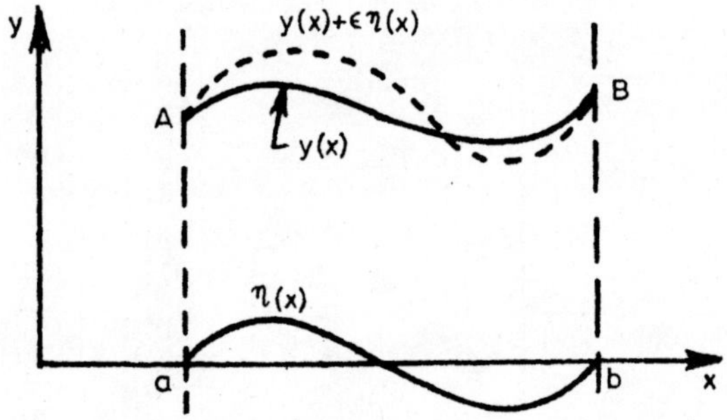

Fig. 13.45

This integral takes minimum values when ε is possible only if

$$\frac{dI(\varepsilon)}{d\varepsilon} = 0, \text{ when } \varepsilon = 0 \qquad ...(13.106)$$

Let $$F_e = F(x, y + \varepsilon\eta, y' + \varepsilon\eta') \qquad ...(13.107)$$

and $$Y(x) = y(x) + \varepsilon\eta(x) \qquad ...(13.108)$$

Then $$\frac{dY}{dx} = Y' = y'(x) + \varepsilon\eta'(x) \qquad ...(13.109)$$

$$\frac{dF_e}{d\varepsilon} = \frac{\partial F_e}{\partial Y}\eta(x) + \frac{\partial F_e}{\partial Y'}\eta'(x) \qquad ...(13.110)$$

Therefore, from Eqs. (13.105) and (13.106), we have

$$\frac{dI(\varepsilon)}{d\varepsilon} = \int_b^a \left[\frac{\partial F_e}{\partial Y} \eta(x) + \frac{\partial F_e}{\partial Y'} \eta'(x) \right] dx = 0 \quad ...(13.111)$$

$$\text{Now} \int_a^b \frac{\partial F_e}{\partial Y'} \eta'(x) dx = \left[\frac{\partial F_e}{\partial Y'} \eta(x) \right]_a^b - \int_a^b \eta(x) \frac{d}{dx}\left(\frac{\partial F_e}{\partial Y'} \right) dx$$

...(13.112)

Since $\eta(a) = \eta(b) = 0,$

$$\left[\frac{\partial F_e}{\partial Y'} \eta(x) \right] = 0, \quad ...(13.113)$$

and Eq. (11) becomes :

$$\frac{dI(\varepsilon)}{d\varepsilon} = \int_a^b \left[\frac{\partial F_e}{\partial Y} \eta(x) - \frac{d}{dx}\left(\frac{\partial F_e}{\partial Y'} \right) \eta(x) \right] dx = 0 \quad ...(13.114)$$

According to Eq. (13.106), this derivative must vanish when $\varepsilon \to 0$. But as $\varepsilon \to 0$, $F_e \to 0\ F$, $Y \to y$, and $Y' \to y'$. Therefore, Eq. (13.114) becomes

$$\int_a^b \left[\frac{\partial F}{\partial y} - \frac{d}{dx} \frac{\partial F}{dy'} \right] \eta(x)\, dx = 0 \quad ...(13.115)$$

If Eq. (13.115) is true for any function $\eta(x)$, then the coefficient of $\eta(x)$ in the integrand must be zero everywhere in (a, b). Thus

$$\frac{\partial F}{\partial y} - \frac{d}{dx}\left(\frac{\partial F}{\partial y'} \right) = 0 \quad ...(13.116)$$

Eq. (13.116) is called Euler's equation and is a necessary condition for $y(x)$ to make I a minimum or a maximum. Now

$$\frac{d}{dx}\left(\frac{\partial F}{\partial y'} \right) = \frac{\partial^2 F}{\partial x \partial y'} + \frac{\partial^2 F}{\partial y \partial y'} \frac{dy}{dx} + \frac{\partial^2 F}{\partial y'^2} \frac{d^2 y}{dx^2}$$

Hence Eq. (13.116) becomes

$$\frac{\partial F}{\partial y} - \frac{\partial^2 F}{\partial x \partial y'} - \frac{\partial^2 F}{\partial y \partial y'} \frac{dy}{dx} - \frac{\partial^2 F}{\partial y'^2} \frac{\partial^2 y}{dx^2} = 0 \quad ...(13.117)$$

Similarly for $I(y) = \int_a^b F(x, y, y'\ y'')\, dx.$

Euler's equation becomes

$$\frac{\partial F}{\partial y} - \frac{d}{dx} \frac{\partial F}{\partial y'} + \frac{d^2}{dx^2} \frac{\partial F}{\partial y''} = 0 \quad ...(13.118)$$

In Eq. (13.108), the change $\varepsilon\eta(x)$ in $y(x)$ is called the variation of y, and is denoted by δy,

$$\delta y = \varepsilon\eta(x) \qquad ...(13.119)$$

Corresponding to the change δy in the function y, F changes by an amount ΔF, where

$$\Delta F = F(x, y, + \varepsilon\eta, y' + \varepsilon\eta') - F(x, y, y') \qquad ...(13.120)$$

Expanding right-hand side in powers of ε, we get

$$\Delta F = \frac{\partial F}{\partial y}\varepsilon\eta + \frac{\partial F}{\partial y'}\varepsilon\eta' + \text{(terms with higher power of } \varepsilon) \qquad ...(13.121)$$

By analogy with the definition of the differential,

$$\delta F = \frac{\partial F}{\partial y}\varepsilon\eta + \frac{\partial F}{\partial y'}\varepsilon\eta' \qquad ...(13.122)$$

where

$$F = y', \delta y' = \varepsilon\eta' \qquad ...(13.123)$$

so that Eq. (13.122) may be written as

$$\delta F = \frac{\partial F}{\partial y}\delta y + \frac{\partial F}{\partial y'}\delta y' \qquad ...(13.124)$$

For a complete analogy with definition of the differential,

$$\delta F = \frac{\partial F}{\partial x}\delta x + \frac{\partial F}{\partial y}\delta y + \frac{\partial F}{\partial y'}\delta y' \qquad ...(13.125)$$

But, in the manipulations of F, x is not varied, so that $\delta x = 0$. From the definition, it follows that the laws of variation of sums, products, ratios and so forth, are completely analogous to those of differentiation. Thus, for example

$$\delta(F_1F_2) = F_1\,\delta F_2 + F_2\,\delta F_1 \qquad ...(13.126)$$

$$\frac{\delta F_1}{F_2} = \frac{F_2\delta F_1 - F_1\delta F_2}{F_2^{\,2}} \qquad ...(13.127)$$

where F_1 and F_2 are two different functions of x, y and y'. From Eqs. (13.119) and (13.123), we deduce that

$$\frac{d}{dx}(\delta y) = \partial\left(\frac{dy}{dx}\right) \qquad ...(13.128)$$

that is, the operators δ and d/dx are commutative.

The necessary condition for I to be stationary is that its first variation vanishes, that is

$$\delta I = \int_a^b \delta F(x, y, y')\,dx = 0 \qquad ...(13.129)$$

$$\text{Indeed} \int_a^b \delta F\ dx = \int_a^b \left[\frac{\partial F}{\partial y}\delta y + \frac{\partial F}{\partial y}\partial y'\right] dx \qquad ...(13.130)$$

Integrating the second term by parts, we have

$$\int_a^b \frac{\partial F}{\partial y'} \delta \left(\frac{dy}{dx}\right) dx = \int_a^b \frac{\partial F}{\partial y'} \frac{d}{dx} (\delta y)\, dx$$

$$= \left[\frac{\partial F}{\partial y'} \delta y\right]_a^b - \int_a^b \frac{d}{dx}\left(\frac{\partial F}{\partial y'}\right) \delta y\, dx \quad ...(13.131)$$

$$\text{Thus, } \int_a^b \delta F\, dx = \int_a^b \left[\frac{\partial F}{\partial y} - \frac{d}{dx}\frac{\partial F}{\partial y'}\right] \delta y\, dx + \left[\frac{\partial F}{\partial y'} \delta y\right]_a^b = 0$$

...(13.132)

Therefore, a stationary function for an integral function is one for which the variation of that integral is zero.

EXERCISES

13.1 (*a*) Determine an expression for the total strain energy in terms of components of stress and strain.

(*b*) Reduce the above expressions for the case of

(*i*) plane stress, (*ii*) simple tension

(*iii*) symmetrical bending and

(*iv*) torsion.

13.2 (*a*) State and prove the Maxwell-Betti reciprocal theorem.

(*b*) For a cantilever beam of length l loaded at the end by a force P, the deflection of the cantilever is given by :

$$\delta = -\frac{P}{6EI}(2l^3 - 3l^2x + x^3)$$

By using Maxwell-Betti reciprocal theorem, determine the deflection of the cantilever beam due a load Q applied at a distance a from the tip.

13.3 Determine the total strain energy of deformation of a beam of rectangular cross-section 60 mm × 80 mm, simply supported at the ends, 3 m long and carrying two concentrated loads of 4.5 kN and 6.5 kN at a distance of 1 m and 2 m respectively from the left end support. Assume

$$E = 2 \times 10^5 \text{ N/mm}^2, \nu = 0.30.$$

13.4 The stress components at a point are

$$\sigma_x = 100, \quad \sigma_y = 200$$
$$\sigma_z = -100, \quad \tau_{xy} = -50$$
$$\tau_{yz} = 200, \quad \tau_{xz} = -40 \text{ N/m}^2$$

Determine the strain energy per unit volume.

$$E = 2.1 \times 10^5 \text{ N/mm}^2.$$
$$G = 1.05 \times 10^5 \text{ N/mm}^2.$$

13.5 The strain components at a point are

$$\varepsilon_x = 0.001,\ \varepsilon_y = 0.002,\ \varepsilon_z = -0.001,$$
$$\gamma_{xy} = -0.001,\ \gamma_{yz} = -0.004,\ \gamma_{xz} = 0.006.$$

Determine the strain energy per unit volume

$$G = 1.05 \times 10^5\ \text{N/mm}^2,\ \nu = 0.30.$$

13.6 (*a*) State and prove Castigliano's theorem.

(*b*) A rectangular beam 80 mm × 100 mm is simply supported at the ends. The beam is 4 m long and carries a load of 2 kN at a distance of 1 m from the left end support. By using Castigliano's theorem, determine the deflection of the beam at mid-span. Take

$$E = 2 \times 10^2\ \text{N/mm}^2.$$

13.7 (*a*) State and prove the principle of virtual work.

(*b*) Determine the horizontal and vertical deflection of the point *B* by virtual work principle (Fig. 13.46).

13.8 (*a*) State and prove the principle of strain energy.

(*b*) Using the principle of strain energy, determine the deflection of point *B* in Fig. 13.46.

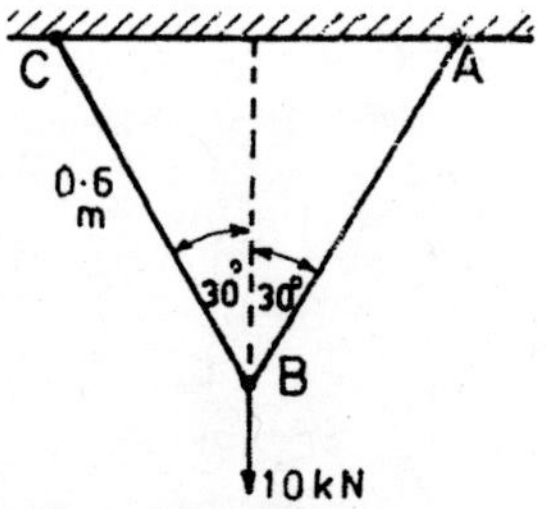

Fig. 13.46

13.9 (*a*) State and prove the complementary energy theorem.

(*b*) Using the complementary energy theorem, determine the horizontal and vertical deflections of point *B* in Fig. 13.46.

13.10 (*a*) Explain the Rayleigh-Ritz method.

(*b*) Assume $y = \sum_{n=1}^{n} C_n \sin\left(\frac{n\pi x}{l}\right)$.

Determine C_n if a beam is under the action of a concentrated load *P* at $x = a$. The beam is simply supported at the ends.

13.11 By assuming the deflection curve to be $y = ax^2 + bx^3$ for a cantilever beam bent by an end load *P*, determine the parameters *a* and *b* which satisfies the boundary conditions

$$y = \frac{dy}{dx} = 0 \text{ at } x = 0.$$

13.12 For the following equations, derive the corresponding finite difference equations.

(*i*) $\nabla^2 f = 0$ (*ii*) $\nabla^2 f = -2$ and (*iii*) $\nabla^4 f = 0$.

13.13 Explain the following :

(*i*) Relaxation patterns.

(*ii*) Block relaxation.

(*iii*) Finite difference approximation near irregular boundaries.

13.14 Explain the finite element method in brief and explain the basic steps involved.

13.15 What are the different steps involved in the displacement approach ? Explain with the help of suitable equations.

13.16 (*a*) What are the various type of elements used for plane stress analysis ?

(*b*) Derive an equation for the stiffness matrix of a triangular element.

13.17 Derive the basic equations for solving axisymmetric elasticity problems by using triangular elements.

13.18 Derive the basic equations for solving three dimensional elasticity problems by using tetrahedral elements.

13.19 (*a*) What do you understand by interpolation function or shape function for an element ? Expain with the help of suitable examples.

(*b*) What are isoparametric elements ? Determine shape functions for a quadrilateral of the linear and quadratic type.

13.20 What are the different type of elements used for three-dimensional stress analysis ? Determine their important properties.

14

Thermal Stresses

14.1 Introduction

Thermal stresses are developed in engineering materials both due to severe temperature gradients and elevated temperature levels. Thermal stresses are frequently used as an important factor in determining materials life. The formulation of elasticity problems including the effect of temperature variation was studied by Duhamel as early as 1835. In this chapter, we shall study the methods of formulation of thermoelastic problems and the solutions for some simple problems will be presented.

14.2 Formulation of Thermoelastic Problems

14.2.1 Basic Equations

(i) Stress-strain relations

$$\left.\begin{aligned}
\varepsilon_x &= \frac{1}{E}[\sigma_x - \nu(\sigma_y + \sigma_z)] + \alpha T \\
\varepsilon_y &= \frac{1}{E}[\sigma_y - \nu(\sigma_z + \sigma_x)] + \alpha T \\
\varepsilon_z &= \frac{1}{E}[\sigma_z - \nu(\sigma_x + \sigma_y)] + \alpha T \\
\gamma_{xy} &= \frac{\tau_{xy}}{G},\ \gamma_{yz} = \frac{\tau_{yz}}{G},\ \gamma_{zx} = \frac{\tau_{zx}}{G}
\end{aligned}\right\} \quad \text{...(14.1)}$$

where α is the coefficient of linear expansion and T is the temperature rise.

Now $$G = \frac{E}{2(1+\nu)}$$

Also $$e = \frac{\Theta}{3K} + 3\alpha T$$

where $$K = \frac{E}{3(1-2\nu)}$$ is the bulk modulus.

$$e = e_x + e_y + e_z = \text{dilatation}$$

$$\Theta = \sigma_x + \sigma_y + \sigma_z$$

Hence

$$\left.\begin{aligned} \sigma_x &= \lambda e + 2G\varepsilon_x - (3\lambda + 2G)\alpha T \\ \sigma_y &= \lambda e + 2G\varepsilon_y - (3\lambda + 2G)\alpha T \\ \varepsilon_z &= \lambda e + 2G\varepsilon_z - (3\lambda + 2G)\alpha T \\ \tau_{xy} &= G\gamma_{xy}, \tau_{yz} = G\gamma_{yz}, \tau_{zx} = G\gamma_{zx} \end{aligned}\right\} \quad \text{...(14.2)}$$

where $\lambda = \dfrac{\nu E}{(1+\nu)(1-2\nu)}$

(ii) Equilibrium Equations

$$\left.\begin{aligned} \frac{\partial \sigma_x}{\partial x} + \frac{\partial \tau_{xy}}{\partial x} + \frac{\partial \tau_{xz}}{\partial z} + B_x = 0 \\ \frac{\partial \tau_{xy}}{\partial x} + \frac{\partial \sigma_y}{\partial y} + \frac{\partial \tau_{yz}}{\partial z} + B_y = 0 \\ \frac{\partial \tau_{xz}}{\partial x} + \frac{\partial \tau_{yz}}{\partial y} + \frac{\partial \sigma_z}{\partial z} + B_z = 0 \end{aligned}\right\} \quad \text{...(14.3)}$$

(iii) Strain Displacement Relations

$$\left.\begin{aligned} \varepsilon_x &= \frac{\partial u_x}{\partial x}, \varepsilon_y = \frac{\partial u_y}{\partial y}, \varepsilon_z = \frac{\partial u_z}{\partial z} \\ \gamma_{xy} &= \frac{\partial u_x}{\partial y} + \frac{\partial u_y}{\partial x} \\ \gamma_{yz} &= \frac{\partial u_y}{\partial z} + \frac{\partial u_z}{\partial y} \\ \gamma_{zx} &= \frac{\partial u_z}{\partial x} + \frac{\partial u_x}{\partial z} \end{aligned}\right\} \quad \text{...(14.4)}$$

(iv) Boundary Conditions

(a) Traction boundary conditions.

$$\left.\begin{aligned} T_x &= \sigma_x a_{nx} + \tau_{xy} a_{ny} + \tau_{xz} a_{nz} \\ T_y &= \tau_{xy}.a_{nx} + \sigma_y a_{ny} + \tau_{yz}.a_{nz} \\ T_z &= \tau_{xz} a_{nx} + \tau_{yz}.a_{ny} + \sigma_z.a_{nz} \end{aligned}\right\} \quad \text{...(14.5)}$$

(b) Displacement boundary conditions. The boundary conditions for this case are expressed through the following equations to be satisfied at every point P of the bounding surface.

$$\left.\begin{aligned} u_x &= f_1(P) \\ u_y &= f_2(P) \\ u_z &= f_3(P) \end{aligned}\right\} \quad \text{...(14.6)}$$

where f_1, f_2 and f_3 are prescribed functions.

14.2.2 Displacement Formulation

Equilibrium equations (14.3) may be expressed in terms of displacements by using Eqs. (14.2) and (14.4) as :

$$\left.\begin{aligned}(\lambda + G)\frac{\partial e}{\partial x} + G\nabla^2 u_x - (3\lambda + 2G)\alpha\frac{\partial T}{\partial x} + B_x = 0\\(\lambda + G)\frac{\partial e}{\partial y} + G\nabla^2 u_y - (3\lambda + 2G)\alpha\frac{\partial T}{\partial y} + B_y = 0\\(\lambda + G)\frac{\partial e}{\partial z} + G\nabla^2 u_z - (3\lambda + 2G)\alpha\frac{\partial T}{\partial z} + B_z = 0\end{aligned}\right\} \quad ...(14.7)$$

These equations, when solved with the displacement boundary conditions (14.6), yield the functions u_x, u_y and u_z for all points in the body. Eqs. (14.4) and (14.2) may then be used to determine the strains and stresses. Similar results may be written in cylindrical and spherical co-ordinates.

14.2.3 Stress Formulation

By using Eqs. (14.1), the compatibility equations can be expressed in terms of stresses as :

$$\left.\begin{aligned}(1 + \nu)\nabla^2\sigma_x + \frac{\partial^2\Theta}{\partial x^2} + \alpha E\left(\frac{1+\nu}{1-\nu}\nabla^2 T + \frac{\partial^2 T}{\partial x^2}\right) = 0\\(1 + \nu)\nabla^2\sigma_y + \frac{\partial^2\Theta}{\partial y^2} + \alpha E\left(\frac{1+\nu}{1-\nu}\nabla^2 T + \frac{\partial^2 T}{\partial y^2}\right) = 0\\(1 + \nu)\nabla^2\sigma_z + \frac{\partial^2\Theta}{\partial z^2} + \alpha E\left(\frac{1+\nu}{1-\nu}\nabla^2 T + \frac{\partial^2 T}{\partial z^2}\right) = 0\\(1 + \nu)\nabla^2\tau_{xz} + \frac{\partial^2\Theta}{\partial x\partial z} + \alpha E\frac{\partial^2 T}{\partial x\partial z} = 0\\(1 + \nu)\nabla^2\tau_{yz} + \frac{\partial^2\Theta}{\partial y\partial z} + \alpha E\frac{\partial^2 T}{\partial y\partial z} = 0\\(1 + \nu)\nabla^2\tau_{xy} + \frac{\partial^2\Theta}{\partial x\partial y} + \alpha E\frac{\partial^2 T}{\partial x\partial y} = 0\end{aligned}\right\} \quad ...(14.8)$$

This formulation requires the solution of Eqs. (14.3) and (14.8) together with Eqs. (14.5).

14.3 Two-Dimensional Formulation

Two types of two-dimensional thermostatic problems may be considered for temperature distributions of the form.

$$T = T(x, y)$$

For long prismatic bodies the concept of plane strain may be used, defined by

$$u_x = u_x(x, y),\ u_y = u_y(x, y),\ u_z = 0$$

or

$$\varepsilon_z = \gamma_{yz} = \gamma_{zx} = 0$$

For thin prismatic bodies the concept of plane stress may be used, defined by

$$\sigma_z = \tau_{yz} = \tau_{zx} = 0$$

The eight quantities σ_x, σ_y, τ_{xy} ; ε_x, ε_y, γ_{xy} ; u_x, u_y satisfy in either case the following eight equations :

2 equilibrium equations (for the case of no body forces)

$$\left.\begin{aligned} \frac{\partial \sigma_x}{\partial x} + \frac{\partial \tau_{xy}}{\partial y} = 0 \\ \frac{\partial \tau_{xy}}{\partial x} + \frac{\partial \sigma_y}{\partial y} = 0 \end{aligned}\right\} \qquad ...(14.9)$$

3 stress-strain relations.

$$\left.\begin{aligned} \varepsilon_x &= \frac{1}{E}[\sigma_x - \nu\sigma_y] + \alpha T \\ \varepsilon_y &= \frac{1}{E}[\sigma_y - \nu\sigma_x] + \alpha T \\ \gamma_{xy} &= \tau_{xy}/G \end{aligned}\right\} \qquad ...(14.10)$$

3 strain-displacement relations.

$$\left.\begin{aligned} \varepsilon_x &= \frac{\partial u_x}{\partial x},\ \varepsilon_y = \frac{\partial u_y}{\partial y} \\ \gamma_{xy} &= \frac{\partial u_x}{\partial y} + \frac{\partial u_y}{\partial x} \end{aligned}\right\} \qquad ...(14.11)$$

For the plane-strain formulation the constants E, ν and α are replaced by E_1, ν_1 and α_1 respectively, where

$$E_1 = \frac{E}{1-\nu^2},\ \nu_1 = \frac{\nu}{1-\nu}$$

$$\alpha_1 = \alpha(1+\nu) \text{ while}$$

$$G = \frac{E}{2(1+\nu)} = \frac{E_1}{2(1+\nu_1)}$$

remains unchanged.

These stresses and strains are accompanied by a stress component σ_z given by

$$\sigma_z = \nu(\sigma_x + \sigma_y) - \alpha E T \text{ for plane stress}$$

and

$$\sigma_z = \frac{1}{1+\nu_1}[\nu_1(\sigma_x + \sigma_y) - \alpha_1 E_1 T] \qquad ...(14.12)$$

for the case of plane strain, and by a strain component ε_z given by

$$\varepsilon_z = -\frac{\nu}{E}(\sigma_x + \sigma_y) + \alpha T \qquad \text{...(14.13)}$$

for the case of plane stress.

The traction boundary conditions become

$$\left.\begin{aligned} \sigma_x . a_{nx} + \tau_{xy}\, a_{ny} &= 0 \\ \tau_{xy} . a_{nx} + \sigma_y . a_{ny} &= 0 \end{aligned}\right\} \qquad \text{...(14.14)}$$

and the displacement boundary conditions :

$$\left.\begin{aligned} u_x &= f_1(P) \\ u_y &= f_2(P) \end{aligned}\right\} \qquad \text{...(14.15)}$$

where P is a generic point of the bounding curve.

For displacement formulation, the equilibrium equations become

$$\left.\begin{aligned} &\frac{E}{2(1-\nu)}\frac{\partial}{\partial x}\left(\frac{\partial u_x}{\partial x} + \frac{\partial u_y}{\partial y}\right) + \frac{E}{2(1+\nu)}\nabla^2 u_x \\ &\qquad - \frac{\alpha E}{1-\nu}\frac{\partial T}{\partial x} + B_x = 0 \\ &\frac{E}{2(1-\nu)}\frac{\partial}{\partial y}\left(\frac{\partial u_x}{\partial x} + \frac{\partial u_y}{\partial y}\right) + \frac{E}{2(1+\nu)}\nabla^2 u_y . \\ &\qquad - \frac{\alpha E}{1-\nu}\frac{\partial T}{\partial y} + B_y = 0 \end{aligned}\right\} \qquad \text{...(14.16)}$$

where $\nabla^2 = \dfrac{\partial^2}{\partial x^2} + \dfrac{\partial^2}{\partial y^2}$

For stress formulation, the compatibility equation (14.8) reduces to

$$\left.\begin{aligned} &(1+\nu)\nabla^2\sigma_x + \frac{\partial^2(\sigma_x + \sigma_y)}{\partial x^2} + \alpha E\left(\frac{1+\nu}{1-\nu}\nabla^2 T + \frac{\partial^2 T}{\partial x^2}\right) = 0 \\ &(1+\nu)\nabla^2\sigma_y + \frac{\partial^2(\sigma_x + \sigma_y)}{\partial y^2} + \alpha E\left(\frac{1+\nu}{1-\nu}\nabla^2 T + \frac{\partial^2 T}{\partial y^2}\right) = 0 \\ &(1+\nu)\nabla^2\tau_{xy} + \frac{\partial^2(\sigma_x + \sigma_y)}{\partial x\,\partial y} + \alpha E\frac{\partial^2 T}{\partial x\,\partial y} = 0 \end{aligned}\right\} \qquad \text{...(14.17)}$$

Eqs. (14.17) may also be written as

$$\nabla^2(\sigma_x + \sigma_y) + \alpha E \nabla^2 T = 0 \qquad \text{...(14.18)}$$

or in the presence of body forces, we have

$$\nabla^2(\sigma_x + \sigma_y + \alpha E T) + (1+\nu)\left(\frac{\partial B_x}{\partial x} + \frac{\partial B_y}{\partial y}\right) = 0 \qquad \text{...(14.19)}$$

The equilibrium equations in the absence of the body forces are satisfied if the stresses are derived from a stress function Φ, the so called Airy stress function, in the following manner :

$$\sigma_x = \frac{\partial^2 \Phi}{\partial y^2}, \sigma_y = \frac{\partial^2 \Phi}{\partial x^2}, \tau_{xy} = -\frac{\partial^2 \Phi}{\partial x \partial y} \quad ...(14.20)$$

Substituting Eqs. (14.20) in (14.18), we get

$$\nabla^4 \Phi = -\alpha E \nabla^2 T \quad ...(14.21)$$

where $$\nabla^4 \Phi = \Delta^2 \nabla^2 \Phi = \frac{\partial^4 \Phi}{\partial x^4} + 2\frac{\partial^4 \Phi}{\partial x^2 \partial y^2} + \frac{\partial^4 \Phi}{\partial y^4}$$

14.4 Free Plate with Temperature Variation Through the Thickness

Consider a plate of arbitrary profile and of constant thickness $2h$ as shown in Fig. 14.1 The plate is completely free of surface traction and the temperature varies through the thickness only, *i.e.* $T = T(z)$. Under these conditions it is reasonable to assume that

$$\sigma_x = \sigma_y = f(z);$$

$$\sigma_z = \tau_{xy} = \tau_{yz} = \tau_{zx} = 0 \quad ...(14.22)$$

Eqs. (14.22) satisfy the equilibrium equations (14.3). Of the six compatibility equations (14.8), the last three are identically satisfied and the first three will be satisfied provided that

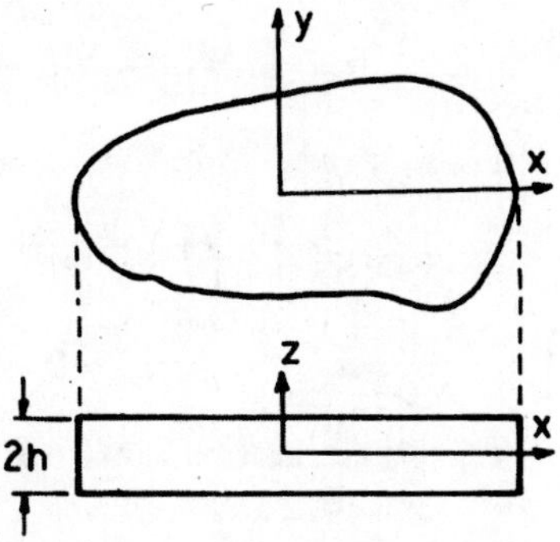

Fig. 14.1

$$\frac{d^2}{dz^2}\left\{f + \frac{\alpha E}{1-\nu}T\right\} = 0 \quad ...(14.23)$$

The form of the non-zero stress components is then

$$\sigma_x = \sigma_y = f = -\frac{\alpha E}{1-\nu}T + C_1 + C_2 z \quad ...(14.24)$$

where the constants C_1 and C_2 are to be determined from the boundary conditions of zero tractions on the edges of the plate. If we choose the constants C_1 and C_2 so that for any temperature $T(z)$ the resultant force and moment per unit of length produced by σ_x and σ_y are zero over the edges of the plate, that is, so that

$$\left.\begin{aligned}\int_{-h}^{+h} \sigma_x \, dz &= \int_{-h}^{+h} \sigma_x \, z dz = 0 \\ \text{and} \qquad \int_{-h}^{+h} \sigma_y \, dz &= \int_{-h}^{+h} \sigma_y \,.\, z dz = 0\end{aligned}\right\} \qquad \text{...(14.25)}$$

The solution then becomes

$$\sigma_x = \sigma_y = \frac{\alpha E}{1-\nu}\left\{-T + \frac{1}{2h}\int_{-h}^{h} T dz + \frac{3z}{2h^3}\int_{-h}^{h} T z dz\right\} \qquad \text{...(14.26)}$$

Now let

$$N_T = \alpha E \int_{-h}^{h} T dz$$

$$M_T = \alpha E \int_{-h}^{h} T z dz \qquad \text{...(14.27)}$$

Then, stress components become

$$\left.\begin{aligned}\sigma_x = \sigma_y &= \frac{1}{1-\nu}\left\{-\alpha E T + \frac{1}{2h} N_T + \frac{3z}{2h^3} M_T\right\} \\ \sigma_z &= \tau_{xy} = \tau_{yz} = \tau_{zx} = 0\end{aligned}\right\} \qquad \text{...(14.28)}$$

The strain components are

$$\left.\begin{aligned}\varepsilon_x = \varepsilon_y &= \frac{1}{E}\left\{\frac{1}{2h} N_T + \frac{3z}{2h^3} M_T\right\} \\ \varepsilon_z &= -\frac{2\nu}{(1-\nu)E}\left\{\frac{1}{2h} N_T + \frac{3z}{2h^3} M_T\right\} + \left(\frac{1+\nu}{1-\nu}\right)\alpha T \\ \gamma_{xy} &= \gamma_{yz} = \gamma_{zx} = 0\end{aligned}\right\} \qquad \text{...(14.29)}$$

Displacement components are

$$u_x = \frac{x}{E}\left\{\frac{1}{2h} N_T + \frac{3z}{2h^3} M_T\right\}$$

$$u_y = \frac{y}{E}\left\{\frac{1}{2h} N_T + \frac{3z}{2h^3} M_T\right\} \qquad \text{...(14.30)}$$

$$u_z = -\frac{3M_T}{4h^3E}(x^2+y^2) + \frac{1}{(1-\nu)E}\left\{(1+\nu)\alpha E\int_0^z T\,dz - \frac{\nu z}{h}N_T - \frac{3\nu z^2}{2h^3}M_T\right\}$$

14.5 Rectangular Beam with Temperature Variation Through the Depth only

Consider a beam (Fig. 14.2) of depth $2h$ and width b such that both b and $2h$ are small compared to the length l. Since the beam

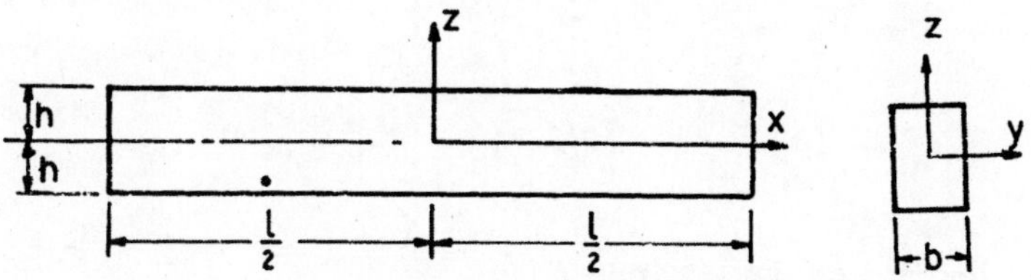

(a) (b)

Fig. 14.2

is thin, the plane stress assumption gives

$$\sigma_y = \tau_{xy} = \tau_{yz} = 0 \qquad ...(14.31)$$

Using the semi-inverse method and assuming that

$$\sigma_z = \tau_{xz} = 0,\ \sigma_x = \sigma_x(z) \qquad ...(14.32)$$

the equations of equilibrium are satisfied identically, and the compatibility equation in terms of stresses reduces to

$$\frac{d^2}{dz^2}(\sigma_x + \alpha ET) = 0 \qquad ...(14.33)$$

$$\therefore \qquad \sigma_x = -\alpha ET + C_1 + C_2 z \qquad ...(14.34)$$

All bounding surfaces are free of traction in the solution found thus far with the exception of the end faces $x = \pm\frac{l}{2}$. The conditions on the latter faces may be replaced by the requirements that the resultant force and moment be zero. The constants C_1 and C_2 are then again determined by Eqs. (14.25), and we get

$$\sigma_x = \alpha E\left\{-T + \frac{1}{2h}\int_{-h}^{h} T\,dz + \frac{3z}{2h^3}\int_{-h}^{h} Tz\,dz\right\} \qquad ...(14.35)$$

Thus the solution becomes,

Stress components

$$\left.\begin{aligned}\sigma_x &= -\alpha ET + \frac{1}{2h}N_T + \frac{3z}{2h^3}M_T \\ &= -\alpha ET + \frac{bN_T}{A} + \frac{z(bM_T)}{I} \\ \sigma_z &= \tau_{xz} = 0\end{aligned}\right\} \quad \text{...(14.36)}$$

where $$A = 2bh, I = \frac{2bh^3}{3}.$$

Strain components,

$$\left.\begin{aligned}\varepsilon_x &= \frac{1}{E}\left\{\frac{(bN_T)}{A} + \frac{z}{I}(bM_T)\right\} \\ \varepsilon_z &= -\frac{\nu}{E}\left\{\frac{(bN_T)}{A} + \frac{z}{I}(bM_T)\right\} + \left(\frac{1+\nu}{E}\right)\alpha T \\ \gamma_{xz} &= 0\end{aligned}\right\} \quad \text{...(14.37)}$$

Displacement components.

$$u_x = \frac{x}{E}\left\{\frac{(bN_T)}{A} + \frac{z}{I}(bM_T)\right\} \quad \text{...(14.38)}$$

$$u_z = -\frac{(bM_T)}{2EI}x^2 - \frac{\nu}{E}\left\{\frac{(bN_T)}{A}z + \frac{z^2}{2I}(bM_T)\right\} + \alpha\left(\frac{1+\nu}{E}\right)\int_0^z T\,dz.$$

Example 14.1 *Cooling of a clamped pipe.*

Solution. Consider a pipe held by two fixed supports as shown in Fig. 14.3. When mounted, let the temperature of the pipe be 20°C. If the pipe is cooled to 0°C by a cold fluid passing through the pipe, then away from the end supports, we can assume for the stress due to temperature change a one dimensional stress distribution in the x-direction. Thus

$$\sigma_y = \sigma_z = \tau_{xy} = \tau_{yz} = \tau_{zx} = 0 \quad \text{...(1)}$$

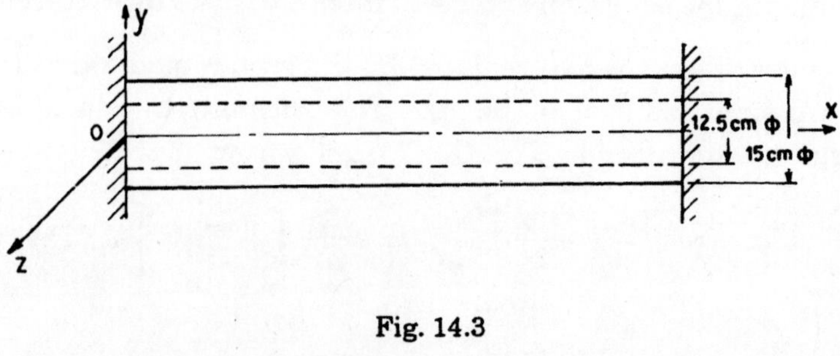

Fig. 14.3

This stress distribution satisfies the boundary conditions. Now

$$\left.\begin{aligned} \varepsilon_x &= \frac{\sigma_x}{E} + \alpha T \\ \varepsilon_y &= -\nu\frac{\sigma_x}{E} + \alpha T \\ \varepsilon_z &= -\nu\frac{\sigma_x}{E} + \alpha T \\ \gamma_{xy} &= \gamma_{yz} = \gamma_{xz} = 0 \end{aligned}\right\} \quad \text{...(2)}$$

Because the ends of the tube are fixed, hence

$$\varepsilon_x = 0$$

or
$$\sigma_x = -E\alpha T \quad \text{...(3)}$$

$$\therefore \quad \left.\begin{aligned} \varepsilon_y &= \alpha T(1+\nu) \\ \varepsilon_z &= \alpha T(1+\nu) \end{aligned}\right\} \quad \text{...(4)}$$

The strains are thus all constant. Thus the compatibility equations are satisfied identically. The tensile force on the supports is given by

$$P = \sigma_x A = -E\alpha TA.$$

Let
$$E = 2 \times 10^6 \text{ N/mm}^2$$
$$\alpha = 1.5 \times 10^{-6}/^\circ\text{C}$$

$$\therefore \quad \sigma_x = -2 \times 10^6 \times 1.5 \times 10^{-6} \times (-20) = 6 \text{ N/mm}^2$$

$$P = 6 \times \frac{\pi}{4}(15^2 - 12.5^2) \times 10^2 = 6 \times \frac{\pi}{4} \times 68.75 \times 10^2$$
$$= 323976 \text{ N}$$

Example 14.2 *Heating of a clamped square plate.*

Solution. Consider a square flat plate clamped rigidly on all sides as shown in Fig. 14.4. The plate is initially at a uniform temperature T_0. Now imagine that edges AB and CD are exposed to a heat source so that the temperature rises linearly along these edges to a value $(T_0 + e)$ at B and C. The edge AD is maintained at temperature T_0 while edge BC is heated to a uniform temperature $(T_0 + e)$.

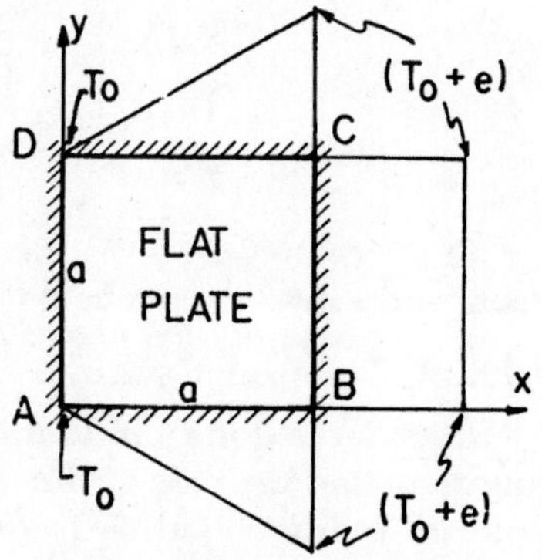

Fig. 14.4

The steady state temperature distribution in the plate is a function only of x in the form

$$T(x, y, z) = T_0 + \frac{x}{a} e \qquad \text{...(1)}$$

This is then clearly a plane stress problem. Neglecting body forces, the equilibrium equations are :

$$\left.\begin{aligned} \frac{\partial \sigma_x}{\partial x} + \frac{\partial \tau_{xy}}{\partial y} = 0 \\ \frac{\partial \tau_{xy}}{\partial x} + \frac{\partial \sigma_y}{\partial y} = 0 \end{aligned}\right\} \qquad \text{...(2)}$$

From Hooke's law

$$\left.\begin{aligned} \varepsilon_x &= \frac{1}{E}[\sigma_x - \nu\sigma_y] + \frac{x}{a} e \\ \varepsilon_y &= \frac{1}{E}[\sigma_y - \nu\sigma_x] + \frac{x}{a} e \\ \gamma_{xy} &= \frac{\tau_{xy}}{G} \end{aligned}\right\} \qquad \text{...(3)}$$

The boundary conditions require that

$$\int_0^a \varepsilon_x \, dx = 0 \, ; \int_0^a \varepsilon_y \, dy = 0 \qquad \text{...(4)}$$

or

$$\int_0^a \left[\frac{\sigma_x - \nu\sigma_y}{E} + \frac{x}{a} e \right] dx = 0$$

$$\int_0^a \left[\frac{\sigma_y - \nu\sigma_x}{E} + \frac{x}{a} e \right] dy = 0$$

or

$$\left.\begin{aligned} \frac{1}{E}\int_0^a (\sigma_x - \nu\sigma_y)\, dx = \frac{ae}{2} \\ \frac{1}{E}\int_0^a (\sigma_y - \nu\sigma_x) dy = xe \end{aligned}\right\} \qquad \text{...(5)}$$

To determine σ_x and σ_y, Eqs. (5) are to be solved simultaneously and then τ_{xy} can be determined from Eqs. (2).

14.6 Thick Circular Cylinder

Consider a long circular cylinder as shown in Fig. 14.5. Taking z-axis as the axis of the cylinder, the stresses acting on an element at radius r, taken from wall of circular cylinder are as shown in Fig. 14.5.

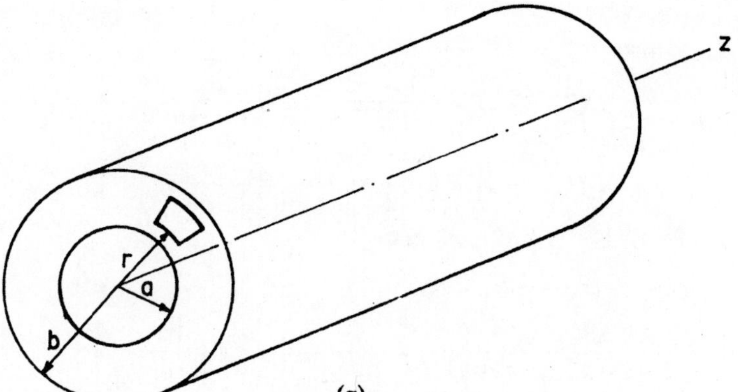

(a)

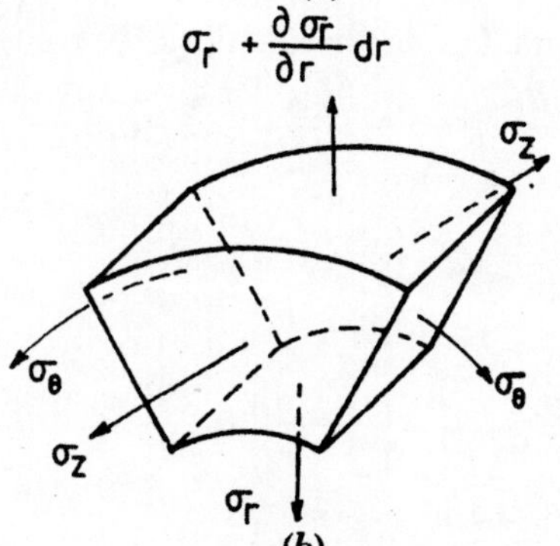

(b)

Fig. 14.5

(*i*) *Any steady state temperature gradient.* For radial force equilibrium, we get

$$\frac{d\sigma_r}{dr} + \frac{\sigma_r - \sigma_\theta}{r} = 0 \qquad \text{...(14.39)}$$

The strains at a point in the cylinder wall are given by

$$\left.\begin{aligned} \varepsilon_r &= \frac{1}{E}[\sigma_r - \nu(\sigma_\theta + \sigma_z)] + \alpha T \\ \varepsilon_\theta &= \frac{1}{E}[\sigma_\theta - \nu(\sigma_z + \sigma_r)] + \alpha T \\ \varepsilon_z &= \frac{1}{E}[\sigma_z - \nu(\sigma_\theta + \sigma_r)] + \alpha T \end{aligned}\right\} \qquad \text{...(14.40)}$$

For a long cylinder, sufficiently away from the ends, plane strain conditions may be assumed to exist, *i.e.* $\varepsilon_z = 0$.

$$\therefore \qquad \sigma_z = \nu(\sigma_r + \sigma_\theta) - \alpha ET \qquad \text{...(14.41)}$$

Hence

$$\left.\begin{aligned}\varepsilon_r &= (1+\nu)\,\alpha T + \left(\frac{1-\nu^2}{E}\right)\left(\sigma_r - \frac{\nu}{1-\nu}\sigma_\theta\right)\\ \varepsilon_\theta &= (1+\nu)\,\alpha T + \left(\frac{1-\nu^2}{E}\right)\left(\sigma_\theta - \frac{\nu}{1-\nu}\sigma_r\right)\end{aligned}\right\} \quad ...(14.42)$$

From Eq. (14.42), we get

$$\left.\begin{aligned}\sigma_r - \sigma_\theta &= \frac{E}{1+\nu}(\varepsilon_r - \varepsilon_\theta)\\ \text{and}\quad \sigma_r &= \frac{E(1-\nu)}{(1-2\nu)(1+\nu)}\left[\left(\varepsilon_r + \frac{\nu}{1-\nu}\varepsilon_\theta\right) - \alpha T\left(\frac{1+\nu}{1-\nu}\right)\right]\end{aligned}\right\} \quad ...(14.43)$$

Substituting Eqs. (14.43) in Eq. (14.39), we get

$$\left(\frac{1-\nu}{1-2\nu}\right)\left[\frac{d\varepsilon_r}{dr} + \frac{\nu}{1-\nu}\frac{d\varepsilon_\theta}{dr} - \frac{1+\nu}{1-\nu}\alpha\frac{dT}{dr}\right] + \frac{\varepsilon_r - \varepsilon_\theta}{r} = 0 \quad ...(14.44)$$

Now $\quad \varepsilon_r = \dfrac{du_r}{dr}, \; \varepsilon_\theta = \dfrac{u_r}{r}$.

Hence Eq. (14.44) reduces to

$$\frac{d^2u_r}{dr^2} + \frac{1}{r}\frac{du_r}{dr} - \frac{u_r}{r^2} = \left(\frac{1+\nu}{1-\nu}\right)\alpha\frac{dT}{dr}$$

or

$$\frac{d}{dr}\left[\frac{1}{r}\frac{d(ru_r)}{dr}\right] = \left(\frac{1+\nu}{1-\nu}\right)\alpha\frac{dT}{dr} \quad ...(14.45)$$

Integrating

$$\frac{1}{r}\frac{d(ru_r)}{dr} = \frac{u_r}{r} + \frac{du_r}{dr} = \left(\frac{1+\nu}{1-\nu}\right)\alpha T + C_1$$

$$ru_r = \left(\frac{1+\nu}{1-\nu}\right)\alpha\int Tr\,dr + \int C_1 r\,dr + C_2$$

or

$$u_r = \frac{\alpha}{r}\left(\frac{1+\nu}{1-\nu}\right)\int_a^r Tr\,dr + \frac{C_1 r}{2} + \frac{C_2}{r} \quad ...(14.46)$$

$$\varepsilon_\theta = \frac{u_r}{r} = \frac{\alpha}{r^2}\left(\frac{1+\nu}{1-\nu}\right)\int_a^r Tr\,dr + \frac{C_1}{2} + \frac{C_2}{r^2}$$

and

$$\varepsilon_\theta + \varepsilon_r = \left(\frac{1+\nu}{1-\nu}\right)\alpha T + C_1.$$

Substituting in Eqs. (14.43), we get

$$\sigma_r = \frac{E}{1+\nu}\left[\frac{C_1}{2(1-2\nu)} - \frac{C_2}{r^2}\right] - \frac{\alpha E}{(1-\nu)r^2}\int_a^r Tr\,dr \quad ...(14.47)$$

$$\sigma_r - \sigma_\theta = \frac{E\alpha T}{1-\nu} - \frac{2\alpha E}{r^2(1-\nu)}\int_a^r Tr\,dr - \frac{2C_2E}{r^2(1+\nu)}$$

$$\sigma_\theta = \frac{E}{(1+\nu)}\left[\frac{C_1}{2(1-2\nu)} + \frac{C_2}{r^2}\right] - \frac{\alpha ET}{(1-\nu)} + \frac{\alpha E}{r^2(1-\nu)}\int_a^r Tr\,dr \qquad \text{...(14.48)}$$

$$\sigma_z = \frac{\nu EC_1}{(1+\nu)(1-2\nu)} - \frac{\alpha ET}{(1-\nu)} \qquad \text{...(14.49)}$$

The boundary conditions of zero tractions on the cylindrical surface are

$$\sigma_r = 0 \text{ at } r = a \text{ and } r = b.$$

Substituting in Eq. (14.47), we get

$$\frac{C_1}{2(1-2\nu)} - \frac{C_2}{a^2} = \left(\frac{1+\nu}{1-\nu}\right)\frac{\alpha}{a^2}\int_a^b Trdr = 0$$

$$\frac{C_1}{2(1-2\nu)} - \frac{C_2}{b^2} = \left(\frac{1+\nu}{1-\nu}\right)\frac{\alpha}{b^2}\int_a^b Trdr$$

Thus $$C_1 = \frac{2(1-2\nu)(1+\nu)}{(1-\nu)} \cdot \frac{\alpha}{(b^2-a^2)}\int_a^b Trdr$$

$$C_2 = \frac{\alpha a^2}{(b^2-a^2)}\left(\frac{1+\nu}{1-\nu}\right)\int_a^b Trdr$$

Hence $$\sigma_r = \frac{\alpha E}{(1-\nu)} \cdot \frac{1}{r^2}\left[\left(\frac{r^2-a^2}{b^2-a^2}\right)\int_a^b Trdr - \int_a^r Trdr\right] \qquad \text{...(14.50)}$$

$$\sigma_\theta = \frac{\alpha E}{(1-\nu)} \cdot \frac{1}{r^2}\left[\left(\frac{r^2+a^2}{b^2-a^2}\right)\int_a^b Trdr + \int_a^r Trdr - Tr^2\right] \qquad \text{...(14.51)}$$

$$\sigma_z = \frac{\alpha E}{(1-\nu)}\left[\left(\frac{2\nu}{b^2-a^2}\right)\int_a^b Trdr - T\right] \qquad \text{...(14.52)}$$

$$u_r = \frac{\alpha}{r}\left\{\frac{(1+\nu)}{(1-\nu)}\int_a^r Trdr + \frac{(1-\nu)r^2 + (1+\nu)a^2}{(b^2-a^2)(1-\nu)}\int_a^b Trdr\right\} \qquad \text{...(14.53)}$$

(ii) Steady state heat flow. Constant temperature difference between the cylinder walls.

If the temperature at $r = a$ is T_i and at $r = b$ is T_0, then if there is a steady state heat flow, the temperature T, at any radius is

$$T = \lambda \ln \frac{b}{r}$$

where
$$\lambda = \frac{T_i - T_0}{\ln \frac{b}{a}}$$

Now $\displaystyle\int_a^r r \ln \frac{b}{r}\, dr = \frac{r^2}{2} \ln \frac{b}{r} - \frac{a^2}{2} \ln \frac{b}{a} + \frac{(r^2 - a^2)}{4}$

Substituting in Eqs. (14.50) to (14.52), we get

$$\left.\begin{aligned}
\sigma_r &= \frac{\alpha E \lambda}{2(1-\nu)} \left[-\ln \frac{b}{r} - \frac{a^2}{(b^2 - a^2)} \left(1 - \frac{b^2}{r^2}\right) \ln \frac{b}{a} \right] \\
\sigma_\theta &= \frac{\alpha E \lambda}{2(1-\nu)} \left[1 - \ln \frac{b}{r} - \frac{a^2}{b^2 - a^2} \left(1 + \frac{b^2}{r^2}\right) \ln \frac{b}{a} \right] \\
\sigma_z &= \frac{\alpha E \lambda}{2(1-\nu)} \left[\nu - 2 \ln \frac{b}{r} - \frac{2\nu a^2}{b^2 - a^2} \ln \frac{b}{a} \right]
\end{aligned}\right\} \quad \text{...(14.54)}$$

14.7 Circular Disc with Radial Temperature Variation

Consider a circular disc of inner radius a and outer radius b. If the thickness h of the disc is such that $\frac{h}{b}$ is very small as compared to unity and the end faces are free of tractions, then the problem is one of plane stress type. In this case, the radial displacement is the solution of the equation

$$\frac{d}{dr}\left[\frac{1}{r}\frac{d(ru_r)}{dr}\right] = \alpha(1+\nu)\frac{dT}{dr} \quad \text{...(14.55)}$$

The general solution of this equation is

$$u_r = (1+\nu)\frac{\alpha}{r}\int_a^r Trdr + C_1 r + \frac{C_2}{r} \quad \text{...(14.56)}$$

Now
$$\varepsilon_\theta = \frac{u_r}{r} = (1+\nu)\frac{\alpha}{r^2}\int_a^r Trdr + C_1 + \frac{C_2}{r^2}$$

$$\varepsilon_\theta + \varepsilon_r = \frac{du_r}{dr} + \frac{u_r}{r} = (1+\nu)\alpha T + C_1$$

$$\therefore \quad \varepsilon_r = (1+\nu)\alpha T - (1+\nu)\frac{\alpha}{r^2}\int_a^r Trdr - \frac{C_2}{r^2}$$

$$\left.\begin{aligned}
\therefore \quad \sigma_r &= -\frac{E\alpha}{r^2}\int_a^r Trdr + \frac{EC_1}{1-\nu} - \frac{EC_2}{(1+\nu)r^2} \\
\sigma_\theta &= \frac{\alpha E}{r^2}\int_a^r Trdr - \alpha ET + \frac{EC_1}{1-\nu} + \frac{EC_2}{(1+\nu)r^2} \\
\tau_{r\theta} &= 0
\end{aligned}\right\} \quad \text{...(14.57)}$$

The boundary conditions of zero tractions on the cylindrical surfaces are then

$$\sigma_r = 0 \text{ at } r = a \text{ and } r = b$$

and thus the constants C_1 and C_2 may now be determined. The final expressions for the non-zero stress and displacement component are

$$\left.\begin{aligned}
\sigma_r &= \frac{\alpha E}{r^2}\left[\left(\frac{r^2 - a^2}{b^2 - a^2}\right)\int_a^b Trdr - \int_a^r Trdr\right] \\
\sigma_\theta &= \frac{\alpha E}{r^2}\left[\left(\frac{r^2 + a^2}{b^2 - a^2}\right)\int_a^b Trdr + \int_a^r Trdr - Tr^2\right] \\
u_r &= \frac{\alpha}{r}\left[(1+\nu)\int_a^r Trdr + \frac{(1-\nu)r^2 + (1+\nu)a^2}{(b^2 - a^2)}\int_a^b Trdr\right]
\end{aligned}\right\} \quad \text{...(14.58)}$$

When the temperature is constant then all stress components are zero and the radial displacement is

$$u_r = \alpha Tr \quad \text{...(14.59)}$$

For a solid disc $a = 0$ and we get

$$\left.\begin{aligned}
\sigma_r &= \alpha E\left[\frac{1}{b^2}\int_0^b Trdr - \frac{1}{r^2}\int_0^r Trdr\right] \\
\sigma_\theta &= \alpha E\left[\frac{1}{b^2}\int_0^b Trdr + \frac{1}{r^2}\int_0^r Trdr - T\right] \\
u_r &= \frac{\alpha}{r}\left[(1+\nu)\int_0^r Trdr + (1-\nu)\left(\frac{r}{b}\right)^2\int_0^b Trdr\right]
\end{aligned}\right\} \quad \text{...(14.60)}$$

These expressions becomes indeterminate for $r = 0$. However, if the temperature is finite at $r = 0$, then by using l'Hospital's rule, we get

$$\left.\begin{aligned} \lim_{r \to 0} \frac{1}{r^2}\int_0^r Trdr &= \frac{1}{2}T(0) \\ \lim_{r \to 0} \frac{1}{r^2}\int_0^r Trdr &= 0 \end{aligned}\right\} \qquad \ldots(14.61)$$

In this case $(u_r)_{r=0} = 0$ and the stresses become

$$(\sigma_r)_{r=0} = (\sigma_\theta)_{r=0} = \alpha E\left[\frac{1}{b^2}\int_0^b Trdr - \frac{1}{2}T(0)\right] \qquad \ldots(14.62)$$

14.8 Solid or Hollow Sphere under Radial Temperature Variation

By using spherical co-ordinates (r, θ, φ) and omitting all displacement components and all derivatives in the θ and φ direction, the equilibrium equations in terms of displacements reduce to

$$\frac{d}{dr}\left[\frac{1}{r^2}\frac{d(r^2u_r)}{dr}\right] = \alpha\left(\frac{1+\nu}{1-\nu}\right)\frac{dT}{dr} \qquad \ldots(14.63)$$

The general solution of this equation is

$$u_r = \alpha\left(\frac{1+\nu}{1-\nu}\right)\frac{1}{r^2}\int_a^r Tr^2dr + C_1r + \frac{C_2}{r^2} \qquad \ldots(14.64)$$

For a hollow sphere of inner radius a and outer radius b the constants C_1 and C_2 are determined from the condition that

$$\sigma_r = 0 \text{ at } r = a \text{ and } r = b \qquad \ldots(14.65)$$

From the strain-displacement relations and Hooke's law the non-zero stress components are :

$$\sigma_r = \lambda\left(\frac{du_r}{dr} + \frac{2u_r}{r}\right) + 2G\frac{du_r}{dr} - (3\lambda + 2G)\alpha T$$

$$= -\frac{4G\alpha}{r^3}\left(\frac{3\lambda + 2G}{\lambda + 2G}\right)\int_a^r Tr^2dr + (3\lambda + 2G)C_1 - \frac{4GC_2}{r^2} \qquad \ldots(14.66)$$

$$\sigma_\theta = \sigma_\varphi = \lambda\left(\frac{du_r}{dr} + \frac{2u_r}{r}\right) + 2G\frac{u_r}{r} - (3\lambda + 2G)\alpha T$$

$$= \frac{2G\alpha}{r^3}\left(\frac{3\lambda + 2G}{\lambda + 2G}\right)\int_a^r Tr^2dr + (3\lambda + 2G)C_1$$

$$+ 2G\frac{C_2}{r^3} - 2G\left(\frac{3\lambda + 2G}{\lambda + 2G}\right)\alpha T \qquad \ldots(14.67)$$

Using boundary conditions (12.65), we get for a hollow sphere

$$u_r = \frac{\alpha}{(b^3 - a^3)}\left(\frac{3\lambda + 2G}{\lambda + 2G}\right)\left[\frac{a^3}{r^3}\int_a^b Tr^2 dr + \frac{b^3}{r^2}\int_a^r Tr^2 dr + \left(\frac{4G}{3\lambda + 2G}\right) r \int_a^b Tr^2 dr\right] \quad ...(14.68)$$

$$\left.\begin{aligned} \sigma_r &= \frac{4G\alpha}{(a^3 - b^3)}\left(\frac{3\lambda + 2G}{\lambda + 2G}\right)\left[\frac{a^3}{r^3}\int_r^b Tr^2 dr + \frac{b^3}{r^3}\int_a^r Tr^2 dr - \int_a^b Tr^2 dr\right] \\ \sigma_\theta = \sigma_\varphi &= \frac{2G\alpha}{(b^3 - a^3)}\left(\frac{3\lambda + 2G}{\lambda + 2G}\right)\left[\frac{a^3}{r^3}\int_a^b Tr^2 dr + \frac{b^3}{r^3}\int_a^r Tr^2 dr + 2\int_a^b Tr^2\, dr - (b^3 - a^3)T\right] \end{aligned}\right\} \quad ...(14.69)$$

In the special case in which T is constant, these equations show that all stress components are zero, and that

$$u_r = \alpha T r \quad ...(14.70)$$

For a solid sphere $a = 0$, and we get

$$u_r = \alpha\left(\frac{3\lambda + 2G}{\lambda + 2G}\right)\left[\frac{1}{r^2}\int_0^r Tr^2 dr + \frac{4G}{(3\lambda + 2G)}\left(\frac{r}{b^3}\right)\int_0^b Tr^2 dr\right]$$

$$\sigma_r = 4G\alpha\left(\frac{3\lambda + 2G}{\lambda + 2G}\right)\left[\frac{1}{b^3}\int_0^r Tr^2 dr - \frac{1}{r^3}\int_0^r Tr^2 dr\right] \quad ...(14.71)$$

$$\sigma_\theta = \sigma_\varphi = 2G\alpha\left(\frac{3\lambda + 2G}{\lambda + 2G}\right)\left[\frac{1}{r^3}\int_0^r Tr^2 dr + \frac{2}{b^3}\int_0^b Tr^2 dr - T\right]$$

Now

$$\left.\begin{aligned} \lim_{r \to 0} \frac{1}{r^3}\int_0^r Tr^2 dr &= \frac{1}{3}T(0) \\ \lim_{r \to 0} \frac{1}{r^2}\int_0^r Tr^2 dr &= 0 \end{aligned}\right\} \quad ...(14.72)$$

Hence

$$(u_r)_{r=0} = 0$$

$$(\sigma_r)_{r=0} = (\sigma_\theta)_{r=0} = (\sigma_\varphi)_{r=0}$$

$$= 4G\alpha\left(\frac{3\lambda + 2G}{\lambda + 2G}\right)\left[\frac{1}{b^3}\int_0^b Tr^2dr - \frac{1}{3}T(0)\right]$$

For an infinite body $b \to \infty$ and, we get

$$\left.\begin{aligned} u_r &= \frac{\alpha}{r^2}\left(\frac{3\lambda + 2G}{\lambda + 2G}\right)\int_0^r Tr^2dr \\ \sigma_r &= -\frac{4G\alpha}{r^3}\left(\frac{3\lambda + 2G}{\lambda + 2G}\right)\int_0^r Tr^2dr \\ \sigma_\theta &= \frac{2G\alpha}{r^3}\left(\frac{3\lambda + 2G}{\lambda + 2G}\right)\left[\int_0^r Tr^2dr - Tr^3\right] \end{aligned}\right\} \quad ...(14.74)$$

14.9 Circular Rectangular Beam with Radial Temperature Variation

Consider a rectangular circular beam as shown in Fig. 14.6, whose median line describes an arc of a circle. All bounding surfaces are assumed to be free of tractions and $T = T(r)$.

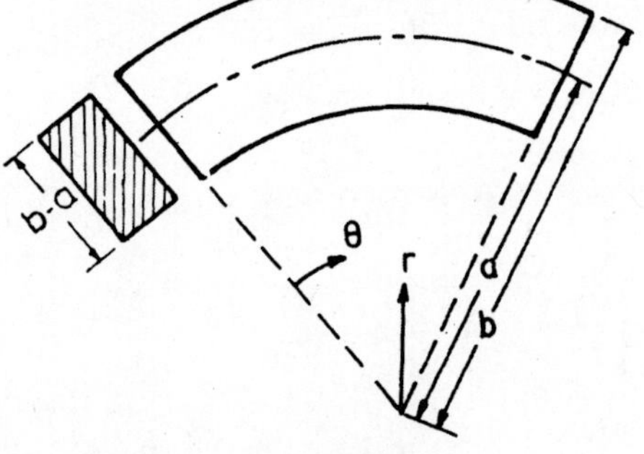

Fig. 14.6

The biharmonic equation is

$$\nabla^4\Phi = -\alpha E\,\Delta^2 T$$

where $$\nabla^2 = \frac{\partial^2}{\partial r^2} + \frac{1}{r}\frac{\partial}{\partial r} + \frac{1}{r^2}\frac{\partial^2}{\partial \theta^2}$$

Since the temperature varies only in the radial direction, therefore the variation with respect to θ is omitted. Hence we get

$$\frac{1}{r}\frac{d}{dr}\left(r\frac{d}{dr}\right)\left\{\frac{1}{r}\frac{d}{dr}\left(r\frac{d\Phi}{dr}\right) + \alpha ET\right\} = 0 \quad ...(14.75)$$

Integrating, we get

$$\Phi = C_1 \log\left(\frac{r}{a}\right) + C_2\left(\frac{r}{a}\right)^2 \log\left(\frac{r}{a}\right) + C_3\left(\frac{r}{a}\right)^2 + C_4 - E\alpha\int_a^r\left[\frac{1}{r_2}\int_a^{r_2} Tr_1 dr_1\right] dr_2 \qquad ...(14.76)$$

The boundary conditions are

$$\Phi = \frac{d\Phi}{dr} = 0 \text{ at } r = a \text{ and } r = b \qquad ...(14.77)$$

The resultant force and moment on the end faces are then also zero. The constants C_1, C_2, C_3 and C_4 are then

$$C_1 = \frac{E\alpha}{N}\left[\left\{2\left(\frac{b}{a}\right)^2 \log\frac{b}{a}\left(2\log\frac{b}{a} - 1\right) + \left(\frac{b}{a}\right)^2 - 1\right\}\int_a^b Trdr - 4\left(\frac{b}{a}\right)^2 \log\frac{b}{a}\int_a^b Tr\log\frac{r}{a}\,dr\right]$$

$$C_2 = \frac{E\alpha}{N}\left[\left\{2\left(\frac{b}{a}\right)^2 \log\frac{b}{a} - \left(\frac{b}{a}\right)^2 + 1\right\}\int_a^b Trdr - 2\left\{\left(\frac{b}{a}\right)^2 - 1\right\}\int_a^b Tr\log\frac{r}{a}\,dr\right]$$

$$C_3 = -C_4 = \frac{E\alpha}{N}\left[-2\left(\frac{b}{a}\right)^2\left(\log\frac{b}{a}\right)^2\int_a^b Trdr + \left\{2\left(\frac{b}{a}\right)^2 \log\frac{b}{a} + \left(\frac{b}{a}\right)^2 - 1\right\}\int_a^b Tr\log\frac{r}{a}\,dr\right]$$

where $N = 4\left(\frac{b}{a}\right)^2\left(\log\frac{b}{a}\right)^2 - \left[\left(\frac{b}{a}\right)^2 - 1\right]^2$

The stress components are

$$\left.\begin{aligned}
\sigma_r &= \frac{1}{r}\frac{\partial\Phi}{\partial r} = \frac{C_1}{r^2} + \frac{C_2}{a^2}\left(2\log\frac{r}{a} + 1\right) + \frac{2C_3}{a^2} - \frac{E\alpha}{r^2}\int_a^r Trdr \\
\sigma_\theta &= \frac{\partial^2\Phi}{\partial r^2} = -\frac{C_1}{r^2} + \frac{C_2}{a^2}\left(2\log\frac{r}{a} + 3\right) + \frac{2C_3}{a^2} - E\alpha\left[T - \frac{1}{r^2}\int_a^r Tr\,dr\right] \\
\tau_{r\theta} &= -\frac{\partial}{\partial r}\left(\frac{1}{r}\frac{\partial\Phi}{\partial\theta}\right) = 0
\end{aligned}\right\} \qquad ...(14.78)$$

14.10 Thermal Stresses in Plates

14.10.1 Basic Plate Equations

A thin plate of thickness h, whose median plane lies in the xy plane, with z denoting the distance from this plane is considered. The displacements, in the x, y and z directions, of points on the median plane are denoted by u_x, u_y and u_z respectively. The strain components are calculated on the basis of the assumptions of plane stress and that sections which are plane and perpendicular to the middle surface remain so after heating. The strain components are :

$$\left.\begin{aligned} \varepsilon_x &= \frac{\partial u_x}{\partial x} - z\frac{\partial^2 u_z}{\partial x^2} \\ \varepsilon_y &= \frac{\partial u_y}{\partial y} - z\frac{\partial^2 u_z}{\partial y^2} \\ \gamma_{xy} &= \frac{\partial u_z}{\partial y} + \frac{\partial u_y}{\partial x} - 2z\frac{\partial^2 u_z}{\partial x\,\partial y} \end{aligned}\right\} \qquad \text{...(14.79)}$$

The curvatures $K_x = \dfrac{1}{r_x}, K_y = \dfrac{1}{r_y}$ of the middle surface, in planes parallel to the xz and yz planes respectively are :

$$K_x = -\frac{\partial^2 u_z}{\partial x^2}, \; K_y = -\frac{\partial^2 u_z}{\partial y^2} \qquad \text{...(14.80)}$$

and the twist of the middle surface is

$$K_{xy} = \frac{1}{r_{xy}} = \frac{\partial^2 u_z}{\partial x\,\partial y} \qquad \text{...(14.81)}$$

From Hooke's law the non-zero stress components are :

$$\left.\begin{aligned} \sigma_x &= \frac{E}{1-\nu^2}[\varepsilon_x + \nu\varepsilon_y - (1+\nu)\alpha T] \\ \varepsilon_y &= \frac{E}{1-\nu^2}[\varepsilon_y + \nu\varepsilon_x - (1+\nu)\alpha T] \\ \tau_{xy} &= \frac{E}{2(1+\nu)}\gamma_{xy} \end{aligned}\right\} \qquad \text{...(14.82)}$$

Now the forces and moments per unit of length are

$$N_x = \int_{-h/2}^{h/2} \sigma_x \, dz, \; N_y = \int_{-h/2}^{h/2} \sigma_y \,.\, dz \qquad \text{...(14.83a)}$$

$$N_{xy} = \int_{-h/2}^{h/2} \tau_{xy} \, dz \qquad \text{...(14.83b)}$$

$$M_x = \int_{-h/2}^{h/2} \sigma_x \,.\, z dz, M_y = \int_{-h/2}^{h/2} \sigma_y\, z dz \qquad \text{...(14.83c)}$$

$$M_{xy} = -\int_{-h/2}^{h/2} \tau_{xy} \,.\, z dz \qquad \text{...(14.83d)}$$

Using Eqs. (14.79) and (14.82), Eqs. (14.83) become

$$\left.\begin{aligned}
N_x &= \frac{Eh}{1-\nu^2}\left(\frac{\partial u_x}{\partial x} + \nu \frac{\partial u_y}{\partial y}\right) - \frac{N_T}{1-\nu} \\
N_y &= \frac{Eh}{1-\nu^2}\left(\frac{\partial u_y}{\partial y} + \nu \frac{\partial u_x}{\partial x}\right) - \frac{N_T}{1-\nu} \\
N_{xy} &= \frac{Eh}{2(1+\nu)}\left(\frac{\partial u_x}{\partial y} + \frac{\partial u_y}{\partial x}\right) \\
M_x &= -D\left(\frac{\partial^2 u_z}{\partial x^2} + \nu \frac{\partial^2 u_z}{\partial y^2}\right) - \frac{M_T}{1-\nu} \\
M_y &= -D\left(\frac{\partial^2 u_z}{\partial y^2} + \nu \frac{\partial^2 u_z}{\partial x^2}\right) - \frac{M_T}{1-\nu} \\
M_{xy} &= (1-\nu) D \frac{\partial^2 u_z}{\partial x \partial y}
\end{aligned}\right\} \qquad \text{...(14.84)}$$

where

$$\left.\begin{aligned}
D &= \frac{Eh^2}{12(1-\nu^2)} \\
N_T &= \alpha E \int_{-h/2}^{h/2} T dz, \quad M_T = \alpha E \int_{-h/2}^{h/2} T z dz
\end{aligned}\right\} \qquad \text{...(14.85)}$$

Now expressing stresses directly in terms of forces and moments, we get

$$\left.\begin{aligned}
\sigma_x &= \frac{1}{1-\nu}\left[-\alpha E T + \frac{1}{h}\left\{(1-\nu) N_x + N_T\right\} + \frac{12z}{h^3}\left\{(1-\nu) M_x + M_T\right\}\right] \\
\sigma_y &= \frac{1}{1-\nu}\left[-\alpha E T + \frac{1}{h}\left\{(1-\nu) N_y + N_T\right\} + \frac{12z}{h^3}\left\{(1-\nu) M_y + M_T\right\}\right] \\
\tau_{xy} &= \frac{1}{h} N_{xy} - \frac{12z}{h^3} M_{xy}
\end{aligned}\right\} \qquad \text{...(14.86)}$$

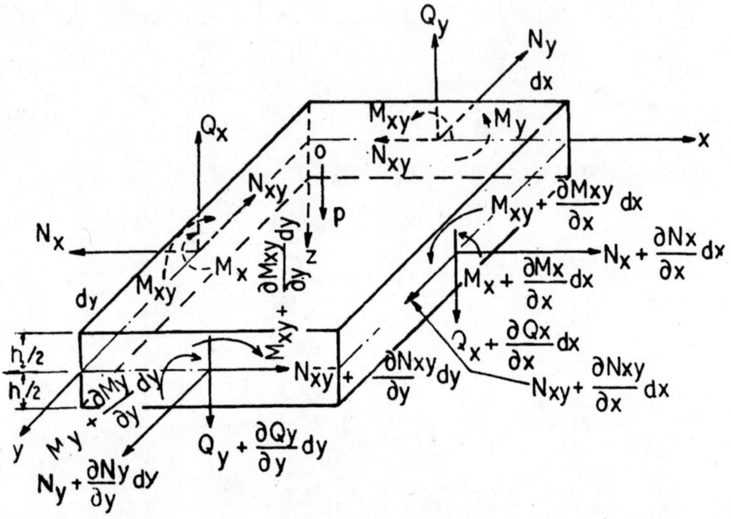

Fig. 14.7

The equations of equilibrium in the plane of the plate are (Fig. 14.7),

$$\left.\begin{aligned}\frac{\partial N_x}{\partial x} + \frac{\partial N_{xy}}{\partial y} = 0\\ \frac{\partial N_{xy}}{\partial x} + \frac{\partial N_y}{\partial y} = 0\end{aligned}\right\} \qquad \ldots(14.87)$$

If we define a stress function $F(x, y)$ such that

$$N_x = \frac{\partial^2 F}{\partial y^2}, \qquad N_y = \frac{\partial^2 F}{\partial x^2}$$

$$N_{xy} = \frac{\partial^2 F}{\partial x \partial y} \qquad \ldots(14.88)$$

then the equation of compatibility for thermoelastic plane stress problem

$$\frac{\partial^2}{\partial y^2}\left[\frac{N_x - \nu N_y + N_T}{h}\right] - 2(1+\nu)\frac{\partial^2}{\partial x \partial y}\left[\frac{N_{xy}}{h}\right]$$

$$+ \frac{\partial^2}{\partial x^2}\left[\frac{N_y - \nu N_x + N_T}{h}\right] = 0 \qquad \ldots(14.80)$$

can be written as

$$\nabla^4 F = -\nabla^2 N_T \qquad \ldots(14.90)$$

where $\nabla^2 = \dfrac{\partial^2}{\partial x^2} + \dfrac{\partial^2}{\partial y^2}$

Thus by solving Eq. (14.90) the forces N_x, N_y and N_{xy} can be determined.

Now the equilibrium equations of forces in the z-direction and moments about the x and y-axes as shown in Fig. 14.7 are :

$$\left.\begin{aligned}\frac{\partial Q_x}{\partial x}+\frac{\partial Q_y}{\partial y}+p=0\\ \frac{\partial M_{xy}}{\partial x}-\frac{\partial M_y}{\partial y}+Q_y=0\\ \frac{\partial M_{xy}}{\partial y}-\frac{\partial M_x}{\partial x}+Q_x=0\end{aligned}\right\} \quad ...(14.91)$$

where Q_x and Q_y are the shear forces per unit of length and $p = p(x, y)$ is the distributed transverse loading per unit of area. Substituting for Q_y and Q_x from the last two equations into the first equation of Eqs. (14.91), we get

$$\frac{\partial^2 M_x}{\partial x^2}-\frac{2\partial^2 M_{xy}}{\partial x\partial y}+\frac{\partial^2 M_y}{\partial y^2}=-p \quad ...(14.92)$$

Substituting for moments from Eqs. (14.84), we get

$$\frac{\partial^2}{\partial x^2}\left[D\left(\frac{\partial^2 u_z}{\partial x^2}+\nu\frac{\partial^2 u_z}{\partial y^2}\right)\right]+2(1-\nu)\frac{\partial^2}{\partial x\partial y}\left[D\frac{\partial^2 u_z}{\partial x\partial y}\right]$$

$$+\frac{\partial^2}{\partial y^2}\left[D\left(\frac{\partial^2 u_z}{\partial y^2}+\nu\frac{\partial^2 u_z}{\partial x^2}\right)\right]=p-\frac{1}{1-\nu}\nabla^2 M_T$$

or

$$D\nabla^4 u_z=p-\frac{1}{1-\nu}\nabla^2 M_T \quad ...(14.93)$$

If $p = 0$, then

$$D\nabla^4 u_z=-\frac{1}{1-\nu}\nabla^2 M_T \quad ...(14.94)$$

Eq. (14.93) or (14.94) may be solved for u_z and then the moments can be determined.

14.10.2 Boundary Conditions

The most frequently used boundary conditions with respect to the coordinate system (s, n) as shown in Fig. 14.8 are :

(*i*) *Clamped edge*

$$u_z=0,\ \frac{\partial u_z}{\partial n}=0$$

Also

$$\frac{\partial u_z}{\partial n}=\frac{\partial u_z}{\partial x}\cos\alpha+\frac{\partial u_z}{\partial y}\sin\alpha \quad ...(14.95)$$

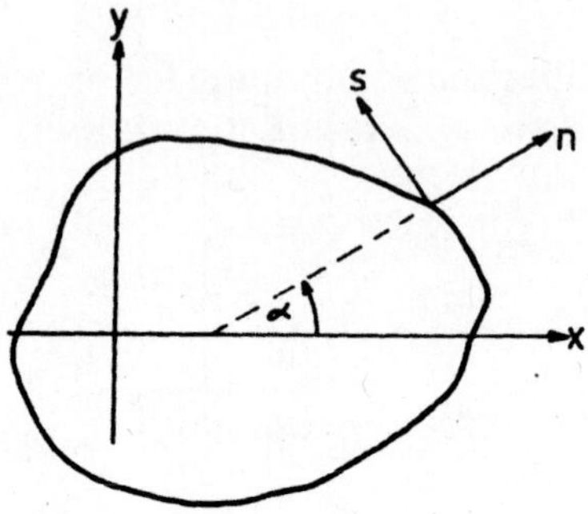

Fig. 14.8

(ii) Simply supported edge

$$u_z = 0, M_n = 0 \quad ...(14.96)$$

$$\left.\begin{aligned} \text{Now}\quad M_n &= M_x \cos^2\alpha + M_y \sin^2\alpha - 2M_{xy}\sin\alpha\cos\alpha \\ M_s &= M_x \sin^2\alpha + M_y \cos^2\alpha + 2M_{xy}\sin\alpha\cos\alpha \\ M_{ns} &= (M_x - M_y)\sin\alpha\cos\alpha + M_{xy}(\cos^2\alpha - \sin^2\alpha) \end{aligned}\right\} \quad ...(14.97)$$

or
$$M_n = -D\left[\left(\frac{\partial^2 u_z}{\partial x^2}\cos^2\alpha + 2\frac{\partial^2 u_z}{\partial x \partial y}\sin\alpha\cos\alpha + \frac{\partial^2 u_z}{\partial y^2}\sin^2\alpha\right) + \nu\left(\frac{\partial^2 u_z}{\partial y^2}\cos^2\alpha - 2\frac{\partial^2 u_z}{\partial x\partial y}\sin\alpha\cos\alpha + \frac{\partial^2 u_z}{\partial x^2}\sin^2\alpha\right)\right] - \frac{M_T}{1-\nu}$$

$$M_s = -D\left[\left(\frac{\partial^2 u_z}{\partial y^2}\cos^2\alpha - \frac{2\partial^2 u_z}{\partial x\partial y}\sin\alpha\cos\alpha + \frac{\partial^2 u_z}{\partial x^2}\sin^2\alpha\right) + \nu\left(\frac{\partial^2 u_z}{\partial x^2}\cos^2\alpha + \frac{2\partial^2 u_z}{\partial x\partial y}\sin\alpha\cos\alpha + \frac{\partial^2 u_z}{\partial y^2}\sin^2\alpha\right)\right] - \frac{M_T}{1-\nu}$$

$$M_{ns} = (1-\nu)D\left[\frac{\partial^2 u_z}{\partial x\partial y}(\cos^2\alpha - \sin^2\alpha) + \left(\frac{\partial^2 u_z}{\partial y^2} - \frac{\partial^2 u_z}{\partial x^2}\right)\sin\alpha\cos\alpha\right] \quad ...(14.98)$$

Alternatively

$$\left.\begin{aligned} M_n &= D(K_n + \nu K_s) - \frac{M_T}{1-\nu} \\ M_s &= D(K_s + \nu K_n) - \frac{M_T}{1-\nu} \\ M_{ns} &= D(1-\nu)K_{ns} \end{aligned}\right\} \quad ...(14.99)$$

Also

$$\left.\begin{aligned} K_n &= K_x \cos^2 \alpha - 2K_{xy} \sin \alpha \cos \alpha + K_y \sin^2 \alpha \\ K_s &= K_y \cos^2\alpha + 2K_{xy} \sin \alpha \cos \alpha + K_x \sin^2 \alpha \\ K_{ns} &= K_{xy} (\cos^2 \alpha - \sin^2 \alpha) + (K_x - K_y) \sin \alpha \cos \alpha \end{aligned}\right\} \quad ...(14.100)$$

For a circular plate, we set $\alpha = \theta$ and $x = r \cos \theta$, $y = r \sin \theta$, $n = r$, $ds = rd\theta$. Thus

$$\left.\begin{aligned} M_r &= -D\left[\frac{\partial^2 u_z}{\partial r^2} + \nu\left(\frac{1}{r^2}\frac{\partial^2 u_z}{\partial \theta^2} + \frac{1}{r}\frac{\partial u_z}{\partial r}\right)\right] - \frac{M_T}{1-\nu} \\ M_\theta &= -D\left[\frac{1}{r^2}\frac{\partial^2 u_z}{\partial \theta^2} + \frac{1}{r}\frac{\partial u_z}{\partial r} + \nu\frac{\partial^2 u_z}{\partial r^2}\right] - \frac{M_T}{1-\nu} \\ M_{r\theta} &= (1-\nu)\,D\left[\frac{1}{r}\frac{\partial^2 u_z}{\partial r \partial \theta} - \frac{1}{r^2}\frac{\partial u_z}{\partial \theta}\right] \end{aligned}\right\} \quad ...(14.101)$$

(iii) Free edge

$$M_n = 0,\ Q_n = 0,\ M_{ns} = 0 \quad ...(14.102)$$

Now M_{ns} is statically equivalent to a distributed shear force of intensity $\left[-\left(\frac{\partial M_{ns}}{\partial_s}\right)_{n=0}\right]$ and, in addition, to concentrated forces R at the corners (if any) of the plate of magnitude

$$R = [(M_{ns})_1 - (M_{ns})_2]_{n=0}$$

where the subscripts 1 and 2 indicate the values of M_{ns} on opposite sides of the corner. Thus at a free edge, we have

$$Q_n - \frac{\partial M_{ns}}{\partial s} = 0 \quad ...(14.103)$$

Now $\quad Q_n = \dfrac{\partial M_n}{\partial n} - \dfrac{\partial M_{ns}}{\partial s}$

Hence at a free edge, then

$$M_n = 0,\ \frac{\partial M_n}{\partial n} - 2\frac{\partial M_{ns}}{\partial s} = 0$$

Now

$$\frac{\partial M_n}{\partial n} - 2\frac{\partial M_{ns}}{\partial s} = -D\left[\frac{\partial}{\partial n}(\nabla^2 u_z) + (1-\nu)\frac{\partial}{\partial s}\left\{\frac{\partial^2 u_z}{\partial x \partial y}\right.\right.$$

$$\left.\left.\times\ (\cos^2\alpha - \sin^2\alpha) + \left(\frac{\partial^2 u_z}{\partial y^2} - \frac{\partial^2 u_z}{\partial x^2}\right)\sin\alpha\cos\alpha\right\}\right]$$

$$-\frac{1}{1-\nu}\frac{\partial M_T}{\partial n} \quad \text{...(14.104)}$$

where

$$\left.\begin{aligned}\frac{\partial}{\partial n} &= \cos\alpha\,.\frac{\partial}{\partial x} + \sin\alpha\,.\frac{\partial}{\partial y}\\ \frac{\partial}{\partial s} &= \cos\alpha\,.\frac{\partial}{\partial y} - \sin\alpha\,.\frac{\partial}{\partial x}\end{aligned}\right\} \quad \text{...(14.105)}$$

For a rectangular plate with $\alpha = 0$, $n = x$, $s = y$, Eq. (14.104) becomes

$$D\left[\frac{\partial^3 u_z}{\partial x^3} + (2-\nu)\frac{\partial^3 u_z}{\partial x \partial y^2}\right] + \frac{1}{1-\nu}\frac{\partial M_T}{\partial x} = 0 \quad \text{...(14.106)}$$

and for a circular plate, we get

$$D\left[\frac{\partial}{\partial r}\left(\frac{\partial^2 u_z}{\partial r^2} + \frac{1}{r}\frac{\partial u_z}{\partial r} + \frac{1}{r^2}\frac{\partial^2 u_z}{\partial \theta^2}\right)\right.$$

$$\left. + (1-\nu)\left(\frac{1}{r^2}\frac{\partial^3 u_z}{\partial r \partial \theta^2} - \frac{1}{r^3}\frac{\partial^2 u_z}{\partial \theta^2}\right)\right] + \frac{1}{1-\nu}\frac{\partial M_T}{\partial r} = 0 \quad \text{...(14.107)}$$

Example 14.3 *Simply-supported rectangular plate.*

Solution. Consider a free simply supported rectangular plate defined by

$$0 \le x \le a,\ 0 \le y \le b,\ -\frac{h}{2} \le z \le \frac{h}{2} \quad \text{...(1)}$$

under an arbitrary temperature distribution. The boundary conditions are

$$u_z(0, y) = u_z(a, y) = u_z(x, 0) = u_z(x, b) = 0 \quad \text{...(2)}$$

$$\left.\begin{aligned}\frac{\partial^2 u_z}{\partial x^2} + \frac{M_T}{(1-\nu)D} &= 0 \text{ at } x = 0, a\\ \frac{\partial^2 u_z}{\partial y^2} + \frac{M_T}{(1-\nu)D} &= 0 \text{ at } y = 0, b\end{aligned}\right\} \quad \text{...(3)}$$

Now let $M_T = M_T(x, y)$. The Eq. (14.94) is equivalent to

$$\left.\begin{aligned} D\nabla^2 u_z + \frac{M_T}{1-\nu} &= f(x, y) \\ \Delta^2 f &= 0 \end{aligned}\right\} \quad \text{...(4)}$$

On the boundary $u_z = 0$, hence $\dfrac{\partial^2 u_z}{\partial s^2} = 0$

Thus $\quad D\,\nabla^2 u_z = -\left(\dfrac{M_T}{1-\nu}\right)$ on the boundary ...(5)

Hence the solution of $\nabla^2 f = 0$ is

$$f = 0 \quad \text{...(6)}$$

and the problem reduces to the solution of

$$D\,\nabla^2 u_z = -\left(\frac{M_T}{1-\nu}\right) \quad \text{...(7)}$$

Under the condition that $u_z = 0$ on the boundary. Eq. (7) can be solved in terms of double trigonometric series given by

$$u_z(x, y) = \sum_{m=1}^{\infty} \sum_{n=1}^{\infty} a_{mn} \sin\left(\frac{m\pi x}{a}\right) \sin\left(\frac{n\pi y}{b}\right) \quad \text{...(8)}$$

and $\quad M_T(x, y) = \displaystyle\sum_{m=1}^{\infty} \sum_{n=1}^{\infty} b_{mn} \sin\left(\frac{m\pi x}{a}\right) \sin\left(\frac{n\pi y}{b}\right)$ ...(9)

where the Fourier coefficients b_{mn} are

$$b_{mn} = \frac{4}{ab}\int_0^a \int_0^b M_T(x, y) \sin\left(\frac{m\pi x}{a}\right) \sin\left(\frac{n\pi y}{b}\right) dx dy \quad \text{...(10)}$$

Substituting these two series in Eqs. (7), we get

$$a_{mn} = \frac{b_{mn}}{(1-\nu)\pi^2 D}\left[\frac{1}{\left(\frac{m}{a}\right)^2 + \left(\frac{n}{b}\right)^2}\right] \quad \text{...(11)}$$

14.10.3 Plates with Temperature Distributions Varying Through the Thickness Only

(a) *Free unsupported plate of arbitrary profile.* The solution of this problem can be obtained by adding the deflections in the plane of the plate and the transverse deflections.

In the first case $N_T = 0$ and Eq. (14.90) becomes

$$\nabla^4 F = 0 \quad \text{...(14.108)}$$

An appropriate solution of this equation satisfying the free boundary conditions is then

$$F = 0 \qquad \text{...(14.109)}$$

and, therefore, $$N_x = N_y = N_{xy} = 0 \qquad \text{...(14.110)}$$

Using first three of Eqs. (14.84), we get

$$\left.\begin{aligned} \frac{\partial u_x}{\partial x} + \nu \frac{\partial u_y}{\partial y} = \frac{\partial u_y}{\partial y} + \nu \frac{\partial u_x}{\partial x} = \frac{(1+\nu)N_T}{Eh} \\ \frac{\partial u_x}{\partial y} + \frac{\partial u_y}{\partial x} = 0 \end{aligned}\right\} \qquad \text{...(14.111)}$$

Solving for u_x and u_y, we get

$$\left.\begin{aligned} u_x(x, y) &= \frac{N_T}{Eh} x + (a + by) \\ u_y(x, y) &= \frac{N_T}{Eh} y + (c - bx) \end{aligned}\right\} \qquad \text{...(14.112)}$$

where the terms in the brackets represent rigid-body motions.

The transverse deflection u_z is a solution of the equation

$$\nabla^4 u_z = 0 \qquad \text{...(14.113)}$$

The solution of this equation can be obtained by introducing the semi-inverse assumption that

$$M_x = M_y = M_{xy} = 0 \qquad \text{...(14.114)}$$

and by then verifying that the deflections calculated on this basis satisfy all the required conditions. With this assumption, the last three of Eqs. (14.84) become

$$\left.\begin{aligned} \frac{\partial^2 u_z}{\partial x^2} + \nu \frac{\partial^2 u_z}{\partial y^2} = \frac{\partial^2 u_z}{\partial y^2} + \nu \frac{\partial^2 u_z}{\partial x^2} = -\frac{M_T}{(1-\nu)D} \\ \frac{\partial^2 u_z}{\partial x \partial y} = 0 \end{aligned}\right\} \qquad \text{...(14.115)}$$

whose general solution is

$$u_z(x, y) = -\frac{6M_T}{Eh^3}(x^2 + y^2) + (d + ex + fy) \qquad \text{...(14.116)}$$

where the last term in brackets represents rigid-body motions. The stresses then become,

$$\left.\begin{aligned} \sigma_x = \sigma_y &= \frac{1}{1-\nu}\left[-\alpha ET + \frac{N_T}{h} + \frac{12M_T}{h^3} z\right] \\ \tau_{xy} &= 0 \end{aligned}\right\} \qquad \text{...(14.117)}$$

(*b*) *Free circular simply supported plate.* In this case the additional boundary condition that $u_z = 0$ at the plate boundary defined by $x^2 + y^2 = b^2$ be also satisfied, giving

$$u_z = \frac{6M_T}{Eh^3}(b^2 - x^2 - y^2) \qquad \text{...(14.118)}$$

instead of Eq. (14.116). The remainder of the solution is valid as in (*a*) above.

(*c*) *Rectangular simply supported plate*

Let $$u_z = \sum_{m=1}^{\infty} Y_m(y) \sin\left(\frac{m\pi x}{a}\right) \qquad \text{...(14.119)}$$

and $$M_T = \left(\frac{4M_T}{\pi}\right) \sum_{m=1,3,5}^{\infty} \frac{1}{m} \sin\left(\frac{m\pi x}{a}\right) \qquad \text{...(14.120)}$$

Substituting in $D\,\nabla^4 u_z = -\left(\frac{M_T}{1-\nu}\right)$

We get

$$\frac{d^2Y_m}{dy^2} - \left(\frac{m\pi}{a}\right)^2 Y_m = \begin{cases} -\dfrac{4M_T}{(1-\nu)Dm\pi} & \text{for } m \text{ odd} \\ 0 & \text{for } m \text{ even} \end{cases} \qquad \text{...(14.121)}$$

Boundary conditions are

$$Y_m(0) = Y_m(b) = 0 \qquad \text{...(14.122)}$$

Solving for Y_m and substituting in Eq. (14.119), we find

$$u_z = \frac{4M_T}{(1-\nu)\pi D} \sum_{m=1,3,5}^{\infty} \left(\frac{1}{m\beta_m^2}\right) \times \left[1 - \frac{\sinh \beta_m y + \sinh \beta_m (b-y)}{\sinh \beta_m b}\right] \sin\left(\frac{m\pi x}{a}\right) \qquad \text{...(14.123)}$$

where $$\beta_m = \frac{m\pi}{a}$$

Let $y' = y - \dfrac{b}{2}$, then

$$u_z = \frac{4M_T}{(1-\nu)\pi D} \sum_{m=1,3,5}^{\infty} \left(\frac{1}{m\beta_m^2}\right)\left[1 - \frac{\cosh(\beta_m y')}{\cosh(\beta_m b/2)}\right] \sin\left(\frac{mx\pi}{a}\right) \qquad \text{...(14.124)}$$

The expressions for moments then become,

$$M_x = -\frac{4M_T}{\pi} \sum_{m=1,3,5}^{\infty} \left(\frac{1}{m}\right) \left[\frac{\cosh\left(\frac{m\pi y'}{a}\right)}{\cosh\left(\frac{m\pi b}{2a}\right)} \right] \sin\left(\frac{m\pi x}{a}\right)$$

$$M_y = -\frac{4M_T}{\pi} \sum_{m=1,3,5}^{\infty} \left(\frac{1}{m}\right) \left[1 - \frac{\cosh\left(\frac{m\pi y'}{a}\right)}{\cosh\left(\frac{m\pi b}{2a}\right)} \right] \sin\left(\frac{m\pi x}{a}\right)$$

$$M_{xy} = -\frac{4M_T}{\pi} \sum_{m=1,3,5}^{\infty} \left(\frac{1}{m}\right) \left[\frac{\sinh\left(\frac{m\pi y'}{a}\right)}{\sinh\left(\frac{m\pi b}{2a}\right)} \right] \cos\left(\frac{m\pi x}{a}\right) \qquad \text{...(14.125)}$$

(*d*) *Free plate with fixed edges.* For a free plate of arbitrary profile, we have at the edges

$$u_z = 0$$

The stresses then become

$$\left.\begin{aligned} \sigma_x &= -\frac{\alpha ET}{1-\nu} + \frac{N_x + \left(\frac{N_T}{1-\nu}\right)}{h} \\ \sigma_y &= -\frac{\alpha ET}{1-\nu} + \frac{N_y + \left(\frac{N_T}{1-\nu}\right)}{h} \\ \tau_{xy} &= \frac{N_{xy}}{h} \end{aligned}\right\} \qquad \text{...(14.126)}$$

Example 14.4 *A bare steam pipe 100 mm inside diameter and 2.5 mm thick, carries dry and saturated steam at 1 MPa. Calculate radial and hoop stresses induced in the pipe at its inner and outer periphery. Take coefficient of thermal expansion as 10×10^{-6} / C° and E = 200 GPa. Ambient temperature is 30°C.*

Solution. For $T = 0$°C

$$\sigma_r = A - \frac{B}{r^2}$$

$$\sigma_\theta = A + \frac{B}{r^2}$$

At $r = 50$ mm, $\sigma_r = -1$ MPa

$r = 52.5$ mm, $\sigma_r = 0$

$$-1 = A - \frac{B}{2500}$$

$$0 = A - \frac{B}{2756.25}$$

$$A = \frac{B}{2756.25}$$

$$B = +26890$$

$$A = +9.756$$

$$\sigma_r = A - \frac{B}{r^2} - \frac{\alpha ET}{r^2}\left(\frac{b^2 - a^2}{2}\right)$$

$$\sigma_\theta = A + \frac{B}{r^2} - \alpha ET + \frac{\alpha ET}{r^2}\left(\frac{b^2 - a^2}{2}\right)$$

At $r = a = 50$ mm

$$\sigma_r = +9.756 - \frac{26890}{2500} - \frac{10 \times 10^{-6} \times 200 \times 10^3 \times 70}{2500 \times 2}(2756.25 - 2500)$$

$$= +9.756 - 10.756 - 7.175 = -8.175 \text{ MPa}$$

$$\sigma_\theta = 9.756 + \frac{26890}{2500} - 10 \times 10^{-6} \times 2 \times 10^5 \times 70$$

$$+ \frac{10 \times 10^{-6} \times 2 \times 10^5 \times 70}{2500}(2756.25 - 2500)$$

$$= 9.756 + 10.756 - 140 + 7{,}175 = -112.3 \text{ MPa}$$

At $r = b = 52.5$ mm

$$\sigma_r = 9.756 - \frac{26890}{2756.25}$$

$$- \frac{10 \times 10^{-6} \times 200 \times 10^3 \times 70(2756.25 - 2500)}{2756.25 \times 2}$$

$$= 9.756 - 9.756 - 6.508 = -6.508 \text{ MPa}$$

$$\sigma_\theta = 9.756 + 9.756 - 140 + 6.508 = -113.98 \text{ MPa}$$

Example 14.5 *A steel disc 150 mm diameter and rotating at 12000 rpm is subjected to temperature rise from 0°C at the centre to 50°C at the outer periphery of the disc. Calculate the maximum stresses set up in the disc. Take E = 200 GPa, $\nu = 0.3$, $\alpha = 12.5 \times 10^{-6}$/°C and $\rho = 7470$ kg/m^3.*

Solution.

$$\sigma_r = \left(\frac{3 + \nu}{8}\right)\rho\omega^2 (b^2 - r^2) - \frac{\alpha E}{r^2}\int Tr\, dr$$

$$\sigma_\theta = \left(\frac{1+3\nu}{8}\right)\rho\omega^2\left[\left(\frac{3+\nu}{1+3\nu}\right)b^2 - r^2\right] - \alpha ET + \frac{\alpha E}{r^2}\int Tr\,dr$$

$T \propto r$

$= kr$, where k is a constant of proportionality.

$$\int Tr\,dr = \int kr^2\,dr = \frac{kr^3}{3}$$

At $\quad r = b,\ T = 50°\text{C}$

$50 = kb$

$k = 50/b$ °C/mm

$$\sigma_r = \left(\frac{3+\nu}{8}\right)\rho\omega^2(b^2 - r^2) - \frac{\alpha Erk}{3}$$

$$\sigma_\theta = \left(\frac{1+3\nu}{8}\right)\rho\omega^2\left[\left(\frac{3+\nu}{1+3\nu}\right)b^2 - r^2\right] - \alpha ET + \frac{\alpha Ekr}{3}$$

$$(\sigma_r)_{max} = \left(\frac{3.3}{8}\right)\times 7470 \times \left(\frac{2\pi\times 12000}{60}\right)^2 (72^2 - 0)\times 10^{-6}$$

$= 27.37$ MPa

$$(\sigma_\theta)_{max} = \left(\frac{1.9}{8}\right)\times 7470 \times \left(\frac{2\pi\times 12000}{60}\right)^2 \times 75^2 \times 10^{-6} \times \left(\frac{3.3}{1.9}\right)$$

$= 27.37$ MPa

EXERCISES

14.1 A thin uniform disc of radius b is enclosed in a heavy ring of the same material into which it just fits when the disc and the ring are at a uniform temperature. If the heat is supplied over the faces of the disc and it is rejected at the circumference, the temperature rise at a distance r from the centre is given by

$$T = (T_1 - T_0) - (T_1 - T_0)\frac{r^2}{b^2}$$

Assume that the ring takes up a uniform temperature T_0 but undergoes no appreciable strain due to the stresses set up. Show that the radial compressive stress in the disc at a radius r is

$$\frac{1}{4}E\alpha(T_1 - T_0)\left(\frac{3-\nu}{1-\nu} - \frac{r^2}{b^2}\right).$$

14.2 The temperature rise of a solid conductor of radius b due to a uniform heat source is given by the formula $T = \lambda(b^2 - r^2)$, where λ is a constant. If there are no external forces resisting longitudinal or radial expansion, determine the stress distribution in the conductor.

14.3 Obtain expressions for the thermal stresses in a ring of rectangular cross-section, inside radius a, outside radius b and height h under linearly varying temperature between T_i (inside) and T_0 (outside). The temperature is uniform in the axial direction. Treat the ring (*i*) as a thin disc (*ii*) as a thick cylinder.

14.4 A rectangular beam 30 mm wide and 40 mm deep is 1 m long. The top surface of the beam is heated to a temperature of 60°C. Determine the thermal stresses set up in the beam at the top and bottom surfaces. Assume $E = 2 \times 10^5$ N/mm^2, $\nu = 0.30$, $\alpha = 12.5 \times 10^{-6}$ per C°, room temperature = 15°C.

14.5 A thick circular cylinder of internal radius 100 mm and external radius 200 mm is heated at the inside surface so that the rise in temperature is 45°C. If there is a steady state heat flow from the inside to the outside of the cylinder, determine the thermal stresses developed in the cylinder wall at the inside and outside surfaces. Assume $E = 2 \times 10^5$ N/mm^2, $\nu = 0.28$, $\alpha = 15 \times 10^{-6}$ per C°.

14.6 A thin circular disc of inside radius 50 mm and outside radius 150 mm is heated at the inside surface to a temperature of 40°C. If the temperature at any radius is given by $T = \lambda\,(225 - r^2)$, where $\lambda = \dfrac{25}{\ln 3}$, determine the stress distribution in the disc. Assume $E = 2 \times 10^5$ N/mm^2, $\nu = 0.30$, $\alpha = 11.5 \times 10^{-6}$ per C°, room temperature = 15°C.

14.7 A hollow sphere of inside radius 50 mm and outside radius 100 mm is heated at the outside surface so that the temperature rise is 40°C. If the variation in the radial temperature distribution is linear and the inside surface temperature is 20°C, determine the thermal stresses in the sphere wall. Assume $E = 2 \times 10^5$ N/mm^2, $\nu = 0.3$, $\alpha = 12.5 \times 10^{-6}$ per C°.

14.8 A rectangular beam of unit width is bent into a circular form of inner radius 400 mm and outer radius 500 mm. Its outer surface temperature is raised to 70°C. If the inside surface temperature is maintained at room temperature of 15°C, determine the stresses at a radius of 450 mm. Assume radial temperature variation only and $E = 2 \times 10^5$ N/mm^2, $\nu = 0.30$, $\alpha = 12.5 \times 10^{-6}$ per C°.

14.9 Derive the equation of equilibrium for a thin plate heated on the lateral surface. State the assumptions made.

14.10 Derive an expression for the deflection of a thin circular plate of radius a heated on the lateral surface.

14.11 Calculate the maximum thermal stress in a cast iron cylinder for which the inner temperature is – 8°C and the outer temperature is zero. Outside and inside radii of cylinder are 15 mm and 10 mm respectively, $E = 90$ GPa, $\alpha = 11.4 \times 10^{-6}$ per °C and $\nu = 0.3$.

14.12 An unloaded closed cylinder has an inner radius of 100 mm and an outer radius of 250 mm. The cylinder is made of steel for which $\alpha = 11.7 \times 10^{-6}$ per °C, $E = 200$ GPa, and $\nu = 0.29$. Determine the stress components at the inner radius for steady-state temperature change with the temperature at the inner radius 100° C greater than the temperature at the outer radius.

15

Hertz's Contact Stresses

15.1 Introduction

Contact-stress problems occur where forces are transmitted through contact from one machine element to another, *e.g.* gearing, roller bearing, cams and pin joints in linkages. There are two major types of contact problems, line contact and point contact.

In determining the contact stresses, it is assumed that the material of each of the two bodies is homogeneous isotropic and elastic in accordance with Hooke's law, but the two bodies may not necessarily be made of the same material.

15.2 Pressure Between two Bodies in Contact

Consider two bodies in contact as shown in Fig. 15.1. Before the load P is applied, the bodies will have a point contact at 0. The distances from the tangent plane at 0 of points M_1 on a meridian section of the body 1 at a very small distance from the point 0 can be written as

$$z_1 = x_1 \tan\frac{\beta}{2} = \frac{1}{2}x_1\beta$$

since angle β is small. From triangle HM_1D (Fig. 15.1 *a*).

$$\tan\beta = \beta = \frac{M_1D}{HD} = \frac{x_1}{R'_1}$$

Since $\qquad HD \simeq HO = R'_1$

Hence $\qquad z_1 = \dfrac{x_1^2}{2R'_1}$

In a similar manner the distance z_1 to the points lying in the plane of radius R''_1 is found to be (Fig. 15.1 *b*)

$$z_1 = \frac{y_1^2}{2R''_1}$$

It is assumed that the distance z_1 to any point not lying in either plane of principal curvature is given by

$$z_1 = \frac{x_1^2}{2R'_1} + \frac{y_1^2}{2R''_1} \qquad ...(15.1)$$

Similarly the distance z_2 from the tangent plane to any point in the surface of body 2 near the point of contact can be obtained as

$$z_2 = \frac{x_2^2}{2R'_2} + \frac{y_2^2}{2R''_2} \qquad ...(15.2)$$

where (x_1, y_1) and (x_2, y_2) are the coordinates with respect to the axes lying in the tangent plane and also in the planes of the principal radii of curvature R'_1, R''_1 and R'_2, R''_2 respectively. The location of the axes x_1, y_1 and x_2, y_2 are shown in Fig. 15.1 (*c*). The axes x_1 and x_2 are separated by a small angle α.

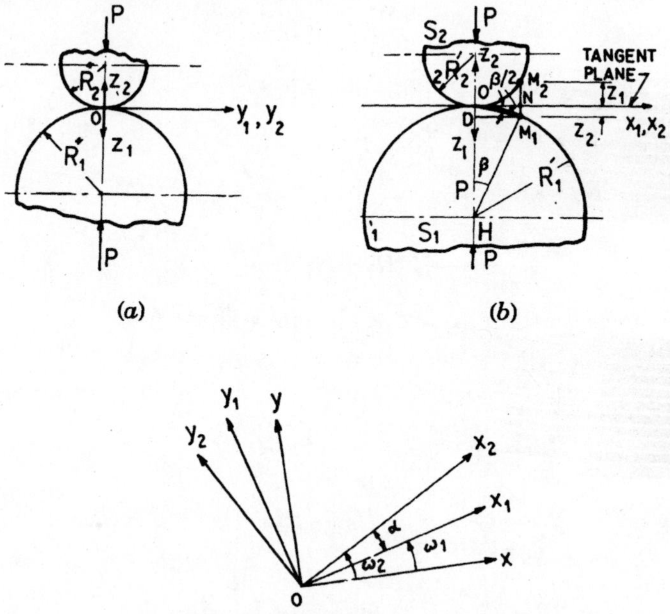

Fig. 15.1

The distance z between points on the two surfaces near the point of contact is the numerical sum of z_1 and z_2. Hence

$$z = z_1 + z_2 = \frac{x_1^2}{2R'_1} + \frac{y_1^2}{2R''_1} + \frac{x_2^2}{2R'_2} + \frac{y_2^2}{2R''_2} \qquad ...(15.3)$$

Referring to Fig. 15.1 (*c*), we have

$$x_1 = x \cos \omega_1 + y \sin \omega_1$$

$$y_1 = -x \sin \omega_1 + y \sin \omega_1$$

$$x_2 = x \cos \omega_2 + y \sin \omega_2$$

$$y_2 = -x \sin \omega_2 + y \cos \omega_2$$

Substituting in Eq. (15.3), we get

$$z = \frac{x^2}{2R_1} + \frac{y^2}{2R_2} + \frac{1}{2}xy(g_1 \sin 2\omega_1 + g_2 \sin 2\omega_2)$$

where
$$\frac{1}{R_1} = \frac{\cos^2 \omega_1}{R'_1} + \frac{\sin^2 \omega_1}{R''_1} + \frac{\cos^2 \omega_2}{R'_2} + \frac{\sin^2 \omega_2}{R''_2}$$

$$\frac{1}{R_2} = \frac{\sin^2 \omega_1}{R'_1} + \frac{\cos^2 \omega_1}{R''_1} + \frac{\sin^2 \omega_2}{R'_2} + \frac{\cos^2 \omega_2}{R''_2}$$

$$g_1 = \frac{1}{R'_1} - \frac{1}{R''_1} \quad \text{and} \quad g_2 = \frac{1}{R'_2} - \frac{1}{R''_2}$$

Let $\omega_1 = \omega - \dfrac{\alpha}{2}$ and $\omega_2 = \omega + \dfrac{\alpha}{2}$

where $\alpha = \omega_2 - \omega_1$

and $2H_1 = \dfrac{1}{R'_1} + \dfrac{1}{R''_1}$; $2H_2 = \dfrac{1}{R'_2} + \dfrac{1}{R''_2}$

$$\therefore \quad \frac{1}{R_1} + \frac{1}{R_2} = 2(H_1 + H_2)$$

If the coordinates are so chosen so that

$$g_1 \sin 2\omega_1 + g_2 \sin 2\omega_2 = (g_1 + g_2) \sin 2\omega \cos \alpha - (g_1 - g_2) \cos 2\omega \sin \alpha = 0$$

$$z = \frac{x^2}{2R_1} + \frac{y^2}{2R_2} = Ax^2 + By^2 \qquad \text{...(15.4)}$$

where
$$2A = \frac{1}{R_1} = H_1 + H_2 + \tfrac{1}{2}[(g_1 + g_2) \cos 2\omega \cos \alpha + (g_1 - g_2) \sin 2\omega \sin \alpha]$$

$$2B = \frac{1}{R_2} = H_1 + H_2 - \tfrac{1}{2}[(g_1 + g_2) \cos 2\omega \cos \alpha + (g_1 - g_2) \sin 2\omega \sin \alpha]$$

Now $(g_1 + g_2) \sin 2\omega \cos \alpha - (g_1 - g_2) \cos 2\omega \sin \alpha = 0$

and
$$(g_1 + g_2) \cos 2\omega \cos \alpha + (g_1 - g_2) \sin 2\omega \sin \alpha = 2\left(\frac{1}{R_1} - H_1 - H_2\right)$$

Hence
$$\sin 2\omega = \frac{2}{\eta^2}\left(\frac{1}{R_1} - H_1 - H_2\right)(g_1 - g_2) \sin \alpha$$

$$\cos 2\omega = \frac{2}{\eta^2}\left(\frac{1}{R_1} - H_1 - H_2\right)(g_1 + g_2) \cos \alpha$$

where
$$\eta^2 = (g_1 + g_2)^2 \cos^2 \alpha + (g_1 - g_2)^2 \sin^2 \alpha = g_1^2 + g_2^2 + 2g_1g_2 \cos 2\alpha.$$

If R_1 is the bigger radius then

$$\frac{1}{R_1} = H_1 + H_2 - \frac{1}{2}\eta$$

$$\frac{1}{R_2} = H_1 + H_2 + \frac{1}{2}\eta$$

and
$$\sin 2\omega = -\frac{g_1 - g_2}{\eta}\sin\alpha$$

$$\cos 2\omega = -\frac{g_1 + g_2}{\eta}\cos\alpha$$

$$\text{Now } \eta^2 = \left(\frac{1}{R'_1} - \frac{1}{R''_1}\right)^2 + \left(\frac{1}{R'_2} - \frac{1}{R''_2}\right)^2 + 2\left(\frac{1}{R'_1} - \frac{1}{R''_1}\right) + \left(\frac{1}{R'_2} - \frac{1}{R''_2}\right)\cos 2\alpha$$

$$H_1 + H_2 = \frac{1}{2}\left[\frac{1}{R'_1} + \frac{1}{R''_1} + \frac{1}{R'_2} + \frac{1}{R''_2}\right] = A + B$$

$$\therefore \quad A = \frac{1}{2R_1} = \frac{1}{4}\left[\frac{1}{R'_1} + \frac{1}{R''_1} + \frac{1}{R'_2} + \frac{1}{R''_2}\right] - \frac{1}{4}\left[\left(\frac{1}{R'_1} - \frac{1}{R''_1}\right)^2 + \left(\frac{1}{R'_2} - \frac{1}{R''_2}\right)^2 + 2\left(\frac{1}{R'_1} - \frac{1}{R''_1}\right)\left(\frac{1}{R'_2} - \frac{1}{R''_2}\right) \times \cos 2\alpha\right]^{1/2} \quad \ldots(15.5)$$

$$B = \frac{1}{2R_2} = \frac{1}{4}\left[\frac{1}{R'_1} + \frac{1}{R''_1} + \frac{1}{R'_2} + \frac{1}{R''_2}\right] + \frac{1}{4}\left[\left(\frac{1}{R'_1} - \frac{1}{R''_1}\right)^2 + \left(\frac{1}{R'_2} - \frac{1}{R''_2}\right)^2 + 2\left(\frac{1}{R'_1} - \frac{1}{R''_1}\right)\left(\frac{1}{R'_2} - \frac{1}{R''_2}\right) \times \cos 2\alpha\right]^{1/2} \quad \ldots(15.6)$$

The constants A and B depend upon the principal radii of curvature of the two bodies at the point of contact and upon the angle α between the corresponding planes of the minimum (or maximum) principal curvatures. If $\alpha = 0°$ then

$$\left.\begin{aligned} A &= \frac{1}{2}\left[\frac{1}{R''_1} + \frac{1}{R''_2}\right] \\ B &= \frac{1}{2}\left[\frac{1}{R'_1} + \frac{1}{R'_2}\right] \end{aligned}\right\} \qquad ...(15.7)$$

If $\alpha = 90°$ then

$$\left.\begin{aligned} A &= \frac{1}{2}\left[\frac{1}{R'_1} + \frac{1}{R'_2}\right] \\ B &= \frac{1}{2}\left[\frac{1}{R''_1} + \frac{1}{R''_2}\right] \end{aligned}\right\} \qquad ...(15.8)$$

When the load P is applied to the bodies, their surfaces deform elastically near the point of contact so that a small area of contact is formed. It is assumed that as this small area of contact is formed, the points which come into contact simultaneously will be points on two surfaces which were originally equal distances apart. Equidistant points on the two surfaces lie on an ellipse. Hence the boundary line of the area of contact is assumed to be an ellipse whose equation is $\frac{x^2}{a^2} + \frac{y^2}{b^2} = 1$.

Let δ_1 and δ_2 be the amounts by which the two bodies move towards each other respectively as shown in Fig. 15.2, which means that the distance between points on two bodies not affected by the local deformation near 0 is shortened by an amount $\delta = \delta_1 + \delta_2$. Points such as M_1 and M_2 move toward each other at distance z given by Eq. (15.4), then meet each other. If they did not meet within the contact area, they might move the whole distance δ. Therefore, the deformation of the two bodies within the contact area is $\delta - z$. If W_1 and W_2 represent the deformations of each of the two bodies, at any point within the contact area, the sum of these deformations is given by

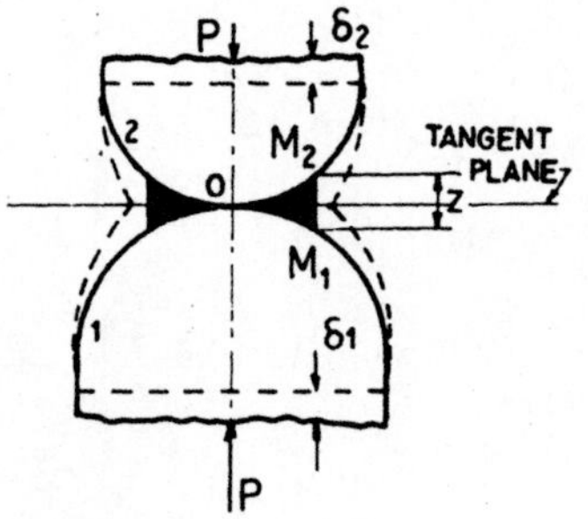

Fig. 15.2

$$W_1 + W_2 = \delta - z$$

Using Eq. (15.4), we get

$$W_1 + W_2 = \delta - Ax^2 - By^2 \qquad ...(15.9)$$

where δ is called the approach of the two bodies or the distance they move toward each other. H. Hetrz showed that this requirement is satisfied by assuming that the intensity of pressure over the surfaces of contact is represented by the ordinates of a semi-ellipsoid constructed on the surface of contact, thus

$$p = p_0 \sqrt{1 - \frac{x^2}{a^2} - \frac{y^2}{b^2}} \qquad ...(15.10)$$

This pressure distribution is illustrated in Fig. 15.3.

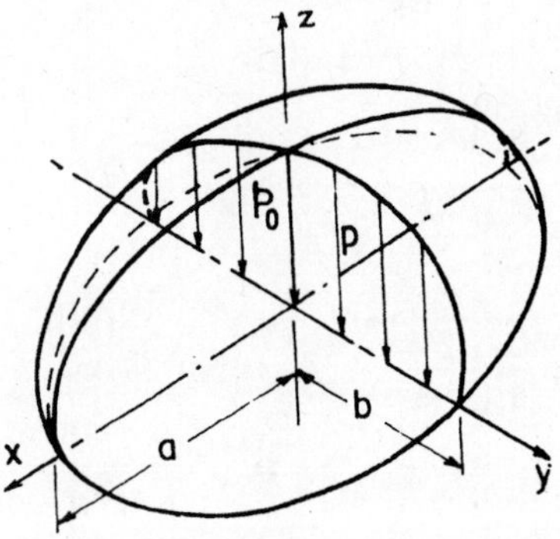

Fig. 15.3

Since total load P is equal to the volume of the semi-ellipsoid, we can write

$$P = \frac{2}{3} \pi\, ab p_0 \qquad ...(15.11)$$

or the maximum pressure

$$p_0 = \frac{3}{2} \frac{P}{\pi ab} \qquad ...(15.12)$$

which is 1.5 times the average pressure on the surface of contact. To calculate p_0, we must know a and b, both of which depend upon the shape, applied load, and the material properties of the two bodies.

Now for two semi-infinite bodies in contact

$$W_1 + W_2 = \left(\frac{1 - \nu_1^2}{\pi E_1} + \frac{1 - \nu_2^2}{\pi E_2} \right) \iint \frac{p dA}{r} \qquad ...(15.13)$$

where pdA is the pressure acting on an infinitely small element of the surface of contact and r is the distance of this element from the point under consideration. The integration must be extended over the entire surface of contact.

From Eqs. (15.13) and (15.9), we get

$$\left(\frac{1-\nu_1^2}{\pi E_1}+\frac{1-\nu_2^2}{\pi E_2}\right)\iint\frac{p\,dA}{r}=\delta-Ax^2-By^2$$

Now $P=\iint p\,dA=\frac{2}{3}\pi abp_0.$

Following the procedure analogous to that used for spherical bodies (see art. 15.3), we find that

$$a=m\sqrt{\frac{3\pi}{4}\frac{P(k_1+k_2)}{(A+B)}}$$

$$=m\sqrt[3]{\frac{3}{4}\frac{P\Delta}{(A+B)}} \qquad ...(15.14)$$

$$b=n\sqrt{\frac{3\pi}{4}\frac{P(k_1+k_2)}{(A+B)}}$$

$$=n\sqrt[3]{\frac{3}{4}\frac{P\Delta}{(A+B)}} \qquad ...(15.15)$$

where $$\Delta=\frac{1-\nu_1^2}{E_1}+\frac{1-\nu_2^2}{E_2}=\frac{1}{\pi}(k_1+k_2) \qquad ...(15.16)$$

and coefficients m and n are constants depending on the ratio $(B-A)/(B+A)$. Using the notation

$$\cos\theta=\frac{B-A}{B+A} \qquad ...(15.17)$$

The values of m and n are given in Table 15.1.

Table 15.1 Values of m and n

θ	30°	35°	40°	45°	50°	55°	60°
m	2.731	2.397	2.136	1.926	1.754	1.611	1.486
n	0.493	0.530	0.567	0.604	0.641	0.678	0.717

θ	65°	70°	75°	80°	85°	90°
m	1.378	1.284	1.202	1.128	1.061	1.000
n	0.759	0.802	0.846	0.893	0.944	1.000

Knowing the distribution of pressure, the stresses at any point can be calculated.

Considering points on the elliptical surface of contact and taking the x and y-axes in the direction of the semi-axes a and b respectively, the principal stresses at the center of the surface of contact are

$$\left.\begin{aligned}\sigma_x &= -2\nu p_0 - (1-2\nu)p_0\left(\frac{b}{a+b}\right)\\ \sigma_y &= -2\nu p_0 - (1-2\nu)p_0\left(\frac{a}{a+b}\right)\\ \sigma_z &= -p_0\end{aligned}\right\} \quad ...(15.18)$$

15.3 Pressure Between Two Spherical Bodies in Contact

Consider two spherical bodies in contact as shown in Fig. 15.4. Let R_1 and R_2 be the radii of these surfaces. If there is no pressure between the bodies, we have contact at point 0. The distances from the tangent plane at 0 of points such as M_1 and M_2, on a meridian section of the spheres at a very small distance r from the axes z_1 and z_2 can be represented by

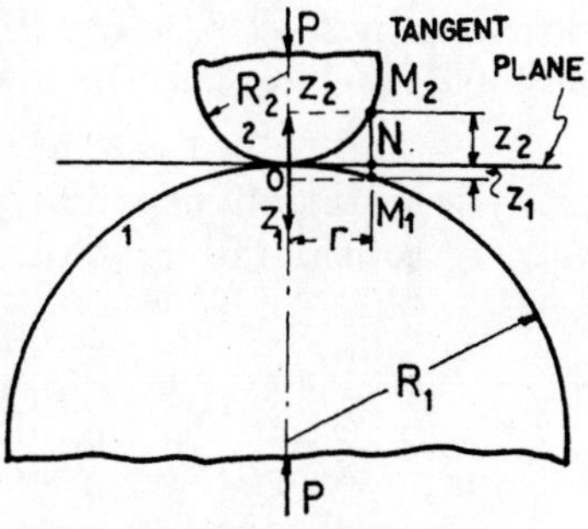

Fig. 15.4

$$z_1 = \frac{r^2}{2R_1}$$

$$z_2 = \frac{r^2}{2R_2}$$

or

$$z_1 + z_2 = \frac{r^2}{2}\left(\frac{1}{R_1} + \frac{1}{R_2}\right) = \frac{r^2(R_1 + R_2)}{2R_1R_2} \quad ...(15.19)$$

In the particular case of contact between a sphere and a plane $R_1 = \infty$ and

$$z_1 + z_2 = \frac{r^2}{2R_2} \quad ...(15.20)$$

In the case of a contact between a ball and a spherical seat R_1 is negative and

$$z_2 - z_1 = \frac{r^2(R_1 - R_2)}{2R_1R_2} \quad ...(15.21)$$

If the bodies are pressed together along the normal by a force P, there will be a local deformation near the point of contact producing contact over a small surface with a circular boundary, called the surface of contact. If the radii of curvature R_1 and R_2 are very large in comparison with the radius of the boundary of the surface of contact, then the local deformations around the point of contact can be obtained by considering the two bodies to be as semi-infinite bodies.

Let W_1 denote the displacement due to the local deformation in the direction z_1 of a point such as M_1 on the surface of the body 1 and W_2 denote the same displacement in the direction z_2 for a point such as M_2 of the body 2. Assuming that the tangent plane and 0 remains immovable during local compression, then due to this compression, any two points will approach each other by a certain amount δ, and the distance between two points such as M_1 and M_2 will decrease by $\delta - (W_1 + W_2)$. If finally due to local compression, the points M_1 and M_2 come inside the surface of contact, we have

$$\delta - (W_1 + W_2) = z_1 + z_2 = \beta r^2 \qquad ...(15.22)$$

in which β is a constant depending on the radii R_1 and R_2. Thus from geometrical consideration we find for any point of the surface of contact.

$$W_1 + W_2 = \delta - \beta r^2 \qquad ...(15.23)$$

Let us now consider local deformations. From the conditions of symmetry the intensity p of pressure between the bodies in contact and the corresponding deformation are symmetrical with respect to the centres 0 of the surface of contact.

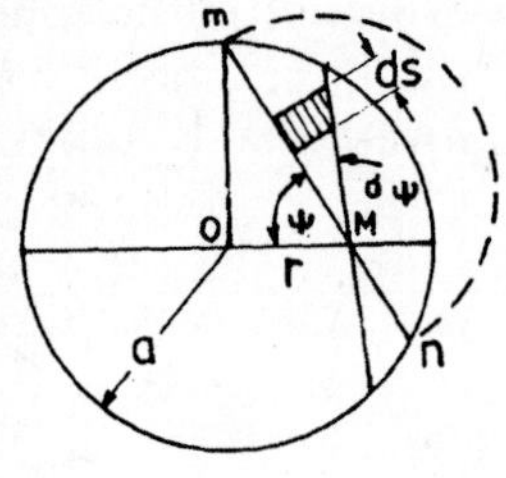

Fig. 15.5

Now the displacement W_1 of point M_1 can be written as (Fig. 15.5)

$$W_1 = \frac{(1 - \nu_1^2)}{\pi E_1} \iint p\,ds\,d\psi$$

The integration is to be carried out over the entire area of contact.

Similarly for the body 2, the displacement W_2 of point M_2 is

$$W_2 = \frac{(1 - \nu_2^2)}{\pi E_2} \iint p\,ds\,d\psi$$

$$\therefore \quad W_1 + W_2 = \left(\frac{1 - \nu_1^2}{\pi E_1} + \frac{1 - \nu_2^2}{\pi E_2} \right) \iint p\,ds\,d\psi$$

Let $\quad k_1 = \dfrac{1 - \nu_1^2}{\pi E_1}, \; k_2 = \dfrac{1 - \nu_2^2}{\pi E_2}$

$$\therefore \qquad W_1 + W_2 = (k_1 + k_2)\iint p\,ds\,d\psi$$

or
$$(k_1 + k_2)\iint p\,ds\,d\psi = \delta - \beta r^2$$

It is assumed that the distribution of pressure p over the contact surface is represented by the ordinates of a hemisphere of radius a constructed on the surface of contact. If p_0 is the pressure at the centre O of the surface of contact, then

$$p_0 = ka$$

in which $k = p_0/a$ is a constant factor representing the scale of our representation of the pressure distribution. Along the chord mn the pressure p varies, as indicated in Fig. 15.5 by the dotted semicircle. Integrating along this chord, we find

$$\int p\,ds = \frac{p_0 A}{a}$$

in which A is the area of the semicircle indicated by the dotted lines and is equal to $\frac{1}{2}\pi\,(a^2 - r^2 \sin^2 \psi)$. Hence we get

$$\frac{\pi(k_1 + k_2)}{a}\,p_0\int_0^{\pi/2}(a^2 - r^2\sin^2\psi)\,d\psi = \delta - \beta r^2$$

or
$$\pi^2(k_1 + k_2)\frac{p_0}{4a}(2a^2 - r^2) = \delta - \beta r^2 \qquad ...(15.24)$$

This equation will be fulfilled for any value of r, and hence the assumed pressure distribution is the correct one if the following relations exist for displacement δ and the radius a of the surface of contact :

$$\delta = (k_1 + k_2)\,p_0\frac{\pi^2 a}{2}$$

$$a = (k_1 + k_2)\frac{\pi^2 p_0}{4\beta}$$

Now for the hemispherical pressure distribution

$$P = \frac{2}{3}\pi a^3 \cdot \frac{p_0}{a}$$

from which
$$p_0 = \frac{3P}{2\pi a^2}$$

i.e. the maximum pressure is 1.5 times average pressure on the surface of contact.

Hence we get $\beta = \dfrac{R_1 + R_2}{2R_1 R_2}$

$$a = \sqrt[3]{\frac{3\pi}{4} \frac{P(k_1 + k_2) R_1 R_2}{(R_1 + R_2)}} \qquad ...(15.25)$$

$$\delta = \sqrt[3]{\frac{9\pi^2}{16} \frac{P^2 (k_1 + k_2)^2 (R_1 + R_2)}{R_1 R_2}} \qquad ...(15.26)$$

If we put $\Delta = \dfrac{1 - \nu_1^2}{E_1} + \dfrac{1 - \nu_2^2}{E_2}$

then
$$a = \sqrt[3]{\frac{3}{4} \frac{P \Delta R_1 R_2}{(R_1 + R_2)}}$$

$$= 0.908 \sqrt[3]{\frac{P \Delta R_1 R_2}{(R_1 + R_2)}} \qquad ...(15.27)$$

and
$$p_0 = \sqrt[3]{\frac{P(R_1 + R_2)^2}{\Delta^2 R_1^2 R_2^2}} \qquad ...(15.28)$$

Assuming that both the spheres have the same elastic properties and taking

$\nu = 0.3$, we find

$$\left.\begin{aligned} a &= 1.109 \sqrt[3]{\frac{P}{E} \frac{R_1 R_2}{(R_1 + R_2)}} \\ \delta &= 1.23 \sqrt[3]{\frac{P^2}{E^2} \frac{(R_1 + R_2)}{R_1 R_2}} \\ p_0 &= 0.388 \sqrt[3]{PE^2 \frac{(R_1 + R_2)^2}{R_1^2 R_2^2}} \end{aligned}\right\} \qquad ...(15.29)$$

(*i*) For a sphere pressed into a plane surface, and the same elastic constants for both bodies, we find by substituting $R_1 = \infty$,

$$\left.\begin{aligned} a &= \sqrt[3]{\frac{3}{4} PR_2 \Delta} = 1.109 \sqrt[3]{\frac{PR_2}{E}} \\ \delta &= 1.23 \sqrt[3]{\frac{P^2}{E^2 R_2}} \\ p_0 &= 0.388 \sqrt[3]{\frac{PE^2}{R_2^2}} \end{aligned}\right\} \qquad ...(15.30)$$

(*ii*) For a sphere in a spherical seat, R_2 is negative and we get

$$\left.\begin{aligned} a &= \sqrt[3]{\frac{3}{4}\frac{P\,\Delta\,R_1\,R_2}{(R_2 - R_1)}} \\ p_0 &= 0.578\sqrt[3]{\frac{P(R_1 - R_2)^2}{\Delta^2\,R_1^{\,2}\,R_2^{\,2}}} \end{aligned}\right\} \qquad \text{...(15.31)}$$

15.4 Contact Pressure Between Two Parallel Cylinders

Two surfaces of contact may be cylindrical and be arranged so that they are in contact approximately along a straight line element before the loads are applied. In the cylinders, radii R_1'' and R_2'' which lie in plane perpendicular to the paper, are each indefinitely large so that

$$\frac{1}{R''_1} = \frac{1}{R''_2} = 0 \text{ and also the angle } \alpha = 0.$$

Hence from Eqs. (15.7) we find that

$$A = 0 \text{ and } B = \frac{1}{2}\left[\frac{1}{R'_1} + \frac{1}{R''_2}\right]$$

Let $R_1' = R_1$ and $R_2' = R_2$ as drawn in Figs. 15.6. Figs. 15.6 (*a*) and (*b*) show two cylinders in contact before and after load P is applied.

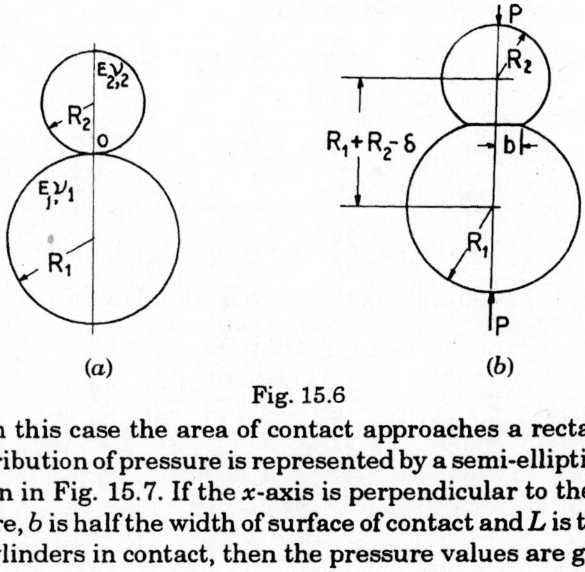

Fig. 15.6

In this case the area of contact approaches a rectangle and the distribution of pressure is represented by a semi-elliptical prism as shown in Fig. 15.7. If the x-axis is perpendicular to the plane of the figure, b is half the width of surface of contact and L is the length of the cylinders in contact, then the pressure values are given by

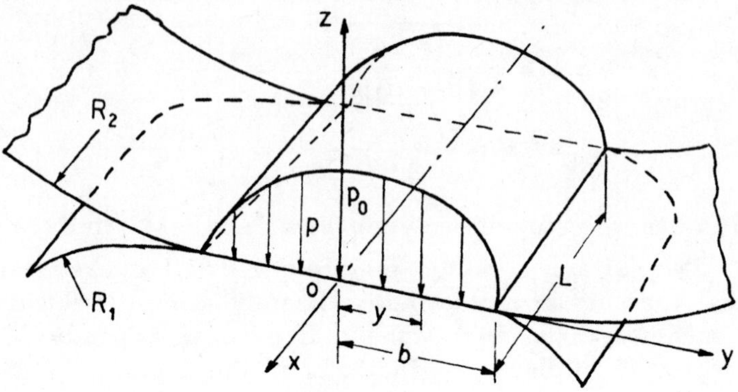

Fig. 15.7

$$p = p_0 \sqrt{1 - \frac{y^2}{b^2}} \qquad \text{...(15.32)}$$

The volume of the semi-elliptical prism is equal to the total load P. Hence

$$P = \frac{\pi L b\, p_0}{2}$$

and

$$p_\theta = \frac{2P}{\pi L b} \qquad \text{...(15.33)}$$

The expression for b is given by

$$b = \sqrt{\frac{4P(k_1 + k_2) R_1 R_2}{L(R_1 + R_2)}}$$

$$= 1.13 \sqrt{\frac{P \Delta R_1 R_2}{L(R_1 + R_2)}} \qquad \text{...(15.34)}$$

and

$$p_0 = 0.564 \sqrt{\frac{P(R_1 + R_2)}{L \Delta R_1 R_2}} \qquad \text{...(15.35)}$$

If both cylinders are of the same material and $\nu = 0.3$, then

$$b = 1.52 \sqrt{\frac{P R_1 R_2}{EL(R_1 + R_2)}} \qquad \text{...(15.36)}$$

In the case of two equal radii

$$R_1 = R_2 = R$$

$$b = 1.08 \sqrt{\frac{PR}{EL^3}} \qquad \text{...(15.37)}$$

For the case of contact of a cylinder with a plane surface so that $\frac{1}{R_1} = 0$.

$$b = 1.13\sqrt{\frac{PR_2\Delta}{L}} \qquad ...(15.38)$$

$$p_0 = 0.564\sqrt{\frac{P}{L\Delta R_2}} \qquad ...(15.39)$$

For the case of the contact of cylinder and cylindrical groove, by taking R_2 as negative, we find

$$b = 1.13\sqrt{\frac{P\Delta R_1 R_2}{L(R_2 - R_1)}} \qquad ...(15.40)$$

and

$$p_0 = 0.564\sqrt{\frac{P(R_2 - R_1)}{L\Delta R_1 R_2}} \qquad ...(15.41)$$

15.5 Stresses Along the Load Axis

(*i*) *Two bodies in contact.* The following expressions have been obtained by Thomas and Hoersch* for the principal stresses σ_x, σ_y and σ_z at a point on the z-axis, the point is at the distance z from the origin which lies in the surface of contact of the two elastic bodies.

$$\sigma_x = [M(\Omega_x + \nu\Omega_x')]\frac{b}{\Delta}(A + B) \qquad ...(15.42)$$

$$\sigma_y = [M(\Omega_y + \nu\Omega_y')]\frac{b}{\Delta}(A + B) \qquad ...(15.43)$$

$$\sigma_z = -\left[\frac{M}{E}\left(\frac{1}{n} - n\right)\right]\frac{b}{\Delta}(A + B) \qquad ...(15.44)$$

in which

$$M = \frac{2k}{k'^2 E(k')}, \quad n = \sqrt{\frac{k^2 + k^2(z/b)^2}{1 + k^2(z/b)^2}}$$

$$\Delta = \frac{1 - \nu_1^2}{E_1} + \frac{1 - \nu_2^2}{E_2}$$

$$k = \frac{b}{a} = \cos\theta\,;\ k \le 1$$

$$k' = \sqrt{1 - k^2} = \sin\theta$$

*Thomas, H.R. and Hoersch. V.A., "Stresses due to the pressure of one elastic solid on another," Bulletin 212, Engineering Experimentation Centre, University of Illinois, June 15, 1930.

$k\left(\frac{z}{b}\right)$ = Relative depth below surface of contact to point on z-axis at which stresses are to be calculated

$= \cot \phi$

A, B = Constant given by Eqs. (15.5) and (15.6)

$$\Omega_x = -\frac{1-n}{2} + k\frac{z}{b}[F(\phi, k') - H(\phi, k')]$$

$$\Omega_x' = -\frac{n}{k^2} + 1 + k\frac{z}{b}\left[\left(\frac{1}{k^2}\right)H(\phi, k') - F(\phi, k')\right]$$

$$\Omega_y = \frac{1}{2n} + \frac{1}{2} - \frac{n}{k^2} + k\frac{z}{b}\left[\frac{1}{k^2}H(\phi, k') - F(\phi, k')\right]$$

$$\Omega_y' = -1 + n + k\frac{z}{b}[F(\phi, k') - H(\phi, k')]$$

where $$F(\phi, k') = \int_0^{\phi} \frac{d\theta}{\sqrt{1 - k'^2 \sin^2 \theta}}$$

$$H(\phi, k') = \int_0^{\phi} \sqrt{1 - k'^2 \sin^2 \theta}\, d\theta$$

are the complete elliptic integrals.

Also $$\frac{B}{A} = \frac{\left(\frac{1}{k^2}\right)E(k') - K(k')}{K(k') - E(k')} \quad ...(15.45)$$

where $$K(k') = F\left(\frac{\pi}{2}, k'\right) = \int_0^{\pi/2} \frac{d\theta}{\sqrt{1 - k'^2 \sin^2 \theta}}$$

$$E(k') = H\left(\frac{\pi}{2}, k'\right) = \int_0^{\pi/2} \sqrt{1 - k'^2 \sin^2 \theta}\, d\theta$$

$$b = \sqrt{\frac{3kE(k')}{2\pi} \frac{P\Delta}{(A+B)}} \quad ...(15.46)$$

(*ii*) *Two spherical bodies in contact.* The equations for stresses at points along the z-axis due to contact of two spheres have been computed by Thomas and Hoersch as :

$$\sigma_x = \sigma_y = \frac{2a\left(\frac{1}{R_1} + \frac{1}{R_2}\right)}{\pi\Delta}\left[(1 + \nu)\left(\frac{z}{a}\cot^{-1}\frac{z}{a} - 1\right) + \frac{1}{2}\frac{a^2}{(a^2 + z^2)}\right] \quad ...(15.47)$$

$$\sigma_z = -\frac{2a\left(\frac{1}{R_1}+\frac{1}{R_2}\right)}{\pi\Delta}\left(\frac{a^2}{a^2+z^2}\right) \qquad \text{...(15.48)}$$

Since $\sigma_x = \sigma_y$, hence τ_{xz} (45°) = τ_{yz} (45°) = $\frac{1}{2}(\sigma_x - \sigma_z)$, and τ_{xy} (45°) = 0. The magnitudes of the shear stresses τ_{xz} and τ_{yz} are greatest at a depth of about half the radius of the contact surface. For $\nu = 0.3$, the magnitude of this stress is approximately 0.33 p_0. Fig. 15.8 shows the distribution of stresses along the load axis for one of a pair of normally loaded spheres. The greatest stress is the compressive stress at the surface, σ_z, where $\sigma_x = -p_0$. The other two principal stresses at the same point are given by the expression

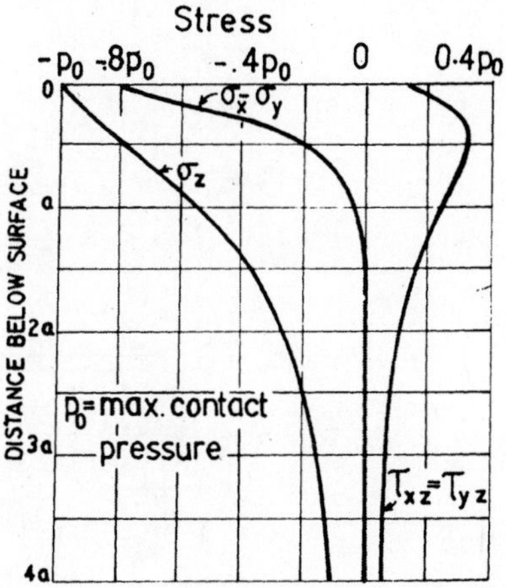

Fig. 15.8

$$\sigma_x = \sigma_y = -\left(\frac{1+2\nu}{2}\right)p_0 \qquad \text{...(15.49)}$$

(*iii*) *Two cylindrical bodies in contact.* When two bodies in contact are cylindrical having radii R_1 and R_2 so that they are in contact approximately along a straight line element before loads are applied, then

$$A = 0 \text{ and } B = \frac{1}{2}\left[\frac{1}{R_1}+\frac{1}{R_2}\right]$$

Hence the value of the ratio $\frac{B}{A}$ is indefinitely large, and from Eq. (15.45) it is seen that the corresponding value of k approaches zero. But k is the ratio of semi-minor axis b of the area of contact to semi-major axis a and, therefore, a must be indefinitely large, which is the case of contact along a line between two bodies. The area of contact when a load of p kg per unit length is applied is a long narrow rectangle of width $2b$ in the y-direction and length $2a = L$ in the x-direction. When $k = 0$, Eqs. (15.42) to (15.44) for the stresses at points on the z-axis at various distances $\frac{z}{b}$ below the contact surface become :

$$\sigma_x = -2\nu\left[\sqrt{1+\left(\frac{z}{b}\right)^2} - \frac{z}{b}\right]\frac{b}{\Delta}.B \qquad ...(15.50)$$

$$\sigma_y = -\left[\frac{\left\{\sqrt{1+\left(\frac{z}{b}\right)^2} - \frac{z}{b}\right\}^2}{\sqrt{1+\left(\frac{z}{b}\right)^2}}\right]\frac{b}{\Delta}.B \qquad ...(15.51)$$

$$\sigma_z = -\left[\frac{1}{\sqrt{1+\left(\frac{z}{b}\right)^2}}\right]\frac{b}{\Delta}.B \qquad ...(15.52)$$

where $b = \sqrt{\frac{2p\Delta}{\pi B}}$

$$\Delta = \frac{1-\nu_1^2}{E_1} + \frac{1-\nu_2^2}{E_2}$$

$$\left.\begin{aligned} |\tau_{xy}| &= |\tfrac{1}{2}(\sigma_x - \sigma_y)| \\ |\tau_{yz}| &= |\tfrac{1}{2}(\sigma_y - \sigma_z)| \\ |\tau_{zx}| &= |\tfrac{1}{2}(\sigma_z - \sigma_x)| \end{aligned}\right\} \qquad ...(15.53)$$

These stresses have been shown in Fig. 15.9.

At the surface *i.e.* $z = 0$, these stresses are :

$$\left.\begin{aligned} \sigma_x &= -2\nu\frac{b}{\Delta}B = -2\nu p_0 \\ \sigma_y &= -\frac{b}{\Delta}B = -p_0 \\ \sigma_z &= -\frac{b}{\Delta}B = -p_0 \end{aligned}\right\} \qquad ...(15.54)$$

Now $|\tau_{yz}| = |\tfrac{1}{2}(\sigma_y - \sigma_z)|$

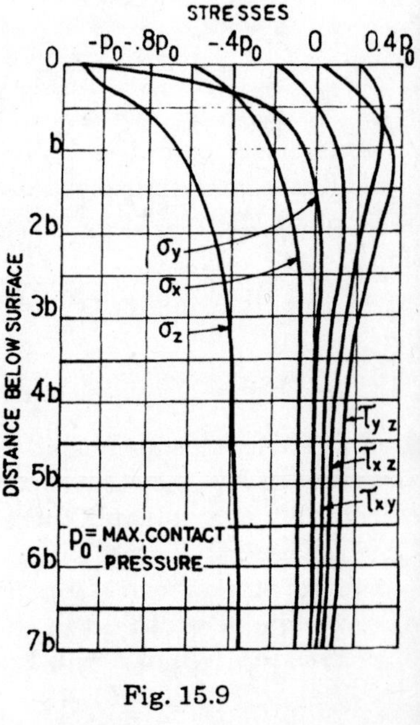

Fig. 15.9

Substituting the values of σ_y and σ_z from Eqs. (15.51) and (15.52) and taking the first derivative with respect to z, and equating to zero, we get the value of $\frac{z}{b} = 0.7861$ at which the shear stress τ_{yz} has maximum value. At this point the principal stresses become :

$$\left.\begin{aligned} \sigma_x &= -0.9718\,\nu\frac{b}{\Delta}\,.\,B \\ \sigma_y &= -0.1856\frac{b}{\Delta}\,.\,B \\ \sigma_z &= -0.7861\frac{b}{\Delta}\,.\,B \end{aligned}\right\} \qquad \text{...(15.55)}$$

and hence for $\nu = 0.3$

$$(\tau_{yz})_{max} = 0.30\frac{b}{\Delta}B = 0.30\,p_0 \qquad \text{...(15.56)}$$

When the two cylinders are crossed at $\alpha = 90°$, for this case, Thomas and Hoersch have developed an empirical equation for the maximum shear stress, given by

$$\tau_{max} = \frac{822.5}{(R_1/R_2)^{0.271}}\sqrt[3]{\frac{P}{R_2^{\,2}}}\ \text{kg/cm}^2 \qquad \text{...(15.57)}$$

where $\nu = 0.3$, $E = 2.1 \times 10^6$ kg/cm^2

and $1 \le \frac{R_1}{R_2} \le 8$

15.6 Stresses for Two Bodies in Line Contact

(*i*) *Normal loading.* For a concentrated load p acting on the surface of contact as shown in Fig. 15.10, the stresses as obtained by Smith and Liu* are

$$\left.\begin{aligned} \sigma_y &= -\frac{2p}{\pi z}\cos^2\theta\sin^2\theta \\ \sigma_z &= -\frac{2p}{\pi z}\cos^4\theta \\ \tau_{yz} &= -\frac{2p}{\pi z}\sin\theta\,.\,\cos^3\theta \end{aligned}\right\} \qquad \text{...(15.58)}$$

where $\theta = \cos^{-1}\frac{z}{\sqrt{z^2 + y^2}}$

*Smith, J.O. and Liu, C.K., "Stresses due to tangential and normal loads on an elastic solid with application to some contact stress problems". J. Appl. Mech., Vol. 20, No. 2, March 1953, pp. 157-166.

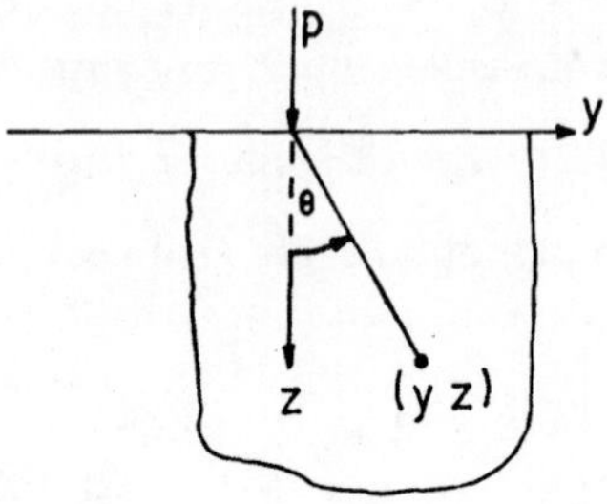

Fig. 15.10

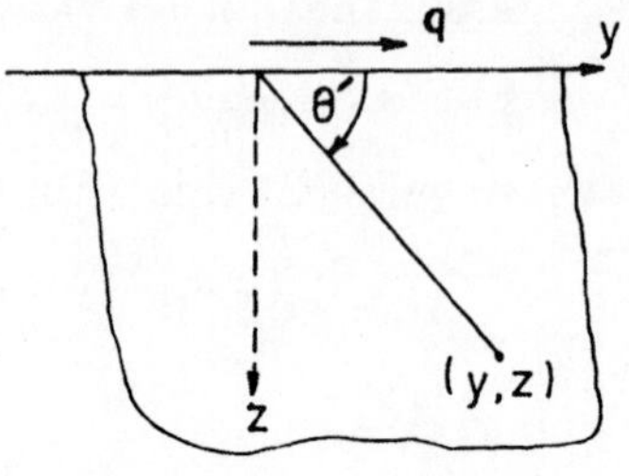

Fig. 15.11

(*ii*) *Tangential loading.* For tangential loading as shown in Fig. 15.11, the stresses as obtained by Smith and Liu are given by

$$\left.\begin{aligned}\sigma_y &= -\frac{2q}{\pi z}\cos^3\theta'\sin\theta' \\ \sigma_z &= -\frac{2q}{\pi z}\cos\theta'\sin^3\theta' \\ \tau_{yz} &= -\frac{2q}{\pi z}\sin^2\theta'\cos^2\theta'\end{aligned}\right\} \quad ...(15.59)$$

where $\theta' = \cos^{-1}\dfrac{y}{\sqrt{y^2+z^2}}$

(*ii*) *Normal load elliptically distributed.* For elliptically distributed normal load as shown in Fig. 15.12, the expressions for stresses as obtained by Smith and Liu are :

$$\left.\begin{aligned}\sigma_{yn} &= -\frac{p_0}{\pi}z\left[\frac{b^2+2y^2+2z^2}{b}\overline{\psi}-\frac{2\pi}{b}-3y\psi\right] \\ \sigma_{zn} &= -\frac{p_0}{\pi}z[b\,\overline{\psi}-y\phi] \\ \tau_{yzn} &= -\frac{p_0}{\pi}z^2\psi\end{aligned}\right\} \quad ...(15.60)$$

where $p_0 = \dfrac{2P}{\pi b},\ b = \sqrt{\dfrac{4P\Delta R_1R_2}{(R_1+R_2)}}$

$$\Delta = \frac{1-{\nu_1}^2}{E_1} + \frac{1-{\nu_2}^2}{E_2}$$

$$\psi = \frac{\pi}{K_1}\frac{1-\sqrt{K_2/K_1}}{\sqrt{\dfrac{K_2}{K_1}}\sqrt{2\sqrt{K_2+\left(\dfrac{K_1+K_2-4a^2}{K_1}\right)}}}$$

$$\bar{\psi} = \frac{\pi}{K_1} \frac{1 + \sqrt{K_2/K_1}}{\sqrt{\frac{K_2}{K_1}} \sqrt{2\sqrt{\frac{K_2}{K_1}} + \left(\frac{K_1 + K_2 - 4a^2}{K_1}\right)}}$$

$$K_1 = (b+y)^2 + z^2$$
$$K_2 = (b-y)^2 + z^2$$

At $z = 0$

$$\sigma_{yn} = \begin{cases} -p_0\sqrt{1-\frac{y^2}{b^2}}, & |y| \le b \\ 0, & y \ge b, y \le -b \end{cases}$$

$$\sigma_{zn} = \begin{cases} -p_0\sqrt{1-\frac{y^2}{b^2}}, & |y| \le b \\ 0, & y \ge b, y \le -b \end{cases}$$

$\tau_{yzn} = 0$ for all points.

(*iv*) *Tangential load elliptically distributed.* **For tangential load elliptically distributed as shown in Fig. 15.13, the expressions for stresses as obtained by Smith and Liu are**

$$\left.\begin{aligned} \sigma_{yt} &= -\frac{q_0}{\pi}\left[(2y^2 - 2b^2 - 3z^2)\psi + 2\pi\frac{y}{b} + 2(b^2 - y^2 - z^2)\frac{y}{b}\bar{\psi}\right] \\ \tau_{zt} &= -\frac{q_0}{\pi} z^2\psi \end{aligned}\right\} \quad ...(15.61)$$

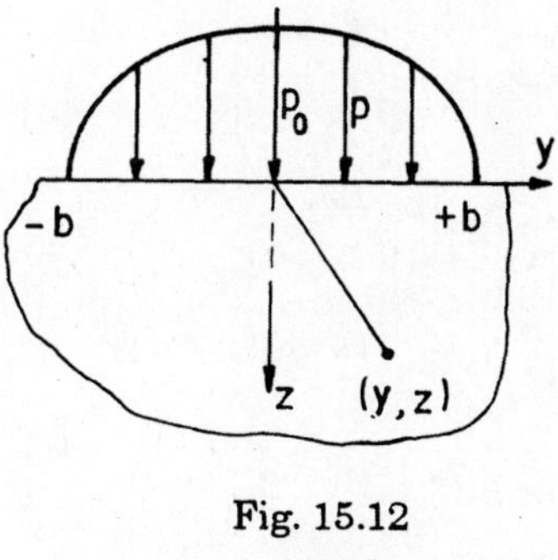

Fig. 15.12

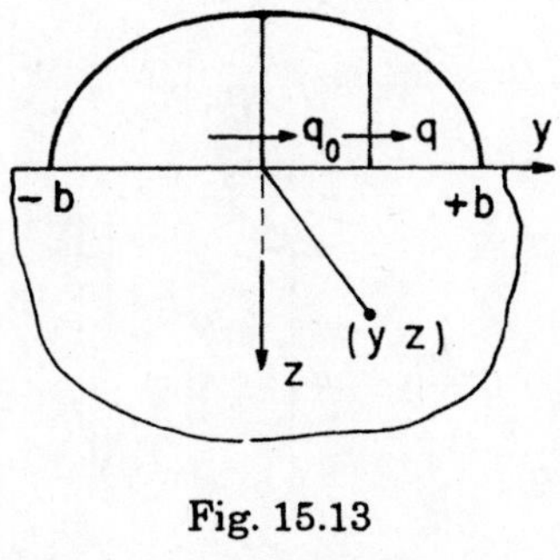

Fig. 15.13

$$\tau_{yzt} = -\frac{q_0}{\pi}\left[(b^2 + 2y^2 + 2z^2)\frac{z}{b}\bar{\psi} - 2\pi\frac{z}{b} - 3yz\psi\right]$$

At $z = 0$

$$\sigma_{yt} = \begin{cases} -2q_0\left[\frac{y}{b} - \sqrt{\frac{y^2}{b^2} - 1}\right], y \le b \\ -2q_0\left[\frac{y}{b} + \sqrt{\frac{y^2}{b^2} - 1}\right], \; y \le -b \\ -2q_0\frac{y}{b}, \quad |\,y\,| \le -b. \end{cases}$$

$$\tau_{yzt} = \begin{cases} -q_0\sqrt{1 - \frac{y^2}{b^2}} \;, \quad |\,y\,| \le b \\ \quad 0 \quad , \qquad y \ge b, 0 \le -b. \end{cases}$$

$\sigma_{zt} = 0$ for all points.

(*v*) *Combined normal and Tangential Loading.* Let $q_0 = fp_0$, where f is the coefficient of friction. Assuming $f = \frac{1}{3}$, Smith and Liu has obtained the following expressions for the stresses under the combined action of normal and tangential loading.

$$\sigma_y{}^* = -\frac{p_0}{\pi}\left\{(b^2 + 2y^2 + yz^2)\frac{z}{b}\overline{\psi} - 2\pi.\frac{z}{b} - 3yz\,\psi\right.$$
$$\left. + \frac{1}{3}\left[(2y^2 - 2b^2 - 3z^2)\,\psi + 2\pi\,\frac{y}{b} + 2(b^2 - y^2 - z^2)\frac{y}{b}\overline{\psi}\right]\right\}$$

$$\left.\begin{aligned} \sigma_z{}^* &= -\frac{p_0}{\pi}z\left[b\,\overline{\psi} - y\psi + \frac{z}{3}\,\psi\right] \\ \tau_{yz}{}^* &= -\frac{p_0}{\pi}\left\{z^2\psi + \frac{1}{3}\left[(b^2 + 2y^2 + 2z^2)\frac{z}{b}\overline{\psi}\right.\right. \\ &\qquad\qquad \left.\left. - 2\pi\,\frac{z}{b} - 3yz\,\psi\right]\right\} \end{aligned}\right\} \quad ...(15.62)$$

At $z = 0$, we find that

$$\sigma_y{}^* = \begin{cases} -\frac{2}{3}p_0\left[\frac{y}{b} - \sqrt{\frac{y^2}{b^2} - 1}\right], y \ge b \\ -p_0\left[\sqrt{1 - \frac{y^2}{b}} + \frac{2}{3}\frac{y}{b}\right], \; |\,y\,| \le b \\ -\frac{2}{3}p_0\left[\frac{y}{b} + \sqrt{\frac{y^2}{b^2} - 1}\right], y \le -b \end{cases}$$

$$\sigma_z^* = \begin{cases} -p_0\sqrt{1-\dfrac{y^2}{b^2}}, & |\, y \,| \le b \\ 0 \quad , & y \ge a, y \le -b \end{cases}$$

$$\tau_{yz}^* = \begin{cases} -\dfrac{1}{3}p_0\sqrt{1-\dfrac{y^2}{b^2}}, & |\, y \,| \le b \\ 0 \quad , & y \ge b, y \le -b \end{cases}$$

Example 15.1 *A railway car wheel rolls on a rail. Both rail and wheel are made of steel for which E = 210 GPa and ν = 0.3. The wheel has a radius of R_1 = 0.4 m, and the cross radius of the rail top surface is R_2 = 0.3 m (Fig. 15.14). Determine the size of the contact area and the maximum contact pressure if compressive load P is 90 kN.*

Solution.

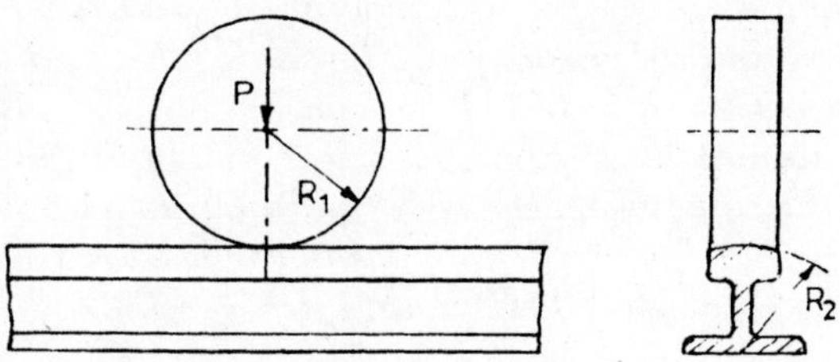

Fig. 15.14

$$\frac{1}{R_1'} = \frac{1}{R_2'} = 0$$

$$\alpha = \pi/2$$

$$\Delta = \frac{4}{\dfrac{1}{R_1}+\dfrac{1}{R_2}} = \frac{4}{\dfrac{1}{0.4}+\dfrac{1}{0.3}} = 0.6857$$

$$D = \frac{4E}{3(1-\nu^2)} = \frac{4\times 210\times 10^9}{3(1-0.09)} = 3.07692\times 10^{11}$$

$$\cos\theta = \frac{\dfrac{1}{0.4}-\dfrac{1}{0.3}}{\dfrac{1}{0.4}+\dfrac{1}{0.3}} = 0.1428 \text{ or } \theta = 81.79°$$

For 0 = 81.79°, from Table 15.1, we have

$$m = 1.1040 \text{ and } n = 0.9112$$

Semi axes of elliptical contact area are :

$$a = m\sqrt[3]{\frac{P\Delta}{D}} = 1.1040\left[\frac{90 \times 10^3 \times 0.6857}{3.07692 \times 10^{11}}\right]^{1/3} = 0.00646 \text{ m}$$

$$b = n\sqrt[3]{\frac{P\Delta}{D}} = 0.9112\left[\frac{90 \times 10^3 \times 0.6857}{3.07692 \times 10^{11}}\right]^{1/3} = 0.00533 \text{ m}$$

Maximum contact pressure,

$$p_{max} = \frac{3}{2}\frac{P}{\pi ab}$$

$$= \frac{1.5 \times 90 \times 10^3}{\pi \times 0.00646 \times 0.00533} = 1248 \text{ MPa}$$

Example 15.2 *The radii of curvature of the two surfaces of semi circular discs at the point of contact are $R'_1 = 60$ mm, $R_1'' = 130$ mm, $R_2' = 80$ mm, and $R_2'' = 200$ mm. The angle α between the planes of minimum curvature is 60°. If a load of $P = 4.5$ kN is applied, determine the maximum principal stress, maximum shearing stress, and locate the point where each of these stresses occur. Also calculate the distance that the two discs move towards each other because of load P. Take $E_1 = E_2 = 200$ GPa, and $\nu_1 = \nu_2 = 0.29$.*

Solution.

$$A = \frac{1}{4}\left[\frac{1}{R_1'} + \frac{1}{R_1''} + \frac{1}{R_2'} + \frac{1}{R_2''}\right] - \frac{1}{4}\left[\left(\frac{1}{R_1'} - \frac{1}{R_1''}\right)^2 + \left(\frac{1}{R_2'} - \frac{1}{R_2''}\right)^2 + 2\left(\frac{1}{R_1'} - \frac{1}{R_1''}\right)\left(\frac{1}{R_2'} - \frac{1}{R_2''}\right)\cos 2\alpha\right]^{1/2}$$

$$= \frac{1}{4}\left[\frac{1}{60} + \frac{1}{130} + \frac{1}{80} + \frac{1}{200}\right] - \frac{1}{4}\left[\left(\frac{1}{60} - \frac{1}{130}\right)^2 + \left(\frac{1}{80} - \frac{1}{200}\right)^2 + 2\left(\frac{1}{60} - \frac{1}{130}\right)\left(\frac{1}{80} - \frac{1}{200}\right)\cos 120°\right]^{1/2}$$

$$= 0.00838 \text{ mm}^{-1}$$

$$B = \frac{1}{4}\left[\frac{1}{R_1'} + \frac{1}{R_1''} + \frac{1}{R_2'} + \frac{1}{R_2''}\right] + \frac{1}{4}\left[\left(\frac{1}{R_1'} - \frac{1}{R_1''}\right)^2 + \left(\frac{1}{R_2'} - \frac{1}{R_2''}\right)^2 + 2\left(\frac{1}{R_1'} - \frac{1}{R_1''}\right)\left(\frac{1}{R_2'} - \frac{1}{R_2''}\right)\cos 2\alpha\right]^{1/2}$$

$$= 0.01255 \text{ mm}^{-1}$$

$$A + B = 0.02093 \text{ mm}^{-1}$$

$$\Delta = \frac{2(1 - \nu^2)}{E} = \frac{2\,[1 - (0.29)^2]}{200 \times 10^3} = 9.159 \times 10^{-6} \text{ mm}^2/\text{N}$$

$$\cos\theta = \frac{B - A}{B + A} = \frac{0.00417}{0.02093} = 0.199236$$

$$\theta = 78.5^\circ$$

From Table 15.1, for $\theta = 78.5^\circ$, $m = 1.1502$, $n = 0.8789$

$$a = m\sqrt[3]{\frac{3}{4}\frac{P\Delta}{(A + B)}}$$

$$= 1.1502\sqrt[3]{\frac{3}{4} \times \frac{4.5 \times 10^3 \times 9.159 \times 10^{-6}}{0.02093}} = 1.30985 \text{ mm}$$

$$b = n\sqrt[3]{\frac{3}{4}\frac{P\Delta}{(A + B)}} = 1.00098$$

$$p_0 = \frac{3}{2}\frac{P}{\pi ab} = \frac{3}{2} \times \frac{4.5 \times 10^3}{\pi \times 1.30985 \times 1.00098} = 1638.728 \text{ N/mm}^2$$

The principal stresses at the centre of the surface of contact are :

$$\sigma_x = -2\nu p_0 - (1 - 2\nu)\, p_0 \left(\frac{b}{a + b}\right)$$

$$= -2 \times 0.29 \times 1638.728 - (1 - 2 \times 0.29) \times 1638.728 \times \frac{1.0098}{2.31083}$$

$$= -950.46 - 298.13 = -1248.59 \text{ N/mm}^2$$

$$\sigma_y = -2\nu p_0 - (1 - 2\nu)\, p_0 \times \left(\frac{a}{a + b}\right)$$

$$= -2 \times 0.29 \times 1638.728 - (1 - 2 \times 0.29) \times 1638.728 \times \frac{1.30985}{2.31083}$$

$$= -950.46 - 390.13 = -1340.59 \text{ N/mm}^2$$

$$\sigma_z = -p_0 = -1638.728 \text{ N/mm}^2$$

$$\tau_{max} = \frac{1638.728 - 1248.59}{2} = 195.07 \text{ N/mm}^2$$

EXERCISES

15.1 A 500 mm radius wheel supporting a 200 kN load rests on a rail 100 mm radius. The material of both is steel with $E = 2 \times 10^3$ N/mm^2 and $\nu = 0.3$. Determine the maximum contact pressure and the total contact area.

15.2 A hardened steel roller ($E = 2 \times 10^5$ N/mm^2, $\nu = 0.29$) runs in an annular groove of a cast-iron member ($E = 1.05 \times 10^5$ N/mm^2, $\nu = 0.26$) with a total load of 20 kN as shown in Fig. 15.15. Determine the maximum contact pressure.

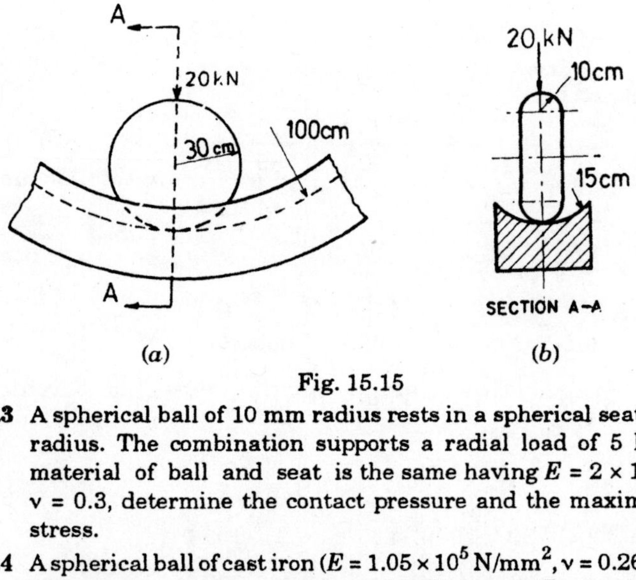

Fig. 15.15

15.3 A spherical ball of 10 mm radius rests in a spherical seat of 20 mm radius. The combination supports a radial load of 5 kN. If the material of ball and seat is the same having $E = 2 \times 10^5$ N/mm^2 $\nu = 0.3$, determine the contact pressure and the maximum shear stress.

15.4 A spherical ball of cast iron ($E = 1.05 \times 10^5$ N/mm^2, $\nu = 0.26$) of radius 100 mm is resting on a spherical ball of steel ($E = 2 \times 10^5$ N/mm^2, $\nu = 0.30$) of radius 50 mm. If they are to support a load of 50 kN, determine the contact pressure and the total area of contact.

15.5 A flat cam of steel ($E = 2 \times 10^5$ N/mm^2, $\nu = 0.30$) having nose radius 50 mm is to actuate a valve through a flat follower of cast iron ($E = 1.05 \times 10^5$ N/mm^2, $\nu = 0.28$). If the force required to open the valve is 600 N, determine the contact pressure between the cam and the follower.

15.6 A uniform line load of intensity 20 N/mm is acting normally to a semi-infinite plate. Determine the stresses in the plate at a depth of 5 mm and 30° angular location from the line of action of load.

15.7 A uniform load of intensity 30 N/mm is acting tangentially to a semi-infinite plate. Determine the stresses in the plate at a depth of 2.5 mm and 30° angular location from the plate surface.

15.8 In a single row ball bearing assembly, the ball diameter is 50 mm ; the groove radius, 30 mm ; the diameter of the outer race, 250 mm ; and the highest compressive force on the ball, $P = 1.8$ kN. Determine the maximum principal stress. Take $E = 200$ GPa and $\nu = 0.3$.

[Ans. 1038.46 MPa]

15.9 A steel railway car wheel may be considered to be approximately a cylinder with a radius of 440 mm. The wheel rolls on a steel rail whose top surface may be considered to be approximately another cylinder with a radius of 300 mm. For the wheel and rail, $E = 200$ GPa and $\nu = 0.29$. If the wheel load is 110 kN, determine the maximum principal stress and maximum shear stress. Also determine the vertical displacement of the centre of the wheel due to deflections in the region of contact. **[Ans.** 0.116 mm**]**

15.10 A cast iron push rod ($E = 117$ GPa, $\nu = 0.20$) in a valve gear is operated by a steel cam ($E = 200$ GPa, $\nu = 0.29$) as shown in Fig. 15.16. The cam is cylindrical in shape and has a radius of curvature of 5 mm at its tip. The surface of the push rod that contacts the cam is spherical in shape with a radius of curvature of 4 m so that the rod and cam are in point contact. If the allowable maximum principal stress for cast iron is – 1400 MPa, determine the maximum load P that may act on the rod. **[Ans.** 5.58 kN**]**

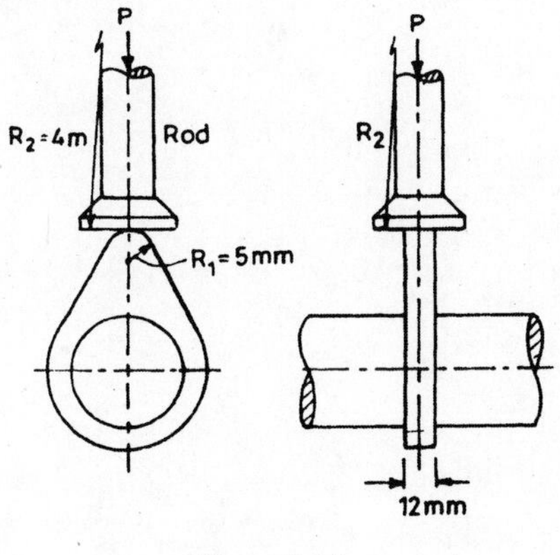

Fig. 15.16

15.11 A hard steel ball (E = 200 GPa, ν = 0.29) of diameter 50 mm is pressed against a thick aluminium plate (E = 72 GPa, ν = 0.33). Determine the maximum load the assembly can carry if the yield strength of aluminium is 150 MPa.

15.12 A steel ball bearing (Fig. 15.17) consists of an inner race an outer race, and 12 balls. The rated load for the bearing is 4.2 kN when operating at 300 rpm. The largest load on a ball is 1.75 kN. At the region of contact between the inner race and this ball, determine the maximum principal stress, the maximum shearing stress, and the distance from the point of contact to the point where these stresses occur. R_1' = 4.76 mm, R_2' = – 4.86 mm, E = 200 GPa, ν = 0.29.

[**Ans.** – 2315 MPa, 695 MPa, 2.099 mm, 0.1574 mm]

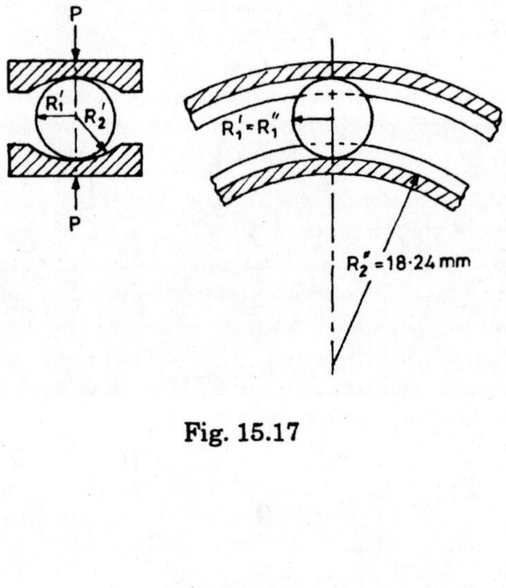

Fig. 15.17

16

Stress Concentration Problems

16.1 Introduction

The conditions that cause the stresses to be greater than those given by the ordinary stress equations of strength of materials are frequently called *discontinuities* and *stress raisers*. The term *stress gradient* is used to indicate the rate of increase of stress as a stress raiser is approached. These localized stresses are called *stress concentrations*. The conditions that may cause the stress at a point in a simple member to be radically different from the value calculated from ordinary formulas include effects such as :

(*i*) abrupt changes in section such as holes, keyways, fillet radii etc.

(*ii*) pressure at the points of application of the external forces such as points of contact of gear teeth, or of ball bearings on the races etc.

(*iii*) material discontinuities such as non-metallic inclusions in steel, air holes in concrete etc.

(*iv*) initial stresses due to over straining and cold working of metals during fabrication, or heat treatment, or to residual stresses due to welding etc., and

(*v*) cracks developed during welding, cold working and grinding etc.

The solutions to the stress concentration problems has been obtained by the theory of elasticity, numerical methods, such as finite elements and experimental methods, such as photoelasticity, strain gauges, brittle coating, plaster model and analogous methods etc.

In this chapter, we consider the stress-concentration problem for simple cases. For more advanced solutions the reader may refer to specialized work.

16.2 Stress-Concentration Factor

The effect of a stress raiser is defined by the stress-concentration factor, which is the ratio of the maximum stress to the nomial stresss, *i.e.*

$$K_t = \sigma_{max}/\sigma_{nom} \quad ...(16.1)$$

The nominal stress is based on the knowledge from strength of materials. The nominal stress may be either based upon the gross or net area of cross-section. Correspondingly, we define, stress-concentration factor, K_{tg} based upon the gross area of cross-section as :

$$K_{tg} = \sigma_{max}/\sigma_{gross} \quad ...(16.2)$$

The stress-concentration factor, K_{tn}, based upon the net area of cross-section is defined as :

$$K_{tn} = \sigma_{max}/\sigma_{net} \quad ...(16.3)$$

These two factors may be combined as follows :

$$K_{tn} = K_{tg}(1 - \lambda) \quad ...(16.4)$$

where λ is a fraction by which area is decreased.

16.3 Fatigue Stress-Concentration Factor

The fatigue stress-concentration factor K_f, is defined as the ratio of fatigue limit of unnotched specimen to the fatigue limit of notched specimen under axial-or bending loads. The effect of a notch on the fatigue strength of a part varies considerably with material and notch geometry. This phenomenon is called "notch sensitivity", q. It may be considered as a measure of the degree to which the theoretical effect is obtained, and is defined as :

$$q = (K_f - 1)/(K_t - 1) \quad ...(16.5)$$

or

$$K_f = 1 + q(K_t - 1) \quad ...(16.6)$$

Values of q ranges from zero to 1.

16.4 Theoretical Stress-Concentration Factors

Here we discuss the values of calculated stress-concentration factors for different discontinuities, based upon the theory of elasticity. The cases considered are not exhaustive but rather to be illustrative of the effects of different discontinuities.

16.4.1 Circular Hole in an Infinite Plate under Uniaxial Tension

This problem was considered in Example 3.5. With respect to polar co-ordinates (r, θ), the plane stress components at any point are given by (Fig. 16.1)

$$\left.\begin{aligned}\sigma_r &= \frac{\sigma_o}{2}\left(1-\frac{a^2}{r^2}\right)+\frac{\sigma_o}{2}\left(1-\frac{4a^2}{r^2}+\frac{3a^4}{r^4}\right)\cos 2\theta \\ \sigma_\theta &= \frac{\sigma_o}{2}\left(1+\frac{a^2}{r^2}\right)-\frac{\sigma_o}{2}\left(1+\frac{3a^4}{r^4}\right)\cos 2\theta \\ \tau_{r\theta} &= -\frac{\sigma_o}{2}\left(1+\frac{2a^2}{r^2}-3\frac{a^4}{r^4}\right)\sin 2\theta\end{aligned}\right\} \quad \text{...(16.7)}$$

where a is the radius of the hole. For $r = a$

$$\sigma_\theta = \sigma_o\,(1 - 2\cos 2\theta) \quad \text{...(16.8)}$$

Hence, for $\theta = \frac{\pi}{2}, \frac{3\pi}{2}$, σ_θ attains its maximum value of $3\sigma_o$. For $\theta = 0, \pi$, σ_θ attains a compressive value of $-\sigma_\theta$.

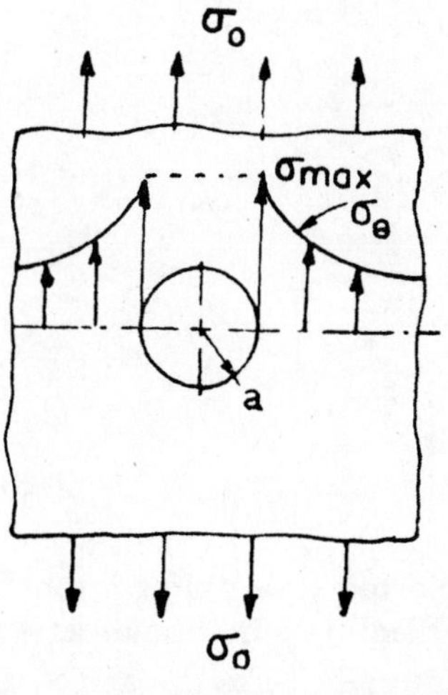

Fig. 16.1

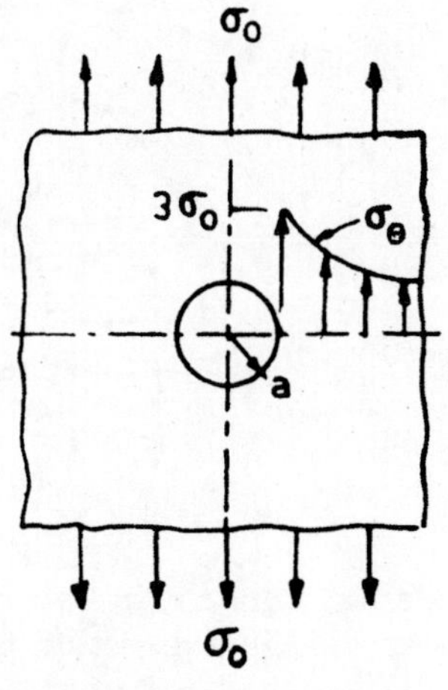

Fig. 16.2. Distribution for σ_o at $\pi/2$, $3\pi/2$.

For $\theta = \frac{\pi}{2}, \frac{3\pi}{2}$ and $r > a$, σ_θ becomes :

$$\sigma_\theta = \frac{\sigma_o}{2}\left(2+\frac{a^2}{r^2}+\frac{3a^4}{r^4}\right) \quad \text{...(16.9)}$$

Fig. 16.2 shows that σ_o decreases rapidly. Thus, the high stress gradient is quite localized.

For $r \simeq 10\,a$, $\sigma_\theta = \sigma_o$

When the diameter of the hole is comparable to the width of the plate, Eqs. (16.7) are considerably in error. Several authors have studied the problem of a large circular hole in a strip by theoretical and experimental methods. The results are summarized by the formula :

$$K_i = \frac{\sigma_{max}}{\sigma_{net}} = \frac{3k - 1}{k + 0.3} \qquad ...(16.10)$$

where $$k = \frac{\text{width of strip}}{\text{dia of hole}}$$

16.4.2 Elliptic Hole in an Infinite Plate

The problem of an elliptic hole in a plate under tension was considered in chapter 4. For an infinite plate with elliptic hole of

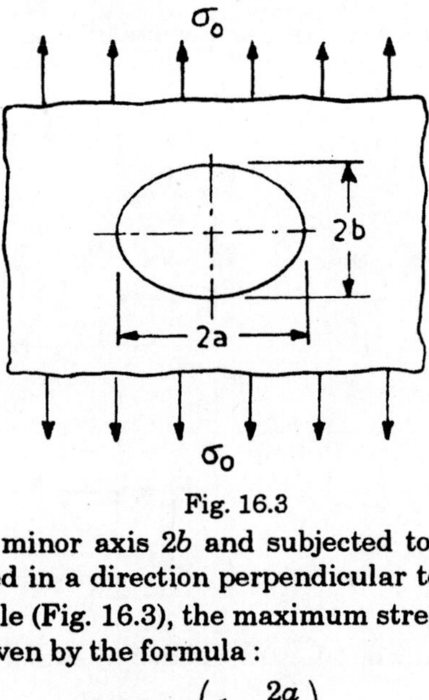

Fig. 16.3

major axis $2a$, minor axis $2b$ and subjected to a uniform tensile stress σ_0 applied in a direction perpendicular to the major axis of the elliptical hole (Fig. 16.3), the maximum stress at the end of the major axis is given by the formula :

$$\sigma_{max} = \sigma_0 \left(1 + \frac{2a}{b}\right) \qquad ...(16.11)$$

When the uniaxial tension is applied perpendicular to the minor axis of the hole, then

$$\sigma_{max} = \sigma_0 \left(1 + \frac{2a}{b}\right) \qquad ...(16.12)$$

By geometry, the radius of curvature of an ellipse at the end of the major axis is :

$$\rho = \frac{b^2}{a} \qquad ...(16.13)$$

Hence, Eq. (16.11) yields :

$$\sigma_{max} = \left(1 + 2\sqrt{\frac{a}{b}}\right) \quad ...(16.14)$$

As $b \to 0$, the elliptic hole in an infinite plate becomes very and approaches the shape of a line crack. The maximum value stress may become very large, depending upon the nature of the load.

16.4.3 Grooves and Holes

The values of theoretical stress-concentation factors for grooves may be obtained for various types of loads as follows :

Let the SCF be K_{ts} when the groove is very shallow. Then

$$K_{ts} = 1 + 2\sqrt{\frac{t}{\rho}} \quad ..(16.15)$$

where t is the depth of notch and ρ its radius.

Let the SCF be K_{td} when the groove is very deep, then

$$K_{td} = \frac{2\left(\frac{b}{\rho} + 1\right)\sqrt{\frac{b}{\rho}}}{\left(\frac{b}{\rho} + 1\right)\tan^{-1}\sqrt{\frac{b}{\rho}} + \sqrt{\frac{b}{\rho}}} \quad ...(16.16)$$

where $2b$ is the effective width (or dia) for a symmetric notch and b the effective width for unsymmetric notch.

Then, according to Neuber, an approximate value of SCF for any depth of groove is given by

$$K_t = 1 + \frac{(K_{ts} - 1)(K_{td} - 1)}{\sqrt{(K_{ts} - 1)^2 + (K_{td} - 1)^2}} \quad ...(16.17)$$

16.4.4 Concentrated Load on the Edge of an Infnite Plate

This problem was considered in Example 3.6. The stresses at a point are given by

$$\sigma_x = \sigma_r \cos^2\theta = -\frac{2P}{\pi r}\cos^3\theta = -\frac{2Px^3}{\pi\,(x^2 + y^2)^2} \quad ...(16.18)$$

$$\sigma_y = \sigma_r \sin^2\theta = -\frac{2P}{\pi r}\sin^2\theta \,.\, \cos\theta = -\frac{2Pxy^2}{\pi(x^2 + y^2)^2} \quad ...(16.19)$$

$$\tau_{xy} = \sigma_r \sin\theta \,.\, \cos\theta = -\frac{2P}{\pi r}\sin\theta\cos^2\theta = -\frac{2Px^2y}{\pi(x^2 + y^2)^2} \quad ...(16.20)$$

Solutions to some more problems may be seen in chapter 3.

16.5 Strain-Concentration Factor

A strain-concentration factor may be defined as

$$K_\varepsilon = \frac{\text{maximum strain at the notch}}{\text{nominal strain at the notch}} = \frac{\varepsilon_{max}}{\varepsilon_{nom}} \qquad ...(16.21)$$

whereas the stress-concentration factor, K_σ, has been defined as

$$K_\sigma = \sigma_{max}/\sigma_{nom} \qquad ...(16.22)$$

The value of K_ε increases as the value of K_σ decreases. For relatively sharp notches, the following relation between K_σ, K_ε and K_t exists :

$$K_\sigma . K_\varepsilon = K_t^{\,2} \qquad ...(16.23)$$

Eq. (16.23) states the "Neuber's Rule". It can be used to calculate the value of K_t in the plastic range also.

Combining Eqs. (16.21), (16.22) and (16.23), we get

$$\sigma_{max} \cdot \varepsilon_{max} = K_t^{\,2} . \sigma_{nom} \cdot \varepsilon_{nom} \qquad ...(16.24)$$

BIBLIOGRPHY

(i) Books

1. *S.P. Timoshenko* and *J.N. Goodier,* 'Theory of Elasticity', McGraw-Hill Kogakusha, Ltd., 3rd. Ed., 1970.
2. *S.P. Timoshenko* and *S. Woinowsky—Krieger,* 'Theory of Plates and Shells', McGraw-Hill Kogakusha, Ltd., 2nd Ed., 1959.
3. *Chi-Teh Wang,* 'Applied Elasticity', McGraw-Hill Book Company, Inc., New York, 1953.
4. *I.H. Shames,* 'Mechanics of Deformable Solids,' Prentice-Hall of India (Pvt.) Ltd., New Delhi, 1965.
5. *N.I. Muskhelishvili,* 'Some Basic Problems of the Mathematical Theory of Elasticity', P. Noordhoff, Ltd. Groningen, Holland, 1953.
6. *I.S. Sokolnikoff,* 'Mathematical Theory of Elasticity', McGraw-Hill Book Company, New York, N.Y. 1956.
7. *Y.C. Fung,* 'Foundations of Solid Mechanics', Prentice-Hall, Inc., Englewood Cliffs, N.J. 1965.
8. *A.E. Green and W. Zerna,* 'Theoretical Elasticity', Oxford University Press, London, England, 1954.
9. *A.E.H. Love,* 'A Treatise on Mathematical Theory of Elasticity', 4th Ed., Dover Publications, Inc., New York N.Y., 1944.
10. *E.E. Sechler,* 'Elasticity in Engineering', John Wiley & Sons, Inc., New York, N.Y., 1962.
11. *H.L. Langhaar,* 'Energy Methods in Applied Mechanics' John Wiley & Sons, Inc., New York, N.Y., 1962.
12. *R.V. Southwell,* Relaxation Methods in Engineering Science', Oxford University Press, 1940.
13. *R.V. Southwell,* 'Relaxation Methods in Theoretical Physics', Oxford At The Clarendon Press, 1946.
14. *M. Hetenyi,* 'Beams on Elastic Foundation', The University of Michigan Press, Ann Arbor, 1970.
15. *S.P. Timoshenko* and *J.M. Gere,* 'Theory of Elastic Stability', McGraw-Hill Kogakusha Ltd., 1961.

16. *R.C. Juvinall,* 'Stress, Strain and Strength', McGraw-Hill, New York, 1967.
17. *E. Volterra* and *J.H. Gaines,* 'Advanced Strength of Materials', Prentice-Hall, Englewood Cliffs. N.J. 1971.
18. *A.H. England,* 'Complex Variable Methods in Elasticity', Wiley-Interscience, London, 1971.
19. *R.G. Budynas,* 'Advanced Strength and Applied Stress Analysis', McGraw-Hill Book Company, 1977.
20. *G. Murphy,* 'Advance Mechanics of Materials,' McGraw-Hill Book Company, New York, N.Y., 1964.
21. *F.B. Seely and J.O. Smith,* 'Advanced Mechanics of Materials,' John Wiley & Sons, Inc., New York, N.Y., 1952.
22. *O.C. Zienkiewicz,* 'The Finite Element Method', 3rd Ed., McGraw-Hill Book Company (U.K.) Limited, 1977.
23. *C.S. Desai* and *J.F. Abel,* 'Introduction to the Finite Element Method', Van Nostrand Reinhold Company, New York, N.Y.
24. *B.A. Boley* and *J.H. Weiner,* 'Theory of Thermal Stresses', John Wiley & Sons, Inc., New York, 1960.
25. *A.I. Lure,* 'Three Dimensional Problems of the Theory of Elasticity', Interscience New York, 1964.
26. *S. Flugge,* 'Hand-Book of Physics', Vol. VI Springer Verlag, 1958.
27. *S.N. Savin,* 'Stress Concentration Around Holes', Pergamon Press, London (1961).
28. *J.P. Den Hartog,* 'Advanced Strength of Materials', McGraw-Hill Book Co., Inc., (1952).
29. *W. Flugge,* 'Stresses in Shells', Springer Varlag, Berlin (1960).
30. *B.E. Gatewood,* 'Thermal Stresses', McGraw-Hill Book Co., Inc., New York (1957).
31. *D.S. Dugdale,* 'Elements of Elasticity', Pergamon Press, Oxford (1968).
32. *P.P. Benham and R.D. Hoyle,* 'Thermal Stresses', Pitman & Co., London (1964).
33. *D.J. Peery,* 'Aircraft Structures', McGraw-Hill Book Co., Inc., New York (1950).

(ii) **Journals**

1. Journal of Applied Mechanics.
2. Quarterly of Applied Mathematics.

3. International Journal of Engineering Science.
4. International Journal of Fracture Mechanics.
5. Applied Mechanics Reviews.
6. International Journal on Numerical Methods in Engineering.
7. Computers and Structures.
8. Bulletin of Japan Society of Mechanical Engineers.
9. International Journal of Mechanical Science.
10. International Journal of Solids and Structures.
11. Journal of Elasticity.
12. Journal of Mechanical Engineering Science.
13. Journal of Engineering Mechanics.
14. Journal of Institution of Engineers (India), Mechanical Engineering Division.
15. PMM : Journal of Applied Mathematics and Mechanics.
16. Quarterly Journal of Mechanics and Applied Mathematics.
17. Engineering Index.
18. AIAA Journal.
19. Journal of Engineering for Industry.
20. Proceedings of the Cambridge Philosophical Society Mathematical and Physical Sciences.

AUTHOR INDEX

H

I

K

L

M

N

P

R

S

T

U

V

W